T0293059

Machine Learning and Data Sciences for Financial Markets

Leveraging the research efforts of more than 60 experts in the area, this book reviews cutting-edge practices in machine learning for financial markets. Instead of seeing machine learning as a new field, the authors explore the connection between knowledge developed in quantitative finance over the past 40 years and modern techniques generated by the current revolution in data sciences and artificial intelligence.

The text is structured around three main areas: "Interacting with investors and asset owners," which covers robo-advisors and price formation; "Towards better risk intermediation," which discusses derivative hedging, portfolio construction, and machine learning for dynamic optimization; and "Connections with the real economy," which explores nowcasting, alternative data, and ethics of algorithms.

Accessible to a wide audience, this invaluable resource will allow practitioners to include machine learning driven techniques in their day-to-day quantitative practices, while students will build intuition and come to appreciate the technical tools and motivation behind the theory.

AGOSTINO CAPPONI is Associate Professor in the Department of Industrial Engineering and Operations Research at Columbia University. He conducts research in financial technology and market microstructure. His work has been recognized with the NSF CAREER Award, and a JP Morgan AI Research award. Capponi is a co-editor of *Management Science and Mathematics and Financial Economics*. He is a Council member of the Bachelier Financial Society, and recently served as Chair of the SIAM-FME and INFORMS Finance.

CHARLES-ALBERT LEHALLE is Global Head of Quantitative R&D at Abu Dhabi Investment Authority and Visiting Professor at Imperial College London. He has a PhD in machine learning, was previously Head of Data Analytics at CFM, and held different Global Head positions at Crédit Agricole CIB. Recognized as an expert in market microstructure, Lehalle is often invited to present to regulators and policy-makers.

Machine Learning and Data Sciences for Financial Markets

A Guide to Contemporary Practices

Edited by

Agostino Capponi
Columbia University, New York

Charles-Albert Lehalle
Abu Dhabi Investment Authority

CAMBRIDGE
UNIVERSITY PRESS

CAMBRIDGE
UNIVERSITY PRESS

Shaftesbury Road, Cambridge CB2 8EA, United Kingdom

One Liberty Plaza, 20th Floor, New York, NY 10006, USA

477 Williamstown Road, Port Melbourne, VIC 3207, Australia

314–321, 3rd Floor, Plot 3, Splendor Forum, Jasola District Centre, New Delhi – 110025, India

103 Penang Road, #05–06/07, Visioncrest Commercial, Singapore 238467

Cambridge University Press is part of Cambridge University Press & Assessment, a department of the University of Cambridge.

We share the University's mission to contribute to society through the pursuit of education, learning and research at the highest international levels of excellence.

www.cambridge.org
Information on this title: www.cambridge.org/9781316516195
DOI: 10.1017/9781009028943

© Cambridge University Press & Assessment 2023

First published 2023

A catalogue record for this publication is available from the British Library.

ISBN 978-1-316-51619-5 Hardback

Contents

Contributors

Robert Almgren *Quantitative Brokers, New York; and Princeton University.*

Andrea Angiuli *Department of Statistics and Applied Probability, University of California, Santa Barbara.*

Shane Barratt *Stanford University, Department of Electrical Engineering.*

Milo Bianchi *Toulouse School of Economics, TSM; and IUF, University of Toulouse Capitole.*

Paul Bilokon *Department of Mathematics, Imperial College, London.*

Jean-Philippe Bouchaud *Capital Fund Management, Paris.*

Stephen Boyd *Stanford University, Department of Electrical Engineering.*

Ḥaoyang Cao *CMAP, École Polytechnique.*

Marie Brière *Amundi, Paris Dauphine University, and Université Libre de Bruxelles.*

Luca Capriotti *Department of Mathematics, University College London; and New York University, Tandon School of Engineering.*

René Carmona *Department of Operations Research and Financial Engineering, Princeton University.*

Álvaro Cartea *Oxford University, Mathematical Institute, and Oxford–Man Institute of Quantitative Finance.*

Umut Çetin *London School of Economics, Department of Statistics.*

Brian Clark *Lally School of Management, Rensselaer Polytechnic Institute.*

Samuel N. Cohen *Mathematical Institute, University of Oxford.*

Francesco D'Acunto *Carroll School of Management, Boston College.*

Carlo de Franchis *ENS Paris-Saclay, CNRS; and Kayrros, Paris.*

Matthew F. Dixon *Department of Applied Mathematics, Illinois Institute of Technology.*

Sébastien Drouyer *ENS Paris-Saclay, CNRS.*

Gabriele Facciolo *ENS Paris-Saclay, CNRS.*

Laurent Ferrara *Skema Business School, University Côte d'Azur; and QuantCube Technology.*

Michael Fleder *Massachusetts Institute of Technology; and Covariance Labs, New York.*

Jean-Pierre Fouque *Department of Statistics and Applied Probability, University of California, Santa Barbara.*

Maximilien Germain *LPSM, Université de Paris.*

Aitor Muguruza Gonzalez *Imperial College London and Kaiju Capital Management.*

Daniel Giamouridis *Bank of America, Data and Innovation Group, London.*

Adam Grealish *Altruist, Los Angeles.*

Rafael Grompone von Gioi *ENS Paris-Saclay, CNRS.*

Olivier Guéant *Université Paris 1 Panthéon-Sorbonne, Centre d'Economie de la Sorbonne.*

Xin Guo *University of California, Berkeley, Department of Industrial Engineering and Operations Research.*

Igor Halperin *AI Research, Fidelity Investments, Boston.*

Artur Henrykowski *Department of Mathematics, University College London.*

Charles Hessel *ENS Paris-Saclay, CNRS; and Kayrros, Paris.*

Blanka Horvath *Technical University of Munich; Munich Data Science Institute; King's College London; and The Alan Turing Institute.*

Lisa L. Huang *Head of AI Investment Management and Planning, Fidelity.*

Sebastian Jaimungal *University of Toronto, Statistical Sciences.*

Apurv Jain *MacroXStudio, San Francisco.*

Prabhanjan Kambadur *Bloomberg, New York.*

Petter N. Kolm *Courant Institute of Mathematical Sciences, New York University.*

Sophie Laruelle *Université Paris Est Creteil, CNRS, LAMA; and Université Gustave Eiffel, LAMA, Marne-la-Vallée.*

Mathieu Laurière *Department of Operations Research and Financial Engineering, Princeton University.*

Jacky Lee *Department of Mathematics, University College London.*

Fabrizio Lillo *University of Bologna; and Scuola Normale Superiore.*

Gideon Mann *Bloomberg, New York.*

Alberto G. Rossi *McDonough School of Business, Georgetown University.*

Jean-Michel Morel *ENS Paris-Saclay, CNRS.*

Gilles Pagès *LPSM, Sorbonne-Université.*

Mikko S. Pakkanen *Imperial College London.*

Georgios V. Papaioannou *Bank of America, Data and Innovation Group, London.*

Markus Pelger *Stanford University, Department of Management Science & Engineering.*

Huyên Pham *LPSM, Université de Paris.*

Nicholas G. Polson *ChicagoBooth, University of Chicago.*

Michael Recce *CEO, AlphaROC Inc., New York.*

Mathieu Rosenbaum *CMAP, École Polytechnique.*

Brice Rosenzweig *Bank of America, Data and Innovation Group, London.*

Leandro Sánchez-Betancourt *Oxford University, Mathematical Institute.*

Devavrat Shah *Massachusetts Institute of Technology.*

Akhtar Siddique *Economics Department, Office of the Comptroller of the Currency.*

Majeed Simaan *School of Business, Stevens Institute of Technology.*

Anna Simoni *CREST, CNRS, ENSAE, École Polytechnique, Institut Polytechnique de Paris.*

Derek Snow *The Alan Turing Institute.*

Amanda Stent *Colby College.*

Lukasz Szpruch *School of Mathematics, University of Edinburgh.*

Jonathan Tuck *Stanford University, Department of Electrical Engineering.*

Xavier Warin *EDF R&D.*

Dominic Wright *Department of Mathematics, University College London.*

Xun Yu Zhou *Columbia University, Department of Industrial Engineering and Operations Research; and The Data Science Institute, New York.*

Preface

Machine learning, Artificial Intelligence (AI), and data science pervade every aspect of our everyday life. Many of the techniques developed by the Computer Science community are becoming increasingly used in the area of financial engineering, ranging from the use of deep learning methods for hedging and risk management through the exploitation of AI techniques for investment or design of trading systems. These techniques are also having enormous implications on the operations of financial markets. It is thus not surprising to see increasingly the proliferation of AI research groups or recently created "AI Labs" at major banks, centered around topics of key relevance to financial services. Those include, among others, explainable AI, human-machine interaction, and DS methods for extracting information from data and using it to support investment decisions. The integration of AI methods in the decision making process may also have unintended or unanticipated consequences especially in a sector like finance, where bad intermediation of risk can spread over the whole economy. Many of the ethical issues expected from AI systems, including privacy, data manipulation, opacity, and discrimination, can be detrimental to financial markets. For example, data leakage is a key concern for banks; regulatory authorities need to deal with it, and so is fairness in the distribution of debt and issuance of loans. In asset management, the question of bias introduced by a dataset and its stationarity has been known for a long time; the more data dominate decisions, the more important they are. All those issues are getting increasing consideration from major regulatory bodies worldwide.

We should mention that if we come back to the early age of machine learning, the techniques and tools used to provide theoretical grounds to the process of learning from data share their roots with the ones that gave birth to online optimization and stochastic control. They are based on asymptotics of discrete stochastic processes and on stochastic algorithms that support frameworks in which the learned parameters, like the weights of a neural network, are seen as controls that evolve during the learning process. These parameters start at an arbitrary point (they are often randomly initialized) and are meant to follow flows which minimize a criterion usually referred to as a loss function: they are "controls" driving the neural network from a random state to a state where a target task can be performed. These technical tools, designed to capture the behavior of a stochastic system that is driven to a specific state in a noisy environment, evolved in parallel to address important problems arising in financial markets. A prominent example is "hedging", where one needs to hedge a portfolio of derivatives by replicating the risks embedded into the derivative constituents of the portfolio. In such a case, this portfolio is a control driving the balance sheet of an institution towards a state with minimal unhedged risk. Other business needs require the design of a portfolio that captures investment goals stated in a more generic way (with no specification in terms of tradable instruments). Hence financial engineering has exploited these tools from the 1980s and contributed to their improvement. This community did it independently from the machine learning community, which also contributed to improving these tools mostly from an algorithmic perspective. In recent years, the disciplines of data science and AI have started to be seriously involved in the analysis of financial markets. It is important to not forget what academics and practitioners understood about these tools, and especially the way they can improve risk management in markets. Since the dream of replacing reasoning and modelling by data and black boxes is dangerous in the non stationary environment of financial markets, it is important to integrate machine learning practices with the structural knowledge developed by quantitative finance during the last 40 years. "Old" knowledge and new approaches should cross-fertilize, injecting the structural nonlinearities of learning machines and their capability to extract structures from data exactly where more formal methods had a lack of adaptiveness.

Inspired by these considerations, we have decided to collect the most relevant sample of cutting edge research developed in the fields of Machine learning, Data Sciences, and AI with application to finance into a book. Our book project has been strongly supported by the academic community. We have invited active researchers with demonstrated expertise and leadership in their own areas of relevance to contribute a chapter to the book. They have enthusiastically responded to our call, and submitted high quality chapters. Their chapters have been reviewed by a team of qualified referees, who have carefully processed the content and provided excellent feedback for improvement. Our project has also received strong support by the Cambridge University Press (CUP), which has kindly agreed to publish the volume. This book follows the tradition of the financial engineering community started in the last decade to spotlight topics of increasing importance for the community and the broad society overall, and culminating then into the *Handbook on Systemic Risk* published by CUP. The topics of the chapters are highly reflective of the research agenda of the two most prominent financial engineering societies, namely the SIAM-FM Activity group currently chaired by Agostino Capponi, and the Finance and Insurance Reloaded program (FaIR) within the Institute of Louis Bachelier Paris, which Charles-Albert Lehalle started a few years ago. The last two biennial meetings of the SIAM-FM group, held in 2019 and 2021, featured many plenary talks, invited minisymposia, and tutorials in the area of machine learning and data science. Talks given by a mix of academics and industry practitioners, reflected both an algorithmic technical perspective and the integration of ML methodologies against financial markets data. Relatedly, the FaIR transverse program has been a unique occasion to meet researchers involved in the use of new technologies for financial markets. The series of thematic workshops organized by FaIR and the ACPR (French regulator for banking), as well as its kick-off workshop at the Collège de France, have specially been places of intense thinking and brainstorming on how machine learning would influence these industries.

Since starting our effort, it has been our intention to structure the book around three main areas of interest: "Interactions with investors and asset owners" which mainly covers robo-advisors and price formation; "Risk intermediation" which covers portfolio construction, and machine learning for dynamic optimization, including optimal trading; and "Connections with the real economy" covers nowcasting, alternative data and ethics of algorithms. This structure offers a comprehensive and easy to read perspective on the areas of machine learning, AI and data science in financial markets.

We believe that now, more than ever, is now a good time to collect the various efforts made by leading and high profile researchers, including academics, practitioners and policy makers, into a book. We have developed this book with the idea that it becomes a key reference in the field. It will serve as the main reference for experienced researchers with training in quantitative methods, who want to increase their awareness of the cutting edge research being done in the area. We have also paid attention to a pedagogic component, and strived to make each chapter comprehensive enough and understandable by advanced graduate students. Those in search of a new topic to explore for their dissertation at the intersection of machine learning, data science, and finance will be inspired by the methodologies and applications presented in the book.

We expect the handbook to be received well beyond the academic community. Financial institutions and policy makers wishing to bring rigor to their business will be able to leverage upon the methodologies discussed in the book, and integrate them with data. As a result, the book will have a high potential of increasing the collaborations of the academia with the public and private sector, and to educate new generations of scientists who will build the new AI technologies in the financial sector.

The editors, Agostino Capponi and Charles-Albert Lehalle
New York and Abu Dhabi

Acknowledgments of referees.

The editors and contributors would like to thank the referees who took time to read and comment the contributions of this book:

- Agustin Lifschitz, Capital Fund Management, Paris, France.
- Amine Raboun, Euronext Paris, Courbevoie, France.
- Andrea Angiuli, Department of Statistics and Applied Probability, University of California, Santa Barbara.
- Bobby Shackelton, Head of Geospatial, Bloomberg LP.
- Emmanuel Série, Capital Fund Management, Paris, France.
- Frederic Bucci,
- Haoran Wang, CAI Data Science and Machine Learning, The Vanguard Group, Inc., Malvern, PA, USA.
- Haoyang Cao, The Alan Turing Institute.
- Harvey Stein, Head, Quantitative Risk Analytics, Bloomberg and Adjunct Professor, Mathematics Department, Columbia University.
- Ibrahim Ekren, Florida State University, Department of Mathematics, Tallahassee, FL.
- Iuliia Manziuk, Engineers Gate, Quantitative Researcher, London.
- Jiacheng Zhang, Department of Operations Research and Financial Engineering, Princeton University.
- Matthew Dixon, Illinois Institute of Technology, Department of Applied Mathematics.
- Michael Fleder, Massachusetts Institute of Technology and Covariance.AI.
- Michael Reher, University of California San Diego, Rady School of Management.
- Noufel Frikha, Université de Paris, Laboratoire de Probabilités, Statistiques et Modélisation.
- Othmane Mounjid, University of California, Berkeley (IEOR department).
- Renyuan Xu, Industrial and Systems Engineering, University of Southern California.
- Ruimeng Hu, Department of Mathematics, Department of Statistics and Applied Probability, University of California, Santa Barbara.
- Shihao Gu, Booth School of Business, University of Chicago.
- Sveinn Olafsson, Stevens Institute of Technology.
- Sylvain Champonnois, Capital Fund Management, Paris, France.
- Symeon Chouvardas, Independent Researcher.
- Zhaoyu Zhang, Department of Mathematics, USC.

INTERACTING WITH INVESTORS AND ASSET OWNERS

Part I

Robo Advisors and Automated Recommendation

1

Introduction to Part I
Robo-advising as a Technological Platform for Optimization and Recommendations

Lisa L. Huang[a]

It may be a self-evident truth that the financial services industry is driven by data and that data is increasing at an exponential pace. Robo-advisors are technological platforms that help individuals make better financial decisions, i.e., deliver 'advice', at scale using large disparate data sets. Advice may mean anything from investment portfolios, to consumption/savings rates, to financial goals, and to withdrawals in retirement, etc. The prefix 'robo' reflects the fact that advice is given most often algorithmically. This does not mean that there is not a human in the loop, and this is most often the case currently. Robo also implicitly means that advice can be delivered at scale. With this scale, the cost of advising can be lowered substantially, which leads naturally to the democratization of financial advice. With this platform, it's not hard to imagine a world where there is universal access to financial services which breaks down traditional economic, social, gender, and geographical barriers.

My own work helping to build one of the first robo-advisors in the world began in 2012 when I first learned of the mission that Betterment was founded upon. I joined Betterment the following year and built many of the foundational algorithms that deliver financial advice at scale to the many users on its platform during my years there.

The robo-advisor market is enormous, not measured in hundreds of billions, but in trillions of dollars. The robo-advisor market is also global, because the need to access financial services at scale is becoming more critical across the world. At the inception of robo-advising, advice was limited in scope to investment and portfolio management. Indexing can be seen as one of the first examples of robo-advice, which provided a ubiquitous and low-cost way for individuals to invest. The first wave of robo-advisors typically used mathematically sophisticated portfolio optimization tools, such as Modern Portfolio Theory (or extensions of it such as Black–Litterman), to create semi-customized solutions for retail investors. These tools for portfolio optimization were well known but not democratized at a cost that was accessible to the masses. These first robo-advisors helped change

[a] Head of AI Investment Management and Planning, Fidelity
Published in *Machine Learning And Data Sciences For Financial Markets*, Agostino Capponi and Charles-Albert Lehalle © 2023 Cambridge University Press.

that and indeed transformed an entire industry. In their chapter *Robo-Advisory: From Investing Principles and Algorithms to Future Development*, Adam Grealish and Petter N. Kolm give the readers a fantastic insider's view of the detailed blueprint of the traditional robo-advisor. What is striking is the simplicity and the elegance of advice when it is guided by a set of principles, as explained by the authors.

However, traditional robo-advising – that is rooted in investment management – is evolving. Since robo-advisors are, at their core, a technological platform for financial services, the scope of what can be achieved on that platform can broaden substantially. Robo-advisors as a platform can educate their users, correct for human bias, help to conceptualize the entire financial life cycle of an individual, and optimize every financial decision to maximize the 'happiness' of the user. The meaning of 'happiness' is a personal one but can in theory be captured algorithmically. With enough data, and allowing for feedback between users and algorithms, robo-advisors have opportunities to help users optimize every personal financial decision. In *New Frontiers of Robo-Advising: Consumption, Savings, Debt Management, and Taxes*, Francesco D'Acunto and Alberto G. Rossi outline the tantalising vision of the 'holistic robo-advisor'.

There are incredible challenges around realizing the full potential of robo-advisors. The most critical is data that gives a complete and holistic view of the financial life of a user. If partial data is available to the robo-advisor, then the algorithms will not be able to come up with the most optimal solutions for the user. Most users have a variety of financial relationships with different financial institutions. For example, they may have multiple bank accounts, brokerage accounts, retirement accounts, etc. Therefore, seeing a holistic picture is often non-trivial.

While the holy grail of robo-advising is personalization, the measurement of personal parameters that are needed for the robo-advisor is potentially fraught with uncertainty. High uncertainty in input parameters will lead to suboptimal outputs. One of these inputs is the 'risk tolerance' parameter. Loosely speaking, risk tolerance is a measure of the attitude toward investment risk. Different robo-advisors will try to access this number in different ways but most use a questionnaire to collect data from users. This is clearly insufficient because the definition of risk tolerance is unclear to begin with. It could be very customized for each user. In some implementations of robo investment advice, this risk number directly maps to a portfolio. Since the measurement of risk tolerance is imprecise, optimization of the portfolio only leads to a false sense of precision.

The last challenge that I will highlight here is a technical one. Many tasks that are universal in the financial lives of users do not yet have an accepted mathematical solution. One such problem that I helped solve during my time at Betterment was how to optimize the location of assets in a multi-account setting, in order to minimize taxes, given different tax treatments across multiple accounts. Surprisingly, the exact mathematical solution for this was not known when we began the work. We eventually solved this problem by mapping the asset location to the mathematical problem called the knapsack problem. However, the knapsack problem only solves the static allocation problem, but not the dynamic one, which is driven by any cash flow into accounts. The dynamic knapsack problem was one of many unsolved problems in financial planning. Another examples is finding the optimal way to save, given a multi-account setting with different risk tolerances for each account and different horizons with different priorities across those accounts? Most often, heuristics are relied upon to solve these mathematically complex problems. Milo Bianchi and Marie Briere, in their chapter, *Robo-Advising: Less AI and More XAI?*, delve into the nuanced nature of algorithmic advice and explore the challenges of how to generate trust in robo-advisors.

Since robo-advising is a technological platform, the users can be retail or institutional investors. In *Recommender Systems for Corporate Bond Trading*, Dominic Wright, Artur Henrykoswki, Jacky Lee and Luca Capriotti have created an application of robo-advising for corporate bond trading which leverages the recommender algorithms that are ubiquitous in retail businesses like Netflix and Facebook.

I will end here by referencing the title of a chapter, called called the *Investor's Worst Enemy*, from Ashwin B. Chhabra's book *The Aspirational Investor* (2015). This enemy, as many have pointed out, is the investor themselves. The promise of the robo-advisor is that the technology platform will help conquer the investor's worst enemy, to improve their financial decisions, and in turn, their lives.

2

New Frontiers of Robo-Advising: Consumption, Saving, Debt Management, and Taxes

Francesco D'Acunto[a] and Alberto G. Rossi[b]

Abstract

Traditional forms of robo-advice were targeted to help individuals make portfolio allocation decisions. Based on the balance-sheet view of households, the scope for robo-advising has been expanding to many other personal-finance choices, such as households' saving and consumption decisions, debt management, mortgage uptake, tax management, and lending. This sub-chapter reviews existing research on these new functions of robo-advising with a special emphasis on the questions that are still open for researchers across several disciplines. We also discuss the attempts to optimize jointly all personal-finance decisions, which we term "Holistic Robo-Advisors." We conclude by assessing fruitful avenues for research and practice in finance, computer science, marketing, decision science, information systems, law, and sociology.

2.1 Robo-advice and the balance-sheet view of the household

Robo-advice is any form of financial advice provided to human decision makers by algorithms. Even though many early applications of robo-advice were concentrated in the context of helping individual investors make portfolio allocation decisions, no inherent characteristic of algorithmic advice limits its application to that narrowly-specified context. And, indeed, the scope of robo-advice has broadened dramatically across all the areas of personal finance and more broadly to all contexts in which inexpert and often financially illiterate consumers need to make important choices that will affect their life-time wealth.

The breadth of applications of robo-advising are defined through the lens of the "balance-sheet view" of the household, which we depict schematically in Figure 2.1.

Under the balance-sheet view, households run dynamic budgets similar to those of firms: households have assets (left-hand side of Fig. 2.1), which include housing, durable goods, human capital, financial investments, and health. Households

---

[a] Carroll School of Management, Boston College
[b] McDonough School of Business, Georgetown University
   Published in *Machine Learning And Data Sciences For Financial Markets*, Agostino Capponi and Charles-Albert Lehalle © 2023 Cambridge University Press.

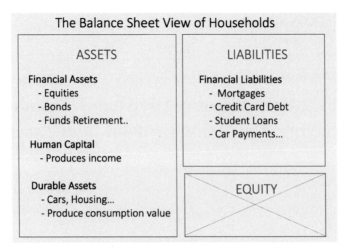

Figure 2.1 Balance Sheet View of the Household

also have to finance liabilities such as mortgages, credit-card debt, student loans, taxes, and insurance premiums. Households need to make decisions about all these budgetary items throughout their lifetime. Many such decisions will have enormous implications for their long-run wealth and financial sustainability.

In contrast to firms, however, the typical household lacks the knowledge and experience needed to make such important choices. For instance, many households only make a decision about purchasing a house and hence borrowing money through mortgages once in a lifetime. Moreover, households usually only face the problem of which form of education to provide to their offspring and how to finance such education once per child. The disconnect between the importance of all these decisions for household budgets and the lack of knowledge and experience in making such decisions stresses the need and scope for advice. Indeed, there is a large literature showing that, when left to their own devices, households make significant and costly mistakes that limit their ability to accumulate wealth over time (see Odean, 1999, Agarwal et al., 2017, and Laibson et al., 1998).

Despite their limitations as economic decision makers, households still need to make decisions that shape their balance sheets both statically and dynamically. For instance, how much and what type of human capital to acquire. Or, what kind of durable goods to purchase – what car to use and what housing condition to live in. All these asset purchases have dramatic implications on the liability side, too. For example, car purchases or leases involve choosing only one out of the very many financing solutions and contracts available. The choice of acquiring human capital – obtaining college and/or graduate education – involves decisions on the ways in which such asset acquisition can be financed, for instance choosing appropriate student loan conditions or even planning on college funds many years before the offspring reaches college age. Also, think about what is possibly the most important choice households make, i.e. the purchase of a house, which requires choosing appropriate mortgage characteristics based on household members income paths and horizons, a decision-making problem under risk and uncertainty that is incredibly hard to solve even for experts.

Historically, whenever choosing how to manage their balance sheets, households had the option of hiring human advisors. This option is less than desirable, however. First, human advisors are relatively costly and have been shown to make suboptimal choices. Suboptimal choices could be due to conflicts of interest in principal-agent relationships with asymmetric information, such as advisors' incentive to propose high-fee financial products to their clients, who are often unaware of the differences across financial products. Behavioral and cognitive biases could drive suboptimal human-advisor decisions as well (Foerster et al., 2017; Linnainmaa et al., 2021). By relying on human advisors, households face at the same time a potentially high cost of advice paired with an often suboptimal quality of advice. Second, supply-side forces might also restrain the availability of human advice to households and especially to lower-income households, who tend to be the most vulnerable when making decisions about managing their balance sheets. Catering to individuals with low net worth might be unpalatable to human advisors due to the low prospective revenues such clients would generate over time (Reher and Sokolinski, 2020).

These severe limitations of human advice in a context in which potential advisees often lack the ability to understand, let alone solve, the decision-making problems they face has represented fertile ground for the swift diffusion of robo-advice, also known as algorithmic advice (see D'Acunto and Rossi, 2021, Rossi and Utkus, 2020a). Robo-advice eliminates the barriers to access advice represented by the cost of human advisers because, in contrast to human advisers, it can be scaled up without virtually any constraints. For this reason, providers of robo-advising services can reduce their fees to a fraction of those commanded by human advisers. Moreover, robo-advice has been shown to make better decisions than humans and experts in several contexts on both the assets and liabilities side of the household balance sheet, such as the allocation of financial investments (e.g., see D'Acunto et al., 2019f; Rossi and Utkus, 2020b) or the take-up of peer-to-peer (P2P) loans (e.g., see D'Acunto et al., 2020a).

In the rest of this chapter, we highlight important recent developments in the evolution of robo-advising services based on the balance-sheet view of the household. We discuss the institutional details of each form of advice as well as the findings of existing research on the characteristics and performance of robo-advice across various domains. In particular, we focus on robo-advice in the domains of households' consumption and savings decisions, borrowing decisions, tax management, and lending choices. Robo-advisors for lending choices are allowing consumers and households who need financing to obtain funds without the need to pay fees to intermediaries. Moreover, they allow households to use their own savings to finance other borrowers and hence reduce the scope for institutional financial intermediaries. For each area, we discuss open questions and opportunities for researchers. We then envision the possibility of forms of robo-advice that optimize jointly households' choices subject to their budget constraint across all the individual parts of households' balance sheets. We term these forms of robo-advice "Holistic Robo-Advisors." Throughout the subchapter, we discuss the challenges and opportunities these recent forms of robo-advice imply and how these challenges and opportunities can translate into fruitful avenues of future research for scholars in as disparate fields as finance, computer science, marketing, decision science, information systems, and sociology.

2.2 Robo-advising for consumption-saving choices

A fundamental factor that determines a household's ability to accumulate wealth throughout the life cycle is the choice of how much to consume and save out of household income in each period in which income is earned. Computing the optimal saving rate requires solving a complicated optimization problem (D'Acunto et al., 2019a) that can prove challenging even for experienced economists. Non-economists are at a further disadvantage, because they often lack a clear understanding of the status of their finances, they cannot assess their own budget constraints, and they do not understand the implications of macroeconomic shocks for their individual consumption-saving decisions (see Agarwal et al., 2009; Agarwal and Mazumder, 2013; Christelis et al., 2010; D'Acunto et al., 2019d). Most households may find it hard to merely conceptualize this problem, even intuitively (see D'Acunto et al., 2019e), let alone to assess the optimal behavior throughout the life-cycle path and subject to budget constraints.

And, indeed, unsurprisingly many households fail to choose a saving rate during their working years that allows them to maintain a lifestyle comparable to the one they enjoyed before retirement (e.g., see Banks et al., 1998; Bernheim et al., 2001; Lusardi and Mitchell, 2007, among many others). This phenomenon represents not only a problem for individual households, but also produces negative externalities for society as a whole as the average tax payer needs to contribute higher taxes to maintain minimal living standards for the undersavers.

Even if potentially less problematic under the societal point of view, the opposite mistake in households' consumption-saving choices has also been detected: several US and European households tend to save large amounts based on perceived rather than actual precautionary savings motives (D'Acunto et al., 2020b). This phenomenon has been detected even during retirement – the phase of their life in which they should be engaging in the process of "decumulation" (See Mitchell and Utkus, 2004) – even when bequest motives are absent. Households' use of rules of thumb based on cultural norms, which substitute for their inability to understand and solve the dynamic optimization problem, have been proposed to explain this type of decisions (e.g., see D'Acunto, 2015). Households' consumption-saving choices are also at the heart of the balance-sheet view of the household discussed above, because the allocation of income across these two alternative types of assets has substantial dynamic implications in terms of long-run net worth.

Pairing the importance of the consumption-saving choice for individual households with the widespread inability of households to conceptualize and optimize such choice represents fertile ground for robo-advising applications. In this context, robo-advising applications might solve two different types of needs. First, they should provide households with information about their own balance sheet, size of assets and liabilities, and budget constraints, in a unified and simple format so that households can understand the parameters of the decision-making problem they face. This information role of robo-advising is especially important for households who have irregular income inflows or those who are self-employed and business owners, and hence whose income streams are irregular and not always easy to forecast.

Second, robo-advising applications to consumption-saving decisions should provide suggestions and advice to households on how to improve their choices as well as easy implementation of such advice. Suggestions can cover several aspects of decision-making such as the choice of which credit card(s) to use, which share of income to save each month based on projections of future values of saved amounts, as well as potential nudges to increase households' incentives to save rather than spend, which would be especially helpful for households who tend to spend more than what the permanent-income hypothesis implies at each point in time.

Real-world applications of robo-advising to the consumption-saving choice based on the criteria discussed above abound. In particular, one class or robo-advisors known as "income aggregators" fulfils this role (e.g., see Olafsson and Pagel, 2017, 2018). As the name suggests, income aggregators are a class of robo-advisors that covers the first scope of robo-advsing in the consumption-saving choice, i.e. providing households with clear and easy-to-grasp information about their own balance sheet and constraints.

Income aggregators require users to provide access to their asset and liability accounts. Asset accounts might include checking, saving, and other forms of financial investment accounts, such as brokerage accounts and retirement accounts. Liability accounts include mortgages, student debt, credit cards, and other forms of debt. In this way, robo-advisors collect information from the households' accounts, typically at the level of the individual transaction. By collecting this large amount of big data across accounts that would otherwise be unlinked, income aggregators are able to construct the balance sheet of each household following the balance-sheet view of the household discussed above. The accuracy of the information income aggregators produce depends on whether users link all their accounts to the robo-advising platform. For this reason, users have a strong incentive to link all their accounts.

The information income aggregators produce has a set of unique characteristics. First of all, income aggregators provide a just-in-time holistic representation of an household's balance sheet, which the household can check at any point in time. This feature is especially compelling for households who have substantial wealth invested in financial markets, the volatility of whose returns might be high. Moreover, income aggregators display information about households' balance sheet and budget constraints vividly in simple graphical forms that are intuitive for households and allow them to grasp basic concepts of household finance even without being trained, such as the balancing of budgets or the sustainability of assets and liabilities accounts. Having access to such intuitive display of information about one's own finances is crucial to create awareness in investors' mind and was shown to have a major impact in helping individuals make better financial decisions (Olafsson and Pagel, 2017, 2018).

However, advising individuals on how much to consume, what items to purchase, and how to split spending between durable and non-durable consumption is more complicated than helping individuals form well-diversified investment portfolios, because an algorithm would need to input specific information regarding individuals' preferences over all possible consumption bundles as well as their beliefs about a large range of future outcomes.

To overcome these limitations, innovative FinTech Apps have proposed alternative ways to help individuals by providing them with simple rules of thumb. A recent example is the US application *Status Money*. Status Money is an income aggregator, and hence as discussed above can compute users' net worth and observe all their transactions, including spending transactions. The unique feature of this App, which provides advice in the form of a rule of thumb, is providing users with information about peers' spending, where peers are defined as individuals observed in a US-representative sample outside the App and who are similar to users based on a set of demographic characteristics. Upon subscribing to the App, users fill in a form about demographic characteristics that include their annual income, age, home-ownership status, location of residence, and location type.

Based on this information Status Money assigns a peer group to each users and provides users with information about the average spending, assets, debts, and net worth of such peers. In this way, users can calibrate their spending to the spending of individuals who look similar to them. This rule of thumb is based on the notion of *the wisdom of the crowd*, whereby agents might obtain valuable signals about their (unknown to the user and to the robo-advisor) optimal spending and saving rate based on the average values of these ratios in a large population of decision makers that look similar to them (Chen et al., 2014; Da and Huang, 2020). Delivering information about crowds through media outlets has been shown to be effective in persuading consumers to change their behavior through the management of their subjective beliefs (Barone et al., 2015). Another channel behind this form of advice is *peer pressure*, whereby it is especially those users who spend substantially more than their peers – and hence are likely to spend more than their own optimal rate – who feel more compelled to react to the peer information and converge to peers' spending than those who spend less than their peers (Rosenberg, 2011). This potential asymmetric reaction to peers' spending information based on users' position relative to their peers would be valuable because overspending, and hence accumulating fewer savings and lower wealth for retirement, is a mistake that creates more issues for individual households and society than underspending.

D'Acunto et al. (2019b) study the effectiveness and the mechanisms behind this form of robo-advice. They find that providing salient peer information through the App has a large effect on users' consumption behavior. Users who were overspending with respect to their peer group at the time of sign-up ended up reducing their spending after signing up for the App. Those individuals who underspent instead, increased their spending but the reaction was much less pronounced for underspenders. D'Acunto et al. (2019b) also show that the informativeness of the peer group plays an important role in explaining users' changes in consumption. The authors conclude that FinTech Apps can provide valuable advice to individuals by collecting and summarizing in an unbiased fashion the decisions made by others and exploiting mechanisms such as the *the wisdom of the crowd* and *peer pressure.*

Another form in which income aggregators provide robo-advice for spending and saving decisions is through nudges, which are based on App notifications and reminders (Acquisti et al., 2017). Notifications and reminders from Apps are becoming ubiquitous and have proven useful in motivating individuals to stay active and eat healthy, among other outcomes. In the context of income aggregators, recent studies have documented the importance and effectiveness of these notifications. For example, Lee (2019) studies individuals' responses to overspending alerts, which are based on the robo-advising algorithm of an income aggregator that compares a users' own spending over time and identifies unusual patterns of spending within their spending history. Lee (2019) finds that users who receive overspending alerts reduce their spending 5.4% more than users who do not receive them. These changes in spending affect long-run cumulative spending. Lee (2019) also finds that the effect of nudges vary across the user population, with older, more financially-savvy, and more educated users adjusting their spending more after receiving overspending notifications, which suggests that more sophisticated users, rather than the least sophisticated, find notifications about their own unusual spending patterns useful. This result encourages further research on how robo-advising could be used to reach to the least sophisticated parts of the population, whose consumption, saving, and education choices tend to be stickier over time than those of the highly educated (D'Acunto, 2014).

Whereas the robo-advising income aggregators discussed so far provide advice on users' spending decisions, another class of robo-advisors target users' saving choices. Consumption and saving choices are obviously strongly interlinked, but the principles extant robo-advisors use to provide advice on these two dimensions are quite different. For example, Apps such as Acorn in the US and Gimme5 in Italy provide robo-advice to their users by helping them to set saving goals and reach such goals using nudges (Gargano and Rossi, 2020).

Goal setting exploits a behavioral mechanism that is not contemplated in standard life-cycle consumption-saving models. According to such models, agents should care about their overall savings but not about the specific objectives for which a certain amount is saved. This is because, for the most part, savings are fungible – they can be used for any purpose at any time (Browning and Crossley, 2001). However, setting budgets and goals is a common feature of agents' daily life, because as a large literature in experimental social psychology shows, agents are intrinsically motivated by goals and work hard to achieve them (Locke and Latham, 1991, 2002, 2006).

Using data from the robo-advisor for saving Gimme5, Gargano and Rossi

(2020) provide the first field analysis of whether goal-setting for specific savings objectives makes individuals save more. They establish a causal effect of goal setting on saving behavior using a formal identification strategy. Overall, Gargano and Rossi (2020) show that goal-setting leads the average user to increase monthly savings by 90%. They find that any goal, as long as it is stated, increases saving propensities, irrespective of the specific purpose of the goal. For instance, users that save for concrete objectives such as a trip or a car achieve their saving goals as often as those who set a generic saving objective without any concrete aims. Whereas goal concreteness seems irrelevant, the feasibility of the time deadline associated with the goal has an important impact on the effectiveness of goal setting on saving choices: users who set long-term deadlines are less likely to achieve their goals relative to users who set short-term deadlines.

The robo-advisors for spending and saving reported in this section focus on consumers' difficulties in computing the optimal spending and saving rate (D'Acunto et al., 2019b). Whereas the robo-advisor cannot compute such optimal ratios for the user, it can provide information, rules of thumb, nudges and reminders, as well as benchmarks in the form of aspirational goals to provide consumers with easy-to-grasp information about their optimal spending and saving rate.

2.2.1 Open areas of inquiry in robo-advising for consumption-saving choices

The potential applications and research questions in the space of robo-advising for spending and saving are many. The extant research discussed in this chapter has analyzed the effectiveness of robo-advice interventions based on the wisdom of the crowd and a set of psychological mechanisms, but many more forms of robo-advisors and mechanisms await to be studied by researchers in several disciplines.

For instance, scholars in finance, economics, marketing, decision science, and social psychology should study how existing mechanisms such as nudges based on one's own past spending behavior could be applied to the fast-growing area of digital-wallet apps (Agarwal and Qian, 2014). Currently, digital wallets, such as WeChat in China or Paytm in India act as instruments for managing households' liquidity. They are helpful insofar as they give households the possibility to engage in electronic payments without the need to rely on credit cards or other high-fee services provided by traditional financial intermediaries (Crouzet et al., 2019). Digital wallets, though, could be transformed into robo-advisors for spending and saving. The digital wallet might warn the user whenever he/she is engaging in anomalous spending or is spending on goods whose price is substantially higher than similar goods the user has purchased in the past.

The principle of social pressure and peer information could also be applied to many other designs of robo-advisors for spending and saving. For instance, developers could create a "FitBit for Finance," whereby users are connected to friends and peers they know in their real life and compete with these peers on achieving goals about spending and saving. This type of robo-advice would add a gamification aspect to the delivery of information about peers (Fitz-Walter et al., 2013; Sailer et al., 2017; Piteira et al., 2018). Gamification might add to peer information and peer pressure in further motivating users to put more effort into maintaining healthy spending and saving rates. This form of robo-advice – which, to the best of our knowledge, has not yet been implemented in the context of consumption and saving decisions – could be studied by scholars in disparate fields in terms of both providing the technical ability to implement such a strategy into apps as well as studying the effects of this form of robo-advice and its mechanisms, both in the laboratory and in the field.

Another direction that begets more research is the deepening of our understanding of the causes of adoption and effects of existing forms of robo-advising for spending and saving. For instance, existing research has not yet been able to assess the extent to which the effects of robo-advising in this context are long-lived. The main limitation to answering this question is that many Apps have only been released in the recent years. Moreover, the structure of Apps often changes over time and hence does not allow researchers to compare the behavior of agents who receive the same exact form of advice repeatedly over time. Also, the churning of users of robo-advising apps is quite high, which implies that within-agent studies of the effects of robo-advising over time are often hard to design with data from Apps in the extant literature. In this vein, further understanding the dimensions that predict adoption is important to ensure that categories that might tend to adopt robo-advising less but for whom the potential benefits from adoption might be high (e.g., see D'Acunto et al., 2021), are specifically targeted. Progress along any of these dimensions would be a crucial contribution to deepen our understanding of the effectiveness of robo-advising for saving and spending.

Finally, one aspect that needs further investigation is the potential pitfalls of robo-advising applications on consumption-saving choices. For instance, suppose that a user observes information about peer spending and saving in a domain in which the median household overspends. The user would infer that overspending is the norm and hence might adjust to that norm, which would worsen her ability to accumualte wealth for retirement. Other pitfalls might relate to the role of "gamification" applications, whereby robo-advisor developers aim to increase users' engagement with the app by creating competitions across users in terms of opening saving and brokerage accounts or new credit card accounts. These competitions might have opposite effects on users' outcomes. On the one hand, they might establish a virtuous circle in which each user tries to improve on the other in terms of saving and reducing spending, thus reinforcing the peer effect mechanism. On the other hand, gamification might bring users to cut on their spending excessively, given that the objective is not reaching a healthy saving share but winning over peers by increasing the saving share more and more than what the peers do in response. Whether gamification interventions improve users' outcomes or produce an excessive treatment effect of robo-advisors is an important open area of future research.

2.3 Robo-advising and durable spending choices

Whereas income aggregator robo-advisors focus on spending on non-durable goods and services, a substantial portion of households' balance sheet consists of durable goods, such as housing, cars, large furniture and electronic items (D'Acunto et al., 2022). Durable spending displays several features that make it different from non-durable spending as far as the scope for robo-advising is concerned. First, because durable goods provide consumption utility over time and often for several years after purchase, they resemble firms' fixed-asset investments and, contrary to non-durable goods, are often financed through consumption loans, credit card debt, or other forms of consumer debt. A robo-advisor that targets durable consumption choices should thus not only advise agents on the types of goods they should purchase but also on the optimal ways to finance such goods.

A second peculiar feature of durable spending that affects the design of robo-advising tools for durables is the fact that the choice of which durable goods to purchase involves more dimensions than the choice of non-durable goods. In the case of non-durable goods, price and quality are the most relevant features consumers consider in their purchase choices. In the case of durable purchases, instead, agents need to consider not only price and quality but also the good's depreciation, the tax implications of usage and depreciation over the years, as well as the costs of maintaining the good over time (Waldman, 2003). A robo-advisor for durable spending thus needs to provide agents with information and/or suggestions about all these aspects that are typically irrelevant for the case of non-durable choices.

In the rest of this section, we focus on existing robo-advising tools for two important durable purchases most households make – houses and cars – and we conclude by suggesting how robo-advising should evolve to adapt to other types of durable goods.

2.3.1 Robo-advising for housing choices

In the absence of robo-advising, house purchases entail multiple days spent with a real estate agent touring homes and discussing budgets. Part of the real estate agents' job is to understand the taste and preferences of their clients and help them navigate housing options that include multiple dimensions to be assessed. Real estate agents thus act as human advisors to prospective home owners.

Over the last few years, robo-advising tools for durable spending decisions have also emerged as an alternative to real-estate agents. For instance, Apps such as REDFIN and ZILLOW in the US fulfill the role of robo-advisors for durable spending based on the two directions discussed above: on the one hand, they provide information about a large set of dimensions agents need to consider when making housing decisions, such as the quality of nearby amenities, the quality of nearby public schools, the extent of walkability of neighborhoods, the crime rates and other characteristics of neighborhoods, and the price trends in various areas (Green and Walker, 2017, Eraker et al., 2017, Gargano and Giacoletti, 2020, Gargano et al., 2020). By providing information on all these dimensions in a concise and easy-to-grasp format, these robo-advisors for housing choices reduce the complexity of the multi-dimensional problem agents have to solve.

Moreover, housing Apps fulfil the second main feature of robo-advisors for housing decisions – they also provide information about the financing choices available to agents for each potential housing solution they might consider. Advice about financing options focuses on two features: (i) it simplifies agents' assessment and computation of the financial needs they might face for each housing option, and (ii) it helps agents compute the estimated monthly payments of mortgages with different characteristics (fixed rate vs. adjustable rate, different maturity options, conforming vs. non-conforming mortgages) (Karch, 2010). Moreover, some Apps also provide direct suggestions on actual options for mortgages from financial institutions for which they agents can apply online (Fuster et al., 2019), thus making the house purchase choice and its financing fully automated. The role of robo-advisors for financing housing solutions through mortgage advice is likely especially important for low- and middle-income households, for whom the supply of mortgage credit by traditional financial institutions has been declining consistently since 2010 (see D'Acunto and Rossi, 2022).

Academic research in the area of robo-advising for housing choices is still in its infancy. More work focusing on the integration of mortgage calculating services with the proposal of actual market offers on mortgages that have the characteristics users require is needed. Moreover, assessing the quality of advice and its effectiveness is crucial to understand the economic and psychological mechanisms behind these forms of robo-advice.

2.3.2 Robo-advising for the purchase of vehicles

The choice of purchasing vehicles and other durable goods, such as big furniture items, can be interpreted as a middle point between non-durable spending choices and durable spending choices as far as the advice to be produced by robo-advising tools is concerned. On the one hand, similar to housing choices, the purchase of vehicles typically needs to be financed. On the other hand, the dimensionality of the sets of characteristics agents have to consider when assessing the purchase of cars or other durables is substantially lower than for the case of housing. Whereas housing requires assessments about amenities, school districts, crime rates, and many other dimensions, cars and other durables are fully movable and hence their quality does not depend on any other dimension.

An example of an extant form of robo-advising for the purchase of vehicles are Apps such as TRUECAR and CARVANA (Garcia III et al., 2018). Similar to the other robo-advising tools we discussed in different domains, the first feature of these Apps is that they provide agents with easy-to-grasp information about otherwise complicated assessments, such as used car valuations as well as distributions of prices of similar cars that have transacted over time across different suppliers. Providing agents with this information abates their search costs and allows them to make informed decisions based on a large-scale number of transactions for similar goods, which would otherwise be impossible to observe given the high costs of obtaining data on individual car transactions for the average US consumer.

Moreover, even in the case of the purchase of vehicles, robo-advisors provide detailed information about financing options. The typical financing option for cars is a lease (Johnson et al., 2014). Robo-advisors for the purchase of vehicles provide agents with estimates and computations of monthly installments based on the maturity and size of the lease.

Despite their rising popularity in the US and abroad, robo-advisors for the purchase of vehicles have been barely studied in terms of their effects on consumers' choices.

2.3.3 *Open areas of inquiry in robo-advising for durable spending*

The study of the characteristics of robo-advising for durable spending is still in its infancy. A set of features that are unique to durable goods make several open questions in this area worth of inquiry by researchers.

First, the choice of durable good investments requires a multi-dimensional assessment of several characteristics at once, and hence is more complex than the choice of non-durable purchases. For this reason, consumers have traditionally relied on human advisors when assessing durable-good investments. An open area of inquiry is thus understanding to what extent robo-advisors for durable spending are complements or substitutes of traditional human advisers. On the one hand, the simplicity and affordability of robo-advisors could often allow agents to automate their choices fully and not resort to human advisers, such as real estate agent. At the same time, though, because the purchase of durable goods requires a substantial investment on the part of consumers as well as financial commitments that bind the consumer for years, consumers might prefer to still resort to a human adviser to at least check on the suggestions of the robo-advisor and validate its choices. The second option might be especially compelling if consumers displayed forms of distrust towards algorithms and finance (for instance, see Dietvorst et al., 2015; D'Acunto, 2020; but also Logg et al., 2019).

A second broadly open question for economists is the extent to which robo-advisors might modify the structure of market prices in markets where information about transaction prices is suddenly made easy to access and analyze on the part of retail consumers. Whereas the prices of housing transactions as well as those of vehicle transactions are in principle, in many cases, public, access to this information is prohibitively costly for the average US consumer. Sellers could therefore assess transaction prices more easily and meaningfully than buyers prior to the advent of robo-advising tools. This advantage of sellers has virtually disappeared since when any interested buyer, even those who have has never experienced the purchase of a durable good before, can simply and cheaply access a large amount of information about historical transaction prices from their phone.

2.4 Robo-advising and consumers' lending decisions

One of the most studied innovations associated with FinTech is the potential for banking disintermediation associated with peer-to-peer (P2P) lending. P2P lending fosters disintermediation in that consumers do not borrow from brick-and-mortar banks or online financial institutions, but from one individual or a pool of individuals who participate in a syndicated loan. P2P platforms connect borrowers and lenders directly and, depending on the borrowers' characteristics, set the rates and terms of the loans.

There are a number of P2P lending firms in the US, including Prosper, Lending-Club and Peerform among others. These platforms differ somewhat in the terms of the loans, eligibility criteria, but the underlying idea is to connect directly borrowers and lenders without the need of a traditional banking intermediary.

In its base implementation, P2P lending is unlikely to disrupt the banking system for a number of reasons (Balyuk and Davydenko, 2019). First, while the P2P platforms screen borrowers, individual lenders may not know how to construct a well-diversified portfolio of loans. Banks are able to diversify away the idiosyncratic risk of individual borrowers by issuing thousand of loans. On the other hand, investors lending a couple of thousand dollars on a lending platform may not realize that it is sub-optimal to lend to just a few borrowers, because of the relatively high probability of losing a large part of the investment. On the other hand, wealthy individuals that are willing to lend hundreds of thousands of dollars may find themselves in the impractical situation of having to manually select hundreds of loans to contribute to. Also, individual lenders may be subject to a number of biases and may therefore lend to individuals rather than others not because of their creditworthiness, but because of dimensions that affect their trust in the borrower (D'Acunto et al., 2020c), which might also include demographic characteristics such as their gender, race or other observable characteristics on the platform (Duarte et al., 2012).

Because of these limitations, P2P platforms have started to introduce automated lending portfolios for their investors who do not want to pick their investments manually. An example in the US is Lending Club, where its investors can choose a fully automated investment portfolio and a customized semi-automated investment portfolio rather than picking loans automatically. As the platform advertises, this allows its investors to generate well-diversified lending portfolios with the click of a button.

Another example is Faircent, a leading Indian P2P lending platform, which gives its investors access to a robo-advising tool named "Auto Invest." Lenders can adopt Auto Invest at any time. At the time of adoption, lenders choose how much of their wealth they want to allocate manually and how much they want to invest using Auto Invest. In addition, for the funds allocated to Auto Invest, lenders can allocate their wealth across six risk-based categories of borrowers.

The intent of choosing risk-based categories of borrowers is to mimic lenders' manual choices, because the six risk categories among which lenders allocate their funds on Auto Invest are the same risk categories they see as attached to borrowers once they appear in the pool.

D'Acunto et al. (2020a) provide a comprehensive analysis of the difference in performance between investors that lend manually on the platform and those who adopt the robo-advisor. They show that, before using the robo-advisor, investors tend to make rather poor investment decisions. For example, they lend to individuals of their own religion and shy away from lending to borrowers of different religions. They also lend to borrowers of higher social castes, such as the Brahmins, Kshatriyas, and Vaishyas at the expense of members of the Shudra caste, which traditionally were at the bottom of the social pyramid. The adoption of Auto Invest corrects these biases, evidenced by the fact that the proportion of loans issued by robo-advised investors across borrowers of different religions and castes reflect the respective proportions on the platform. Finally, D'Acunto et al. (2020a) show that correcting for these cultural biases improves investors' lending performance: robo-advised investors face 32% lower default rates and 11% higher returns on the loans they issue to borrowers who belong to favored demographic groups relative to available borrowers in discriminated groups.

The results in D'Acunto et al. (2020a) emphasize an important and often neglected role of robo-advising tools: they can eliminate biases in decision-makers' choices even in cases in which such biases are implicit, as is the case with ingrained cultural biases that affect decision-makers' choices under the form of rules of thumb in unfamiliar decision contexts (for instance, see D'Acunto et al., 2019c)

2.5 Areas of consumer finance with a scarce presence of robo-advising

So far, we have discussed several areas of consumer finance in which the use of robo-advising has been diffusing swiftly. Reviewing the peculiar features of each setting, which are often tailored to the characteristics of the decision-making problem agents need to solve, helps to take stock of our existing knowledge as well as to pave the way for future research endeavors in this area.

At the same time, the balance-sheet view of the household also includes several more types of households' assets and liabilities for which, so far, robo-advising applications are quite rare. In this section, we discuss these areas as well as the reasons why the problems households need to solve in these areas might also benefit from robo-advising applications. We hope that this discussion can influence both the development of robo-advising tools in these areas as well as the study of the adoption and effects of such tools on the quality of households' choices and their decision-making mechanisms.

2.5.1 Robo-advising and consumer credit management

A fundamental portion of household liabilities, especially in the US, is represented by consumer credit (Agarwal et al., 2007). Consumer credit is an important source of financing for non-durable and durable consumption for many US households and its take up follows some empirical regularities in observational data. First, typically less sophisticated households and low-income households tend to accumulate consumer credit debt on their balance sheets (Melzer, 2011; Chang, 2019). Second, the costs of this form of debt are typically substantial and often not fully transparent to non-expert households, which raises the issue of whether households that rely heavily on this form of debt understand its current and future costs fully (Brown et al., 2010).

Because of these two peculiar features of consumer credit, this area should represent an obvious arena in which robo-advising tools can be applied: for the first feature – high-cost debt taken up by unsophisticated households – robo-advising tool could provide simple rules of thumb to help households understand the trade offs of higher current consumption and higher future debt. For instance, inspired by the tools of robo-advising for the purchase of durable goods discussed above, robo-advising for consumer credit would provide automated calculators that allow households to assess the present value of their future debt debentures as well as the horizon of repayment based on the features of the consumer's credit card at hand, when considering whether to engage in a certain expense and after providing the maximum monthly payment the consumer is willing to face. Moreover, a robo-advising tool for consumer credit might monitor the credit options available on the market, e.g. credit card characteristics across financial institutions, and suggest that agents switch to alternative cards to reduce their APRs and annual fees. Robo-advising features similar to this last one have started to appear in some income-aggregating robo-advisors for spending, e.g. on Status Money.

The second peculiar feature of consumer credit is the lack of information and understanding about the costs of this form of credit, especially on the part of less sophisticated households. Even in this respect, robo-advising tools could provide more vivid information about, for instance, credit cards' APRs as well as shrouded attributes of credit cards and payday loan contracts. This function is likely to have a relevant impact on a household's understanding of the characteristics of consumer credit contracts, because research finds that, despite the mandated disclosure of credit-card characteristics, many consumers do not understand the implications of such features in terms of the cost of debt and the relationship between principal and interest in debt repayment (Salisbury and Zhao, 2020).

Despite representing such an obvious potential application for robo-advising tools, the extent to which such tools have been developed so far is scant. One obvious difference between this potential application of robo-advising and the applications that have obtained more diffusion so far lies in the incentives that supply-side actors have to provide the two forms of robo-advising. When it comes to investment allocation, financial institutions have a clear incentive to enlarge their pool of advisees – who pay fees on such advice, invest money in an institution's products, and are a target of cross-selling of other products by the institutions – to agents who would otherwise not participate in financial investments. Robo-advising for investment allocation thus provides financial institutions with a means to acquire customers that would have otherwise not been using an institution's services.

When it comes to consumer credit management, based on the features of this form of debt discussed above, financial institutions lack incentives to provide robo-advising services. Because of the high costs of this form of debt, their opacity, and the fact that it is often low-income and unsophisticated households who take up this type of product, financial institutions would only reduce their margins by providing borrowers with more transparent information on the costs of consumer credit and/or with strategies that would reduce the costs agents pay to access this form of debt (which represent financial institutions' revenues in this case). The lack of strong incentives on the supply side is likely to help explain why this component of households' balance sheets has seen fewer applications of robo-advising to date.

A potential solution to the misalignement of incentives in the introduction of robo-advising tools between consumers and financial institutions is the intervention of regulators. In terms of providing more easily accessible information about the characteristics of consumer credit contracts, regulators are already imposing disclosure requirements to financial institutions. However, those households who appear to rely substantially on credit card debt also happen to be households that do not understand the information disclosed to them. Because the objective of regulators is that information is understood by households and not just delivered in an incomprehensible format to households, a natural perspective for robo-advising is that regulators mandate financial institutions to provide information in the format of a robo-advising tool. For instance, instead of reporting the structure of interest rates households are required to pay if they accumulate debt (i.e., the typical structure of zero introductory APR and high APRs after a certain period of time) regulators could require institutions to introduce an automatic calculator that allows the household to input the amounts they want to borrow and the maximum monthly payment they are willing/able to make and delivers the present value of the debentures to the institution as well as the timing of full repayment of this debt based on households' inputs.

A second way to enhance consumers' understanding of their debt-management decisions with robo-advising is the introduction of "hints" that act as substitutes for households' (lack of) financial literacy. Indeed, most households lack an understanding of basic financial concepts such as the compounding of interest rates or the optimality of paying down debts subject to higher interest rates before other debt, all else equal. Whereas one solution to this problem would be requiring households to sit in financial literacy classes, this solution is extremely costly on the side of both households (both economically and cognitively) and regulators. Robo-advising would be a natural solution to this problem, because when households face several debts with different characteristics, they could be provided with rules of thumb based on basic financial principles. For instance, households would see a hint suggesting that "debts for which you pay the highest interest rate should always be paid before others."

Studying the design and effects of robo-advising for debt management and comparing the costs and benefits of this form of robo-advising with the costs and benefits of providing financial literacy content to consumers are wide open areas for future research and policy assessment.

2.5.2 *Robo-advising and human capital Investments*

A second area in which robo-advising applications are surprisingly lacking is the choice of financing human-capital investments. A notable example being the choice of how to finance children's college education.

Decisions about the financing of human capital investments have three peculiar features that make robo-advising applications particularly beneficial to households' decision-making. First, because in countries like the US the cost of higher education (e.g., college, professional schools, and master-level programs) is quite substantial, households who do not belong to the highest portions of the wealth distribution need to plan on financing options many years before the actual expense is incurred. Second, contrary to all other households' investments whose timing is endogenous and can be moved by households over time, the timing of expense of college tuitions and other college costs is pre-specified at the time of birth of the child and corresponds to the age of graduation from high school. A third aspect is that households face very different options to finance this expense. Whenever thinking about consumer credit contracts, agents would typically compare credit card accounts based on details of their characteristics, but the decision is among financial contracts that are similar. Instead, for the case of financing the higher education of their offspring, households need to compare as disparate options as college saving accounts that need to be built up for decades before the child becomes of age for college with, for instance, student loans that would need to be taken up at the time in which households face the actual college expense. And, adding to the complication of this problem, because most governments around the world recognize a positive externality to society from the increase of education levels of their citizens, some of these options are subsidized by governments but others are not.

As of the time of drafting this chapter, we are not aware of robo-advising applications that help households consider the alternative options for the financing of their child's higher education and to simplify the complex dynamic problem they need to solve to choose the best option. One direction robo-advising in this area could experiment is to simplify the comparison of choices across different horizons in terms of vividly representing these choices in terms of present value, so as to make the costs of each option easily comparable even though these costs would be paid by households at very different horizons.

2.5.3 Robo-advising and tax management

The last component of household balance sheets we discuss explicitly in this section, and which would represent an important potential area of application of robo-advising, is household tax management. Direct taxes, and especially income, wealth, utility, and estate taxes are an important liability that households face at pre-specified dates.

The peculiar characteristic of tax management that makes it viable for robo-advising applications is that minimizing tax debentures requires substantial and detailed institutional knowledge of the tax code which would be too costly for most households to build, both financially and cognitively. At the same time, because the institutional features of the tax code are pre-specified and do not require judgement based on preferences (they are not risky) or beliefs (they are not uncertain), optimization could be done virtually instantaneously by a well-designed algorithm.

And, indeed, tax-planning robo-advisors such as Turbotax, H&R Block, and TaxAct have quickly gained very large market shares around the world over the past two decades. Goolsbee (2004) provides an early assessment of the Turbotax from 2004, where the firm was not as sophisticated as it is today. Using a sample of 90,000 users, Goolsbee (2004) shows that Turbotax users have incomes 40% higher than non-users. They are also much more likely to have a retirement and a brokerage account. Finally, they are more technologically sophisticated than non-users.

Because the extent of digital literacy of the broader population has increased dramatically over the last two decades, the characteristics of users and non-users of robo-advisors for tax planning might be completely different at the present day. Updated research on these aspects is thus warranted.

Other open questions about this application of robo-advising relate to the quantification of the monetary and non-monetary benefit of using an robo-advisor like Turbotax, as opposed to having households filing their own taxes or through the more expensive services of a human accountants. Outcome variables that would provide a broad view of these pros and cons include the overall tax amount paid by otherwise similar households who use different types of services, the incidence of mistakes in tax reporting and fines households have to pay conditional on the occurrence of such mistakes, and the overall costs of using these different services, including the labor cost of the time households employ when filing their taxes individually.

Moreover, existing robo-advisors for tax management, especially in the US, are mainly focused on helping households decide whether they should file full requests of various forms of deductions or just pay the alternative minimum tax. At the same time, though, a broader robo-advisor for tax management would not only be used by households at the time of filing. Rather, such robo-advisor could be consulted by households throughout the fiscal year so as to obtain information about how certain expenses, if incurred, might be deducted as well as compare spending and investment options households face in their daily lives based on their tax implications at the end of the fiscal year. Indeed, most consumers tend to focus their attention on tax management only at the time of tax filing and robo-advisors might instead make consumers recognize that all their consumption and investment choices have implications in terms of direct taxation. We are not aware of robo-advisors for tax planning of this type and welcome their design, implementation, and study of their characteristics and effects of households' choices.

2.6 E pluribus unum: Is the Holistic Robo-Advisor the future of robo-advising?

So far, we have discussed a set of independent applications of robo-advising to various areas of households' balance sheets. Because most robo-advising applications are independently provided by supply-side operators, such as financial institutions and tax-management companies, the fact that different areas of households' balance sheet are the focus of separate and independent robo-advising tools and applications seems natural. Moreover, as we have argued in each of the previous sections, most components of households' balance sheets have peculiarities in terms of type of decision-making problems to be solved. We should thus not be surprised that the first attempts to introduce robo-advising tools would focus on providing tailored information and solutions for each of these problems separately.

And, yet, the balance-sheet view of the household we introduced at the beginning of this survey stresses an obvious route for the future of robo-advising: households do not need dozens of applications and tools to assist with one specific type of choice at once. Rather, the ideal robo-advisor for households, and especially for those with lower levels of sophistication, is a robo-advisor that provides a holistic approach to the dynamic optimization of households' balance sheets. Households need a "Holistic Robo-Advisor" that allows them to jointly optimize all the decision-making problems discussed in this survey.

Obviously, the conceptualization and realization of such a Holistic Robo-Advisor is extremely complicated. Although economic theory proposes rich models of dynamic optimization of households' consumption-saving life cycle choices, a unifying theory of households' balance-sheet management is still missing and perhaps will never exist. Whereas economic theory and psychology might thus inspire the design of robo-advising tools targeted at specific decision-making problems, as we have also emphasized when discussing each individual area of application of robo-advising, the possibility that economic theory or psychology might provide a unifying treatment of all these problems to inspire the design of a Holistic Robo-Advisor seems out of reach.

A direction that instead might be more promising for the realization of a Holistic Robo-Advisor is relying on data-based methods. In particular machine-learning techniques might allow robo-advising researchers and developers to train algorithms based on the observed joint choices of millions of households in the field across different parts of their balance sheets. Researchers and developers would first need to set criteria to assess the quality of joint households' choices; for instance, a bundle of life-time choices might be assessed based on the difference between the wealth accumulated up to retirement with the wealth needed for maintaining a household's standards of living for a certain period of time. Once the criteria are set, the choices of those households who perform better based on such criteria could be analyzed as desirable choices, whereas the choices of those households who perform worse as undesirable choices. Ultimately, this theory-free analysis of the data would thus allow to isolate "optimal" joint choices across various areas of households' balance sheets, which would represent the guiding principles for the design of a Holistic Robo-Advisor to households.

Although some recent commercial applications argue that they are already able to provide such a form of holistic robo-advice – for instance, see PEFIN in the US – the viability and effectiveness of these platforms has not yet been assessed empirically. And, yet, robo-advising will not be able to fully replace more expensive reliance on human advisors unless Holistic Robo-Advisors are produced that can fully replace the global financial-planning services human advisors currently provide.

2.7 Conclusions

This chapter argues that robo-advising, also knows as algorithmic advice, involves every aspect of the balance sheet of households, including, among others, consumption-saving choices, debt management, tax management and financing of human capital investments.

So far, the term robo-advisor has been mainly used to label the first form of robo-advice that has been developed in personal finance – robo-advisors for the management of household financial assets. The fact that this specific component of households' balance sheet has been the first target of robo-advising is not surprising, because the returns to providing advice in this realm, which mainly involves wealthy households, are higher than the returns to providing robo-advising for the optimization of other household-finance related choices.[1]

And, yet, all forms of algorithmic advice applied to household finance choices of any type represent robo-advising applications. Because different components of households' balance sheet require households to solve different types of optimization problems, this subchapter has reviewed the existing applications of robo-advising to alternative problems as well as the peculiar features of each robo-advising application based on the underlying mechanisms and common mistakes in households' decision making as documented in the field and the laboratory across several areas of research such as economics, finance, marketing, accounting, social psychology, and information systems.

---

[1] Moreover, note that many early applications of robo-advising to a household's financial portfolio allocation do not provide households with advice, which households can decide to follow, but manage a household's portfolios directly without barely any involvement on the side of the household. For this reason, a more appropriate label for what industry participants label robo-advisors would be "robo-managers."

D'Acunto, Francesco, Rauter, Thomas, Scheuch, Christoph K., and Weber, Michael. 2020b. Perceived precautionary savings motives: Evidence from fintech. National Bureau of Economic Research Tech. Report. Available at: `https://www.nber.org/papers/w26817`.

D'Acunto, Francesco, Xie, Jin, and Yao, Jiaquan. 2020c. Trust and contracts: empirical evidence. Available at SSRN 3728808.

D'Acunto, Francesco, Malmendier, Ulrike, and Weber, Michael. 2021. Gender roles produce divergent economic expectations. *Proceedings of the National Academy of Sciences*, **118**(21), 1–10.

D'Acunto, Francesco, Hoang, Daniel, and Weber, Michael. 2022. Managing households' expectations with unconventional policies. *Review of Financial Studies*, **35**(4), 1597–1642.

Dietvorst, Berkeley J, Simmons, Joseph P, and Massey, Cade. 2015. Algorithm aversion: People erroneously avoid algorithms after seeing them err. *Journal of Experimental Psychology: General*, **144**(1), 114.

Duarte, Jefferson, Siegel, Stephan, and Young, Lance. 2012. Trust and credit: The role of appearance in peer-to-peer lending. *Review of Financial Studies*, **25**(8), 2455–2484.

Eraker, David, Dougherty, Adam Michael, Smith, Edward M, and Eraker, Stephen. 2017 (Sept. 12). User interfaces for displaying geographic information. US Patent 9,760,237.

Fitz-Walter, Zachary, Tjondronegoro, Dian, and Wyeth, Peta. 2013. Gamifying everyday activities using mobile sensing. Pages 98–114 of: *Tools for Mobile Multimedia Programming and Development*. IGI Global.

Foerster, Stephen, Linnainmaa, Juhani T., Melzer, Brian T., and Previtero, Alessandro. 2017. Retail financial advice: does one size fit all? *Journal of Finance*, **72**(4), 1441–1482.

Fuster, Andreas, Plosser, Matthew, Schnabl, Philipp, and Vickery, James. 2019. The role of technology in mortgage lending. *Review of Financial Studies*, **32**(5), 1854–1899.

Garcia III, Ernest C., Behrens, Nicole, Swofford, Adam, and Adams, William. 2018 (July 19). Methods and Systems For Online Transactions. US Patent App. 15/924,084.

Gargano, Antonio, and Giacoletti, Marco. 2020. Cooling auction fever: Underquoting laws in the housing market. Working Paper. Available at SSRN 3561268.

Gargano, Antonio, and Rossi, Alberto G. 2020. Goal setting and saving in the fintech era. Working Paper. Available at SSRN 3579275.

Gargano, Antonio, Giacoletti, Marco, and Jarnecic, Elvis. 2020. Local experiences, attention and spillovers in the housing market. Working Paper. Available at SSRN 3519635.

Goolsbee, Austan. 2004. The turbotax revolution: Can technology solve tax complexity? Pages 124–147 in: *The Crisis in Tax Aadministration*, Brookings Institution Press.

Green, Joanna, and Walker, Russell. 2017. *Neighborhood watch: The rise of zillow*. Working Paper. Available at: `https://sk.sagepub.com/cases/neighborhood-watch-the-rise-of-zillow`.

Jansen, Mark, Nguyen, Hieu, and Shams, Amin. 2020. Human vs. machines: underwriting decisions in finance. Fisher College of Business Working Paper. Available at SSRN 3664708.

Johnson, Justin P., Schneider, Henry S., and Waldman, Michael. 2014. The role and growth of new-car leasing: Theory and evidence. *Journal of Law and Economics*, **57**(3), 665–698.

Karch, Marziah. 2010. Specialized apps for professionals. Pages 233–254 of: *Android for Work*. Springer.

Laibson, David I., Repetto, Andrea, Tobacman, Jeremy, Hall, Robert E., Gale, William G., and Akerlof, George A. 1998. Self-control and saving for retirement. *Brookings Papers on Economic Activity*, **1998**(1), 91–196.

Lee, Sung K. 2019. Fintech nudges: Overspending messages and personal finance management. NYU Stern School of Business Working Paper. Available at SSRN 3390777.

Linnainmaa, Juhani T., Melzer, Brian, and Previtero, Alessandro. 2021. The misguided beliefs of financial advisors. *Journal of Finance*, **76**(2), 587–621.

Locke, Edwin A., and Latham, Gary P. 1991. A theory of goal setting & task performance. *Academy of Management Review*, **16**(2), 480–483.

Locke, Edwin A., and Latham, Gary P. 2002. Building a practically useful theory of goal setting and task motivation: A 35-year odyssey. *American Psychologist*, **57**(9), 705.

Locke, Edwin A., and Latham, Gary P. 2006. New directions in goal-setting theory. *Current Directions in Psychological Science*, **15**(5), 265–268.

Logg, Jennifer M., Minson, Julia A., and Moore, Don A. 2019. Algorithm appreciation: People prefer algorithmic to human judgment. *Organizational Behavior and Human Decision Processes*, **151**, 90–103.

Lusardi, Annamaria, and Mitchell, Olivia S. 2007. Baby boomer retirement security: The roles of planning, financial literacy, and housing wealth. *Journal of Monetary Economics*, **54**(1), 205–224.

Melzer, Brian T. 2011. The real costs of credit access: Evidence from the payday lending market. *Quarterly Journal of Economics*, **126**(1), 517–555.

Mitchell, Olivia S., and Utkus, Stephen P. 2004. Lessons from behavioral finance for retirement plan design. Pages 3–41 in: *Pension Design and Structure: New Lessons from Behavioral Finance*, Oxford University Press.

Odean, Terrance. 1999. Do investors trade too much? *American Economic Review*, **89**(5), 1279–1298.

Olafsson, Arna, and Pagel, Michaela. 2017. *The ostrich in us: Selective attention to financial accounts, income, spending, and liquidity*. Working Paper. Available at SSRN 3057176.

Olafsson, Arna, and Pagel, Michaela. 2018. The liquid hand-to-mouth: Evidence from personal finance management software. *Review of Financial Studies*, **31**(11), 4398–4446.

Piteira, Martinha, Costa, Carlos J., and Aparicio, Manuela. 2018. Computer programming learning: how to apply gamification on online courses? *Journal of Information Systems Engineering and Management*, **3**(2), 11.

Reher, Michael, and Sokolinski, Stanislav. 2020. Automation and inequality in wealth management. Working Paper. Available at SSRN 3515707.

Rosenberg, Tina. 2011. *Join the Club: How Peer Pressure Can Transform the World*. WW Norton & Company.

Rossi, Alberto G., and Utkus, Stephen P. 2020a. The needs and wants in financial advice: Human versus robo-advising. Available at SSRN 3759041.

Rossi, Alberto G., and Utkus, Stephen P. 2020b. Who benefits from robo-advising? Evidence from machine learning. Working Paper. Available at SSRN 3552671.

Sailer, Michael, Hense, Jan Ulrich, Mayr, Sarah Katharina, and Mandl, Heinz. 2017. How gamification motivates: An experimental study of the effects of specific game design elements on psychological need satisfaction. *Computers in Human Behavior*, **69**, 371–380.

Salisbury, Linda Court, and Zhao, Min. 2020. Active choice format and minimum payment warnings in credit card repayment decisions. *Journal of Public Policy & Marketing*, **39**(3), 284–304.

Waldman, Michael. 2003. Durable goods theory for real world markets. *Journal of Economic Perspectives*, **17**(1), 131–154.

Zheng, X. 2020. How can innovation screening be improved? A machine learning analysis with economic consequences for firm performance. Working Paper. Available at SSRN 3845638.

3

Robo-Advising: Less AI and More XAI? Augmenting Algorithms with Humans-in-the-Loop

Milo Bianchi[a] and Marie Brière[b]

Abstract

We start by revisiting some key reasons behind the academic and industry interest in robo-advisors. We discuss how robo-advising could potentially address some fundamental problems in investors' decision making as well as in traditional financial advising by promoting financial inclusion, providing tailored recommendations based on accountable procedures, and, ultimately, by making investors better off. We then discuss some open issues in the future of robo-advising. First, what role Artificial Intelligence plays and should play in robo-advising. Second, how far should we go into personalization of robo-recommendations. Third, how trust in automated financial advice can be generated and maintained. Fourth, whether robots are perceived as complements to or substitutes for humans. We conclude with some thoughts on what the next generation of robo-advisors may look like. We highlight the importance of recent insights in Explainable Artificial Intelligence and how new forms of AI applied to financial services would benefit from importing insights from economics and psychology to design effective human-robo interactions.

Acknowledgments.

Milo Bianchi acknowledges funding from LTI@Unito, TSE Sustainable Finance Center and ANR (ANR-17-EURE-0010 grant). We have no material interest that relates to the research described in the chapter.

3.1 Introduction

Automated portfolio managers, often called robo-advisors, are attracting a growing interest both in academia and in the industry. In this chapter, we aim first at reviewing some of the reasons behind such a growing interest. We emphasize

---

[a] Toulouse School of Economics, TSM, and IUF, University of Toulouse Capitole
[b] Amundi, Paris Dauphine University, and Université Libre de Bruxelles
Published in *Machine Learning And Data Sciences For Financial Markets*, Agostino Capponi and Charles-Albert Lehalle © 2023 Cambridge University Press.

how robo-advising can be seen in the broader context of the so-called Fintech revolution. We also emphasize some more specific reasons of interest in automated financial advice, building on fundamental problems that individual investors face in taking financial decisions, and on the limits often observed in traditional financial advising.

We then discuss how robo-advising could potentially address these fundamental problems and highlight robots' main promises. First, promote financial inclusion by reaching under-served investors; second, provide tailored recommendations based on accountable procedures; and finally, make investors better off. For each of these, we revisit the reasons why some hope can be placed on robots and we take a stand on what the academic literature has shown so far.

In the third part of the chapter, we address what we believe are fundamental open issues in the future of robo-advising. First, we discuss what role Artificial Intelligence (AI) plays and should play. We stress constraints both in terms of regulatory challenges and in terms of conceptual advances of portfolio theory that may limit how much AI can be placed into robo-advising. We also stress how the quest for simplicity and explainablity in recommendations could make AI not desirable even if feasible. Second, we discuss how far we should go into personalization of the robo-recommendations, highlighting the trade-off between aiming at bringing the portfolio closer to the specific individual needs and the risks related to possible measurement errors of relevant individual characteristics (say, risk aversion) and to the sensitivity of algorithms to parameter uncertainty. Third, we discuss how robo-advising can shed light on the broader issues of the human-robo interactions and on the mechanics of trust in automated financial services. We revisit the arguments of algorithm aversion, and the possible ways to reduce it, and how those can be applied in the context of automated financial advice. Finally, we discuss some evidence of whether robots are perceived as complements or substitutes to human decisions.

We conclude with some thoughts on how the next generation of robo-advisors may look like. Rather than continuing on the trend of using more data, more complex models, and more automated interactions, we define an alternative path building on the key premises of robo-advisers in terms of increased accountability and financial inclusion, and on the key challenges of developing trust in financial technology. We highlight the importance of recent insights on XAI (i.e., Explainable Artificial Intelligence) and stress how new forms of AI applied to financial services can benefit from importing insights from social sciences such as economics and psychology.

This chapter does not aim at being exhaustive. Rather, it should be seen as complementary to existing reviews (such as D'Acunto and Rossi, 2020) and to the other chapters in this book.

3.2 Why so popular?

Robo-advisors use automated procedures, ranging from relatively simple algorithms using limited information on the client to artificial intelligence systems built on big data, with the purpose of recommending how to allocate funds across different types of assets. First, a client profiling technique is used to assess investors' characteristics (risk aversion, financial knowledge, horizon, ...) and goals. Second, an investment universe is defined and, third, a portfolio is proposed by taking into account investment goals and the desired risk level. As documented in Beketov et al. (2018), in most cases, the optimal portfolio builds on modern portfolio theory, dating back to Markowitz in 1952. In addition to recommending an initial allocation of funds, algorithms can be designed to continuously monitor portfolios and detect deviations from the targeted profile. Whenever deviations are identified, the client is alerted and/or the portfolio is automatically rebalanced. The portfolio can also be automatically rebalanced to reduce risks as time goes by or when the investor changes her risk tolerance or investment goals. Some robots also propose to implement "tax harvesting" techniques: selling assets that experience a loss and using the proceeds to buy assets with similar risk, which decreases capital gains and so taxable income without affecting the exposure to risk. Apart from the portfolio allocation, the robot can display statistics of interest to the client, such as the expected annual return and volatility, often by showing historical performances and Monte Carlo simulations of possible future realizations of the portfolio allocation.

The market is growing rapidly. Most practitioners estimate that the global market is today around $400–500bn, as compared to $100bn in 2016 (S&P Global Market Intelligence, Backend Benchmarking, Aite Group – see Buisson, 2019). Assets under management in the robo-advisors segment worldwide are projected to reach between $1.7trn and $4.6trn in 2022 (Statista, BI Intelligence). The number of users is expected to amount to 436M by 2024 (Statista 2020). This growth is driven by the entry of large incumbents in the digital service arena (for example, JPMorgan and Goldman Sachs announced the launch of a digital wealth management services for 2020) and the migration of assets managed by large financial institutions to their robo-advisors, which amounts to 8% of their AUM and to one-quarter of the assets in accounts with less than $1m. At the same time, clients have increased their demand for digital investment tools, and in particular for low-cost portfolio management and adjacent services such as financial planning. If the United States remains, by far, the leading market for robo-advising (more than 200 robo-advisors registered), the number of robo advisors is growing rapidly in Europe (more than 70), but also in Asia, driven by an emerging middle class and high technological connectivity (Abraham et al., 2019).[1]

We refer to Grealish and Kolm (2022) for more details on the functioning of robo-advisors and on recent market trends, and stress a few reasons which may motivate such a rapid market growth and increased interest in academia and policy circles.

---

[1] Robo-advisors are already present in China, India, Japan, Singapore, Thailand and Vietnam.

3.2.1 Fintech revolution

Part of the interest in robo-advising comes from the broader trend of applying new technologies and novel sources of data in the financial domain, a phenomenon often dubbed as fintech. The word has played a central role in many academic and policy debates in the past few years. Enthusiasts about fintech talk about a revolution that promises to disrupt and reshape the financial service industry.[2]

Buchanan (2019) discusses the global growth of the AI industry and of its application to the finance industry. Quoting a 2017 report, she mentions that 5,154 AI startups have been established globally during the past five years, representing a 175% increase relative to the previous 12 years. This impressive growth has been driven by the advances in computing power, leading to a decline in the cost of processing and storing data, and secondly, and the same time by the availability of data of increased size and scope. Similarly, AI related patent publications (denoted by the AI keyword) in the US have grown from around 50 in 2013 to around 120 in 2017. In China, such growth has been even more dramatic, with about in 120 patents in 2013 and 640 patents in 2018. Buchanan (2019) also discusses the broad range of ways in which AI is changing the financial services industry, not only in terms of robo-advising but also for fraud detection and compliance, chatbots, and algo trading.

In academic circles, the increased attention can be seen for example from the exponential growth in finance academic primarily centered around AI. Bartram et al. (2020) analyze the number of AI-related keywords in the title, abstract, or listed keywords of all working papers posted in the Financial Economics Network (FEN) between 1996 and 2018.[3] In 1996, no working paper with any AI-related keyword was uploaded, in 2018 the number of posted papers including such keywords were 410, accounting for 3% of all papers posted in 2018.

Robo-advisors promise to apply new technologies and procedures to improve financial decision making, as we discuss below, and as such they can be seen as a piece of the broader fintech revolution, just like digital currencies promise to redefine the role of traditional money and platform lending promises to redefine the role of traditional access to credit.

3.2.2 Fundamental problems with investors

From a more specific perspective, one key interest in robo-advising is that it is now commonly understood that many investors face ample margins of improvement in their financial decisions. In the past decades, the literature has documented various ways in which investors' decisions may deviate, sometimes in a fundamental way, from the standard premises of a fully rational economic agent, who knows the entire set of possible alternatives, the associated outcomes in a probabilistic sense,

---

[2] See e.g. The Economist (2015) on The Fintech Revolution or The World Economic Forum (2017) on Beyond FinTech: A pragmatic assessment of disruptive potential in financial service.

[3] Keywords included artificial intelligence, machine learning, cluster analysis, genetic algorithm or evolutionary algorithm, lasso, natural language processing, neural network or deep learning, random forest or decision tree, and support vector machine.

and can correctly match all her information in order to maximize her life-time utility.

Restricting to the investment domain, which has also been so far the typical focus of robo-advisors, investors have been found to display low participation (Mankiw and Zeldes, 1991), underdiversification (Grinblatt et al., 2011; Goetzmann and Kumar, 2008; Bianchi and Tallon, 2019), default bias (Benartzi and Thaler, 2007), portfolio inertia (Agnew et al., 2003; Bilias et al., 2010), excessive trading (Odean, 1999), trend chasing (Greenwood and Nagel, 2009), poor understanding of matching mechanism (Choi et al., 2009). Many of those investment behaviors are associated to a poor understanding of basic financial principles (Lusardi et al., 2017; Lusardi and Mitchell, 2014; Bianchi, 2018).

Several surveys provide a comprehensive list of biases and associated trading mistakes (see e.g. Guiso and Sodini, 2013, Barber and Odean, 2013, Beshears et al., 2018). For the purpose of this chapter, two points are worth stressing. First, these mistakes are not small; on the contrary, their welfare implications can be substantial (Campbell, 2006, Campbell et al., 2011). Second, they do not cancel out in equilibrium; rather, they have important effects on the functioning of financial markets and on broader macroeconomic issues such as wealth inequality (Vissing-Jorgensen, 2004; Lusardi et al., 2017; Bach et al., 2020; Fagereng et al., 2020).

Motivated by this evidence, it is clear that improving financial decision making can be seen as a major goal of financial innovation, and part of the interest in robo advising lies in its promise to help investors in these dimensions.

3.2.3 Fundamental problems with advisors

A natural response to investors' poor financial decision making is to delegate the task to professional experts, who have the time and skills to serve investors' best interest. The argument relies on a few important assumptions, which may sometimes be difficult to meet in practice. First, it is required that advisors are able to recognize and adapt their strategies to match their clients' preferences and needs. This is far from obvious, and recent evidence in fact suggests that investors may themselves have misguided beliefs. Foerster et al. (2017) analyze trading and portfolio decisions of about 10,000 financial advisors and 800,000 clients in four Canadian financial institutions. They show that clients' observable characteristics (risk tolerance, age, income, wealth, occupation, financial knowledge) jointly explain only 12% of the cross-sectional variation in clients' risk exposure. This is remarkably low, especially as compared to the effect of just being served by a given advisor, which explains 22% of the variation in a client's risk exposure. In terms of incremental explanatory power, adding advisor effects to a model in which investors' risk exposure is explained by their observable characteristics improves the adjusted R^2 goes from 12% to 30%. This evidence suggests that, in some cases, financial recommendations are closer to "one size fits all" than being fully tailored to a client's specific preferences and needs. Furthermore, Linnainmaa et al. (2021) show that some advisors, when trading with their own money, display very similar trading biases as their clients: they prefer active management, they chase returns, they are not well diversified.

A second key aspect is that advisors need to have the incentives to act in clients' best interests, rather than pursuing their own goals. Again, recent evidence suggests this need not be the case. Mullainathan et al. (2012) conducted a study by training auditors, posing as customers of various financial advisors, and (randomly) asking them to represent different investment strategies and biases. They show that advisors display a significant bias towards active management, they initially support clients' requests but their final recommendations are orthogonal to clients' stated preferences. At the end, advisors fail to correct client biases and even make clients worse off. Similarly, Foà et al. (2019) document banks' strategic behaviors in their offer of mortgage contracts. A more extensive review of advisors' conflicted advice is provided in Beshears et al. (2018).

A third key aspect is that, even abstracting from the previous concerns, financial advising is costly, and a significant part of the cost has a fixed component (say, the advisor's time). This implies that financial advice may not be accessible to investors with lower levels of wealth, who may in fact be those who need it the most.

3.3 Promises

3.3.1 Accountable procedures and tailored recommendations

Robo-advisors' services offer accountable procedure to allocate an individual's portfolio across various asset classes and different types of funds, depending on her individual characteristics. Two stages of the process are crucial to this: (1) client profiling; and (2) asset allocation. While tailored recommendations are offered to clients, there is considerable heterogeneity in the recommended allocations, and the exact algorithm used by robo-advisors is typically not transparent.

Client profiling

Robo-advisors typically use an online questionnaire to assess investor's financial situation, characteristics and investment goals. This questionnaire is a regulatory requirement under SEC guidelines in the US (SEC, 2006; SEC, 2019). A "suitability assessment" is also mandatory under MiFID (Markets in Financial Instruments Directive) regulation in Europe.[4]

Individual characteristics, such as age, marital situation, net worth, investment horizon, risk tolerance are used to assess the investor's situation. Interestingly, a large variety of questions can be used to estimate one particular characteristic. For

---

[4] In Europe, the MiFID regulation has set the objective of increased and harmonized individual investor protection, according to their level of financial knowledge. MiFID I (2004/39/3C), implemented in November 2007, requires investment companies to send their clients a questionnaire to determine their level of financial knowledge, their assets and their investment objectives. MiFID I has been replaced in January 2018 by MiFID II (2014/65/EU), which has demanded a strengthening of legislation in several areas, in particular in the requirements of advice independence and transparency (on costs, available offering, etc.).

example, if you consider risk tolerance, most of the robo-profilers use subjective measures of risk aversion based on a self-assessment. Some robo-profilers use risk capacity metrics (measuring the ability to bear losses), estimated from portfolio loss constraints, financial obligations or expenses, balance sheet information, etc. In Europe, under MiFID II, advisors should also assess the clients' "experience and knowledge" to understand the risks involved in the investment product or service offered.[5] Robo-advisors thus ask questions about the clients' financial literacy and reduce the individuals' risk tolerance when financial literacy is low.

Robo-advisors typically propose that clients pick a goal (for example, retirement, buying a house, a bequest to family members, a college/education fund, a safety net) among several possibilities during the risk profiling questionnaire. This goal can define the investment horizon or the risk capacity in the optimal portfolio allocation. Other robo-advisors allow their clients to name their goal before or outside of the risk profiling process, and do not necessarily incorporate it into the portfolio allocation. Finally, a few robo-advisors permit their clients to set multiple goals, thus offering their clients the ability to explicitly put their portfolio in a mental account (Das et al., 2010). One of their limitations is that robo-advisors frequently lack a global view on investor's overall financial situation, as savings outside of the robo platform are rarely taken into account. Some of them have a broader view of the client's financial situation through partnerships with financial account aggregators or digital platforms of investment.[6]

Asset allocation

In a second step, the robo-advisor proposes structuring a portfolio by taking into account investment goals and the desired risk level. Beketov et al. (2018) analyzed a set of 219 robo advisors from 28 countries (30% in the US, 20% in Germany, 14% in the UK), that were founded between 1997 and 2017. As shown in Figure 3.1, representing the word count of the occurrence of different methods within robo advisors, a large variety of portfolio construction techniques are used. Beketov et al. (2018) show that most robo advisors use simple Markowitz optimization or a variant of it such as Black–Litterman (40%), sample portfolios applying a pre-defined grid (27%) or constant portfolio weights (14%). A minority of robo advisors are using alternative portfolio construction techniques such as liability driven investment, full-scale optimization, risk parity, constant proportion portfolio insurance.

If most robo-advisors perform asset allocation with a mean–variance analysis or a variant of it, they rarely disclose information on how they chose their asset class investment universe or how they estimate variances and correlations between asset classes. They even more rarely disclose these expected return and risk parameters explicitly. Among the dominant players in the US, Wealthfront is probably one of the few exceptions. They disclose on their website their portfolio optimization method (Black–Litterman), but also their expected returns,

---

[5] see Article 25(3) and 56.

[6] For example, Wealthfront recently featured direct integrations with digital platforms of investment (Venmo, Redfin, Coinbase), lending (Lending Club) and tax calculation (turbotax).

Figure 3.1 Word count of the occurrence of different methods within the existing robo-advisors. *Source: Beketov et al. (2018)*

volatilities and correlation matrices and the way they estimated it.[7] Betterment is also relatively transparent. They provide justification and detail on the choice of their investment universe, their portfolio optimization method (Black–Litterman) and the way they calculated expected returns and risk, without disclosing them explicitly.[8] Schwab Intelligent Portfolios also disclose the portfolio optimization method, a variant of the Markowitz approach (using Conditional Value at Risk instead of the variance). However, they are less transparent on their Monte-Carlo simulation method and expected returns hypotheses.[9]

---

[7] https://research.wealthfront.com/whitepapers/investment-methodology/

[8] On the investment universe, they excluded asset classes such as private equity, commodities and natural resources, since "estimates of their market capitalization is unreliable and there is a lack of data to support their historical performance". Expected returns are derived from market weights, through a classical reverse optimization exercise that uses the variance–covariance matrix between all asset classes. An estimation of this covariance matrix is made using historical data, combined with a target matrix, and using the Ledoit and Wolf (2003) shrinkage method to reduce estimation error. Portfolios can also be tilted towards Fama and French (1993) value and size factors, the size of the tilt being freely parametrized by the confidence that Betterment has in these views. See https://www.betterment.com/resources/betterment-portfolio-strategy/#citations.

[9] They simulate 10,000 hypothetical future realizations of returns, using fat-tailed assumptions for the distribution of asset returns, also allowing for changing correlations modeled with a Copula approach. See https://intelligent.schwab.com/page/our-approach-to-portfolio-construction.

Heterogeneity in the proposed asset allocations

In theory, these rigorous procedures and their systematic nature should make it possible to overcome the shortcomings of human advisers, by reducing unintentional biases, and by simplifying the interaction with the client. Rebalancing is for example made easier through robo-advising platforms that implement it automatically or require a simple validation by the client. Also, if individual characteristics are measured with sufficient precision, robo-advising services should make it possible to offer investment recommendations that are tailored to each investor's situation.

In practice, a large disparity in the proposed asset allocations has been documented, for the same investor's profile. For example, Boreiko and Massarotti (2020) analyses 53 robo-advisors operating in the US and Germany in 2019. They show that a "moderate" profile invests in average 56% in equity, but the standard deviation of the proposed equity exposure is large (23%). Equity exposure can go from 14% to 100%, depending on the robo-advisor. Aggressive or conservative asset allocations have similar features, with an average equity exposure of 73% and 35% respectively, but a range between 18% and 100% for aggressive allocations, and from 0 to 100% for conservative allocations.

This disparity in the proposed allocations can have several sources. It could perhaps come from different portfolio construction methods or different expected risk/return hypotheses. It may also reflect robo-advisors' conflicts of interest. Boreiko and Massarotti (2020) show that the asset managers' expertise in a given asset class (proxied as the percentage of funds of a given asset class in the total universe of funds proposed by the robo-advisor) is the main driver. Conflicts of interests were also demonstrated in the case of Schwab Intelligent Portfolio, recommending that a significant portion of the clients' portfolio being invested in money market funds. Lam (2016) argued that this unusually large asset allocation to cash allowed Schwab Intelligent Portfolios to delegate cash management to Schwab Bank, allowing the firm to profit from the interest rate difference between lending rates and the paid rate of return (Fisch et al., 2019).

3.3.2 *Make investors better off*

As for many innovative financial services, a key promise of robo-advising is to make investors better off. This claim is obviously difficult to test, it requires having a good understanding of investors' preferences, constraints, and outside opportunities (say, how they would otherwise use the capital invested with the robo-advisor), as well as a complete picture on investors' assets. Moreover, even if one can have reasonable approximations on how investors trade off risk and returns, investors may care about other dimensions. For example, some investors may just use financial advice to acquire peace of mind. Gennaioli et al. (2015) propose a model in which a financial advisor acts as "money doctor" and allows investors to effectively decrease their reluctance to take risk. Rossi and Utkus

(2019a) document that acquiring peace of mind of one of the key driver of the demand for financial advice.

Most academic studies do not venture into developing a fully fledged welfare analysis, they take a more limited view and check whether having access to the robot increases investors returns, after having controlled for some measures of portfolio risk. Improvement along risk-adjusted returns can come from static changes in portfolio choices, for example by improving diversification and so allowing to reduce risk for a given level of expected returns. Or they may occur over time, by allowing investors to rebalance their portfolios in a way to stay closer to their target risk-return profile.

Recent academic studies document that robo-advising services tend to improve investors' diversification and risk-adjusted performance. For example, D'Acunto et al. (2019) study a portfolio optimizer targeting Indian equities, and find that robo-advice was beneficial to *ex ante* under diversified investors, by increasing their portfolio diversification, reducing their risk and increasing their *ex post* mean returns. However, the robo-advisor did not improve the performance of already-diversified investors. Rossi and Utkus (2019b) study the effects of a large US robo-advisor on a population of previously self-directed investors. They find that, across all investors, robo-takers reduced their money market investment and increased their bond holdings. Robo-introduction also reduced idiosyncratic risk by lowering the holdings of individual stocks and active mutual funds, and raising exposure to low-cost indexed mutual funds. It also reduced home bias by significantly increasing international equity and fixed income diversification. The introduction of the robot increased individuals' overall risk-adjusted performance. In a different sample, Reher and Sun (2019) also pointed to a diversification improvement of robo-takers generated by a large US robo-advisor. Bianchi and Brière (2020) study the introduction of a large French robo-advisor on employee savings's plans. They find that relative to self-managing, accessing the robo-services is associated to an increase in individuals' investment and risk-adjusted returns. Investors bear more risk, and rebalance their portfolio in a way to keep their allocation closer to the target. This increased risk taking is also found by Hong et al. (2020), studying a Chinese robo-advisor, and using unique account-level data on consumption and investments from Ant Group. Robo-adoption helped households to move toward optimal risk-taking, reducing their consumption volatility.

3.3.3 Reach under-served investors

One the most important promise of the fintech revolution is linked to financial inclusion. As mentioned, offering financial services often involves substantial fixed costs, which can make it unprofitable to serve poorer consumers. New technologies allow a dramatic decrease of transaction costs (Goldfarb and Tucker, 2019, identify various ways through which this could happen). By reducing these costs, new technologies may allow reaching those who have been traditionally under-served (Philippon, 2019).

Robo-advisors can be seen as part of this promise. First, they typically require lower initial capital to open an account. For example, Bank of America requires US$25,000 to open an account with a private financial advisor, but only US$5,000 to open an account with their robo-advisor. Some robo-advisors, such as Betterment, do not require a minimum investment at all. Second, they typically charge lower fees than human advisors. The automation of the advising process allows to reduce the advising fixed costs. For example, a fully automated robo-advisor typically charge a fee between 0.25% and 0.50% of assets managed in the US (between 0.25% and 0.75% in Europe),[10] whereas the fees for traditional human advisors hardly fall short of 0.75% and can even reach 1.5% (Lopez et al., 2015; Better Finance, 2020).

Academic studies on robo advising and financial inclusion are scarce, but the first results seem in line the above claims. Hong et al. (2020) show that the adoption of a popular fintech platform in China is associated with increased risk taking, and the effect is particularly large for households residing in ares with low financial service coverage. Reher and Sokolinski (2020) exploit a shock in which a major US robo-advisor reduced its account minimum from $5,000 to $500. They show that, thanks to this reduction, there is a 59% increase in the share of "middle class" participants (with wealth between $1,000 and $42,000), but no increase in participation by households with wealth below $1,000. The majority of new middle-class robo participants are also new in the stock market and, relative to upper class participants, they increase their risky share by 13 pps and their total return by 1.2 pps. Bianchi and Brière (2020) also show that robo participants increase their risk exposure and their risk adjusted returns. Importantly, the increase in risk exposure is larger for investors with smaller portfolio and lower equity exposure at the baseline, and the increase in returns is larger for smaller investors and for investors with lower returns at the baseline. These results suggest that having access to a robo advisor may be particularly important for investors who are less likely to receive traditional advice, and as such it can be seen as an important instrument towards financial inclusion.

3.4 Open questions

3.4.1 Why not more AI/big data?

As mentioned, most robo-advising today build on rather simple procedures both in terms of the information employed to profile the client and on how this information is used to construct the optimal portfolio. As emphasized in Beketov et al. (2018), modern portfolio theory remains dominant, forms of artificial intelligence are hardly employed. This may seem surprising given the increased interest in AI and Big Data mentioned above, and given that robo-advisors are often presented as incorporating those latest trends. One may wonder why we fail to see more AI built in robo-advising.

A first reason may be that, while such inclusion would be desirable, it is not

---

[10] We consider here management fees only, not underlying ETFs or funds' fees.

feasible due to technological or knowledge constraints. That is, finance theory has not advanced enough to be able to give recommendations on how to incorporate AI into finance models. Some scholars would not agree. Bartram et al. (2020) summarize the shortcomings of classical portfolio construction techniques and highlight how AI techniques improve the practice. In particular, they argue that AI can produce better risk-return estimates, solve portfolio optimization problems with complex constraints, and yield to better out-of-sample performance compared to traditional approaches.

A second reason may be that including more AI would violate regulatory constraints. According to the current discipline, as a registered investment advisor, a robo-advisor has a fiduciary duty to its clients. As discussed by Grealish and Kolm (2022), the fiduciary duty in the U.S. builds on the 1940 Advisers Act and it has been adapted by the SEC in 2017 so as to accommodate the specifics of robo-advising. In particular, robo-advisors are required to elicit enough information on the client, use properly tested and controlled algorithms, and fully disclose the algorithms' possible limitations.

Legal scholars debate on how much a robo-advisor can and should be subject to a fiduciary duty. Fein (2017) argues that robo-advisors cannot be fully considered as fiduciaries since they are programmed to serve a specific goal of the client, as opposed to considering her broader interest. As such, they cannot meet the standard of care of the prudent investor required to human advisers. Similarly, Strzelczyk (2017) stresses that robo-advisors cannot act as a fiduciary since they do not provide individualized portfolio analysis but rather base their recommendations on a partial knowledge of the client. On the other hand, Ji (2017) argues that robo-advisors can be capable of exercising the duty of loyalty to their clients so as to meet Advisers Act's standards. In a similar vein, Clarke (2020) argues that the fiduciary duty can be managed by basing recommendations on finance theory and by fully disclosing any possible conflict of interest.

A third reason may be that having more AI into robo-advising is simply not desirable. Incorporating AI would at least partly make robots a black-box, it would make it harder to provide investors with explanations of why certain recommendations are given. Patel and Lincoln (2019) identify three key sources of risk associated to AI applications: first, opacity and complexity; second, distancing of humans from decision making and third, changing incentive structures for example in data collection efforts. They consider the implications of these sources of risk on several domains, ranging from damaging trust in financial services, propagating biases, harming certain group of customers possibly in an unfair way. They also consider market level risks ranging from financial stability, cybersecurity, and new regulatory challenges.

Algorithm complexity could be particularly problematic in bad times. Financial Stability Board (2017) argues that the growing use of AI in financial services can threaten financial stability. One reason is that AI can create new forms of interconnectedness between financial markets and institutions, since for example various institutions may employ previously unrelated data sources. Moreover, the opacity of AI learning methods could become a source of macro-level risk due to their possibly unintended consequences.

Algorithm complexity is also particularly problematic for those with lower financial capabilities. Complex financial products have been shown to be particularly harmful for less sophisticated investors (see e.g. Bianchi and Jehiel (2020) for a theoretical investigation, Ryan et al. (2011) and Lerner and Tufano (2011) for historical evidence, and Célérier and Vallée (2017) for more recent evidence). As for many (financial) innovations, the risk is that they do not reach those who would need it the most, or that they end up being misused.

In this way, some key promises of robo-advising, notably on improved financial inclusion and accountability, can be threatened by the widespread use of opaque models.

3.4.2 *How far shall we go into personalization?*

The promise of robo-advisors is to combine financial technology and artificial intelligence and offer to each investor a personalized advice based on her objectives and preferences. One important difficulty lies in the precise measurement of investor characteristics. A second issue is related to the sensitivity of the optimal asset allocation to these characteristics, which can be subject to a large uncertainty. This can lead the estimated optimal portfolio to be substantially different from the true optimal one, with dramatic consequences for the investor.

Difficulty in measuring individual's characteristics

Lo (2016) calls for the development of smart indices, that would be tailored to individual circumstances and characteristics. If we are not there yet, robo-advisors can make a step in that direction, by helping to precisely define the investor's financial situation and goals (Gargano and Rossi, 2020). As has been demonstrated by considerable academic research, optimal portfolio choices rely on various individual characteristics such as human capital (Cocco et al., 2005; Benzoni et al., 2007; Bagliano et al., 2019), housing market exposure (Kraft and Munk, 2011), time preference, risk aversion, ambiguity aversion (Dimmock et al., 2016; Bianchi and Tallon, 2019), etc. The possibilities for personalization are much wider than what is currently implemented in robo-advisor services.

However, some individual characteristics are difficult to measure and subject to a large uncertainty. Risk aversion is one of them. Different types of methods have been developed by economists and psychologists to measure an individual's risk aversion. Most of them are experimental measurements based on hypothetical choices. For example, the lotteries of Barsky et al. (1997) offer individuals to choose between an employment situation with a risk-free salary, and a higher but risky salary. Other work (Holt and Laury, 2002; Kapteyn and Teppa, 2011; Weber et al., 2013) measure preferences based on a series of risk/return tradeoffs. The choice between a certain gain and a risky lottery is repeated, gradually increasing the prize until the subject picks one lottery.

One reason for this difficulty of measuring risk aversion might be that people interpret outcomes as gains and losses relative to a reference point and are more sensitive to losses than to gains. Kahneman et al. (1990) and Barberis et al. (2001) report experimental evidence of loss aversion. Loss aversion can also explain why many investors enjoy portfolio insurance products offering capital guarantees (Calvet et al., 2020).

In practice, robo-advisors frequently assess a client's risk tolerance based on a self-declaration. People are asked to rate themselves in their ability to take risks on a scale of 1 to 10 (Dohmen et al., 2005). These measures have the disadvantage of not being very comparable across individuals. Scoring techniques are also frequently used by robo-advisors. They propose to the individual a large number of questions of all kinds, covering different aspects of life (consumption, leisure, health, financial lotteries, work, retirement, family). Global scores are obtained by adding the scores on various dimensions, keeping only the questions which prove to be the most relevant ex-post to measure the individual's risk aversion, a statistical criterion eliminating the questions that contribute least (Arrondel and Masson, 2013).

In Europe, the implementation of MiFID regulation led to several academic studies assessing the risk profiling questionnaires. The European regulation does not impose a standardized solution, each investment company remains free to develop its questionnaire as it wishes, which explains the great heterogeneity of the questionnaires distributed in practice to clients. Marinelli and Mazzoli (2010) sent three different questionnaires used by banks to 100 potential investors to verify the consistency of the client risk profiles. Only 23% of individuals were profiled in a consistent way across the three questionnaires, a likely consequence of the differences in the contents and scoring methods of the questionnaires. Other work carried out in several European countries (De Palma et al., 2009; Marinelli and Mazzoli, 2010; Linciano and Soccorso, 2012) reach the same conclusion.

Algorithm sensitivity to parameter uncertainty

Optimal allocations are usually very sensitive to parameters (expected returns, covariance of asset returns) which are hard to estimate. They also depend crucially on an investor's characteristics (financial wealth, human capital, etc.) often known with poor precision. On the one hand, there is a cost for suboptimal asset allocations (one size does not fit all) and substantial gains to individualize (see Dahlquist et al., 2018; Warren, 2019). On the other hand, there is a risk of overreaction to extreme/time-varying individual characteristics, potentially leading to "extreme" asset allocations, as has been shown in the literature on optimization with parameter uncertainty (see for example Garlappi et al., 2007). Blake et al. (2009) claim that some standardization is needed, as in the aircraft industry, to guarantee investors' security. How much customization is needed depends largely on the trade-off between the gains in bringing the portfolio closer to the needs of individuals and the risks of estimating an individual's characteristics with a large error.

How stable are individual characteristics in practice also remains an open question. Capponi et al. (2019) show that if these risk profiles are changing through time (depending on idiosyncratic characteristics, market returns, or economic conditions), the theoretical optimal dynamic portfolio of a robo-advisors should adapt to the client's dynamic risk profile, by adjusting the corresponding intertemporal hedging demands. The robo-advisor faces a trade-off between receiving client information in a timely manner and mitigating behavioral biases in the risk profile communicated by the client. They show that with time-varying risk aversion, the optimal portfolio proposed by the robo-advisor should counters the client's tendency to reduce market exposure during economic contractions.

3.4.3 Can humans trust robots?

In the interaction between humans and robo-advisors, a key ingredient is trust, determining the individual's willingness to use the service and to follow the robo recommendations. We review what creates trust in algorithms and discuss the impact of trust on financial decisions.

Trust is key for robo adoption

Trust has been demonstrated to be a key driver of financial decisions (see Sapienza et al., 2013 for a review). For example, trusting investors are significantly more likely to invest in the stock market (Thakor and Merton, 2018). Trust is also a potential key driver of robo-advisor adoption. As stated by Merton (2017), "What you need to make technology work is to create trust."

Trust has been studied in a variety of disciplines, including sociology, psychology and economics, in order to understand how humans interact with other humans, or more recently with machines. Trust is a "multidimensional psychological attitude involving beliefs and expectations about the trustee's trustworthiness, derived from experience and interactions with the trustee in situations involving uncertainty and risk" (Abbass et al., 2018). One can also see trust as a transaction between two parties: if A believes that B will act in A's best interest, and accepts vulnerability to B's actions, then A trusts B (Misztal, 2013). Importantly, trust exists to mitigate uncertainty and risk of collaboration by enabling the trustor to anticipate that the trustee will act in the trustor's best interests.

While trust has both cognitive and affective features, in the automation literature, cognitive (rather than affective) processes seem to play a dominant role. Trust in robots is multifaceted. It has been shown to depend on robot reliability, robustness, predictability, understandability, transparency, and fiduciary responsibility (Sheridan, 1989; Sheridan, 2019; Muir and Moray, 1996). One key feature of robo-advisors is their reliance on more or less complicated algorithms in several steps of the advisory process. An algorithm is used to profile the investor, and then to define the optimal asset allocation. A client delegating the decision to the robot bears the risk that a wrong decision by the robot will lead to poor performance of her savings. Trust in these algorithms is thus key for robo-advisor adoption.

Algorithm aversion

Survey evidence (HSBC, 2019) shows that there is a general lack of trust in algorithms. While most people seem to trust their general environment and the technology (68% of the survey respondents said they will trust a person until proved otherwise, 48% believe the majority of people are trustworthy and 76% that they feel comfortable using new technology), artificial intelligence is not yet trusted. Only 8% of the respondents would trust a robot programmed by experts to offer mortgage advice, compared to 41% trusting a mortgage broker. As a comparison, 9% would be likely to use a horoscope to guide investment choices! 14% would trust a robot programmed by leading surgeons to conduct open heart surgery on them, while 9% would trust a family member to do operation supported by a surgeon. Only 19% declare they would trust a robo-advisor to help make choices in investment. There are large differences across countries however. The percentage of respondents trusting robo-advisors rises to 44% and 39% in China and India respectively, it is only 9% and 6% in France and Germany.

Some academic studies have shown that decision makers are often averse to using algorithms, most of the time preferring less accurate human judgments. For example, professional forecasters have been shown to fail to use algorithms or give them insufficient weight (Fildes and Goodwin, 2007). Dietvorst et al. (2015) gave participants the choice of either exclusively using an algorithm's forecasts or exclusively using their own forecasts during an incentivized forecasting task. They found that most participants chose to use the algorithm exclusively when they had no information about the algorithm's performance. However, when the experimenter told them it was imperfect, they were much more likely to choose the human forecast. This effect persisted even when they had explicitly seen the algorithm outperform the human's forecasts. This tendency to irrationally discount advice that is generated and communicated by computer algorithms has been called "algorithm aversion". In a later experimental study (Dietvorst et al., 2018), participants were given the chance to modify the algorithm. Participants were considerably more likely to choose the imperfect algorithm when they could modify its forecasts, even if they were severely restricted in the modifications they could make. This suggests that algorithm aversion can be reduced by giving people some control over an imperfect algorithm's forecast.

Recent experimental evidence shows less algorithm aversion. Niszczota and Kaszás (2020) tested if people exhibited algorithm aversion when asked to decide whether they would use human advice or an artificial neural network to predict stock price evolution. Without any prior information on the human vs robot performance, they find no general aversion towards algorithms. When it was made explicit that the performances of the human advisor was similar to that of the algorithm, 57% of the participants showed a preference for the human advice. In another experiment, subjects were asked to choose a human or robo-advisor to exclude stocks that were controversial. Interestingly, people perceived algorithms as being less effective than humans when the tasks require to make a subjective judgment, such as morality.

Germann and Merkle (2019) also find no evidence of algorithm aversion. In a laboratory experiment (mostly based on business or economics' students), they asked participants to choose between a human fund manager and an investment algorithm. The selection process was repeated 10 times, which allowed to study the reaction to the advisor's performance. With equal fees for both advisors, 56% of participants decided to follow the algorithm. When fees differed, most participants (80%) chose the advisor with the lower fees. Choices were strongly influenced by the cumulative past performance. But investors did not lose confidence in the algorithm more quickly after seeing forecasting errors. An additional survey provided interesting qualitative explanations to their results. Participants believed in the ability of the algorithm to be better able to learn than humans. They viewed humans as having a comparative advantage in using qualitative data and dealing with outliers. All in all, the algorithms are viewed as a complement rather than a competitor to a human advisor.

What creates trust in algorithm?

Jacovi et al. (2020) distinguish two sources of trust in algorithms: intrinsic and extrinsic. Intrinsic trust can be gained when the observable decision process of the algorithm matches the user priors. Explanations of the decision process of the algorithm can help creating intrinsic trust.[11] Additionally, an algorithm can become trustworthy through its actual behavior: in this case, the source of trust is not the decision process of the model, but the evaluation of its output.

The European (Commission, 2019) recently listed a number of requirements for trustworthy algorithms. Related to intrinsic trust are the requirements of (1) user's agency and human oversight, (2) privacy and data governance, (3) transparency and explainability of the algorithm. Extrinsic trust can be increased by (4) the technical robustness and safety of the algorithm, (5) the interpretability of its output, (6) its accountability and auditability. In addition, ethical and fairness considerations such as (7) avoiding discrimination, promoting diversity and fairness or (8) encouraging societal and environmental well-being are also considered as being key components of trust.

Trust in the algorithms also crucially depends on the perception of the expertise and reliability of the humans or institutions offering the service (Prahl and Van Swol, 2017). "Technology doesn't create trust on its own" (Merton, 2017). People trust humans certifying a technology, not necessarily the technology itself. In the specific case of robo advice, Lourenço et al. (2020) study consumer decision to adopt the service and show that this decision is clearly influenced by the for-profit vs. not-for-profit orientation of the firm offering the service (for example private insurance and investment management firm vs. pension fund or government sponsored institution). Transparency, explainability and interpretability may not by itself be sufficient for enhancing decisions and increasing trust. However, informing about key hypothesis and potential shortcomings of the

---

[11] For example, a robo-advisor may disclose its risk profiling methodology, its optimization method and risk/return hypotheses, or reveal the signals leading to portfolio rebalancing.

algorithms when making certain decisions, might be an fundamental dimension to be worked on.

Trust in robots and financial decisions

Not everyone trusts robot-advisors. In a sample of 34,000 savers in French employee savings plans, Bianchi and Brière (2020) document that individuals who are young, male, and more attentive to their saving plans (measured by the time spent on the savings plan website), have a higher probability of adopting a robo-advising service. The probability of taking up the robot is also negatively related to the size of the investor portfolio, which suggests that the robo-advisor is able to reach less wealthy investors,[12] a result also confirmed by Brenner and Meyll (2020). Investors with smaller portfolios are also more likely to assign a larger fraction of their assets to the robot.

A unique feature of the robo-service analyzed by Bianchi and Brière (2020) allows them to analyze both "robo-takers" and "robo-curious," i.e., individuals who get to the point of observing the robot's recommendation without eventually subscribing to it. Interestingly, the further away is the recommendation of the robot relative to the current allocation, the larger is the probability that the investor subscribes to the robot. This finding can be contrasted with the observation that human advisers tend to gain trust from their clients by being accommodating with clients (Mullainathan et al., 2012). Moreover, investors who are younger, female, those who have larger risk exposure and lower past returns as well as less attentive investors are more likely to accept a larger increase in their exposure to risky assets, such as equities.

Trust can have a large impact on investor decisions. Bianchi and Brière (2020) and Hong et al. (2020) provide evidence of increased risk taking, a result consistent with increased trust. For example, Bianchi and Brière (2020) document a 7% increase in equity exposure after robo adoption (relative to an average 16% exposure). Hong et al. (2020) document a 14% increase (relative to an average risky exposure of 37% on their sample of 50,000 Chinese consumers clients of Alibaba). Interestingly, Hong et al. (2020) additionally show that this result is likely not driven by an increase in an individual's risk tolerance driven by robo support. Rather, it seems to reflect a better alignment of the investment portfolio with the actual risk tolerance of the individual. In particular, they show that after robo adoption, exposure to risky assets is more in line with an individual's risk tolerance estimated from the individual's consumption growth volatility (Merton, 1971), measured from Alibaba's Taobao online shopping platform. The robo-advisor seems to help individuals move closer to their optimal alignment of risk-taking and consumption. These results should however be taken with caution, as both studies concentrate on a relatively short period of investment (absent very serious market crash) and lack a global view on the individual's overall portfolios. More work would need to be done to document a long term impact.

---

[12] Conversely, wealthier investors are more likely to acquire information about the robot without subscribing to the service.

3.4.4 Do robots replace or complement human decisions?

Autonomous systems are developing in large areas of our everyday life. Under-standing how humans will interact with them is a key issue. In particular, should we expect that robots will become substitutes for humans or rather complements? In the special case of financial advice, are they likely to replace human advisors?

Using a representative sample of US investors, Brenner and Meyll (2020) investigate whether robo-advisors reduce investor demand for human financial advice offered by financial service providers. They document a large substitution effect and show that this effect is driven by investors who fear being victimized by investment fraud or worried about potential conflicts of interest. In practice however, a number of platforms that were entirely digital decided to reintroduce human advisors. For example, Scalable Capital, the European online robo-advice company backed by BlackRock, or Nutmeg, reintroduced over-the-phone and face-to-face consultations after finding that a number of clients preferred talking to human advisors rather than answering online questionnaires alone.

Another related question is about understanding how people will interact with robots. Will they delegate the entire decision to the robot or will they keep an eye on it, monitoring the process and intervening if necessary? In certain experiments, users put too much faith in robots. Robinette et al. (2016) designed an experiment where participants were asked to choose whether or not to follow the robot's instructions in an emergency. All participants followed the robot in the emergency, even if half of them observed the same robot perform poorly in a non-emergency navigation guidance task just a few minutes before. Even when the robot pointed to a dark room with no discernible exit the majority of people did not choose to safely exit the way they entered. Andersen et al. (2017) expand on this work and show that such an excess of trust can also affect human–robot interactions that are outside an emergency setting.

In the context of financial decisions, Bianchi and Brière (2020) document that robo-advisor adoption leads to a significant increase in attention in the savings plan, in the months following the adoption. Individuals are in general more attentive to their saving plan, and in particular when they receive their variable remuneration and need to take an investment decision. This seems to indicate that people do not take the robot as a substitute for their own attention.

3.5 The next generation of robo-advisors

It is not clear which generation of robo-advisors we are currently facing. Beketov et al. (2018) focus on robots of third and fourth generation, which differ from earlier generations as they use more automation and more sophisticated methods to construct and rebalance the portfolios. One possibility is just that the next generation of robots would continue on the trend of using more data and more complex models.

One may however imagine an alternative path. As discussed above, incorpo-rating more AI into robo-advising (and more generally into financial services) faces three key challenges. First, while highly personalized asset allocations have the great potential of accommodating individuals' needs, they are also more exposed to measurement errors of relevant individual characteristics and to parameter uncertainty. Second, to the extent that increased AI would be associated to increased opacity, we should be careful of not missing some key promises of inclusion. Third, trust is key for technology adoption, even more so in the domain of financial advice. These challenges call in our view for devising algorithms that can be easily interpreted and evaluated. Toreini et al. (2020) discusses how developing trust in (machine learning) technologies requires them to be fair, explainable, accountable, and safe (FEAS).

Under this perspective, recent advances in so-called XAI, i.e., explainable artificial intelligence, can be particularly useful when thinking about the future of robo-advisors. Explainability refers first to the possibility to explain a given prediction or recommendation, even if based on a possibly very complicated model, for example by evaluating the sensitivity of the prediction when changing one of the inputs. Second, it refers to how much a given model can itself be explained. Explanations can help humans in performing a given task and at the same time in evaluating a given model (see e.g. Biran and Cotton, 2017 for a recent survey). As discussed in Doshi-Velez and Kim (2017), explainability can be considered a desiderata both in itself, in relation to the issues of trust and accountability expressed above, and also as a tool to assess whether other desiderata, such as fairness, privacy, reliability, robustness, causality, usability, are met.

There is a large academic literature testing whether explainable artificial intelligence can improve human decision-making. How much explanation is needed of the actual functioning of an automated system remains an open question, and it is especially debated for example in the context of self-driving cars. On the one hand, psychological research on decision-making suggests that when decisions involve complex reasoning, ignoring part of the available information and using heuristics, can help in dealing more robustly with uncertainty than simply relying on resource-intensive processing strategies (Gigerenzer and Brighton, 2009). On the other hand, experimental studies show that providing the driver with information on why and how an autonomous vehicle acts, is important for safe driving (Koo et al., 2015). This information is particularly key in emergency situations. Drivers receiving such information tend to trust the car less and are quicker to take control of the car when a dangerous situation occurs (Helldin et al., 2013). One should also be particularly attentive to the risk of information overload. An algorithm is easier to interpret and to use when it focuses on a few features, it is also easier to correct in instances of error (Poursabzi-Sangdeh et al., 2018).

In the context of robo-advisors, explainability is not an easy task. Evaluating the performance of a robo-recommendation is not straightforward, especially if one uses AI to move towards fully personalized allocations to be confronted against fully personalized benchmarks (as described in Lo (2016)). Even more difficult for the client is to build counterfactuals on performance. And probably even more difficult is to appreciate the underlying finance model which governs the algorithm, especially if one wishes to serve less experienced investors.

In that respect, the quest is not for full transparency on the potentially complicated algorithm underlying the robo-advising process, disclosing for example all the details on the portfolio optimization method or the covariance matrix estimates. It would probably be more effective to disclose for example in which economic scenarios the algorithm might perform less accurately, possibly proving ex-post sub-optimal, and informing clients about the potential limitations of the algorithm.

Another potentially interesting development would be to strengthen the interactions with clients. For example, some robo-advisors send alerts when a client's portfolio deviates significantly from the target asset allocation (see e.g. Bianchi and Brière, 2020). These alerts could be seen also as an opportunity to interact with the client. Alerts could for example be used to explain why the deviation occurred (market movements, change in personal characteristics, etc.) and a given rebalancing is recommended. As another example, one could elicit customer perceptions on the quality of the response provided by the algorithm and integrating this feedback as part of the evaluation of the robo-service (Dupont, 2020).

These issues are not new in AI. Biran and Cotton (2017) discusses earlier approaches of explainability of decisions in expert systems in the 1970's and more recently in recommender systems. One may argue, however, that probably today models are more complex, more autonomous, and they span a larger set of decisions on a larger set of agents (including possibly less sophisticated one), which make these issues particularly relevant in current debates. Indeed, improving transparency is central also in the policy domain, such as in the recent EU regulation on data protection (GDPR). As discussed in Goodman and Flaxman (2017), the law defines a right to explanation, whereby users can inquire about the logic involved in an algorithmic decision affecting them (say, through profiling), and this calls for algorithms which are as much explainable as they are efficient.

Some prominent scholars argue that the AI revolution has not happened yet. Instead of better mimicking human interactions or most sophisticated human thinking, the AI revolution will happen when new forms of intelligence will be considered (Jordan, 2019a). In this effort, importing insights from social sciences seems crucial. AI needs psychology to capture how humans actually think and behave, or, to say it with Lo (2019), to include forms of "artificial stupidity." Insights from philosophy, psychology, and cognitive sciences are also key in informing how explanations are and should be communicated. Miller (2019) reviews the large literature in these fields and emphasizes the importance of providing selective explanations, based on causal relations and counterfactuals rather than on likely statistical relations, and of allowing a social dimension in which explainers and "explainees" may interact. AI also needs economics not only to help addressing causality and discussing counterfactuals, but also to help designing new forms of collective intelligence. These new forms may go beyond a purely anthropocentric approach, and build on some understanding of how markets functions and how they may fail (Jordan, 2019b). We share the enthusiasm of these scholars when imagining advances in these directions, we look forward to seeing more social sciences in the next generation of robo-advisors!

References

Abbass, Hussein A., Scholz, Jason, and Reid, Darryn J. 2018. *Foundations of Trusted Autonomy*. Springer Nature.

Abraham, Facundo, Schmukler, Sergio L., and Tessada, Jose. 2019. Robo-advisors: Investing through machines. *World Bank Research and Policy Briefs*.

Agnew, Julie, Balduzzi, Pierluigi, and Sunden, Annika. 2003. Portfolio choice and trading in a large 401(k) plan. *American Economic Review*, **93**(1), 193–215.

Andersen, Kamilla Egedal, Köslich, Simon, Pedersen, Bjarke Kristian Maigaard Kjær, Weigelin, Bente Charlotte, and Jensen, Lars Christian. 2017. Do we blindly trust self-driving cars. Pages 67–68 of: *Proc. Companion of the 2017 ACM/IEEE International Conference on Human-Robot Interaction*.

Arrondel, Luc, and Masson, André. 2013. Measuring savers' preferences how and why? `https://halshs.archives-ouvertes.fr/halshs-00834203`.

Bach, Laurent, Calvet, Laurent E., and Sodini, Paolo. 2020. Rich pickings? Risk, return, and skill in household wealth. *American Economic Review*, **110**(9), 2703–47.

Bagliano, Fabio C., Fugazza, Carolina, and Nicodano, Giovanna. 2019. Life-cycle portfolios, unemployment and human capital loss. *Journal of Macroeconomics*, **60**, 325–340.

Barber, Brad M., and Odean, Terrance. 2013. The behavior of individual investors. Pages 1533–1570 of: *Handbook of the Economics of Finance*, vol. 2. Elsevier.

Barberis, Nicholas, Huang, Ming, and Santos, Tano. 2001. Prospect theory and asset prices. *Quarterly Journal of Economics*, **116**(1), 1–53.

Barsky, Robert B., Juster, F. Thomas, Kimball, Miles S., and Shapiro, Matthew D. 1997. Preference parameters and behavioral heterogeneity: An experimental approach in the health and retirement study. *Quarterly Journal of Economics*, **112**(2), 537–579.

Bartram, Söhnke M., Branke, Jürgen, and Motahari, Mehrshad. 2020. Artificial Intelligence in Asset Management. `https://www.cfainstitute.org/-/media/documents/book/rf-lit-review/2020/rflr-artificial-intelligence-in-asset-management.pdf`

Beketov, Mikhail, Lehmann, Kevin, and Wittke, Manuel. 2018. Robo advisors: quantitative methods inside the robots. *Journal of Asset Management*, **19**(6), 363–370.

Benartzi, Shlomo, and Thaler, Richard. 2007. Heuristics and biases in retirement savings behavior. *Journal of Economic perspectives*, **21**(3), 81–104.

Benzoni, Luca, Collin-Dufresne, Pierre, and Goldstein, Robert S. 2007. Portfolio choice over the life-cycle when the stock and labor markets are cointegrated. *Journal of Finance*, **62**(5), 2123–2167.

Beshears, John, Choi, James J., Laibson, David, and Madrian, Brigitte C. 2018. Behavioral household finance. Pages 177–276 of: *Handbook of Behavioral Economics: Applications and Foundations*, vol. 1. Elsevier.

Bianchi, Milo. 2018. Financial literacy and portfolio dynamics. *Journal of Finance*, **73**(2), 831–859.

Bianchi, Milo, and Brière, Marie. 2020. Robo-advising for small investors. *Amundi Working Paper*. `https://research-center.amundi.com/article/robo-advising-small-investors-evidence-employee-savings-plans`.

Bianchi, Milo, and Jehiel, Philippe. 2020. Bundlers' dilemmas in financial markets with sampling investors. *Theoretical Economics*, **15**(2), 545–582.

Bianchi, Milo, and Tallon, Jean-Marc. 2019. Ambiguity preferences and portfolio choices: Evidence from the field. *Management Science*, **65**(4), 1486–1501.

Bilias, Yannis, Georgarakos, Dimitris, and Haliassos, Michael. 2010. Portfolio inertia and stock market fluctuations. *Journal of Money, Credit and Banking*, **42**(4), 715–742.

Biran, Or, and Cotton, Courtenay. 2017. Explanation and justification in machine learning: A survey. Pages 8–13 of: *Proc. IJCAI-17 Workshop on Explainable AI (XAI)*.

Blake, David, Cairns, Andrew, and Dowd, Kevin. 2009. Designing a defined-contribution plan: What to learn from aircraft designers. *Financial Analysts Journal*, **65**(1), 37–42.

Financial Stability Board. 2017. *Artificial Intelligence and machine learning in financial services: Market developments and financial stability implications*. https://www.fsb.org/wp-content/uploads/P011117.pdf.

Boreiko, Dmitri, and Massarotti, Francesca. 2020. How risk profiles of investors affect robo-advised portfolios How risk profiles of investors affect robo-advised portfolios. *Frontiers in Artificial Intelligence*, **3**, 60.

Brenner, Lukas, and Meyll, Tobias. 2020. Robo-advisors: A substitute for human financial advice? *Journal of Behavioral and Experimental Finance*, **25**, 100275.

Buchanan, Bonnie. 2019. Artificial intelligence in finance. Available at http://doi.org/10.5281/zenodo.2612537.

Buisson, Pascal. 2019. Pure robo-advisors have become viable competitors in the US. *Amundi Digibook*.

Calvet, Laurent E., Celerier, Claire, Sodini, Paolo, and Vallee, Boris. 2020. Can security design foster household risk-taking? Available at SSRN 3474645.

Campbell, John Y. 2006. Household finance. *Journal of Finance*, **61**(4), 1553–1604.

Campbell, John Y., Jackson, Howell E., Madrian, Brigitte C., and Tufano, Peter. 2011. Consumer financial protection. *Journal of Economic Perspectives*, **25**(1), 91–114.

Capponi, Agostino, Olafsson, Sveinn, and Zariphopoulou, Thaleia. 2019. Personalized robo-advising: Enhancing investment through client interaction. ArXiv:1911.01391.

Célérier, Claire, and Vallée, Boris. 2017. Catering to investors through security design: Headline rate and complexity. *Quarterly Journal of Economics*, **132**(3), 1469–1508.

Choi, James J., Laibson, David, and Madrian, Brigitte C. 2009. Mental accounting in portfolio choice: evidence from a flypaper effect. *American Economic Review*, **99**(5), 2085–95.

Clarke, Demo. 2020. Robo-advisors-market impact and fiduciary duty of care to retail investors. Available at SSRN 3539122.

Cocco, Joao F. Gomes, Francisco J., and Maenhout, Pascal J. 2005. Consumption and portfolio choice over the life cycle. *Review of Financial Studies*, **18**(2), 491–533.

Commission, European. 2019. Ethics Guidelines for Trustworthy AI. Available at https://ec.europa.eu/futurium/en/ai-alliance-consultation.

D'Acunto, Francesco, and Rossi, Alberto G. 2020. Robo-advising. CESifo Working Paper 8225.

D'Acunto, Francesco, Prabhala, Nagpurnanand, and Rossi, Alberto G. 2019. The promises and pitfalls of robo-advising. *Review of Financial Studies*, **32**(5), 1983–2020.

Dahlquist, Magnus, Setty, Ofer, and Vestman, Roine. 2018. On the asset allocation of a default pension fund. *Journal of Finance*, **73**(4), 1893–1936.

Das, Sanjiv, Markowitz, Harry, Scheid, Jonathan, and Statman, Meir. 2010. Portfolio optimization with mental accounts. *Journal of Financial and Quantitative Analysis*, **45**(2), 311–334.

De Palma, André, Picard, Nathalie, and Prigent, Jean-Luc. 2009. Prise en compte de l'attitude face au risque dans le cadre de la directive MiFID. Available at HAL-00418892.

Dietvorst, Berkeley J., Simmons, Joseph P., and Massey, Cade. 2015. Algorithm aversion: People erroneously avoid algorithms after seeing them err. *Journal of Experimental Psychology: General*, **144**(1), 114.

Dietvorst, Berkeley J., Simmons, Joseph P., and Massey, Cade. 2018. Overcoming algorithm aversion: People will use imperfect algorithms if they can (even slightly) modify them. *Management Science*, **64**(3), 1155–1170.

Dimmock, Stephen G, Kouwenberg, Roy, Mitchell, Olivia S,, and Peijnenburg, Kim. 2016. Ambiguity aversion and household portfolio choice puzzles: Empirical evidence. *Journal of Financial Economics*, **119**(3), 559–577.

Dohmen, Thomas J., Falk, Armin, Huffman, David, Sunde, Uwe, Schupp, Jürgen, and Wagner, Gert G. 2005. Individual risk attitudes: New evidence from a large, representative, experimentally-validated survey.

Doshi-Velez, Finale, and Kim, Been. 2017. Towards a rigorous science of interpretable machine learning. ArXiv:1702.08608.

Dupont, Laurent. 2020. Gouvernance des algorithmes d'intelligence artificielle dans le secteur financier: Analyse des réponses à la consultation. `https://acpr.banque-france.fr/gouvernance-des-algorithmes-dintelligence-artificielle-dans-le-secteur-financier`

Fagereng, Andreas, Guiso, Luigi, Malacrino, Davide, and Pistaferri, Luigi. 2020. Heterogeneity and persistence in returns to wealth. *Econometrica*, **88**(1), 115–170.

Fama, E. F., and French, K. R. 1993. Common risk factors in the returns on stocks and bonds. *Journal of Financial Economics*, **33**(1), 3–56.

Fein, Melanie L. 2017. Are robo-advisors fiduciaries? Available at SSRN 3028268.

Fildes, Robert, and Goodwin, Paul. 2007. Against your better judgment? How organizations can improve their use of management judgment in forecasting. *Interfaces*, **37**(6), 570–576.

Finance, Better. 2020. Robo-advice 5.0: Can consumers trust robots? `https://betterfinance.eu/publication/robo-advice-5-0-can-consumers-trust-robots/`.

Fisch, Jill E., Labourě, Marion, and Turner, John A. 2019. The emergence of the robo-advisor. Pages 13–37 of: *The Disruptive Impact of FinTech on Retirement Systems*, J. Agnew and O. Mitchell (eds). Oxford University Press.

Foà, Gabriele, Gambacorta, Leonardo, Guiso, Luigi, and Mistrulli, Paolo Emilio. 2019. The supply side of household finance. *Review of Financial Studies*, **32**(10), 3762–3798.

Foerster, Stephen, Linnainmaa, Juhani T., Melzer, Brian T., and Previtero, Alessandro. 2017. Retail financial advice: does one size fit all? *Journal of Finance*, **72**(4), 1441–1482.

Gargano, Antonio, and Rossi, Alberto G. 2020. There's an app for that: Goal-setting and saving in the FinTech era. Available at SSRN 3579275.

Garlappi, Lorenzo, Uppal, Raman, and Wang, Tan. 2007. Portfolio selection with parameter and model uncertainty: A multi-prior approach. *Review of Financial Studies*, **20**(1), 41–81.

Gennaioli, Nicola, Shleifer, Andrei, and Vishny, Robert. 2015. Money doctors. *Journal of Finance*, **70**(1), 91–114.

Germann, Maximilian, and Merkle, Christoph. 2019. Algorithm aversion in financial investing. Available at SSRN 3364850.

Gigerenzer, Gerd, and Brighton, Henry. 2009. Homo heuristicus: Why biased minds make better inferences. *Topics in Cognitive Science*, **1**(1), 107–143.

Goetzmann, William N., and Kumar, Alok. 2008. Equity portfolio diversification. *Review of Finance*, **12**(3), 433–463.

Goldfarb, Avi, and Tucker, Catherine. 2019. Digital economics. *Journal of Economic Literature*, **57**(1), 3–43.

Goodman, Bryce, and Flaxman, Seth. 2017. European Union regulations on algorithmic decision-making and a "right to explanation". *AI Magazine*, **38**(3), 50–57.

Grealish, Adam, and Kolm, Petter N. 2022. Robo-advisory: From investing principles and algorithms to future developments. In: *Machine Learning in Financial Markets: A Guide to Contemporary Practice*, A. Capponi and C.-A. Lehalle (eds). Cambridge University Press.

Greenwood, Robin, and Nagel, Stefan. 2009. Inexperienced investors and bubbles. *Journal of Financial Economics*, **93**(2), 239–258.

Grinblatt, Mark, Keloharju, Matti, and Linnainmaa, Juhani. 2011. IQ and stock market participation. *Journal of Finance*, **66**(6), 2121–2164.

Guiso, Luigi, and Sodini, Paolo. 2013. Household Finance: An Emerging Field. Pages 1397–1532 of: *Handbook of the Economics of Finance*, vol. 2. Hans R. Stoll, George M. Constantinides, Milton Harris, and R.M.Stulz (eds). Elsevier.

Helldin, Tove, Falkman, Göran, Riveiro, Maria, and Davidsson, Staffan. 2013. Presenting system uncertainty in automotive UIs for supporting trust calibration in autonomous driving. Pages 210–217 of: *Proc. 5th International Conference on Automotive User Interfaces and Interactive Vehicular applications*.

Holt, Charles A., and Laury, Susan K. 2002. Risk aversion and incentive effects. *American Economic Review*, **92**(5), 1644–1655.

Hong, Claire Yurong, Lu, Xiaomeng, and Pan, Jun. 2020. FinTech adoption and household risk-taking. *NBER Working Paper No. 28063*.

HSBC. 2019. Trust in technology.

Jacovi, Alon, Marasović, Ana, Miller, Tim, and Goldberg, Yoav. 2020. Formalizing trust in artificial intelligence: Prerequisites, causes and goals of human trust in AI. ArXiv:2010.07487.

Ji, Megan. 2017. Are robots good fiduciaries: Regulating robo-advisors under the Investment Advisers Act of 1940. *Colum. L. Rev.*, **117**, 1543.

Jordan, Michael I. 2019a. Artificial intelligence – the revolution hasn't happened yet. *Harvard Data Science Review*, **1**(1).

Jordan, Michael I. 2019b. Dr. AI or: How I learned to stop worrying and love economics. *Harvard Data Science Review*, **1**(1).

Kahneman, Daniel, Knetsch, Jack L., and Thaler, Richard H. 1990. Experimental tests of the endowment effect and the Coase theorem. *Journal of Political Economy*, **98**(6), 1325–1348.

Kapteyn, Arie, and Teppa, Federica. 2011. Subjective measures of risk aversion, fixed costs, and portfolio choice. *Journal of Economic Psychology*, **32**(4), 564–580.

Koo, Jeamin, Kwac, Jungsuk, Ju, Wendy, Steinert, Martin, Leifer, Larry, and Nass, Clifford. 2015. Why did my car just do that? Explaining semi-autonomous driving actions to improve driver understanding, trust, and performance. *International Journal on Interactive Design and Manufacturing (IJIDeM)*, **9**(4), 269–275.

Kraft, Holger, and Munk, Claus. 2011. Optimal housing, consumption, and investment decisions over the life cycle. *Management Science*, **57**(6), 1025–1041.

Lam, J.W. (2016). Robo-Advisers: A Portfolio Management Perspective. Senior Thesis, Yale College.

Ledoit, Olivier, and Wolf, Michael. 2003. Improved estimation of the covariance matrix of stock returns with an application to portfolio selection. *Journal of Empirical Finance*, **10**(5), 603–621.

Lerner, Josh, and Tufano, Peter. 2011. The consequences of financial innovation: a counterfactual research agenda. *Annu. Rev. Financ. Econ.*, **3**(1), 41–85.

Linciano, Nadia, and Soccorso, Paola. 2012. Assessing investors' risk tolerance through a questionnaire.

Linnainmaa, Juhani T., Melzer, Brian, and Previtero, Alessandro. 2021. The misguided beliefs of financial advisors. *Journal of Finance*, **76**(2), 587–621.

Lo, Andrew W. 2016. What is an index? *Journal of Portfolio Management*, **42**(2), 21–36.

Lo, Andrew W. 2019. Why artificial intelligence may not be as useful or as challenging as artificial stupidity. *Harvard Data Science Review*, **1**(1).

Lopez, Juan C., Babcic, Sinisa, and De La Ossa, Andres. 2015. Advice goes virtual: how new digital investment services are changing the wealth management landscape. *Journal of Financial Perspectives*, **3**(3).

Lourenço, Carlos J.S., Dellaert, Benedict G.C., and Donkers, Bas. 2020. Whose algorithm says so: The relationships between type of firm, perceptions of trust and expertise, and the acceptance of financial Robo-advice. *Journal of Interactive Marketing*, **49**, 107–124.

Lusardi, Annamaria, and Mitchell, Olivia S. 2014. The economic importance of financial literacy: theory and evidence. *Journal of Economic Literature*, **52**(1), 5–44.

Lusardi, Annamaria, Michaud, Pierre-Carl, and Mitchell, Olivia S. 2017. Optimal financial knowledge and wealth inequality. *Journal of Political Economy*, **125**(2), 431–477.

Mankiw, N. Gregory, and Zeldes, Stephen P. 1991. The consumption of stockholders and nonstockholders. *Journal of Financial Economics*, **29**(1), 97–112.

Marinelli, Nicoletta, and Mazzoli, Camilla. 2010. Profiling investors with the MiFID: current practice and future prospects. Research Paper. Available at `https://www.ascosim.it/public/19_ric.pdf`.

Merton, Robert. 1971. Optimal portfolio and consumption rules in a continuous-time model. *Journal of Economic Theory*, **3**(4), 373–413.

Merton, Robert C. 2017. The future of robo-advisors. Video available at `https://www.cnbc.com/2017/11/05/mit-expert-robert-merton-on-the-future-of-robo-advisors.html`.

Miller, Tim. 2019. Explanation in artificial intelligence: Insights from the social sciences. *Artificial Intelligence*, **267**, 1–38.

Misztal, Barbara. 2013. *Trust in Modern Societies: The Search for the Bases of Social Order*. John Wiley & Sons.

Muir, Bonnie M, and Moray, Neville. 1996. Trust in automation. Part II. Experimental studies of trust and human intervention in a process control simulation. *Ergonomics*, **39**(3), 429–460.

Mullainathan, Sendhil, Noeth, Markus, and Schoar, Antoinette. 2012. The market for financial advice: An audit study. Tech. rept. National Bureau of Economic Research.

Niszczota, Paweł, and Kaszás, Dániel. 2020. Robo-investment aversion. *PLOS One*, **15**(9), e0239277.

Odean, Terrance. 1999. Do investors trade too much? *American Economic Review*, **89**(5), 1279–1298.

Patel, Keyur, and Lincoln, Marshall. 2019. It's not magic: Weighing the risks of AI in financial services. Report available at `https://www.european-microfinance.org/publication/its-not-magic-weighing-risks-ai-financial-services`.

Philippon, Thomas. 2019. On fintech and financial inclusion. *NBER Working Paper No. 26330*.

Poursabzi-Sangdeh, Forough, Goldstein, Daniel G., Hofman, Jake M., Vaughan, Jennifer Wortman, and Wallach, Hanna. 2018. Manipulating and measuring model interpretability. ArXiv:1802.07810.

Prahl, Andrew, and Van Swol, Lyn. 2017. Understanding algorithm aversion: When is advice from automation discounted? *Journal of Forecasting*, **36**(6), 691–702.

Reher, Michael, and Sokolinski, Stanislav. 2020. Does finTech democratize investing? Available at SSRN 3515707.

Reher, Michael, and Sun, Celine. 2019. Automated financial management: Diversification and account size flexibility. *Journal of Investment Management*, **17**(2), 1–13.

Robinette, Paul, Li, Wenchen, Allen, Robert, Howard, Ayanna M., and Wagner, Alan R. 2016. Overtrust of robots in emergency evacuation scenarios. Pages 101–108 of: *Proc. 11th ACM/IEEE International Conference on Human–Robot Interaction*.

Rossi, A., and Utkus, S. 2019a. The needs and wants in financial advice: Humans versus robo-advising. Working Paper, George Washington University. Available at SSRN 3759041.

Rossi, A., and Utkus, S. 2019b. Who benefits from robo-advising? Evidence from machine learning. Working Paper, George Washington University. Avalaible at SSRN 3552671.

Ryan, Andrea, Trumbull, Gunnar, and Tufano, Peter. 2011. A brief postwar history of US consumer finance. *Business History Review*, **85**(03), 461–498.

Sapienza, Paola, Toldra-Simats, Anna, and Zingales, Luigi. 2013. Understanding trust. *Economic Journal*, **123**(573), 1313–1332.

SEC. 2006. Questions advisers should ask while establishing or reviewing their compliance programs. Available at `https://www.sec.gov/info/cco/advise\_compliance\_questions.htm`.

SEC. 2019. Commission interpretation regarding standard of conduct for investment advisors. Available at `https://www.sec.gov/info/cco/adviser\_compliance\_questions.htm`.

Sheridan, T.B. 1989. Trustworthiness of command and control systems. Pages 427–431 of: *Analysis, Design and Evaluation of Man–Machine Systems 1988*. Elsevier.

Sheridan, Thomas B. 2019. Individual differences in attributes of trust in automation: measurement and application to system design. *Frontiers in Psychology*, **10**, 1117.

Strzelczyk, Bret E. 2017. Rise of the machines: the legal implications for investor protection with the rise of robo-advisors. *DePaul Bus. & Comm. LJ*, **16**, 54.

Thakor, Richard T., and Merton, Robert C. 2018. Trust in lending. Tech. rept. National Bureau of Economic Research.

Toreini, Ehsan, Aitken, Mhairi, Coopamootoo, Kovila, Elliott, Karen, Zelaya, Carlos Gonzalez, and van Moorsel, Aad. 2020. The relationship between trust in AI and trustworthy machine learning technologies. Pages 272–283 of: *Proc. Conference on Fairness, Accountability, and Transparency*.

Vissing-Jorgensen, Annette. 2004. Perspectives on behavioral finance: Does "irrationality" disappear with wealth? Evidence from expectations and actions. Pages 139–208 of: *NBER Macroeconomics Annual 2003, Volume 18*, Mark Gertler and Kenneth Rogoff (eds). MIT Press.

Warren, Geoffrey J. 2019. Choosing and using utility functions in forming portfolios. *Financial Analysts Journal*, **75**(3), 39–69.

Weber, Martin, Weber, Elke U., and Nosić, Alen. 2013. Who takes risks when and why: Determinants of changes in investor risk taking. *Review of Finance*, **17**(3), 847–883.

4

Robo-Advisory: From Investing Principles and Algorithms to Future Developments

Adam Grealish[a] and Petter N. Kolm[b]

Abstract

Advances in financial technology have led to the development of easy-to-use online platforms known as *robo-advisors* or *digital-advisors*, offering automated investment and portfolio management services to retail investors. By leveraging algorithms embodying well-established investment principles and the availability of exchange traded funds (ETFs) and liquid securities in different asset classes, robo-advisors automatically manage client portfolios that deliver similar or better investment performance at a lower cost compared with traditional financial retail services.

In this chapter we explore how robo-advisors translate core investing principles and best practices into algorithms. We discuss client onboarding and algorithmic approaches to client risk assessment and financial planning. We review portfolio strategies available on robo-advisor platforms and algorithmic implementations of ongoing portfolio management and risk monitoring. Since robo-advisors serve individual retail investors, tax management is a focal point on most platforms. We devote substantial attention to automated implementations of a number of tax optimization strategies, including tax-loss harvesting and asset location. Finally, we explore future developments in the robo-advisory space related to goal-based investing, portfolio personalization, and cash management.

4.1 From investing principles to algorithms

Robo-advisors use automation to systematically implement best practices in portfolio management, tax management, and financial planning. For example, most robo-advisors offer a mean-variance optimized portfolio framework that rebalances client portfolios automatically. Given the algorithmic nature of their services and access to client data, robo-advisors are able to augment traditional financial planning and asset management practices with machine learning (ML) techniques to provide more personalized portfolio management and financial advice. Robo-advisors' automated systems are capable of running these processes at scale, potentially across millions of accounts.

[a] Altruist, Los Angeles
[b] Courant Institute of Mathematical Sciences, New York University
Published in *Machine Learning And Data Sciences For Financial Markets*, Agostino Capponi and Charles-Albert Lehalle © 2023 Cambridge University Press.

A core innovation of robo-advisors is the automation of wealth management processes that traditionally have been manual. Traditional financial advisors may spend a significant amount of time on mechanical tasks such as rebalancing a portfolio to its target weights or identifying securities at a loss to harvest for tax purposes. Automating these tasks removes the risk of manual error and allows such processes to be run across many accounts in parallel at low cost in real-time. By using automation, robo-advisors are able to realize economies of scale and provide high quality investment management at a low cost to clients. In comparison with traditional financial advisors, the costs of using robo-advisors are lower (Kitces, 2017; Uhl and Rohner, 2018). Advisory fees at most robo-advisors generally range from 0.25% to 0.50% compared to approximately 1% for a traditional advisor (Bol et al., 2020), making them attractive for retail investors.

Moreover, robo-advisors' usage of automation and passive investment strategies reduce risks of internal agency conflicts and conflicts of interest that traditionally could arise between financial advisors and their clients (Inderst and Ottaviani, 2009).

Beyond automation of rote tasks, robo-advisors have integrated ML into numerous practice areas, including portfolio construction, financial planning, and behavioral interventions. For example, clustering and supervised learning algorithms are deployed to identify segments of the client base that would benefit from behavioral interventions during down markets. Timely and tailored algorithmic client interaction can improve investor outcomes (Capponi et al., 2019). While not as widely adopted, reinforcement learning (RL) can be used to uncover investor risk preferences (Alsabah et al., 2020) and generate optimal strategies for multi-period deposit and withdrawal behavior (Das and Varma, 2020; Dixon and Halperin, 2020; Charpentier et al., 2021). Generative adversarial nets (GAN) can generate synthetic data with complex properties similar to that of real financial data for use in portfolio simulations (Takahashi et al., 2019). ML techniques can also be used to solve mixed integer programming optimizations, a class of problems that can arise when selecting a portfolio of individual securities to track a broader index (Khalil, 2016; Bengio et al., 2020). Needless to say, some aspects of investment management processes can be automated more easily than others. Managing complex financial circumstances, estate planning, and personal coaching generally fall outside the scope of robo-advisor capabilities. As technology advances, it is likely that robo-advisors will find ways to automate even some of these investment services.

Since automated systems do not have the flexibility and perspective of human judgement, fully automated wealth management processes present a new set of challenges. For instance, a traditional advisor may notice that a tax rate assumption does not coincide with their client's taxable income and make the necessary adjustment for planning and tax management purposes. An automated system must incorporate internal consistency checks to avoid acting on incomplete or incorrect information.

Whether novice or seasoned retail investor, there are several core investing principles that are important:

- **Establish an investment plan with clear objectives.** Establishing an investment plan involves much more than picking a few stocks, mutual funds or ETFs to invest in. For instance, it is important to consider one's current financial situation and goals, investment timeline and risk appetite.
- **Seek broad diversification.** First coined by Miguel de Cervantes, the author of Don Quixote, in the early 1600s, "don't put all your eggs in one basket" is perhaps the most well-known metaphor expressing the investment advice of the benefits of diversification. Many decades of academic research and centuries of practical experience have shown the importance of diversifying across investment themes and asset classes. Seeking broad diversification should be one of the main goals of any retail investment strategy. Economic theory states that the market portfolio is the most diversified portfolio.
- **Weigh investment cost and value added.** Any investment process needs to weigh the cost of exposure to the assets with their added value. While this is common sense, it is "easier said than done" as costs appear in many different forms such as transaction and trading costs, management fees, and taxes.
- **Account for taxes.** For retail investors tax considerations and tax optimization are crucial as a significant part of their investments are held in taxable accounts.

Robo-advisors have incorporated these core investment principles in their services and product offerings, and it is perhaps not surprising that investors who are less experienced, trade frequently, hold high-cost funds, or large cash positions particularly benefit from robo-advisory services (Rossi and Utkus, 2020). Many robo-advisors have integrated these core principles as a three-step process:

1. Client assessment and onboarding,
2. Implementation of the investment strategy, and
3. Ongoing management of the investment strategy,

where each step is fully, or at least to a large extent, automated. Figure 4.1 provides a schematic view of this process and its elements. In the next sections, we examine each one separately.

4.1.1 Client assessment and onboarding

Most commonly, a robo-advisor provides a new client with an automated online survey to collect general and investment specific information, including age, net worth, investment goals, risk capacity and risk tolerance (Andrus, 2017). Investment goals may include generating income, saving for retirement, planning for large future expenditures (such as the purchase of a house or car) and establishing a financial safety net for emergencies. The robo-advisor inquires about a client's investing experience, level of sophistication and level of risk tolerance. The specific questions and formats to elicit this information can vary considerably across robo-advisory services. We discuss some of the developments in goals-based

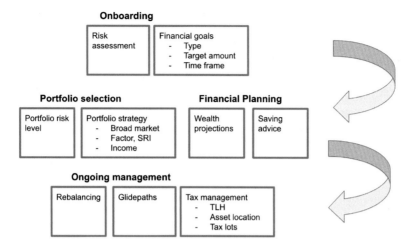

Figure 4.1 Schematic view of the robo-advisory process and its elements.

investing in Section 4.4.1. The data collected through the survey is used to automatically setup the new client's account and provide initial investment settings and recommendations.

This automated assessment represents a big change in client onboarding as compared to traditional wealth management services that rely on an initial in-person consultation. While less personal, and perhaps not as comprehensive, the online assessments are less time-consuming and offer great cost-savings over the traditional approach. It is no wonder that many clients prefer the simplicity and ease of those robo-advisors who have onboarding processes that take no more than fifteen minutes to complete (Lo et al., 2018).

Client risk assessment

SEC guidelines outline that a financial advisor should make a reasonable inquiry into a client's financial situation, risk tolerance, and financial goals (SEC, 2006, 2019). Robo-advisors accomplish this through client questionnaires, goal-based investing frameworks, and user interaction which highlight potential downside portfolio performance. Approaches to estimating risk tolerance include the one-question Survey of Consumer Finances (SCF) risk-tolerance item, adapted versions of the Gable-Lytton risk tolerance assessment (Grable and Lytton, 1999), and gauging reactions to hypothetical or simulated portfolio performance. The variety in approaches reflects the different techniques in risk tolerance literature (Callan and Johnson, 2002).

Risk tolerance is unobservable and can only be measured indirectly and imperfectly. The shortcomings of survey-based approaches are well documented. Klement (2015) find that risk questionnaires explain less than 15% of variation in risky assets between investors. Guillemette and Finke (2014) find that risk tolerance scores are heavily influenced by market events. Schooley and Worden (2016) show that perceived and realized risk tolerance are not always well connected.

Some robo-advisors address these challenges with *experience sampling*, where investors can see how their wealth would change under different market conditions before investing. Market conditions "experienced" by investors are usually pulled from history or generated through simulation. Investors exposed to experience sampling tend to have better alignment between portfolio performance and expectations (Faloon and Scherer, 2017; Bradbury et al., 2015).

While not widely used in robo-advisors today (Faloon and Scherer, 2017), ML algorithms, particularly RL frameworks, can also infer investor preferences based on their behavior and existing portfolios (Yu et al., 2020; Alsabah et al., 2020). In the future, such insights may augment and potentially de-bias survey-based risk assessments.

Robo-advisors consider the client's financial goal. The type of financial goal (i.e., retirement or major purchase) can play an important part in recommending an appropriate level of risk as different types of goals may have different liquidation profiles. For example, compare the liquidation strategies for a retiree versus someone who is saving towards a down payment on a house. The retiree expects to liquidate their portfolio with smaller, recurring withdrawals over the subsequent thirty years. By contrast, the investor saving for a down payment will prefer to liquidate the full value of their portfolio all at once when purchasing their home.

Each financial objective and its associated liquidation plan impacts the recommended portfolio risk level. Consequently, even for investors with the same risk tolerance and investment horizon, their appropriate portfolio risk levels will differ based on the type of goal and its associated liquidation profile.

Robo-advisors update client risk recommendations with time as appropriate. For example, in Section 4.4.1 on glide paths, we will see that risk levels are automatically decreased as goals are approaching the end of their horizon.

Financial planning and investment advice

Through apps and websites provided by the robo-advisor, their clients receive financial planning advice and other investment updates in an ongoing fashion. The advice may include scenario analysis, projections of potential asset growth through time, recommended savings advice, feedback on portfolio risk levels, and specific retirement planning alerts.

Investor risk capacity is closely associated with household balance sheet items (Scherer, 2017). As such, many robo-advisors incorporate externally held assets into their recommendations. To present a view of the full household balance sheet, some robo-advisors will incorporate and synchronize with external assets and accounts when authorized by the client. For example, in order to track their full net worth, a client may add home value, outstanding mortgage balance, and investments held outside of the robo-advisory platform. Some robo-advisors can incorporate these external sources towards a specific goal. For example, an employer 401(k) account could be assigned to a retirement goal monitored by the robo-advisor. Clearly, with this additional information, a robo-advisor can provide a more holistic view of a client's risk and more accurate growth projections. Additionally, a robo-advisor can alert clients if fees in external investments are becoming abnormally high, or if a fund is holding an unusually high amount of cash.

A challenge for a robo-advisor that seeks to incorporate a client's full financial holdings is the accurate identification of non-public assets. For instance, a fund company may have negotiated pricing terms for a 401(k) plan, in which case, individual investors would have non-public share classes of mutual funds in their 401(k) accounts. Such non-public assets may not have risk and performance data easily available. Finding a suitable proxy asset or set of assets becomes the task of the robo-advisor. Some robo-advisors employ natural language processing techniques to descriptive textual data available for non-public assets in order to identify close proxies. Robo-advisors may use third-party services to provide estimates of less liquid assets such as home valuations.

Wealth projections and advice

A core aspect of robo-advisory is helping clients plan their investment goals and understand the tradeoffs between portfolio risk, growth, and savings rate. For this purpose, a robo-advisor projects portfolio growth trajectories for the remaining life of the goals. These projections may include growth projections under average, good and poor market conditions. Wealth projections may also include planned changes to the portfolio through the life of the goal, such as planned future deposits or allocation changes. Figure 4.2 shows an example of a financial planning page that includes projected growth of a portfolio over an eight-year horizon under average, good and poor market performance.

By assuming asset return distributions are multivariate normal, wealth projections can be efficiently computed in closed-form. However, to model more complex market dynamics (i.e., fat-tails, skewness, serial correlation) and cash in- and outflow behaviors, some robo-advisors use Monte Carlo simulations. While more flexible, Monte Carlo simulation is computationally intensive, which can make servicing real-time requests at scale challenging.

A robo-advisor may also recommend the savings rate, often represented as a monthly deposit amount, that would be needed to likely reach the target goal amount. In the example in Figure 4.2 the client has set the monthly deposit amount to the recommended savings rate of $825.07. Visualizing the savings rate, portfolio risk and growth together allows clients to better understand their tradeoffs.

The probability of reaching, subceeding or exceeding the goal target can be particularly instructive in assisting clients in managing longer term goals. As the investment horizon increases, the probability of achieving the goal becomes less sensitive to immediate changes in portfolio value. Figure 4.3 depicts the change in probability of achieving a hypothetical goal subject to different portfolio drawdown scenarios. We observe that for a goal with investment horizons of 20 year or more, larger portfolio drawdowns have less impact on the likelihood of the portfolio reaching or exceeding the target amount. By illustrating the likelihood of

Figure 4.2 A financial planning page showing projections of a client goal with monthly deposits of $825 over an eight-year horizon under average, good and poor market performance.

success, clients can get a better perspective of the impact of market drawdowns, especially at longer investment horizons.

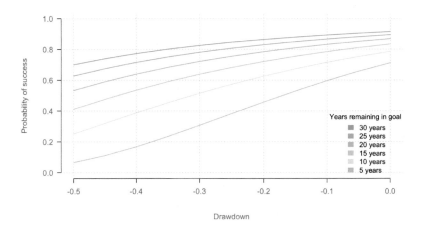

Figure 4.3 Probability of reaching or exceeding a target dollar amount at the end of the investment horizon.

4.1.2 Implementation of the investment strategy

Robo-advisors offer investment strategies that seek global diversification across equity and fixed income asset classes. Portfolios are most often comprised of low cost, index tracking ETFs. Next, we address several aspects of portfolio construction and investment selection amongst robo-advisors.

Portfolio construction

Robo-advisors use algorithms embodying well-established investment principles to automatically construct and periodically rebalance a diversified portfolio of liquid assets. Most commonly, robo-advisors build and manage portfolios based on mean-variance optimization (MVO) from modern portfolio theory (MPT) (Markowitz, 1952). In its basic form, MVO provides a framework for constructing a portfolio by choosing the amount to invest in each asset such that the expected return of the resulting portfolio is maximized at a prespecified level of risk as measured by portfolio volatility. Some robo-advisors use various extensions of MVO to incorporate additional features in their models, such as transaction costs, tax lots, other risk specifications and the ability to incorporate subjective views in the portfolio construction process (see, for example, Kolm et al., 2014, 2021). Some robo-advisors employ other portfolio construction methodologies, such as balancing portfolio weights across sectors (equal weight strategies) or balancing risk contributions across asset classes (see, Section 4.4.5 on risk parity).

Common ML techniques, such as regularization and cross-validation, can be applied to make the portfolio optimization process more robust to real world data (Ban et al., 2018). RL provides an alternative to MVO that relaxes model assumptions. RL approaches can perform multi-period optimization while incorporating complex trading costs structures, investor preferences and other constraints (Kolm and Ritter, 2020; Neuneier, 1998; Benhamou et al., 2020).

While automation and ML can improve outcomes, investors can harbor a degree of skepticism about an entirely computer-driven investment process. The avoidance of algorithms even as they produce better results is referred to as *algorithm aversion* (Dietvorst et al., 2015). Bianchi and Briere (2022) suggest that algorithm aversion may be mitigated with richer explanations of the factors driving investment decisions.

Investment selection

Generally, through investment selection robo-advisors aim to gain broad asset class exposure while keeping overall investment costs low. Robo-advisors predominantly use low cost, index tracking funds, which tend to outperform actively managed funds. Over a 15-year period ending in December 2019, active global equity funds underperformed a passive global index by 0.5%, with 83% of the funds trailing the index. A similar pattern holds for shorter time periods as well (Liu, 2019). While ongoing portfolio management is entirely automated, investment selection commonly involves a level of human oversight from investment professionals who evaluate fund suitability.

Robo-advisors predominantly use ETFs, as opposed to mutual funds or individual securities, in their offerings. ETFs give a robo-advisor's clients broad asset class exposure at a low cost. In addition, ETFs confer certain tax advantages over mutual funds beneficial to individual investors. Because of their legal structure, ETFs tend to avoid passing capital gains incurred while managing the fund through to the shareholder, especially when compared to mutual funds (Gastineau, 2010).

As a consequence of the wide adoption of ETFs in the market place, ETFs can be traded at low cost throughout the day and be used efficiently for tax-loss harvesting (TLH). As we discuss in more detail in Section 4.2, the purpose of TLH is to realize current losses that can later be offset against future gains.

Investment selection lends itself to the application of ML methods, such as clustering and classification algorithms, for identifying similar securities to be evaluated for inclusion in portfolios or consideration as alternative investment options to be used in a TLH strategy. For example, support vector machines (SVM) can classify credit quality of securities from market observables such as credit spread and duration. Figure 4.4 depicts an SVM classifying bond ETFs into high, medium and low credit quality based on weighted average option-adjusted spread (OAS) and weighted average life. Once similar funds are identified, a robo-advisor may examine the funds' expense ratios, tracking errors, trading volumes, efficiency, and their securities lending policies before deciding to include them in client portfolios.

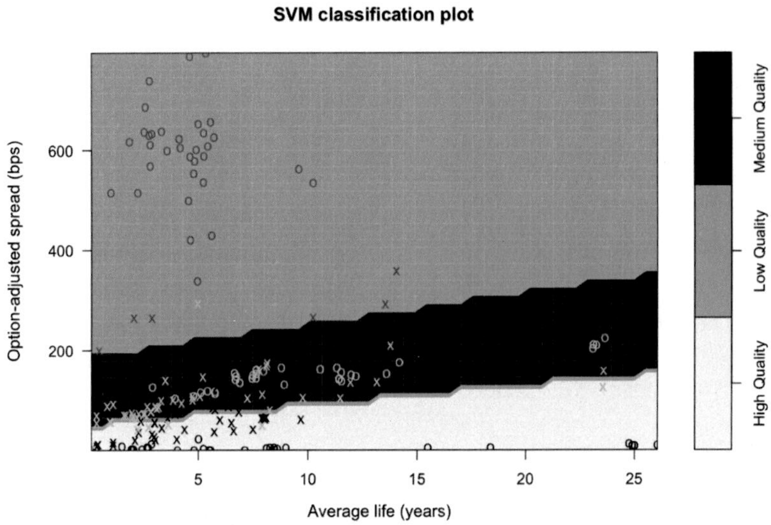

Figure 4.4 SVM classifier of bond ETFs into high, medium and low credit quality based on weighted average option-adjusted spread (OAS) and weighted average life.

ETFs versus individual stocks

Some robo-advisors build portfolios using individual stocks. Most often a robo-advisor will seek to replicate the performance of a generic index. For example, the robo-advisor may seek to track the performance of the 500 largest U.S. stocks. Buying stocks to track index performance, referred to as *direct indexing*, offers potential tax benefits and greater flexibility for client personalization compared to an ETF pursuing a similar strategy.

Because of fund structure, ETFs and mutual funds must pass capital gains through to shareholders. However, capital losses stay within the fund and are not passed on. By contrast, in a direct indexing strategy, where individual stocks that comprise the index are held in the client's name, capital losses can be offset against capital gains from other investments.

Direct indexing allows for more TLH opportunities by taking advantage of the cross-sectional dispersion in returns of individual index constituents. For example, an index could be up over the year, but some stocks in the index may have experienced negative returns. An investor in the index tracking fund will only have the opportunity to tax-loss harvest when the fund itself is down. However, a direct index investor can harvest losses in stocks that performed poorly even when the overall index return for the year was positive.

Direct investing in stocks underlying an index may result in cost savings, as compared to an index ETF, as fund expense ratios are avoided. However, expense ratios for many index funds are already quite low.

Needless to say, individual stock portfolios allow for an added level of personalization. For example, some robo-advisors allow clients to specify a "do not buy" list of stocks, or assist clients in constructing portfolios of individual stocks based on their preferences and views. Frequently, this approach is used to tailor socially responsible investment portfolios, where client preferences often are heterogeneous (see, Section 4.4.3).

4.1.3 Ongoing management of the investment strategy

An important aspect of robo-advisory services is the ongoing monitoring and rebalancing of client portfolios. We address several aspects in this section including portfolio rebalancing, risk management, ongoing automated advice and glide paths.

Portfolio rebalancing and risk management

There are four main situations in which portfolio management algorithms rebalance: (a) when, because of market movements, portfolio holdings drift too far away from their desired target allocations; (b) when tax-loss harvesting opportunities are identified; (c) when clients update their preferences; and (d) when there are cashflows, such as deposits, dividends, or withdrawals. Robo-advisors may also rebalance a portfolio to reduce risk as an investing goal gets closer to the end of its horizon. In general, annual turnover of robo-portfolios are low as robo-advisors design their investment strategies for the long-term.

While rebalancing is primarily a risk control strategy, disciplined rebalancing from automated software may also provide some systematic performance enhancements by selling overweight securities and buying underweight securities (Bouchey et al., 2012). Using simulated data, Huss and Maloney (2017) find that a rebalanced portfolio has higher median performance, a narrower range of final wealth outcomes and smaller drawdowns compared to a portfolio with no rebalancing.

Algorithms are particularly well suited for the task of rebalancing, which is often mechanical in nature with set rebalancing triggers based on the deviation from portfolio target weights or based on a regular calendar schedule. An automated system is able to monitor thousands of accounts as market conditions change throughout the day and rebalance when necessary.

If possible, a robo-advisor will rebalance a client portfolio toward target weights using cash flows by buying underweight securities with deposits and dividend proceeds and selling overweight securities for withdrawals. Rebalancing with deposits is particularly tax efficient since no selling occurs, avoiding any potential capital gains.

If a robo-advisor needs to sell overweight assets in order to rebalance, it may choose to sell specific lots in order to minimize the tax impact of the sale. Tax lot management, discussed in more detail in Section 4.2.3, is offered by many robo-advisors. A robo-advisory offering becomes particularly valuable when its numerous features work in concert. For example, considering and minimizing tax impacts is important in all rebalancing activities.

Ongoing automated advice

As markets fluctuate, robo-advisors update their wealth projections and planning advice. As these updates are ongoing and occur in real time, clients are able to see the implications that market changes may have on their financial plan. For example, after a rally in the equity market, a client may log into their account and find that they can take less risk and still achieve their investment target. Or, after a market drop, a client may find that they need to increase their savings rate in order to reach their goal with sufficient certainty.

Many robo-advisors will take market changes into account when recommending a safe withdrawal amount for retirement. Scott et al. (2009) show that static withdrawals strategies can leave a surplus during good markets and suffer shortfalls when markets underperform. By updating as the market changes, dynamic withdrawal recommendations avoid these shortcomings. However, a dynamic strategy lacks the predictable withdrawal amount associated with withdrawal strategies that do not update with market conditions, like the "4% rule" (Bengen, 1994).

4.2 Automated tax Management for retail investors

Besides diversification and automatic rebalancing, one of the most valuable services robo-advisors provide is automated tax management. Constantinides (1984) showed that it is optimal to realize capital losses in stocks immediately and to defer capital gains for as long as possible.

Individual investors often overlook the impact of taxes on performance (Horan and Adler, 2009); however, taxes represent a meaningful drag on after-tax returns for the individual investor and can be actively managed for better outcomes

(Jeffrey and Arnott, 1993; Stein, 1998). Tax-aware investment strategies consider the impact of taxes during portfolio construction and in decisions to rebalance the portfolio (Apelfeld et al., 1996; Brunel, 1997, 2001). Additionally, the taxable investor can benefit from active portfolio management to capture tax losses and shield tax-inefficient investments (Reichenstein, 2004; Stein et al., 2008). A common approach is to take the strategic asset allocation as given and overlay "tax algorithms" either at the account or trading level to generate so-called *tax alpha* through tax-loss harvesting, asset location, and other tax minimization approaches.

It is important to consider how a robo-advisor's different investment management algorithms interact with each other. Typically, these interactions are accretive to the overall value of the features. For example, after selling a security for tax-loss harvesting, instead of simply using the full proceeds to buy a security with similar market exposure, a rebalancing algorithm may use some of the proceeds to rebalance the overall portfolio. Alternatively, certain algorithms may marginally reduce the stand-alone efficiency of others. For example, an asset allocation algorithm may prefer to hold tax-inefficient high-growth assets, like a REIT, in a tax-exempt account. However, by allocating a REIT to a tax-exempt account, a TLH algorithm no longer has the opportunity to harvest a volatile asset. Rollén (2019) accounts for such interactions by evaluating a robo-advisor's investment management algorithms together, as opposed to studying individual features in isolation.

In this section we elaborate on tax optimization approaches deployed by robo-advisors such as tax-loss harvesting, asset location, tax lot management, and cash flow-based rebalancing.

4.2.1 Tax-loss harvesting

Tax-loss harvesting (TLH) is a strategy where assets held in the portfolio at a loss are sold opportunistically to generate tax savings (Jeffrey and Arnott, 1993; Stein and Narasimhan, 1999; Wilcox et al., 2006). When losses are realized, they can be used to offset capital gains in the current or future tax years, or offset taxable income. While TLH traditionally has been performed manually on a quarterly or annual frequency, algorithms can continuously monitor and act on opportunities as they arise. Hence, robo-advisors can significantly increase the number of TLH opportunities relative to traditional investment managers. Arnott et al. (2001) find a 0.50% annualized benefit to tax-loss harvesting over a 25 year holding period assuming a 35% tax rate, using simulated data for US stocks. Berkin and Ye (2003) consider the impact of cash flows and lot-level accounting treatment and find a persistent benefit to TLH of 0.40% annualized.

The so-called *wash sale rule* can make TLH a complex endeavor. A wash sale occurs when an investor sells an asset at a loss and then buys that same asset or "substantially identical" assets within a window of thirty days before or after the sale date. The wash sale rule was designed to discourage investors from selling an asset at a loss to claim a tax benefit, again and again.

One can navigate the wash sale rule in a few different ways. A common approach is to sell an asset at a loss and buy an alternate asset that replicates the exposure of the original asset. In this context, selling one asset and buying a replicating asset is known as a *dual ticker strategy*. For instance, one might sell the holdings in an ETF tracking the Russell 1000 index and later buy back an ETF tracking the S&P 500. Most robo-advisors that offer a TLH program will use a dual ticker strategy.

TLH algorithms need to determine when to sell an asset and buy the alternate, and vice versa. As part of this decision, the robo-advisor must consider that it will likely be blocked from harvesting again in that security for the 30 days following. This results in an opportunity cost of harvesting too soon, whereby additional losses may go unharvested. Frequently, TLH algorithms set loss thresholds that must be met before harvesting in order to balance the tradeoff between harvesting too soon, or waiting too long and missing an opportunity. Asset volatility will impact the effectiveness of TLH. Higher volatility assets offer more opportunities for an asset's price to drop below its cost basis and thus results in more harvesting opportunities.

Robo-advisors must carefully consider actions that could impair the tax efficiency of the TLH program. For example, realizing a short-term capital gain in order to harvest a long-term capital loss can impair efficiency since short-term gains are taxed at a higher rate. Additionally, an effective TLH program will take into account necessary holding periods for dividends to receive *qualified dividend income* (QDI) treatment. Many TLH algorithms consider the interaction with tax-advantaged accounts and seek to avoid the permanent wash sale rule related to IRA and 401(k) accounts.

Importantly, the wash sale rule does not just apply to the activity in accounts held by an individual investor but for all accounts in which the investor has a beneficial interest. This means that accounts held outside of the robo-advisor, such as spousal accounts must be considered. Therefore, some robo-advisors use their online platforms to coordinate tax-loss harvesting activity between beneficial accounts. In such cases, harvesting decisions navigate wash sales by considering all the activity across beneficial accounts.

Naturally, TLH strategies incur additional turnover compared to a buy and hold strategy. The availability of highly liquid ETFs allows robo-advisors to implement TLH strategies with minimal transaction costs. Some robo-advisors offer TLH strategies on liquid individual stock portfolios as part of a direct indexing strategy.

Because TLH is primarily a tax deferral strategy, the after-tax benefit from TLH depends on the investment horizon and liquidation strategy. Most harvesting happens in the first years of an investment since the market value and cost basis are usually closer to one another. However, the longer the overall investment horizon, the more time deferred tax dollars have to grow, which increases the overall value of TLH. Of course, when positions are finally liquidated, taxes (which were previously deferred) must be paid.

4.2.2 Asset location

Asset location is a tax overlay strategy where tax *inefficient* assets (often bonds) are placed in tax *efficient* accounts (for example, an IRA or 401(k) account) in order to mitigate the tax drag on the overall portfolio. The strategy is based on the fact that the after-tax return of an asset can be very different if held in a tax-deferred, tax-exempt or taxable account. For instance, the coupon payment on a bond held in a taxable account is taxed as ordinary income. In this situation, if possible, an investor should instead hold the bond in a tax-exempt account.

For retail investors who have a number of taxable, tax-deferred and tax-exempt accounts, robo-advisors can assist clients by optimally allocating assets preferentially to the different accounts so as to maximize the after-tax return while the overall strategic allocation is maintained (Huang et al., 2016; Huang and Kolm, 2019). Figure 4.5 shows a stylized depiction of account preference for an asset based on its tax efficiency.

Figure 4.5 A stylized example of a balanced stock and bond portfolio with and without asset location. Note that the overall portfolio allocation is the same in both treatments.

Asset location algorithms consider expected annual taxes and taxes at portfolio liquidation. For this purpose, a robo-advisor will weigh how a security's expected dividends, amount of qualified dividends, and expected growth rate may impact after-tax return.

The after-tax benefit from asset location strategies depends on the account types available to an investor, the balances in their accounts, the overall asset mix, and of course, the investor's tax rate. Asset location strategies are most effective when an investor has roughly equal balances in taxable and tax-advantaged accounts, the overall asset mix is balanced between stocks and bonds, and when the investor has a high tax rate. Under these conditions, the after-tax benefit is estimated to be roughly 0.75% of additional annualized return (Kinniry Jr. et al., 2014; Huang et al., 2016).

4.2.3 *Tax lot management*

When selling securities, either as part of rebalancing or withdrawals, the seller is faced with a choice of what specific tax lots to sell. In the absence of other instructions, a broker will generally use the first in, first out (FIFO) rule for selecting lots. Frequently, a FIFO lot selection strategy will be tax inefficient; resulting in selling the most appreciated shares, as they have had the most time to increase in value.

Robo-advisors will often automate lot selection, seeking to maximize capital losses or minimize capital gains in a given security. Such algorithms may go beyond sorting by cost basis and consider whether the capital gains (losses) would receive long- or short-term tax treatment. Additionally, algorithms may consider how long the lot has been held in order to meet QDI holding period criteria.

4.3 Investor interaction

Robo-advisors permit clients to override their algorithms in several ways, including changing portfolio allocations, modifying their risk profile, and updating their preferences. Unnecessary client overrides can trigger significant trading that results in tax consequences. Additionally, changes to portfolio allocations introduce an element of market timing that on average leads to underperformance (Montier, 2002; Richards, 2012).

Unlike a human advisor, the option to call a client and "talk them off the ledge" before making a potentially unwise investment decision is not practical for robo-advisors. Given their scale, instead they must ensure proper investor behavior through their client interfaces and other forms of electronic communication. Robo-advisors employ a number of methods to positively influence investor behavior, including notifications, nudges, smart defaults, and thoughtful use of color and animation. Robo-advisors often use smart default settings and automation to make good investing behavior easier. For example, many robo-advisors provide clients with tools to set up automatically recurring deposits to help them save for different goals.

4.3.1 *Investor education*

A key benefit of a robo-advisor is that pertinent information that an investor needs to make an informed decision is presented at the time the investor is faced with that decision. For example, potential tax impact might be surfaced to an investor before they complete a security sale, or potential upside and downside performance information is displayed when a client is choosing the risk level of their portfolio. In addition, many robo-advisors make vast resource libraries available to clients and the general public.

4.3.2 Data collection and split testing

Robo-advisor platforms collect large amounts of data on client behavior and transactions. This data can highlight common patterns in behavior and inform potential interventions to improve client outcomes. Given the scale and amount of data collected, robo-advisory platforms present a somewhat unique opportunity in the financial advisory space to experiment with interventions and learn from the results.

Robo-advisors may test a new intervention against a control group to understand its efficacy, a practice commonly referred to as *split testing* or *A/B testing*. For example, during a period of higher volatility, a robo-advisor may observe clients changing their allocations at higher rates. To assuage fears, robo-advisors may use email communication to inform clients and put the recent volatility into context. However, instead of sending the email to all clients at once, the robo-advisor may hold out a control group and study the impact of the communication strategy on login rates and allocation changes.

4.4 Expanding service offerings

The robo-advisory landscape is evolving rapidly, with main trends including greater personalization, improved integration of services and platforms, and increased automation. *Autonomous finance* is quickly becoming the new norm, where algorithms assist us in making more disciplined financial decisions such as investing for long-term retirement goals and managing cash flows for future liabilities. A recent study suggests that close to 60% of the US population will be using robo-advisors by 2025 (Schwab, 2018).

In this section we examine developments of service offerings by robo-advisors in goals-based investing, retirement planning, responsible investing, smart beta and factor investing, risk parity, user-defined portfolios, and cash management.

4.4.1 Goals-based investing

It is well-known that individuals do not treat all of their investments the same, but rather practice what is known as *mental accounting* (Thaler, 1985, 1999). In particular, individuals assign different risk-return preferences to their savings, depending on how they see each "chunk" of money being used in the future. *Goals-based wealth management* is an investment and portfolio management approach that focuses directly on investors' financial goals.

Shefrin and Statman (2000) suggest, in behavioral portfolio theory (BPT), that investors behave as if they have multiple mental accounts. Each mental account has varying levels of aspiration, depending on its goals. BPT results in a portfolio management framework where investors are goal-seeking (aspirational) while remaining concerned about downside risk. For example, an individual may view their homeownership differently from that of their stock portfolio. They may tolerate a larger loss in their stock portfolio, but may not be willing to risk losing their home. Specifically, rather than to trade off return versus risk as in MVO, investors should trade off goals versus safety (Brunel, 2003; Nevins, 2004; Chhabra, 2005; Brunel, 2015). As one would expect, BPT leads to normatively different statements about the optimal portfolio than those based on modern portfolio theory (see, for example, Das et al., 2010; Parker, 2016; Das et al., 2018).

Based on its intuitive appeal and ability to model individual investor's financial goals in a flexible and customizable way, goals-based wealth management principles are emerging as the predominant approach in retail investment management and have been adopted by a number of robo-advisors.

Glide paths

For many investment goals, it is prudent for the investor to reduce risk as they approach the end of their investment horizon. Most robo-advisors provide portfolio risk advice that considers the investment horizon during the creation of a new investment goal. Many will also automatically adjust target portfolio allocations according to a *glide path* as the end of the investment horizon nears (Gomes et al., 2008; Mladina, 2014).

Automatically adjusting target portfolio allocations helps clients stay closer to their recommended risk level as it changes over time. In the absence of automation, an investor is unlikely to make the necessary portfolio adjustments with appropriate frequency. Instead they may prefer to revisit their portfolio quarterly or annually. This can result in an investor portfolio taking too much risk, particularly towards the end of a goal's investment term where glide paths can be particularly steep. Figure 4.6 illustrates that adjusting a goal portfolio's target stock allocation annually results in higher risk level for much of the year as compared to more frequent monthly adjustments.

Glide path automation becomes particularly powerful in the presence of periodic deposits. As a portfolio's target allocation updates, new deposits can be used to rebalance the portfolio. This can reduce or eliminate the need to sell assets to rebalance towards the new allocation. With fewer sales, rebalancing has lower transaction costs and less potential for realizing capital gains.

4.4.2 Retirement planning

One of the central goals for the individual investor is retirement. Therefore, many robo-advisors provide wealth projections and financial advice to meet the complexities that arise in retirement planning.

Determining the amount of money needed for retirement can prove challenging to individual investors. Robo-advisors help clients ascertain an appropriate target

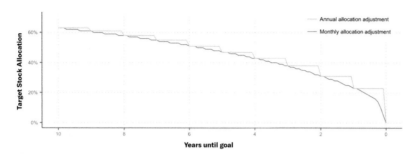

Figure 4.6 Glide path for a major purchase goal with monthly and annual adjustments to the portfolio target allocation. Annual adjustments result in higher risk level for much of the year as compared to more frequent monthly adjustments.

retirement balance needed to replace a desired income level. These projections need to account for inflation, cost of living in the retirement location, Social Security benefits, tax rates and longevity.

Investors planning for retirement easily find themselves overwhelmed by the various accounts that are available to them such as 401(k)s, Roth IRAs, Traditional IRAs, and taxable accounts. In fact, the choice between a Roth or Traditional IRA/401(k) is perplexing to many individual investors. Each account has different tax treatments and contribution limits. For 401(k)s, potential employer matching can vary. Robo-advisors seek to simplify these decisions with clear advice on which accounts to contribute to and in what order of priority to maximize the after-tax value over the life of the goal.

Once retirement is reached, a robo-advisor can provide advice on how much to safely withdraw from the retirement account. Wealth projections in the presence of recurring withdrawals allow the client to understand how long their retirement balance might last under various market conditions and withdrawal patterns.

The complexity of retirement planning is another ripe area for the flexibility of reinforcement learning: it is capable of handling the many facets of retirement planning, including tax effects, complex return dynamics, time-varying bond yield curves, and uncertain life expectancies. Irlam (2020) find that for simple scenarios, RL solutions perform similarly to known optimal solutions. For more complex scenarios, where an optimal solution is unknown, machine learning is found to outperform other common approaches.

4.4.3 Responsible investing

Increasingly, individuals wish to consider the environmental, social and governance (ESG) practices underlying their investments. There are several challenges for robo-advisors here. First is to find liquid ETFs that can be used for constructing ESG-aware portfolios. Second, as investor preferences related to responsible investing can vary considerably, robo-advisors must provide new levels of customization.

Robo-advisors address heterogeneous investor preferences towards responsible investing through a number of different approaches. Some robo-advisors offer portfolios that seek to balance environment, social, and governance factors. However, choosing which factors to emphasize and by how much can prove challenging without additional investor input. Portfolio construction becomes more challenging when the investor also seeks to enhance a dimension with no clear monetary correspondence.

To address this challenge, some robo-advisors have allowed for greater flexibility in constructing an ESG portfolio. These approaches have been implemented with ETFs and by constructing individual stock portfolios. To increase flexibility, certain robo-advisors will elicit client preferences on ESG issues, then alter the securities held in the client portfolio to reflect those preferences. This approach may include overweighting or substituting funds with specific ESG focuses. Alternatively, individual stock portfolios may be constructed to overweight companies that align with client preferences and underweight or avoid those that do not. Commonly, ESG scoring methodologies from third party data providers are used to quantify responsible investing criteria and serve as an additional input to match ESG exposures with client preferences during portfolio construction.

4.4.4 Smart beta and factor investing

It is common that in investment management innovative products are first introduced in institutional contexts, and only after significant delay are they later, gradually made available in the retail space. Such has been the case with smart beta offerings. Smart beta is a set of investment strategies that aim at capturing market inefficiencies and risk premia in a rules-based and transparent way. Today, many smart beta strategies are available as liquid ETFs. Perhaps somewhat surprisingly, there are few smart beta products available in the robo-advisory space.

In the institutional space, smart beta ETFs have seen increased interest due to improved technology, reduced costs and an evergrowing body of empirical evidence of what drives underlying risk premia. Because of this evolution, the retail market today has a strong foundation to build upon when implementing smart beta solutions. Huang and Kolm (2019) argue that smart beta is ripe for the retail audience and discuss some of the challenges in implementing smart beta in robo-advisory offerings.

Most often, exposure to smart beta factors is at the fund level. Portfolios are constructed using smart beta ETFs, where implementation of the strategy is managed by the fund. However, some robo-advisors extend the direct indexing framework to include factor investing by building portfolios of individual stocks with exposures to factors such as value and momentum. Here, the robo-advisor needs to select the factors, score stocks on each factor, and manage the factor exposures at the portfolio level. Smart beta implementations with individual stocks share the same tax and cost advantages as other direct indexing strategies.

Agather and Gunthorp (2018) suggest that smart beta products are increasingly popular amongst financial advisors across Canada, UK and the US for the purpose of diversifying client portfolios and to express strategic views. Continued increase in liquid smart beta ETFs will provide robo-advisors an opportunity to offer a broader suite of smart beta options to the retail audience.

4.4.5 Risk parity

Risk parity is another strategy that originated in the institutional space and has been made available to retail investors (Roncalli, 2013). Some robo-advisors make this strategy available on their platforms. Risk parity offers clients an alternative way to manage risk in their portfolios where risk contributions from various asset classes are balanced.

Because risk parity seeks to balance risk when determining asset weights, it does not require any assumptions about the future growth rate of assets. Instead, only estimates of future asset variances and covariances are required for portfolio construction. This is appealing as volatility (or its squared form, variance) is generally more stable to estimate compared to returns.

Risk parity strategies tend to have large allocations to bonds, due to lower volatility of bonds compared to stocks. Thus, in order to reach investor return targets, risk parity may require the use of leverage through futures contracts or total return swaps. A levered investment strategy is not commonly employed by individual investors. By offering risk parity on their platform, a robo-advisor needs to advise clients on the appropriateness of the strategy and its proper amount of leverage. In practice, risk parity only makes up a small portion of client portfolios, with most clients' accounts being managed predominantly using MVO.

4.4.6 User-defined portfolios

Some clients may want to define their own portfolios in order to express specific market views or to account for investments held outside of a robo-advisor's platform. For instance, consider an investor who has a 401(k) account at their current employer. This individual may wish to employ an asset location strategy across their 401(k) account and their taxable account with their robo-advisor. In this case, the investor would (a) hold fixed income securities, which are more tax-inefficient, in the 401(k); and (b) hold more stocks, which are relatively tax-efficient, in the taxable account. In order to maintain the desired total portfolio mix across both accounts, the investor would deviate from the balanced portfolio recommended by their robo-advisor by overweighting equities assets, since they will be balanced by the bonds in the 401(k) account.

Some robo-advisors offer this flexibility while still allowing the client to benefit from automated portfolio management, such as automated rebalancing and tax optimization features. As a client customizes their portfolio, a robo-advisor provides immediate feedback by calling attention to the risk and return profile of the overall portfolio and its level of diversification.

4.4.7 Cash management

Robo-advisor clients would like advice and management of their entire financial lives. The most common financial transactions by individuals involve cash moving to and from their checking and saving accounts. Consequently, many robo-advisors have expanded their mandate and provide advice across both investment and cash accounts.

Several robo-advisors offer sweep accounts for cash management, which allow greater flexibility and higher interest than a traditional savings account. Commonly, a robo-advisor will deposit cash with multiple banks, affording them greater flexibility in allocating client monies at higher interest rates and providing higher FDIC insurance limits.

Robo-advisors extend their financial advice and goal-based investing capabilities to cash management, allowing customers to create goals for their cash savings, similar to an investing goal. The benefits of goal-based investing – increased accuracy of projections and mental accounting – are extended to clients' cash positions and incorporated into their financial plans. The growth of the cash account can be projected based on current and expected changes in interest rates and the client's planned future deposits.

The addition of daily financial transactions, either from a synced external checking account or a checking account provided by the robo-advisor, presents additional information to help clients manage their cash positions. The wealth of data from daily spending transactions provides opportunities for robo-advisors to deploy machine learning techniques to understand spending and income patterns and make recommendations about appropriate saving and spending levels. For example, natural language processing (NLP) techniques may be used on a transaction's memo field in order to identify similar transaction types. Clustering algorithms may also be used to identify similar transactions and detect predictable patterns in cash flows.

Robo-advisors use algorithms to make recommendations on optimal levels of liquidity in checking accounts to safely cover immediate and expected expenses. These often rely on structured and semi-structured user expense data to predict future spending patterns. Additional automation can be built on top of client cash flow predictions including automatic sweeps of excess funds from a checking account to savings or investment accounts with higher risk-adjusted returns.

In this new world where robo-advisors and fintech companies, more broadly, may become the gatekeepers of the access to banking services, an expansion of their services is crucial to compete with traditional banks. Many robo-advisors have moved strategically in this direction and are offering a suite of services including cash and checking accounts, debit cards, lending and retirement services (McCann, 2020). Recognizing that automation cannot replace human touch everywhere, some robo-advisors are offering customers financial advice from financial planners on staff who can assist in making decisions such as how to start investing; address significant life events (changing job, having a child, purchasing a home, etc.); and plan for college, marriage and retirement, to name a few.

4.5 Conclusion

Robo-advisors are playing an important role in offering institutional investment services to the individual investor. Few individual investors have the resources, time or expertise to build or manage portfolios with optimization software, monitor positions day to day, or optimize taxes. More than ten years in the making, robo-advisors continue to grow both in size and product offerings. Key reasons contributing to their success include:

Low cost. Fully automatic algorithm-driven management of client portfolios that significantly lowers the cost of financial advice and wealth management.

Personalization and customization. By providing a general investment management framework and a suite of financial services, robo-advisors can in a highly scalable fashion customize investment strategies and provide a digital banking experience that suits the specific needs of each individual investor.

Anywhere, anytime convenience. People have gotten used to accessing their digital lives and beyond through mobile apps on their smartphones and laptops. Robo-advisors provide their clients with this convenience for their investment portfolios and other aspects of their financial life, anywhere and anytime.

Wealth management services for the masses. Robo-advisors are making many sophisticated investment and financial advisory services, that in the past were only accessible to high net worth individuals, available to anyone at low cost.

The automated nature of their investment processes and their access to client data make robo-advisors well positioned to take advantage of the latest advances in ML, particularly as they provide more adaptive and individualized investment plans.

References

Agather, Rolf, and Gunthorp, Peter. 2018. Smart beta: 2018 global survey findings from asset owners. Tech. Rept. FTSE Russell. https://www.ftserussell.com/research/smart-beta-2018-global-survey-findings-asset-owners.

Alsabah, Humoud, Capponi, Agostino, Ruiz Lacedelli, Octavio, and Stern, Matt. 2020. Robo-advising: learning investors' risk preferences via portfolio choices. *Journal of Financial Econometrics*, **19**(2), 369–392.

Andrus, Danielle. 2017. 4 ways robo-advisors improve client onboarding, https://www.thinkadvisor.com/2017/06/16/4-ways-robo-advisors-improve-client-onboarding/?slreturn=20211128113846.

Apelfeld, Roberto, Fowler Jr., Gordon B., and Gordon Jr., James P. 1996. Tax-aware equity investing. *Journal of Portfolio Management*, **22**(2), 18.

Arnott, Robert D., Berkin, Andrew L., and Ye, Jia. 2001. Loss harvesting: What's its worth to the taxable investor? *Journal of Wealth Management*, **3**(4), 10–18.

Ban, Gah-Yi, El Karoui, Noureddine, and Lim, Andrew E.B. 2018. Machine learning and portfolio optimization. *Management Science*, **64**(3), 1136–1154.

Bengen, William P. 1994. Determining withdrawal rates using historical data. *Journal of Financial Planning*, **7**(4), 171–180.

Bengio, Yoshua, Lodi, Andrea, and Prouvost, Antoine. 2020. Machine learning for combinatorial optimization: a methodological tour d'horizon. *European Journal of Operational Research*, **290**(2), 405–421.

Benhamou, Eric, Saltiel, David, Ungari, Sandrine, and Mukhopadhyay, Abhishek. 2020. Bridging the gap between Markowitz planning and deep reinforcement learning. ArXiv:2010.09108.

Berkin, Andrew L., and Ye, Jia. 2003. Tax management, loss harvesting, and HIFO accounting. *Financial Analysts Journal*, **59**(4), 91–102.

Bianchi, Milio, and Brière, Marie. 2022. Robo-advising: Less AI and more XAI? Augmenting algorithms with humans-in-the-loop. Pages 33-58 in: *Machine Learning and Data Sciences For Financial Markets*, A. Capponi and C.-A. Lehalle (eds). Cambridge University Press.

Bol, Kieran, Kennedy, Patrick, and Tolstinev, Dmitry. 2020. *The State of North American Retail Wealth Management*. 9th Annual PriceMetrix Report. `https://www.mckinsey.com/industries/financial-services/our-insights/the-state-of-north-american-retail-wealth-management`.

Bouchey, Paul, Nemtchinov, Vassilii, Paulsen, Alex, and Stein, David M. 2012. Volatility harvesting: Why does diversifying and rebalancing create portfolio growth? *Journal of Wealth Management*, **15**(2), 26–35.

Bradbury, Meike A.S., Hens, Thorsten, and Zeisberger, Stefan. 2015. Improving investment decisions with simulated experience. *Review of Finance*, **19**(3), 1019–1052.

Brunel, Jean L.P. 1997. The upside-down world of tax-aware investing. *Trusts And Estates (Atlanta)*, **136**, 34–42.

Brunel, Jean L.P. 2001. A tax-efficient portfolio construction model. *Journal of Wealth Management*, **4**(2), 43–49.

Brunel, Jean L.P. 2003. Revisiting the asset allocation challenge through a behavioral finance lens. *Journal of Wealth Management*, **6**(2), 10–20.

Brunel, Jean L.P. 2015. *Goals-Based Wealth Management: An Integrated and Practical Approach to Changing the Structure of Wealth Advisory Practices*. John Wiley & Sons.

Callan, Victor, and Johnson, Malcolm. 2002. Some guidelines for financial planners in measuring and advising clients about their levels of risk tolerance. *Journal of Personal Finance*, **1**, 31–44.

Capponi, Agostino, Olafsson, Sveinn, and Zariphopoulou, Thaleia. 2019. Personalized robo-advising: Enhancing investment through client interaction. ArXiv:1911.01391.

Charpentier, Arthur, Elie, Romuald, and Remlinger, Carl. 2021. Reinforcement learning in economics and finance. *Computational Economics*, `https://doi.org/10.1007/s10614-021-10119-4`, 38 pages.

Chhabra, Ashvin B. 2005. Beyond Markowitz: A comprehensive wealth allocation framework for individual investors. *Journal of Wealth Management*, **7**(4), 8–34.

Constantinides, George M. 1984. Optimal stock trading with personal taxes: Implications for prices and the abnormal January returns. *Journal of Financial Economics*, **13**(1), 65–89.

Das, Sanjiv, Markowitz, Harry, Scheid, Jonathan, and Statman, Meir. 2010. Portfolio optimization with mental accounts. *Journal of Financial and Quantitative Analysis*, **45**(2), 311–334.

Das, Sanjiv R., and Varma, Subir. 2020. Dynamic goals-based wealth management using reinforcement learning. *Journal of Investment Management*, **18**(2).

Das, Sanjiv R., Ostrov, Daniel, Radhakrishnan, Anand, and Srivastav, Deep. 2018. A new approach to goals-based wealth management. *Journal of Investment Management*, **16**(3), 1–27.

Dietvorst, Berkeley J, Simmons, Joseph P, and Massey, Cade. 2015. Algorithm aversion: People erroneously avoid algorithms after seeing them err. *Journal of Experimental Psychology: General*, **144**(1), 114.

Dixon, Matthew, and Halperin, Igor. 2020. G-Learner and GIRL: Goal Based wealth management with reinforcement learning. ArXiv:2002.10990.

Faloon, Michael, and Scherer, Bernd. 2017. Individualization of robo-advice. *Journal of Wealth Management*, **20**(1), 30–36.

Gomes, Francisco J., Kotlikoff, Laurence J., and Viceira, Luis M. 2008. Optimal life-cycle investing with flexible labor supply: A welfare analysis of life-cycle funds. *American Economic Review*, **98**(2), 297–303.

Grable, John, and Lytton, Ruth H. 1999. Financial risk tolerance revisited: The development of a risk assessment instrument. *Financial Services Review*, **8**(3), 163–181.

Guillemette, Michael, and Finke, Michael. 2014. Do large swings in equity values change risk tolerance. *Journal of Financial Planning*, **27**(6), 44–50.

Horan, Stephen M., and Adler, David. 2009. Tax-aware investment management practice. *The Journal of Wealth Management*, **12**(2), 71–88.

Huang, Lisa, and Kolm, Petter N. 2019. Smart beta investing for the masses: The case for a retail offering. In *Equity Smart Beta and Factor Investing for Practitioners*, K. Ghayur, R.G. Heaney, and S.C. Platt (eds). Wiley.

Huang, Lisa, Khentov, Boris, and Vaidya, Rukun. 2016. Asset location methodology. `https://www.betterment.com/resources/asset-location-methodology`

Huss, John, and Maloney, Thomas. 2017. Portfolio rebalancing: Common misconceptions. `https://www.aqr.com/Insights/Research/White-Papers/Portfolio-Rebalancing-Common-Misconceptions`

Inderst, Roman, and Ottaviani, Marco. 2009. Misselling through agents. *American Economic Review*, **99**(3), 883–908.

Irlam, Gordon. 2020. Machine learning for retirement planning. *Journal of Retirement*, **8**(1), 32–39.

Jeffrey, Robert H., and Arnott, Robert D. 1993. Is your alpha big enough to cover its taxes? *Journal of Portfolio Management*, **19**(3), 15–25.

Khalil, Elias B. 2016. Machine learning for integer programming. Pages 4004–4005 of: *Proc. 25th IJCAI*.

Kinniry Jr., Francis M., Jaconetti, Colleen M., DiJoseph, Michael A., Zilbering, Yan, and Bennyhoff, Donald G. 2014. Putting a value on your value: Quantifying Vanguard advisor's alpha. Vanguard Research, `https://www.vanguard.co.uk/content/dam/intl/europe/documents/en/quantifying-vanguards-advisers-alpha.pdf`.

Kitces, Michael. 2017. Financial advisor fees comparison: All-in costs for the typical financial advisor? `https://www.kitces.com/blog/independent-financial-advisor-fees-comparison-typical-aum-wealth-management-fee/`.

Klement, Joachim. 2015. Investor risk profiling: an overview. `https://www.cfainstitute.org/en/research/foundation/2015/investor-risk-profiling-an-overview`

Kolm, Petter N., and Ritter, Gordon. 2020. Modern perspectives on reinforcement learning in finance. *Journal of Machine Learning in Finance*, **1**(1), 28 pages.

Kolm, Petter N., Tütüncü, Reha, and Fabozzi, Frank J. 2014. 60 Years of Portfolio Optimization: Practical Challenges and Current Trends. *European Journal of Operational Research*, **234**(2), 356–371.

Kolm, Petter N., Ritter, Gordon, and Simonian, Joseph. 2021. Black–Litterman and beyond: The Bayesian paradigm in investment management. *Journal of Portfolio Management*, **47**(5), 91–113.

Liu, Berlinda. 2019.   SPIVA US scorecard. `https://www.spglobal.com/spdji/en/documents/spiva/spiva-us-year-end-2019.pdf`.

Lo, Joseph, Campfield, Darrell, and Brodeur, Michael. 2018.   Onboarding: The imperative to improve the first experience. `https://www.broadridge.com/white-paper/onboarding-the-imperative-to-improve-the-first-experience`

Markowitz, Harry M. 1952. Portfolio selection. *Journal of Finance*, **7**(1), 77–91.

McCann, Bailey. 2020. Robo advisers keep adding on services. *Wall Street Journal*, March 8, 2020.

Mladina, Peter. 2014.   Dynamic asset allocation with horizon risk: Revisiting glide path construction. *Journal of Wealth Management*, **16**(4), 18–26.

Montier, James. 2002. *Behavioral Finance: Insights into Irrational Minds and Markets*. John Wiley & Sons.

Neuneier, Ralph. 1998. Enhancing Q-learning for optimal asset allocation. Pages 936–942 of: *Advances in Neural Information Processing Systems*.

Nevins, Daniel. 2004. Goals-based investing: Integrating traditional and behavioral finance. *Journal of Wealth Management*, **6**(4), 8–23.

Parker, Franklin J. 2016. Goal-based portfolio optimization. *Journal of Wealth Management*, **19**(3), 22–30.

Reichenstein, William R. 2004. Tax-aware investing: Implications for asset allocation, asset location, and stock management style. *Journal of Wealth Management*, **7**(3), 7–18.

Richards, Carl. 2012. *The Behavior Gap: Simple Ways to Stop Doing Dumb Things with Money*. Penguin.

Rollén, Sebastian. 2019.   How we estimate the added value of using Betterment. `https://www.betterment.com/`.

Roncalli, Thierry. 2013. *Introduction to Risk Parity and Budgeting*. Chapman & Hall/CRC Financial Mathematics Series.

Rossi, Alberto G., and Utkus, Stephen P. 2020. Who benefits from robo-advising? Evidence from machine learning. Working paper, available at SSRN 3552671.

Scherer, Bernd. 2017. Algorithmic portfolio choice: Lessons from panel survey data. *Financial Markets and Portfolio Management*, **31**(1), 49–67.

Schooley, Diane K., and Worden, Debra Drecnik. 2016. Perceived and realized risk tolerance: Changes during the 2008 financial crisis. *Journal of Financial Counseling and Planning*, **27**(2), 265–276.

Schwab, Charles. 2018.   The rise of robo: Americans' perspectives and predictions on the use of digital advice. `https://content.schwab.com/web/retail/public/about-schwab/charles_schwab_rise_of_robo_report_findings_2018.pdf`.

Scott, Jason S., Sharpe, William F., and Watson, John G. 2009. The 4% rule: At what price? *Journal of Investment Management*, **7**(3), 31–48.

SEC. 2006 (May).   Questions advisers should ask while establishing or reviewing their compliance programs. `https://www.sec.gov/info/cco/adviser_compliance_questions.htm#:~:text=Annual%20review&text=Does%20or%20did%20the%20review,Are%20any%20changes%20under%20consideration%3F`.

SEC. 2019 (July). Commission interpretation regarding standard of conduct for investment advisers. `https://www.sec.gov/rules/interp/2019/ia-5248.pdf`.

Shefrin, Hersh, and Statman, Meir. 2000. Behavioral portfolio theory. *Journal of Financial and Quantitative Analysis*, **35**(2), 127–151.

Stein, David M. 1998. Measuring and evaluating portfolio performance after taxes. *Journal of Portfolio Management*, **24**(2), 117–124.

Stein, David M., and Narasimhan, Premkumar. 1999. Of passive and active equity portfolios in the presence of taxes. *Journal of Wealth Management*, **2**(2), 55–63.

Stein, David M., Vadlamudi, Hemambara, and Bouchey, Paul W. 2008. Enhancing active tax management through the realization of capital gains. *Journal of Wealth Management*, **10**(4), 9–16.

Takahashi, Shuntaro, Chen, Yu, and Tanaka-Ishii, Kumiko. 2019. Modeling financial time-series with generative adversarial networks. *Physica A: Statistical Mechanics and its Applications*, **527**, 121261.

Thaler, Richard H. 1985. Mental accounting and consumer choice. *Marketing Science*, **4**(3), 199–214.

Thaler, Richard H. 1999. Mental accounting matters. *Journal of Behavioral Decision Making*, **12**(3), 183–206.

Uhl, Matthias W., and Rohner, Philippe. 2018. Robo-advisors versus traditional investment advisors: An unequal game. *Journal of Wealth Management*, **21**(1), 44–50.

Wilcox, Jarrod W., Horvitz, Jeffrey E., and DiBartolomeo, Dan. 2006. Investment Management for Taxable Private Investors. Research Foundation of CFA Institute. `https://www.cfainstitute.org/-/media/documents/book/rf-publication/2006/rf-v2006-n1-3933-pdf.ashx`.

Yu, Shi, Chen, Yuxin, and Dong, Chaosheng. 2020. Learning time varying risk preferences from investment portfolios using inverse optimization with applications on mutual funds. ArXiv:2010.01687.

5

Recommender Systems for Corporate Bond Trading

Dominic Wright[a], Artur Henrykowski[a], Jacky Lee[a]
and Luca Capriotti[b]

Abstract

In this chapter, we illustrate how market makers in the corporate bond business can effectively employ machine learning based recommender systems. These techniques allow them to filter the information embedded in Requests for Quote (RFQs) to identify the set of clients most likely to be interested in a given bond, or, conversely, the set of bonds that are most likely to be of interest to a given client. We consider several approaches known in the literature and ultimately suggest the so-called *latent factor collaborative filtering* as the best choice. We also suggest a scalable optimization procedure that allows the training of the system with a limited computational cost, making collaborative filtering practical in an industrial environment. Finally, by combining the collaborative filtering approach with more standard content-based filtering, we propose a methodology that allows us to provide some narrative to the recommendations provided.

5.1 Introduction

Market makers, also known as dealers, play the role of liquidity providers in the financial markets by quoting both buy and sell prices for many different financial assets and trading on their own account. Market makers are compensated for the service they provide (and the risk they take in holding inventory) by charging for an asset at any given time a higher (ask) price than the one they are willing to pay (bid). In some situations, the dealer is able to match pairs of clients willing to buy and sell the same asset, thus monetizing the full bid-ask spread instantly. More frequently, to satisfy a client's requests, market makers have to enter in outright positions and hold an inventory. The value of such inventory is typically subject to variation in prices due to market dynamics. Some of the market risk can be hedged by trading appropriate assets or derivatives. However, market makers have to deal with the residual risk that they may close their position at a loss because of adverse market moves. In addition to increasing balance sheet costs, the longer

[a] Department of Mathematics, University College London
[b] Department of Mathematics, University College London; Columbia University; and New York University, Tandon School of Engineering
Published in *Machine Learning And Data Sciences For Financial Markets*, Agostino Capponi and Charles-Albert Lehalle© 2023 Cambridge University Press.

Client	Bond Id	Quantity	Side
Client1	Bond1	400K	Buy

Figure 5.1 An example RFQ.

an open position sits on the inventory, the higher the risk market makers will incur a loss by the time they close it. It is therefore of paramount importance to turn around inventory as efficiently as possible.

In the corporate bond business, market makers need to handle large amounts of requests from clients, typically in the form of electronic inquiries – or Requests for Quote (RFQs). An example of RFQ is shown in Fig. 5.1. If the quote offered by the dealer is accepted by the client, the market maker enters into a position (either long or short the bond), bearing market risk until it is closed out. When a position needs to be closed out, sales teams contact clients who may be interested in taking over that position. However, since it is generally possible to contact only a very small fraction of the dealer's clients, it is of paramount importance for salespeople to be intimately familiar with the clients' trading preferences.

This is particularly challenging because, at any given time, most of the market activity is concentrated on a small number of bonds while trading on the majority of the inventory happens fairly infrequently. This is known as the *long-tail* problem. In this situation an effective *recommender system* (RS) – that is, an algorithm able to identify the small population of clients that are most likely to be interested in a given bond – could bring substantial value to the dealer and, by virtue of providing a better service, to its clients.

Similar problems are not uncommon in many other industries. A common challenge of e-commerce websites is helping customers sort through a large variety of offered products to easily find the ones they are most interested in. Music and video streaming services, like Netflix or Spotify, are equipped with algorithms which aim at personalized recommendations to their users to improve their experience. One of the tools commonly employed for these tasks are RS (Goldberg et al., 1992; Linden et al., 2003).

In this chapter, we investigate the application to corporate bond trading of RS based on machine-learning techniques able to use the information embedded in RFQs. Two main categories of models are described: content-based filtering and collaborative filtering, along with approaches to training and testing that we trialed on example data. We also suggest a few practical optimizations that are essential for reducing the time necessary to train the algorithms at a level that makes their usage viable in an industrial setting.

5.2 Bond recommender systems

Broadly speaking, RS fall into two categories: *content-based* and *collaborative* filtering, differing in their interactions with *users* (the agents we would like to

Company	Currency	Coupon	Rating	Maturity	Industry	Region	Callable
Issuer1	USD	7.5/Variable	Ba3/BB-	Perpetual	Financials	EMEA	True

Figure 5.2 An example for bond static data.

make recommendations to) and *items* (the set of objects we need to recommend). Content-based filtering (CBF) methods (Lops et al., 2011) create profiles for users and items in order to characterize their nature and then try to match the user–item pairs via metrics based on the similarity between profiles. In the context of the bond market making business, each bond can be characterized by a set economic features and each client by the features of the bonds that have historically interested them. Collaborative filtering (CF) (Goldberg et al., 1992), instead, only employs past user behavior in order to detect users with similar preferences over items. For example, in the specific context, by knowing what bonds clients have historically inquired, one can infer the interdependencies among clients and bonds and thus find potential associations for new client–bond pairs.

5.2.1 Content-based filtering

In general, CBF models assume that clients are looking for bonds with certain economic characteristics or features. For example, some clients are more likely to trade long-dated bonds within certain industries. Based on this idea, the profile of the clients can be represented by the features of previously traded bonds. If bond i has similar features to those traded by client u, then it makes sense to recommend bond i to client u. This can be formalized as follows.

Each bond is characterized by a set of *categorical* features (e.g., region, industry, coupon type, callability) and *numerical* features (e.g. maturity, yield). An example of bond static data is shown in Fig. 5.2. We indicate with $y^i = [C^i; N^i]^t$ the set of features for bond i, $i = 1, \ldots, N$, where $C^i = [C^i_1, \ldots, C^i_{n_c}]^t$ is the vector of categorical features and $N^i = [N^i_1, \ldots, N^i_{n_n}]^t$, with $N^i_k = [N^i_{k1}, \ldots, N^i_{kn_b}]^t$, is the matrix of numerical features. Here n_c and n_n are the number of categorical and numerical features, respectively, and n_b is the number of intervals in which the domain of numerical features is discretized into. For each bond i, the entries of the vector C^i correspond to the categorical features characterizing the bond (e.g., region: European, industry: financial, coupon type: fixed), encoded as strings. This gives a more concise representation compared to transforming categorical data to a numerical binary representation (Huang, 1997, 1998). For each vector N^i_k, $k = 1, \ldots, n_n$, only the single entry corresponding to the interval in which the bond's kth feature falls into is equal to one, while the remaining $n_b - 1$ components are set to zero.

Similarly, we indicate with $x^u = [C^u; N^u]^t$ the set of features for client u, $u = 1, \ldots, M$. In this case, for each vector N^u_k, $k = 1, \ldots, n_n$, the entry N^u_{kl} is set to the frequency of bonds with the kth feature falling in the lth interval, as

	Item 1	Item 2	Item 3	Item 4	Item 5	Item 6
User 1	X		X		X	
User 2		X	X			
User 3				X		X
User 4					X	
User 5	X	X		X		X
User 6			X	X		
User 7	X	X	X		X	X
User 8		X		X		
User 9			X			

Figure 5.3 User-item interaction matrix.

observed in the historical sample of RFQs. Each entry of the vector C^u is set to the most commonly observed categorical feature.

A simple way to use such data to make recommendations is to compute the *predicted* preference of client u for bond i, $\hat{p}_{ui}$, as a pseudo inner product of the client profile vector x^u and the bond features vector y^i. We define this pseudo inner product as

$$\hat{p}_{ui} \equiv \sum_{k=1}^{n_c} C_k^u C_k^i + \sum_{k=1}^{n_n} N_k^u \cdot N_k^i , \qquad (5.1)$$

where we have used the notation

$$C_k^u C_k^i \equiv \delta(C_k^u, C_k^i) , \qquad (5.2)$$

with $\delta(a, b)$ the generalized Kronecker delta. This is equal to 1 if $a = b$ (for any type a and b, including strings as in this case), and zero otherwise.

The estimator above assumes all the features are of equal importance. A more accurate estimator can be obtained by weighting each of the features in (5.1) and computing the following *weighted* pseudo inner product:

$$(x^u \circ y^i)^t \cdot w^u \equiv \sum_{k=1}^{n_c} w_k^u C_k^u C_k^i + \sum_{k=1}^{n_n} w_{n_c+k}^u N_k^u \cdot N_k^i, \qquad (5.3)$$

where $\circ$ is the element-wise product and $w^u = [w_1^u, \ldots, w_{n_f}^u]^t$, with $n_f = n_n + n_c$, is the feature weight vector for client u. Given the clients' and the bonds' features, the objective is to find the optimal weights w_k^u for each client u. This leads to the ridge regression (Hastie et al., 2009) based content filtering:

$$\min_{\boldsymbol{w}^u} \sum_{u,i} c_{ui}(p_{ui} - (\boldsymbol{x}^u \circ \boldsymbol{y}^i) \cdot \boldsymbol{w}^u)^2 + \lambda_{\text{reg}}||\boldsymbol{w}^u||^2 \ . \tag{5.4}$$

Here, following Hu et al. (2008) the *preference* of client u for bond i, given the historical sample of RFQs in a given time horizon, p_{ui}, is defined as the binary variable

$$p_{ui} = \begin{cases} 1 \text{ if client } u \text{ traded bond } i \\ 0 \text{ otherwise.} \end{cases} \tag{5.5}$$

This is also known as user-item interaction matrix. A pictorial representation is given in Fig. 5.3. The *confidence* we have in such preference, c_{ui}, is defined as

$$c_{ui} = 1 + \alpha r_{ui}, \tag{5.6}$$

where r_{ui} is given by the total notional traded by client u in bond i, and α is an adjustable parameter.

By differentiation, the set of weights minimizing Eq. (5.4) reads:

$$\boldsymbol{w}^u = (A^u)^{-1} \boldsymbol{B}^u \ , \tag{5.7}$$

where

$$A^u_{lm} = x^u_l x^u_m \sum_{i=1}^N c_{ui} y^i_l y^i_m + \lambda_{\text{reg}} \delta_{lm} \ ,$$

$$B^u_l = x^u_l \sum_{i=1}^N c_{ui} p_{ui} y^i_l \ ,$$

and N is the number of bonds. After obtaining $\boldsymbol{w}^u$, the preference of client u for bond i can be computed by:

$$\hat{p}^{\text{CBF}}_{ui} = (\boldsymbol{x}^u \circ \boldsymbol{y}^i) \cdot \boldsymbol{w}^u \ . \tag{5.8}$$

For any client u, the larger $\hat{p}_{ui}$, the more likely client u is to be interested in bond i.

5.2.2 Collaborative filtering

In contrast to CBF, collaborative filtering (CF) can be performed using only the information contained in the user-item observations matrix (Hu et al., 2008). The entries in this matrix can be either user ratings for explicit feedback data or built from the preference and indicator matrices, Eqs. (5.5) and (5.6), for implicit data. Given the observed entries in such a matrix, different methods can be used to compute the missing ones. At the core, these methods infer similarity between clients from their expression of interests in the same set of bonds, or, vice versa similarity between bonds from having attracted the interest of the same set of clients. This is pictorially illustrated in Fig. 5.4.

	Bond 1	Bond 2	Bond 3	Bond 4	Bond 5
Client 1	1	0	0	1	1
Client 2	0	1	0	0	1
Client 3	1	1	1	1	0
Client 4	1	1	1	0	0
Client 5	0	0	0	1	0

Figure 5.4 The basic idea of collaborative filtering: Clients 3 and 4 are similar; Bonds 1,2 and 3 are similar.

Neighborhood models

The most common approach to CF is based on neighborhood models (Hastie et al., 2009), which usually have two forms: user-oriented and item-oriented. User-oriented neighborhood CF (U-NCF) models try to estimate the unknown preference of a client for a bond given the preferences of similar clients. Conversely, item-oriented neighborhood CF (I-NCF) models use the information about a client's preference for similar bonds.

Given the user-item observation matrix p_{ui} in Eq. (5.5) for all client–bond pairs, the similarity between two bonds i and j can be computed as the following 'cosine' similarity:

$$s_{ij} = \frac{\sum_u p_{ui} p_{uj}}{\sqrt{\sum_u p_{ui}^2}\sqrt{\sum_u p_{uj}^2}}. \tag{5.9}$$

Likewise, the similarity between two clients u and v can be computed as:

$$s_{uv} = \frac{\sum_i p_{ui} p_{vi}}{\sqrt{\sum_i p_{ui}^2}\sqrt{\sum_i p_{vi}^2}}. \tag{5.10}$$

After computing the pairwise similarity s_{uv} or s_{ij} for all clients and bonds, the missing preference of client u over bond i can be decided by finding either the top k most similar clients or most similar bonds. For example, denoting the set of the top k most similar bonds to bond i by $S^k(i)$, the preference for client u over all bonds is:

$$\hat{p}_{ui}^{\text{I–NCF}} = \frac{\sum_{j \in S^k(i)} s_{ij} p_{uj}}{\sum_{j \in S^k(i)} s_{ij}}. \tag{5.11}$$

Similarly, denoting the set of the top k most similar clients to client u by $\tilde{S}^k(u)$, the preference for client u over all bonds can also be estimated as

$$\hat{p}_{ui}^{\text{U–NCF}} = \frac{\sum_{v \in \tilde{S}^k(u)} s_{uv} p_{uj}}{\sum_{u \in \tilde{S}^k(v)} s_{uv}}. \tag{5.12}$$

Latent factor models

The basic idea underlying latent factor models is the factorization of the client–bond observation matrix, p_{ui}, into a product of smaller matrices, which can be interpreted as the latent features for clients and bonds respectively, as depicted

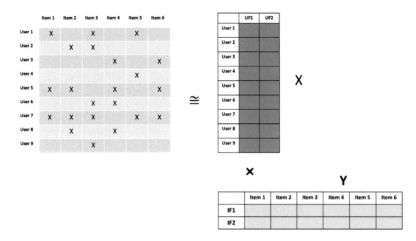

Figure 5.5 Illustration of matrix factorization for CF.

in Fig. 5.5. Following Hu et al. (2008), this can be formulated as the following (non-convex) optimization problem

$$\min_{\boldsymbol{x},\boldsymbol{y}} \sum_{u,i} c_{ui}(p_{ui} - \boldsymbol{x}^u \cdot \boldsymbol{y}^i)^2 + \lambda_{\text{reg}}(||\boldsymbol{x}^u||^2 + ||\boldsymbol{y}^i||^2), \qquad (5.13)$$

where $\boldsymbol{x}^u = [x_1^u, \ldots, x_K^u]^t$ and $\boldsymbol{y}^u = [y_1^i, \ldots, y_K^i]^t$ are the K latent factors vectors for client u and the bond i, respectively and c_{ui}, p_{ui} and λ_{reg} are defined as in Eq. (5.4).

A common approach to this optimization is the so-called alternating-least-squares (ALS) (Hu et al., 2008), where the optimal user-factors are computed assuming that the item-factors are fixed and vice versa until convergence. In this case:

$$\boldsymbol{y}^i = (X^t C^i X + \lambda_{\text{reg}} I)^{-1} X^t C^i \boldsymbol{p}^i \qquad (5.14)$$

$$\boldsymbol{x}^u = (Y^t \tilde{C}^u Y + \lambda_{\text{reg}} I)^{-1} Y^t \tilde{C}^u \tilde{\boldsymbol{p}}^u \qquad (5.15)$$

where $X_{lm} = x_m^l$, $Y_{lm} = y_m^l$, $C_{lm}^i = \delta_{lm} c_{li}$, $\tilde{C}_{lm}^u = \delta_{lm} c_{ul}$, $p_l^i = p_{li}$, $\tilde{p}_l^u = p_{ul}$ and I is the identity matrix in $\mathbb{R}^K$. After computing $\boldsymbol{x}^i$ and $\boldsymbol{y}^u$ for a number of iterations until the desired degree of convergence is achieved, recommendations can be made using the metric

$$\hat{p}_{ui}^{\text{LF}} = \boldsymbol{x}^u \cdot \boldsymbol{y}^i. \qquad (5.16)$$

Implementation

The latent factor CF is significantly more computationally demanding than the other methods. As a result, to make the approach practical, it is important to optimize the computation of Eqs. (5.14) and (5.15). First, one can avoid the matrix inversion and compute the solution of the linear systems by means of the conjugate gradient method (Hastie et al., 2009). This lowers the computational complexity per client or user (when using a standard matrix inversion) from $O(K^3)$ to $O(mn_I)$, where m is the number of non-zero entries in the matrix and n_I is the number of iterations for convergence. Second, a further optimization can be obtained by factorizing the matrices $X^t C^i X$ and $Y^t C^u Y$. As explained in Hu et al. (2008) this lowers the overall computational complexity per bond for calculating $X^t C^i X$ (resp. $Y^t C^u Y$) from $O(K^2 M)$ (resp. $O(K^2 N)$) to $O(K^2(1 + N_u))$ (resp. $O(K^2(1 + M_u)))$, where N_u (resp., M_i) is the number of nonzero elements in the matrix p_{ui} for client u (resp., for bond i). When applied across all bonds and clients this lowers the computational complexity of Eqs. (5.14) and (5.15) from

$$O(K^2 MN + K^3(M + N) + KMN + K^2(M + N))$$

to

$$O(K^2((N + M) + \sum_{u=1}^{M} N_u + \sum_{i=1}^{N} M_i) + mn_I(M + N) + KMN)$$

per iteration. Finally, as seen from Eqs. (5.14) and (5.15), the calculations for each client and each bond latent factors can be performed in parallel in a multi-threaded environment so that the training cost can be reduced by the number of threads available, which is currently of order 10 on a standard desktop computer. Our Cython-based[1] Python implementation was able to train the latent factor CF on our dataset within a few seconds on a desktop computer with commercially standard specifications.

5.3 Testing

A proportion of the RFQ data must be reserved for testing performance, we refer to this as the validation data set. For each item (user) the recommender system gives a list of users (items) ordered by preference, with the most highly recommended at the top. We step though this list of users (items) and check whether it is present in the validation data set. If so, we label it as a correct recommendation. Starting from the top of this list, the false positive (FPR) and true positive rate (TPR) are calculated for each item (user). The TPR is the proportion of correct recommendations so far in the ordered list of users (items) relative to the total number of correct recommendations. Similarly, the FPR is the proportion of incorrect recommendations relative to the total number of incorrect recommendations. Plotting the TPR on the y-axis versus the FPR on the x-axis gives a curve that is referred to as the receiver operating characteristic (ROC) curve. The ROC curve starts at $(0,0)$ and, after going through the entire list of users (items) in order, ends at $(1, 1)$. Each correct recommendation increases the TPR while the FPR remains constant, similarly each incorrect recommendation increases the FPR while the TPR remains constant. Calculating the area under this curve gives the *area under ROC curve (AUC) score* (Hastie et al., 2009), which is shown in grey in Fig. 5.6. We use this metric to compare the performance of

---

[1] http://cython.org/

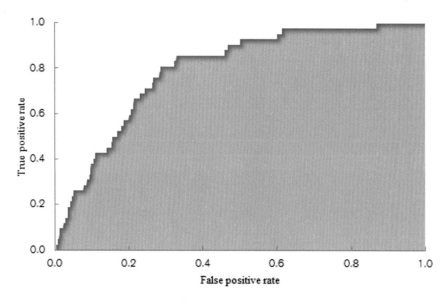

Figure 5.6 Example of ROC curve in red and AUC in grey. (While the plot has been generated with simulated data, the results are indicative of the performance that can be expected from recommender systems in practice.)

our models. An area of 1 represents a perfect performance and an area of 0.5 is equivalent to a random guess.

5.3.1 Hyperparameter optimization

Before performing the evaluation, the model 'hyperparameters' must be decided. These are α and λ_{reg} in Eq. (5.4) for the CBF; the number of 'nearest neighbors' k in Eqs. (5.11) and (5.12) for the neighborhood CF; and α, λ_{reg} and the number of latent factors K in Eq. (5.13) for the latent factor CF. A simple grid-based optimization approach with AUC as metrics and standard k-fold cross-validation (Hastie et al., 2009) can be used for this. For example, one could use 80% of the data for training the model for each combination of hyperparameters, 10% for validation (namely choosing the set of hyper-parameters providing the largest AUC on the validation test), and 10% for the actual back-testing (see Fig. 5.7). Similarly, for the latent factor CF a 3D grid search can be performed for the three hyperparameters α, λ_{reg} and K. For the neighborhood CF, only the number of neighbors k needs to be chosen.

5.3.2 Testing results

Our testing in a practical setting has shown that the collaborative filtering techniques perform best in terms of AUC score on corporate bond data. In particular, the latent factor collaborative filter gives the best performance.

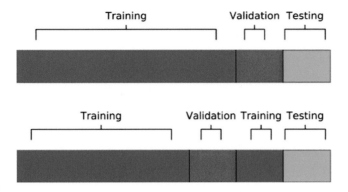

Figure 5.7 Cross validation.

5.4 Explaining recommendations

One drawback of the latent factor collaborative filter is that it does not provide an intuitive explanation of the recommendations, apart from a generic one such as *clients like you have demonstrated a preference for this bond*. In the following, we describe a possible way to build a content-based explain of the recommendations provided by a latent factor collaborative filter.

The idea is to run the content-based optimization of Eq. (5.4) after replacing the standard user-item interaction matrix in Eq. (5.5) with one in which the zero-entries are replaced with the implicit interest as defined by the recommendation score, Eq. (5.16), produced by the latent factor collaborate filter. As before, the recommendation score is given by

$$\hat{p}_{ui}^{\text{Exp}} = (\boldsymbol{x}^u \circ \boldsymbol{y}^i) \cdot \boldsymbol{w}^u, \tag{5.17}$$

with the pseudo inner product defined as in Eq. (5.3), while the resulting weights, appropriately normalized,

$$R_k^u = \frac{w_k^u}{\sum_{k=1}^{n_f} w_k^u}, \tag{5.18}$$

provide a measure of how relevant any bond feature is for each client (see Fig. 5.8).

More directly, the following quantity can be used to rank the contribution of each feature of bond i in the recommendation provided to client u:

$$C_k(u, i) = \frac{(\boldsymbol{x}^u \circ \boldsymbol{y}^i)_k w_k^u}{\hat{p}_{ui}^{\text{Exp}}}. \tag{5.19}$$

These two quantities provide useful insight in a client's past behavior that sales personnel can use to build a narrative around the recommendation (for a graphical representation, see Fig. 5.9).

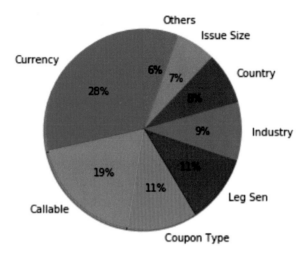

Figure 5.8 The preference demonstrated by a client for different bonds features.

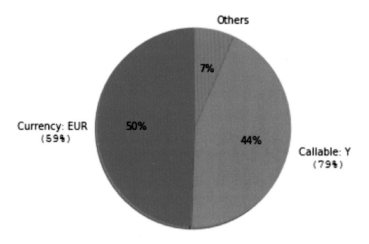

Figure 5.9 The contribution of each bond feature in a recommendation to a certain client.

5.5 Conclusions

In this chapter, we have presented an application of recommender systems, ubiquitous in eCommerce and streaming services, to the corporate bond market-making business, based on RFQ data. We outlined two sets of approaches: content-based filtering, which identifies similarities between bonds based on their features, and collaborative filtering, which identifies similarities based on user preferences. Based on the examples we studied, we found that the collaborative filtering techniques perform best in terms of AUC score. In particular, the latent factor collaborative filter gave the best performance.

An advantage of collaborative filtering (in addition to improved performance), which makes it well suited for the large variety of products available in the financial industry, is that it does not require the expert knowledge of product features required in content-based filtering. A disadvantage is that it is less transparent in providing insight into what features are attractive for a given client, thus possibly making approaching a client with a recommendation more awkward. Another drawback compared to content-based filtering, instead, is that it requires items to be in existence for a certain amount of time in order for some user-item interactions to be recorded, and to become eligible for recommendation.

While the latent factor collaborative filter is the most computationally intensive approach, we suggested computational optimizations that reduce the computational time required for training to a few minutes, thus making the approach practical in an industrial environment.

By combining the collaborative filtering approach with more standard content-based filtering, we proposed a methodology that allows us to provide some narrative to the recommendations provided.

Acknowledgments

We are grateful to Jodie Humphreys for initial work on this topic and Fengrui Shi for his help with the implementation. We are grateful to James Roberts and Nicholas Holgate for their continued collaboration. The views and opinions expressed in this article are those of the authors and do not represent the views of their employers. Analysis and examples discussed are based only on publicly available information.

References

Goldberg, David, Nichols, David, Oki, Brian M., and Terry, Douglas. 1992. Using collaborative filtering to weave an information tapestry. *Commun. ACM*, **35**(12), 61–70.

Hastie, Trevor, Tibshirani, Robert, and Friedman, Jerome. 2009. *The Elements of Statistical Learning*. Springer.

Hu, Yifan, Koren, Yehuda, and Volinsky, Chris. 2008. Collaborative filtering for implicit feedback datasets. Pages 263–272 of: *IEEE International Conference on Data Mining (ICDM 2008)*.

Huang, Zhexue. 1997. Clustering large data sets with mixed numeric and categorical values. Pages 21–34 of: *Proceedings of the 1st Pacific–Asia Conference on Knowledge Discovery and Data Mining, (PAKDD)*.

Huang, Zhexue. 1998. Extensions to the k-means algorithm for clustering large data sets with categorical values. *Data Mining and Knowledge Discovery*, **2**(3), 283–304.

Linden, Greg, Smith, Brent, and York, Jeremy. 2003. Amazon.com recommendations: Item-to-item collaborative filtering. *IEEE Internet Computing*, **7**(1), 76–80.

Lops, Pasquale, de Gemmis, Marco, and Semeraro, Giovanni. 2011. Content-based recommender systems: State-of-the-art and trends. Pages 73–105 of: *Recommender System Handbook*. Springer.

Part II

How Learned Flows Form Prices

6

Introduction to Part II
Price Impact: Information Revelation or Self-Fulfilling Prophecies?

Jean-Philippe Bouchaud[a]

6.1 Liquidity hide-and-seek

Why do prices move and why do markets participants trade? Most theoretical attempts at answering these questions distinguish between two different types of trades in financial markets:

- *Informed trades* are attributed to sophisticated traders with some information about the future price of an asset, which these traders buy or sell to eke out a profit.
- *Uninformed trades* are attributed either to unsophisticated traders with no access to (or the inability to correctly process) information, or to liquidity trades (e.g., trades triggered by a need for immediate cash, a need to reduce portfolio risk, or a need to offload an inventory imbalance). These trades are often called *noise trades,* because from an outside perspective they seem to occur at random: they do not correlate with long-term future price changes and they are not profitable on average.

In most cases, however, this seemingly intuitive partitioning of trades as informed or uninformed suffers from a problem: information is difficult to measure – and even to define. For example, is the observation of, say, a buy trade itself information? If so, how much? And how strongly might this impact subsequent market activity? For most large-cap US stocks, about 0.5% of the market capitalisation changes hands every day. Given that insider trading is prohibited by law, and that managers systematically over-performing their benchmark are scarce, it is highly implausible that a significant fraction of this vast market activity can be attributed to informed trades.

Still, as argued by Giamouridis et al. in their chapter, *some* trades can *sometimes* be informed, to the detriment of liquidity providers who bear the risk of being picked off by a truly informed trader. To minimise this risk, and perhaps even

[a] Capital Fund Management, Paris
Published in *Machine Learning And Data Sciences For Financial Markets*, Agostino Capponi and Charles-Albert Lehalle© 2023 Cambridge University Press.

to bait informed traders and to out-guess their intentions, liquidity providers only offer relatively small quantities for trade. This creates a kind of hide-and-seek game in financial markets: buyers and sellers want to trade, but both avoid showing their hands and revealing their true intentions. As a result, markets operate in a regime of small *revealed liquidity* but large *latent liquidity*. This simple observation leads to many empirical microstructure regularities that are discussed in the chapter by Lillo; see also Bouchaud et al. (2018). For example, the scarcity of available liquidity has an immediate and important consequence: large trades must be fragmented. More precisely, market participants who wish to buy or sell large volumes of a given asset must chop up their orders into smaller pieces, and execute them incrementally over time. Therefore, even an inside trader with clear information about the likely future price of an asset cannot use all of this information immediately, lest he or she scares the market and gives away the private information – this is the crucial insight behind Kyle's model, which is reviewed in depth by U. Cetin in his chapter.

But Kyle's framework misses a crucial point: it is unable to explain why the sign of the order flow (+ for buys, – for sells) is empirically found to be *long range correlated*, see Lillo (2022) and Bouchaud et al. (2018). The long memory of the market order signs is a striking stylised fact in market microstructure. At first sight, the effect is extremely puzzling, because it appears to contradict the near-absence of predictability in price series. And in fact, for this very reason, Kyle's theory predicts that the sign of the order flow must be uncorrelated. A recent attempt to reconcile Kyle's model with long range correlated order flow has recently been proposed in Vodret et al. (2021).

6.2 Information efficiency vs. statistical efficiency

From a conceptual viewpoint, the most important consequence of the chronic dearth of liquidity in financial markets is that prices cannot be in equilibrium, in the traditional sense that supply and demand are matched at some instant in time. Since transactions must be fragmented, the instantaneous traded volume is much smaller than the underlying "true" supply and demand waiting to be matched. Part of the imbalance is necessarily latent, and can only be *slowly* digested by markets, as emphasized in Bouchaud et al. (2009).

But if prices cannot be in equilibrium (except perhaps on long enough time scales), can we trust theories built on the postulate that a fundamental price exists, and to which the traded price is strongly anchored? Does this provide the correct foundation to understand how order flows impact prices? Can such theories account for the volatility observed in real markets?

Consider the case of a typical US large-cap stock, say Apple. Each second, one observes on average 6 transactions and of the order of 100 order book events for this stock alone. Compared to the typical time between news arrivals that could potentially affect the price of a the company (which are on the scale of one every few days, or perhaps hours), these frequencies are extremely high, suggesting that market activity is not only driven by news.

Indeed, the number of large price jumps is found to be much higher than the number of relevant news arrivals (Cutler et al., 1989; Fair, 2002; Joulin et al., 2008). In other words, most large price moves seem to be unrelated to news, but rather to arise endogenously from trading activity itself. As emphasised by Cutler, Poterba & Summers: *The evidence that large market moves occur without identifiable major news casts doubts on the view that price movements are fully explicable by news.* It is as if price changes *themselves* were the main source of news, and induce a feedback that creates excess volatility and, most probably, those price jumps that occur without any news at all (Fosset et al., 2020). Interestingly, all quantitative volatility/activity feedback models (such as ARCH-type models or Hawkes processes: Bacry et al., 2015) suggest that at least 80% of the price variance is induced by self-referential effects. This adds credence to the idea that the lion's share of the short- to medium-term activity of financial markets is unrelated to any fundamental information or economic effects. The reason why prices are close to a random walk at high frequencies is not because fundamental value is a martingale but rather because of competition between liquidity providers and/or high frequency statistical arbitrage strategies, which remove any exploitable price pattern induced by the autocorrelation of the order flow. This is essentially the content of the "propagator model" (Bouchaud et al., 2004), in which impact decay is fine-tuned to compensate the long memory of order flow, and causes the price to be close to a martingale – as explained in Lillo (2022) and in Bouchaud et al. (2018). This mechanism makes prices *statistically efficient* without necessarily being *fundamentally efficient*. In other words, competition at high frequencies is enough to whiten the time series of returns, but not necessarily to ensure that prices reflect fundamental values.

Such a scenario is also vindicated by the ever more pervasive (machine) "learning" approach to high-frequency trading and market-making, for which what matters most is not the fundamental price but any detectable statistical pattern in the intertwined order flow and price dynamics. Reinforcement learning algorithms devised to take advantage of such statistical regularities end up whitening the sequence of price returns.

In fact, the Kyle model can itself be rephrased as a learning mechanism which produces white returns. Now suppose that there is no fundamental price at all, like in zero-intelligence models of the order book (Daniels et al., 2003; Bouchaud et al., 2018), but that the market maker believes that there is one, with a volatility calibrated on past price changes. In the Kyle model, the market maker will set the price as to make returns unpredictable, with precisely the assumed volatility! Because there is in reality no "terminal time" when a "true price" is revealed, the price will wander off endlessly, entirely driven by order flow. Interestingly, this self-fulfilling picture may offer a natural framework to understand how volatility can feedback on itself and generate wild, intermittent fluctuations of prices (Bouchaud, 2011; Fosset et al., 2020).

6.3 Price "discovery" vs. price "formation"

Echoing the discussion in Section 6.1 above, there are two strands of interpretation for the correlation between price and order flow, which reflect the great divide between efficient-market proponents (who believe that the price is always close to its fundamental value) and skeptics (who believe that the dynamics of financial markets are primarily governed by order flow). At the two extremes of this spectrum are the following stories, as also mentioned in Lillo (2022):

1. *Agents successfully forecast short-term price movements, and trade accordingly.* This clearly results in a positive correlation between the sign of the trade and the subsequent price change(s), even if the trade by itself has no effect on prices. In this framework, a noise-induced trade that is based on no information at all should have no long-term impact on prices. This is the case within the Kyle model, where noise traders do not contribute to the price volatility. By this interpretation, if the price was meant to move due to information, it would eventually do so even *without* any trades.
2. *Price impact is a reaction to order-flow imbalance.* This view posits that the fundamental value is irrelevant, at least on short time scales, and that even if a trade reflected no information in any reasonable sense, then price impact would still occur.

Although both of the above scenarios result in a positive correlation between trade signs and price movements, they are conceptually very different. In the first story, trades reveal private information about the fundamental value, creating a so-called *price discovery* process. In a very Platonic way, the fundamental price exists *in abstracto* and trading merely reveals it.

In the second story, the act of trading itself impacts the price. In this case, one should remain agnostic about the information content of the trades, and should therefore speak of *price formation* rather than price discovery. If market participants believe that the newly established price is the "right" price and act accordingly, "information revelation" might simply be a self-fulfilling prophecy, as we argued in section 2 above in the context of the Kyle model.

The zero-intelligence Santa Fe model (Daniels et al., 2003) provides another illustration of the second story. In this model, the mechanism that generates impact can be traced back to the modelling assumption that at any given time, agents submitting orders always use the current mid-price as a reference point. Any upwards (resp. downwards) change in mid-price therefore biases the subsequent order flow in an upwards (resp. downwards) direction. This causes the model to produce a diffusive mid-price in the long run, resulting from the permanent impact of a purely random order flow, in a purely random market.

Whether prices are "formed" or "discovered" remains a topic of much debate. At this stage, there is no definitive answer, but because the line between real information and noise is so blurry, reality probably lies somewhere between these two extremes. Since some trades may contain real private information, and since other market participants do not know which trades do and do not contain such information, it follows that all trades must (on average) impact the price, at least temporarily – but maybe also permanently, as recently argued in the context of the "Inelastic Market Hypothesis" (see Gabaix and Koijen, 2020, van der Beck and Jaunin, 2021, and also Bouchaud, 2021). The question of how much real information is revealed by trades is obviously crucial in determining whether markets are closer to the first picture or the second picture. This is why the kind of data analyzed by Giamouridis et al. (2022) is absolutely fascinating. While empirical results using anonymous order flow suggest that the short term impact of random trades is similar to that of putative informed trades (Toth et al., 2017), data broken down by categories of market participants seem to reveal a much richer structure.

Such studies could at last reveal what really goes on in financial markets, and help formulating a consistent theory of prices and order flow. This is extremely important from many standpoints, including regulation and market stability. Indeed, if order flow turns out to be the dominant cause of price changes, all sorts of destabilising feedback loops can emerge (see e.g. Bouchaud, 2011; Fosset et al., 2020). Markets may not be stabilized by a strong anchor to an elusive fundamental value, but rather by carefully engineered market design and smart regulation.

References

Bacry, E., Mastromatteo, I., and Muzy, J. F. 2015. Hawkes processes in finance. *Market Microstructure and Liquidity*, **1**.

Bouchaud, J. P. 2011. The endogenous dynamics of markets: Price impact, feedback loops and instabilities. In *Lessons from the Credit Crisis*, Arthur M. Berd (ed). Risk Publications.

Bouchaud, J.P. 2021. The inelastic market hypothesis: A microstructural interpretation. Available at SSRN 3896981.

Bouchaud, J. P., Gefen, Y., Potters, M., and Wyart, M. 2004. Fluctuations and response in financial markets: the subtle nature of 'random' price changes. *Quantitative finance*, **4**(2), 176–190.

Bouchaud, J. P., Farmer, J. D., and Lillo, F. 2009. How markets slowly digest changes in supply and demand. In Pages 57–160 of: *Handbook of Financial Markets: Dynamics and Evolution*, Thorsten Hens and Klaus Reiner Schenk-Hoppé (eds). North-Holland.

Bouchaud, J. P., Bonart, J., Donier, J., and Gould, M. 2018. *Trades, Quotes and Prices: Financial Markets under the Microscope*. Cambridge University Press.

Cetin, U. 2022. Price formation and learning in equilibrium under asymmetric information. In *Machine Learning and Data Sciences for Financial Markets: A Guide to Contemporary Practice*, A. Capponi and C-A. Lehalle (eds). Cambridge University Press.

Cutler, D. M., Poterba, J. M., and Summers, L. H. 1989. What moves stock prices? *Journal of Portfolio Management*, **15**(3), 4–12.

Daniels, M. G., Farmer, J. D., Gillemot, L., Iori, G., and Smith, E. 2003. Quantitative model of price diffusion and market friction based on trading as a mechanistic random process. *Physical Review Letters*, **90**(10), 108102.

Fair, R. C. 2002. Events that shook the market. *Journal of Business*, **75**(4), 713–731.

Fosset, A., Bouchaud, J. P., and Benzaquen, M. 2020. Endogenous liquidity crises. *Journal of Statistical Mechanics: Theory and Experiment*, **2020**(6), 063401.

Gabaix, X., and Koijen, R. S. 2020. In search of the origins of financial fluctuations: The inelastic markets hypothesis. Available at SSRN 3686935.

Giamouridis, Daniel, Papaioannou, Georgios V., and Rosenzweig, Brice. 2022. Deciphering how investors' daily flows are forming prices. In *Machine Learning and Data Sciences for Financial Markets: A Guide to Contemporary Practice*, A. Capponi and C-A. Lehalle (eds). Cambridge University Press.

Joulin, A., Lefevre, A., Grunberg, D., and Bouchaud, J. P. 2008. Stock price jumps: news and volume play a minor role. *Wilmott Magazine*, Sept/Oct 2008, 1–7.

Lillo, F. 2022. Order flow and price formation. In *Machine Learning and Data Sciences for Financial Markets: A Guide to Contemporary Practice*, A. Capponi and C-A. Lehalle (eds). Cambridge University Press.

Toth, B., Eisler, Z., and Bouchaud, J. P. 2017. The short-term price impact of trades is universal. *Market Microstructure and Liquidity*, **3**(02), 1850002.

van der Beck, P., and Jaunin, C. 2021. The equity market implications of the retail investment boom. Available at SSRN 3776421.

Vodret, M., Mastromatteo, I., Tóth, B., and Benzaquen, M. 2021. A stationary Kyle setup: microfounding propagator models. *Journal of Statistical Mechanics: Theory and Experiment*, **2021**(3), 033410.

7

Order Flow and Price Formation

Fabrizio Lillo[a]

Abstract

I present an overview of some recent advancements on the empirical analysis
and theoretical modeling of the process of price formation in financial markets
as the result of the arrival of orders in a limit order book exchange. After discussing critically the possible modeling approaches and the observed stylized
facts of order flow, I consider in detail market impact and transaction cost of
trades executed incrementally over an extended period of time, by comparing
model predictions and recent extensive empirical results. I also discuss how the
simultaneous presence of many algorithmic trading executions affects the quality
and cost of trading.

7.1 Introduction

Understanding the price formation process in markets is of paramount importance
both from an academic and from a practical perspective. Markets can be seen
as a collective evaluation system where the 'fair' price of an asset is found by
the aggregation of information dispersed across a large number of investors. Non
informed investor (roughly speaking, intermediaries and market makers) also
participate to the process, in the attempt of profiting from temporal or 'spatial'
(i.e. across market venues or assets) local imbalance between supply and demand,
thus acting as counterparts when liquidity is needed.

Order submission and trading constitute the way aggregation of information
is obtained. The process through which this information is impounded into price
is highly complex and might depend on the specific structure of the investigated
market. Prices emerge as the consequence of the arrival of orders, which in turn
are affected, among other things, by the recent dynamics of prices. Despite the
fact this feedback process is of paramount importance, the complexity of the
process is only partial understood and many different models are able to provide
only a partial description of it.

The two main components of the price formation process are order flow and

[a] University of Bologna and Scuola Normale Superiore, Italy
Published in *Machine Learning And Data Sciences For Financial Markets*, Agostino Capponi and
Charles-Albert Lehalle © 2023 Cambridge University Press.

market impact. The former refers to the dynamical process describing the arrival of buy and sell orders to the market. As detailed below, this is in general a complicated process whose modelization is challenging because of the high dimensionality and the presence of strong temporal correlations. Market impact is, broadly speaking, the correlation between an incoming order and the subsequent price change. Since in each trade there is a buyer and a seller, it is not a priori obvious whether a given trade should move on average the price up or down. Considering the role of information on prices, one can advance few alternative explanations on the origin of market impact (for a more detailed discussion on this point, see Bouchaud et al., 2009):

- **Trades convey a signal about private information.** The arrival of new private information causes trades, which cause other agents to update their valuations, which changes prices.
- **Agents successfully forecast short-term price movements and trade accordingly.** Thus there might be market impact even if these agents have absolutely no effect on prices. In the words of Hasbrouck 'orders do not impact prices. It is more accurate to say that orders forecast prices'.
- **Random fluctuations in supply and demand.** Fluctuations in supply and demand can be completely unrelated to information, but the net effect regarding market impact is the same. In this sense impact is a completely mechanical (or statistical) phenomenon.

In the first two explanations, market impact is a friction but it is also the mechanism that let prices adjust to the arrival of new information. In the third explanation, instead, market impact is unrelated to information and may merely be a self-fulfilling prophecy that would occur even when the fraction of informed traders is zero. Identifying the dominating mechanism in real markets is therefore of fundamental importance to understand price formation.

Price formation and market impact are very relevant also from the practitioner perspective of minimizing transaction costs. For medium and large size investors, the main source of trading costs is the one associated with market impact, since by executing progressively an order in response to a given trading decision, the price is moved in a direction adverse to the trader and the later trades/orders become more and more expensive. Minimizing market impact cost by designing optimal execution strategies is an active field of research in academia and industry (Almgren and Chriss, 2001).

Market impact is thus a critical quantity to understand the informativeness of a trade as well as the cost for the trader, but its nature and properties are still vigorously debated. Also the empirical analysis and characterization of price formation and order flow is challenging, despite the availability of very high-resolution market data. This is due in part to the difficulty of controlling several potential confounding effects and biases and in part to the fact that market data are often not sufficient to answer some fundamental questions. However recent years have witnessed a booming increase in the number of empirical studies of market impact and transaction cost analysis of algorithmic executions, and we are now able to model them with a great level of accuracy and to dissect the problem under different conditioning settings.

In this chapter I will review of some of these recent advancements in the modeling and empirical characterization of order flow, price formation, and market impact. I will focus on a specific, yet widespread, market mechanism namely the Limit Order Book, which is presented in section 7.2. In Section 7.3 I will present an overview of the different modeling approaches to order flow and price formation, clarifying the different choices that the modeler has and why and when some should be preferred to others. Section 7.4 reviews some results on order flow modeling and in Section 7.5 I will consider cross-impact, i.e. how the price of an asset responds to trades (and orders) executed on a different asset. The study of cross-impact is important when a portfolio of assets is liquidated, since cross-asset effects can deteriorate the quality of the trade if not properly included in the optimal execution scheme. Section 7.6 presents empirical evidences and theoretical results on the market impact of metaorders, i.e. sequences of orders sent by the same trader as a consequence of a unique trading decision. Section 7.7 discusses the problem of the simultaneous presence of many metaorders and how market impact behaves under aggregation. The response of price to multiple simultaneous metaorders has been recently termed co-impact and its characterization and modeling is important to study the effect of crowding on price dynamics and cost analysis. Finally, in Section 7.8 I will briefly present some open problems in the field of order flow and price formation and I will delineate few possible research avenues.

7.2 The limit order book

Market microstructure is, by definition, very specific about the actual mechanism implemented in the investigated market, because it can affect the price formation process. Financial markets are characterized by a variety of structures, and attempting to make a classification is outside the scope of this subchapter. In the following the focus will be on the most popular market mechanism, namely the Limit Order Book (LOB). A LOB, used actively also outside finance, is a mechanism for double auction and it is essentially a queuing system. Traders can decide to send their order (to buy or to sell) in two different ways: either they require to buy or sell a certain amount of shares at the best available price or they specify also the worst price at which they are willing to trade, thus the highest price for a buy or the lowest price for a sell. In the first case they send a *market order* (or, equivalently, a crossing limit order) and, unless there is no one on the opposite case, the order is executed and leads to a *transaction*. In the second case they send a *limit order*, where the specified price is called the *limit price*, and, if no one is on the opposite side with the same (or more favorable) price, the limit order is stored in a queue of orders at the limit price. An agent can decide to cancel a limit order at any time, for example if the price moves in an adverse direction. At any time, the highest standing limit price to buy (sell) is called bid (ask) or best bid (best ask). The mean price between the bid and the ask is the *midprice* and the difference is the *spread*. Orders arrive and are canceled asynchronously in the market and what is normally called 'the price' is something in between the best ask and the best bid. However, from the above description it is clear that at a certain time there is not a unique price in the market.

Broadly speaking three modeling approaches have been pursued: (i) econometric models, fitting for example large dimensional linear models on market data (queues, prices, order arrivals); (ii) statistical models of the LOB, where orders arrive in the market as a random point process and the resulting properties of the price is studied; (iii) computational agent based models, where a set of heterogeneous agents trade in a realistic environment, such as a LOB. I will mostly focus on the first two approaches, despite the fact the third approach often provide important insights, especially for testing alternative policy measures.

7.3 Modeling approaches

Modeling order flow and price formation is a challenging task because of the complexity of the system and the large number of variables potentially involved. The modeler has different choices to make, which in part depend on the available data and methods, but more often depend on the objectives of the model.

The first choice is whether to work in continuous or in discrete time. The first option is the most complete, i.e. it does not discard any information of the process of price formation. Inter-event times can, in fact, provide relevant information on the event is going to occur. For example, the price change triggered by a trade can depend on the time elapsed from the last trade. The modeling in discrete time disregards this information but allows to use all the machinery of discrete time series analysis (ARMA, VAR, etc). Discrete time modeling can be deployed either by advancing the clock by one unit any time a specific event occurs, for example a trade or an order arrival, or by considering a finite interval of physical time, say one second, and by considering aggregated quantities (e.g., average or end-of-period LOB, total order flow, one second price return, etc).

Let us consider first the continuous time approach and let K the number of available limit prices[1]. Denoting by p_t^i and q_t^i, with $i = 1, \ldots, K$, the price and the number, respectively, of shares on the ith limit price at time t, the LOB dynamics is described by the continuous-time process $\mathcal{L}_t = (p_t^i, q_t^i : i = 1, \ldots, K)$. The order flow is described by the multivariate marked point process whose components are the intensity of limit orders (λ_t^i), cancellations (v_t^i), and buy and sell market orders (μ_t^b and μ_t^s). The marks correspond to the volumes of the order, but for expositional simplicity we will assume that all the orders have unitary volume. In general the rates are not constant in time but can depend on the past history of the order flow, on the state of the order book ($\mathcal{L}_{\{s<t\}}$), and possibly on other covariates. Let us call O_t the multivariate point process generated by the intensities $(\lambda_t^i, v_t^i, \mu_t^b, \mu_t^s : i = 1, \ldots, K)$ and fully describing the order flow.

[1] Following Cont et al. (2010), we consider K large enough that it is unlikely that in the considered period orders are placed outside the grid.

It is important to stress that the state of the LOB at a given time is *completely determined* by the past order flow, plus some initial condition. In other words, once we choose an observable price p_t as reference (for example the midprice, the microprice, the ask), there exists a deterministic function F such that

$$\Delta p_t \equiv p_t - p_{t-\tau} = F(\mathcal{L}_{t-\tau}, O_{s \in (t-\tau, t)}) \tag{7.1}$$

Thus, from a purely econometric point of view, one could simply model the point process process O_t. This type of models is often analytically tractable and, for this reason, it has been explored in the past twenty years in several papers. The Zero Intelligence (or Santa Fe) model of Daniels et al. (2003) and the model in Cont et al. (2010), for example, consider independent Poisson processes for the different components of O_t. In order to include memory of the past order flow, Abergel and Jedidi (2015) considered instead a multivariate Hawkes processes able to describe auto- and lagged cross-correlation between the different components of the order flow.

The observable reference price in the LOB might not the fully reflect the economic conditions of the firm. For this reason, many models postulate the existence of an unobservable *efficient* price, which typically follows a semimartingale dynamics. Market data (e.g., trade or mid price) are a noisy version of the efficient price and a lot of econometric effort is devoted to remove the microstructure noise either to filter it or to estimate from ultra high frequency data some of its statistical properties (for example the volatility) useful in applications such as option pricing or risk management.

Although order flow determines uniquely (observable) price changes, it is possible that a better model (in terms, for example, of explained variance) is obtained by considering the order flow intensities as dependent on LOB state $\mathcal{L}_t$ or of a function of it, such as the reference price p_t. The reason is that, in general, the relation between intensities and past order flow is strongly non-linear and high dimensional. On the contrary, simpler and easier to estimate parametric models can be chosen by identifying the drivers that supposedly influence real traders decision to submit a specific type of order[2]. For example, real traders likely decide when and where to place an order taking into account the LOB state and the price. Thus one could use a model, which instead of modeling autonomously the order flow, makes the intensities dependent on the state of the LOB or of part of it (Huang et al., 2015).

Choosing to build models using functions of the order flow could be also useful when deciding to restrict the dimensionality of the problem and restricting it to a subpart of the order flow (and of LOB). The reasons for this choice are manyfold: either for data availability (especially in the old times), for purely statistical reasons (dimensionality reduction and improved estimation), because

---

[2] A recent alternative approach is to use modeling approach suited for high-dimensional non linear models, such as Deep Neural Networks. Even in these cases however it might be better to use LOB state rather than past order flow to forecast the LOB state at a future time. For example, Deep Learning has been used to forecast short term price movement from LOB state and recent order flow (see, for example, Sirignano, 2019).

one believes that some parts of the order flow (e.g., trades) might be more informative on price dynamics, or because we are interested in modeling a part of the order flow and its effect on price (for example our order flow in a real trading problem). In these cases the reduced model giving the price as a function of the (sub)order flow becomes stochastic and the randomness describes the effect of the unmodeled part of order flow. Following this line, one can take two approaches:

(1)

Treat the order flow as exogenous to the price. In this case, the model connects the considered part of the order flow to the price, but neglects the reverse effect, i.e. how price dynamics can affect order flow. Classical market microstructure models following this approach are the Roll model (and its generalization) and the Madhavan–Richardson–Roomans (Madhavan et al., 1997) model. More recently, the Transient Impact Model (TIM, see Bouchaud et al., 2004, 2009) and its generalizations with multiple propagators have been proposed to describe the relation between order flow and price. In a nutshell, the general TIM can be written in discrete time (see below for the continuous time version) as

$$p_t = \sum_{s<t} G_{\pi_s}(t-s)f(v_s) + \xi_t + p_{-\infty} \tag{7.2}$$

where v_s is the signed volume of the order at time s, $f(x) = \text{sign}(x)h(|x|)$ with $h(\cdot)$ a concave function[3], as observed empirically (Lillo et al., 2003), π_s indicates the type of event at time s (e.g., market order, limit order at a given price, etc), $G_{\pi_s}(t-s)$ is a function, termed *kernel* or *propagator*, quantifying the lagged effect of the event π_s at time s on the price at time t, and ξ_t is a noise term describing the effect on price of all the orders which are not considered in the model. If the functions G_{π_s} are not constant, Eq. (7.2) describes the *transient* nature of impact of event π_s, i.e. the fact that the effect of an order on price is not permanent, but declines with time. Many empirical analyses show that G_{π_s} are slowly decaying functions, typically well fitted asymptotically by a power law function. The transient nature of impact can be related to the very persistent autocorrelations of order flow (see next section) and to the diffusivity and efficiency of prices (see Bouchaud et al., 2009, for an extensive discussion). While the original TIM was considering only one type of events, namely market orders, subsequent works have included also limit orders and cancellations, while others have discriminated more finely between orders changing and not changing the price, since the (lagged) effect on price is shown to be different in these cases (Eisler et al., 2012; Taranto et al., 2018).

The TIM describes trades that impact prices, but with a time dependent, decaying impact function $G(t)$. One can interpret the same model slightly differently. Considering the model with one propagator associated with trades and taking

---

[3] To avoid dealing with volume fluctuations, strongly dependent on LOB state, often it is chosen $f(x) = \text{sign}(x)$. By sign(x) we denote the sign function.

$f(v_t) = \text{sign}(v_t) \equiv \epsilon_t$, one can rewrite the model as

$$\Delta p_t = G(1)(\epsilon_t - \hat{\epsilon}_t) + \tilde{\xi}_t, \tag{7.3}$$

$$\hat{\epsilon}_t = -\sum_{s>0} \frac{G(s+1) - G(s)}{G(1)} \epsilon_{t-s} \tag{7.4}$$

with $\tilde{\xi}_t = \Delta \xi_t$. The quantity $\hat{\epsilon}_t$ can be seen as the (linear) predictor of trade sign given the past history of the signs and the model tells us that the deviation of the realized sign ϵ_t from an expected level $\hat{\epsilon}_t$ impacts the price linearly and permanently. If $\hat{\epsilon}_t$ is the best possible predictor of ϵ_t, then the above equation leads by construction to an exact martingale for the price process. This model has been termed History Dependent Impact Model (HDIM) (Lillo and Farmer, 2004; Bouchaud et al., 2009; Taranto et al., 2018) and in the simple setting above is mathematically equivalent to the TIM when the best predictor is linear in the past order signs. Taranto et al. (2018) showed that as soon as one attempts to generalize the model to multiple event types, TIM and HDIM become no longer equivalent. In fact, the HDIM with different events can be rewritten as

$$\Delta p_t = G_{\pi_t}(1) \left[\epsilon_t + \sum_{s<t} \frac{\kappa_{\pi_s,\pi_t}(t-s)}{G_{\pi_t}(1)} \epsilon_s \right] + \tilde{\xi}_t \tag{7.5}$$

where $\kappa_{\pi_s,\pi_t}(t-s)$ is an influence kernel that depends on both the past event type π_s and the current event π_t. The matrix of two point kernels makes the model more complicated to estimate (see Taranto et al., 2018) while clearly HDIM reduces to TIM when $\kappa_{\pi_s,\pi_t}(t-s)$ is a function only of the triggering event π_s.

The approach taking as exogenous the order flow includes also other models connecting the order flow in a given time interval with the simultaneous price change. A paradigmatic example was given by Cont et al. (2014) who introduced a stylized model of the order book predicting a *contemporaneous* linear relation between the price change in a given time interval and a linear combination of level-I order flow components (the Order Flow Imbalance or OFI). The goodness of the model (for large tick stocks) is testified by the high R^2 empirically obtained in the linear regression between Δp_t and OFI.

(2)

The above approach, however, leaves the order flow as completely exogenous. The limits of this approach are evident for example when considering the negative lag response, i.e. the lagged cross correlation between past price returns and future order flow, i.e.

$$R(\tau) = \mathbb{E}[f(v_t)(p_{t+\tau} - p_t)] \tag{7.6}$$

with $\tau < 0$. When the simple TIM model with only one propagator for trades is calibrated using $R(\tau)$ (or other equivalent methods) with $\tau > 0$, it is observed that the predicted negative lag response is smaller than the empirical one, indicating, as intuitive, that a declining (increasing) price attracts in the future more buy (sell) trades (Taranto et al., 2018). To overcome this problem, one jointly models the dynamics of price and order flow. The seminal work in this context was

Hasbrouck (1991) who proposed a discrete time structural VAR model for the vector $\boldsymbol{x}_t = (\Delta p_t, f(v_t))'$, v_t being the volume of the market order at time t, of the form

$$A_0\boldsymbol{x}_t = \sum_{i=1}^{p} A_i\boldsymbol{x}_{t-i} + \boldsymbol{\xi}_t \qquad (7.7)$$

and $A_0 = \begin{pmatrix} 1 & g \\ 0 & 1 \end{pmatrix}$ and A_i are other 2×2 matrices to be estimated. The parameter g describes the immediate impact of a trade on price.

This model has been generalized in several directions. First, instead of considering only market orders and a single reference price (Hasbrouck used the midprice), Hautsch and Huang (2012) considered a vector containing bid and ask prices, the queue volume at the first three quotes on either sides of the LOB, and two dummy variables indicating the occurrence of buy and sell trades. They modeled this ten dimensional vector $\boldsymbol{y}_t$ with the Vector Error Correction Model

$$\Delta\boldsymbol{y}_t = \boldsymbol{\mu} + \alpha\boldsymbol{\beta}'\boldsymbol{y}_{t-1} + \sum_{i=1}^{p} \Gamma_i\Delta\boldsymbol{y}_{t-i} + \boldsymbol{\xi}_t \qquad (7.8)$$

where $\boldsymbol{\mu}$ is a constant vector, α and β denote the loading and cointegrating matrices and Γ_i are parameter matrices. Using impulse response functions, they measured impact, separately on bid and ask prices, of the arrival of a limit order on a queue (for an approach based on TIM, see Eisler et al., 2012), as well of course the impact of the arrival of a market order.

The second generalization considered instead continuous-time models. For example, Bacry and Muzy (2014) introduced a Hawkes process for the four-dimensional counting process $\boldsymbol{P}_t = (T_t^+, T_t^+, N_t^+, N_t^-)'$, where the first two components describe the arrival of buy and sell market orders and the last two the upward/downward movements of the reference price. The model for the intensity $\boldsymbol{\lambda}_t$ reads

$$\boldsymbol{\lambda}_t = \boldsymbol{\mu} + \int_{-\infty}^{t} \Phi(t-s)d\boldsymbol{P}_s \qquad (7.9)$$

where $\boldsymbol{\mu}$ is a constant baseline intensity, $\Phi(t-s)$ is a 4×4 matrix of kernels describing the lead–lag effects. Some care must be taken to model the immediate impact (the g term in the Hasbrouck model above) by introducing a Dirac delta component in some elements of $\Phi(t-s)$. While this model can be directly put in correspondence with the Hasbrouck's VAR, its generalizations can easily include other components of the order flow, such as limit orders and cancellations, or even to take into account order volume (see, for example, Rambaldi et al., 2017 discussed below). Moreover the continuous time approach allows to consider in the modeling the time between events, which are clearly neglected in discrete time approach á la Hasbrouck.

Understanding the relation between order flow and price is important for many reasons, such as to create realistic LOB simulators, to study the stability of markets under different rules, etc. However, it is often very relevant to study how price reacts to a specific sequence of orders generated by a specific trading decision, i.e. what we called a metaorder, because this is related to transaction cost (mainly due to market impact) and to the release of private information into the prices. It is evident that, since we are neglecting a very large fraction of orders, those due to all other traders, the relation between price dynamics and order flow of a single metaorder will become very noisy and large samples are required to obtain clean measures. Section 7.6 presents some empirical evidences on the market impact of metaorders and the price dynamics during their execution.

7.4 Order flow

Order flow is the process describing the arrival of orders in the market. If this works with a LOB, then the order flow is the multivariate point process describing the arrival of market orders, limit orders, and cancellations. Since limit orders (and cancellations) are also characterized by a limit price, a component of the multivariate process should be associated with each limit price, making immediately the problem high-dimensional.

Different point process models of order flow have been proposed, ranging from (compound) Poisson processes (Daniels et al., 2003) to self-exciting Hawkes processes (Abergel and Jedidi, 2015; Rambaldi et al., 2017). Here we first review some of the empirical evidences and stylized facts observed in order flow which make challenging the development of a realistic model of the LOB.

The first empirical evidence (even chronologically) is the so-called *diagonal effect* (Biais et al., 1995) i.e. orders of a specific type are more likely to be observed just after orders of the same type. Interestingly, Rambaldi et al. (2017) extended this analysis by using a Hawkes process approach and shows that, by including volume into the analysis, the diagonal effect is markedly stronger for same-type same-size orders (see below).

The diagonal effect is the manifestation of a more significant regularity observed in real LOBs: components of the order flow are extremely persistent, i.e. long range autocorrelated in time. To present a specific example, consider only market orders, where volume is neglected and time is discretized in such a way that it increases by one unit each time a new market order arrives. Denoting with ϵ_t the sign of the t-th market order, being equal to $+1$ (-1) for a buy (sell) order, it has been empirically shown (Lillo and Farmer, 2004; Bouchaud, 2011) that its autocorrelation function behaves asymptotically as $C(\tau) \equiv Cor[\epsilon_t, \epsilon_{t+\tau}] \sim \tau^{-\gamma}$ with $\gamma \in [0, 1]$. The empirical value of $\gamma \simeq 0.5$ shows that the market order sign is a long memory process, i.e. it lacks a typical time scale, with Hurst exponent $H = 1 - \gamma/2 \simeq 0.75$. A similar behavior has been observed for the other components of the order flow.

Several explanations have been proposed for this stylized fact, empirically observed in many different markets, asset classes, and time periods. The theories can be clustered in two classes: the first states that this is the effect of *herding*, i.e. several investors share the same view on the asset around the same time and trade accordingly. The second explanation is instead related to the fact that each trader creates an autocorrelated order flow and this is due to the practice of *order splitting*. Despite the fact that it is possible to create agent based models with either of the two mechanisms reproducing a correlated order flow, the assessment of the mechanism mainly responsible for this observation should be based on empirical evidences. Tóth et al. (2015) proposed a method to disentangle the herding and splitting contributions to the autocorrelation. The idea to use labeled data, i.e. data where the identity of the trader sending the order is known (even if anonymized). The autocorrelation function of order flow can be exactly decomposed as $C(\tau) = C_{split}(\tau) + C_{herd}(\tau)$, where the first (second) term is the contribution to the correlation considering only cases when the two market orders at time t and $t + \tau$ were placed by the same (different) trader(s). To measure the relative importance of the two components, Tóth et al. (2015) uses brokerage data. Some exchanges provide data where each order contains the coded identity of the broker who sent the order. An extensive investigation of LSE data shows unambiguously that $C_{split}(\tau)$ explains always more than 75% of $C(\tau)$ and, except for very short τ (one or two trades) the value is above 85%. This empirical finding strongly indicates that order splitting is the main driver of the correlated order flow. Similar results are obtained when using data with agents rather than brokers.

Market orders describe only part of the order flow dynamics. Among the several approaches to describe the full correlation structure of order flow (and price) we mention here the one using Hawkes processes. Generalizing a pioneering paper (Large, 2007), Bacry et al. (2016) modeled level-I order book data by using 8-dimensional Hawkes process whose components are market, limit, cancel order (buy and sell), and mid-price changes (up and down). Using a non-parametric estimation method, their main finding is that the dominating driver of the process is self-excitation (i.e., once more, the diagonal effect). The only exceptions are the mid-price components for which cross-excitation effects are strongly dominating. Moreover there is a significant mean reversion of price, since present price changes trigger price changes in the opposite direction. Interestingly most of the estimated diagonal kernels of the Hawkes process are slowly decreasing and well described by a power-law behavior, consistent with the long memory described above.

This type of approach can be generalized in several directions. For example, Muni Toke (2010) considered a full order book modeling using Hawkes process to disentangle the role and interaction between liquidity takers and providers. Another generalization considers the fact that orders are also characterized by a volume. Mathematically one can treat volumes as marks of the multivariate point process. Alternatively, when only few levels are considered in the analysis, one can bin the volume in D groups and consider the volume process as the superposition of D unmarked point processes, each of which corresponds to one of the possible D values that volume can take (Rambaldi et al., 2017). It is found that order size does matter, since kernels for different volume bins are quite different. Moreover large orders trigger cascade of small orders and small limit orders and cancellations strongly cross-excite, indicating hectic order repositioning by market makers.

Despite the fact one can decide to model the (marked) multivariate process autonomously, obtaining as a 'byproduct' the state of the LOB and, as a consequence, the price dynamics, it is likely that LOB state (but also recent price dynamics) describes better, or in a more parsimonious way, the local intensities of the orders arrival. The intuition is also related to the fact that traders often condition the decision of submitting an order to the state of the LOB. This suggests a class of models where intensities are a function of the LOB state. This approach was pioneered in Huang et al. (2015) where orders arrivals are modeled as Poisson processes whose intensity is a function of the current state of the LOB. Thus the empirically observed autocorrelation of the order flow is seen as a 'consequence' of the persistence of the queue size, but, conditionally on them, the arrival of orders follows a Poisson process. A natural extension of this model considers the order arrival intensity as a function both of the LOB state and of the past order flow. By using an Hawkes model with a kernel depending on both these variables, the State Dependent or Queue Reactive Hawkes models (Morariu-Patrichi and Pakkanen, 2018; Wu et al., 2019) have been proposed.

7.5 Cross impact

Up to now we have considered the market impact of trades and orders from a single asset. However, institutional investors rebalancing their portfolio very often trade simultaneously many assets. Both the optimal execution problem and the assessment of transaction costs of metaorders should therefore take into account possible interactions between assets.

Generically, there are three sources of interaction: (i) statistical dependence in asset prices, i.e. the well-known fact that returns of different assets are correlated; (ii) commonality in liquidity across assets (Chordia et al., 2000), i.e. the fact that, for example, the arrival rate of signed market (or limit) orders is correlated across assets, and (iii) quote revision effects, i.e. a trade in an asset can lead market makers to modify the bid and ask price in another related asset. The (lagged) correlation between price and order flow is termed *cross-impact*. As in the single asset case, the entire order flow completely determines the (reference) prices of the assets, thus one can trivially explain cross-impact (as well as self-impact) as a mere consequence of order flow dynamics and correlations. However, when conditioning to a subset of the order flow (for example a market order or the child orders of a metaorder), or when the *future* price evolution is of interest, the dynamics becomes stochastic, because of the unmodeled part of the order flow and suitably modeling cross-impact becomes critical for predictions or ex-ante cost estimation. Under this conditioning, cross-impact can be dissected as the result of the three sources described above (Benzaquen et al., 2017).

Cross-impact has been empirically studied recently, see e.g., Benzaquen et al. (2017), and Schneider and Lillo (2019), and its role in optimal execution has been highlighted in Mastromatteo et al. (2017) and Tsoukalas et al. (2019). We review here some results obtained when considering the market order flow. First, there is a measurable cross asset effect between order flow and price as can be measured by the cross response function, which generalizes Eq. 7.6 as

$$R^{ij}(\tau) = \mathbb{E}[f(v_t^i)(p_{t+\tau}^j - p_t^j)] \tag{7.10}$$

between an order on asset i at time t and the price change of asset j in $[t, t + \tau]$. $R^{ij}(\tau)$ is found to be different from zero and smaller than $R^{ii}(\tau)$ by a factor ~ 5 (Benzaquen et al., 2017; Schneider and Lillo, 2019). To investigate the source of this lagged correlation (Schneider and Lillo, 2019), by investigating empirically the high frequency dynamics of Italian sovereign bonds traded in an double auction market, find evidence that both lagged correlations of orders across assets and quote revisions play a role in forming cross-impact. This result is obtened by investigating the effect on price of bonds of isolated trades, i.e. trades on a bond such that no other trade is observed in other bonds a time window around it. This results indicates that both commonality in liquidity taking and price revision across assets are responsible for cross impact effects.

The TIM can be easily extended to the multi asset case. Considering the continuous time version of the TIM, the price of asset i at time t is

$$p_t^i = p_0^i + \sum_j \int_0^t f^{ij}(\dot{x}_s^j) G^{ij}(t - s) ds + \int_0^t \sigma_s^i dW_s^i \tag{7.11}$$

where $f^{ij}(\dot{x}_s^j)$ is the (instantaneous) impact on the price of asset i of trading asset j at a rate $\dot{x}_s^j$, $G^{ij}(\cdot)$ is the decay kernel describing the lagged effect of trading on price, σ_s^i is the volatility of asset i and W_s^i is a Wiener process. This model can be estimated on real data and it is found that: (i) f^{ij} is non-linear and well described by a power law function with an exponent smaller than 1 as for f^{ii}; (ii) the kernels G^{ij} also display a power law behavior similar to G^{ii}, but with a significantly smaller amplitude; (iii) the matrix $\{G^{ij}\}_{i,j=1,N}$ has a strong sectorial structure, similar to the one observed for returns (Benzaquen et al., 2017; Schneider and Lillo, 2019). These regularities and the modeling can be successfully used to design optimal portfolio executions (Mastromatteo et al., 2017).

Another important question is whether a model like (7.11) is always well posed or if there are trading strategies $\Pi = \{x_t\}_{t \in [0.T]}$ allowing for price manipulation. More precisely, we remind that a *round-trip trade* is a sequence of trades whose sum is zero, i.e. a trading strategy Π with $\int_0^T \dot{x}_t dt = 0$. A *price manipulation* is a round-trip trade Π whose expected cost $C(\Pi)$ is negative and the principle of no-dynamic-arbitrage states that such a price manipulation is impossible. For the multi asset TIM this implies that

$$C(\Pi) = \sum_{i,j} \int_0^T \dot{x}_t^i dt \int_0^t f^{ij}(\dot{x}_s^j) G^{ij}(t - s) ds \geq 0 \tag{7.12}$$

Schneider and Lillo (2019) proved a series of theorems constraining the form of f and G in order to avoid price manipulation. In particular authors showed that for bounded decay kernels instantaneous cross-impact f must be an odd and linear function of trading intensity and cross-impact from asset i to asset j must be equal to the one from j to i. When a non vanishing bid-ask spread is considered, some inequalities between spread, maximum trading speed, and cross-impact asymmetry must be verified to avoid price manipulation.

7.6 Market impact of metaorders

While the above models generically describe the relation between order flow and price, it is often of practical and academic interest to study the price dynamics when conditioning this relation to the execution of a (large) order by a specific trader following a single trading decision (a metaorder). In a seminal paper Kyle (1985) showed that for a trader with insider information it is optimal to split the volume to be executed in many transactions to be executed incrementally over an extended period of time.

Apart from the practical problem of minimizing transaction costs, the relation between metaorder execution and price dynamics is relevant to understand how information is incorporated into price. In fact, a metaorder by definition corresponds to a trading decision, which in general is the response of the trader to a piece of information. For this reason, it is important to understand, not only how the price changes during the execution of the metaorder, but also the long term level reached by the price when the transient effects due to the imbalance between supply and demand are dissipated.

Note again the difference between the relation between order flow and price dynamics when one considers all market participants or only the order flow generated by the trading decision of a single trader. As said above, price dynamics is a *deterministic* function of the order flow, while, when conditioning on the order flow of a specific trader, we expect a very noisy relation between signed volume and price change. However the objective here is not to have an high R^2 between them, but to answer the question: how much my trading activity consequent to a trading decision is going to affect *on average* the price?

Measuring market impact of metaorders is typically quite complicated because it requires suitable data that cannot be inferred from public (e.g., market) data. In fact, it is necessary to have access to data where one can track the activity of a single trader (broker or investor) following a given trading decision. For this reason, most of the empirical researches on this topic has been performed by using trading data from a given institution or trading desk (Torre and Ferrari, 1998; Almgren et al., 2005; Tóth et al., 2011). A part from the difficulty of accessing such data, this type of analyses runs the risk of being biased, since the sample is limited to a specific fund, which might have an idiosyncratic trading style. Market wide investigations of market impact of metaorders have been conducted by following two approaches. First, some exchanges exceptionally provide data where the coded identity of the market member is disclosed; thus by using suitable statistical methods, one can infer metaorders as sequences of trades/orders by the same member on the same asset with the same sign (see for example, Moro et al., 2009; Vaglica et al., 2010; Tóth et al., 2010). The other approach requires the access to databases collected by specialized institutions and containing information about the metaorder executions of a large set of investors. The most important example is probably the dataset provided by ANcerno Ltd, a transaction cost analysis firm for institutional investors. According to some estimates, it accounts for more than 10% of CRSP volume in US markets, thus providing a wide coverage of metaorder trading activity from many different institutional investors.

Methodologically there are two main problems in measuring market impact of metaorders. First, impact might depend on several conditioning variables, such as the market conditions at the time of the trade, the execution algorithm, etc., thus different conclusions might be drawn depending on the choice made. Second, market impact of metaorders is typically very noisy (see above), and, as a consequence, large datasets are required to obtain small error bars on the estimated impact. It is important to stress that market impact contributes as a *drift* term to the unperturbed dynamics of price. For this reason, in order to measure market impact it is fundamental to take into account the sign of the trade of the metaorder.

The main quantity of interest is the *metaorder impact* defined as

$$\mathcal{I}(Q,T) \equiv \mathbb{E}[\epsilon \Delta \log p | Q, T] \tag{7.13}$$

where $\Delta \log p$ is the logprice change between the end and the start of the metaorder, Q is the size of the metaorder (in shares), T is the metaorder duration (in seconds or in volume time, to minimize possible intraday effects), and ϵ is the sign of the metaorder (i.e $\epsilon = +1$ for a buy and $\epsilon = -1$ for a sell order). Notice that $\mathcal{I}(Q,T)$ is directly related to the average impact cost of a metaorder execution. In fact, for an execution described by $\Pi = \{x_t\}_{t \in [0,T]}$, where x_t is the asset position at time t, the expected implementation shortfall cost, i.e. the difference between the expected cost and the theoretical cost obtained by marking to market the trade with the initial price, is

$$C(\Pi) = \int_0^T \dot{x}_t \mathcal{I}(x_t, t) \, dt \tag{7.14}$$

where $\dot{x}_t$ is the time derivative of x_t (i.e. the trading speed). Market impact is better described in terms of normalized quantities which also allows to consider different assets and different time periods in the same analysis. The first key quantity is the daily (or volume) fraction, defined as $\phi = Q/V$, where V is the average daily traded volume[4]. The second quantity is the participation rate η, i.e the ratio between Q and the volume traded in the market during the execution. The third one is the metaorder duration T, which can be obtained from $T = \phi/\eta$.

Remarkably, many empirical studies (for example, Torre and Ferrari, 1998; Moro et al., 2009; Zarinelli et al., 2015; Tóth et al., 2011; Bershova and Rakhlin, 2013; Waelbroeck and Gomes, 2015 seem to agree on the validity of the 'square-root impact law', obtained when conditioning on the volume fraction of the

---

[4] V and σ (see below) are typically estimated over the past $10 \div 25$ trading days, excluding the day when the metaorder is executed.

metaorder

$$\mathcal{I}(Q,T) \approx Y\sigma\sqrt{\phi} \qquad (7.15)$$

where $Y \simeq 1$ is a numerical constant and σ is the daily volatility of the asset. Eq. 7.15 has been empirically shown also for disparate asset classes as options (Tóth et al., 2016) and Bitcoin (Donier and Bonart, 2015). This empirical relation is at first sight surprising: it indicates that the style of trading (for example using limit orders or market orders), the duration T of the execution, the trading speed (i.e. the number of shares traded per unit time), etc, are not relevant! These observations indicate that there must be some limitations to the validity of this 'law'. For example, the prefactor Y might depend on the trading algorithm.

More recent and extensive empirical analyses (Zarinelli et al., 2015) clarify the limits of the square-root impact law and highlight some deviations. Specifically:

- Considering a power law dependence on T and η, Zarinelli et al. (2015) investigated the regression

$$\mathcal{I}(Q,T) = A\, T^{\delta_T}\eta^{\delta_\eta} \cdot noise \qquad (7.16)$$

  to measure the dependence of metaorder impact *separately* on participation rate and duration. The fitted exponents are $\delta_T = 0.54 \pm 0.01$ and $\delta_\eta = 0.52 \pm 0.01$, and $A = 0.207 \pm 0.005$. The fact that both exponents are very close to $1/2$ indicates that $\mathcal{I}(Q,T) \approx \sqrt{\phi}$, at least as a first approximation, even when considering the effect of participation rate and duration.
- By considering $\mathcal{I}(Q,T)$ as a function only of $\phi = Q/V$, it is clear that a logarithmic function fits the data better than a power law function; this indicates a linear behavior of impact for small volumes and an extra concavity (likely due to a selection bias) for very large volumes. Below we will present two possible explanations for the linear behavior of the impact for small ϕ.
- By considering $\mathcal{I}(Q,T)$ as a function of both variables, Zarinelli et al. (2015) introduced the *market impact surface* and showed that a double logarithmic function outperforms the power law form of Eq. 7.16.

Interestingly, Eq. 7.16 can be predicted from the execution of a metaorder with constant participation rate in the continuous time TIM model with $f(v) = \text{sign}(v)|v|^\delta$ and $G(t) = t^{-\gamma}$ with $\delta_T = 1 - \gamma$ and $\delta_\eta = \delta$, thus $\delta = \gamma = 1/2$ (Gatheral, 2010).

Notice that the square root impact law is not related to the fact that volatility scales as the square root of (execution) time, which, for a fixed participation rate, is proportional to metaorder size. First, according to definition 7.13, market impact is a drift term and the inclusion of the metaorder sign ϵ is critical in the definition, while neglecting ϵ simply highlights the relation between volatility and volume. Second, the result of the regression of Eq. 7.16 indicates that, by controlling for both T and η, market impact is mainly dependent on $\sqrt{T\eta} = \sqrt{\phi}$. Third, as shown explicitly in Bucci et al. (2019b), market impact curves of metaorders with $\phi \gtrsim 5 \cdot 10^{-4}$ (roughly 80% of those in the ANcerno database) are independent of T and consistent with a square-root dependence on ϕ. Once more, impact of the remaining small metaorders are better described by a linear relation. Bucci et al. (2019b) also shows that the variance of impact depends linearly on T, as expected by the diffusivity of price, and this price uncertainty largely exceeds the average reaction impact contribution (which in turn explains why the R^2 in the market impact estimation is typically very small).

From a modeling perspective, the square root impact law and its deviations are well described by the Locally Linear Order Book (LLOB) model for the coarse-grained dynamics of latent liquidity (Donier et al., 2015). In a nutshell, LLOB is a limit order book model whose quantity of interest is the density $\varphi(x,t)$ of latent orders around price x at time t. Conventionally, one can choose φ to be positive for buy latent[5] orders (corresponding to $x < p(t)$, where $p(t)$ is here the current transaction price) and negative for sell latent orders (corresponding to $x > p(t)$). The coarse-grained dynamics of the latent liquidity is well described by

$$\partial_t \varphi = D\partial_{xx}\varphi - \nu\varphi + \lambda \ \text{sign}(y) + m \ \delta(y), \qquad (7.17)$$

where $y \equiv p(t) - x$, and ν describes order cancellation, λ new order deposition and $D\partial_{xx}$ limit price reassessments. The final "source" term corresponds to a metaorder of size Q executed at a constant rate $m = Q/T$, corresponding to a flux of orders localized at the transaction price $p(t)$. In the absence of a metaorder ($m = 0$), Eq. (7.17) admits a stationary solution in the price reference frame, which is linear when y is small, i.e. $\varphi_{st}(y) = \mathcal{L}y$ where $\mathcal{L} = \lambda/\sqrt{D\nu}$ is a measure of liquidity. The total transaction rate J is simply given by the flux of orders through the origin, i.e. $J \equiv D\partial_y\varphi_{st}|_{y=0} = D\mathcal{L}$.

In the limit of a slow latent order book (i.e. $\nu T \ll 1$), the price trajectory $p_m(t)$ during the execution of the metaorder (obtained as the solution of $\varphi(p_m,t) = 0$) is given by the self-consistent expression (Donier et al., 2015)

$$p_m(t) = p_0(t) + y(t), \qquad (7.18)$$

$$y(t) = \frac{m}{\mathcal{L}} \int_0^t \frac{ds}{\sqrt{4\pi D(t-s)}} \exp\left[-\frac{(y(t)-y(s))^2}{4D(t-s)}\right], \qquad (7.19)$$

where $p_0(t)$ is the price trajectory in the absence of the metaorder that starts at $t = 0$ and ends at $t = T$. Interestingly, when impact is small, i.e. if $\forall t, s$ it is $|y(t) - y(s)| \ll D(t-s)$, the above expression for the price dynamics coincides with the TIM with $\delta = 1$ and $\gamma = 1/2$.

Price impact of a metaorder in the LLOB model is then defined as $\mathcal{I}(Q,T) = y(T)$, and is found to be given by

$$\mathcal{I}(Q,T) = \sqrt{\frac{DQ}{J}} \mathcal{F}(\eta), \quad \text{with} \quad \eta \equiv \frac{Q}{JT}, \qquad (7.20)$$

where η is the participation rate and the scaling function $\mathcal{F}(\eta) \approx \sqrt{\eta/\pi}$ for $\eta \ll 1$ and $\approx \sqrt{2}$ for $\eta \gg 1$. Hence, $\mathcal{I}(Q,T)$ is linear in Q for small Q at fixed T, and crosses over to a square-root for large Q. Note that in the square-root regime,

---

[5] The LLOB model was originally developed for describing the latent liquidity, not necessarily the visible one, however close to the spread the two liquidities should coincide.

impact is predicted to be independent of the execution time T, as approximately observed empirically (see the discussion above).

The theoretical predictions of LLOB model have been empirically tested in Bucci et al. (2019a) where, using a large dataset of more than 8 million metaorders from the ANcerno database, there was shown a remarkable qualitative agreement between the data and the model. However the original model in Donier et al. (2015) predicts the crossover of impact from the linear to the square root regime at $\eta^* = 1$, while empirical data shows that this value is much closer to 10^{-3}. Benzaquen and Bouchaud (2018) generalized the model of Donier et al. (2015) by introducing (at least) two types of liquidity providers, acting on two different time scales: slow and persistent agents are able to resist the impact of the metaorder and fast agents who lubricate the high-frequency activity of markets. The introduction of two types of agents modifies the value of the crossover participation rate η^*. Bucci et al. (2019a) showed that the LLOB model with two types of agents fits quantitatively extremely well the shape of $\mathcal{I}(Q, T)$ as a function of ϕ and η when tested on the ANcerno database.

Besides the total impact of a metaorder, it is interesting to investigate the properties of the average price dynamics during and after the execution of a metaorder, because this analysis gives insightful information on the price impact dynamics and the role of information in trading. The first problem was investigated in Zarinelli et al. (2015) by computing the average price path during metaorder execution by considering subsets of metaorders with different duration T and participation rate η. Again, large samples are required due to the high level of noise in this type of data and ANcerno dataset allows to perform robust statistical analyses. One of the investigated question is whether, given two metaorders with the same participation rate η and different durations T_1 and T_2 ($T_1 < T_2$), the market impact reached at time T_1 is the same for the two metaorders. The empirical answer is clearly negative: The market impact trajectories deviate from the market impact surface. For small participation rates, this effect is stronger and price trajectories are well above the immediate impact. Moreover, in most cases the price reverts before the end of the metaorder (see also Bacry et al., 2015), while for larger η, the price trajectories become closer and closer to the values of the impact surface. The observation of non-overlapping trajectories might be explained in terms of executions with variable participation rate. A front-loaded execution, i.e., an execution with a decreasing participation rate, produces a strong impact at the beginning and a milder impact toward the end, as observed in real data. This choice might be due to risk aversion (Almgren and Chriss, 2001) or to the attempt to catch as much liquidity on the book as possible. It is quite interesting to observe that the TIM model with a front-loaded execution is able to reproduce the observed fact that price impact trajectories revert during the execution of the metaorder. On the contrary, a model with permanent impact, such as the Almgren–Chriss model (Almgren and Chriss, 2001), always gives monotonic price trajectories if the sign of the trades is uniform.

The behavior of price after the end of the metaorder is more complicated to estimate, in part because the noise level is even larger than during the metaorder execution. The observed average price dynamics is consistent with a reversion of the price with respect to the value reached at the end of the execution. This is another confirmation of the transient nature of market impact as described, for example, by the TIM. The long term value of the price is even more complicated to estimate for several reasons. First, the very slow decay of impact requires to measure impact on a long time horizon, when volatility effects become dominant. Second, end of day effect and overnight returns could make difficult to estimate permanent impact if the decay continues the days after the metaorder execution. Third, metaorders are sometimes split over multiple days creating an autocorrelation of metaorders, which makes hard to estimate the 'bare' decay of price impact.

From a theoretical point of view, in the 'fair pricing' theory of Farmer et al. (2013) an equilibrium condition is derived between liquidity providers and a broker aggregating informed metaorders from several funds. The theory predicts that the average price payed during the execution is equal to the price at the end of the reversion phase. If metaorder size distribution is a power law with tail exponent $3/2$ (as empirically observed), the impact is predicted to decay towards a plateau value whose height is $2/3$ of the peak impact, i.e. the impact reached exactly when the metaorder execution is completed.

Interestingly, several empirical studies reports results compatible with the $2/3$ factor (see Moro et al., 2009; Zarinelli et al., 2015; Bershova and Rakhlin, 2013; Waelbroeck and Gomes, 2015 although the last of these papers notes that the impact of uninformed trades appears to relax to zero. On the other hand, Brokmann et al. (2015) underlined the importance of metaorders split over many successive days, as this may strongly bias upwards the apparent plateau value. After accounting for metaorder autocorrelations (from a single fund), the paper concludes that impact decays as a power-law over several days, with no clear asymptotic value. A more extensive analysis has been performed using the ANcerno database in Bucci et al. (2018) which shows that while at the end of the same day the average price is on average close to $2/3$ of the peak impact, the decay continues the next days, following a power-law function at short time scales, and apparently converges to a non-zero asymptotic value at long time scales (roughly 50 days) close to $1/3$ of the peak impact. For such long time lags, however, market noise becomes dominant and makes it difficult to conclude on the asymptotic value of impact, which is a proxy for the (long time) information content of the trades.

7.7 Co-impact

In the previous section, market impact of a metaorder is defined by conditioning only on its properties (size and duration). However, in a given day there is typically a large number of funds simultaneously trading the same stock. As empirically observed in Zarinelli et al. (2015) by investigating the ANcerno database, there is a clear tendency of traders to send metaorders with the same sign (buy or sell) on the same asset. The reason for this coherent behavior are manyfold, but probably the most important one is related to the similarity of trading strategies among institutional investors. One can thus ask how the presence of other metaorders, modifies market impact and the associated transaction cost of a given metaorder. This crowding effect on market impact was termed *co-impact* in Bucci et al. (2020). We are thus changing the conditioning variables in the definition of market impact by considering a vector of simultaneously present metaorders. We will then averaging this quantity by using their joint distribution, keeping as conditioning variable the metaorder whose impact we are interested in.

Bucci et al. (2020) investigated how the expected open-to-close daily logreturn $\Delta p^{(d)} \equiv \log p_{\text{close}}/p_{\text{open}}$ depends on the order flow generated by the ANcerno metaorders. Consider a day when N metaorders are simultaneously present, each described by $\tilde{\phi}_i \equiv \epsilon_i Q_i/V$, $(i = 1, \ldots, N)$, where V is again the average daily volume and ϵ_i and Q_i are, respectively, the sign and the size of the i-th metaorder. Defining the vector $\tilde{\varphi}_N = (\tilde{\phi}_1, \ldots, \tilde{\phi}_N)$, the quantity of interest is

$$I(\tilde{\varphi}_N) \equiv \mathbb{E}[\Delta p^{(d)}|\tilde{\varphi}_N] \qquad (7.21)$$

This is however a function of N variables and some parametric restriction must be made to estimate it from data. Bucci et al. (2020) empirically found that the above quantity is well described by $I(\tilde{\varphi}_N) = Y \cdot f_\delta(\Phi)$, where $\tilde{\Phi} = \sum_{i=1}^N \tilde{\phi}_i$ and $f_\delta(v) = \text{sign}(v)|v|^\delta$ with $\delta \simeq 1/2$. Thus the average price mainly reacts with a square root law to the total net order flow of ongoing metaorders. This means that the market, due also to the fact that trading is anonymous, is unable to individually distinguish them. Despite the insensitivity of the price to individual metaorders is quite intuitive, it also raises some issues on how the square root impact can hold. Let consider a simple example where there is a buy metaorder with order flow $\tilde{\phi} > 0$, which is traded simultaneously with other metaorders with total order flow $\tilde{\phi}_m > 0$. Assuming that the square root law applies for the total order flow, the observed impact is

$$\mathbb{E}[\Delta p^{(d)}|\tilde{\phi}, \tilde{\phi}_m] \propto \sqrt{\tilde{\phi} + \tilde{\phi}_m} \qquad (7.22)$$

Keeping $\tilde{\phi}_m$ fixed, when $\tilde{\phi} \to 0$ market impact tends to a constant, for $\tilde{\phi} \ll \tilde{\phi}_m$, instead, $\mathbb{E}[\Delta p|\tilde{\phi}, \tilde{\phi}_m]$ is linear in $\tilde{\phi}$, and only when $\tilde{\phi} \gg \tilde{\phi}_m$ a square root behavior is expected. Thus, how can a non-linear impact law survive in the presence of a large number of simultaneously executed metaorders?

The argument can be made mathematically more precise by asking what is the expected impact of a metaorder, labeled with k, when other $N - 1$ are simultaneously being executed. Given the evidence above, this impact can be written as[6]

$$\mathcal{I}_N(\tilde{\phi}) \equiv \mathbb{E}[\Delta p^{(d)}|\tilde{\phi}_k = \tilde{\phi}, N] = Y \int d\tilde{\phi}_1 \cdots d\tilde{\phi}_N P(\tilde{\varphi}_N|\tilde{\phi}_k = \tilde{\phi}) f_\delta\left(\tilde{\phi}_k + \sum_{i \neq k} \tilde{\phi}_i\right)$$
$$(7.23)$$

---

[6] Note that we are conditioning on the signed volume fraction $\tilde{\phi}_k$, which, under buy-sell symmetry, is equivalent to compute the expectation of $\epsilon_k \Delta p^{(d)}$ conditional to absolute volume fraction ϕ. In other words, also here we are measuring a drift term.

One can then obtain the unconditional impact by averaging $\mathcal{I}_N(\tilde{\phi})$ over the distribution $P(N)$ of the number of metaorders per day. Thus $\mathcal{I}_N(\tilde{\phi})$ depends on the joint distribution of order flows $P(\tilde{\varphi}_N)$ and Bucci et al. (2020) derived the analytical expression for $\mathcal{I}_N(\tilde{\phi})$ under different specification for it (for example multivariate Gaussian). A crossover from a linear to a square root behavior is predicted and the transition point depends on the number of metaorders N and on their correlation (more generally, statistical dependence). When N is small, a small investor will observe linear impact with a non-zero intercept $\mathcal{I}_0$, crossing over to a square-root law at larger $\tilde{\phi}$. The intercept $\mathcal{I}_0$ grows with the correlation between the signs of the metaorders and can be interpreted as the average impact of all the other metaorders. When the number of metaorders is large and the investor has no correlation with their average sign, one should expect on a given day a square-root impact randomly shifted upwards or downwards by $\mathcal{I}_0$. Averaged over all days, a pure square-root law emerges, which explains why such behavior has been reported in many empirical papers.

Calibrating such model on real data requires to make some assumptions on the joint distribution $P(\tilde{\varphi}_N)$. Bucci et al. (2020) showed that the correlation of absolute volume fractions $\phi_i = |\tilde{\phi}_i|$ is negligible, while correlation between metaorder signs plays an important role. By calibrating a simple heuristic model where a single factor drives the metaorder signs, Bucci et al. (2020) reproduced to a good level of precision the different regimes of the empirical market impact curves as a function of $\tilde{\phi}$, N, and the correlation of their signs. In particular, for a metaorder uncorrelated with the rest of the market, the impacts of other metaorders cancel out on average. Conversely, any intercept of the impact law can be interpreted as a non-zero correlation with the rest of the market.

It is interesting to make a comparison with what simple models of market impact predict on price impact when many informed agents are simultaneously present. Bagnoli et al. (2001) investigated the equilibrium in a one-period Kyle model (Kyle, 1985). N symmetrically and informed agents trade one asset in a market where uninformed agents and market makers are also present. Bagnoli et al. (2001) shows that the Kyle's lambda, i.e. the proportionality factor between price impact and aggregated order flow, scales as $N^{-1/\alpha}$, where α is the exponent of the stable law describing the price and uninformed order flow distribution. Moreover if the second moment of both variables is finite, Bagnoli et al. (2001) shows that the Kyle's lambda scales as $1/\sqrt{N}$. Interestingly, Figure 3 of Bucci et al. (2020) shows that market impact of a ANcerno metaorder decreases with the number of metaorders simultaneously present.

From a practical perspective, the model and the empirical observations are important for traders to estimate (pre- and post-execution) the cost of their trades, and thus to help them deciding when is the right moment to trade. For example, Briere et al. (2020), investigating the ANcerno database, found an approximately linear relation between the implementation shortfall of a metaorder and the net trading imbalance due to the other metaorders simultaneously traded. When the trade is in the same direction as the net order flow imbalance, one could expect to pay a significant trading cost up to 0.4 points of price volatility, while one could expect to benefit from a price improvement of 0.3 points of volatility when the trader is almost alone in front of his competitors aggregate flow. In a normal trading situation, the information on the ongoing metaorders is not available, thus statistical and machine learning methods could be used to infer, at least partly, this information from the visible order flow.

7.8 Conclusion

As should be clear from this short review, in the last twenty years we have made huge progress in understanding the important and fascinating problem of how price is formed in financial markets as the result of order flow and trading activity. This advancement is due to the availability of very detailed and rich datasets and to the development of sophisticated models able to capture, at least partly, the strong dependencies and feedbacks between orders and prices. Still much remains to do. For example, most models are inherently stationary and with fixed parameters, while liquidity, as many market variables, are highly dynamic and latent. Methods from econometrics (filtering, score-driven models) and machine learning (reinforcement learning) can provide the tools for tackling this important aspect of market dynamics. Combining these models with optimal execution or optimal market making solutions available in real time would certainly provide a great addition for the industry.

References

Abergel, F., and Jedidi, A. 2015. Long-time behavior of a Hawkes process-based limit order book. *SIAM Journal of Financial Mathematics*, **6**, 1026–1043.

Almgren, R., and Chriss, N. 2001. Optimal execution of portfolio transactions. *Journal of Risk*, **3**, 5–40.

Almgren, Robert, Thum, Chee, Hauptmann, Emmanuel, and Li, Hong. 2005. Direct estimation of equity market impact. *Risk*, **18**(7), 5862.

Bacry, E., and Muzy, J.-F. 2014. Hawkes model for price and trades high-frequency dynamics. *Quantitative Finance*, **14**, 1–20.

Bacry, E., Iuga, A., Lasnier, M., and Lehalle, C.-A. 2015. Market impacts and the life cycle of investors orders. *Market Microstructure and Liquidity*, **1**, 1550009.

Bacry, E., Jaisson, T., and Muzy, J.F. 2016. Estimation of slowly decreasing Hawkes kernels: application to high-frequency order book dynamics. *Quantitative Finance*, **16**, 1179–1201.

Bagnoli, Mark, Viswanathan, S., and Holden, Craig. 2001. On the existence of linear equilibria in models of market making. *Mathematical Finance*, **11**(1), 1–31.

Benzaquen, M., and Bouchaud, J.-P. 2018. Market impact with multi-timescale liquidity. *Quantitative Finance*, **18**, 1781.

Benzaquen, Michael, Mastromatteo, Iacopo, Eisler, Zoltan, and Bouchaud, Jean-Philippe. 2017. Dissecting cross-impact on stock markets: An empirical analysis. *Journal of Statistical Mechanics*, **2017**, 023406.

Bershova, Nataliya, and Rakhlin, Dmitry. 2013. The non-linear market impact of large trades: Evidence from buy-side order flow. *Quantitative Finance*, **13**(11), 1759–1778.

Biais, B, Hillion, P, and Spatt, C. 1995. An empirical analysis of the limit order book and order flow in the Paris bourse. *Journal of Finance*, **50**, 11655–1689.

Bouchaud, J.-P., Gefen, Y., Potters, M., and Wyart, M. 2004. Fluctuations and response in financial markets: the subtle nature of 'random' price changes. *Quantitative Finance*, **4**(2), 176–190.

Bouchaud, Jean-Philippe, Farmer, J. Doyne, and Lillo, Fabrizio. 2009. How markets slowly digest changes in supply and demand. Pages 57–160 of: *Handbook of Financial Markets: Dynamics and Evolution*. Elsevier.

Briere, M., Lehalle, C.-A., Nefedova, T., and Raboun, A. 2020. Modeling Transaction Costs When Trades May Be Crowded: A Bayesian Network Using Partially Observable Orders Imbalance. Pages 387–430 of: *Machine Learning for Asset Management: New Developments and Financial Applications*, Emmanuel Jurczenko (ed), Wiley. .

Brokmann, X, Serie, E, Kockelkoren, J, and Bouchaud, J-P. 2015. Slow decay of impact in equity markets. *Market Microstructure and Liquidity*, **1**(02), 1550007.

Bucci, F., Mastromatteo, I., Eisler, Z., Lillo, F., Bouchaud, J.-P., and Lehalle, C.-A. 2020. Co-impact: crowding effects in institutional trading activity. *Quantitative Finance*, **20**(2), 193–205.

Bucci, Frédéric, Benzaquen, Michael, Lillo, Fabrizio, and Bouchaud, Jean-Philippe. 2018. Slow Decay of Impact in Equity Markets: Insights from the ANcerno Database. *Market Microstructure and Liquidity*, **4**(3), 1950006.

Bucci, Frédéric, Benzaquen, Michael, Lillo, Fabrizio, and Bouchaud, Jean-Philippe. 2019a. Crossover from linear to square-root market impact. *Physical Review Letters*, **122**(10), 108302.

Bucci, Frédéric, Mastromatteo, Iacopo, Bouchaud, Jean-Philippe, and Benzaquen, Michael. 2019b. Impact is not just volatility. *Quantitative Finance*, **19**(11), 1763–1766.

Chordia, Tarun, Roll, Richard, and Subrahmanyam, Avanidhar. 2000. Commonality in liquidity. *Journal of Financial Economics*, **56**(1), 3 – 28.

Cont, R., Stoikov, S., and Talreja, R. 2010. A stochastic model for order book dynamics. *Operations Research*, **58**(3), 549–563.

Cont, R., Kukanov, A., and Stoikov, S. 2014. The price impact of order book events. *Journal of Financial Econometrics*, **12**, 47–88.

Daniels, Marcus G., Farmer, J. Doyne, Gillemot, László, Iori, Giulia, and Smith, Eric. 2003. Quantitative model of price diffusion and market friction based on trading as a mechanistic random process. *Physical Review Letters*, **90**(10), 108102–4.

Donier, J., and Bonart, J. 2015. A million metaorder analysis of market impact on the bitcoin. *Market Microstructrure and Liquidity*, **1**, 1550008.

Donier, J., Bonart, J., Mastromatteo, I., and Bouchaud, J.-P. 2015. A fully consistent, minimal model for non-linear market impact. *Quantitative Finance*, **15**, 1109.

Eisler, Z., Bouchaud, J.-P., and Kockelkoren, J. 2012. The price impact of order book events: Market orders, limit orders and cancellations. *Quantitative Finance*, **12**, 1395–1419.

Farmer, J.D., Gerig, A., Lillo, F., and Waelbroeck, H. 2013. How efficiency shapes market impact. *Quantitative Finance*, **13**, 1743–1758.

Gatheral, Jim. 2010. No-dynamic-arbitrage and market impact. *Quantitative Finance*, **10**(7), 749–759.

Hasbrouck, J. 1991. Measuring the information content of a trade. *The Journal of Finance*, **46**, 179–207.

Hautsch, N., and Huang, R. 2012. Measuring the information content of a trade. *Journal of Economic Dynamics & Control*, **36**, 501–522.

Huang, W., Lehalle, C.-A., and Rosenbaum, M. 2015. Simulating and analyzing order book data: The queue-reactive model. *Journal of the American Statistical Association*, **110**, 107–122.

Kyle, Albert S. 1985. Continuous auctions and insider trading. *Econometrica*, **53**, 1315–1335.

Large, J. 2007. Measuring the resiliency of an electronic limit order book. *Journal of Financial Markets*, **10**, 1–25.

Lillo, F., and Farmer, J.D. 2004. The long memory of the efficient market. *Studies in nonlinear dynamics & econometrics*, **8**, 3.

Lillo, Fabrizio, Farmer, J. Doyne, and Mantegna, Rosario N. 2003. Econophysics: Master curve for price-impact function. *Nature*, **421**, 129–130.

Madhavan, A., Richardson, M., and Roomans, M. 1997. Why do security prices change? A transaction-level analysis of NYSE stocks. *Review of Financial Studies*, **10**, 1035–1064.

Mastromatteo, Iacopo, Benzaquen, Michael, Eisler, Zoltan, and Bouchaud, Jean-Philippe. 2017. Trading lightly: Cross-impact and optimal portfolio execution. *Risk*, **30**, 82–87.

Morariu-Patrichi, M., and Pakkanen, M.S. 2018. State-dependent Hawkes processes and their application to limit order book modelling. Arxiv.org/pdf/1809.08060.

Moro, Esteban, Vicente, Javier, Moyano, Luis G., Gerig, Austin, Farmer, J. Doyne, Vaglica, Gabriella, Lillo, Fabrizio, and Mantegna, Rosario N. 2009. Market impact and trading profile of hidden orders in stock markets. *Physical Review E*, **80**(6), 066102.

Muni Toke, I. 2010. Market making in an order book model and its impact on the bid-ask spread. Pages 49–64 of: *Econophysics of Order-Driven Markets*. " F. Abergel, B.K. Chakrabarti, A. Chakraborti, M. Mitra (eds). Springer.

Rambaldi, M., Bacry, E., and Lillo, F. 2017. The role of volume in order book dynamics: a multivariate Hawkes process analysis. *Quantitative Finance*, **17**(7), 999–1020.

Schneider, Michael, and Lillo, Fabrizio. 2019. Cross-impact and no-dynamic-arbitrage. *Quantitative Finance*, **19**(1), 137–154.

Sirignano, J.A. 2019. Deep learning for limit order books. *Quantitative Finance*, **19**(4), 549–570.

Taranto, D.E., Bormetti, G., Bouchaud, J.-P., Lillo, F., and Tóth, B. 2018. Linear models for the impact of order flow on prices I. Propagators: transient vs. history-dependent impact. *Quantitative Finance*, **18**, 903–915.

Torre, Nicolo G., and Ferrari, Mark J. 1998. The market impact model. *Horizons, The Barra Newsletter*, **165**.

Tóth, B, Lillo, F., and Farmer, J.D. 2010. Segmentation algorithm for non-stationary compound Poisson processes. With an application to inventory time series of market members in a financial market. *European Physical Journal B*, **78**, 235–243.

Tóth, B., Palit, I., Lillo, F., and Farmer, J.D. 2015. Why is equity order flow so persistent? *Journal of Economic Dynamics & Control*, **51**, 218–239.

Tóth, B., Eisler, Z., and Bouchaud, J.-P. 2016. The square-root impact law also holds for option markets. *Wilmott*, **85**, 70.

Tóth, Bence, Lemperiere, Yves, Deremble, Cyril, De Lataillade, Joachim, Kockelkoren, Julien, and Bouchaud, J-P. 2011. Anomalous price impact and the critical nature of liquidity in financial markets. *Physical Review X*, **1**(2), 021006.

Tsoukalas, Gerry, Wang, Jiang, and Giesecke, Kay. 2019. Dynamic portfolio execution. *Management Science*, **65**(5), 2015–2040.

Vaglica, Gabriella, Lillo, Fabrizio, and Mantegna, Rosario N. 2010. Statistical identification with hidden Markov models of large order splitting strategies in an equity market. *New Journal of Physics*, **11**, 075031.

Waelbroeck, Henri, and Gomes, Carla. 2015. Is market impact a measure of the information value of trades? Market response to liquidity vs. informed trades. *Quantitative Finance*, **15**, 773–793.

Wu, P., Rambaldi, M., Muzy, J.-F, and Bacry, E. 2019. Queue-reactive Hawkes models for the order flow. Arxiv.org/pdf/1901.08938.

Zarinelli, Elia, Treccani, Michele, Farmer, J. Doyne, and Lillo, Fabrizio. 2015. Beyond the square root: Evidence for logarithmic dependence of market impact on size and participation rate. *Market Microstructure and Liquidity*, **1**(02), 1550004.

8

Price Formation and Learning in Equilibrium under Asymmetric Information

Umut Çetin[a]

Abstract

This chapter studies the financial equilibrium and its properties among asymmetrically informed market participants starting with the seminal work of Kyle (1985). Using its continuous-time formulation by Back (1992) as the underlying framework, equilibrium strategies of informed traders and market makers will be derived in the original model as well as in a number of key extensions including the models that account for competition among multiple insiders, default risk and dynamic information acquisition. Moreover, the interplay between the batch auction model of Kyle and the sequential arrival model of Glosten and Milgrom (1985) will be discussed. The mathematical analysis will rely on the combination of stochastic filtering and Markovian bridge techniques that are tailored for this equilibrium framework. Finally, by incorporating risk averse market makers to the model we will obtain an equilibrium that simultaneously exhibits price reversal and permanent impact, and thereby bridging the gap between the earlier and more recent market microstructure models.

8.1 Introduction

One of the goals of Market Microstructure (MS) models is to understand the *temporary* and *permanent* impacts of the trades on the asset price and how the price-setting rules evolve in time. In real markets bid and ask prices are announced by *specialists* or *dealers*, whom we will henceforth collectively call *market makers*. The early literature on market microstructure (Garman, 1976; Stoll, 1978; Amihud and Mendelson, 1980; Ho and Stoll, 1981) have started with the simple observation that the trades could involve some implicit costs due to the need for immediate execution, which is provided by the market makers. At the same time, the market makers take into account their inventory level when making pricing decisions. These works have concluded that the market makers adjust the prices in order to keep their inventories around a certain level in the long run: they lower the price when their inventory levels are too high and raise

[a] London School of Economics
Published in *Machine Learning And Data Sciences For Financial Markets*, Agostino Capponi and Charles-Albert Lehalle© 2023 Cambridge University Press.

the prices when they are short large quantities. As the market makers want to keep their inventories around a fixed level, the impact of trades are *transitory* since the prices are also expected to mean revert.

The MS research have later shifted its focus to models with asymmetric information, which account for permanent changes in the price. The canonical model of markets with asymmetric information is due to Kyle (1985). He studied a market for a single risky asset whose price is determined in equilibrium in discrete time. The key feature of this model is that the market makers cannot distinguish between the informed and uninformed trades and compete to fill the net demand. In this model market makers 'learn' from the net demand by 'filtering' what the informed trader knows, which is 'corrupted' by the demand of the uninformed traders. The market makers learn from the order flow and they update their pricing strategies as a result of this learning mechanism.

In contrast to the batch arrival model of Kyle (1985), Glosten and Milgrom (1985) study the equilibrium pricing in a model where market makers quote bid and ask prices and market orders of unit size arrive sequentially. Nevertheless, the market makers do not know whether the arriving order is informed or not. Thus, a similar learning mechanism has to take place in order to price the risky asset efficiently.

This chapter will give a brief discussion on the fundamentals of the original Kyle model with risk neutral market makers as well as its extensions to include dynamic information flow, multiple informed traders, and defult risk. Moreover, a suitable version of the Glosten–Milgrom model will be presented and its connection to the Kyle model will be discussed.

The empirical studies on the inventories of market makers demonstrate mean reversion, which is a sign of risk aversion. In Section 8.8 we shall study the impact of market makers being risk averse on equilibrium. Consistent with empirical studies such a change will result in mean reverting inventories for market makers. From another perspective having risk averse market makers in the Kyle model bridges the earlier MS literature with that following Kyle's framework.

Surveying, even only listing, all the relevant literature in this limited space is impossible. The last section nevertheless is devoted to brief remarks on some other works that are closely related to the topics discussed in earlier pages.

8.2 The Kyle model

8.2.1 A toy example

To get a flavour of the Kyle model suppose that there is an asset whose value V will be revealed at time 1. Assume further the existence of an insider who knows the value of V at time 0. To simplify the matters the insider will be allowed to trade once at time 0 and liquidate her position at time 1.

At time 0 there are also *noise traders* who are not strategic and their cumulative demand for the asset is given by $v \sim N(0, \sigma_v^2)$. Consistent with the term 'noise' v is assumed to be independent of V.

If the insider trades θ many shares, the market makers observe the net demand $Y := \theta + n$ and take the opposite side to clear the market by setting a price. They know the distribution of V but no other relevant information regarding its value. The market makers are *risk neutral* and compete in a Bertrand fashion to fill the aggregate order Y. That is, the price $h(y)$ chosen by the market makers for $Y = y$ is such that their expected profit is 0. Since they will also liquidate their position at time 1 at price V, this implies

$$h(y) = E[V|Y = y]. \tag{8.1}$$

Given this *pricing rule* of market makers the insider finds her optimal trading amount based on her private information. In this idealisation of the market the market price of the traded asset will be determined in a Bayesian Nash-type equilibrium:

The pair (θ, h) will constitute and equilibrium if

1. Given h, θ maximises the expected profit of the insider;
2. Given θ, h satsifies (8.1).

Suppose further that $V \sim N(0, \sigma^2)$. Let us next observe that a *linear equilibrium* in which $h(y) = a + \lambda y$ and $\theta = \alpha + \beta V$ exists. First of all, if $h(y) = a + \lambda y$, the insider's optimisation problem given $V = v$ is

$$\max_{\alpha, \beta} E[(\alpha + \beta v)(v - a - \lambda(v + \alpha + \beta v))].$$

The profit/loss is quadratic in parameters and the first order condition yields:

$$\alpha + \beta v = \frac{v - a}{2\lambda}. \tag{8.2}$$

On the other hand, (8.1) requires

$$a + \lambda Y = E[V|Y].$$

Now, since (V, v) is a Gaussian vector, the conditional distribution of V given Y is also Gaussian, which can be determined by Bayes' rule. Formally,

$$P(V \in dv|Y = y) \sim \frac{P(Y \in dy|V = v)}{dy} P(V \in dv).$$

Moreover, given $V = v$, $Y := v + \theta \sim N(\alpha + \beta v, \sigma_v^2)$. Thus, $P(Y \in dy \mid V = v)$ is proportional to

$$\exp\left(-\frac{(y - \alpha - \beta v)^2}{2\sigma_v^2}\right).$$

Hence,

$$P(V \in dv|Y = y) \sim \exp\left(-\frac{(v - \hat{\mu})^2}{2\Sigma^2}\right),$$

where

$$\frac{1}{\Sigma^2} = \frac{1}{\sigma^2} + \frac{\beta^2}{\sigma_v^2}, \quad \hat{\mu} = \beta(y - \alpha)\frac{\Sigma^2}{\sigma_v^2}.$$

That is, V is Gaussian with mean $\hat{\mu}$ and variance Σ^2 given $Y = y$. Thus,

$$a + \lambda y = \beta(y - \alpha)\frac{\Sigma^2}{\sigma_v^2} = \beta(y - \alpha)\frac{\sigma^2}{\beta^2\sigma^2 + \sigma_v^2},$$

which in turn yields

$$\lambda = \frac{\beta\sigma^2}{\beta^2\sigma^2 + \sigma_v^2} \text{ and } a = -\frac{\alpha}{\beta}\lambda.$$

Recall that (8.2) implies $2\lambda\beta = 1$. Therefore,

$$\beta = \frac{\sigma_v}{\sigma} \text{ and } \lambda = \frac{\sigma}{2\sigma_v}.$$

The remaining two equations for a and α are satisfied only if $a = \alpha = 0$.

Kyle's lambda:
A widely used metric for the amount of liquidity available in a given market is the so-called *Kyle's lambda*. It is a measurement of the sensitivity of prices to the volume and is roughly defined as the inverse of the volume needed to move the prices by one unit. More precisely, it is the derivative of the function h defined above with respect to y, which is given by λ! As such, a low λ is a sign of low liquidity costs. Given the above description of λ a liquid market requires a sufficiently large volume of noise trading in the presence of asymmetric information. This is quite reasonable: the higher the adverse selection faced by the market makers, the higher the level of compensation they require to clear the market.

The value of information:
Information acquisition is costly. Although how the informed trader has obtained her private information is not modelled in the Kyle model, it is possible to compute the value of private information. Given the above explicit characterisation of equilibrium the equilibrium level of wealth of the insider is given by

$$(1 - \lambda\beta)\beta v^2 = \beta\frac{v^2}{2}$$

It is also not difficult to see that an uninformed strategic trader will make 0 expected profit in this model as the prices evolve as a martingale for the uninformed traders. Thus, the value of information equals *ex ante*, i.e. unconditional, profit, which is given by

$$\beta\frac{\sigma^2}{2} = \frac{\sigma\sigma_v}{2}. \tag{8.3}$$

8.2.2 The Kyle model in continuous time

If a trader has some private information regarding the future value of the asset, she would like to take advantage of this and trade dynamically, not just once as above. The continuous time version of the Kyle model is formalised in Back

(1992). Although in the literature it is usually assumed that the informed investor knows the future asset value perfectly, this is not a necessary assumption as we shall soon see.

Let us suppose that the time-1 value of the traded asset is given by some random variable V, which will become public knowledge at $t = 1$ to all market participants.

We shall work on a filtered probability space $(\Omega, \mathcal{G}, (\mathcal{G}_t)_{t \in [0,1]}, \mathbb{Q})$.

Three types of agents trade in the market. They differ in their information sets, and objectives, as follows.

- *Noise/liquidity traders* trade for liquidity reasons, and their total demand at time t is given by a standard $(\mathcal{G}_t)$-Brownian motion B independent of V. This normalisation of the variance of the noise trades is without loss of generality as long as the variance process is perfect knowledge among all market participants.
- *Market makers* observe only the total demand

$$Y = \theta + B,$$

where θ is the demand process of the informed trader. The admissibility condition imposed later on θ will entail in particular that Y is a semimartingale.

They set the price of the risky asset via a *Bertrand competition* and clear the market. We assume that the market makers set the price as a function of the total order process at time t, i.e. we consider pricing functionals $S\left(Y_{[0,t]}, t\right)$ of the following form

$$S\left(Y_{[0,t]}, t\right) = H(t, Y_t), \qquad \forall t \in [0,1). \tag{8.4}$$

Moreover, a pricing rule H has to be admissible in the sense of Definition 8.1. In particular, $H \in C^{1,2}$ and, therefore, S will be a semimartingale as well.

- *The informed trader (insider)* observes the price process $S_t = H(t, Y_t)$ and her private signal, Z, which is possibly time varying Markov process and is independent of B. Based on her signal, she makes an educated guess about V. We shall assume a Markovian framework in the sense that

$$E[V | \sigma(Z_t; t \leq 1)] = E[V | Z_1].$$

Thus, there exists a measurable function f such that

$$f(Z_1) = E[V | \sigma(Z_t; t \leq 1)].$$

We assume that Z_t is a continuous random variable for each $t > 0$ and f is continuous. Moreover, f can be taken strictly increasing. This entails in particular that the larger the signal Z_1 the larger the value of the risky asset for the informed trader.

She is assumed to be risk-neutral, her objective is to maximize the expected final wealth.

$$\sup_{\theta \in \mathcal{A}(H)} E^{0,z} \left[W_1^{\theta} \right], \text{ where}$$

$$W_1^{\theta} = (V - S_{1-}))\theta_{1-} + \int_0^{1-} \theta_{s-} dS_s.$$

However, using the tower property of conditional expectations, the above problem is equivalent to

$$\sup_{\theta \in \mathcal{A}(H)} E^{0,z} \left[W_1^{\theta} \right], \text{ where} \tag{8.5}$$

$$W_1^{\theta} = (f(Z_1) - S_{1-})\theta_{1-} + \int_0^{1-} \theta_{s-} dS_s. \tag{8.6}$$

In above $\mathcal{A}(H)$ is the set of admissible trading strategies for the given pricing rule[1] H, which will be defined in Definition 8.3. Moreover, $E^{0,z}$ is the expectation with respect to $P^{0,z}$, which is the probability measure on $\sigma(Y_s, Z_s; s \leq 1)$ generated by (Y, Z) with $Y_0 = 0$ and $Z = z$.

The informed trader and the market makers not only differ in their information sets but also in their probability measures. To precisely define the probability measure of the market makers consider $\mathcal{F} := \sigma(B_t, Z_t; t \leq 1)$ and let $Q^{0,z}$ be the probability measure on $\mathcal{F}$ generated by (B, Z) with $B_0 = 0$ and $Z_0 = z$. Next introduce the probability measure $\mathbb{P}$ on $(\Omega, \mathcal{F})$ by

$$\mathbb{P}(e) = \int_{\mathbb{R}} Q^{0,z}(e) \mathbb{Q}(Z_0 \in dz), \tag{8.7}$$

for any $e \in \mathcal{F}$. This is the probability measure used by the uninformed market makers in this model. Note that the probability measure of the informed can be *singular* with respect to that of the market makers. Indeed, if Z_0 has a continuous distribution, $P^{0,z}(Z_0 = z) = 1$ while $\mathbb{P}(Z_0 = z) = 0$.

Due to the discrepancies in the null sets of the market makers and those of the informed trader there are also delicate issues regarding the completion of filtration. As such a technical discussion will muddle the presentation and won't have a significant contribution to the understanding of the fundamentals of the model, the interested reader is referred to Section 6.1 in Çetin and Danilova (2018a). What is important to know at this point is that the insider's filtration $\mathcal{F}^I$ is generated by (Z, S) while the market makers' filtration is generated by the observation of Y only.

We can now define the rational expectations equilibrium of this market, i.e. a pair consisting of an *admissible* pricing rule and an *admissible* trading strategy such that: *a)* given the pricing rule the trading strategy is optimal, *b)* given the trading strategy, the pricing rule is *rational* in the following sense:

$$H(t, Y_t) = S_t = \mathbb{E}\left[V | \mathcal{F}_t^M \right] = \mathbb{E}\left[f(Z_1) | \mathcal{F}_t^M \right], \tag{8.8}$$

---

[1] Note that this implies the insider's optimal trading strategy takes into account the *feedback effect*, i.e. that prices react to her trading strategy.

where $\mathbb{E}$ corresponds to the expectation operator under $\mathbb{P}$. Note that the last equality follows from the tower property of conditional expectations and the independence of B from V and Z as

$$\mathbb{E}\left[V|\mathcal{F}_t^M\right] = \mathbb{E}\left[\mathbb{E}\left[V|\sigma(B_s, Z_s; s \le t)\right]|\mathcal{F}_t^M\right] = \mathbb{E}\left[\mathbb{E}\left[V|\sigma(Z_s; s \le t)\right]|\mathcal{F}_t^M\right].$$

Observe that in view of (8.8) what is important is not the exact value of V but its valuation by the informed trader. That is, the informed trader does not have to be an insider.

To formalize the above notion of equilibrium, we first define the sets of admissible pricing rules and trading strategies.

Definition 8.1 An *admissible pricing rule* is any function H fulfilling the following conditions:

1. $H \in C^{1,2}([0,1) \times \mathbb{R})$.
2. $x \mapsto H(t,x)$ is strictly increasing for every $t \in [0,1)$;

Remark 8.2 The strict monotonicity of H in the space variable implies H is invertible prior to time 1, thus, the filtration of the insider is generated by Y and Z. This in turn implies that $(\mathcal{F}_t^{S,Z}) = (\mathcal{F}_t^{B,Z})$, i.e. the insider has full information about the market.

In view of the above one can take $\mathcal{F}_t^I = \mathcal{F}_t^{B,Z}$ for all $t \in [0,1]$.

Definition 8.3 An $\mathcal{F}^{B,Z}$-adapted θ is said to be an admissible trading strategy for a given pricing rule H if

1. θ is adapted and absolutely continuous on $(\Omega, \mathcal{F}, (\mathcal{F}_t^{B,Z}), Q^{0,z})$; that is, $d\theta_t = \alpha_t dt$ for some adapted and integrable α.
2. and no doubling strategies are allowed, i.e. for all $z \in \mathbb{R}$

$$E^{0,z}\left[\int_0^1 H^2(t, X_t)\, dt\right] < \infty. \tag{8.9}$$

The set of admissible trading strategies for the given H is denoted with $\mathcal{A}(H)$.

The hypothesis of absolutely continuity is standard in the literature. It was proved by Back (1992) that this restriction was without loss of generality when the insider's signal is static, i.e. $Z_t = Z_0, t \le 1$. That it suffices to consider only the absolutely continuous strategies in the dynamic case has been recently proved in Çetin and Danilova (2018b).

Definition 8.4 A couple $(H^*\theta^*)$ is said to form an equilibrium if H^* is an admissible pricing rule, $\theta^* \in \mathcal{A}(H^*)$, and the following conditions are satisfied:

1. *Market efficiency condition:* given θ^*, H^* is a rational pricing rule, i.e. it satisfies (8.8).
2. *Insider optimality condition:* given H^*, θ^* solves the insider optimization problem:

$$\mathbb{E}[W_1^{\theta^*}] = \sup_{\theta \in \mathcal{A}(H^*)} \mathbb{E}[W_1^\theta].$$

8.3 The static Kyle equilibrium

In this section we consider the case when the private signal of the informed trader is unchanged during the trading period, i.e. $Z_t = Z_1, t \leq 1$. That is, we are considering the extension of the toy example to the case of continuous trading. We shall also assume without loss of generality that Z_1 is standard normal. Before finding the optimal strategy of the insider let us formally deduce the Hamilton-Jacobi-Bellmann (HJB) equation associated to the value function of the insider.

Let H be any rational pricing rule and suppose that $d\theta_t = \alpha_t dt$. First, notice that a standard application of integration-by-parts formula applied to W_1^θ gives

$$W_1^\theta = \int_0^1 (f(Z_1) - S_s)\alpha_s \, ds. \tag{8.10}$$

Furthermore,

$$E^{0,z}\left[\int_0^1 (f(Z_1) - S_s)\alpha_s ds\right] = E^{0,z}\left[\int_0^1 (f(z) - S_s)\alpha_s ds\right]. \tag{8.11}$$

In view of (8.10) and (8.11), insider's optimization problem becomes

$$\sup_\theta E^{0,z}[W_1^\theta] = \sup_\theta E^{0,z}\left[\int_0^1 (f(z) - H(s, Y_s))\alpha_s ds\right]. \tag{8.12}$$

Let us now introduce the value function of the insider:

$$\phi(t, y, z) := \operatorname{ess\,sup}_\alpha E^{0,z}\left[\int_t^1 (f(z) - H(s, Y_s))\alpha_s ds | Y_t = y, Z = z\right], \quad t \in [0, 1].$$

Applying formally the dynamic programming principle, we get the following HJB equation:

$$0 = \sup_\alpha \left(\left[\phi_y + f(z) - H(t, y)\right]\alpha\right) + \phi_t + \frac{1}{2}\phi_{yy}. \tag{8.13}$$

Thus, for the finiteness of the value function and the existence of an optimal α we need

$$\phi_y + f(z) - H(t, y) = 0 \tag{8.14}$$

$$\phi_t + \frac{1}{2}\phi_{yy} = 0. \tag{8.15}$$

Differentiating (8.14) with respect to y and since from (8.14) it follows that $\phi_y = H(t, y) - f(z)$, we get

$$\phi_{yy} = H_y(t, y), \quad \phi_{yyy} = H_{yy}. \tag{8.16}$$

Since differentiation (8.14) with respect to t gives

$$\phi_{yt} = H_t(t, y),$$

(8.16) implies after differentiating (8.15) with respect to y

$$H_t(t, y) + \frac{1}{2}H_{yy}(t, y) = 0. \tag{8.17}$$

Thus, the equations (8.15) and (8.17)seem to be necessary to have a finite solution to the insider's problem.

Before presenting a solution of the equilibrium let's briefly observe one immediate consequence of (8.17). First recall that

$$dY_t = dB_t + \alpha_t dt,$$

where α_t is the rate of trade of the informed trader. Since the market makers only observe the batched order and cannot differentiate between the informed and the uninformed, the decomposition of the total order into a martingale component and a drift component will be different for market makers. The theory of non-linear filtering comes to the rescue here and one can write

$$dY_t = dB_t^Y + \hat{\alpha}_t dt,$$

where B^Y is the so-called innovation process, i.e. a Brownian motion with respect to the filtration of the market makers, and $\hat{\alpha}_t = \mathbb{E}[\alpha_t | \mathcal{F}_t^M]$.

Next set $S_t = H(t, Y_t)$ and observe in view of (8.17) that

$$dS_t = H_y(t, Y_t) dB_t^Y + H_y(t, Y_t)\hat{\alpha}_t dt.$$

Since S must be a martingale in equilibrium and $H_y > 0$, we expect to have $\hat{\alpha} \equiv 0$ in equilibrium. That is, the informed trader should hide her trades among the noise traders so that she gives the impression that she does not trade (well, locally)! We shall see that this is indeed the case in equilibrium.

Theorem 8.5 *Let H be an admissible pricing rule satisfying (8.17) and assume that $Z_t = Z_1$, $t \le 1$, where Z_1 is a standard normal random variable. Then $\theta \in \mathcal{A}(H)$ is an optimal strategy if $H(1-, Y_{1-}) = f(Z_1)$, $P^{0,z}$-a.s..*

Proof   Using Itô's formula we obtain

$$dH(t, Y_t) = H_t(t, Y_t)dt + H_y(t, Y_t)dY_t + \frac{1}{2}H_{yy}(t, Y_t)d[Y, Y]_t$$
$$= H_y(t, Y_t)dY_t.$$

Also recall that

$$W_1^\theta = f(Z_1)\theta_1 - \int_0^1 H(t, Y_t))d\theta_t. \tag{8.18}$$

Consider the function

$$\Psi^a(t, x) := \int_{\xi(t,a)}^x (H(t, u) - a)du + \frac{1}{2}\int_t^1 H_y(s, \xi(s, a))ds \tag{8.19}$$

where $\xi(t, a)$ is the unique solution of $H(t, \xi(t, a)) = a$. Direct differentiation with respect to x gives that

$$\Psi_x^a(t, x) = H(t, x) - a. \tag{8.20}$$

Differentiating above with respect to x gives

$$\Psi_{xx}^a(t, x) = H_x(t, x). \tag{8.21}$$

Direct differentiation of $\Psi^a(t,x)$ with respect to t gives

$$\Psi^a_t(t,x) = \int_{\xi(t,a)}^x H_t(t,u)du - \frac{1}{2}H_x(t,\xi(t,a))$$

$$= -\frac{1}{2}H_x(t,x).$$

Combining the above with (8.21) gives

$$\Psi^a_t + \frac{1}{2}\Psi^a_{xx} = 0. \tag{8.22}$$

Applying Ito's formula we thereby deduce

$$d\Psi^a(t,Y_t) = (H(t,Y_t) - a)\,dY_t.$$

The above implies

$$\Psi^a(1-,Y_{1-}) = \Psi^a(0,0) + \int_0^{1-} H(t,Y_t)(dB_t + d\theta_t) - a(B_1 + \theta_1).$$

Combining the above and (8.18) yields

$$E^{0,z}\left[W^\theta_1\right] = E^{0,z}\left[\Psi^{f(Z_1)}(0,0) - \Psi^{f(Z_1)}(1-,Y_1) - f(Z_1)B_1 + \int_0^{1-} H(t,Y_t)dB_t\right]$$

$$= E^{0,z}\left[\Psi^{f(Z_1)}(0,0) - \Psi^{f(Z_1)}(1-,Y_{1-})\right].$$

Moreover, $\Psi^{f(Z_1)}(1-,Y_{1-}) \geq 0$ with an equality if and only if $H(1-,Y_{1-}) = f(Z_1)$. Therefore, $E^{0,z}\left[W^\theta_1\right] \leq E^{0,z}\left[\Psi^{f(Z)}(0,0)\right]$ for all admissible θs, and equality is reached if and only if $H(1-,Y_{1-}) = f(Z_1)$, $P^{0,z}$-a.s.. □

The above result shows that the insider will drive the market prices to her own valuation at time 1. We will see that this will be the case in many other extensions.

Let us now compute the equilibrium in the case of bounded asset value.

Theorem 8.6 *Suppose f is bounded. Define θ by setting $\theta_0 = 0$ and*

$$d\theta_t = \frac{Z_1 - Y_t}{1 - t}dt.$$

Let H be the unique solution of

$$H_t + \frac{1}{2}H_{yy} = 0, \quad H(1,y) = f(y).$$

Then, (H,θ) is an equilibrium. In particular, Y is a Brownian motion in its own filtration and $Y_1 = Z_1$.

Proof First note that since f is bounded, H is bounded by the same constant due to its Feynman–Kac representation. Thus, to show that θ is admissible it suffices to show that it is a semimartingale. Indeed, given $Z_1 = z$

$$Y_t = B_t + \theta$$

is a Brownian bridge converging to z. Thus, Y is a $P^{0,z}$-semimartingale for each

z. Consequently, θ is a $P^{0,z}$-semimartingale for each z. Moreover, $H(1, Y_1) = f(Z_1)$, $P^{0,z}$-a.s.. Thus, θ is optimal given H.

Therefore, it remains to show that H is a rational pricing rule. Note that if Y is a Brownian motion in its own filtration,

$$H(t, Y_t) = \mathbb{E}[f(Y_1) \mid \mathcal{F}_t^Y]$$

due to the Feynman–Kac representation of H, which in turn implies H is a rational pricing rule.

Let us next show that Y is a Brownian motion in its own filtration. This requires finding the conditional distribution of Z given Y. This is a classical Kalman–Bucy filtering problem on $[0, T]$ for any $T < 1$. It is well-known (see, e.g., Theorem 3.4 in Çetin and Danilova, 2018a) the conditional distribution of Z given $\mathcal{F}_t^Y$ is Gaussian with mean $\widehat{X}_t := \mathbb{E}[Z | \mathcal{F}_t^Y]$ and variance $v(t)$, where

$$(1 - t)^2 v'(t) + v^2(t) = 0,$$

and

$$\widehat{X}_t = \int_0^t \frac{v(s)}{1 - s} dN_s,$$

where N is the innovation process.

The unique solution of the ODE with $v(0) = 1$ is given by $v(t) = 1 - t$. Consequently, $\widehat{X} = N$, i.e. $\widehat{X}$ is an $\mathcal{F}^Y$-Brownian motion. Let us now see that $\widehat{X} = Y$.

Indeed,

$$d\widehat{X}_t = dN_t = dY_t - \frac{\widehat{X}_t - Y_t}{1 - t} dt.$$

In other words, $\widehat{X}$ solves an SDE given Y. Since this is a linear SDE, it has a unique solution, which is given by Y itself. Hence Y is an $\mathcal{F}^Y$-Brownian motion.   □

Some remarks are in order. First of all, the boundedness assumption is only imposed for the brevity of the proof of admissibility and is easily satisfied for many natural boundary conditions.

The total order process Y is a Brownian motion in its own filtration. Thus, the distribution of Y is the same as that of noise trades. That is, the insider hides her orders among the noise traders and, thereby, the *inconspicuous trade theorem* holds.

The *Kyle's lambda* or the market impact of trades is given by

$$\lambda(t, y) := H_y(t, y).$$

Thus, the flatter f the more liquid is the market. Also note that the insider is indifferent among all bridge strategies that bring the market price to $f(Z_1)$ at time 1. One such bridge is when

$$dY_t = dB_t + k \frac{Z_1 - Y_t}{1 - t} dt$$

for some k while H is still as in Theorem 8.6. Although this is optimal for the insider, it cannot make an equilibrium when combined with H since $H(t, Y_t)$ will not be a martingale when Y is as above.

8.4 The static Kyle model with multiple insiders

The equilibrium in the previous section shows that the informed trader trades moderately in the sense that she reveals her private information slowly. In fact she only reveals her hand fully at the end of the trading period. This is crucially dependent on the fact that the informed trader has a monopoly over the meaningful information on the future asset price. Indeed, Holden and Subrahmanyam (1992) conjectured by taking the continuous-time limit of their discrete-time model that the insiders reveal their information immediately in case of two or more insiders possessing the same information.

This conjecture was later proven by Back et al. (2000) in the setting of the previous section under the assumption that V is normally distributed with mean 0. Moreover, they have also considered the case of multiple insiders when their private information are not perfectly correlated and have established the existence of equilibrium in a special case.

To present their results let's denote the number of insiders by $N \geq 2$ and assume that

$$V = \sum_{i=1}^{N} Z^i,$$

where Z_i is the private signal of insider i. It is assumed that the private information is symmetric; that is, the joint distribution of Z^is is invariant to permutations.

They also limit themselves to linear equilibria given the Gaussian structure, where the rate of trade of insider i at time t is of the form

$$\alpha_i(t)S_t + \beta_i(t)Z^i$$

and α and β are deterministic functions, and the price changes is given by

$$dS_t = \lambda(t) \left\{ dB_t + \sum_{i=1}^{N} \left(\alpha_i(t)S_t + \beta_i(t)Z^i \right) dt \right\}$$

with λ a deterministic function.

Note that the equilibrium rate of trade for the insider and the equilibrium price process obtained in Section 8.3 when f is affine is of this form.

Despite all these simplifying assumptions the solution of the individual insider's optimisation problem is still a difficult task. However, Back et al. obtain a clever resolution of this stochastic control problem. Their main result is the following theorem that describes the equilibrium in this setting.

Theorem 8.7 *Let*

$$\phi := \frac{\mathrm{Var}(V)}{\mathrm{Var}(NZ^i)}$$

and consider the constant

$$\kappa = \int_1^\infty x^{2(N-2)/N} e^{-2x(1-\phi)/N\phi} dx.$$

If $N > 1$ and the Z^i are perfectly correlated, i.e. $\phi = 1$, there is no equilibrium. Otherwise, there is a unique linear equilibrium. Set $\Sigma(0) = \mathrm{Var}(V)$ and define $\Sigma(t)$ for each $t < 1$ by

$$\int_1^{\frac{\Sigma(0)}{\Sigma(t)}} x^{2(N-2)/N} e^{-2x(1-\phi)/N\phi} dx = \kappa t.$$

An equilibrium is

$$\beta(t) = \left(\frac{\kappa}{\Sigma(0)}\right)^{\frac{1}{2}} \left(\frac{\Sigma(t)}{\Sigma(0)}\right)^{\frac{N-2}{N}} \exp\left(\frac{1}{N}\frac{1-\phi}{\phi}\frac{\Sigma(0)}{\Sigma(t)}\right),$$

$$\alpha(t) = -\frac{\beta(t)}{N},$$

$$\lambda(t) = \beta(t)\Sigma(t).$$

Furthermore, $\Sigma(t)$ is the conditional variance of V given market makers' information at time t.

It is easy to see that $\Sigma(t) \to 0$ and, thus, $\beta(t) \to \infty$ as $t \to 1$. This implies,

$$S_1 = \sum_{i=1}^N Z^i = V,$$

establishing that the prices converge to the true value at the end of trading.

In case $N = 1$ the above characterisation yields $\phi = \kappa = 1$. Thus, $\Sigma(t) = (1 - t)\Sigma(0)$. Therefore,

$$\beta(t) = \frac{1}{(1-t)\sqrt{\Sigma(0)}}, \quad \alpha(t) = -\frac{1}{(1-t)\sqrt{\Sigma(0)}}, \text{ and } \lambda(t) = \sqrt{\Sigma(0)},$$

coinciding with the findings of Theorem 8.6.

Given the competition among insiders one naturally wonders whether they trade more or less aggressively compared to the monopolist insider of Kyle. Fortunately, using the explicit form of equilibrium it is easy to analyse the impact of competition among the informed traders. Back et al. measures the intensity of informed trading by the coefficient of $V^i - S$ in the rate of trade for insider i, where V^i is the private valuation of the traded asset by insider i, and is shown to be a linear combination of market price and initial signal as follows:

$$V^i = (1 - \delta(t))S_t + \delta(t)NZ^i,$$

where

$$\delta(t) = \frac{\phi\Sigma(t)}{(1-\phi)\Sigma(0) + \phi\Sigma(t)}.$$

If $N = 2$ and the signals are not perfectly correlated, trade intensity is easily less than or equal to $\frac{1}{1-t}$, which is the corresponding intensity for the monopolist insider. Thus, the insiders reveals less in the presence of competition.

This should lead one to conjecture that the markets are informationally less efficient when there is a competition among insiders. Indeed, the residual uncertainty at time t as measured by $\Sigma(t)$ is greater than $1 - t$ when $N = 2$ and V is standard nnormal. It is a straightforward exercise in Gaussian filtering to conclude from Theorem 8.6 that in case of a monopolist insider the conditional variance of V given market makers' information at time t equals $1 - t$.

Another important metric is, of course, the market depth as measured by Kyle's lambda. In case of monopolistic insider $\lambda = 1$ once we assume that V is normally distributed. Back et al. show that

$$\lim_{t \to 1} \frac{1}{\lambda(t)} = 0.$$

In other words, the market approaches to complete illiquidity as the date of public announcement of V approaches.

In summary, the competition leads to relatively low informed trading intensity, lower level of informational efficiency, and lower liquidity.

8.5 Dynamic Kyle equilibrium

Section 8.3 assumes that the informed trader receives a private information only at the beginning of the trading period. In this section we shall relax this assumption by considering the case of a single informed trader receiving a continuous signal converging to Z_1 as time approaches to the public announcement date of the value of the traded asset.

Following Back and Pedersen (1998) we assume that the private signal of the insider is the following Gaussian process:

$$Z_t = Z_0 + \int_0^t \sigma(s)dW_s,$$

where Z_0 is a mean-zero Normal random variable, W is a Brownian motion independent of B, and $\text{Var}(Z_1) = 1$. This normalisation is for the sake of easy comparison with the static equilibrium from Section 8.3. Back and Pedersen placed a certain restriction on the mapping $t \mapsto \text{Var}(Z_t)$, which have been relaxed by Danilova (2010). The following assumption on σ follows Danilova (2010) (see also Section 5.1 in Çetin and Danilova, 2018a).

Assumption 8.8 Let $c = \text{Var}(Z_0)$ and define $\Sigma(t) = c + \int_0^t \sigma^2(s)ds$. Then Σ satisfies the following conditions:

1. $\Sigma(t) > t$ for every $t \in (0,1)$, and $\Sigma(1) = 1$.
2. $\int_0^t \frac{1}{(\Sigma(s)-s)^2}ds < \infty$ for all $t \in [0,1)$.
3. $\lim_{t \to 1} s^2(t)S(t)\log S(t) = 0$, where $s(t) = \exp\left(-\int_0^t \frac{1}{\Sigma(s)-s}ds\right)$ and $S(t) = \int_0^t \frac{1+\sigma^2(r)}{s^2(r)}dr$.

4. σ is bounded.

Although the third condition above seems involved, it is satisfied in practical situations. For instance, it is always satisfied if $S(1) < \infty$ since $s(1) = 0$ under the first condition. Also, an application of L'Hôpital rule shows its validity when σ is constant.

In this case the optimal strategy of the insider is still to bring the market price to her own time-1 valuation gradually. More precisely, the equilibrium total order process is given by

$$Y_t = B_t + \int_0^t \frac{Z_s - Y_s}{\Sigma(s) - s} ds.$$

As in static case, the informed traders' trades are inconspicuous, i.e. Y is a Brownian motion in its own filtration.

There is no change in the equilibrium pricing rule. Indeed, it is given by the solution to the same boundary value problem:

$$H_t + \frac{1}{2}H_{yy} = 0, \qquad H(1, y) = f(y). \tag{8.23}$$

This in particular implies the Kyle's lambda, i.e. $H_y(t, Y_t)$, have the same properties, too.

Thus, whether the information flow is dynamic or static does not have any impact on the qualitative properties of the equilibrium when there is a single informed trader.

One can also consider more general Markovian information flows. The reader is referred to Campi et al. (2011) for the details and, in particular, the concept of general dynamic Markovian bridges (see also Çetin and Danilova, 2018a).

8.6 The Kyle model and default risk

In earlier sections we have considered the pricing of a default-free risky asset. However, it is also possible to use a similar framework when the risky asset is also subject to default. This was analysed by Campi and Çetin (2007) in a static setting and by Campi et al. (2013) in a dynamic one.

Suppose for simplicity that the normalised cash balances of the firm is modelled by $1 + \beta_t$, where β is a standard Brownian motion, and the default occurs at time T_0, where

$$T_0 = \inf\{t > 0 : 1 + \beta_t = 0\}.$$

If the insider has perfect knowledge of T_0, analogous to the case studied in Section 8.3, the problem can be treated within the paradigm of static information flow as done in Campi and Çetin (2007). The other possibility is that the insider receives a dynamic information flow that gradually reveals the default time. What is assumed in Campi et al. (2013) is that the insider's signal is given by

$$Z_t = 1 + \beta_{\Sigma(t)},$$

where Σ is a continuously differentiable function with $\Sigma(0) = 0$, $\Sigma(1) = 1$ and $\Sigma(t) > t$ for $t \in (0, 1)$. This in particular implies that

$$Z_t = 1 + \int_0^t \sigma(s)dW_s,$$

where $\sigma(t) = \sqrt{\Sigma'(t)}$, as well as $T_0 = \Sigma(\tau)$ with

$$\tau = \inf\{t > 0 : Z_t = 0\}.$$

Now let us consider the pricing of defaultable asset whose value at time-1 is given by $1_{[T_0 > 1]} f(Z_1)$ for some continuous and strictly increasing f. The information structure is almost identical to the previous cases except that the market makers not only observe the total order process Y but also whether the default has occurred or not, i.e the default indicator process $D_t := 1_{[T_0 > t]}$.

In earlier default-free models the insider's goal was to bring the market valuation of the risky asset to her own valuation using a Brownian bridge strategy. A similar phenomenon occurs here as well. However, the insider now must convey relevant information not only about the value of Z_1 but also regarding the default time. And the analogue of the Brownian bridge in this setting is the Bessel bridge.

Following Chapter 8 of Çetin and Danilova (2018a) we shall consider two cases: (1) the static case that comprises the insider knowing τ and Z_1 in advance; and (2) the dynamic case in which the insider observes the process Z only, where Σ is satisfying Assumption 8.8 with $c = 0$.

In the static case the insider's strategy in the equilibrium is given by

$$d\theta_t = \left(\frac{q_x(1 - t, Y_t, Z_1)}{q(1 - t, Y_t, Z_1)} 1_{[T_0 > 1]} + \frac{\ell_a(T_0 - t, Y_t)}{\ell(T_0 - t, Y_t)} 1_{[\tau \leq 1]} \right) dt$$

where

$$q(t, x, z) = \frac{1}{\sqrt{2\pi t}} \left(\exp\left(-\frac{(x - z)^2}{2t} \right) - \exp\left(-\frac{(x + z)^2}{2t} \right) \right) \quad \text{and}$$

$$\ell(t, a) = \frac{a}{\sqrt{2\pi t^3}} \exp\left(-\frac{a^2}{2t} \right).$$

Therefore, if the insider knows that the default will not happen before time 1, she will bring the total demand to the same level as Z_1 as she did in earlier models. However, if the default is going to take place before time 1, she will drive the total demand to 0 at the time of default.

A similar but more complicated trading strategy is employed in the dynamic case. Note that since $\Sigma(t) > t$ for $t \in (0, 1)$, if $T_0 < 1$, the insider will receive the news of default a bit earlier, more precisely, at time τ, than T_0 since $\tau = \Sigma^{-1}(T_0) < T_0$. How much in advance depends on the structure of Σ and how significantly it differs from the identity function. Such difference can indeed happen. It is documented that there is a difference between the recorded default time and the economic default time (see Guo et al., 2014).

In the dynamic case the trading strategy of the informed trader in equilibrium is given by

$$d\theta_t = \left(\frac{q_x(\Sigma(t) - t, Y_t, Z_t)}{q(\Sigma(t) - t, Y_t, Z_t)} 1_{[t \leq \tau \wedge 1]} + \frac{\ell_a(T_0 - t, Y_t)}{\ell(T_0 - t, Y_t)} 1_{[\tau \wedge 1 < t \leq T_0 \wedge 1]} \right) dt.$$

Again, the total order process is driven to Z_1 in case of no-default whereas it converges to 0 when default happens before time 1.

In both cases the pricing rule is given by the solution of a boundary value akin to the one given by (8.23) with the extra side condition that H vanishes at 0. The solution to this boundary value problem is given by a Feynman–Kac representation in terms of a *killed* Brownian motion (see Campi et al., 2013, for details).

As in the default-free case, there is no qualitative difference between the equilibrium with a dynamical signal and the one with a static signal.

8.7 Glosten–Milgrom model

Glosten and Milgrom (1985) study a model in which competitive risk-neutral market makers quote bid and ask prices to trade a single unit of an asset with a trader who submits a market order. The market order can be informed or coming from a *noise* trader who trade for liquidity reasons endogenous to the model. We shall be using the version of the Glosten–Milgrom model studied in Çetin and Xing (2013) that is a formalisation of the version considered by Back and Baruch (2004).

In this model the cumulative demand of the noise traders is given by the difference of two jump process X^B (representing buy orders) and X^S (representing sell orders). Each order is of fixed magnitude of δ and the arrival times of buy and sell orders are following two independent Poisson processes of constant intensity β.

The value of the risky asset V is either 0 or 1. This value will become public information at time 1 but is already known to the insider at time 0.

As usual, the market makers only observe the total order flow, and the insider is assumed to observe the noise orders as well. Note that in this model the insider will never trade of size different than δ or trade at the same time in the same direction since such actions will immediately reveal the presence of the insider and whether she is buying or selling.

Çetin and Xing show that, differently from the Kyle model, the insider uses a mixed strategy. That is, the trades of the insider not only depend on the total order and her private information but also on an extra randomisation. She achieves this by randomly meeting the orders of the noise traders by submitting a market order in the opposite direction and, thus, in a way acting like a market maker.

The techniques for establishing the equilibrium in this model is quite different than the ones discussed in earlier sections and relies on enlargement of filtration arguments for point process (see Çetin and Xing, 2013, for details). However, the equilibrium strategy of the insider is still a bridge strategy: The equilibrium price converges to V at the end of the trading horizon.

Çetin and Xing also study the asymptotics of Glosten–Milgrom equilibria by setting $\beta = (2\delta^2)^{-1}$ and letting $\delta \to 0$. It is shown therein that the limiting equilibrium is that of a Kyle model where V is Bernoulli random variable taking values in $\{0, 1\}$. Thus, the continuous-time Kyle model can be viewed as an idealisation of a Glosten–Milgrom model with high trading activity.

8.8 Risk aversion of market makers

Whereas the risk-neutrality of the market makers makes the model tractable, it is not consistent with the observed market behaviour. Indeed, there is vast empirical evidence that the market makers are risk averse and quote prices in a way to ensure their inventories mean revert around a target level at a speed determined by their risk aversion (see Huang and Stoll, 1997, and Madhavan and Smidt, 1993, for New York Stock Exchange, Hansch et al., 1998, for London Stock Exchange, Bjønnes and Rime, 2005, for Foreign Exchange; for a survey of related literature and results, see Sections 1.2 and 1.3 in Biais et al., 2005).

Although relaxing the assumption of market makers' risk neutrality is natural and has been prompted by empirical evidence, there have been limited attempts in the literature for a theoretical investigation of its impact due to the technical complexity of the model. Subrahmanyam (1991) considered a one-period Kyle model where market makers with identical exponential utilities set the price assuming autarky utility, i.e. their mark-to-market utilities are martingales.

Çetin and Danilova (2016) developed and solved a continuous-time version of the problem introduced by Subrahmanyam. The model assumes N identical market makers quoting prices assuming autarky utilities. The market makers are risk averse and have exponential utility with risk aversion coefficient γ. Moreover, the total demand is assumed to be split equally in case of draw, which will be the case in equilibrium as the market makers are identical. The information flow of the insider is static, i.e. the insider knows V from the beginning. Note that the model presented below is for an insider, and not for an informed trader with an unbiased estimator for V. This is due to the fact that the market makers are risk-averse, thus the argument leading to (8.8) no longer holds as one needs to work with certainty equivalents due to risk aversion.

Given this warning the characterisation of the equilibrium in this market is as follows: The market makers choose the pricing rule H so that

$$H_t + \frac{\sigma^2}{2}H_{yy} = 0$$

and the total demand for the asset in its own filtration has the decomposition

$$dY_t = \sigma \, dB_t^Y - \frac{\sigma^2 \gamma}{2N} Y_t H_y(t, Y_t) \, dt,$$

where B^Y is an $\mathcal{F}^Y$-Brownian motion. On the other hand the optimality condition of the informed trader requires that $H(1, Y_1) = f(V)$. It turns out that the equilibrium dynamics in this framework is given by the forward-backward system

$$H_t + \frac{\sigma^2}{2} H_{yy} = 0;$$

$$dY_t = \sigma d\beta_t - \frac{\sigma^2 \gamma}{2N} Y_t H_y(t, Y_t) \, dt; \tag{8.24}$$

$$H(1, Y_1) \overset{d}{=} f(V),$$

where the last equality is equality in distribution, β is a given Brownian motion, and the solution of the SDE is required to be strong. Note that the terminal condition of the backward PDE is determined by the time-1 distribution of the solution of the forward SDE, which itself depends on the solution of the PDE.

Under the assumption that f is bounded with a continuous derivative, it is shown in Çetin and Danilova (2016) via a Schauder's fixed point argument the existence of a solution to the above system. Moreover, the solution to the SDE given in (8.24) has a smooth transition density $q(s, y; t, z)$ implying that the equilibrium level of demand in this economy is given by

$$Y_t = \sigma B_t - \int_0^t \frac{\sigma^2 \gamma}{2N} Y_s H_y(s, Y_s) \, ds + \sigma^2 \int_0^t \frac{q_y}{q}(s, Y_s; 1, H^{-1}(1, f(V))) \, ds, \tag{8.25}$$

while the price is similarly given by $H(t, Y_t)$. Since the price is always a strictly increasing function of the demand, the solution to (8.24) is mean reverting as predicted by the empirical studies.

Interestingly the price chosen by the market makers in the above model is a solution to a particular backward stochastic differential equation (BSDE). If one denotes by S the price set by the market makers, then given any (not necessarily the optimal) trading strategy of the informed trader, S satisfies

$$dS_t = -\frac{\sigma^2 \gamma}{2N} Y_t \lambda_t^2 \, dt + \sigma \lambda_t dB_t^Y, \tag{8.26}$$

where B^Y is a Brownian motion in the natural filtration of Y, and S_1 is determined via the terminal condition

$$\exp(\gamma Y_1 S_1) = \mathbb{E}\left[\exp(\gamma Y_1 V) \middle| \mathcal{F}_1^Y \right]. \tag{8.27}$$

If one can find a solution (S, λ) to the above BSDE, then S determines the price in this market while λ can be considered as the *price impact* of the trades given that the martingale part of Y is σB^Y extending the notion of price impact in the previous Markovian equilibria where it is given by $H_y(t, Y_t)$.

Although at the first sight the above BSDE seems to be quadratic in λ, its coefficient is proportional to Y, which is in general unbounded being a Brownian motion. In the Markovian setting Y will be the solution to a forward SDE

$$dY_t = \sigma dB_t^Y + \hat{\alpha}(t, Y_t, S_t, Z_t) dt. \tag{8.28}$$

Even if one tries to handle this difficulty via localisation, the terminal condition (8.27) is highly non-standard and depends possibly on the whole history of Y.

The equilibrium found in Çetin and Danilova (2016) in fact shows that when the informed trader is behaving optimally, there is a Markovian solution to the above BSDE when Y is defined by the equilibrium demand process.

In the Kyle model and its extensions discussed in earlier sections, the Kyle's lambda is a martingale. In fact, it was conjectured in Kyle (1985) that:

[. . .] neither increasing nor decreasing depth is consistent with behavior by the informed trader which is "stable" enough to sustain an equilibrium. If depth ever increases, the insider wants to destabilize prices (before the increase in depth) to generate unbounded profits. If depth ever decreases, the insider wants to incorporate all of his private information into the price immediately.

However, when the market makers are risk averse, the Kyle's lambda is no longer a martingale, while the insider still has bounded profits. This is due to the risk sharing between the market makers and the insider. Indeed, if the trader attempts to follow the strategy outlined by Kyle, she would be moving the total order away from its mean, leaving the market makers exposed to the risk of large orders. This would violate the risk sharing mechanism in equilibrium and cause the market makers to adjust the prices eliminating favourable gains for the insider.

It is also shown in Çetin and Danilova (2016) that the sensitivity of prices to the total order can be a submartingale for certain model parameters. This implies that the execution costs are, on average, increasing toward the end of a trading period, which is consistent with the empirical results obtained in Madhavan et al. (1997).

8.9 Conclusion and further remarks

In all the models discussed so far the informed traders reveal their private information slowly and make sure that the market prices converge to their own valuation by the end of trading period. However, all models considered assumed a single traded asset. Two notable extensions of the single-period Kyle model to multiple assets are Caballé and Krishnan (1994) and Garcia del Molino et al. (2020). A more recent paper (Cocquemas et al., 2020) uses methods from optimal transport to study equilibrium with multiple assets. Back studies in a continuous-time setting an extension of the Kyle model to allow trading in an option on the stock and shows that possibility of trading in the option introduces a stochastic volatility component to the stock (Back, 1993). Stochastic volatility of equilibrium prices is also obtained in the setting considered by Collin-Dufresne and Fos (2016).

Choi et al. (2019) study a dynamic Kyle model in discrete time in which there are two strategic traders: one is the informed trader as above and the other is an uninformed trader with a target amount in the traded stock to liquidate, which is unknown to the others. The model therefore combines the informed trading model of Kyle with the literature on optimal execution for uninformed traders with liquidity motives. Although it cannot be solved in closed form, the equilibrium can be computed numerically and the model has testable implications.

Risk aversion of the insider is studied in a one-period setting by Subrahmanyam

(1991) and in continuous-time but restrictive assumptions by Cho (2003). In an unpublished manuscript that constitutes a part of the dissertation of P. Shi, Danilova and Shi established the equilibrium in a fairly general setting for a risk-averse insider with static information flow (Danilova and Shi, 2014). More recently, Bose and Ekren (2020) studied the equilibrium with a risk-averse insider receiving a static signal using methods of optimal transport.

Risk aversion of the market makers and its impact on market risk premium is also the subject of Ying (2020).

Back et al. (2018a) propose a model of informed trading in Kyle's framework that allows for the detection of information events based on market data.

In all these models the terminal value of the asset V is exogenous. Back et al. study activist trading in Kyle's model in which the terminal value of the traded asset depends on the trades of the activist (Back et al., 2018b), and, thus, is endogenously determined in equilibrium.

An important aspect of the research literature that has not been touched upon so far in this chapter is that on limit order markets and the equilibria therein. This is mostly due to the technical difficulties involved in modelling and the scarcity of solvable models in contrast to the Kyle model. Moreover, the competition among market makers submitting limit orders are fundamentally different. Bernhardt and Hughson (1997) show that the limit order traders makes positive expected gains if there are only finitely many of them, while only two market makers are enough to drive the profits to zero in the Kyle model. Glosten (1994) assumes infinitely many limit order traders to get around this issue. Çetin and Waelbroeck (2020) propose a setting to combine the Glosten model of limit order trading and the Kyle model in a single equilibrium framework, albeit in a single-period model!

References

Amihud, Yakov, and Mendelson, Haim. 1980. Dealership market: Market-making with inventory. *Journal of Financial Economics*, **8**(1), 31–53.

Back, Kerry. 1992. Insider trading in continuous time. *Review of Financial Studies*, **5**(3), 387–409.

Back, Kerry. 1993. Asymmetric information and options. *Review of Financial Studies*, **6**(3), 435–472.

Back, Kerry, and Baruch, Shmuel. 2004. Information in securities markets: Kyle meets Glosten and Milgrom. *Econometrica*, **72**(2), 433–465.

Back, Kerry, and Pedersen, Hal. 1998. Long-lived information and intraday patterns. *Journal of Financial Markets*, **1**(3-4), 385–402.

Back, Kerry, Cao, C. Henry, and Willard, Gregory A. 2000. Imperfect competition among informed traders. *Journal of Finance*, **55**(5), 2117–2155.

Back, Kerry, Crotty, Kevin, and Li, Tao. 2018a. Identifying information asymmetry in securities markets. *Review of Financial Studies*, **31**(6), 2277–2325.

Back, Kerry, Collin-Dufresne, Pierre, Fos, Vyacheslav, Li, Tao, and Ljungqvist, Alexander. 2018b. Activism, strategic trading, and liquidity. *Econometrica*, **86**(4), 1431–1463.

Bernhardt, Dan, and Hughson, Eric. 1997. Splitting orders. *Review of Financial Studies*, **10**(1), 69–101.

Biais, Bruno, Glosten, Larry, and Spatt, Chester. 2005. Market microstructure: A survey of microfoundations, empirical results, and policy implications. *Journal of Financial Markets*, **8**(2), 217–264.

Bjønnes, Geir Høidal, and Rime, Dagfinn. 2005. Dealer behavior and trading systems in foreign exchange markets. *Journal of Financial Economics*, **75**(3), 571–605.

Bose, Shreya, and Ekren, Ibrahim. 2020. Kyle–Back models with risk aversion and non-Gaussian beliefs. ArXiv:2008.06377.

Caballé, Jordi, and Krishnan, Murugappa. 1994. Imperfect competition in a multi-security market with risk neutrality. *Econometrica*, **62**(3), 695–704.

Campi, Luciano, and Çetin, Umut. 2007. Insider trading in an equilibrium model with default: a passage from reduced-form to structural modelling. *Finance and Stochastics*, **11**(4), 591–602.

Campi, Luciano, Çetin, Umut, and Danilova, Albina. 2011. Dynamic Markov bridges motivated by models of insider trading. *Stochastic Processes and their Applications*, **121**(3), 534–567.

Campi, Luciano, Çetin, Umut, and Danilova, Albina. 2013. Equilibrium model with default and dynamic insider information. *Finance and Stochastics*, **17**(3), 565–585.

Çetin, Umut, and Danilova, Albina. 2016. Markovian Nash equilibrium in financial markets with asymmetric information and related forward–backward systems. *Annals of Applied Probability*, **26**(4), 1996–2029.

Çetin, Umut, and Danilova, Albina. 2018a. *Dynamic Markov Bridges and Market Microstructure: Theory and Applications*. Springer.

Çetin, Umut, and Danilova, Albina. 2018b. On pricing rules and optimal strategies in general Kyle-Back models. ArXiv:1812.07529.

Çetin, Umut, and Waelbroeck, Henri. 2020. Informed trading, limit order book and implementation shortfall: equilibrium and asymptotics. ArXiv:2003.04425.

Çetin, Umut, and Xing, Hao. 2013. Point process bridges and weak convergence of insider trading models. *Electronic Journal of Probability*, **18**.

Cho, Kyung-Ha. 2003. Continuous auctions and insider trading: uniqueness and risk aversion. *Finance and Stochastics*, **7**(1), 47–71.

Choi, Jin Hyuk, Larsen, Kasper, and Seppi, Duane J. 2019. Information and trading targets in a dynamic market equilibrium. *Journal of Financial Economics*, **132**(3), 22–49.

Cocquemas, François, Ekren, Ibrahim, and Lioui, Abraham. 2020. A general solution method for insider problems. ArXiv:2006.09518.

Collin-Dufresne, Pierre and Fos, Vyacheslav. 2016. Insider trading, stochastic liquidity, and equilibrium prices. *Econometrica*, **84**(4), 1441–1475.

Danilova, A., and Shi, P. 2014. Insider trading when information is static: impact of insider's risk aversion on equilibrium. *Preprint*; see http://etheses.lse.ac.uk/3156/.

Danilova, Albina. 2010. Stock market insider trading in continuous time with imperfect dynamic information. *Stochastics*, **82**(1), 111–131.

Garcia del Molino, Luis Carlos, Mastromatteo, Iacopo, Benzaquen, Michael, and Bouchaud, Jean-Philippe. 2020. The multivariate Kyle model: More is different. *SIAM Journal on Financial Mathematics*, **11**(2), 327–357.

Garman, Mark B. 1976. Market microstructure. *Journal of Financial Economics*, **3**(3), 257–275.

Glosten, Lawrence R. 1994. Is the electronic open limit order book inevitable? *Journal of Finance*, **49**(4), 1127–1161.

Glosten, Lawrence R., and Milgrom, Paul R. 1985. Bid, ask and transaction prices in a specialist market with heterogeneously informed traders. *Journal of Financial Economics*, **14**(1), 71–100.

Guo, Xin, Jarrow, Robert A., and de Larrard, Adrien. 2014. The economic default time and the arcsine law. *Journal of Financial Engineering*, **1**(03), 1450025.

Hansch, Oliver, Naik, Narayan Y., and Viswanathan, S. 1998. Do inventories matter in dealership markets? Evidence from the London Stock Exchange. *Journal of Finance*, **53**(5), 1623–1656.

Ho, Thomas, and Stoll, Hans R. 1981. Optimal dealer pricing under transactions and return uncertainty. *Journal of Financial Economics*, **9**(1), 47–73.

Holden, Craig W., and Subrahmanyam, Avanidhar. 1992. Long-lived private information and imperfect competition. *Journal of Finance*, **47**(1), 247–270.

Huang, Roger D., and Stoll, Hans R. 1997. The components of the bid–ask spread: A general approach. *Review of Financial Studies*, **10**(4), 995–1034.

Kyle, Albert S. 1985. Continuous auctions and insider trading. *Econometrica*, **53**(6), 1315–1335.

Madhavan, Ananth, and Smidt, Seymour. 1993. An analysis of changes in specialist inventories and quotations. *Journal of Finance*, **48**(5), 1595–1628.

Madhavan, Ananth, Richardson, Matthew, and Roomans, Mark. 1997. Why do security prices change? A transaction-level analysis of NYSE stocks. *Review of Financial Studies*, **10**(4), 1035–1064.

Stoll, Hans R. 1978. The supply of dealer services in securities markets. *Journal of Finance*, **33**(4), 1133–1151.

Subrahmanyam, Avanidhar. 1991. Risk aversion, market liquidity, and price efficiency. *Review of Financial Studies*, **4**(3), 417–441.

Ying, Chao. 2020. The pre-FOMC announcement drift and private information: Kyle meets macro-finance. Available at SSRN 3644386.

9

Deciphering How Investors' Daily Flows are Forming Prices

Daniel Giamouridis[a], Georgios V. Papaioannou[a]
and Brice Rosenzweig[a]

Abstract

Different types of participants have distinct impacts on prices. We employ well-suited machine learning techniques and a unique set of data with explicit information about the participant type and side of order. We find that net Quant flow has positive price impact, while net Hedge Fund and Broker Dealer flows have negative price impact, consistent with liquidity provision. We additionally find that Quants, Hedge Funds, and Institutions have positive co-impact when we examine their intra-sector trade impact. Our findings are intuitive and robust to various parameter checks. By contrast, when we explore whether similar patterns can be detected with linear model specifications, we get neither intuitive nor robust results; for example, that order flow does not have any price impact.

9.1 Introduction

In this chapter, we investigate the relation between order flow and prices, and in particular, the effect that different type of participants might have on prices. Cogniscent of the fact that prices are affected by potentially multiple variables, which may also interact non-linearly, we employ well suited machine learning techniques to identify participant type related flow-prices relations, directionally but also by order of importance. Our modelling approach allows us to study not only stock-level relations between order flow and prices but also the effect of stock-group order flow on the group constituents' stock price, a phenomenon most commonly referred to as 'co-impact'.

More specifically, we are using a unique dataset of order flow for European stocks, characterized by the nature of the participant type investment activities into *Quant*, *Institutional*, *Hedge Fund*, and *Broker Dealer*. We are able to sign order flow explicitly, based on the originator's direction of order, i.e. buy/sell. We then calculate measures of daily flow and order imbalance by participant type for each company in our sample. As in Boehmer and Wu (2008), we conjecture that

[a] Bank of America, Data and Innovation Group, London. Contact Author: Georgios V. Papaioannou
(george.papaioannou@bofa.com)
Published in *Machine Learning And Data Sciences For Financial Markets*, Agostino Capponi and
Charles-Albert Lehalle © 2023 Cambridge University Press.

these types of market participants are likely to differ in the quantity and quality of their information, how they use it for executions, and what relation to expect between the respective order flow and prices.

We further condition the flow based on other characteristics such as the style factor, sector, country, and currency. Those can be thought of as non-orthogonal dimensions to multisect the flow. To better address the non-linear interactions as well as the variance introduced by the increased dimensionality, we apply machine learning techniques including artificial neural networks (ANN), tree and ensemble methods, gradient boosting trees and XGBoost, and support vector machine regressors. These are non-linear in nature and along with regularization methods, are better suited for identifying potentially complex, informative, patterns of the flow at the more granular level. The second main question of this chapter is therefore to explore how machine learning techniques can help us advance our empirical understanding of market microstructure.

Our main empirical finding is that different types of participants do affect prices differently and machine learning collectively improves our understanding of this relation. We find that order flow information is a significant determinant of price. We also find that granular information on the originator of the flow explains price formation even further. By contrast, comparable linear models provide mixed conclusions, for example that while aggregate order flow imbalance affects prices, when it is decomposed into its individual components, it actually does not. By means of importance, *Quant* is the most important driver of price, followed by *Broker Dealer*. *Institutional* and *Hedge Fund* order imbalances are amongst the top-ten highest important drivers, albeit not as significant. More specifically *Quant* order imbalance is positively associated with contemporaneous returns due to either superior information and/or large trading pressure. In terms of how stock-group order flow affects the group constituents' prices, we find that the impact varies depending on the originator and hence, when it is viewed on aggregate, may actually dilute the magnitude of co-impact effects.

Our work extends the empirical literature on flow imbalance and price formation. Earlier work has documented strong correlation between flow and returns and risk in the time series or cross-section. Andrade et al. (2008) and Chordia et al. (2002) are indicative such works, whereas, closer to ours, Griffin et al. (2003), and Boehmer and Wu (2008) distinguish order flow imbalance by participant type. More recently, Bianchi et al. (2019) and Ha and Hu (2017) have also introduced the notion of participant type into their analysis. An additional dimension in the study of order flow imbalance and price formation was introduced by Giamouridis et al. (2017). The motivation of their approach lay in the empirical observation that increasingly so institutional investors use common inputs in their investment process, i.e. styles. They investigated style-trading behaviour and style returns for different type of investors and found that style-level imbalances are significantly positively related to future style returns. Overall, this literature focuses on the US equity market – except Andrade et al. (2008) – and typically conducts time-series regressions of stock or broad market portfolio returns on a small number of predictor or explanatory variables.

More generally, we are also adding to a significant body of literature that has studied the asset pricing implications of institutional holdings and changes in holdings of specific different investor groups. Earlier work has focused on the trading behavior of institutional investors and in particular mutual funds (see for example Frazzini and Lamont, 2008). Beyond that, linking investor trading behaviour to asset prices has been extended to other types of institutional investors such as pension funds (Lakoshinok et al., 1992), hedge funds (e.g., Akbas et al., 2015; Cao et al., 2016; Chen et al., 2020), and retail investors (e.g., Kelley and Tetlock, 2013). Along these lines Froot and Teo (2008) document that institutional investors reallocate capital across styles, and find that style inflows positively predict returns for stocks in the same style.

Our modelling approach allows us also to provide empirical insights into recent literature on 'co-impact' or 'cross-impact'. Giamouridis (2017) discussed the increasing institutional investing focus on systematic strategies, the paradigm shift to coordinated trading – as more assets are traded as index, sector, factor portfolio or other, and the implications for asset prices, risk management, and market microstructure. As to the last, it is argued that systematic investment strategies naturally lead to coordinated portfolio trade lists' natural order flow. The market impact from executing these portfolio trade lists increases the correlations of intraday stock returns and affects the covariance of stock returns. The coordinated order flow also suggests an increase in the covariance of intraday traded volume, which could itself be a source of the variability of execution costs. A number of recent studies (see, e.g., Benzaquen et al., 2016; Bucci et al., 2020; Min et al., 2018) provide a detailed account of this phenomenon. In our analysis we attempt to further our understanding of co-impact through the lens of more granular flow information.

Finally, this chapter adds to the growing literature of machine learning studies in empirical finance. Recent papers apply machine learning methods to pursue research questions in asset pricing (see, e.g., Bianchi et al., 2021; Chen et al., 2019; Feng et al., 2020; Gu et al., 2020; Holley et al., 2021); empirical corporate finance (see, e.g., Li et al., 2020); market microstructure (see, e.g., Briere et al., 2020; Easley et al., 2020) as well as in other areas such as portfolio optimization (see, e.g., Ban et al., 2018) and trade idea classification (see Papaioannou and Giamouridis, 2020). We document the benefits of machine learning over conventional, linear, methods in the context of our research question. In particular, we are able to provide consistent and intuitive interpretations that are, as we document, not attainable with a linear model.

The chapter is organized as follows. In Section 9.2 we discuss and explore the data that the analysis was based on. Section 9.3 discusses the modeling techniques applied. In Section 9.4 we present the results and findings of our analysis focusing on the fundamental questions this chapter is trying to address: Are aggregate flows contributing to daily price formation, or returns, and what is their role in that? At the more granular level, what is the effect of different participant-type flows at the daily aggregate level? How and why are machine learning methods beneficial to the analysis? Finally, in Section 9.5, we summarize and conclude the chapter.

9.2 Data description and exploratory statistics

Our sample consists of the total daily trading flow executed through BofA Securities for all stocks in the STOXX 600 index for the sample period running from March 2019 to March 2021. For each stock and day the sample contains the buy and sell US Dollar notional executed through BofA Securties by each of the four following participant types: *Quant, Institutional, Hedge Fund*, and *Broker Dealer*. The assignment of participants to participant types follows an internal scheme based on the nature of the participants' investment activities:

- *Institutional* includes traditional, long-only asset managers, pension funds, insurance, etc.
- *Hedge Fund* include alternative investment managers implementing long-short strategies, and multi-strategies.
- *Quant* includes fully systematic funds and proprietary money managers.
- Finally *Broker Dealer* are intermediary firms buying and selling on behalf of clients.

The last includes part of the retail category often found in the literature. BofA Securities is a major counterparty in terms of equity flow market share. Our dataset aggregates more than one thousand individual participants. Given the comprehensive coverage of stocks and clients by BofA Securties, the dataset is not expected to exhibit significant idiosyncrasies relative to the entire equity market flow.

We obtain stock level information from the Axioma risk model database. The stock level information comprises of aggregate and idiosyncratic risk data, as well as exposures to Axioma's fundamental risk factors, industry and country participation, and also currency. Additional stock level data include liquidity, market capitalization, and other return-based metrics we compute. Table 9.1 tabulates the definitions of Axioma's fundamental risk factors.

To get a better understanding of the composition and time variation of the order flow data we use in our analysis, we present in Figure 9.1 the quarterly variations of the gross and net flow by participant type. The most represented participant type in our data is the *Institutional*. It represents about 41–46% of the total gross annually. *Hedge Fund* is the second largest with 25–30% annually. *Quant* gross flow ranges between 19–28% of the gross annually and *Broker Dealer* is the smallest category, with about 5.3–6.4% of the total gross order flow. A noteworthy empirical feature of our data relates to the side of the order flow. When we look at the net values part of Figure 9.1 we observe that, at this level of aggregation, *Quant* flow seems, at least qualitatively and almost consistently, to be on the opposite side of all other flows. We will revisit this observation in the context of interpreting the relation between order imbalances and price formation.

We further report the variation and composition of our flow by sector in Figures 9.2 and 9.3, respectively. We observe that *Financials*, *Industrials* and *Consumer Discretionary* represent the largest three, and *Energy*, *Utilities* and

Table 9.1 Style factors

Dividend Yield	Ratio of sum of the dividends paid over the most recent year to average market capitalization
Earnings Yield	Earnings-to-price and estimated earning-to-price
Foreign Exchange Sensitivity	1-year weekly beta to returns of a basket of major currencies
Growth	Realized sales growth, forecast sales growth, realized earnings growth, forecast earnings growth
Leverage	Total debt to total assets and total debt to equity
Liquidity	Natural logarithm of the ratio of 1-month average daily volume and 1-month average market capitalization, inverse of 3-month Amihud illiquidity ratio and proportion of returns traded over the last calendar year.
Market Sensitivity	1-year weekly β
Medium-Term Momentum	Cumulative return over past year excluding the most recent months
Profitability	Return-on-equity, return-on-assets, cash flow to assets, cash flow to income, gross margin, and sales-to-assets
Size	Natural logarithm of market capitalization
Short-Term Momentum	Cumulative return over past month
Value	Book-to-price
Volatility	3-month average of absolute returns over cross-sectional standard deviation, fully orthogonalized to market sensitivity

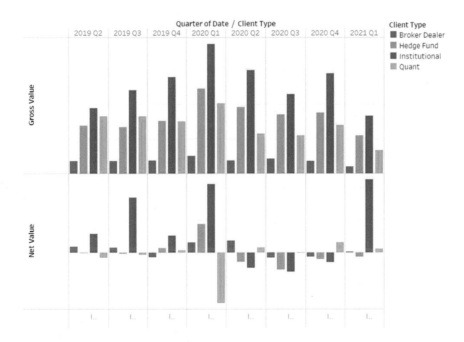

Figure 9.1 Quarterly flow by participant type

Telecommunications the bottom three sector order flows. Apparently this is expected to also be driven by the participant type representation in our sample as well as the contribution of the respective sectors in major market benchmarks.

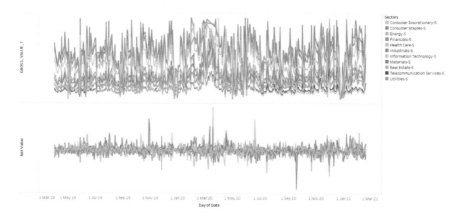

Figure 9.2 Gross and Net flow time series by sector.

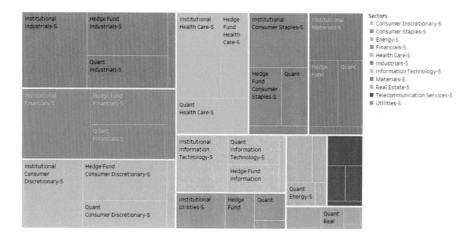

Figure 9.3 Aggregate gross notional flow by participant type and style.

In our empirical analysis, we add multiple dimensions to the order flow information. We express the multisected flow by $F_{k,s:(l,m,n,p)}^{t,d}$ where t denotes the date, d denotes 'Buy' or 'Sell', or the transformed variables 'Gross' and 'Net', s is the security and k denotes the participant type. The security s maps uniquely to l, m, n, p, where l is the style and takes values in Table 9.1, m denotes the sector, n the country and p the currency. In Section 9.4, among others, we compare two sets of results. In one we include the direct stock level flow features $F_{k,s}^{t,d}$ along with stock information, such as categorical information about the stock such as sector, country, style, etc, and allow the task of the model to create the relations between the flow of stocks corresponding to each category. In another, we explicitly use features $F_{k,s:(l,m,n,p)}^{t,d}$ where we have aggregated the multisected flow at different levels and mapped it to the corresponding stock level information. The market participant type split is at the highest level of the split hierarchy. A pictorial view to help understand the bisection of flow by participant-type and sector is shown in Figure 9.3. Figure 9.4 shows correlation heatmaps across market participant

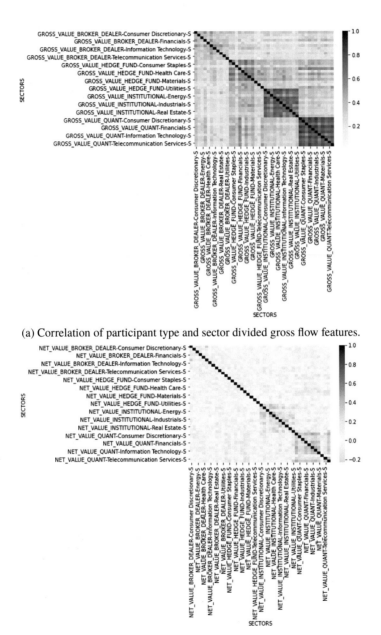

(a) Correlation of participant type and sector divided gross flow features.

(b) Correlation of participant type and sector divided net flow features.

Figure 9.4 Correlation across different participant type flows per sector.

type and sector grouped as gross and net flows. It is observed that in the *Quant* category, the cross-correlations across different sectors are quite high. This effect is present, although weaker, in the *Institutional* category, and weakest in the *Broker Dealer* category. Focusing on the net flow, Figure 9.4, we see negative correlations between *Quant* and *Hedge Fund*. The lowest correlations in the net flow matrix, with correlation coefficients less or equal to −0.2, are observed between: *Institutional* flow in Utilities and *Quant* flow in Utilities; *Institutional* flow in Financials and *Institutional* flow in IT; and *Institutional* flow in Utilities and *Quant* flow in Health Care. This shows that, within a sector, different participants may take distinct positions, and that within each participant category sector positions may be negatively correlated, motivating the feature engineering with flow granularity proposed as a means of helping the learner identify those patterns among a large number of variables and a somewhat limited number of historic dates.

9.3 Modeling and methodology

To establish what the role is of flows, and in particular different participant type flows, in price formation, we use contemporaneous analysis on the flows; that is, we try to explain the returns, or rather log-returns, of a stock at time t based on: flow data up to and including time t; categorical information of the stock, namely style, sector, country, and currency; as well as other risk and market information of the security up to time $t - 1$. This can be expressed as

$$\mathcal{Y}_s^t = \log\left(r_s^{(t)}\right) = \mathcal{G}\left(\mathcal{B}_{s,j}^{(t-1)}, \mathcal{H}_i\left(F_{k,s:(l,m,n,p)}^{t,d}\right)\right), \tag{9.1}$$

where $\mathcal{B}_{s,j}^{(t-1)}$, with $j \in \{1,\ldots,J\}$, representing a number of stock-related variables, s, at time $t - 1$, and $\mathcal{H}_i\left(F_{k,s:(l,m,n,p)}^{t,d}\right)$, with $i \in \{1,\ldots,I\}$. a set of flows related variables at time t. More specifically for each security s we can present aggregations of the flow sharing a common style, sector, etc. For example:

$$\mathcal{H}_{i,s} = \sum_{s_j:l(s_j)=l(s)} F_{s:(l,m,n,p)}^{t,d} \tag{9.2}$$

We are trying to approximate the function $\mathcal{G}$ using different machine learning models trained at subsets of the data corresponding to time segments of the history. The backtest is performed in a rolling-window fashion as shown in Figure 9.5 to avoid forward looking bias. We use daily data for each of the months in the period October 2020 to February 2021, as testing months. Each month's prediction model is trained with data from the preceding 18 months. This is a validation method suitable for time series data. Linear regression was used as a benchmark.

To quantify the importance of participant flow to the predictability of the model we follow two approaches. In the first, more naive, approach we run with and without sets of features and compare the resulting R^2, which is the coefficient of

Figure 9.5 Rolling Window Backtests. 30-day predictions preceded by 540 days of training. Approximately 95% to 5% split between training and testing

determination describing the proportion of the variance in the dependent variable that is predictable from the independent variables and defined as:

$$R^2 = 1 - \frac{\sum (y_i - \hat{y}_i)^2}{\sum (y_i - \bar{y})^2},$$ (9.3)

where y_i are the data to be fitted, $\bar{y} = 1/N \sum_i^N y_i$ and $\hat{y}_i$ the predictions.

In particular we are adding features incrementally as follows: In baseline case we only use the stock-related information at $t - 1$, i.e. $\mathcal{Y}_s^t = \log\left(r_s^{(t)}\right) = \mathcal{G}\left(\mathcal{B}_{s,j}^{(t-1)}\right)$. Then some, but not all the multisected, flow features are introduced: $\mathcal{Y}_s^t = \log\left(r_s^{(t)}\right) = \mathcal{G}\left(\mathcal{B}_{s,j}^{(t-1)}, \mathcal{H}_i\left(F_{k,s}^{t,d}\right)\right)$.

In the next instance we introduce more complexity in the flows data by including the multisected flow as features to individual stock level returns helping the model discover the connections between the flows of stocks that belong to the same style, sector, country etc.

This last approach increases the number of variables significantly so finally we use autoencoders, and in particular a variant of variational autoencoders, called β-VAE (Kingma and Welling, 2019), to 'compress' or encode some of the multisected flow information into latent states and use those as features for the model, instead of the full multisected flow. To avoid forward-looking bias, the encoding proceeds with time, capturing flow information up to time t and compressing it to a reduced number of latent states as shown in Figure 9.6. The encoding can be done on different supersets of the flow subdivision or at the most granular level. The more granular multi-section of the flow, the more variables have to be encoded. Considering the history of flows is limited, variance is an issue, particularly for the early time periods we apply the encoding. In such a setting, applying variational autoencoders has benefits compared with applying autoencoders. The β-VAE additionally results in more disentangled latent states, as explained in Burgess et al. (2017). In the present form we use the granular slicing of the gross, net and daily change to gross, into participant types, styles, sectors, countries and currencies as variables into the variational autoencoder. We discover the latent states corresponding to each time t and use those then as predictive features along with stock level characteristics.

In the second, more systematic approach we calculate the SHAP factors, which

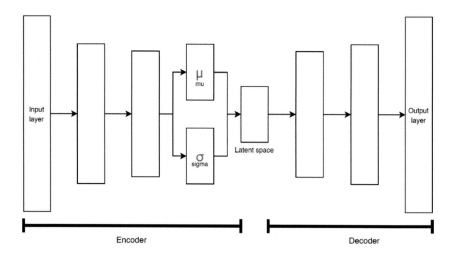

Figure 9.6 Variational Autoencoder Schematic

stand for Shapley additive explanations and is a game-theoretical approach to determining feature importance. Reproducing the outcome of the model is the 'game' and the features of the model are the 'players' in this game. Essentially what the method does is it quantifies the contribution of each player to the prediction made by the model. Each observation is one game. The interested reader may refer to Lundberg and Lee (2017).

A suite of machine learning techniques were tested on the data, including artificial neural networks (ANN), a comprehensive treatise of which can be found in Goodfellow et al. (2016). Tree and ensemble methods (Efron, 1979) such as random forests (Ho, 1995), gradient boosting trees and XGBoost (Chen and Guestrin, 2016), and support vector machine regressors (Drucker et al., 1997). In all cases the mean squared error of the log-returns was selected as loss function to minimize.

9.4 Empirical results

In this section we present the results of the methodology explained in Section 9.3 applied to the data presented in Section 9.2.

9.4.1 Aggregate flow and price formation

Our first empirical objective is to establish the link between order flow and price in our data and to also demonstrate the benefits of our proposed model, discussed in Section 9.3. At this level of investigation, we will be relying on the examination of R^2 across different models and feature sets. Table 9.2 shows the results of our stepwise approach in introducing flow related features, as well as comparison between machine learning models and linear regression, that previous literature has heavily relied on. Feature number refers to the cardinality of explicit or latent variables used in the analysis per the notation of $F_{k,s:(l,m,n,p)}^{t,d}$ discussed earlier.

In this context, we establish two interesting empirical results. The first is that we are not able to establish the link between flows and equity returns in the linear regression case. The linear model, with no information about flows, has a better R^2 of –0.0447 compared to 0.0095 when incorporating flow variables. As we introduce more granularity to the linear model, furthering our understanding of the price formation process and its nuances, the R^2 for the multisected flow data specification is –0.0413 , and the R^2 for the more parsimonious partly encoded data specification is –0.0696 .

This pattern reverses with machine learning models. Introducing flows improves over the no-flows results. Additionally, the multisected mapped flows improve performance of the machine learning models further, by making more 'explicit' the links between pieces of flow corresponding to stocks that are part of the same sector, style, and others. The last row of the table shows the effect of compressing some of the multisected flow features with encoded latent states obtained using the β-VAE methodology. We see that the achieved R^2 is somewhat worse than the multisected flow results, but better than the coarse flow data information. Although we run some robustness tests with different number of states $z = 8$ and $z = 10$ yielding similar results, a full hyper-parameter optimization has not been performed. It should also be noted that as the telescopic window is used to train the encoder with data up to each time t some of the early times in the data set had a limited number of prior points on which to train.

In short in this subsection we have provided evidence that flow information is related to price variation, but the benefit of further analysis and granularity of insights seems to only be feasible with machine learning methods. The latter are superior to linear regression in the case where we have a large number of variables; linear regression may cause over-fitting and may happen to violate some of the linear regression conditions such as independence and normality or that interact in non-linear ways. Machine learning methods employ regularization to reduce variance, are non-linear by nature, thus having the ability to capture non-linear interactions and are more tolerant to possible co-linearities in the inputs.

9.4.2 *Participant type flow imbalance and price formation*

Our second empirical objective is to understand the nature of the relations among different types of participants' flows and prices. Our empirical investigation aims to understand not only which market participants are more important but also how different participants contribute to the price formation process. To this end, we use two approaches. The first is to split flow by participant type and fit a different model for each. In the second, we construct a unified model to also explore the impact of individual order flow information in the presence of all the others. We base our inference on out-of-sample R^2 and feature importance analysis.

Table 9.3 shows the results of each participant type flow specification estima-

Table 9.2 R^2 comparison among different models and different feature sets.

Run	Feature Number	R^2 Lin. Reg.	R^2 GBT	R^2 SVM	R^2 XGBoost
No Flow Data	29	0.0095	0.0073	0.0111	0.0038
Flow Data	92	−0.0447	0.0300	0.0061	0.0061
Multisected Flow Data	188	−0.0413	0.0502	0.0124	0.0208
Partly Encoded Flow Data ($z = 5$)	143	−0.0696	0.0280	0.0051	0.0022

Table 9.3 R^2 comparison among different models. Institutional, Broker Dealer, Quant, and Hedge Fund splits the flow and fits models separately for each category.

	No-Flow		Flow	
Client type	Linear	ML	Linear	ML
Institutional	0.0140	0.0136	−0.0884	0.0119
Broker dealer	0.0116	0.0109	−0.0040	0.0182
Hedge fund	−0.0002	0.0014	−0.0071	0.0091
Quant	0.0101	0.0115	-0.0297	0.0285

tion. The results suggest that the model fitted with *Quant* flows information is the most explanatory of equity returns followed by that including *Broker Dealer* flow information with R^2 of 0.0285 and 0.0182 respectively. To give more context however we report the feature importance analysis of the unified-flow model based on the SHAP values in Table 9.4, which reports the ten highest ranked features as well as a brief explanation for each one of them. Interestingly, eight out of these ten are order flow related. This is consistent with the increase in the R^2 when flow features are included in Table 9.2. That said, we should reiterate that the flow related variables are contemporaneous vs. all other variables in the model that are lagged by one day. Net *Quant* flow is the second most important feature of the model. *Hedge Fund* and *Broker Dealer* Net flow are ranked ninth and tenth, less important than Net *Quant* flow at $t − 1$. Aggregate gross order flow is found to be the most important feature, but also interestingly different participant gross flow is found highly important in the price formation process. Feature ranking is robust to the parameters of the run while possibly swapping positions 'locally' in the ordering. Our most important non-flow related features are in line with Gu et al. (2020) who found as most prominent features liquidity, volatility and price trends, even though the forecasting horizons they looked at where longer and is likely that relative feature importance changes with forecasting horizon.

We now turn to the key question of our empirical analysis, which is how different market participants affect prices. More generally, we attempt to further understand how particular variables or features affect the model predictions. We

Table 9.4 Feature importance. Top 10 ranking features for unified flow model. $G_{\text{TOT}}^{(t)} = \sum_s \sum_k G_{ks}^t$

		Unified flow model
Rank	Feature	Explanation
1	GROSS T	Total day Gross $ across stocks: $\sum_s G_s^{(t)}$
2	NET QUANT T FRAC	Net $ flow of Quant clients on day t: $\sum_s N_{s,k=Q}^t / G_{\text{TOT}}^{(t)}$
3	1 DAY RETURN L001	$(\text{Return})_s^{(t-1)}$
4	TOTAL RISK L001	$(\text{Total Risk})^{(t-1)}$
5	NET QUANT DL01 T FRAC	Difference to previous day total Net Quant flow $\left(\Delta_{(t-1)}^{(t)} \sum_s N_{s,k=Q}^{(t)}\right) / G_{\text{TOT}}^{(t)}$
6	GROSS BROKER DEALER DL01 T FRAC	Difference to previous day total Gross Broker Dealer flow $\left(\Delta_{(t-1)}^{(t)} \sum_s G_{s,k=B}^{(t)}\right) / G_{\text{TOT}}^{(t)}$
7	GROSS INSTITUTIONAL T FRAC	Gross $ flow of Institutional clients on day t: $\sum_s G_{s,k=I}^t / G_{\text{TOT}}^{(t)}$
8	GROSS QUANT DL01 T FRAC	Difference to previous day total Gross Quant flow $\left(\Delta_{(t-1)}^{(t)} \sum_s G_{s,k=Q}^{(t)}\right) / G_{\text{TOT}}^{(t)}$
9	NET HEDGE FUND T FRAC	Net $ flow of Hedge Fund clients on day t: $\sum_s N_{s,k=H}^t / N_{\text{TOT}}^{(t)}$
10	NET BROKER DEALER T FRAC	Net $ flow of Hedge Fund clients on day t: $\sum_s N_{s,k=B}^t / N_{\text{TOT}}^{(t)}$

start off with a discussion of the aggregate flow and then proceed with more granular order flow information. Our discussion is based on plots of the impact of changes in the feature's value on returns, while keeping all other features constant.

Figure 9.7 depicts the impact of changes of gross order flow on equity returns. The pattern suggests that the gross notional value traded on a certain stock on a single day is inversely related to the price movement on that day. That is, higher gross volume of a stock is traded on days when the stock price drops. The corresponding net notional plot shows an interesting non-linear, almost piecewise linear, pattern, suggesting a rather sharp increase from negative net values up to the value of 0 and a rather flat profile after that. In other words, sell out is strongly correlated with drops in price of a stock, indicating possible over-reaction or fear, while for positive returns other factors may dominate, with the daily net flow playing a smaller role on its own right. To gain more insight into this pattern we plot in Figure 9.8 positive and negative returns for binned daily gross flow, and

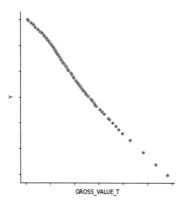 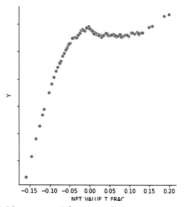

(a) Total gross notional $ value of the day   (b) Net notional $ value of the day / total daily gross

Figure 9.7

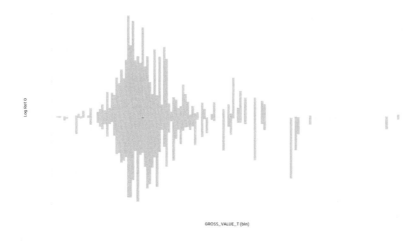

Figure 9.8 Positive and negative log-returns for binned daily gross flow. Bias is observed to negative returns for the high daily gross bins.

verify the tilt of negative returns for the very high gross volume bins, that are driving the behavior of the model.

Figure 9.9 shows the interplay of different participant-type Gross and Net daily flows as well as their change from the previous day. The latter is motivated by the high ranking of lagged Net *Quant* flow; see Table 9.4. We observe a stark difference between how different participants affect prices. This is inferred from the variation in the slopes of the curves in their sign but also in their magnitude. *Quant* net flow is positively related to contemporaneous returns. *Broker Dealer* and *Hedge Fund* net flow show negative impact. The effect of *Institutional* is fairly flat. Lagged net flow effect is positive for both *Quant* and *Hedge Fund* participants, and negative for *Institutional* and *Broker Dealer* participants.

These results shed more light on the conclusions from Tables 9.4 and 9.3

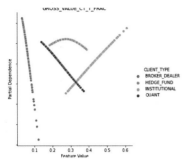

(a) Effect of gross participant type notional $ of the day/total daily gross

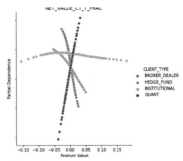

(b) Effect of net participant type notional $ value of the day/total daily gross

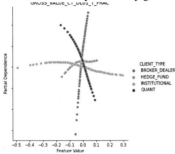

(c) Effect of daily change of participant type gross notional $ of the day/total daily gross

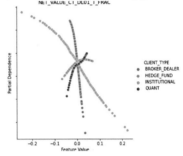

(d) Effect of daily change of participant type net notional $ value of the day/total daily gross

Figure 9.9 Comparing participant types' effect of of day, and day change in gross and net notional.

with regard to the different participant type roles in price formation. The distinct role of *Quant* flow is consistent with the observation made earlier in Figure 9.1 where we often saw the net aggregate quant flow being of opposite sign to the other client types. We rely on Boehmer and Wu (2008) to interpret these findings, with the caveat that our categorizations differ largely, and hence are not directly comparable. Our evidence suggests that *Quant* flow can be interpreted as informed due to positive price impact, and *Broker Dealer* and *Hedge Fund* as liquidity provision flow. Retail flow with negative price impact in Boehmer and Wu (2008) is within *Broker Dealer* category, whereas Institutional flow that has positive impact is partly included in our *Hedge Fund* flow.

Collectively, our analysis implies that different participants play different roles in the price formation process, at different times. Suitable machine learning techniques allow us to distil the information embedded in participant order flow. Consistently, *Quant* flow is the most important contemporaneous and lagged feature. Our intuition suggests that this relation is most likely due to superior information.

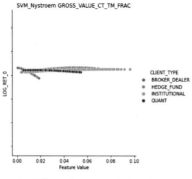

(a) Effect of gross participant type
notional $ of the day on corresponding
sector/total daily gross

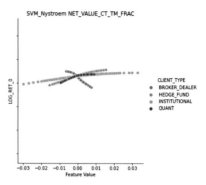

(b) Effect of net participant type notional
$ value of the day on corresponding
sector/total daily gross

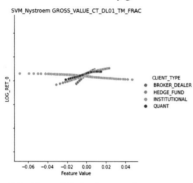

(c) Effect of daily change of participant
type gross notional $ of the day on
corresponding sector/total daily gross

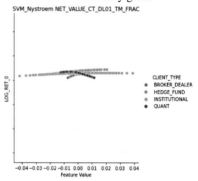

(d) Effect of daily change of participant
type net notional $ value of the day on
corresponding sector/total daily gross

Figure 9.10 Comparing participant types' effect of day, and day change in gross and net notional on corresponding sector.

9.4.3 Co-impact

Our third objective is to further our empirical understanding about how flows, and specifically different participant flows, on a group of stocks affect individual stocks of that group. Rather than exploring every possible group of related stocks that we can within our modelling framework, we present results for sector groupings only and leave style, country, currency groupings for future research. As in Section 9.4.2 our discussion is based on plots of the impact of changes in the feature's value on returns, while keeping all other features constant. The additional element here is that we can further condition the flow to the respective sector.

In Figure 9.10 we show the effect, on the returns of the stock, of the flow traded by each client type on the whole sector to which the stock belongs. This information is visible for all available slicings of the flow, but it is more relevant for those features that come up high in the feature importance ranks. The most interesting result of this analysis is that the effect of the aggregate flow on the same sector is smaller than the effect we document when we consider different participant type order flow. When we inspect Figure 9.10 closely this is not a surprising outcome given the nature of the relation that is revealed. A noteworthy outcome of this analysis is the contrast with the effect illustrated in Figure 9.9. We see that while Figure 9.9 documents adversarial impact between *Quant* net flow and *Hedge Fund* and *Institutional* net flows, the co-impact is actually in the same direction and of not so different magnitude.

To summarize, we conclude that there is more price formation information in the sector flow under the prism of the different participant types trading it than if seen on its own. And that *Quant*, *Hedge Fund*, and *Institutional* net flow co-impact is in the same direction.

9.5 Summary and conclusions

In this chapter we investigate the relation between flows and prices and, in particular, the effect that different type of participants might have on prices. We use a unique dataset of flows for European stocks, classified by participant type, country, sector, style particpation and other characteristics and appropriate machine learning techniques to effectively identify potentially complex, informative, patterns of the order flow at the more granular level.

We find that flow is related to prices, but the benefit of further analysis and granularity of insights seems to only be feasible with machine learning methods. We also find that different participants play different roles in the price formation process at different times. With our approach we are able to conclude that *Quant* flow is the most important contemporaneous and lagged participant flow feature in the price formation process. Finally we found that the co-impact of trades is important primarily when it is studied in the context of different participant types.

References

Akbas, F., Armstrong, W.J., and Sorescu, S. 2015. Smart money, dumb money, and capital market anomalies. *Journal of Financial Economics*, **118**(2), 335–382.

Andrade, S.C., Chang, C., and Seasholes, M.S. 2008. Trading imbalances, predictable reversals, and cross-stock price pressure. *Journal of Financial Economics*, **88**, 406–423.

Ban, G.Y, El Karoui, N., and Lim, A.E.B. 2018. Machine learning and portfolio optimization. *Management Science*, **64**(3), 1136–1154.

Benzaquen, M., Mastromatteo, I., Eisler, Z., and Bouchaud, J.P. 2016. Dissecting cross-impact on stock markets: An empirical analysis. *Journal of Statistical Mechanics Theory and Experiment*, **2017**.

Bianchi, D., Buchner, M., and Kozhan, R. 2019. Predictability of order imbalance, market quality and equity cost of capital. Available at SSRN 3297233.

Bianchi, D., Buchner, M., and Tamoni, A. 2021. Bond risk premiums with machine learning. *Review of Financial Studies*, **34**(2), 1046—1089.

Boehmer, E., and Wu, J. 2008. Order flow and prices. *AFA 2007 Chicago Meetings Paper*. Available at SSRN 891745.

Briere, M., Lehalle, C.A., Nefedova, T., and Raboun, A. 2020. Modelling transaction costs when trades may be crowded: A Bayesian network using partially observable orders imbalance. Pages 387–430 of: *Machine Learning for Asset Management: New Developments and Financial Applications*, E. Jurczenko (ed). Wiley.

Bucci, F., Iacopo, M., Eisler, Z., Lillo, F., Bouchaud, J.P., and Lehalle, C.A. 2020. Co-impact: Crowding effects in institutional trading activity. *Quantitative Finance*, **20**(2), 193–205.

Burgess, C., Higgins, I., Pal, A., Matthey, L., Watters, N., Desjardins, G., and Lerchner, A. 2017. Understanding disentangling in beta-VAE. *NIPS Workshop on Learning Disentangled Representations*. Available at `arXiv:1804.03599`.

Cao, C., Chen, Y., Goetzmann, W. N., and Liang, B. 2016. The role of hedge funds in the security price formation process. *Financial Analysts*, **61**(12).

Chen, L., Pelger, M., and Zhu, J. 2019. Deep learning in asset pricing. Available at SSRN 3350138.

Chen, T., and Guestrin, C. 2016. XGBoost: A scalable tree boosting system. In: *Proc. 22nd ACM SIGKDD International Conference*. Available at ArXiv:1603.02754.

Chen, Y., Kelly, B., and Wu, W. 2020. Sophisticated investors and market efficiency: Evidence from a natural experiment. *Journal of Financial Economics*, **138**(2), 316–341.

Chordia, T., Roll, R., and Subrahmanyam, A. 2002. Order imbalance and individual stock returns. *Journal of Financial Economics*, **65**(1), 111–130.

Drucker, H., Burges, H.C., Kaufman, L., Smola, A., and Vapnik, V. 1997. Support vector regression machines. *Advances in Neural Information Processing Systems*, **9**, 155–161.

Easley, D., de Prado, M.L., O'Hara, M., and Zhang, Z. 2020. Microstructure in the machine age. *Review of Financial Studies*, **34**(7), 3316–3363.

Efron, B. 1979. Bootstrap methods: Another look at the jackknife. *Annals of Statistics*, **7**(1), 1–26.

Feng, G., Giglio, S., and Xiu, D. 2020. Taming the factor zoo: A test of new factors. *Journal of Finance*, **75**, 1327–1370.

Frazzini, A., and Lamont, O. 2008. Dumb money: Mutual fund flows and the cross-section of stock returns. *Journal of Financial Economics*, **88**(2), 299–322.

Froot, K., and Teo, M. 2008. Style investing and institutional investors. *Journal of Financial and Quantitative Analysis*, **43**(4), 883–907.

Giamouridis, D. 2017. Systematic investment strategies. *Financial Analysts Journal*, **73**(4), 10–14.

Giamouridis, D., Neumann, M., and M., Steliaros. 2017. Go with the Flow or Hide from the Tide? Trading flow as a signal in style investing. In *Factor Investing*, E. Jurczenko (ed). Elsevier.

Goodfellow, I., Bengio, Y., and Courville, A. 2016. *Deep Learning*. MIT Press.

Griffin, J.M., Harris, J.H., and Topaloglu, S. 2003. The dynamics of institutional and individual trading. *Journal of Finance*, **58**(6), 2285–2320.

Gu, S., Kelly, B., and Xiu, D. 2020. Empirical asset pricing via machine learning. *Review of Financial Studies*, **33**(5), 2223–2273.

Ha, J.G., and Hu, J. 2017. How smart is institutional trading. SSRN 2907612.

Ho, T.K. 1995. Random decision forests. *Proc. 3rd International Conference on Document Analysis and Recognition*. `10.1109/ICDAR.1995.598994`.

Holley, J.R., Papaioannou, G., and Giamouridis, D. 2021. Cross-asset risk premia prediction with recurrent GANs and disentangled feature encoding using beta-VAE. NVIDIA GTC21.

Kelley, E., and Tetlock, P.C. 2013. How wise are crowds? Insights from retail orders and stock returns. *Journal of Finance*, **68**(3), 1229–1265.

Kingma, D.P., and Welling, M. 2019. An introduction to variational autoencoders. *Foundations and Trends in Machine Learning*, **12**(4), 307–392.

Lakoshinok, J., Shleifer, A., and Vishny, R. 1992. The impact of institutional trading on stock prices. *Journal of Financial Economics*, **32**, 23–34.

Li, K., Mai, F., Shen, R., and Yan, X. 2020. Measuring corporate culture using nachine learning. *Review of Financial Studies*, **34**(7), 3265–3315.

Lundberg, M.S., and Lee, S.I. 2017. A unified approach to interpreting model predictions. In *Proc. Neural Information Processing Systems*. Available at arXiv:1705.07874.

Min, S., Maglaras, C., and Moallemi, C. 2018. Cross-sectional variation of intraday liquidity, cross impact, and their effect in portfolio execution. *Columbia Business School Research Paper*, **19**(4). Available at arXiv:1811.05524.

Papaioannou, G., and Giamouridis, D. 2020. Enhancing alpha signals from trade ideas data using supervised learning. Pages 167–189 of: *Machine Learning for Asset Management: New Developments and Financial Applications*, E. Jurczenko (ed). Wiley.

TOWARDS BETTER RISK INTERMEDIATION

Part III

High Frequency Finance

10

Introduction to Part III

Robert Almgren[a]

Machine learning for trading has a seductive appeal. Financial markets are complicated beyond the understanding of any human. There seems to be an ample amount of historical data, especially at the high-frequency scale considered in this chapter. The economic value of even a small ability to predict future prices and to direct trading is immense. Machine learning offers the hope of taking in all this data, and developing nearly optimal actions without the need to build a statistical model. Given the success of computer learning in other fields such as image identification, face recognition, self-driving cars, and an increasing range of games, surely there must be good ways to use machine learning for trading.

The chapters in this Part illustrate some of the approaches currently in progress, and highlight some of the challenges that must be overcome become these techniques can meet the success that they have in other areas. Chief among these challenges is the low signal-to-noise ratio of financial markets. This is a reflection of the inexorable efficient market hypothesis, which says that nearly all information available to anyone has already been incorporated in current prices, and hence future price changes depend to a very large extent on information which by definition is not currently available.

10.1 Chapters in this Part

These three chapters highlight these issues from different points of view. They all step around the statistical modeling aspects of giving a description of market data, and proceed directly to the challenge of determining optimal trade actions. Thus, they might all be considered versions of reinforcement learning in contrast to supervised or unsupervised, though their approaches are very different.

Olivier Guéant discusses the challenges in directly applying the traditional framework of reinforcement learning to high-frequency trading. As he points out, one of the central difficulties is that "Finance is not a game." The state space is vast and choices must be made to reduce the problem to anything near tractable. The objective function is not entirely clear, since trading does not have a fixed end time, and risk usually must be considered as an intrinsic part of

[a] Quantitative Brokers, New York and Princeton University
Published in *Machine Learning And Data Sciences For Financial Markets*, Agostino Capponi and Charles-Albert Lehalle © 2023 Cambridge University Press.

the problem. The traditional formulation of reinforcement learning involves a sequence of transitions among discrete states at discrete times; in trading, real time is continuous but must be discretized to fit within the classic framework.

Guéant emphasizes the importance of calibrating learning models on real market data, as well as the limitations of real data. Although the quantity of available data may seem immense – modern databases can contain every trade, quote, or order book update for a large number of assets across many years – it is surprisingly usually not enough to train reinforcement learning problems. The usable horizon is limited because market dynamics are nonstationary, and if multiple assets are included the data scarcity becomes even worse.

Nonetheless, Guéant cites numerous examples of partial successes in applying machine learning techniques to algorithmic execution and market making. Future advances are likely to come from combination of improved computational power and computational techniques, with the mathematical and domain-specific insights that are steadily improving.

Sophie Laruelle presents the mathematical underpinnings of a commonly used method: that of iteratively finding a minimum, a maximum, or a zero crossing of a function that can be computed only as the expected value of more complex function. This is a common case in trading problems, where we are interested in minimizing an expected cost or maximizing an expected gain, although we do not know how to write or calculate this expected value in terms of the observable and controllable parameters.

The mathematics of these stochastic optimization problems is quite highly developed, and rigorous results are available on conditions of convergence and rates of convergence to the true optimum. Furthermore, these results are posed in ways that are directly applicable to problems of real practical interest.

After summarizing the state of theoretical knowledge, Laruelle considers two such practical problems: the choice of allocations to a collection of dark pools, and optimal placement of a limit order. An especially interesting aspect of these examples is the need to test them on "pseudo-real data." Since the exact probabilistic distribution of the underlying variables is not known, a reasonable simulation must be constructed using market traded volumes as proxies for the dark pool fill volumes. For the limit order placement, a Poisson process is calibrated using real high-frequency data.

Álvaro Cartea, Sebastian Jaimungal, and Leandro Sánchez-Betancourt present a detailed construction of a reinforcement learning strategy to devise optimal trades in foreign currency triples. They use an increasingly sophisticated collection of networks to model the state function, the transition probabilities, and the rewards, in order to determine optimal strategies given extensive historical observations of the price processes and the results of trades. The state includes the relative prices of the three currencies, the holdings of the trader, and the number of steps remaining before the position must be liquidated.

They first use a deep Q-learning model (DQN) to model the Q-function, using a fully connected feed-forward network with two layers of 64 nodes. They then implement a Reinforced Deep Market Model (RDMM), in which the observed state is a noisy version of a latent state, a generalization of a Kalman filter model.

They calibrate both of these models on a simulated market model, in which all three currency pairs undergo a mean-reverting random walk. The networks are calibrated using one million steps for the DQN, and 100,000 steps for the RDMM, across 60 simulations. Both algorithms develop very reasonable trading strategies, in which they take short positions in overvalued currencies, long positions in undervalued currencies, and reduce the positions to zero as the end time approaches.

These three chapters cover a cross-section of the current state of machine learning in high frequency trading: the challenges of applying standard methods of reinforcement learning, applications of sophisticated mathematics, and detailed implementation of neural networks.

10.2 State of the field and future prospects

In the view of this Introduction, there are two big challenges that face the use of machine learning methods in high frequency trading: the insatiable needs of such methods for training data, and the need to properly represent the game-like aspects of trading in real markets.

10.2.1 Data needs and simulation

Machine learning algorithms are staggeringly inefficient compared not only with human learning methods but even with simple animals: as Simões et al. (2021) observe in comparing learning of a fruit fly to an electronic vehicle, "fly heat avoidance involves decision-making, relies on rapid learning, and is robust to new conditions, features generally associated with more complex behavior." We need learning algorithms that can do as well as a fly.

In the absence of generalized learning ability, machine learning algorithms must be trained on vast quantities of data, all of which is drawn from exactly the same population on which the data will be used in practice. Unfortunately, the available quantity of real historical data is finite, and for this reason many algorithms are trained on simulated data. This data is typically constructed either to embody specific signals that can be detected by the algorithm, or to mimic the statistical properties of real data to the extent that can be characterized by the researcher. The advantage of simulated data is that arbitrarily large quantities can be generated for algorithm training.

The chapter by Cartea, Jaimungal, and Sánchez-Betancourt is an excellent article in the former category (see Ritter, 2017 for another example). It demonstrates that if the exchange-rate triple contains a mean-reverting signal of the type implemented in the model, then the algorithm will be able to discover and to exploit this pattern. Left open is the question of whether the exchange rates contain such a pattern, or how the algorithm will behave on real data. In real market data, predictable price signals are typically very weak and have very subtle structures. Of course it is useful to begin with simple structures, but it is then necessary to compare with real data.

The second category consists of constructing artificial market data using a Markov process whose parameters have been tuned with reference to real market data. The chapter by Laruelle takes this approach, or see for example Huang et al. (2015). The main advantage of this approach is that it makes direct use of real market properties, and thus may be considered closer to the real market than an artificially imposed signal. The disadvantage is that the properties that are reproduced in the simulated data are only those that are known to the experimenter. Real market data has structures, at all time scales, that are not easily captured by any Markov model that can be written down in simple terms. Thus they provide a useful building block but again it must ultimately be translated to real data.

There are many fewer studies that calibrate machine learning models only against real market data, Zhang et al. (2019) being one such. These models typically use not only the historical time series of top-of-book price and volume, but full depth information in order to extract the extremely weak signals that are characteristic of real markets. Such models are extremely computationally intensive to calibrate and while promising, are not yet in widespread use.

To summarize, the reason for these challenges is the extreme "fragility" of machine learning algorithms. A human being would be able to trade a new market, or a fly would quickly learn to react to new stimuli, based on approximate analogies to other situations that the person or the fly has been in before. After all, most markets work in more or less the same way; they are traded by other human beings with similar motivations, and broadly speaking the dynamics are comparable. We are still very far from having computational methods that have this degree of adaptability, and this lack is one of the biggest challenges in using them in markets and trading.

10.2.2 Game formulation

The second challenge is the fact that markets are not an engineering system that can be modeled with arbitrary precision if our models are sophisticated enough. Rather, they are a game, in which all participants have partial information, and all are trying to learn the information possessed by others. The market is not an "it:" it is a "them."

Viewed in this way, even using historical data for training is only an approximate solution. To take an extreme analogy, it would be like calibrating a chess program by replaying the record of historical championship games: of course the actual play would have been completely different if either player had been different, for example if one of them were the algorithm being developed. Successful development of systems that can play chess or Go rely on simulation of the game itself, developing strategies by trading against other competitive agents.

Market impact models are an attempt to capture this effect in a tractable way, but they are an extremely approximate description of the real flow of information in markets. Similarly, the calibrated Markov models incorporate some of this strategic interaction, but in a very stylized way. None of these approaches gets to the heart of how markets really work.

In this light, an important way to develop trading algorithms would be to implement an adversarial game, in which certain participants are given information about the future prices or cash flows of the asset, and others attempt to learn this information. A summary of the history of this approaches is given by Cliff and Rollins (2020), and an example of using supervised learning in a simulated environment is used by Wray et al. (2020). Such a model would still be highly stylized, like the ones described above, since the information flows in markets, like everything else in markets, are much more complex and subtle than can be captured in any simple model structure. Nonetheless, this approach would capture the essential aspects of what makes financial markets such a fascinating laboratory for human goal-driven interaction.

10.2.3 Conclusion

It is clear that the topic of machine learning for high frequency trading is immensely appealing and has a lot of promise, but challenges remain. The three chapters in this part outline different views of the current state of the field: rigorous mathematics, calibration of realistic models on plausible price models, and an overview of reinforcement learning for trading. But fundamental gaps remain in our ability to apply these models. One cause of these gaps is the fragility of these models which cause them to require vast amounts of homogeneous data for calibration. Another cause is the need to incorporate the information flows which underlie essentially all dynamics in real markets. As work progresses on these challenges, machine learning methods for high-frequency trading will become more and more powerful.

References

Cliff, Dave, and Rollins, Michael. 2020 (December). Methods Matter: A Trading Agent with No Intelligence Routinely Outperforms AI-Based Traders. Pages 392–399 of: *2020 IEEE Symposium Series on Computational Intelligence (SSCI2020)*.

Huang, Weibing, Lehalle, Charles-Albert, and Rosenbaum, Mathieu. 2015. Simulating and Analyzing Order Book Data: The Queue-Reactive Model. *J. Amer. Statist. Assoc.*, **110**(509), 107–122.

Ritter, Gordon. 2017. Machine Learning for Trading. *Risk*, October.

Simões, José Miguel, Levy, Joshua I., Zaharieva, Emanuela E., Vinson, Leah T., Zhao, Peixiong, Alpert, Michael H., Kath, William L., Para, Alessia, and Gallio, Marco. 2021. Robustness and Plasticity in *Drosophila* Heat Avoidance. *Nat. Commun.*, **12**, 2044.

Wray, Aaron, Meades, Matthew, and Cliff, Dave T. 2020 (December). Automated Creation of a High-Performing Algorithmic Trader via Deep Learning on Level-2 Limit Order Book Data. Pages 1067–1074 of: *2020 IEEE Symposium Series on Computational Intelligence (SSCI2020)*.

Zhang, Zihao, Zohren, Stefan, and Roberts, Stephen. 2019. DeepLOB: Deep Convolutional Neural Networks for Limit Order Books. *IEEE Trans. Signal Process.*, **67**(11), 3001–3012.

11

Reinforcement Learning Methods in Algorithmic Trading

Olivier Guéant[a]

Abstract

This chapter is dedicated to the third paradigm of machine learning alongside supervised and unsupervised learning: reinforcement learning (RL). RL methods have recently been successful in solving complex dynamic optimization problems in domains such as robotics, video games, and board games. Being flexible in terms of modelling and scalable to high dimensions, they are often regarded as good candidates to solve many financial problems, especially in the field of algorithmic trading. The goal of this subchapter is multifold: presenting the main ideas and concepts of RL, discussing their relevance for addressing algorithmic trading problems, reviewing the existing applications, and discussing the future. In particular, our view is that the range of problems that could be addressed with RL techniques is narrower than what most people think, but that RL-based trading programs could be competitive in execution and market making if traditional quants, computer scientists, and engineers united forces.

11.1 Introduction

11.1.1 The recent successes of reinforcement learning

Since the middle of the 2010s, all fields of science have been impacted by machine learning techniques. Finance, and in particular algorithmic trading, is, of course, no exception. Many techniques of supervised and unsupervised learning have indeed become fashionable challengers to existing statistical and econometrical techniques and have been tested by both academics and practitioners; sometimes with great success!

Alongside supervised and unsupervised learning, reinforcement learning (RL) – the third machine learning paradigm – also came into the limelight as RL-based computer programs have recently been successful in playing video games and board games. Researchers at DeepMind (see Mnih et al., 2015) have indeed built a deep Q-network agent playing a long list of Atari 2600 games at human level

[a] Université Paris 1 Panthéon-Sorbonne

or above. What stunned most researchers was that this agent (i) only learned with the pixels and the game score as inputs, and (ii) used a single algorithm, network architecture, and set of hyperparameters for all games. A few months later, another RL-based computer program called AlphaGo made the headlines after defeating several professional Go players. AlphaGo itself was later soundly defeated by AlphaGo Zero. The latter is another RL-agent that learned to master the game of Go from self-play with no initial knowledge but the game rules; and with the position of the stones on the board as its input instead of hand-engineered features.[1] Interestingly, these programs are often regarded as "creative" as they developed unconventional strategies.

Although RL is not a new field, the buzz surrounding its recent successes has led to new research efforts and new hopes in domains as varied as robotics, self-driving cars, healthcare, and, of course, finance.

11.1.2 Finance, it might be your go

RL is aimed at solving problems involving an agent interacting with an environment – possibly stochastic – so as to maximize an expected numerical reward. Therefore, it is no surprise that the financial community has recurrently been curious about RL tools.

In fact, as we shall see in Section 11.4, there has been for at least two decades some research works here and there using RL techniques in the domain of finance, especially in algorithmic trading. However, RL tools have settled over the last couple of years at the forefront of the financial scene. Of course the initial trigger was the successes of DeepMind discussed above, but the new popularity of RL has also been due to (i) the communication of some banks announcing the advent of execution trading robots based on deep RL (see for instance Noonan, 2017, in the Financial Times), and (ii) the publication of the now famous paper "Deep Hedging" (see Buehler et al., 2019).

"Deep Hedging" aimed at showing that it is possible to find the optimal strategy to hedge a European contingent claim in any model thanks to neural networks and a direct policy search algorithm. The authors proposed more precisely recurrent and semi-recurrent neural network architectures in order to approximate the strategy that minimizes the risk borne by a hedger[2] – the risk being measured through several risk measures including CVaR.[3]

Although it only deals with option pricing (in fact option hedging), "Deep Hedging" has influenced the whole quantitative finance community beyond

---

[1] For more details, see Silver et al. (2016) and Silver et al. (2017). See also Silver et al. (2018) for another version called AlphaZero that learned (at the same time!) to master chess and shōgi as well as Go from self-play. Very recently, a new step forward was made with the MuZero algorithm that learned to play the above board games alongside Atari video games without being told the rules – see Schrittwieser et al. (2020).

[2] The use of neural networks for the pricing and hedging of options is not a new topic and we refer the interested readers to the thorough review work that was recently carried out in Ruf and Wang (2020).

[3] In particular, by using a famous trick due to Rockafellar and Uryasev (see Rockafellar and Uryasev, 2002), they have showed that CVaR is a great risk measure for some RL models that need to account for risk (at least for problems solved using a direct policy search approach).

derivatives. First, it has forced a lot of quant practitioners to consider optimization tools with fresh eyes and to recall that pricing derives from hedging – and not the other way round, contrary to what the usual computation of Greeks could let think. Second, in addition to bringing optimization tools into the limelight, the paper has exemplified the flexibility of RL methods based on simulated data.[4] The direct policy search method proposed in "Deep Hedging" can indeed use, for training, data simulated from any model or even from a mix of models.[5] This second point is important as it enlarges the range of models that can be used; and this may be crucial in an industry that increasingly aims to manage model risk. Third, the paper has conveyed the message that RL methods could be used to solve a wide range of high-dimensional problems, far beyond those (often linear) traditionally addressed with Monte-Carlo simulations.[6]

In this context, this chapter aims to discuss the interest of RL techniques for algorithmic trading. We start by presenting the main concepts and ideas traditionally associated with RL. We then highlight the numerous differences between (i) most of the toy examples of the RL community or even video games and board games, and (ii) the real-life problems of algorithmic trading. Further, we review the existing research works applying RL ideas to algorithmic trading problems.[7] Finally, we discuss future perspectives and insist on the fact that, if the quantitative finance community wishes to see RL algorithms implemented on a large scale, then the involvement of computer scientists and engineers is of utmost importance.

11.2 A brief introduction to reinforcement learning

RL encompasses a wide range of methods aimed at maximizing the expected reward of an agent interacting with a deterministic or stochastic environment. In quantitative finance, this type of problems is often addressed by using the

[4] We intentionally avoid the use of the expression "model-free" because it is ambiguous, at least for financial applications (see also Section 11.3).

[5] Historical data can also be used but RL methods usually require more data points than what historical data can provide (see Section 11.3 for more details).

[6] In recent years, in parallel to this renewed interest for RL as a way to approximate the optimal solution of high-dimensional stochastic optimal control problems, some other approaches have been proposed, in particular to numerically approximate the solutions of Hamilton–Jacobi–Bellman equations in high dimension. In particular, in a series of papers including for instance E et al. (2017) and Han et al. (2018), a group of researchers used the representation of a linear or nonlinear parabolic partial differential equation (PDE) with a backward stochastic differential equation (BSDE) in order to build what they call a BSDE solver that approximates the solution of the PDE and – in fact, through – its gradient. In addition to the above papers, we refer to Becker et al. (2019), Henry-Labordere (2017), Huré et al. (2019) and Pham et al. (2019) for additional discussions and extensions, especially to the case of optimal stopping problems leading to variational inequalities or more complex PDE. Some authors classify these methods as RL because of (i) the use of machine learning techniques (neural networks, stochastic gradient descent, etc.), and (ii) the central use of the gradient of the solution of the PDE, which, in the case of a Hamilton–Jacobi–Bellman equation, is intimately related to the optimal control (or optimal action in the vocabulary of RL). This classification is in fact questionable and we do not consider these approaches RL ones. In particular, it should be noted that these methods do not "explore" unlike many RL methods.

[7] The readers with wider economic or financial interests can read the recent reviews carried out in Charpentier et al. (2020), Fischer (2018), and Kolm and Ritter (2020).

mathematical tools of deterministic and stochastic optimal control. Most techniques of optimal control, in particular those based on the dynamic programming principle, are part of RL techniques, but RL methods often go beyond optimal control/dynamic programming methods in at least two aspects. First, many RL methods are not based on grids,[8] but instead on function approximations. Therefore, they do not suffer (or at least suffer less) from the curse of dimensionality.[9] Second, many RL techniques use data samples and do not require to know the transition kernel (that defines the dynamics of the state variables and the distribution of the rewards).

Let us now start our brief introduction to RL. Our goal is not to be exhaustive, but rather to present the main ideas and concepts. As in many RL seminal references (we recommend in particular Bertsekas and Tsitsiklis, 1996, Bertsekas, 2019, Powell, 2011, Sutton and Barto, 2018, and Szepesvári, 2010, and refer to them for a more detailed introduction), we consider a discrete-time approach.[10] We therefore start with Markov Decision Processes (MDP)[11] and present the different types of optimization problems addressed in RL. This presentation is followed by a list of concepts that are commonly employed in the RL literature. Then we present and discuss the different families of RL algorithms.

11.2.1 *Markov decision processes and optimization problems*

Formally, a MDP is a triplet $(\mathcal{S}, \mathcal{A}, \mathcal{P})$ where:

- $\mathcal{S}$, called the state space, describes the different states in which the system can be – it is typically either a finite or countable set or a subset of a finite-dimension space;
- $\mathcal{A}$, called the action space, describes the different actions the agent (or decision-maker) can choose – it is typically either a finite or countable set or a subset of a finite-dimension space;
- $\mathcal{P}$ is a probability kernel that maps a couple $(s, a) \in \mathcal{S} \times \mathcal{A}$ to a probability measure $\bar{p}(\cdot, \cdot | s, a)$ on $\mathcal{S} \times \mathbb{R}$ (or $\mathcal{S} \times (\mathbb{R} \cup \{-\infty\})$) where $\bar{p}(\cdot, \cdot | s, a)$ models for a given state s and action a the distribution of the next state and the associated reward.

In practice, it may be more convenient to replace the transition kernel by two concepts. First, a state transition kernel that maps a couple $(s, a) \in \mathcal{S} \times \mathcal{A}$ to a probability measure $p(\cdot | s, a)$ on $\mathcal{S}$ modelling for a given state s and action a the distribution of the next state s'. Second, a probability distribution for the reward

---

[8] The usual grid methods based on the dynamic programming principle are part of the so-called tabular methods in RL.

[9] In order to beat the curse of dimensionality, it is possible, when the dimension remains reasonable, to use quantization methods (see Pagès et al., 2004). This is for instance what was done in Abergel et al. (2020) dealing with market making in a limit order book.

[10] Those readers used to continuous-time optimal control may find it more natural to start with the work of Munos, for instance his *habilitation* (Munos, 2004) and the references therein.

[11] We do not cover the case of Partially Observable MDP (POMDP). We refer to Bäuerle and Rieder (2011) for a detailed introduction to MDP (with applications to finance) that covers POMDP.

given (s, a, s') or, as it is often enough, a reward function r that maps a triplet (s, a, s') of current state, action, and next state to the expected reward, or a variant of it where one takes the expectation over all the possible next states s' given (s, a).

MDP are essential for modelling sequential decision-making problems. Starting from a given state S_0, one can build recursively a sequence $(S_n, A_n, R_{n+1})_n$ of states, actions, and rewards by assigning at date n, once A_n has been chosen, the distribution $p(\cdot|S_n, A_n)$ to S_{n+1} and setting R_{n+1} to the associated reward or its expectation, i.e. to $r(S_n, A_n, S_{n+1})$ or $r(S_n, A_n)$ – we use the latter to simplify in most of what follows.[12]

Given a MDP, RL methods are aimed at maximizing an objective function or expected score that depends on the choice of actions. Two main types of optimization problems are traditionally considered:[13]

- Finite-horizon problems, in which one maximizes the expected score

$$\mathbb{E}\left[\sum_{n=0}^{N-1} r(S_n, A_n) + R(S_N)\right]$$

  for a given time horizon N and a final payoff function R;
- Infinite-horizon problems, in which one maximizes the expected discounted score

$$\mathbb{E}\left[\sum_{n=0}^{+\infty} \gamma^n r(S_n, A_n)\right],$$

  for a given discount $\gamma \in (0, 1)$.

11.2.2 Basic concepts

Several concepts have been introduced in order to address the above dynamic optimization problems:

- *Policy*: a policy is a function that maps a time and a state to an action (in the case of a deterministic policy) or to a probability measure on the action space

---

[12] A specific case plays an important part in the literature: bandit problems, where $p(\cdot|s, a)$ does not depend on a. In that case, the state space is often a singleton, but it can also be more complex in the case of contextual bandits. In any case, the problem is essentially that of a gambler in front of a set of slot machines who needs to choose the best machine to play (in an online manner). We cover in more details RL methods where $p(\cdot|s, a)$ does depend on a, but it is interesting to notice that the classical approaches (see, for instance, Thompson, 1933, on Thompson sampling and Auer et al., 2002; Auer, 2002, on the upper confidence bound paradigm) are useful in algorithmic finance, for instance to choose between algorithms. In particular multi-armed bandit methods based on the exploration-exploitation trade-off could be very useful when several execution algorithms with the same benchmark (e.g. VWAP) are available and one needs to determine which one is the best in practice.

[13] Other types of problems do exist, such as infinite-time problems with no discount but an absorbing state, average reward (ergodic) problems, etc.

(in the case of a so-called stochastic policy).[14] We call stationary a policy that does not depend on time.

- *Optimal policy*: an optimal policy is one that maximizes the objective function / expected score. Optimal policies are what RL algorithms ultimately look for.
- *Value function (or state value function)*: value functions map states to expected scores in order to evaluate the performance of a policy. In the case of a finite-horizon problem, the value function V^π associated with a policy π is defined as

$$V^\pi : (k, s) \mapsto \mathbb{E}\left[\sum_{n=k}^{N-1} r(S_n, \pi_n(S_n)) + R(S_N)|S_k = s\right].$$

In the case of an infinite-horizon problem, the value function V^π associated with a stationary policy π is defined as

$$V^\pi : s \mapsto \mathbb{E}\left[\sum_{n=0}^{+\infty} \gamma^n r(S_n, \pi(S_n))\middle| S_0 = s\right].$$

- *Optimal value function*: the optimal value function V^* is the value function associated with an optimal policy.[15]
- *State-action value function or Q function*: the state-action value function associated with a policy is a variant of the value function associated with that policy where the first action is prescribed. In the case of a finite-horizon problem, Q^π is defined as[16]

$$Q^\pi : (k, s, a) \mapsto \mathbb{E}\left[r(S_k, A_k) + \sum_{n=k+1}^{N-1} r(S_n, \pi_n(S_n)) + R(S_N)\middle| S_k = s, A_k = a\right].$$

In the case of an infinite-horizon problem (where it is often used), Q^π is defined as[17]

$$Q^\pi : (s, a) \mapsto \mathbb{E}\left[r(S_0, A_0) + \sum_{n=1}^{+\infty} \gamma^n r(S_n, \pi(S_n))\middle| S_0 = s, A_0 = a\right].$$

- *Optimal state-action value function or optimal Q function*: the optimal state-action value function Q^* is the state-action value function associated with an optimal policy.[18]
- *Greedy policy*: in the case of a finite-horizon problem and given a function

---

[14] A deterministic policy is of course a stochastic policy where the probability measure is a Dirac measure.

[15] The readers accustomed with (stochastic) optimal control must note that what they usually call value function is called here optimal value function.

[16] We have of course $Q^\pi(n, s, a) = r(s, a) + \int_{s'} V^\pi(n + 1, s')p(s'|s, a)ds'$ and $V^\pi(n, s) = Q^\pi(n, s, \pi(s))$.

[17] We have of course $Q^\pi(s, a) = r(s, a) + \gamma \int_{s'} V^\pi(s')p(s'|s, a)ds'$ and $V^\pi(s) = Q^\pi(s, \pi(s))$.

[18] We have of course (in the case of infinite-horizon problems) $Q^*(s, a) = r(s, a) + \gamma \int_{s'} V^*(s')p(s'|s, a)ds'$ and $V^*(s) = \max_a Q^*(s, a)$.

$v : \{0, \ldots, N-1\} \times \mathcal{S} \to \mathbb{R}$, a policy is greedy (for v) if for each time n and state s we have

$$\pi_n(s) \in \text{argmax}_a r(s,a) + \int_{s'} v(n+1,s')p(s'|s,a)ds',$$

and the concept is similarly defined in an infinite-horizon problem for a function $v : \mathcal{S} \to \mathbb{R}$ by

$$\forall s \in \mathcal{S}, \quad \pi(s) \in \text{argmax}_a r(s,a) + \gamma \int_{s'} v(s')p(s'|s,a)ds'.$$

It is noteworthy that any policy that is greedy for V^* is an optimal policy.

11.2.3 Main RL methods

In order to discuss the main methods of the RL literature, or at least the main ideas underlying them, it is useful to consider separately the two different objective functions considered above. We start with the case of infinite-horizon problems and go on with the case of finite-horizon ones.

Infinite-horizon problems

Many methods have been proposed for the case of infinite-horizon (stationary) problems.[19] In order to understand the main families of methods let us first introduce Bellman equations and the associated operators.

Bellman equations are functional equations solved by the value functions. For a given policy π, V^π is solution of the linear Bellman equation

$$V^\pi(s) = r(s,\pi(s)) + \gamma \int_{s'} p(s'|s,\pi(s))V^\pi(s')ds',$$

that we can write with a linear operator as $V^\pi = \mathcal{T}^\pi V^\pi$. As far as the optimal value function is concerned, we have another Bellman equation, nonlinear in that case,

$$V^*(s) = \max_a r(s,a) + \gamma \int_{s'} p(s'|s,a)V^*(s')ds',$$

that we can write with a nonlinear operator as $V^* = \mathcal{T}^* V^*$. Similar equations exist for the functions Q^π and Q^*. The interesting point is that value functions are fixed points of contracting operators (because $\gamma \in (0,1)$).

Given the latter remark, it is natural to introduce a method – called Value Iteration – that starts from an initial function V_0 and iteratively builds a sequence

---

[19] It is always possible – although not recommended – to approximate an infinite-horizon problem by a finite-horizon problem by fixing an end date N sufficiently large. The methods of the next section could be used in that case. In all instances, an infinite-horizon problem can also be seen as a finite-horizon problem with a random terminal time (following a geometric distribution in the case of a constant discount rate γ as above).

$(V_k)_k$ by $V_{k+1} = \mathcal{T}^* V_k$.[20] This method converges[21] but it is often infeasible in practice when the state space and the action space are too large. Approximate dynamic programming techniques (which are part of RL techniques) can then help to beat the curse of dimensionality by replacing the above recursive equation by another of the form $V_{k+1} = \tilde{\mathcal{A}} \mathcal{T}^* V_k$ where $\tilde{\mathcal{A}}$ is an approximation operator. In practice this means that value functions are parametrized (they take the form of a neural network or a linear combination of well-chosen features) and that, at step k, values of $\mathcal{T}^* V_k$ are sampled for many points in order to feed a supervised learning algorithm (represented by $\tilde{\mathcal{A}}$) that allows to obtain V_{k+1} in a (similar) parametrized form. This type of approximate dynamic programming methods is interesting but the algorithms do not always converge.

Another family of methods is called Policy Iteration. It does not use the Bellman equation for the optimal value function but rather that for given policies. It starts from an initial (stationary) policy and works through iterations of policy evaluation and policy improvement steps. During policy evaluation, one evaluates the value function of the current policy, while at policy improvement steps, one updates the policy by considering a greedy policy with respect to the value function computed in the evaluation step (or at least moves the policy from the current one towards one that is closer to the greedy one).

In order to evaluate the value function associated with a policy, classical methods include solving the linear Bellman equation on a grid, the use of an iterative method as in the Value Iteration algorithm (with the operator $\mathcal{T}^\pi$ instead of $\mathcal{T}^*$), Monte-Carlo techniques, or the use of temporal differences (TD). The use of TD learning is inspired from stochastic approximation and is really one of the cornerstones of many RL algorithms. In particular, when using TD learning, one does not require to know the transition kernel and instead works with realized or simulated data.

The main idea behind TD learning, when applied to a value function V^π, is that one can build approximations thanks to a realization $(s_n, a_n, r_{n+1})_n$ of the MDP for a policy π by updating (in a synchronous or asynchronous manner) the current approximation $\hat{V}^\pi(s_n)$ of the value function at point s_n in the direction of $r_{n+1} + \gamma \hat{V}^\pi(s_{n+1}) - \hat{V}^\pi(s_n)$. There are of course many variants based on similar ideas. When tabular methods cannot be applied because of the size of the state space and when value functions are instead parametrized, TD learning methods update the parameters "so as" to reduce the gap between $r_{n+1} + \gamma \hat{V}^\pi(s_{n+1})$ and $\hat{V}^\pi(s_n)$ – see for instance Sutton and Barto (2018) and Szepesvári (2010) for a detailed introduction, in particular as semi-gradient methods are often used and cannot be described in depth in this chapter.

In practice, one usually does not wait for a precise evaluation of the value function associated with a policy before changing the policy: the value function, regarded as a "critic", evolves progressively to guide the "actor" in his policy

[20] By definition of the $\mathcal{T}^*$ operator, the use of Value Iteration requires information about the transition kernel, in particular the state transition kernel and the expected rewards.

[21] We mean convergence in terms of value functions. This does not mean that the associated sequence of greedy policies (which is often chosen) converges.

changes. Also, many RL methods are based on a parametrized policy (with neural networks for instance) as is the case for the value function. In that case, the parameters of the policy are updated to move the policy in what we believe is the right direction according to the current value function.

Regarding the policy improvement step, it is noteworthy that determining the greedy policy associated with a value function requires to know the transition kernel. A classical alternative consists, instead of evaluating the value function V^π of the policy π in the policy evaluation step, in evaluating the state-action value function Q^π (using methods similar to those used for V^π). The policy improvement step then boils down to finding for each state s the maximum of $Q^\pi(s, \cdot)$ if one wants to be greedy.

There are in fact many methods based on Q functions. The application of TD learning ideas to Q functions leads to two important standard algorithms: SARSA and Q-learning (see the classical textbooks cited above for detailed presentations). These algorithms are aimed at directly finding the optimal state-action value function and exist in many variants, in particular when Q functions are parametrized with a neural network (hence the expression Q-network agent), and have known great successes in playing games.

Overall, it is noteworthy that the use of TD learning ideas enables to learn without knowledge of the underlying model. Combined with approximation techniques that enable to beat the curse of dimensionality, these ideas are central to modern RL techniques. Of course, the devil often lies in the details and it is clear that beyond the basic ideas presented in this subchapter, experience in the design of RL agents has strongly contributed to recent successes (e.g. the choice of learning rates in TD learning, the use of several neural networks, the use of experience replay, the architecture of neural networks, the parallelization of computations, the use of GPU and TPU, etc.).

Finite-horizon problems

The case of finite-horizon problems has to be considered separately from that of infinite-horizon problems although many ideas are common to both. In what follows we first briefly discuss value function methods and then go on with direct policy search methods.

As in the case of infinite-horizon problems, the value functions of finite-horizon problems are characterized by Bellman equations. For a given policy π, V^π is indeed a solution to the linear Bellman equation

$$V^\pi(n, s) = r(s, \pi_n(s)) + \int_{s'} p(s'|s, \pi_n(s))V^\pi(n+1, s')ds',$$

i.e. $V^\pi(n, \cdot) = \mathcal{T}^\pi V^\pi(n+1, \cdot)$, for $n < N$, and $V^\pi(N, \cdot) = R(\cdot)$. As for the optimal value function, it solves the nonlinear Bellman equation

$$V^*(n, s) = \max_a r(s, a) + \int_{s'} p(s'|s, a)V^*(n+1, s')ds',$$

i.e. $V^*(n, \cdot) = \mathcal{T}^* V^*(n+1, \cdot)$, for $n < N$, and $V^*(N, \cdot) = R(\cdot)$. Similar equations exist for the functions Q^π and Q^* in this case.

Some of the methods presented in the previous section can be adapted to the finite-horizon case. For instance, methods inspired from Policy Iteration are well suited and work almost as if time was part of the state space. For the policy evaluation step, if tabular methods cannot be used because of the dimensionality of the problem, Monte Carlo and TD methods with approximation can be used. For the policy improvement step, it can be done date by date in an independent manner, except if one is looking for a policy in the form of a parametric function that depends on time, as it is sometimes the case.

In the case of finite-horizon problems, it is often interesting to use methods based on the backward induction underlying the dynamic programming principle. Indeed, the problem can be decomposed into N one-step problems, assuming for each date that we know how to solve the tail problem. In addition to the classical tabular method that consists in solving the Bellman equation for V^* step by step – backward in time – on a grid, there are classical approximate dynamic programming methods that approximate sequentially – backward in time – the functions $(V^*(n, \cdot))_n$ or $(Q^*(n, \cdot, \cdot))_n$ using regression techniques (see for instance Bertsekas and Tsitsiklis, 1996, or the recent papers Bachouch et al., 2018; Huré et al., 2018). One of the main difficulties with these methods is that we do not know where to sample points at each step, because we do not know where we are going to need approximations of the value function in the future (that is, at previous time steps of the dynamic optimization problem).

In the case of finite-horizon problems, it is possible to use RL methods that do not rely on value functions, but instead directly search for the optimal policy – hence their name: direct policy search methods. In direct policy search methods, policies π^θ are parametrized by a vector θ (often the coefficients of a neural network or those of a simple linear combination of well-chosen features[22]) and the goal is to maximize over θ the expected score $\mathbb{E}\left[\sum_{n=0}^{N-1} r(S_n, \pi^\theta(S_n)) + R(S_N)\right]$. The problem becomes therefore a pure stochastic optimization problem and many approaches are available, going from gradient methods to non-gradient methods (simulated annealing, evolutionary approaches, etc.). It is noteworthy that it is often interesting to use stochastic policies in this context (see for instance the REINFORCE algorithm and the numerous variants that have been proposed to accelerate convergence).

Direct policy search methods are interesting but they often suffer from the vanishing/exploding gradient phenomenon because of the recurrent nature of the problem (a change of action at date k potentially changes indeed all states and rewards after date k). To avoid this, it is sometimes possible to proceed by backward induction: approximate the optimal decision rules for the last periods using a direct policy search approach and freeze them to approximate the optimal decision rules for the previous periods, and so on (see for instance Bachouch

---

[22] The coordinates of θ can also be the values of the policy if the state space is small.

et al., 2018; Huré et al., 2018). In that case, the same sampling problem as above occurs once again.

The above ideas are general and have to be adapted to each context of application. Let us now discuss what makes finance specific when it comes to the use of RL ideas.

11.3 Finance is not a game

The craze around RL-based computer programs triggered by the successes of DeepMind goes along, in the financial community, with the hope that the progress made in board and video games can be translated into progress for algorithmic trading. However, expectations need to be managed when it comes to finance problems. Although many finance problems are dynamic optimization problems that could be addressed with the tools of RL, they do not have the same characteristics as the toy problems (Cart Pole, Mountain Car, etc.) of the RL community and are different from those recently addressed with success.

11.3.1 States and actions

A first important difference between games or toy problems and finance problems has to do with the definition of the state space. In the former, the state space may be complicated but it is naturally and unambiguously defined because it is imposed by the problem: the place of pieces on the board in the case of chess or draughts, the pixels (with history maybe) in the case of video games, etc. In the case of a finance problem, the state space is open and must definitely be regarded as a modelling choice. In the case of problems involving a limit order book (LOB) for instance, the state of the LOB (which can usually be reduced to a few limits) – and maybe its history – is naturally part of the story but the state space can also include numerous signals related to trends, historical or implied volatilities, market volumes, etc. There is in fact no limit to the size of the state space and one does not know a priori amongst the numerous candidate variables those that are relevant and those that should be discarded.

Regarding the action space, it can be discrete or continuous, sometimes partly discrete and partly continuous (think of the discrete choice – because of the tick size – of the limit price and the continuous – although in practice it involves integers – choice of the order size, in order placement problems), but it is usually easily defined in algorithmic trading problems.

11.3.2 The role of models

One could naively think that, as many RL methods do not require the knowledge of the transition kernel, the latter does not really matter. In fact, the recent successful RL agents have been trained on simulated data. As a consequence the transition kernel was "known" by the simulator, even though it was not used by the learning algorithm. In particular, it is essential to understand that the often-employed expression "model-free" only applies to the learning method. That said, is a model-based simulator essential for addressing algorithmic trading problems, or, more generally, finance problems, with RL tools? As we shall see, the answer is often positive[23] and this has important consequences on what we can expect from RL methods in algorithmic trading.

Many finance problems can be solved using the tools of optimal control. When this is not the case, RL methods need samples. A tempting alternative to a model-based simulator could be the use of historical data as a way to sample data. This is often proposed but raises a lot of concerns. First, historical data is typically not sufficient if one wants to take account of microstructural concerns (for instance priority issues in LOB cannot be addressed with common high-frequency data sets) or essential feedback effects like market impact. Second, many RL algorithms fail when used on a pre-collected batch of data as is the case when one uses historical data (see for instance Fujimoto et al., 2019). Third, historical data is scarce. In the case of a single-asset trading algorithm, training it on (high-frequency) LOB data may work, but when it comes to multi-asset frameworks or lower-frequency data, most RL methods require far more data points than what historical datasets can offer. In fact, the usual rule of thumb stating that the number of data points should exceed by far the number of parameters to estimate applies to RL techniques and often disqualifies the raw use of historical data.[24]

Among the consequences of the above discussion, one is particularly important to highlight: most RL methods (even the so-called "model-free" methods) can only solve a finance problem given a choice of state space and a model of kernel transition. Given that financial market dynamics do not follow specified rules and are non-stationary – unlike what happens with games or other problems – the solution found through RL methods can only be useful as long as the market, as a system, behaves as described by the model/the simulated data. Expecting too much from RL in algorithmic finance is like expecting a RL-based agent to play a new game every day when it has only been trained to play games from a given list.

The craze around RL should not lure financial practitioners. There is indeed no magic: non-stationarity will not disappear thanks to RL techniques. Model-free RL methods are nevertheless very useful in that they can be used upon simulated data based on different models without knowing these models. In other words,

---

[23] The online use of bandit algorithms to choose among a list of algorithms to carry out a given task is of course an exception, by nature.

[24] This remark is particularly important when it comes to multi-asset portfolio construction.

the place of models might be reduced in the future to data generation/simulation, should RL techniques be adopted.[25] [26]

11.3.3 The question of risk

In the RL literature, objective functions are almost always expected rewards. In finance however, maximizing expectation is often not enough as risk must be mitigated. This seems to disqualify RL, but this is of course not the case. In fact, in the design of MDP for financial applications, risk must be embedded in rewards.[27]

A classical approach in algorithmic trading, where the basic measure of performance is Profit & Loss (PnL), consists in maximizing a final reward that accounts for the risk. Maximizing a Von Neumann–Morgenstern expected utility of the PnL is an important example. Maximizing a risk-adjusted performance indicator such as a final Sharpe ratio is another. Differential versions of these approaches allow one to go from a final reward to running rewards, which may be preferable for some learning algorithms.

Another typical (brute force) approach consists in using penalty terms in running rewards in order to penalize for risk at each date. This is for instance a common approach in market making models where large inventories are penalized by a local variance term.

11.3.4 The question of time steps

We previously discussed the construction of the MDP. Other important points could be highlighted when it comes to finance problems.

For example, in most problems, decisions are taken sequentially and the usual MDP setup is well adapted. The question of time may however be important for some specific problems in which the agent only (re)acts upon the occurrence of some specified events while the system evolves on its own between events. In that case, MDP may not be enough and more general frameworks are needed.

11.4 A review of existing works

As discussed above, RL techniques have been proposed for the pricing and hedging of contingent claims. However, option pricing is clearly out of the scope of this chapter on algorithmic trading. Even though portfolio decisions are more

---

[25] Even with good-quality simulators, learning will not be easy. Financial data is indeed often characterized by a low signal-to-noise ratio (compared to toy examples for instance). The only good point with low signal-to-noise ratio data is that exploration is somehow carried out naturally on some financial state variables.

[26] Some market multi-agent-based simulators are based on RL ideas (see for instance Lussange et al., 2019b,a).

[27] A classical approach to transforming mean–variance dynamic optimization problems into classical time-consistent problems is that of Zhou and Li (2000).

and more made algorithmically and at higher and higher frequency and despite numerous research works suggesting the use of RL for optimizing portfolios, we do not cover optimal portfolio choice either. One reason is that this subfield of quantitative finance is traditionally not considered part of the algorithmic trading field. More importantly, another reason is that, in spite of their initial appeal,[28] most approaches proposed in the literature are dubious as they use structures (for instance neural networks) in which the number of degrees of freedom (i.e. the number of parameters) is large given the volume of historical data used for training (in particular when low-frequency data is used and when the number of assets is large).[29] [30]

Let us now review existing works involving RL techniques for addressing three kinds of problems: the design of statistical arbitrage strategies,[31] the optimization of execution strategies, and the optimization of market making strategies.

11.4.1 Statistical arbitrage

The literature on statistical arbitrage is very diverse. Overall, papers in this field have often been written to advocate for the use of particular tools and techniques, or at least to exemplify them.[32] RL tools appeared in the literature on statistical arbitrage more than twenty years ago. The first wave of papers was followed by another one in recent years to advocate again for the use of RL techniques. Our goal here is to shed light on a few representative papers to help the readers in their own research in the field.[33]

In a series of papers including Moody et al. (1998) and Moody and Saffell (2001) – see also the references therein – researchers proposed direct policy search methods to optimize the strategy of a trader. In particular, in Moody and Saffell (2001), the authors built a mid-frequency long/short trading strategy on USD/GBP based on 30-minute data. Their method[34] used a neural network for the position to take with a short history of past returns as inputs, and a gradient ascent in order to optimize various risk-adjusted performance measures such as differential forms of Sharpe and Sortino ratios, while taking into account trans-action costs. Interestingly, they found little evidence for the need of exploration, probably because the signal-to-noise ratio is so small in finance that it naturally induces exploration. Gold (2003) used a very similar approach to design a trad-

---

[28] Merging the estimation/forecasting step with the investment decision step is indeed attractive.

[29] The use of synthetic market data is sometimes proposed to circumvent this (overfitting) problem but the literature is, as of today, limited – see for instance Wiese et al. (2020) or Yu et al. (2019).

[30] Interesting ideas have recently been developed to avoid the above problems while using RL techniques. See for instance Wang (2019) and Wang and Zhou (2020).

[31] The design of statistical arbitrage strategies can be regarded as some form of portfolio choice problem. However, for one- or two-asset problems with high-frequency data, RL techniques may be less prone to overfitting.

[32] People building strategies that work in real financial markets are indeed unlikely to publish papers with full details.

[33] Our goal is not to be exhaustive. Also, we do not really assess the quality of the papers.

[34] They also tested a Q-learning approach and obtained results in favour of direct policy search.

ing strategy on several currency pairs. An extension of the approach with a risk management layer and a dynamic optimization layer was proposed in Dempster and Leemans (2006) with 1-minute EUR/USD market data.

Recently, a similar approach was proposed in Lu (2017) with the additional use of an LSTM network. Approaches based on value functions have also been proposed in a second wave of papers. Cumming et al. (2015) presented a policy iteration approach in order to maximize the expected PnL of a high-to-mid frequency foreign exchange (FX) trader. More precisely, they used 1-minute candlesticks history on several currency pairs to train an algorithm working through LSTD (a classic of TD learning algorithms) for policy evaluation and improving policies with a standard greedy step. Carapuço et al. (2018) proposed more recently an FX trading algorithm based on a deep Q-network (DQN) trained on a dataset that accounts for market microstructure, but only mid-frequency trading decisions were considered.

Most of the papers that use RL tools to design statistical arbitrage strategies are applied to FX markets. Outside of FX markets, we can first cite Deng et al. (2016). Their approach is based on a gradient-based direct policy search where strategies are represented through a neural network. They exemplified their algorithm on 1-minute data for a Chinese stock index futures and two commodity futures.[35] Recently, Théate and Ernst (2020) also proposed a MDP framework for optimizing (with a DQN)[36] algorithmic trading strategies. Their framework is very rich and detailed with open-high-low-close data, macroeconomic indicators, and even news as state variables, together with realistic rewards involving transaction costs, but their application is restricted to simple contexts with daily data. The same researchers, however, along with co-authors, proposed in Boukas et al. (2020) an RL approach for intraday bidding on energy markets.

11.4.2 Optimal execution

The optimal execution of orders raises a lot of issues associated with the trade-off between liquidity costs and volatility, but also with deep market microstructural questions. The literature on optimal execution started around two decades ago with the seminal work of Almgren and Chriss (1999, 2001) tackling the optimal scheduling problem of agents willing to balance, one the one hand, their incentive to trade fast in order to avoid market fluctuations, and, on the other hand, their incentive to trade slowly in order to have as little impact and liquidity-related costs as possible. Since then, many research works have been carried out on the optimal scheduling problem with different modelling assumptions regarding market impact and transaction costs, and different objective functions. Beyond the optimal scheduling problem that focusses on the splitting of a large parent

---

[35] They obtained better results with another approach that is not based on RL in Deng et al. (2015).

[36] The possibility of using Q-learning to build statistical arbitrage strategies is also exemplified in Ritter (2017). In a very simple simulation model where returns are mean reverting, the author built a Q-learning agent that takes into account trading costs, market impact, and a form of risk aversion through a quadratic penalty in rewards.

order into child orders, another strand of research has focussed on the child order placement problem.[37] Some papers focus on the trade-off between (i) posting a liquidity-providing limit order and having no guarantee of execution but a good execution price, and (ii) sending liquidity-taking orders but paying the bid–ask spread. Some others study the optimal routing of orders in a fragmented market with many lit and dark pools.[38]

Most of the papers in the literature are based on optimal control tools. RL techniques could therefore be a way to study models with more state variables and more complicated dynamics.

One of the first papers to advocate the use of RL techniques to solve optimal execution problems was Nevmyvaka et al. (2006). The problem addressed in that paper is that of an agent willing to sell a given number of shares within a short period of time (a few minutes in their case) by posting a single limit order in a LOB – and potentially updating it – or crossing the spread. Their approach is a brute force form of Q-learning in tabular mode trained on NASDAQ high-frequency data. Although it is simplistic with a low-dimensional state space suffering from too much discretization, that paper paved the way to other RL applications to optimal execution. It surprisingly remains, however, one of the only RL papers dealing with limit order placement. An exception is a recent paper (Schnaubelt, 2022) that deals with limit order placement in cryptocurrency markets. Another exception is the very recent paper Karpe et al. (2020) that uses the market simulator of Byrd et al. (2019) to build a double DQN agent that places limit orders or market orders as a function of remaining quantity and time, but also bid–ask spread, market imbalance, and past evolution of prices.

Regarding optimal scheduling,[39] Hendricks and Wilcox (2014) used Q-learning to perform better than the trading curves of Almgren–Chriss models, by making decisions not only based on remaining inventory and time, but also on bid–ask spreads and volumes. The simple model they used relied on discretized variables so as to be able to use tabular methods, and learned on 5-minute bins constructed with granular historical data for South African stocks (ignoring therefore the impact of actions on the market variables). Ning et al. (2018) considered a double DQN approach to train an agent that makes decisions (based on remaining inventory, time, mid-price, and a volatility measure) on the size of the next market order in an Almgren–Chriss-like framework with no information on the price dynamics. Their training set was based on 1-second mid-price historical data and therefore, once again, no feedback of the actions on the market dynamics was taken into account.[40] Their approach probably constitutes one of the most

---

[37] The former problem corresponds to the strategic layer of most execution algorithms while the latter corresponds to their tactical layer.

[38] Optimal execution has been one of the very active fields of quantitative finance in the last decade. We refer to the books of Cartea et al. (2015) and Guéant (2016) for a detailed description of the field.

[39] Optimal execution models that only consider market orders should be regarded as optimal scheduling models.

[40] They argue that historical data is enough to train the model if it is then continuously updated when used in reality. This is an interesting idea, but it requires an intensive usage of the algorithm to be able to account for market impact.

interesting starting points for real RL-based execution algorithms. Dabérius et al. (2019) is another interesting paper that compared the use of a double DQN and that of a policy-based approach for solving problems inspired from the optimal execution models of Cartea et al. (2015). Strangely, they did not test their results on historical data or simulated data backed by historical data.

In the optimal execution field, dark pool exploration has also been addressed using online RL tools. We refer to Ganchev et al. (2010) and Laruelle et al. (2011) and to the chapter by Sophie Laruelle for more details.

11.4.3 Market making

Economists interested in market microstructure have studied the behaviour of market makers/dealers/market specialists for a long time with the aim of understanding market liquidity and the different factors explaining the very existence or the magnitude of bid-ask spreads. The two usual types of model are: (i) those where one or several risk-averse market makers optimize their pricing policy for managing their inventory risk models (see Amihud and Mendelson, 1980, Ho and Stoll, 1980, 1981, 1983, and O'Hara and Oldfield, 1986); and (ii) models focussed on information asymmetries where bid–ask spreads derive from adverse selection (see for instance Copeland and Galai, 1983, or Glosten and Milgrom, 1985). Other classic economic references on market making include Grossman and Miller (1988) and the review Stoll (2003).

In 2008, largely inspired by Ho and Stoll (1981), Avellaneda and Stoikov (2008) proposed a stochastic optimal control model to determine the optimal bid and ask quotes that a single-asset risk-averse market maker should set. The authors paved the way to a new literature on market making that put more focus on the very problem faced by a market maker; unlike earlier work that focussed rather on a general understanding of liquidity. The models of this new research can be divided into two groups: those adapted to the problem of a market maker in a limit order book and those adapted to OTC markets where market-making automation is now commonplace (bonds, FX, etc.). Most of them use stochastic optimal control tools (see the books Cartea et al., 2015, and Guéant, 2016, for detailed discussions) but many RL approaches have been proposed over the recent past.

The use of RL techniques for market-making automation is, however, not a new idea. The earliest paper is indeed Chan and Shelton (2001). The goal of the authors was clearly to advocate the use of RL techniques: they proposed a model inspired by Glosten and Milgrom (1985) for the market with informed and uninformed traders, and used several RL methods (Monte Carlo, SARSA, actor–critic) in order to find the optimal quotes of their market maker. They allow for the use of several relevant state variables (inventory, imbalance, market quality) and several forms of rewards including proxies of the risk borne by the market maker. Of course their model was too simple for any practical use but they clearly anticipated the relevance of RL tools in the field of automated market making.[41]

In recent years, the renewed popularity of RL techniques has been associated with a significant number of new RL papers dealing with market making.

For order-driven markets, the most cited reference is certainly Spooner et al. (2018). The authors of that article used historical LOB data to train a market maker using several RL methods (SARSA, Q-learning, double Q-learning, etc.) in a model where the state space is simplified thanks to tile coding approximation. That paper is interesting as it constitutes a good starting point for future research. Furthermore, it sheds light on the need for a market simulator: the authors indeed acknowledged that priority issues in LOB cannot be addressed by only using common historical LOB data.

For quote-driven markets, Guéant and Manziuk (2019) proposed several actor–critic approaches in which the value function and the policies are approximated with neural networks. Using RL techniques, they showed how to approximate the optimal quotes in one of the multi-asset market making models proposed in Guéant (2017). Their applications concerned a portfolio of 20 corporate bonds.[42] Another interesting paper for OTC markets is Ganesh et al. (2019) in which the authors are able to train a market maker thanks to a policy-search approach in a simulated market with several dealers.

Other recent works include Lim and Gorse (2018) and Kumar (2020) but the version of the papers we had access to did not contain enough details.

11.5 Conclusion and perspectives for the future

In finance, many problems can be modelled with MDP: portfolio choice, hedging in complete and incomplete markets, optimal execution, market making, etc. This mathematical framework being exactly that of RL, the enthusiasm around RL techniques following the recent successes of DeepMind came with the hope of being able to solve most of the problems usually addressed using MDP and / or the tools of (stochastic) optimal control. In particular, academics and quantitative analysts in the financial industry hoped to get alternatives to numerical methods based on grids – which are known to suffer from the curse of dimensionality – in order to solve high-dimensional problems. Another hope was to get rid of the simplifying assumptions on the dynamics of financial variables in most models. RL indeed came with the promise that models were not always necessary to address dynamic optimization problems, i.e. that observations could be enough (a claim that we clarified above for financial applications).

In this short paper, we have put into perspective the use of RL techniques for addressing finance problems. In addition to highlighting what made finance special compared to games and other fields where RL led to successes, we have insisted on the need to carry out important works in order to obtain satisfactory

---

[41] Their work in a Glosten–Milgron information model inspired the recent paper Mani et al. (2019).

[42] Baldacci et al. (2019) also used an actor-tcritic approach to solve a two-stage principal-agent problem involving an exchange (which sets fees) and a market maker.

market simulators – with characteristics that depend on the problem type – for training RL algorithms.

We have also presented examples of articles using RL techniques for building simple statistical arbitrage trading algorithms or for solving optimal execution and market-making problems. These articles should be regarded as proofs of concept and we believe it will soon be the time for building within financial institutions scalable RL-based execution and market-making trading algorithms.

As noted by several renowned scientists, the recent breakthroughs involving RL are mainly technological, not scientific. For instance, Dimitri Bertsekas, one of the greatest specialists of optimal control, claimed that the great success of AlphaZero was due to a "skillful implementation/integration of known ideas, and awesome computational power". Subsequently, a necessary condition for soon seeing RL-based trading agents in many financial institutions is that traditional quants, computer scientists, and engineers unite forces and ride the learning curve together.

References

Abergel, Frédéric, Huré, Côme, and Pham, Huyên. 2020. Algorithmic trading in a microstructural limit order book model. *Quantitative Finance*, **20**(8), 1263–1283.

Almgren, Robert, and Chriss, Neil. 1999. Value under liquidation. *Risk*, **12**(12), 61–63.

Almgren, Robert, and Chriss, Neil. 2001. Optimal execution of portfolio transactions. *Journal of Risk*, **3**, 5–40.

Amihud, Yakov, and Mendelson, Haim. 1980. Dealership market: Market-making with inventory. *Journal of Financial Economics*, **8**(1), 31–53.

Auer, Peter. 2002. Using confidence bounds for exploitation-exploration trade-offs. *Journal of Machine Learning Research*, **3**(Nov), 397–422.

Auer, Peter, Cesa-Bianchi, Nicolo, and Fischer, Paul. 2002. Finite-time analysis of the multi-armed bandit problem. *Machine Learning*, **47**(2-3), 235–256.

Avellaneda, Marco, and Stoikov, Sasha. 2008. High-frequency trading in a limit order book. *Quantitative Finance*, **8**(3), 217–224.

Bachouch, Achref, Huré, Côme, Langrené, Nicolas, and Pham, Huyên. 2018. Deep neural networks algorithms for stochastic control problems on finite horizon, part 2: Numerical applications. ArXiv:1812.05916.

Baldacci, Bastien, Manziuk, Iuliia, Mastrolia, Thibaut, and Rosenbaum, Mathieu. 2019. Market making and incentives design in the presence of a dark pool: a deep reinforcement learning approach. ArXiv:1912.01129.

Bäuerle, Nicole, and Rieder, Ulrich. 2011. *Markov Decision Processes with Applications to Finance*. Springer Science & Business Media.

Becker, Sebastian, Cheridito, Patrick, and Jentzen, Arnulf. 2019. Deep optimal stopping. *Journal of Machine Learning Research*, **20**, 74.

Bertsekas, Dimitri P. 2019. *Reinforcement Learning and Optimal Control*. Athena Scientific.

Bertsekas, Dimitri P, and Tsitsiklis, John N. 1996. *Neuro-dynamic Programming*. Athena Scientific.

Boukas, Ioannis, Ernst, Damien, Théate, Thibaut, Bolland, Adrien, Huynen, Alexandre, Buchwald, Martin, Wynants, Christelle, and Cornélusse, Bertrand. 2020. A deep reinforcement learning framework for continuous intraday market bidding. ArXiv:2004.05940.

Buehler, Hans, Gonon, Lukas, Teichmann, Josef, and Wood, Ben. 2019. Deep hedging. *Quantitative Finance*, **19**(8), 1271–1291.

Byrd, David, Hybinette, Maria, and Balch, Tucker Hybinette. 2019. Abides: Towards high-fidelity market simulation for ai research. ArXiv:1904.12066.

Carapuço, João, Neves, Rui, and Horta, Nuno. 2018. Reinforcement learning applied to Forex trading. *Applied Soft Computing*, **73**, 783–794.

Cartea, Álvaro, Jaimungal, Sebastian, and Penalva, José. 2015. *Algorithmic and High-Frequency Trading*. Cambridge University Press.

Chan, Nicholas Tung, and Shelton, Christian. 2001. An electronic market-maker. AI Memo 2001-005, MIT AI Lab.

Charpentier, Arthur, Elie, Romuald, and Remlinger, Carl. 2020. Reinforcement learning in economics and finance. ArXiv:2003.10014.

Copeland, Thomas E., and Galai, Dan. 1983. Information effects on the bid–ask spread. *Journal of Finance*, **38**(5), 1457–1469.

Cumming, James, Alrajeh, Dalal, and Dickens, Luke. 2015. An investigation into the use of reinforcement learning techniques within the algorithmic trading domain. Preprint, Imperial College London: London, UK.

Dabérius, Kevin, Granat, Elvin, and Karlsson, Patrik. 2019. Deep execution-value and policy-based reinforcement learning for trading and beating market benchmarks. Available at SSRN 3374766.

Dempster, Michael A.H., and Leemans, Vasco. 2006. An automated FX trading system using adaptive reinforcement learning. *Expert Systems with Applications*, **30**(3), 543–552.

Deng, Yue, Kong, Youyong, Bao, Feng, and Dai, Qionghai. 2015. Sparse coding-inspired optimal trading system for HFT industry. *IEEE Transactions on Industrial Informatics*, **11**(2), 467–475.

Deng, Yue, Bao, Feng, Kong, Youyong, Ren, Zhiquan, and Dai, Qionghai. 2016. Deep direct reinforcement learning for financial signal representation and trading. *IEEE Transactions on Neural Networks and Learning Systems*, **28**(3), 653–664.

E, Weinan, Han, Jiequn, and Jentzen, Arnulf. 2017. Deep learning-based numerical methods for high-dimensional parabolic partial differential equations and backward stochastic differential equations. *Communications in Mathematics and Statistics*, **5**(4), 349–380.

Fischer, Thomas G. 2018. Reinforcement learning in financial markets – a survey. Tech. Rept. FAU Discussion Papers in Economics. `https://econpapers.repec.org/paper/zbwiwqwdp/122018.htm`.

Fujimoto, Scott, Meger, David, and Precup, Doina. 2019. Off-policy deep reinforcement learning without exploration. Pages 2052–2062 in: *Proc. International Conference on Machine Learning 2019*.

Ganchev, Kuzman, Nevmyvaka, Yuriy, Kearns, Michael, and Vaughan, Jennifer Wortman. 2010. Censored exploration and the dark pool problem. *Communications of the ACM*, **53**(5), 99–107.

Ganesh, Sumitra, Vadori, Nelson, Xu, Mengda, Zheng, Hua, Reddy, Prashant, and Veloso, Manuela. 2019. Reinforcement learning for market making in a multi-agent dealer market. ArXiv:1911.05892.

Glosten, Lawrence R, and Milgrom, Paul R. 1985. Bid, ask and transaction prices in a specialist market with heterogeneously informed traders. *Journal of Financial Economics*, **14**(1), 71–100.

Gold, Carl. 2003. FX trading via recurrent reinforcement learning. Pages 363–370 of: *Proc. IEEE International Conference on Computational Intelligence for Financial Engineering*.

Grossman, Sanford J., and Miller, Merton H. 1988. Liquidity and market structure. *Journal of Finance*, **43**(3), 617–633.

Guéant, Olivier. 2016. *The Financial Mathematics of Market Liquidity: From Optimal Execution to Market Making*. CRC Press.

Guéant, Olivier. 2017. Optimal market making. *Applied Mathematical Finance*, **24**(2), 112–154.

Guéant, Olivier, and Manziuk, Iuliia. 2019. Deep reinforcement learning for market making in corporate bonds: beating the curse of dimensionality. *Applied Mathematical Finance*, **26**(5), 387–452.

Han, Jiequn, Jentzen, Arnulf, and E, Weinan. 2018. Solving high-dimensional partial differential equations using deep learning. *Proc. National Academy of Sciences*, **115**(34), 8505–8510.

Hendricks, Dieter, and Wilcox, Diane. 2014. A reinforcement learning extension to the Almgren-Chriss framework for optimal trade execution. Pages 457–464 of: *Proc. IEEE Conference on Computational Intelligence for Financial Engineering & Economics*.

Henry-Labordere, Pierre. 2017. Deep primal–dual algorithm for BSDEs: Applications of machine learning to CVA and IM. Available at SSRN 3071506.

Ho, Thomas, and Stoll, Hans R. 1980. On dealer markets under competition. *Journal of Finance*, **35**(2), 259–267.

Ho, Thomas, and Stoll, Hans R. 1981. Optimal dealer pricing under transactions and return uncertainty. *Journal of Financial Economics*, **9**(1), 47–73.

Ho, Thomas S.Y., and Stoll, Hans R. 1983. The dynamics of dealer markets under competition. *Journal of Finance*, **38**(4), 1053–1074.

Huré, Côme, Pham, Huyên, Bachouch, Achref, and Langrené, Nicolas. 2018. Deep neural networks algorithms for stochastic control problems on finite horizon, part I: convergence analysis. ArXiv:1812.04300.

Huré, Côme, Pham, Huyên, and Warin, Xavier. 2019. Some machine learning schemes for high-dimensional nonlinear PDEs. ArXiv:1902.01599.

Karpe, Michaël, Fang, Jin, Ma, Zhongyao, and Wang, Chen. 2020. Multi-agent reinforcement learning in a realistic limit order book market simulation. ArXiv:2006.05574.

Kolm, Petter N., and Ritter, Gordon. 2020. Modern perspectives on reinforcement learning in finance. *Journal of Machine Learning in Finance*, **1**(1).

Kumar, Pankaj. 2020. Deep reinforcement learning for market making. Pages 1892–1894 of: *Proc. 19th International Conference on Autonomous Agents and Multiagent Systems*.

Laruelle, Sophie, Lehalle, Charles-Albert, and Pagés, Gilles. 2011. Optimal split of orders across liquidity pools: a stochastic algorithm approach. *SIAM Journal on Financial Mathematics*, **2**(1), 1042–1076.

Lim, Ye-Sheen, and Gorse, Denise. 2018. Reinforcement learning for high-frequency market making. Pages 521–526 of: *Proc. European Symposium on Artificial Neural Networks, Computational Intelligence and Machine Learning*.

Lu, David W. 2017. Agent inspired trading using recurrent reinforcement learning and lstm neural networks. ArXiv:1707.07338.

Lussange, Johann, Lazarevich, Ivan, Bourgeois-Gironde, Sacha, Palminteri, Stefano, Gutkin, Boris, et al. 2019a. Stock market microstructure inference via multi-agent reinforcement learning. ArXiv:1910.05137

Lussange, Johann, Bourgeois-Gironde, Sacha, Palminteri, Stefano, and Gutkin, Boris. 2019b. Stock price formation: useful insights from a multi-agent reinforcement learning model. ArXiv:1910.05137.

Mani, Mohammad, Phelps, Steve, and Parsons, Simon. 2019. Applications of Reinforcement Learning in Automated Market-Making. In *Proceedings of Games, Agents and Incentives Workshops, May 2019, Montreal, Canada*.

Mnih, Volodymyr, Kavukcuoglu, Koray, Silver, David, Rusu, Andrei A, Veness, Joel, Bellemare, Marc G, Graves, Alex, Riedmiller, Martin, Fidjeland, Andreas K, Ostrovski, Georg, et al. 2015. Human-level control through deep reinforcement learning. *Nature*, **518**(7540), 529–533.

Moody, John, and Saffell, Matthew. 2001. Learning to trade via direct reinforcement. *IEEE Transactions on Neural Networks*, **12**(4), 875–889.

Moody, John, Wu, Lizhong, Liao, Yuansong, and Saffell, Matthew. 1998. Performance functions and reinforcement learning for trading systems and portfolios. *Journal of Forecasting*, **17**(5–6), 441–470.

Munos, Rémi. 2004. *Contributions à l'apprentissage par renforcement et au contrôle optimal avec approximation*. PhD thesis.

Nevmyvaka, Yuriy, Feng, Yi, and Kearns, Michael. 2006. Reinforcement learning for optimized trade execution. Pages 673–680 of: *Proceedings of the 23rd International Conference on Machine Learning*.

Ning, Brian, Ling, Franco Ho Ting, and Jaimungal, Sebastian. 2018. Double deep Q-learning for optimal execution. ArXiv:1812.06600.

Noonan, Laura. 2017. JPMorgan develops robot to execute trades. *Financial Times*, `https://www.ft.com/content/16b8ffb6-7161-11e7-aca6-c6bd07df1a3c`.

O'Hara, Maureen, and Oldfield, George S. 1986. The microeconomics of market making. *Journal of Financial and Quantitative Analysis*, **21**(4), 361–376.

Pagès, Gilles, Pham, Huyên, and Printems, Jacques. 2004. Optimal quantization methods and applications to numerical problems in finance. Pages 253–297 of: *Handbook of Computational and Numerical Methods in Finance*. Springer.

Pham, Huyen, Warin, Xavier, and Germain, Maximilien. 2019. Neural networks-based backward scheme for fully nonlinear PDEs. ArXiv:1908.00412.

Powell, Warren B. 2011. *Approximate Dynamic Programming: Solving the Curses of Dimensionality*. John Wiley & Sons.

Ritter, Gordon. 2017. Machine learning for trading. Available at SSRN 3015609.

Rockafellar, R. Tyrrell, and Uryasev, Stanislav. 2002. Conditional value-at-risk for general loss distributions. *Journal of Banking and Finance*, **26**(7), 1443–1471.

Ruf, Johannes, and Wang, Weiguan. 2020. Neural networks for option pricing and hedging: a literature review. *Journal of Computational Finance*, **24**(1), 1–46.

Schnaubelt, Matthias. 2022. Deep reinforcement learning for the optimal placement of cryptocurrency limit orders. *European Journal of Operational Research*, **296**(3), 993–1006.

Schrittwieser, Julian, Antonoglou, Ioannis, Hubert, Thomas, Simonyan, Karen, Sifre, Laurent, Schmitt, Simon, Guez, Arthur, Lockhart, Edward, Hassabis, Demis, Graepel, Thore et al. 2020. Mastering Atari, Go, Chess and Shogi by planning with a learned model. *Nature*, **588**(7839), 604–609.

Silver, David, Huang, Aja, Maddison, Chris J, Guez, Arthur, Sifre, Laurent, Van Den Driessche, George, Schrittwieser, Julian, Antonoglou, Ioannis, Panneershelvam, Veda, Lanctot, Marc, et al. 2016. Mastering the game of Go with deep neural networks and tree search. *Nature*, **529**(7587), 484–489.

Silver, David, Schrittwieser, Julian, Simonyan, Karen, Antonoglou, Ioannis, Huang, Aja, Guez, Arthur, Hubert, Thomas, Baker, Lucas, Lai, Matthew, Bolton, Adrian, et al. 2017. Mastering the game of Go without human knowledge. *Nature*, **550**(7676), 354–359.

Silver, David, Hubert, Thomas, Schrittwieser, Julian, Antonoglou, Ioannis, Lai, Matthew, Guez, Arthur, Lanctot, Marc, Sifre, Laurent, Kumaran, Dharshan, Graepel, Thore, et al. 2018. A general reinforcement learning algorithm that masters Chess, Shogi, and Go through self-play. *Science*, **362**(6419), 1140–1144.

Spooner, Thomas, Fearnley, John, Savani, Rahul, and Koukorinis, Andreas. 2018. Market making via reinforcement learning. ArXiv:1804.04216.

Stoll, Hans R. 2003. Market microstructure. Pages 553–604 of: *Handbook of the Economics of Finance*, vol. 1. Elsevier.

Sutton, Richard S, and Barto, Andrew G. 2018. *Reinforcement Learning: An Introduction*. MIT Press.

Szepesvári, Csaba. 2010. Algorithms for reinforcement learning. *Synthesis Lectures on Artificial Intelligence and Machine Learning*, **4**(1), 1–103.

Théate, Thibaut, and Ernst, Damien. 2020. An application of deep reinforcement learning to algorithmic trading. ArXiv:2004.06627.

Thompson, William R. 1933. On the likelihood that one unknown probability exceeds another in view of the evidence of two samples. *Biometrika*, **25**(3/4), 285–294.

Wang, Haoran. 2019. Large scale continuous-time mean-variance portfolio allocation via reinforcement learning. Available at SSRN 3428125.

Wang, Haoran, and Zhou, Xun Yu. 2020. Continuous-time mean–variance portfolio selection: A reinforcement learning framework. *Mathematical Finance*, **30**(4), 1273–1308.

Wiese, Magnus, Knobloch, Robert, Korn, Ralf, and Kretschmer, Peter. 2020. Quant gans: Deep generation of financial time series. *Quantitative Finance*, **20**(9), 1419–1440.

Yu, Pengqian, Lee, Joon Sern, Kulyatin, Ilya, Shi, Zekun, and Dasgupta, Sakyasingha. 2019. Model-based deep reinforcement learning for dynamic portfolio optimization. ArXiv:1901.08740.

Zhou, Xun Yu, and Li, Duan. 2000. Continuous-time mean–variance portfolio selection: A stochastic LQ framework. *Applied Mathematics and Optimization*, **42**(1), 19–33.

12

Stochastic Approximation Applied to Optimal Execution: Learning by Trading

Sophie Laruelle[a]

Abstract

In this chapter we introduce the basic tools of stochastic approximation (SA) theory. We first recall the results in the deterministic framework (like the Newton–Raphson algorithm), then we state the theorems on a.s. convergence of the recursive procedure (using the martingale approach, the ODE method and the constrained setting). Afterwards, we give the weak rates of convergence with CLT and non-CLT rates depending on the spectrum of the differential matrix at the target. We end this theoretical part with the averaging principle of Ruppert and Polyak to smoothen the behavior of the algorithm and to reach the optimal asymptotic variance. The second part is dedicated to applications to algorithmic trading: we design the stochastic recursive procedure to minimize the mean execution cost when a trader wants to split a volume across different liquidity pools and when he/she looks for the optimal posting price of a limit order.

12.1 Introduction

Searching for points where a function reaches a certain level, its minimum or its maximum, is a common problem in numerical analysis. Its domain of applications is wide: For instance, in physics, one looks for minimum points of a potential; in economics, one wants to minimize a cost function or maximize a profit or some returns; in finance, one looks for calibrating parameters to price some derivatives. In fact, both problems are linked since we can bring them down to a zero search procedure, across the derivatives for the optimization part.

If we have access to the analytic form of the function, we can use deterministic procedures (like the Newton–Raphson algorithm). In this case, the recursive procedure of zero search can be written as:

$$\text{for all } n \geq 0, \quad \theta_{n+1} = \theta_n - \gamma_{n+1} h(\theta_n), \quad \text{with } 0 < \gamma_n \leq \gamma_0 < +\infty, \quad (12.1)$$

where $\theta_0 \in \mathbb{R}$ (could be random) and $h : \mathbb{R}^d \to \mathbb{R}^d$ is a continuous vector field

[a] Université Paris Est Creteil, and Université Gustave Eiffel, Marne-la-Vallée, France
Published in *Machine Learning And Data Sciences For Financial Markets*, Agostino Capponi and Charles-Albert Lehalle © 2023 Cambridge University Press.

satisfying an assumption of sub-linear growth at infinity. Under some appropriate assumptions of mean-reversion, one can prove that the sequence $(\theta_n)_{n \geq 0}$ is bounded and eventually converges to a zero θ^* of h. For instance, the Newton–Raphson algorithm satisfies (12.1) with $h(x) = Dh(x)^{-1}h(x)$ and $\gamma_n = 1$.

Stochastic approximation is an extension of this method when the function has a representation as an expectation: namely we want to find θ^* such that $h(\theta^*) = 0$ with $h(\theta) = \mathbb{E}[H(\theta, Y)]$, where Y a random vector (either simulateable or built from real data). Thus we design the following stochastic recursive procedure:

$$\theta_{n+1} = \theta_n - \gamma_{n+1} H(\theta_n, Y_{n+1}), \quad (Y_n)_{n \geq 1} \text{ i.i.d. with the same law as } Y,$$

which converges to θ^*, zero of h, under appropriate assumptions on h or H and on the step sequence $(\gamma_n)_{n \geq 1}$. When h is the gradient of a function we want to optimize, this algorithm is usually called "Stochastic Gradient Descent (SGD)".

The study of stochastic approximation began in the 1950s with the works of Robbins and Monro (1951), then Kiefer and Wolfowitz (1952) adapted the procedure by introducing a finite difference method with decreasing step to approximate the gradient. Generalizations with constraints on the parameter θ can be found in Kushner and Clark (1978) and Kushner and Yin (2003). The original work focused on i.i.d. sequences for the innovation process $(Y_n)_{n \geq 1}$, but it has been extended to Markov chains (see Benveniste et al., 1990), mixing processes (see Dedecker et al., 2007), low-discrepancy sequences (see Lapeyre et al., 1990) and averaging processes (see Laruelle and Pagès, 2012).

This chapter will focus on the original work on i.i.d. sequences for the innovation process by giving the results on a.s. convergence and its rate. For the a.s. convergence, we have two different methods: the first one is based on martingale approach with the introduction of a Lyapunov function and the Robbin–Siegmund lemma (this approach is called the martingale method); the second one is linked to the study of the asymptotic behavior of the ODE $\dot{\theta} = -h(\theta)$ (it is called the ODE method). The result on the rate of convergence exhibits three kind of behaviors regarding the spectrum of $Dh(\theta^*)$: either a CLT, or a convergence in distribution with a different rate, or an a.s. convergence towards a random variable. We will end the theoretical part with the averaging principle of Ruppert and Polyak: The original idea was to smooth the behavior of the stochastic algorithm by considering the empirical mean of its past values up to the nth iteration. Surprisingly, with a scale of the step sequence $(\gamma_n)_{n \geq 1}$, we freely reach the optimal convergence rate, namely we obtain the optimal asymptotic variance for our algorithm!

The second part of this chapter will present some applications of stochastic approximation to optimal execution. When you have have a certain volume to buy over a defined period, several strategies are possible. You can just split the volume across different trading venues, or you can try to post the whole volume at the best price (if the period is short) or you can do both. In such a context, we will write the average cost of our execution policy and try to minimize it regarding the split of the volumes across different liquidity pools (see Laruelle et al., 2011), or the posting price (see Laruelle et al., 2013). We will show how to build such a cost function depending on the model we consider and how to design the algorithm to solve the execution problem by referring to theoretical results of the first part of this chapter. All these examples will be illustrated on real market data.

12.2 Stochastic approximation: results on a.s. convergence and its rate

Before introducing the results on stochastic approximation, let us go back to deterministic recursive procedures for zero search and the optimization point of view.

12.2.1 Back to deterministic recursive methods

Zero search of a function.

Let $h: \mathbb{R}^d \to \mathbb{R}^d$ be a vector field and let us consider the following recursive procedure of zero search:

$$\text{for all } n \geq 0, \quad \theta_{n+1} = \theta_n - \gamma_{n+1} h(\theta_n), \ \theta_0 \in \mathbb{R}^d, \tag{12.2}$$

where $(\gamma_n)_{n \geq 1}$ is a sequence of real positive numbers.

Theorem 12.1 (1) *If $\theta \mapsto \theta - \gamma h(\theta)$ is (locally) contracting and $\gamma_n = \gamma > 0$, then*

$$\theta_n \underset{n \to \infty}{\longrightarrow} \theta^* \quad \text{unique zero of } h.$$

(2) *If h is continuous with linear growth, $\theta \mapsto h(\theta)$ is separating in θ^*; i.e.*

$$\text{for all } \theta \neq \theta^*, \quad \langle h(\theta)|\theta - \theta^* \rangle > 0,$$

and

$$\sum_{n \geq 1} \gamma_n = +\infty, \quad \sum_{n \geq 1} \gamma_n^2 < +\infty, \tag{12.3}$$

then

$$\theta_n \underset{n \to \infty}{\longrightarrow} \theta^*, \quad \text{the unique zero of } h.$$

An example of item (1) is the Newton method: we then replace h by $\widetilde{h} = (Dh)^{-1} h$ if $Dh(\theta)$ is invertible in the neighborhood of θ^*.

Optimization point of view.

We can interpret the above result as the search for the minimum of the (convex) function $L: \theta \mapsto |\theta - \theta^*|^2$ which can be seen as a potential.

Theorem 12.2 *Let $L: \mathbb{R}^d \to \mathbb{R}_+$ be an essentially quadratic function, i.e. L is C^1, ∇L is Lipschitz,*

$$|\nabla L|^2 \leq C(1 + L) \quad \text{and} \quad \lim_{|\theta| \to +\infty} L(\theta) = +\infty.$$

If h is continuous with $\sqrt{L}$-linear growth, i.e.

$$|h| \leq C(1 + L)^{\frac{1}{2}},$$

satisfies

$$\text{for all } \theta \neq \theta^*, \qquad \langle \nabla L \mid h \rangle(\theta) > 0,$$

and if the step sequence $(\gamma_n)_{n \geq 1}$ satisfies (12.3), then:

(1) $\{h = 0\} = \{\nabla L = 0\} = \arg\min_\theta L(\theta) = \{\theta^*\};$
(2) *the procedure (12.2) satisfies*

$$\theta_n \underset{n \to \infty}{\longrightarrow} \theta^*.$$

Example 12.3 (1) **Gradient descent:** $h(x) = \nabla L(x)$ or $h(x) = \rho(x)\nabla L(x)$, with $\rho > 0$ bounded.
(2) **Convex framework:** If $V : \mathbb{R}^d \to \mathbb{R}_+$ is a differentiable convex function with a unique minimum at θ^*, then we can take V to be $L(\theta) = |\theta - \theta^*|^2$ since

$$\text{for all } \theta \neq \theta^*, \qquad \langle \nabla V(\theta) \mid \theta - \theta^* \rangle > 0.$$

The advantage is then that L is trivially essentially quadratic.

12.2.2 Stochastic recursive methods

From now on assume that no direct access to numerical values of $h(\theta)$ is possible, but that h admits an integral representation with respect to a $\mathbb{R}^d$-valued random vector Y, namely

$$h(\theta) = \mathbb{E}\,[H(\theta, Y)], \quad H : \mathbb{R}^d \times \mathbb{R}^q \overset{\text{Borel}}{\longrightarrow} \mathbb{R}^d, \quad Y \overset{(d)}{=} \mu, \qquad (12.4)$$

satisfying $\mathbb{E}\,|H(\theta, Y)| < +\infty$ for every $\theta \in \mathbb{R}^d$. If $H(\theta, y)$ is easy to compute for any couple (θ, y) and the distribution μ of Y is simple to simulate, then a first idea could be to randomize the deterministic zero search procedure (12.2) by using at each step a Monte-Carlo simulation to approximate $h(\theta_n)$.

A more sophisticated idea is to try to do both simultaneously by using $H(\theta_n, Y_{n+1})$ instead of $h(\theta_n)$ where $(Y_n)_{n \geq 1}$ is a sequence of i.i.d. random variables with the same law as Y.

Based on this heuristic analysis, we can reasonably hope that the recursive procedure

$$\theta_{n+1} = \theta_n - \gamma_{n+1} H(\theta_n, Y_{n+1}), \quad Y_{n+1} \text{ i.i.d. with law } \mu, \quad n \geq 0, \qquad (12.5)$$

also converges towards a zero θ^* of h, at least under some appropriate assumptions (specified thereafter) on H and on the step sequence $(\gamma_n)_{n \geq 1}$.

a.s. convergence: the martingale approach.

Stochastic approximation provides various theorems which guarantee the a.s. convergence and/or in L^p of the stochastic recursive procedure (12.5). We first state below a general preliminary result known as the Robbins–Siegmund Lemma from which the main convergence results will be easily deduced.

In what follows, the function H and the sequence $(Y_n)_{n \geq 1}$ are defined by (12.4), and h is the vector field from $\mathbb{R}^d$ to $\mathbb{R}^d$ defined by $h(\theta) = \mathbb{E}\,[H(\theta, Y_1)]$.

Theorem 12.4 (Robbins–Siegmund Lemma)   *Let $h\colon \mathbb{R}^d \to \mathbb{R}^d$ and $H\colon \mathbb{R}^d \times \mathbb{R}^q \to \mathbb{R}^d$ be two functions satisfying (12.4). Assume that there exists a continuously differentible function $L\colon \mathbb{R}^d \to \mathbb{R}^+$ satisfying*

$$\nabla L \text{ is Lipschitz continuous and } \quad |\nabla L|^2 \leq C(1+L) \qquad (12.6)$$

such that h satisfies the mean-reverting assumption

$$\langle \nabla L \mid h \rangle \geq 0. \qquad (12.7)$$

Furthermore, assume that H satisfies the following assumption on pseudo-linear growth:

$$\text{for all } \theta \in \mathbb{R}^d, \quad \|H(\theta,Y)\|_2 \leq C\sqrt{1+L(\theta)}, \qquad (12.8)$$

which implies that $|h| \leq C\sqrt{1+L}$.
  Let $\gamma = (\gamma_n)_{n\geq 1}$ be a step sequence satisfying

$$\sum_{n\geq 1} \gamma_n = +\infty \quad \text{and} \quad \sum_{n\geq 1} \gamma_n^2 < +\infty. \qquad (12.9)$$

Finally, assume that θ_0 is independent of $(Y_n)_{n\geq 1}$ and that $\mathbb{E}\left[L(\theta_0)\right] < +\infty$.
  Then, the recursive procedure defined by (12.5) satisfies $\theta_n - \theta_{n-1} \xrightarrow{\text{P-a.s. and } L^2(\mathbb{P})} 0$, $(L(\theta_n))_{n\geq 0}$ is $L^1(\mathbb{P})$-bounded:

$$L(\theta_n) \xrightarrow[n\to\infty]{\text{a.s.}} L_\infty \in L^1(\mathbb{P}) \quad \text{and} \quad \sum_{n\geq 1} \gamma_n \langle \nabla L | h \rangle (\theta_{n-1}) < +\infty \quad \text{a.s.}$$

Remark 12.5   (1) If the function L also satisfies $\lim_{|\theta|\to\infty} L(\theta) = +\infty$, then L is often called a Lyapunov function of the system, as in the ODE theory.
(2) Note that the assumption (12.6) on L implies that $\nabla\sqrt{1+L}$ is bounded, so that $\sqrt{L}$ has at most linear growth, i.e. L has at most quadratic growth.
(3) If we assume that the innovation sequence $(Y_n)_{n\geq 1}$ is not i.i.d. but only $\mathscr{F}_n$-adapted, we then get a stochastic algorithm of the following form:

$$\text{for all } n \geq 0, \quad \theta_{n+1} = \theta_n - \gamma_{n+1} h(\theta_n) + \gamma_{n+1} \left(\Delta M_{n+1} + r_{n+1}\right),$$

where $\Delta M_{n+1} = H(\theta_n, Y_{n+1}) - \mathbb{E}\left[H(\theta_n, Y_{n+1})|\mathscr{F}_n\right]$ is a $\mathscr{F}_n$-maringale increment and $r_{n+1} = \mathbb{E}\left[H(\theta_n, Y_{n+1})|\mathscr{F}_n\right] - h(\theta_n)$ is a $\mathscr{F}_{n+1}$-adapted remainder term. In that case, if we assume that

$$\sum_{n\geq 1} \gamma_n |r_n|^2 < +\infty \quad \text{a.s.},$$

then the conclusions of the Robbins–Siegmund Lemma stay valid.

  The key for the proof of this lemma is the convergence theorem for non-negative super-martingale: If $(S_n)_{n\geq 0}$ is a non-negative super-martingale ($S_n \in L^1(\mathbb{P})$ and $\mathbb{E}\left[S_{n+1}|\mathscr{F}_n\right] \leq S_n$, a.s.), then S_n converges $\mathbb{P}$-a.s. to an integrable (non-negative) random variable S_∞. For a detailed proof of the Robbin–Siegmund Lemma see Pagès (2018).

Optimization point of view: the Kiefer–Wolfowitz approach.

Let us go back to the optimization problem, namely $\min_{\mathbb{R}^d} L$, where $L(\theta) = \mathbb{E}[\Lambda(\theta, Y)]$. If there is no local gradient $\frac{\partial \Lambda}{\partial \theta}(\theta, y)$ or if the computation of $\frac{\partial \Lambda}{\partial \theta}(\theta, y)$ is not competitive with respect to $\Lambda(\theta, x)$ for instance, there exists an alternative to gradient methods, namely finite differences approaches.

The idea is simply to approximate the gradient ∇L by

$$\frac{\partial L}{\partial \theta_i}(\theta) \approx \frac{L(\theta + \eta^i e_i) - L(\theta - \eta^i e_i)}{2\eta^i}, \quad 1 \le i \le d,$$

where $(e_i)_{1 \le i \le d}$ denotes the canonical basis of $\mathbb{R}^d$ and $\eta = (\eta^i)_{1 \le i \le d}$. The term of finite difference admits an integral representation given by

$$\frac{L(\theta + \eta^i e_i) - L(\theta - \eta^i e_i)}{2\eta^i} = \mathbb{E}\frac{\Lambda(\theta + \eta^i e_i, Y) - \Lambda(\theta - \eta^i e_i, Y)}{2\eta^i}.$$

Starting from this representation, we can deduce a stochastic recursive procedure for θ_n as follows:

$$\theta_{n+1}^i = \theta_n^i - \gamma_{n+1}\frac{\Lambda(\theta_n + \eta_{n+1}^i e_i, Y_{n+1}) - \Lambda(\theta_n - \eta_{n+1}^i e_i, Y_{n+1})}{2\eta_{n+1}^i}, \quad 1 \le i \le d.$$

We state below the convergence result for Kiefer–Wolfowitz procedures (which is the natural counterpart of the Robbins–Siegmund Lemma in the stochastic gradient framework).

Theorem 12.6 (Kiefer–Wolfowitz procedure, see Kiefer and Wolfowitz, 1952) *Assume that the function $\theta \mapsto L(\theta)$ is twice differentiable with a Lipschitz Hessian. Assume that*

$$\theta \mapsto \Lambda(\theta, Y) \text{ is Lipschitz in } L^2$$

and that the step sequences satisfy

$$\sum_{n \ge 1} \gamma_n = \sum_{n \ge 1} \eta_n^i = +\infty, \quad \sum_{n \ge 1} \gamma_n^2 < +\infty, \quad \eta_n \to 0$$

and

$$\sum_{n \ge 1} \left(\frac{\gamma_n}{\eta_n^i}\right)^2 < +\infty, \quad 1 \le i \le d.$$

Then θ_n a.s. converges to a connected component of $\{L = \ell\} \cap \{\nabla L = 0\}$ for some level $\ell \ge 0$.

Convergence result for constrained algorithms.

The aim is to determine $\{\theta \in \Theta\colon h(\theta) = \mathbb{E}[H(\theta, Y)] = 0\}$, where $\Theta \subset \mathbb{R}^d$ is a closed convex set, $h\colon \mathbb{R}^d \to \mathbb{R}^d$ and $H\colon \mathbb{R}^d \times \mathbb{R}^q \to \mathbb{R}^d$. For $\theta_0 \in \Theta$, we consider the $\mathbb{R}^d$-valued sequence $(\theta_n)_{n \ge 0}$ defined by

$$\theta_{n+1} = \Pi_\Theta\left(\theta_n - \gamma_{n+1} H(\theta_n, Y_{n+1})\right), \quad n \ge 0, \tag{12.10}$$

where $(Y_n)_{n\geq 1}$ is an i.i.d. sequence with the same distribution as Y and Π_Θ denotes the Euclidian projection on Θ. The recursive procedure (12.10) can be rewritten as follows:

$$\theta_{n+1} = \theta_n - \gamma_{n+1} h(\theta_n) - \gamma_{n+1} \Delta M_{n+1} + \gamma_{n+1} p_{n+1}, \quad n \geq 0, \qquad (12.11)$$

where $\Delta M_{n+1} = H(\theta_n, Y_{n+1}) - h(\theta_n)$ is a martingale increment and

$$p_{n+1} = \frac{1}{\gamma_{n+1}} \Pi_\Theta \left(\theta_n - \gamma_{n+1} H(\theta_n, Y_{n+1}) \right) - \frac{1}{\gamma_{n+1}} \theta_n + H(\theta_n, Y_{n+1}).$$

Theorem 12.7 (see Kushner and Clark, 1978; Kushner and Yin, 2003) *Let $(\theta_n)_{n\geq 0}$ a sequence defined by (12.11). Assume that there exists a unique value $\theta^* \in \Theta$ such that $h(\theta^*) = 0$ and that the mean function h satisifes the mean-reverting assumption in θ, namely*

$$\text{for all } \theta \neq \theta^* \in \Theta, \quad \langle h(\theta) \mid \theta - \theta^* \rangle. \qquad (12.12)$$

Assume that the step sequence $(\gamma_n)_{n\geq 1}$ satisfies

$$\sum_{n\geq 1} \gamma_n = +\infty \quad \text{and} \quad \sum_{n\geq 1} \gamma_n^2 < +\infty. \qquad (12.13)$$

Furthermore, if the function H satisfies

$$\text{for all } \theta \in \Theta, \quad \mathbb{E}\left[|H(\theta, Y)|^2 \right] \leq K(1 + |\theta|^2), \quad K > 0, \qquad (12.14)$$

then

$$\theta_n \xrightarrow[n\to+\infty]{\text{a.s.}} \theta^*.$$

a.s. convergence: the ODE method.

Consider the following recursive procedure defined on a filtered probability space $(\Omega, \mathcal{A}, (\mathcal{F}_n)_{n\geq 0}, \mathbb{P})$ having values in a convex set $C \subset \mathbb{R}^d$,

$$\text{for all } n \geq 0, \quad \theta_{n+1} = \theta_n - \gamma_{n+1} h(\theta_n) + \gamma_{n+1} (\Delta M_{n+1} + r_{n+1}), \qquad (12.15)$$

where $(\gamma_n)_{n\geq 1}$ is a $(0, \bar{\gamma}]$-valued step sequence for some $\bar{\gamma} > 0$, $h \colon C \to \mathbb{R}^d$ is a continuous function with linear growth (the *mean field* of the algorithm) such that

$$(I_d - \gamma h)(C) \subset C \text{ for every } \gamma \in (0, \bar{\gamma}], \qquad (12.16)$$

and θ_0 is an $\mathcal{F}_0$-measurable finite random vector and, for every $n \geq 1$, we have that ΔM_n is an $(\mathcal{F}_n)_n$-martingale increment and r_n is an $(\mathcal{F}_n)_n$-adapted remainder term.

Let us introduce a few additional notions on differential systems. We consider the differential system $\text{ODE}_h \equiv \dot{x} = -h(x)$ associated to the (continuous) *mean field* $h \colon C \to \mathbb{R}^d$. We assume that this system has a C-valued *flow* $\Phi(t, \xi)_{t\in\mathbb{R}_+, \xi\in C}$: For every $\xi \in C$, $(\Phi(t, \xi))_{t\geq 0}$ is the unique solution to ODE_h defined on the whole positive real line. This flow exists whenever h is locally Lipschitz with linear growth.

Let K be a compact connected, flow-invariant subset of C, i.e. such that $\Phi(t, K) \subset K$ for every $t \in \mathbb{R}_+$.

A non-empty subset $A \subset K$ is an *internal attractor* of K for ODE$_h$ if:

(i) $A \subsetneq K$;
(ii) there exists $\varepsilon_0 > 0$ such that $\displaystyle\sup_{x \in K, \mathrm{dist}(x,A) \le \varepsilon_0} \mathrm{dist}(\Phi(t,x),A) \to 0$ as $t \to +\infty$.

A compact connected flow invariant set K is a *minimal attractor* for ODE$_h$ if it contains no internal attractor. This terminology, coming from dynamical systems, may be misleading: Thus any equilibrium point of ODE$_h$ (zero of h) is a minimal attractor by this definition, regardless of its stability (see Claim (b) in Theorem 12.9 below).

Remark 12.8   When the flow does not exist, the above definition should be understood as follows. One replaces the flow $\Phi(x, \cdot)$ by the family of all solutions of ODE$_h$ starting from x at time 0 (whose existence follow from Peano's Theorem). For more details on this natural extension, we refer to Fort and Pagès (2002) (see Appendix "The ODE method without flow"). Up to this extension, the theorem below remains true even when uniqueness of solutions of ODE$_h$ fails.

Theorem 12.9 (a.s. convergence with ODE method, see, e.g., Benveniste et al., 1990, Duflo, 1997, Kushner and Yin, 2003, Fort and Pagès, 1996, Benaïm, 1999) *Assume that $h : C \to \mathbb{R}^d$ satisfies* (12.16) *and that ODE$_h$ has a C-valued flow (for example, because h is a locally Lipschitz function with linear growth). Assume furthermore that*

$$r_n \xrightarrow[n \to +\infty]{\text{a.s.}} 0 \quad \text{and} \quad \sup_{n \ge 0} \mathbb{E}\left[\|\Delta M_{n+1}\|^2 \mid \mathcal{F}_n\right] < +\infty \quad \text{a.s.},$$

and that $(\gamma_n)_{n \ge 1}$ is a positive sequence satisfying ($\gamma_n \in (0, \bar{\gamma}]$, $n \ge 1$) and

$$\sum_{n \ge 1} \gamma_n = +\infty \quad \text{and} \quad \sum_{n \ge 1} \gamma_n^2 < +\infty.$$

On the event $A_\infty = \{\omega : (h(\theta_n(\omega)))_{n \ge 0}$ is bounded$\}$, $\mathbb{P}(d\omega)$-a.s., the set $\Theta^\infty(\omega)$ of the limiting values of $(\theta_n(\omega))_{n \ge 0}$ as $n \to +\infty$ is a compact connected flow invariant minimal attractor for ODE$_h$ (see Proposition 5.3 in Section 5.1 of Benaïm, 1999).

Furthermore:

(a) *Equilibrium point(s) as limiting value(s). If $\mathrm{dist}(\Phi(\theta_0,t), \{h = 0\}) \to 0$ as $t \to +\infty$, for every $\theta_0 \in \mathbb{R}^d$, then $\Theta^\infty(\omega) \cap \{h = 0\} \ne \varnothing$.*
(b) *Single stable equilibrium point. If $\{h = 0\} = \{\theta^*\}$ and $\Phi(\theta_0,t) \to \theta^*$ as $t \to +\infty$ locally uniformly in θ_0, then $\Theta^\infty(\omega) = \{\theta^*\}$ i.e. $\theta_n \xrightarrow{\text{a.s.}} \theta^*$ as $n \to +\infty$.*
(c) *1-dimensional setting. If $d = 1$ and $\{h = 0\}$ is locally finite, then $\Theta^\infty(\omega) = \{\theta_\infty\} \subset \{h = 0\}$ i.e. $\theta_n \xrightarrow{\text{a.s.}} \theta_\infty \in \{h = 0\}$.*

Note also that examples of situation (a) where the algorithm a.s. *does not converge* are developed in Fort and Pagès (1996), Fort and Pagès (2002), and Benaïm (1999) – necessarily with $d \ge 2$ owing to Claim (c).

Rates of convergence.

In standard frameworks, a stochastic algorithm converges to its target with a rate $\sqrt{\gamma_n}$ (which suggests using steps $\gamma_n = \frac{c}{n}$, $c > 0$). To be precise, writing $J_h(\theta^*)$ for the Jacobian of h evaluated at θ^*, under some assumptions on the spectrum of $J_h(\theta^*)$ and on the remainder sequence $(r_n)_{n \geq 1}$, $\dfrac{\theta_n - \theta^*}{\sqrt{\gamma_n}}$ converges in distribution to some Gaussian distribution with covariance matrix depending on $Dh(\theta^*)$. We consider here stochastic algorithms with remainder term defined by (12.15). First, we need to introduce a new notion for our mean function h.

We will say that h is ϵ-differentiable ($\epsilon > 0$) at θ^* if

$$h(\theta) = h(\theta^*) + J_h(\theta^*)(\theta - \theta^*) + o(\|\theta - \theta^*\|^{1+\epsilon}) \quad \text{as} \quad \theta \to \theta^*.$$

Theorem 12.10 (Rate of convergence see Duflo, 1997, Theorem 3.III.14, p. 131, or Zhang, 2016 (for the CLT see also, for example, Benveniste et al., 1990; Kushner and Yin, 2003)) *Let θ^* be an equilibrium point of $\{h = 0\}$ and $\{\theta_n \to \theta^*\}$ the convergence event associated to θ^* (supposed to have a positive probability). Set the gain parameter sequence $(\gamma_n)_{n \geq 1}$ as follows*

$$\text{for all } n \geq 1, \quad \gamma_n = \frac{1}{n}. \tag{12.17}$$

Assume that the function h is differentiable at θ^* and all the eigenvalues of $J_h(\theta^*)$ have positive real parts. Assume that, for a real number $\delta > 0$,

$$\sup_{n \geq 0} \mathbb{E}\left[\|\Delta M_{n+1}\|^{2+\delta} \mid \mathcal{F}_n\right] < +\infty \text{ a.s.},$$

$$\mathbb{E}\left[\Delta M_{n+1}\Delta M_{n+1}^t \mid \mathcal{F}_n\right] \xrightarrow[n \to +\infty]{\text{a.s.}} \Gamma^* \quad \text{on } \{\theta_n \to \theta^*\},$$

where $\Gamma^* \in \mathcal{S}^+(d, \mathbb{R})$ (deterministic symmetric positive matrix) and for an $\varepsilon > 0$ and a positive sequence $(v_n)_{n \geq 1}$ (specified below),

$$n\, v_n \mathbb{E}\left[\|r_{n+1}\|^2 \, \mathbf{1}_{\{\|\theta_n - \theta^*\| \leq \varepsilon\}}\right] \xrightarrow[n \to +\infty]{} 0. \tag{12.18}$$

Let $\lambda_{\min}$ denote the eigenvalue of $J_h(\theta^*)$ with the lowest real part and set $\Lambda := \mathfrak{Re}(\lambda_{\min})$.

(a) *If $\Lambda > \frac{1}{2}$ and $v_n = 1$, $n \geq 1$, then the weak convergence rate is ruled on the convergence event $\{\theta_n \xrightarrow{\text{a.s.}} \theta^*\}$ by the following Central Limit Theorem*

$$\sqrt{n}\,(\theta_n - \theta^*) \xrightarrow[n \to +\infty]{\mathcal{L}_{\text{stably}}} \mathcal{N}(0, \Sigma^*)$$

with

$$\Sigma^* := \int_0^{+\infty} e^{-u\left(J_h(\theta^*)^t - \frac{I_d}{2}\right)} \Gamma^* e^{-u\left(J_h(\theta^*) - \frac{I_d}{2}\right)} du.$$

(b) *If $\Lambda = \frac{1}{2}$, $v_n = \log n$, $n \geq 1$, and h is ϵ-differentiable at θ^*, then*

$$\sqrt{\frac{n}{\log n}}\,(\theta_n - \theta^*) \xrightarrow[n \to +\infty]{\mathcal{L}_{\text{stably}}} \mathcal{N}(0, \Sigma^*) \quad \text{on } \{\theta_n \to \theta^*\},$$

where

$$\Sigma^* = \lim_n \frac{1}{n} \int_0^n e^{-u\left(J_h(\theta^*)^t - \frac{I_d}{2}\right)} \Gamma e^{-u\left(J_h(\theta^*) - \frac{I_d}{2}\right)} du.$$

(c) *If $\Lambda \in \left(0, \frac{1}{2}\right)$, $v_n = n^{2\Lambda - 1 + \eta}$, $n \geq 1$, for some $\eta > 0$, and h is ϵ-differentiable at θ^*, for some $\epsilon > 0$, then $n^\Lambda (\theta_n - \theta^*)$ is a.s. bounded on $\{\theta_n \to \theta^*\}$ as $n \to +\infty$.*

If, moreover, $\Lambda = \lambda_{\min}$ ($\lambda_{\min}$ is real), then $n^\Lambda (\theta_n - \theta^*)$ a.s. converges as $n \to +\infty$ toward a finite random variable.

The *stable convergence in distribution*, denoted by $\mathcal{L}_{\text{stably}}$ in items (a) and (b), means that there exists an extension $(\Omega', \mathcal{A}', \mathbb{P}')$ of $(\Omega, \mathcal{A}, \mathbb{P})$ and $Z : (\Omega', \mathcal{A}', \mathbb{P}') \to \mathbb{R}^d$ with $\mathcal{N}(0, I_d)$ distribution such that, for every bounded continuous function f and every $A \in \mathcal{A}$,

$$\mathbb{E}\left[\mathbf{1}_{A^* \cap A} f\left(\sqrt{n}(\theta_n - \theta^*)\right)\right] \xrightarrow[n \to +\infty]{} \mathbb{E}\left[\mathbf{1}_{A^* \cap A} f\left(\sqrt{\Sigma^*} Z\right)\right],$$

where $A^* = \{\theta_n \xrightarrow{\text{a.s.}} \theta^*\}$.

Averaging principle (Ruppert–Polyak).

In practice, the convergence of a stochastic algorithm ruled by a CLT is chaotic, even in the final convergence phase, except if we optimize the step to have the optimal asymptotic variance. But this optimal step depends on the spectrum of the differential matrix at the target of the algorithm that we are trying to compute.

The original idea of the averaging principle was to "smooth" the behavior of a converging stochastic algorithm by considering the arithmetic mean of the past values up to the nth iteration rather than the nth computed value of the algorithm itself. Surprisingly, if this averaging procedure is combined with a scaling of the step sequence (that will decrease slowly), we obtain freely the best possible rate of convergence!

To be precise: Let $(\gamma_n)_{n \geq 1}$ be a step sequence satisfying

$$\gamma_n \sim \left(\frac{c}{b+n}\right)^\vartheta, \qquad \vartheta \in (1/2, 1), \quad c > 0, \quad b \geq 0.$$

Then we implement the standard recursive procedure

$$\text{for all } n \geq 1, \quad \theta_{n+1} = \theta_n - \gamma_{n+1} H(\theta_n, Y_{n+1})$$

and we set

$$\text{for all } n \geq 1, \quad \bar{\theta}_n := \frac{\theta_0 + \cdots + \theta_{n-1}}{n}.$$

Under natural assumptions (see Pelletier, 1998), we can prove that

$$\bar{\theta}_n \xrightarrow{\text{a.s.}} \theta^*$$

where θ^* is the target of the algorithm and additionally

$$\sqrt{n}(\bar{\theta}_n - \theta^*) \xrightarrow{\mathcal{L}} \mathcal{N}(0; \Sigma_{\min}^*)$$

where $\Sigma^*_{\min}$ is the lowest covariance matrix: thus, if $d = 1$, then

$$\Sigma^*_{\min} = \frac{\text{Var}(H(\theta^*, Y))}{h'(\theta^*)^2}.$$

Theorem 12.11 (Ruppert & Polyak; see e.g. Duflo, 1996)   *Define the recursive procedure*

$$\theta_{n+1} = \theta_n - \gamma_{n+1}(h(\theta_n) + \Delta M_{n+1})$$

where h is a Borel function, continuous at its unique zero θ^*, satisfying

$$\text{for all } \theta \in \mathbb{R}^d, \quad h(\theta) = J_h(\theta^*)(\theta - \theta^*) + O(|\theta - \theta^*|^2)$$

where all the eigenvalues of $J_h(\theta^*)$ have positive real parts. Furthermore, assume that, for some constant $C > 0$,

$$\mathbb{E}[\Delta M_{n+1} \mid \mathcal{F}_n] \mathbf{1}_{\{|\theta_n - \theta^*| \leq C\}} = 0 \quad \text{a.s.}$$

and there exists an exponent $\delta > 0$ such that

$$\mathbb{E}\left[\Delta M_{n+1}(\Delta M_{n+1})^t \mid \mathcal{F}_n\right] \xrightarrow{\text{a.s.}} \mathbf{1}_{\{|\theta_n - \theta^*| \leq C\}} \Gamma > 0 \in \mathcal{S}(d,)$$

$$\sup_n \mathbb{E}\left[|\Delta M_{n+1}|^{2+\delta} \mid \mathcal{F}_n\right] \mathbf{1}_{\{|\theta_n - \theta^*| \leq C\}} < +\infty.$$

Then, if $\gamma_n = \frac{c}{n^\alpha}$, $n \geq 1$, $1/2 < \alpha < 1$, the sequence of arithmetic means

$$\bar{\theta}_n = \frac{\theta_0 + \cdots + \theta_{n-1}}{n}$$

satisfies on $\{\theta_n \xrightarrow{\text{a.s.}} \theta^*\}$, the CLT with the optimal asymptotic variance

$$\sqrt{n}(\bar{\theta}_n - \theta^*) \xrightarrow{\mathcal{L}} \mathcal{N}(0; J_h(\theta^*)^{-1} \Gamma J_h(\theta^*)) \qquad \text{on } \{\theta_n \xrightarrow{\text{a.s.}} \theta^*\}.$$

Remark 12.12   The arithmetic (or empirical) mean satisfies the following recursive procedure

$$\text{for all } n \geq 0, \quad \bar{\theta}_{n+1} = \bar{\theta}_n - \frac{1}{n+1}(\bar{\theta}_n - \theta_n), \qquad \bar{\theta}_0 = 0,$$

which can be used in practice to update recursively the values of the algorithm with the averaging principle.

This last outcome ends this first part about the main results on stochastic approximation. We focused on i.i.d. innovation sequences $(Y_n)_{n \geq 1}$, but convergence results for more general dynamics exist: for Markov chains (see Benveniste et al., 1990), for mixing processes (see Dedecker et al., 2007), for low-discrepancy sequences (see Lapeyre et al., 1990) and for averaging processes (see Laruelle and Pagès, 2012).

When you deal with multiple equilibrium points, some are local attractors (or "targets"), but others are "parasitic" ones (repeller, saddle points, etc.) and a.s. cannot appear as an asymptotic value of the recursive procedure. These terms should be understood in the sense of the ODE attached to the mean field function h, namely $\dot{\theta} = -h(\theta)$. To eliminate the equilibrium points which are parasitic, one can take advantage of a third aspect of SA theory: these are results that ensure the a.s. non-convergence of a stochastic algorithm towards a noisy equilibrium point (called "traps" in Brandière and Duflo, 1996, see also Pemantle, 1990, Lazarev, 1992), but also, in some situations, the a.s. non-convergence towards noiseless equilibrium points (see Lamberton et al., 2004 Lamberton and Pagès, 2008).

12.3 Applications to optimal execution

The trading landscape has seen a large number of evolutions following two regulations: Reg NMS in the US and MiFID in Europe. One of their consequences is the ability to exchange the same financial instrument on different trading venues (called Electronic Communication Network (ECN) in the US and Multilateral Trading Facilities (MTF) in Europe). Each trading venue differentiates from the others at any time because of the fees or rebate it demands to trade and the behavior of the liquidity it offers. From a regulatory viewpoint, these changes have been driven by a run for quality for the price formation process.

Moreover, with the growth of electronic trading in recent years, most of the transactions in the markets occur in Limit Order Books (LOB) with two kinds of orders: passive orders (that is, limit or patient orders) that will not give rise to a trade but will stay in the LOB, and aggressive orders (that is, market or impatient orders) that will generate a trade. When a trader has to buy or sell a large number of shares, it's not optimal for him to just send his large order at once because it would consume all of the available liquidity in the LOB, impacting the price to his disadvantage; instead, he has to schedule his trading rate to strike a balance between market risk and market impact. Several theoretical frameworks have been proposed for optimal scheduling of large orders see Almgren and Chriss (2000); Bouchard et al. (2011); Predoiu et al. (2011); Alfonsi et al. (2010). Once the optimal trading rate is known, the trader has to send smaller orders in the LOB by alternating limit orders and market orders. For patient orders, he has to find the best posting price: not too far in the LOB to ensure an almost full execution, and not too close to the best limits to have a better price.

In this section, we first introduce a stochastic algorithm to split a volume across special liquidity pools, known as dark pools. Then, we consider the problem of optimal posting price of a single limit order on a lit pool.

12.3.1 *Optimal split of an order across liquidity pools*

We will focus here on the splitting order problem in the case of (competing) *dark pools*. The execution policy of a dark pool differs from a primary market: thus a dark pool proposes bid/ask prices with no guarantee of executed quantity at the occasion of an over-the-counter transaction. Usually its bid price is lower than the bid price offered on the regular market (and the ask price is higher). Let us temporarily focus on a buying order sent to several dark pools. One can model the impact of the existence of N dark pools ($N \geq 2$) on a given transaction as follows: let $V > 0$ be the random volume to be executed and let $\theta_i \in (0,1)$ be the *discount factor* proposed by the dark pool $i \in \{1, \ldots, N\}$. We will make the assumption that this discount factor is deterministic or at least known prior to the execution. Let r_i denote the percentage of V sent to the dark pool i for execution and let $D_i \geq 0$ be the quantity of securities that can be delivered (or made available) by the dark pool i at price $\theta_i S$ where S denotes the bid price on the primary market (this is clearly an approximation since on the primary market, the order will be decomposed into slices executed at higher and higher prices following the order book). The rest of the order has to be executed on the primary market, at price S. Then the cost C of the executed order is given by

$$C = S \sum_{i=1}^{N} \theta_i \min(r_i V, D_i) + S \left(V - \sum_{i=1}^{N} \min(r_i V, D_i) \right)$$

$$= S \left(V - \sum_{i=1}^{N} \rho_i \min(r_i V, D_i) \right)$$

where $\rho_i = 1 - \theta_i > 0$, $i = 1, \ldots, N$. At this stage, one may wish to minimize the mean execution cost C, *given the price S*. This amounts to solving the following (conditional) maximization problem

$$\max \left\{ \sum_{i=1}^{N} \rho_i \, \mathbb{E}\left[\min(r_i V, D_i) \mid S \right], \; r \in \mathcal{P}_N \right\}, \qquad (12.19)$$

where $\mathcal{P}_N := \{ r = (r_i)_{1 \leq i \leq n} \in \mathbb{R}_+^N \mid \sum_{i=1}^{N} r_i = 1 \}$. However, since none of the agents are insiders, they do not know the price S when the agent decides to buy the security and when the dark pools answer to their request. This means that one may assume that $(V, D_1, \ldots, D_n)$ and S are independent so that the maximization problem finally reads

$$\max \left\{ \sum_{i=1}^{N} \rho_i \mathbb{E}\left[\min(r_i V, D_i) \right], \; r \in \mathcal{P}_N \right\} \qquad (12.20)$$

where we assume that all the random variables $\min(V, D_1), \ldots, \min(V, D_N)$ are integrable (otherwise the problem is meaningless).

To use the results on stochastic approximation for i.i.d. innovation process, we assume that the sequence $(V^n, D_1^n, \ldots, D_N^n)_{n \geq 1}$ is i.i.d. with distribution $\nu = \mathcal{L}(V, D_1, \ldots, D_N)$, $V \in L^2(\mathbb{P})$ and the (right continuous) distribution function of D_i/V is continuous on $\mathbb{R}_+$, for every $i \in \{1, \ldots, N\}$.

Optimal allocation: a stochastic Lagrangian algorithm.
To each dark pool $i \in \{1, \ldots, N\}$ is attached a (bounded concave) mean execution function $\varphi_i(r_i) = \rho_i \mathbb{E}\left[\min(r_i V, D_i) \right]$. Then for every $r = (r_1, \ldots, r_N) \in \mathcal{P}_N$,

$$\Phi(r_1, \ldots, r_N) := \sum_{i=1}^{N} \varphi_i(r_i). \qquad (12.21)$$

As said, we aim at solving the following maximization problem $\max_{r \in \mathcal{P}_N} \Phi(r)$. Let us have a look at a Lagrangian approach: $r^* \in \operatorname{argmax}_{\mathcal{P}_N} \Phi$ if and only if

$\varphi_i'(r_i^*)$ is constant when i runs over $\{1, \ldots, N\}$ or equivalently if

$$\text{for all } i \in \{1, \ldots, N\}, \qquad \varphi_i'(r_i^*) = \frac{1}{N} \sum_{j=1}^{N} \varphi_i'(r_j^*). \tag{12.22}$$

Design of the stochastic algorithm.

Using the fact that $\varphi_i'(r_i) = \rho_i \mathbb{E}\left[\mathbf{1}_{r_i V \leq D_i} V\right]$ for every $i \in \{1, \ldots, N\}$, it follows (under additional assumptions which are not written here, see Laruelle et al., 2011 for more details) that $r^* \in \operatorname{argmax}_{\mathcal{P}_N} \Phi$ if and only if

$$\text{for all } i \in \{1, \ldots, N\}, \ \mathbb{E}\left[V\left(\rho_i \mathbf{1}_{\{r_i^* V \leq D_i\}} - \frac{1}{N}\sum_{j=1}^{N} \rho_j \mathbf{1}_{\{r_j^* V \leq D_j\}}\right)\right] = 0.$$

Consequently, this leads to devising the following zero-search procedure:

$$r^n = \Pi_{\mathcal{P}_N}\left(r^{n-1} + \gamma_n H(r^{n-1}, V^n, D_1^n, \ldots, D_N^n)\right), \ n \geq 1, \quad r^0 \in \mathcal{P}_N, \tag{12.23}$$

where

$$H_i(r, V, D_1, \ldots, D_N) = V\left(\rho_i \mathbf{1}_{\{r_i V \leq D_i\} \cap \{r_i \in [0,1]\}} - \frac{1}{N}\sum_{j=1}^{N} \rho_j \mathbf{1}_{\{r_j V \leq D_j\} \cap \{r_j \in [0,1]\}}\right).$$

Then, using the results on convergence of constrained stochastic algorithm, it follows, if the step sequence satisfies

$$\sum_{n \geq 1} \gamma_n = +\infty \quad \text{and} \quad \sum_{n \geq 1} \gamma_n^2 < +\infty,$$

and $(V^n, D_1^n, \ldots, D_N^n)_{n \geq 1}$ is an i.d.d. sequence, that

$$r^n \longrightarrow r^* \quad \text{a.s.}$$

Furthermore, assume that $\{h = 0\} = \operatorname{argmax}_{\mathcal{P}_N} \Phi = r^* \in \operatorname{int}(\mathcal{P}_N)$. Assume that $r^n \to r^*$ $\mathbb{P}$-a.s. as $n \to \infty$, $V \in L^{2+\delta}(\mathbb{P})$, $\delta > 0$, and

$$\gamma_n = \frac{c}{n}, \ n \geq 1 \text{ with } c > \frac{1}{2\Re e(\lambda_{\min})}$$

where $\lambda_{\min}$ denotes the eigenvalue of $A^\infty := -Dh(r^*)_{|1^\perp}$ with the lowest real part, where $\mathbf{1}^\perp := \{u \in \mathbb{R}^N \mid \sum_{i=1}^{N} u_i = 0\}$. Then, under some additional assumptions, we can prove that the stochastic recursive procedure satisfies a CLT, namely

$$\sqrt{n}(r^n - r^*) \xrightarrow{\mathcal{L}} \mathcal{N}(0; \sqrt{c}\, \Sigma^\infty)$$

where the asymptotic covariance matrix Σ^∞ is given by

$$\Sigma^\infty = \int_0^\infty e^{u(A^\infty - \frac{\mathrm{Id}}{2c})} C^\infty e^{u(A^\infty - \frac{\mathrm{Id}}{2c})^t} \, du$$

where

$$C^\infty = \mathbb{E}\left[H(r^*, V, D_1, \ldots, D_N)H(r^*, V, D_1, \ldots, D_N)^t\right]_{|1^\perp}$$

and $(A^\infty - \frac{\mathrm{Id}}{2c})^t$ stands for the *transpose operator* of $A^\infty - \frac{\mathrm{Id}}{2c} \in \mathcal{L}(\mathbf{1}^\perp)$.

Remark 12.13 The above claim is consistent since $u \mapsto H(r, v, \delta_1, \ldots, \delta_N)^t u$ preserves $\mathbf{1}^\perp$.

Numerical experiments on pseudo-real data.

Two natural situations of interest can be considered *a priori*: *abundance*, namely when $\mathbb{E}V \leq \sum_{i=1}^N \mathbb{E}D_i$; and *shortage*, namely when $\mathbb{E}V > \sum_{i=1}^N \mathbb{E}D_i$. Then we define a reference strategy, called an "oracle strategy", devised by an insider who knows all the values V^n and D_i^n before making his/her optimal execution requests to the dark pools. It can be described as follows: assume for simplicity that the rebates are ordered; i.e., $\rho_1 > \rho_2 > \cdots > \rho_N$. Then, it is clear that the "oracle" strategy yields the following cost reduction (CR) of the execution at time $n \geq 1$,

$$
CR^{\text{oracle}} := \begin{cases} \displaystyle\sum_{i=1}^{i_0-1} \rho_i D_i^n + \rho_{i_0}\left(V^n - \sum_{i=1}^{i_0-1} D_i^n\right), & \text{if } \displaystyle\sum_{i=1}^{i_0-1} D_i^n \leq V^n < \sum_{i=1}^{i_0} D_i^n \\ \displaystyle\sum_{i=1}^{N} \rho_i D_i^n, & \text{if } \displaystyle\sum_{i=1}^{N} D_i^n < V^n. \end{cases}
$$

Now, we introduce indexes to measure the performances of our recursive allocation procedure.

- *Relative cost reduction (with respect to the regular market):* they are defined as the ratios between the cost reduction of the execution using dark pools and the cost resulting from an execution on the regular market for the three algorithms, i.e., for every $n \geq 1$, CR^{oracle}/V^n for the oracle and $CR^{\text{algo}}/V^n = \sum_{i=1}^N \rho_i \min\left(r_i^n V^n, D_i^n\right)/V^n$ for the stochastic algorithm.
- *Performances (with respect to the oracle):* the ratio between the relative cost reductions of our allocation algorithms and that of the oracle, i.e. for every $n \geq 1$, $CR^{\text{opti}}/CR^{\text{oracle}}$ and $CR^{\text{reinf}}/CR^{\text{oracle}}$.

Firstly we explain how the data have been created. We have considered for V the traded volumes of a very liquid security – namely the asset BNP – during an 11-day period. Then we selected the N most correlated assets (in terms of traded volumes) with the original asset. These assets are denoted S_i, $i = 1, \ldots, N$ and we considered their traded volumes during the same 11-day period. Finally, the available volumes of each dark pool i have been modeled as follows using the mixing function

$$
\text{for all } 1 \leq i \leq N, \quad D_i := \beta_i \left((1 - \alpha_i)V + \alpha_i S_i \frac{\mathbb{E}V}{\mathbb{E}S_i}\right)
$$

where $\alpha_i \in (0, 1)$, $i = 1, \ldots, N$ are the recombining coefficients, β_i, $i = 1, \ldots, N$ are some scaling factors and $\mathbb{E}V$ and $\mathbb{E}S_i$ stand for the empirical mean of the data sets of V and S_i. The shortage situation corresponds to $\sum_{i=1}^N \beta_i < 1$ since it implies $\mathbb{E}\left[\sum_{i=1}^N D_i\right] < \mathbb{E}V$.

The simulations presented here have been made with $N = 4$. The data used here covers 11 days and it is clear that unlike the simulated data, these pseudo-real data are not stationary: in particular they are subject to daily changes of trend and volatility (at least). To highlight these resulting changes in the response of the algorithms, we have specified the days by drawing vertical doted lines. The dark pool pseudo-data parameters are set to

$$\beta = (0.1, 0.2, 0.3, 0.2)', \quad \alpha = (0.4, 0.6, 0.8, 0.2)', \quad \rho = (0.01, 0.02, 0.04, 0.06)'.$$

Firstly, we benchmarked both algorithms on the whole data set (11 days) as though it were stationary without any resetting (step, starting allocation, etc.). In particular, the running means of the satisfactions ratios are computed from the very beginning for the first 1500 data, and by a moving average on a window of 1500 data. As a second step, we proceed on a daily basis by resetting the parameters of both algorithms (initial allocation for both and the step parameter γ_n of the optimization procedure) at the beginning of every day.

Long-term optimization.

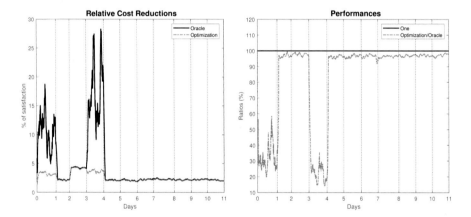

Figure 12.1 *Long term optimization*: Case $N = 4$, $\sum_{i=1}^{N} \beta_i < 1$, $0.2 \le \alpha_i \le 0.8$ and $r_i^0 = 1/N$, $1 \le i \le N$.

This test confirms that the statistical features of the data are strongly varying from one day to another (see Figure 12.1), so there is no hope that our procedures converge in standard sense on a long term period. Consequently, it is necessary to switch to a short term monitoring by resetting the parameters of the algorithms on a daily basis as detailed below.

Daily resetting of the procedure.

We consider now that we reset on a daily basis all the parameters of the algorithm, namely we reset the step γ_n at the beginning of each day and the satisfaction parameters and we keep the allocation coefficients of the preceding day. We obtain the results in Figure 12.2. We observe (see Figure 12.2) that the optimization algorithm reaches more 95 % of the performance of the oracle. Furthermore, although not represented here, the allocation coefficients look more stable.

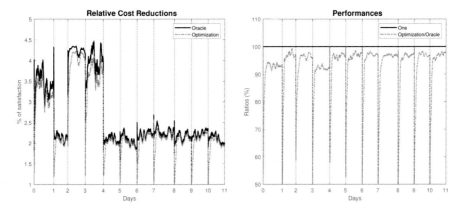

Figure 12.2 *Daily resetting of the algorithms parameters*: Case $N = 4$, $\sum_{i=1}^{N} \beta_i < 1$, $0.2 \leq \alpha_i \leq 0.8$ and $r_i^0 = 1/N$ $1 \leq i \leq N$.

12.3.2 Optimal posting price of limit orders

We focus our work on the problem of optimal trading with limit orders on one security without needing to model the limit order book dynamics. To be more precise, we will focus on buy orders rather than sell orders in all that follows. We only model the execution flow which reaches the price where the limit order is posted with a general price dynamics $(S_t)_{t \in [0,T]}$ since we intend to use real data. However there will be two frameworks for the price dynamics: either $(S_t)_{t \in [0,T]}$ is a process bounded by a constant L (which is obviously an unusual assumption but not unrealistic on a short time scale) or $(S_t)_{t \in [0,T]}$ is ruled by a Brownian diffusion model.

We consider on a short period T, say a dozen seconds, a Poisson process modeling the execution of posted passive *buy* orders on the market, namely

$$\left(N_t^{(\delta)}\right)_{0 \leq t \leq T} \quad \text{with intensity} \quad \lambda(S_t - (S_0 - \delta)) \text{ at time } t \in [0,T], \quad (12.24)$$

where $0 \leq \delta \leq \delta_{\max}$ (here $\delta_{\max} \in (0, S_0)$ is the depth of the order book), $\lambda : [-S_0, +\infty) \to \mathbb{R}_+$ is a non-negative non-increasing function and $(S_t)_{t \geq 0}$ is a stochastic process modeling the dynamics of the "fair price" of a security stock (from an economic point of view). In practice one may regard S_t as representing the best opposite price at time t. It will be convenient to introduce the cumulated intensity defined by

$$\Lambda_t(\delta, S) := \int_0^t \lambda(S_s - (S_0 - \delta))ds. \quad (12.25)$$

We assume that the function λ is defined on $[-S_0, +\infty)$ as a finite non-increasing convex function. Its specification will rely on parametric or non-parametric statistical estimation based on previously obtained transactions. At time $t = 0$, buy orders are posted in the limit order book at price $S_0 - \delta$. Between t and $t + \Delta t$, the probability for such an order to be executed is $\lambda(S_t - (S_0 - \delta))\Delta t$ where $S_t - (S_0 - \delta)$ is the distance to the current fair price of our posted order at time t. The further the order is at time t, the lower is the probability for this order to be executed since λ is decreasing on $[-S_0, +\infty)$. Empirical tests strongly confirm this kind of relationship with a convex function λ (even close to an exponential shape; see Avellaneda and Stoikov, 2008, Guéant et al., 2013, and Bayraktar and Ludkovski, 2011). We then choose the following parametrization $\lambda(x) = Ae^{-ax}$, $A > 0$, $a > 0$.

Over the period $[0, T]$, we aim to execute a portfolio of size $Q_T \in \mathbb{N}$ invested in the asset S. The execution cost for a distance δ is $\mathbb{E}\left[(S_0 - \delta)\left(Q_T \wedge N_T^{(\delta)}\right)\right]$. We add to this execution cost a penalty function depending on the remaining quantity to be executed, namely we want to have Q_T assets in the portfolio at the end of the period T, so we buy the remaining quantity $\left(Q_T - N_T^{(\delta)}\right)_+$ at price S_T.

At this stage, we introduce a *market impact penalty function* $\Phi \colon \mathbb{R} \mapsto \mathbb{R}_+$, non-decreasing and convex, with $\Phi(0) = 0$, to model the additional cost of the execution of the remaining quantity (including the market impact). Then the resulting cost of execution on a period $[0, T]$ reads

$$C(\delta) := \mathbb{E}\left[(S_0 - \delta)\left(Q_T \wedge N_T^{(\delta)}\right) + \kappa\, S_T\, \Phi\left(\left(Q_T - N_T^{(\delta)}\right)_+\right)\right], \qquad (12.26)$$

where $\kappa > 0$ is a free tuning parameter (the true cost is with $\kappa = 1$, but we could overcost the market order due to the bad estimation of the market impact of this order, or conversely). When $\Phi(Q) = Q$, we assume we buy the remaining quantity at the end price S_T. Introducing a market impact penalty function $\Phi(x) = (1 + \eta(x))x$, where $\eta \geq 0$, $\eta \not\equiv 0$, models the market impact induced by the execution of $\left(Q_T - N_T^{(\delta)}\right)_+$ at time T while neglecting the market impact of the execution process via limit orders over $[0, T)$. Our aim is then to minimize this cost by choosing the distance at which to post; namely to solve the following optimization problem:

$$\min_{0 \leq \delta \leq \delta_{\max}} C(\delta). \qquad (12.27)$$

Our strategy for solving (12.27) numerically using a large enough dataset is to take advantage of the representation of C and its first two derivatives as expectations to devise a recursive stochastic algorithm, specifically a stochastic gradient procedure, to find the minimum of the (penalized) cost function (see below). To ensure the well-posedness of our optimization problem and the uniqueness of its solution, we will show that, under natural assumptions on the quantity Q_T to be executed and on the parameter κ, the function C is twice differentiable and *strictly convex* on $[0, \delta_{\max}]$ with $C'(0) < 0$. Consequently,

$$\mathrm{argmin}_{\delta \in [0, \delta_{\max}]} C(\delta) = \{\delta^*\}, \quad \delta^* \in (0, \delta_{\max}]$$

and

$$\delta^* = \delta_{\max} \quad \text{if and only if} \quad C \text{ is non-increasing on } [0, \delta_{\max}].$$

Criteria involving κ and based on both the risky asset S and the trading process especially the execution intensity λ, are established under a co-monotony principle satisfied by the price process $(S_t)_{t \in [0, T]}$ (see Donati et al., 2013 and Laruelle et al., 2013 for more details) and can be stated as follows. We define the two following constants:

$$\bar{a}_0 := \mathbb{P}\text{-esssup}\left(-\frac{\frac{\partial}{\partial\delta}\Lambda_T(0,S)}{\Lambda_T(0,S)}\right) \geq \underline{a}_0 := \mathbb{P}\text{-essinf}\left(-\frac{\frac{\partial}{\partial\delta}\Lambda_T(0,S)}{\Lambda_T(0,S)}\right) \geq \underline{a}_1 > 0.$$

Then $C'(0) < 0$ whenever $Q_T \geq 2T\lambda(-S_0)$ and

$$\kappa \leq \frac{1 + \underline{a}_0 S_0}{\bar{a}_0 \mathbb{E}\left[S_T\right]\left(\Phi(Q_T) - \Phi(Q_T - 1)\right)}.$$

Assume that $s^* := \|\sup_{t\in[0,T]} S_t\|_{L^\infty} < +\infty$. If $Q_T \geq 2T\lambda(-S_0)$ and

$$\kappa \leq \frac{1 + a(S_0 - \delta_{\max})}{a\,s^*} \text{ if } \Phi \neq \text{id}, \quad \kappa \leq \frac{1 + a(S_0 - \delta_{\max})}{a\,s^*(\Phi(Q_T) - \Phi(Q_T - 1))} \text{ if } \Phi = \text{id},$$

then $C''(\delta) \geq 0$, $[0,\delta_{\max}]$, so that C is convex on $[0,\delta_{\max}]$. The same inequality holds for the Euler scheme of $(S_t)_{t\in[0,T]}$ with step $\frac{T}{m}$, for $m \geq m_{b,\sigma}$, or for any $\mathbb{R}^{m+1}$-time discretization sequence $(S_{t_i})_{0\leq i\leq m}$ that satisfies a co-monotony principle.

We specify representations as expectations of the function C and its derivatives C' and C''. In particular, we will exhibit a Borel functional

$$H: [0,\delta_{\max}] \times \mathbb{D}\left([0,T],\mathbb{R}\right) \longrightarrow \mathbb{R},$$

where $\mathbb{D}\left([0,T],\mathbb{R}\right) = \{f: [0,T] \to \mathbb{R}\text{càdlàg}\}$, such that

$$\text{for all } \delta \in [0,\delta_{\max}], \quad C'(\delta) = \mathbb{E}\left[H\left(\delta,(S_t)_{t\in[0,T]}\right)\right].$$

We introduce some notation for the convenience of readers. Let $(N^\mu)_{\mu>0}$ be a family of Poisson distributed random variables with parameter $\mu > 0$. We set

$$\mathbb{P}^{(\delta)}\left(N^\mu > Q_T\right) = \mathbb{P}\left(N^\mu > Q_T\right)_{|\mu=\Lambda_T(\delta,S)} \text{ and } \mathbb{E}^{(\delta)}\left[f(\mu)\right] = \mathbb{E}\left[f(\mu)\right]_{|\mu=\Lambda_T(\delta,S)}.$$

If $\Phi \neq \text{id}$, then

$$H(\delta,S) = -Q_T\mathbb{P}^{(\delta)}\left(N^\mu > Q_T\right)$$
$$+ \left(\frac{\partial}{\partial\delta}\Lambda_T(\delta,S)(S_0 - \delta) - \Lambda_T(\delta,S)\right)\mathbb{P}^{(\delta)}\left(N^\mu \leq Q_T - 1\right)$$
$$- \kappa S_T\frac{\partial}{\partial\delta}\Lambda_T(\delta,S)\varphi^{(\delta)}(\mu), \tag{12.28}$$

where $\varphi^{(\delta)}(\mu) = \mathbb{E}^{(\delta)}\left[\left(\Phi(Q_T - N^\mu) - \Phi(Q_T - N^\mu - 1)\right)\mathbf{1}_{\{N^\mu \leq Q_T-1\}}\right]$ and

$$C''(\delta)$$
$$= \mathbb{E}\Bigg[\left((S_0 - \delta)\frac{\partial^2}{\partial\delta^2}\Lambda_T(\delta,S) - 2\frac{\partial}{\partial\delta}\Lambda_T(\delta,S)\right)\mathbb{P}^{(\delta)}\left(N^\mu \leq Q_T - 1\right)$$
$$- \kappa S_T\frac{\partial^2}{\partial\delta^2}\Lambda_T(\delta,S)\varphi^{(\delta)}(\mu) - (S_0 - \delta)\left(\frac{\partial}{\partial\delta}\Lambda_T(\delta,S)\right)^2\mathbb{P}^{(\delta)}\left(N^\mu = Q_T - 1\right)$$
$$+ \kappa S_T\left(\frac{\partial}{\partial\delta}\Lambda_T(\delta,S)\right)^2\psi^{(\delta)}(\mu)\Bigg],$$

where

$$\psi^{(\delta)}(\mu) = \mathbb{E}^{(\delta)}\left[\Phi\left((Q_T - N^\mu - 2)_+\right) - 2\Phi\left((Q_T - N^\mu - 1)_+\right) + \Phi\left((Q_T - N^\mu)_+\right)\right].$$

If $\Phi = \mathrm{id}$, then

$$H(\delta, S) = -Q_T \mathbb{P}^{(\delta)}\left(N^\mu > Q_T\right)$$
$$+ \left(\left(S_0 - \delta - \kappa S_T\right)\frac{\partial}{\partial\delta}\Lambda_T(\delta, S) - \Lambda_T(\delta, S)\right)\mathbb{P}^{(\delta)}\left(N^\mu \le Q_T - 1\right)$$

and

$$C''(\delta) = \mathbb{E}\left[\left(\left(S_0 - \delta - \kappa S_T\right)\frac{\partial^2}{\partial\delta^2}\Lambda_T(\delta, S) - 2\frac{\partial}{\partial\delta}\Lambda_T(\delta, S)\right)\mathbb{P}^{(\delta)}\left(N^\mu \le Q_T - 1\right)\right.$$
$$\left. -\left(S_0 - \delta - \kappa S_T\right)\left(\frac{\partial}{\partial\delta}\Lambda_T(\delta, S)\right)^2 \mathbb{P}^{(\delta)}\left(N^\mu = Q_T - 1\right)\right].$$

In particular, any quantity $H\left(\delta, (S_t)_{t\in[0,T]}\right)$ can be simulated, up to a natural time discretization, either from a true dataset (of past executed orders) or from the stepwise constant discretization scheme of a formerly calibrated diffusion process modeling $(S_t)_{t\in[0,T]}$ (see below). This will lead us to replace, for practical implementations, the continuous time process $(S_t)_{t\in[0,T]}$ over $[0,T]$ with either a discrete time sample; i.e., a finite-dimensional $\mathbb{R}^{m+1}$-valued random vector $(S_{t_i})_{0\le i\le m}$ (where $t_0 = 0$ and $t_m = T$) or with a time discretization scheme with step $\frac{T}{m}$ (typically the Euler scheme when $(S_t)_{t\in[0,T]}$ is a diffusion).

A theoretical stochastic learning procedure.

Based on the previous representation (12.28) of C', we can formally devise a recursive stochastic gradient descent a.s. converging toward δ^*. However to make it consistent, we need to introduce constraints so that it lives in $[0, \delta_{\max}]$ (see Kushner and Clark, 1978; Kushner and Yin, 2003). This amounts to using a variant *with projection on the "order book depth interval"* $[0, \delta_{\max}]$, namely

$$\delta_{n+1} = \Pi_{[0,\delta_{\max}]}\left(\delta_n - \gamma_{n+1}H\left(\delta_n, \left(S_t^{(n+1)}\right)_{t\in[0,T]}\right)\right), \quad n \ge 0, \ \delta_0 \in (0, \delta_{\max}), \quad (12.29)$$

where

- $\Pi_{[0,\delta_{\max}]} : x \mapsto 0 \vee x \wedge \delta_{\max}$ denotes the projection on (the nonempty closed convex) $[0, \delta_{\max}]$;
- $(\gamma_n)_{n\ge 1}$ is a positive step sequence satisfying (at least) the minimal *decreasing step assumption* $\sum_{n\ge 1}\gamma_n = +\infty$ and $\gamma_n \to 0$;
- the sequence $\left\{(S_t^{(n)})_{t\in[0,T]}, n \ge 1\right\}$, is the "innovation" sequence of the procedure: ideally it is either a sequence of simulable independent copies of $(S_t)_{t\in[0,T]}$ or a sequence sharing some ergodic (or averaging) properties with respect to the distribution of $(S_t)_{t\in[0,T]}$.

The case of independent copies can be understood as a framework where the dynamics of S is typically a Brownian diffusion solution to a stochastic differential equation, which has been calibrated beforehand on a dataset in order to be simulated on a computer. The case of ergodic copies corresponds to a dataset which is directly plugged into the procedure; i.e., $S_t^{(n)} = S_{t-n\Delta t}$, $t \in [0,T]$, $n \geq 1$, where $\Delta t > 0$ is a fixed shift parameter. To make this second approach consistent, we need to make the assumption that at least within a lapse of a few minutes, the dynamics of the asset S (starting in the past) is *stationary* and shares, for example, mixing properties.

The resulting implementable procedure.

In practice, the above procedure cannot be implemented since the full path $(S_t(\omega))_{t\in[0,T]}$ of a continuous process cannot be simulated nor can a functional $H(\delta, (S_t(\omega))_{t\in[0,T]})$ of such a path be computed. So we are led in practice to replace the "copies" $S^{(n)}$ by copies $\bar{S}^{(n)}$ of a time discretization $\bar{S}$ of step, say $\Delta t = \frac{T}{m}$, with $m \in \mathbb{N}^*$. The time discretizations are formally defined in continuous time as follows:

$$\bar{S}_t = \bar{S}_{t_i}, \ t \in [t_i, t_{i+1}), \ i = 0, \ldots, m-1 \quad \text{with } t_i = \frac{iT}{m}, \ i = 0, \ldots, m,$$

where

- $(\bar{S}_{t_i})_{0 \leq i \leq m} = (S_{t_i})_{0 \leq i \leq m}$ when $(S_{t_i})_{0 \leq i \leq m}$ can be simulated exactly at a reasonable cost;
- $(\bar{S}_{t_i})_{0 \leq i \leq m}$ is a time discretization scheme (at times t_i) of $(S_t)_{t\in[0,T]}$, typically an Euler scheme with step $\frac{T}{m}$.

Then, with an obvious abuse of notation for the function H, we can write the *implementable procedure* as follows:

$$\delta_{n+1} = \Pi_{[0,\delta_{\max}]}\left(\delta_n - \gamma_{n+1} H\left(\delta_n, (\bar{S}_{t_i}^{(n+1)})_{0 \leq i \leq m}\right)\right), \ n \geq 0, \ \delta_0 \in [0, \delta_{\max}], \quad (12.30)$$

where $(\bar{S}_{t_i}^{(n)})_{0 \leq i \leq m}$ are copies of $(\bar{S}_{t_i})_{0 \leq i \leq m}$ either independent or sharing "ergodic" properties, namely some averaging properties in the sense of Laruelle and Pagès (2012). In the first case, one will think about simulated data after a calibration process and in the second case to a direct implementation using a historical high frequency database of best opposite prices of the asset S (with, e.g., $\bar{S}_{t_i}^{(n)} = S_{t_i - n\frac{T}{m}}$).

Numerical experiments.

The self-adaptive nature of the recursive procedure (12.30) allows us to implement it on true market data, even if these data do not exactly fulfill our averaging assumptions. In the numerical example, the trader reassesses her order using the optimal posting procedure on true market data on which the parameters of the models have been previously fitted. As a market data set for the fair price process $(S_t)_{t\in[0,T]}$, we use the best bid prices of Accor SA (ACCP.PA) of 11/11/2010. We divide the day into periods of 15 trades with empty intersection which will denote the steps of the stochastic procedure. Let N_{cycles} be the number of these periods. We approximate the cumulated jump intensity of the Poisson process $\Lambda_{T^n}(\delta, \bar{S})$, where $T^n = \sum_{i=0}^{14} t_i$, by a Riemann sum, namely

$$\text{for all } n \in \{1, \ldots, N_{\text{cycles}}\}, \quad \Lambda_{T^n}(\delta, \bar{S}) = A \sum_{i=1}^{14} e^{-a(\bar{S}_{t_i}^{(n)} - \bar{S}_{t_0} + \delta)}(t_i - t_{i-1}).$$

The cumulated intensity function $\Lambda_T(\delta, \bar{S})$ is approximated by the estimator $\bar{\Lambda}(\delta, \bar{S})$ (plotted in Figure 12.3) defined by

$$\bar{\Lambda}(\delta, \bar{S}) = \frac{1}{N_{\text{cycles}}} \sum_{n=1}^{N_{\text{cycles}}} \Lambda_{T^n}(\delta, \bar{S}).$$

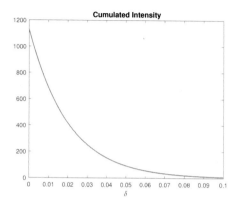

Figure 12.3 Fit of the exponential model on real data (Accor SA (ACCP.PA) 11/11/2010): $A = 1/50$, $a = 50$ and $N_{\text{cycles}} = 220$.

The penalty function has one of the following forms: $\Phi(x) = (1 + A'e^{a'x})x$. The cost function is specified with the following parameter values: $A = 1/50$, $a = 50$, $Q = 100$, $A' = 0.001$ and $a' = 0.0005$. The function C and its derivative are plotted in Figure 12.4.

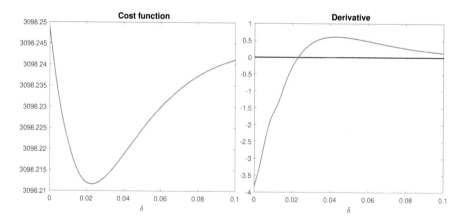

Figure 12.4 Computation of the cost function and its derivative for $\delta = \frac{i}{1000}$, $0 \leq i \leq 100$, with $\Phi \neq \text{id}$, $A = 1/50$, $a = 50$, $Q = 100$, $\kappa = 1$, $A' = 0.001$, $a' = 0.0005$ and $N_{\text{cycles}} = 220$.

Now we present the results of the stochastic recursive procedure: first with no smoothing (see Figure 12.5), then using the Ruppert and Polyak averaging principle (see Figure 12.6).

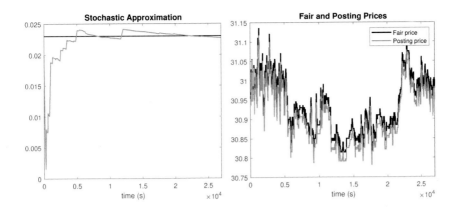

Figure 12.5 Convergence of the *crude algorithm* (left) and comparison between the best bid price and the posting price (right) with $\Phi \not\equiv$ id, $A = 1/50$, $a = 50$, $Q = 100$, $\kappa = 1$, $A' = 0.001$, $a' = 0.0005$, $N_{\text{cycles}} = 220$ and $\gamma_n = \frac{1}{550n}$, $1 \le n \le N_{\text{cycles}}$.

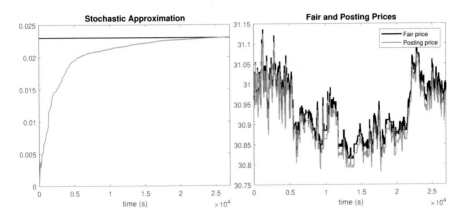

Figure 12.6 Convergence of the *averaged algorithm* (left) and comparison between the best bid price and the posting price (right) with $\Phi \not\equiv$ id, $A = 1/50$, $a = 50$, $Q = 100$, $\kappa = 1$, $A' = 0.001$, $a' = 0.0005$, $N_{\text{cycles}} = 220$ and $\gamma_n = \frac{1}{550n^{0.95}}$, $1 \le n \le N_{\text{cycles}}$.

Remark 12.14 A setting where a trader want to split a volume across different lit pools with limit orders and find simultaneously the optimal distribution and posting prices, with the same frameworks as above, can be found in Laruelle (2014). For a stochastic algorithm with several limit and market orders across different lit pools, see Cont and Kukanov (2017).

References

Alfonsi, A., Fruth, A., and Schied, A. 2010. Optimal execution strategies in limit order books with general shape functions. *Quantitative Finance*, **10**(2), 143–157.

Almgren, R. F., and Chriss, N. 2000. Optimal execution of portfolio transactions. *Journal of Risk*, **3**(2), 5–39.

Avellaneda, M., and Stoikov, S. 2008. High-frequency trading in a limit order book. *Quantitative Finance*, **8**(3), 217–224.

Bayraktar, E., and Ludkovski, M. 2011. Liquidation in Limit Order Books with Controlled Intensity. In *Proc. CoRR*.

Benaïm, M. 1999. Dynamics of stochastic approximation algorithms. Pages 1–68 of: *Séminaire de Probabilités, XXXIII.* Lecture Notes in Math., vol. 1709. Springer.

Benveniste, A., Métivier, M., and Priouret, P. 1990. *Adaptive Algorithms and Stochastic Approximations.* Springer-Verlag.

Bouchard, B., Dang, N.-M., and Lehalle, C.-A. 2011. Optimal control of trading algorithms: a general impulse control approach. *SIAM J. Financial Math.*, **2**, 404–438.

Brandière, O., and Duflo, M. 1996. Les algorithmes stochastiques contournent-ils les pièges? *Ann. Inst. H. Poincaré Probab. Statist.*, **32**(3), 395–427.

Cont, R., and Kukanov, A. 2017. Optimal order placement in limit order markets. *Quantitative Finance*, **17**(1), 21–39.

Dedecker, J., Doukhan, P., Lang, G., León R., J. R., Louhichi, S., and Prieur, C. 2007. *Weak Dependence: With Examples and Applications.* Lecture Notes in Statistics, vol. 190. Springer.

Donati, C., Lejay, A., Pagès, G., and Rouault, A. 2013. A functional co-monotony principle with an application to peacocks and barrier options. Pages 365–400 of: *Séminaire de Probabilités.*

Duflo, M. 1996. *Algorithmes Stochastiques.* Springer-Verlag.

Duflo, M. 1997. *Random Iterative Models.* BSpringer-Verlag.

Fort, J.-C., and Pagès, G. 1996. Convergence of stochastic algorithms: from the Kushner–Clark theorem to the Lyapounov functional method. *Adv. in Appl. Probab.*, **28**(4), 1072–1094.

Fort, J.-C., and Pagès, G. 2002. Decreasing step stochastic algorithms: a.s. behaviour of weighted empirical measures. *Monte Carlo Methods Appl.*, **8**(3), 237–270.

Guéant, O., Fernandez-Tapia, J., and Lehalle, C.-A. 2013. Dealing with the inventory risk. *Mathematics and Financial Economics*, **7**, 477–507.

Kiefer, J., and Wolfowitz, J. 1952. Stochastic estimation of the maximum of a regression function. *Ann. Math. Statistics*, **23**, 462–466.

Kushner, H. J., and Clark, D. S. 1978. *Stochastic Approximation Methods for Constrained and Unconstrained Systems.* Springer-Verlag.

Kushner, H. J., and Yin, G. G. 2003. *Stochastic Approximation and Recursive Algorithms and Applications.* Second edn. Springer-Verlag.

Lamberton, D., and Pagès, G. 2008. How fast is the bandit? *Stoch. Anal. Appl.*, **26**(3), 603–623.

Lamberton, D., Pagès, G., and Tarrès, P. 2004. When can the two-armed bandit algorithm be trusted? *Ann. Appl. Probab.*, **14**(3), 1424–1454.

Lapeyre, B., Pagès, G., and Sab, K. 1990. Sequences with low discrepancy—generalisation and application to Robbins–Monro. *Statistics*, **21**(2), 251–272.

Laruelle, S. 2014. Optimal split and posting price of limit orders across lit pools using stochastic approximation. In: *Research Initiative "Market Microstructure" of Kepler–Cheuvreux, under the supervision of Institut Europlace de Finance.*

Laruelle, S., and Pagès, G. 2012. Stochastic approximation with averaging innovation applied to finance. *Monte Carlo Methods Appl.*, **18**(1), 1–51.

Laruelle, S., Lehalle, C.-A., and Pagès, G. 2011. Optimal split of orders across liquidity pools: A stochastic algorithm approach. *SIAM J. Financial Math.*, **2**(1), 1042–1076.

Laruelle, S., Lehalle, C.-A., and Pagès, G. 2013. Optimal posting price of limit orders: learning by trading. *Mathematics and Financial Economics*, **7**, 359–403.

Lazarev, V. A. 1992. Convergence of stochastic approximation procedures in the case of regression equation with several roots. *Problemy Peredachi Informatsii*, **28**(1), 75–88.

Pagès, G. 2018. *Numerical Probability : An Introduction with Applications to Finance.* Springer.

Pelletier, M. 1998. Weak convergence rates for stochastic approximation with application to multiple targets and simulated annealing. *Ann. Appl. Probab.*, **8**(1), 10–44.

Pemantle, R. 1990. Nonconvergence to unstable points in urn models and stochastic approximations. *Ann. Probab.*, **18**(2), 698–712.

Predoiu, S., Shaikhet, G., and Shreve, S. 2011. Optimal execution of a general one-sided limit-order book. *SIAM J. Financial Math*, **2**, 183–212.

Robbins, H., and Monro, S. 1951. A stochastic approximation method. *Ann. Math. Statistics*, **22**, 400–407.

Zhang, L.-X. 2016. Central limit theorems of a recursive stochastic algorithm with applications to adaptive design. *Ann. Appl. Prob.*, **26**, 3630–3658.

13

Reinforcement Learning for Algorithmic Trading

Álvaro Cartea[a], Sebastian Jaimungal[b] and Leandro Sánchez-Betancourt[c]

Abstract

We employ reinforcement learning (RL) techniques to devise statistical arbitrage strategies in electronic markets. In particular, double deep Q network learning (DDQN) and a new variant of reinforced deep Markov models (RDMMs) are used to derive the optimal strategies for an agent who trades in a foreign exchange (FX) triplet. An FX triplet consists of three currency pairs where the exchange rate of one pair is redundant because, by no-arbitrage, it is determined by the exchange rates of the other two pairs. We use simulations of a co-integrated model of exchange rates to implement the strategies and show their financial performance.

13.1 Learning in financial markets

Continuous-time models and stochastic optimal control techniques are the cornerstone of a great deal of research in algorithmic trading. The design of these models seeks to balance tractability and usefulness. Simple models are ideal to derive trading strategies – in closed-form or as numerical solutions of a system of equations. However, in most cases, these models oversimplify the behaviour of markets to the detriment of the success of the strategies. On the other hand, realistic models are more complex and are often mathematically and computationally intractable with the classical tools of stochastic optimal control.

In this chapter, we show how to leverage on tools and methods of model-based reinforcement learning (RL) to employ more comprehensive market models, while retaining tractability and interpretability of trading strategies. Our framework is general and may be applied in all asset classes where instruments trade in an electronic market.

The prices of financial instruments are heteroskedastic, go through regimes of mean-reversion, momentum, and random walks; among other features. While there is a large body of literature on models of financial environments, it is

[a] Oxford University and Oxford–Man Institute of Quantitative Finance
[b] University of Toronto
[c] Oxford University
Published in *Machine Learning And Data Sciences For Financial Markets*, Agostino Capponi and Charles-Albert Lehalle © 2023 Cambridge University Press.

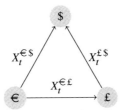

Figure 13.1 FX triplet. When the trader goes long the pair i/j, she purchases the base currency i (start of the arrow) and pays X^{ij} units of the quote currency j (end of the arrow). Similarly, if the trader sells the pair i/j, she sells the base currency i (start of the arrow) and receives X^{ij} units of the quote currency j (end of the arrow).

challenging to represent the key features with simple models. For example, in algorithmic trading, the complex dynamics of the limit order book (LOB), including the liquidity taking orders, are generally represented by simple continuous-time models. These models, however, cannot focus on all relevant features of the supply and demand of liquidity, and benefit from only a fraction of the data available to market participants; see Guéant et al. (2012), Cartea et al. (2014), Cartea et al. (2015), Guéant (2016), Cartea et al. (2017), Guilbaud and Pham (2013), Huang et al. (2015).

There are three building blocks associated with devising a trading strategy: develop a model of the environment, where states are potentially driven by latent factors; specify a performance criterion, which may include constraints such as risk controls; and, finally, employ techniques to solve a high-dimensional optimisation problem.

Each building block poses challenges and tradeoffs between implementability and accuracy. As a whole, due to the complexity of the trader's objectives, it is very challenging to pose and solve a problem with standard tools of stochastic control in realistic market environments – be it through the dynamic programming principle, stochastic Pontryagin maximum principle, or variational analysis techniques.

In this chapter, we show how to use RL-based techniques to solve for optimal trading strategies. To streamline the discussion, we focus on foreign exchange (FX) markets. Specifically, we use deep Q-learning and reinforced deep Markov models to trade in an FX triplet. An FX triplet consists of three currency pairs where the exchange rate of one pair is redundant because, by no-arbitrage in a frictionless market (i.e., no fees and zero spread), it is determined by the exchange rates of the other two pairs.

We consider an agent who trades in a triplet of currency pairs that combine three currencies: EUR, GBP, USD – see Figure 13.1. When the trader goes long the currency pair GBP/USD, which trades with exchange rate $X^{£\$}$, she buys one unit of currency £, known as the base currency in the pair, for which she pays $X_t^{£\$}$ units of currency \$, known as the quote currency in the pair. Similarly, the cost of one unit of base currency € in the pair EUR/USD is $X_t^{€\$}$ units of the quote currency \$; and the cost of one unit of the base currency € in the pair EUR/GBP is $X_t^{\text{\texteuro}\pounds}$ units of the quote currency £. The convention in the FX market is to quote the exchange rate between EUR and USD as EUR/USD, between GBP and USD as GBP/USD, and between EUR and GBP as EUR/GBP. In a frictionless market, one could think of the no-arbitrage exchange rates for the reverse direction of the pairs as $1/X_t^{\$\pounds}$, $1/X_t^{\$\text{\texteuro}}$, $1/X_t^{\pounds\text{\texteuro}}$, but these are not quoted in exchanges. A trader who wishes to exchange GBP into USD will sell the currency pair GBP/USD, i.e., consume liquidity from the bid side of the LOB that trades the pair EUR/USD with rate $X^{\pounds\$}$; and a trader who wishes to exchange USD into GBP will buy the currency pair GBP/USD, i.e., consume liquidity from the ask side of the LOB for the pair GBP/USD – see Cartea et al. (2020b).

Therefore, the key identity of a triplet in a frictionless market is the no-arbitrage relationship:

$$X_t^{\text{\texteuro}\$} \, X_t^{\$\pounds} \, X_t^{\pounds\text{\texteuro}} = 1, \tag{13.1}$$

so, as mentioned above, the exchange rate of one pair in the triplet is redundant, i.e., the rate of two pairs determines the rate of the third pair.

A classical strategy in FX is to take positions in the three pairs of a triplet when there are arbitrage opportunities – this is commonly known as a triangular arbitrage. The execution of this arbitrage is a mechanical application of a rule-of-thumb that consists of three simultaneous trades to arbitrage the misalignment in the exchange rates of a triplet. These unusual arbitrage opportunities rely on speed advantage and produce riskless profits.

Cartea et al. (2020a,b) study the optimal exchange of a foreign currency into a domestic currency when the currency pair is illiquid. The novelty in the strategies they develop is to trade in an FX triplet (which includes the illiquid pair) to compensate the illiquidity of the one currency pair with two other liquid pairs. The authors show that trading in the triplet considerably enhances the performance of the strategy compared with that of an agent who uses only the illiquid pair to exchange the foreign into the domestic currency. There are several other applications of RL for algorithmic trading in the recent literature, including Ning et al. (2018); Casgrain et al. (2019); Guéant and Manziuk (2019); Leal et al. (2020).

In this chapter, we focus on strategies that are based on the statistical properties of the exchange rates of a triplet.

13.2 Statistical arbitrage: trading an FX triplet

We develop a statistical arbitrage strategy for a trader who buys and sells the currency pairs of a triplet over a fixed time horizon T. The success of the strategy stems from taking advantage of co-movements and reversions in the exchange rates of the currency pairs. A statistical arbitrage is not a riskless strategy – the objective of the strategy is to make positive profits (on average), but there is no guarantee that the strategy will not deliver losses.

The trader takes positions in the three currencies of a triplet to maximise a performance criterion given an initial inventory. The trade actions are denoted by $\{a_t^{\pounds\$}, a_t^{\text{\texteuro}\$}, a_t^{\text{\texteuro}\pounds}\}_{t\in\mathbb{T}/\{T\}}$, $\mathbb{T} := \{0, 1, \ldots, T\}$. When a_t^{ij} is positive, the trader purchases a_t^{ij} units of the base currency i, for which she pays $X_t^{ij} \, a_t^{ij}$ units of the quote currency j. When a_t^{ij} is negative, the trader pays a_t^{ij} units of the base currency i and receives $X_t^{ij} a_t^{ij}$ units of the quote currency j. Finally, note that the units of the quantity a^{ij} are those of the currency i.

The inventories in the currencies USD, GBP, EUR, are denoted respectively by $\{q_t^\$, q_t^\pounds, q_t^\euro\}_{t \in \mathbb{T}}$. It follows that at time t, the inventory in each currency pair is

$$q_t^\$ = q_0^\$ - \sum_{j=0}^{t-1} X_j^{\euro\$} a_j^{\euro\$} - \sum_{j=0}^{t-1} X_j^{\pounds\$} a_j^{\pounds\$},$$

$$q_t^\pounds = q_0^\pounds + \sum_{j=0}^{t-1} a_j^{\pounds\$} - \sum_{j=0}^{t-1} X_j^{\euro\pounds} a_j^{\euro\pounds},$$

$$q_t^\euro = q_0^\euro + \sum_{j=0}^{t-1} a_j^{\euro\$} + \sum_{j=0}^{t-1} a_j^{\euro\pounds}.$$

The trader optimises the performance criterion

$$\mathbb{E}\left[q_T^\$ + q_T^\pounds \left(X_T^{\pounds\$} - \alpha^\pounds q_T^\pounds \right) + q_T^\euro \left(X_T^{\euro\$} - \alpha^\euro q_T^\euro \right) \right.$$
$$\left. - \phi^{\pounds\$} \sum_{t=0}^{T-1} \left(a_t^{\pounds\$} \right)^2 - \phi^{\euro\$} \sum_{t=0}^{T-1} \left(a_t^{\euro\$} \right)^2 - \phi^{\euro\pounds} \sum_{t=0}^{T-1} \left(a_t^{\euro\pounds} \right)^2 \right], \tag{13.2}$$

where α^k, for $k \in \{\pounds, \euro\}$, are non-negative liquidation penalty parameters, and ϕ^l, for $l \in \{\pounds\$, \euro\$, \euro\pounds\}$, are non-negative impact parameters that represent the costs, in USD, of walking the limit order book to complete the trade. The units of the terminal penalty parameters α^k are USD/k^2, and that of the cost parameters ϕ^l are USD/l^2.

On the right-hand side of the performance criterion in (13.2), the first line shows the USD value of terminal inventory in the three currency pairs. The first term is the inventory in USD, and the other two terms are the terminal inventories in GBP and EUR, which are exchanged into USD with one market order and the order is subject to a quadratic penalty in the quantity traded with the market order. For example, at the terminal date T, the inventory $q_T^\pounds$ is exchanged into USD at the rate $X_T^{\pounds\$}$ and the strategy pays the penalty $\alpha^\pounds (q_T^\pounds)^2$.

The second line on the right-hand side of (13.2) shows the total costs from walking-the-book that are paid by the market orders to buy and sell the currency pairs at every point in time until $T - 1$. For example, every time the trader sends a market order for $a^{\pounds\$}$ units in the currency pair GBP/USD, the costs (in USD) from walking the book are $\phi^{\pounds\$} (a^{\pounds\$})^2$.

In the trader's performance criterion, the value of the parameters α and ϕ dictate the severity of the penalties and costs paid by orders sent to the exchange. When the value of the terminal penalty parameter and that of the impact parameter is the same (i.e., $\alpha^\pounds = \phi^{\pounds\$}$ and $\alpha^\euro = \phi^{\euro\$}$), the costs of executing the orders during the trading window, including the orders executed at T, are the same. On the other hand, if the values of the terminal penalty parameters α^k are very high, relative to the values of ϕ^l, one interprets the terminal penalty paid by the strategy not as a cost of walking the book, but as a penalty that curbs the strategy to reach T with inventories in EUR and GBP very close to or at zero. Thus, if one objective is to reach time T with no inventory in EUR and GBP, one must implement the trading strategy with an arbitrarily high value of α^k to ensure that for all paths of the strategy the terminal inventory q_T^k is zero both in EUR and GBP to avoid the large terminal penalty.

Equivalently, the performance criterion in terms of changes in the exchange rates is

$$
\begin{aligned}
V^a(S_0) := \mathbb{E}\Bigg[&\sum_{t=0}^{T-1} \left\{ \left(q_t^\pounds + a_t^{\pounds\$} - X_t^{\in\pounds} a_t^{\in\pounds} \right) \left(X_{t+1}^{\pounds\$} - X_t^{\pounds\$} \right) \right. \\
&\left. + \left(q_t^\in + a_t^{\in\$} + a_t^{\in\pounds} \right) \left(X_{t+1}^{\in\$} - X_t^{\in\$} \right) \right\} \\
&- \phi^{\pounds\$} \sum_{t=0}^{T-1} \left(a_t^{\pounds\$} \right)^2 - \phi^{\in\$} \sum_{t=0}^{T-1} \left(a_t^{\in\$} \right)^2 - \phi^{\in\pounds} \sum_{t=0}^{T-1} \left(a_t^{\in\pounds} \right)^2 \\
&- \alpha^\pounds \left(q_T^\pounds \right)^2 - \alpha^\in \left(q_T^\in \right)^2 \Bigg].
\end{aligned}
\tag{13.3}
$$

The equivalence of the two expressions for the performance criterion follows from an inductive argument that we omit for brevity. Below, in the applications, we employ the representation in (13.3) because the step-by-step rewards of the trading strategy are explicit.

The trader's value function is

$$
V(S_0) = \sup_{a \in \mathcal{A}} V^a(S_0)
\tag{13.4}
$$

where $S_0 = (X_0^{\in\$}, X_0^{\pounds\$}, X_0^{\pounds\in}, q_0^\in, q_0^\pounds)$, $a = (a^{\in\$}, a^{\pounds\$}, a^{\in\pounds})$, and $\mathcal{A}$ is the set of admissible actions.

Below, we simulate the exchange rate dynamics and apply deep Q-learning and reinforced deep Markov models to learn the optimal trading strategies in the three currency pairs of the triplet.

13.2.1 Market model

In this chapter, we simulate market data to derive the trading strategies and gain insights into their behaviour. Specifically, we consider the following cointegrated model of FX rates for two of the currency pairs:

$$
X_{t+1}^{\in\$} = X_t^{\in\$} + \kappa_0 \left(\bar{x}^{\in\$} - X_t^{\in\$} \right) + \eta_0 \left(\bar{x}^{\pounds\$} - X_t^{\pounds\$} \right) + \epsilon_t^{\in\$},
\tag{13.5a}
$$

$$
X_{t+1}^{\pounds\$} = X_t^{\pounds\$} + \kappa_1 \left(\bar{x}^{\pounds\$} - X_t^{\pounds\$} \right) + \eta_1 \left(\bar{x}^{\in\$} - X_t^{\in\$} \right) + \epsilon_t^{\pounds\$},
\tag{13.5b}
$$

where $\epsilon_0^{\in\$}, \epsilon_1^{\in\$}, \ldots$ and $\epsilon_0^{\pounds\$}, \epsilon_1^{\pounds\$}, \ldots$ are i.i.d. mean-zero normal random variables. The exchange rate of the third currency pair follows from the no-arbitrage condition in (13.1).

In the subsequent numerical examples, we employ the following parameters:

$$\bar{x}^{\text{€\$}} = 1.1, \ \bar{x}^{\text{£\$}} = 1.3, \ \kappa_0 = \kappa_1 = 0.5, \ \eta_0 = -0.3, \ \eta_1 = -0.3,$$

and the standard deviation of the independent shocks $\epsilon^{\text{€\$}}$ and $\epsilon^{\text{£\$}}$ is 0.01. This represents typical weekly FX rate changes, and we consider each time step to be one week in the subsequent analysis.

Although this model does not incorporate the impact of the agent's actions into exchange rates, one can modify it so that the agent's buy and sell activity affects the rates. For example, one can use a propagator model of transient impact, which assumes that the (unobservable) fundamental exchange rates follow (13.5), while the observed FX rates in the LOB have an additional component due to transient impact. For example, for the currency pair $l \in \{\text{€\$}, \text{£\$}, \text{€£}\}$ assume that the innovation noise in (13.5) is $\epsilon_t^l \overset{\mathbb{P}}{\sim} \mathcal{N}(\eta^\top i_t^l, \sigma^2)$, where $i_{t+1}^{j,l} = \gamma^j \, i_t^{j,l} + \zeta_t^{j,l}$ with $\zeta_0^{j,l}, \zeta_1^{j,l}, \ldots$ i.i.d. mean-zero normal random variables. In the absence of trading, the transient impact factors decay to zero, and the observed FX rates converge to the fundamental ones. Otherwise, exchange rates temporarily drift away from the fundamental rates due to trading actions.

13.3 The reinforcement learning paradigm

The objective of RL is to optimise in environments where there is no a priori assumption on the dynamics of the environment and how it is affected by the actions of the agent. A key feature of RL is to "learn" from the interplay between actions and changes in the environment.

In the simplest setting, the evolution of the environment, action, and reward may be viewed as in Figure 13.2. Then the agent's actions a_t affect the transition from state S_t into state S_{t+1}. The actions depend only on the state S_t, and the triple (S_t, S_{t+1}, a_t) affect the reward R_t. The collection $[S_t, S_{t+1}, a_t, R_t]$ is known as a 4-tuple. We further assume that the system is Markov.[1]

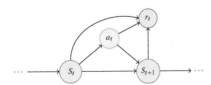

Figure 13.2 Directed graph representation of the stochastic control problem.

The value of an action policy π (which, in the simplest case, maps states uniquely and deterministically into actions) is given by

$$V^\pi(S) := \mathbb{E}\left[\sum_{t=0}^\infty \gamma^t \, R(S_t, S_{t+1}^{a_t^\pi}; a_t^\pi) \ \middle|\ S_0 = S \right], \tag{13.6}$$

---

[1] In the problem formulation above, S_t is the four-dimensional vector $(X_t^{\text{€\$}}, X_t^{\text{£\$}}, q_t^{\text{€}}, q_t^{\text{£}})$. The step-by-step rewards are in (13.3).

where a_t^π denotes the agent's action from following policy π, $R(S_t, S_{t+1}^{a_t^\pi}; a_t^\pi)$ denotes the rewards received using action policy π, $S_{t+1}^{a_t^\pi}$ denotes the one-step ahead state given that the trader takes actions a_t^π when in state S_t, $\gamma \in (0, 1)$ is a discount factor, and the initial state of the system is S.

The *value function* is the value of the best action policy, i.e., the value of the policy that maximises the expectation above, and is written as

$$V(S) := \max_{\pi \in \mathcal{A}} V^\pi(S), \tag{13.7}$$

where $\mathcal{A}$ is the set of admissible policies.

Due to stationarity, there is no time component in the value function. We restrict the admissible set $\mathcal{A}$ to deterministic policies, so that actions are strictly a function of state and identify the policy π with action a. Under mild conditions, the value function then satisfies the Bellman equation

$$V(S) = \max_{a \in \mathcal{A}} \mathbb{E}[\, R(S, S^a; a) + \gamma\, V(S^a)\,], \tag{13.8}$$

where the maximisation is taken over a single-action a taken at the current time, S^a denotes the (random) state the system evolves to under this arbitrary action, and $R(S, S^a; a)$ is the (random) reward received from the action a and corresponding state evolution.

To continue with our exposition, we briefly introduce temporal difference (TD) learning, which is the basis for Q-learning. TD learning has the ability to learn from experience without any modeling from the environment. The simplest TD method is known as TD(0) or one-step TD – see Sutton and Barto (2018). This method, when used for the value function V, follows the update rule

$$V(S_t) \leftarrow V(S_t) + \alpha\,[R(S_t, S_{t+1}; a_t) + \gamma\, V(S_{t+1}) - V(S_t)] \tag{13.9}$$

$$= (1 - \alpha)V(S_t) + \alpha\,[R(S_t, S_{t+1}; a_t) + \gamma\, V(S_{t+1})], \tag{13.10}$$

where $\alpha \in (0, 1]$ is a learning rate parameter. The key idea is to update $V(S_t)$ following a step in the direction of the new estimate $R(S_t, S_{t+1}; a_t) + \gamma\, V(S_{t+1})$: note that (13.10) is a weighted-average between the current and new estimate for $V(S_t)$. TD learning allows updates at each step, and it is at the core of the update rules we discuss next.

A useful concept in learning optimal strategies is the **Q-function**, which measures the *quality* (and hence its name) of an action a at a given point in state space S. The Q-function corresponds to the argument of the max operator in (13.8), more precisely

$$Q(S, a) := \mathbb{E}[\, R(S, S^a; a) + \gamma\, V(S^a)\,]. \tag{13.11}$$

From (13.8), the Q-function also satisfies a Bellman-like equation

$$Q(S, a) = \mathbb{E}\left[\, R(S, S^a; a) + \gamma \max_{a' \in \mathcal{A}} Q(S^a, a')\,\right]. \tag{13.12}$$

RL focuses on learning this function through interaction with the environment. Tabular Q-learning (Watkins and Dayan, 1992) is one of the classical forms of RL, and entails approximating the state/action space by a discrete set, which

---

Algorithm 13.1: Q-learning with experience replay algorithm.

---

1 initialize $Q(s, a)$ and state s ;
2 **for** $i \leftarrow 1$ **to** N **do**
3   select ε-greedy action a from Q;
4   observe $s \xmapsto{\ a\ } (s', R)$;
5   update $Q(s, a) \leftarrow (1 - \beta_k) Q(s, a) + \beta_k \left[R + \gamma \max_{a' \in \mathcal{A}} Q(S', a') \right]$;
6   store (s, a, s', R) in replay buffer $\mathcal{D}$;
7   **for** $j \leftarrow 1$ **to** M **do**
8   select $(\tilde{s}, \tilde{a}, \tilde{s}', \tilde{R})$ from $\mathcal{D}$;
9   update $Q(\tilde{s}, \tilde{a}) \leftarrow (1 - \beta_k) Q(\tilde{s}, \tilde{a}) + \beta_k \left[R + \gamma \max_{\tilde{a}' \in \mathcal{A}} Q(\tilde{S}', \tilde{a}') \right]$;
10   **end**
11   update $s \leftarrow s'$;
12 **end**

---

renders the Q-function into a large table. The methodology uses the Bellman equation in (13.12) to obtain updated estimates of the Q-function from actions and observations. Actions are typically ε-greedy; that is, select a random action with probability ε, otherwise select the optimal action based on the current estimate of the Q-function. This allows agents to explore the state/action space, while exploiting what they know so far. After an action, the update of the Q-function is

$$Q(\tilde{s}, \tilde{a}) \leftarrow (1 - \beta_k) Q(\tilde{s}, \tilde{a}) + \beta_k \left[R + \gamma \max_{\tilde{a}' \in \mathcal{A}} Q(\tilde{S}', \tilde{a}') \right] , \qquad (13.13)$$

where the parameter set $\{\beta_k\}_{k=1,\dots,N}$ is the learning rate, with $\beta_k > 0$ decreasing, $\sum_{k=1}^{\infty} \beta_k = \infty$, and $\sum_{k=1}^{\infty} \beta_k^2 < +\infty$. Typical examples are $\beta_k = A/(B + k)$ for positive constants $A \geq B$.

A simple way to accelerate learning is to add a so-called *replay buffer* which stores historical 4-tuples (S, S', a, R) corresponding to a state, the action taken, the new state the system evolves to, and the reward received. After taking an action, but before sampling from the environment again, the agent randomly selects 4-tuples from the replay buffer and updates the Q-function. This procedure is known as Q learning with experience replay; see Algorithm 13.1.

13.3.1 Deep Q-learning (DQN)

Deep Q-learning (DQN) is conceptually similar to tabular Q-learning, but rather than approximating the state/action space with a discrete space, DQN uses artificial neural nets (ANNs) to approximate the Q-function. The agent assumes that the Q-function is the output of the ANN – the specific architecture of the ANN is arbitrary and should be tuned to the specific problem. Also, the action may be included in the ANN inputs or, if the actions are discrete, the ANN may output the Q-function for all possible action values; see Figure 13.3. We denote the network parameters by θ. The DQN algorithm (Mnih et al., 2015) proceeds as in Algorithm 13.1, but the update rules in steps 5 and 9 are replaced by minimising

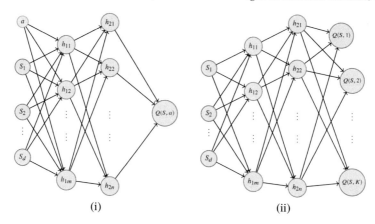

Figure 13.3 DQN takes either (i) states and actions as inputs and outputs the Q-function value, or, when actions are discrete, (ii) states as inputs and outputs the Q-function for all actions. In this work, we employ (i).

the loss function

$$L(\theta; \theta_T) = \sum_{j=1}^{J} \left(\left[R_{(j)} + \gamma \max_{a' \in \mathcal{U}_j} Q\left(S'_{(j)}, a' \,\middle|\, \theta_T \right) \right] - Q\left(S_{(j)}, a_{(j)} \,\middle|\, \theta \right) \right)^2 \quad (13.14)$$

over the network parameters θ. Here, θ_T denotes a target network that is updated to equal the main network θ every M iterations, and $(S_{(j)}, S'_{(j)}, a_{(j)}, R_{(j)})_{j=1,...,J}$ corresponds to a mini-batch of 4-tuples. As usual in deep learning, minimising proceeds using gradient decent through back propagation. The loss in (13.14) uses the optimal actions from the target network rather than the main network θ; this has been shown to result in sub-optimal strategies.

To improve performance, double deep Q network learning (DDQN) uses the optimal action from the main network instead. Thus, we replace the loss in (13.14) by

$$L(\theta; \theta_T) = \sum_{j=1}^{J} \left(\left[R_{(j)} + \gamma \, Q\left(S'_{(j)}, a^*_{(j)}(\theta) \,\middle|\, \theta_T \right) \right] - Q\left(S_{(j)}, a_{(j)} \,\middle|\, \theta \right) \right)^2, \quad (13.15)$$

where

$$a^*_{(j)}(\theta) = \arg\max_{a' \in \mathcal{U}_j} Q\left(S'_{(j)}, a' \,\middle|\, \theta \right). \quad (13.16)$$

Algorithm 13.2 provides an outline of the procedure in DDQN.

We perform 1,000,000 learning steps of the DQN algorithm with the assumptions and model parameters in Section 13.2.1. The remainder of the parameters are: learning rate of the ANN that parameterises the Q-function is $l = 10^{-4}$, replacement frequency $M = 100$, minibatch size of $n_b = 64$, time horizon $T = 10$ weeks, maximum buy/sell quantity is 200,000 units of the base currency at each time step (which we scale down to represent one unit), time step $\Delta_t = 1$ week, discount parameter $\gamma = 0.999$, and the size of the replay buffer is 10,000 previous experiences. The ε-greedy action starts with $\varepsilon = 1$ and decreases by 0.001 with

---

Algorithm 13.2: Double DQN learning algorithm.

1 initialize main and target networks θ, θ_T and state s;
2 **for** $i \leftarrow 1$ **to** N **do**
3   **for** $j \leftarrow 1$ **to** M **do**
4   select ε-greedy action a from $Q(s, a; \theta)$;
5   observe $s \xmapsto{a} (s', R)$ and store in replay buffer;
6   grab mini-batch J from replay buffer;
7   update main network θ using SGD on mini-batch loss

$$L(\theta; \theta_T) = \sum_{j=1}^{J} \left(\left[R_{(j)} + \gamma\, Q\left(S'_{(j)}, a^*_{(j)}(\theta) \,\middle|\, \theta_T \right) \right] - Q\left(S_{(j)}, a_{(j)} \,\middle|\, \theta \right) \right)^2;$$

        update $s \leftarrow s'$;
8   **end**
9   update target network $\theta_T \leftarrow \theta$;
10   repeat until converged ;
11 **end**

---

every epoch, until it reaches a minimum value of 0.01. The action space is taken to be discrete with values in A^3 where $A = \{-1.0, -0.9, \ldots, 0.9, 1.0\}$. The architecture is kept simple and consists of a fully connected feed-forward ANN with two layers with 64 nodes in each layer. We employ the ReLU activation function in both layers and perform update rules according to the Adam optimizer (Kingma and Ba, 2015).

13.3.2 Reinforced deep Markov models

Reinforced Deep Markov models (RDMMs), introduced in Ferreira (2020) for single asset optimal execution over a few seconds, takes a different approach from that in RL. Instead of approximating the Q-function (see (13.11)), RDMM proposes a rich graphical model for the environment, actions, and rewards, to capture a wide range of features observed in real markets. In what follows, we first describe deep Markov models (DMMs) (Krishnan et al., 2015), which models the environment's evolution, and then discuss RDMM.

DMMs may be viewed as stochastic processes that are driven by a latent state which drives the environment and latent states are (randomly) mapped to observable states. Figure 13.4 shows a graphical model representation of a DMM. The conditional latent state $Z_t|_{Z_{t-1}}$ is independent of $Z_{1:t-2}$ and $S_{1:t-1}$, i.e., Z is Markov, and the conditional observable state $S_t|_{Z_t}$ is independent of

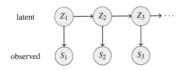

Figure 13.4 Directed graph representation of a DMM

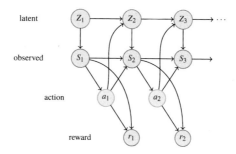

Figure 13.5 Generative model description of the RDMM framework.

all other random variables.[2] This is a direct generalisation of Hidden Markov models, where the latent sate is finite-dimensional. It may also be viewed as a generalisation of Kalman filter models.

In DMMs, the typical conditional one-step evolution of the states is modeled as

$$Z_{t+1}|z_t \overset{\mathbb{P}}{\sim} \mathcal{N}\left(\mu_z^\theta(Z_t)\,;\,\Sigma_z^\theta(Z_t)\right), \qquad \text{latent dynamics,} \qquad (13.17a)$$

$$S_t|z_t \overset{\mathbb{P}}{\sim} \mathcal{N}\left(\mu_s^\theta(Z_t)\,;\,\Sigma_s^\theta(Z_t)\right), \qquad \text{observed data,} \qquad (13.17b)$$

and the means and covariance matrices are parameterised by ANNs with parameters θ. Thus, there are four ANNs, one for each mean vector, and one for each covariance matrix. When the mean ANNs are replaced by affine functions of Z_t, and the covariance matrices are constant, the DMM reduces to a Kalman filter model. The latent state dynamics may be viewed as a time-discretisation of Itô processes, where the instantaneous drift and covariance matrix are given by ANNs: $dZ_t = \mu_z^\theta(Z_t)\,dt + \sigma_z^\theta(Z_t)\,dW_t$, where W is a vector of independent Brownian motions and $\sigma_z^\theta(z)$ is the Cholesky decomposition of $\Sigma_z^\theta(z)$. Given a sequence $S_1, S_2, \ldots$ of data, one maximises the log-likelihood of observing the sequence of states and rewards (given the actions taken) to estimate the ANN parameters. The log-likelihood, however, is intractable because the posterior distribution of the latent states given data, i.e., $\mathbb{P}(Z_{1:T} \mid S_{1:T})$, is not attainable in closed-form, nor is it computationally feasible. Instead, in cases like this, we may use variational inference (VI); see Ormerod and Wand (2010) for an overview. VI uses an approximate posterior and rather than maximising the log-likelihood, it maximises what is known as the evidence lower bound (ELBO). We elaborate on this approach after we introduce the full RDMM below.

RDMMs are generalised versions of DMMs for incorporating actions and rewards. Here, we adopt the graphical model shown in Figure 13.5: observable states affect actions, observable states and actions affect rewards, and actions

---

[2] We use the slice notation $x_{a:b}$ ($a < b$ and integer) to denote the sequence $x_a, x_{a+1}, \ldots, x_b$.

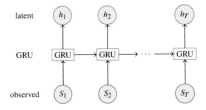

Figure 13.6 GRU for encoding observations.

affect the latent and observable states. Thus, there are additional ANNs that connect the various components in the graphical model representation – this architecture is similar to that in Ferreira (2020). Here, however, the difference is that there are connections between observables and rewards, rather than between latent states and rewards, and there are connections between observables (which is more reflective of how financial markets work). Moreover, we assume that

$$Z_{t+1}|Z_t,a_t \overset{\mathbb{P}}{\sim} \mathcal{N}\left(\mu_z^\theta(Z_t,a_t)\,;\,\Sigma_z^\theta(Z_t,a_t)\right), \qquad \text{latent dynamics,} \quad (13.18a)$$

$$S_t|Z_t,a_{t-1},S_{t-1} \overset{\mathbb{P}}{\sim} \mathcal{N}\left(\mu_s^\theta(Z_t,S_{t-1},a_{t-1})\,;\,\Sigma_s^\theta(Z_t,S_{t-1}a_{t-1})\right), \qquad \text{observed state,} \quad (13.18b)$$

$$r_t = R(S_t,a_t), \qquad \text{observed reward.} \quad (13.18c)$$

We further assume that actions are the output of an ANN ϑ with input features given by the observable state

$$a_t = g_a^\vartheta(S_t). \qquad (13.19)$$

One can replace this assumption with a probabilistic model, e.g.,

$$a_t|_{S_t} \overset{\mathbb{P}}{\sim} \mathcal{N}\left(\mu_a^\vartheta(S_t)\,;\,\Sigma_a^\vartheta(S_t)\right), \qquad (13.20)$$

which allows for built-in exploration; we leave such generalisations to future work.

As mentioned earlier, to estimate the DMM that drives the system from observations, we require the posterior distribution of latent states given observations, i.e., we require $\mathbb{P}(Z_{1:T} \mid S_{1:T}, r_{1:T}, a_{0:T-1})$. This posterior is intractable – both analytically and computationally. Therefore, we adopt the VI approach, and introduce a new probability measure $\mathbb{Q}$ to approximate the posterior distribution as

$$Z_{t+1}|Z_t,S_{1:T},r_{1:T},a_t \overset{\mathbb{Q}}{\sim} \mathcal{N}\left(\mu_z^\phi(Z_t,a_t,h_T)\,;\,\Sigma_z^\phi(Z_t,a_t,h_T)\right). \qquad (13.21)$$

In the above relationship, h_T is a summary of the state sequence $S_{1:T}$ resulting from, e.g., passing $S_{1:T}$ through gated recurrent unit (GRU) Cho et al. (2014) or long-short-term-memory (LSTM) Hochreiter and Schmidhuber (1997) networks. These may also be replaced by their bidirectional versions.

The network θ is often referred to as the decoding network (as it maps latent states to observables), while the network ϕ is often referred to as the encoding network (as it maps the observables to latent states).

To estimate model parameters, we maximise the log-likelihood over the decoding network parameters θ. The log-likelihood is intractable, so instead, we use an approximation to the posterior to compute a lower bound for the log-likelihood as follows:

$$\mathcal{L}(S_{1:T}, r_{1:T} | a_{0:T-1})$$

$$= \log \mathbb{P}(S_{1:T}, r_{1:T} | a_{0:T-1}) \tag{13.22a}$$

$$= \log \int \mathbb{P}(S_{1:T}, r_{1:T} | Z_{1:T}; a_{0:T-1}) \, d\mathbb{P}(Z_{1:T} | a_{0:T-1}) \tag{13.22b}$$

$$= \log \int \left(\frac{\mathbb{P}(S_{1:T}, r_{1:T} | Z_{1:T}; a_{0:T-1}) \mathbb{P}(Z_{1:T} | a_{0:T-1})}{\mathbb{Q}(Z_{1:T} | S_{1:T}, a_{0:T-1})} \right) d\mathbb{Q}(Z_{1:T} | S_{1:T}, a_{0:T-1}) \tag{13.22c}$$

$$\geq \int \log \left(\frac{\mathbb{P}(S_{1:T}, r_{1:T} | Z_{1:T}; a_{0:T-1}) \mathbb{P}(Z_{1:T} | a_{0:T-1})}{\mathbb{Q}(Z_{1:T} | S_{1:T}, a_{0:T-1})} \right) d\mathbb{Q}(Z_{1:T} | S_{1:T}, a_{0:T-1}) \tag{13.22d}$$

$$= \mathbb{E}^{\mathbb{Q}} \left[\log \left(\mathbb{P}(S_{1:T}, r_{1:T} | Z_{1:T}; a_{0:T-1}) \right) \right] - KL \left[\mathbb{Q}(Z_{1:T} | a_{0:T-1}) \| \mathbb{P}(Z_{1:T} | a_{0:T-1}) \right], \tag{13.22e}$$

where the function KL denotes the Kullback-Leibler divergence (also called relative entropy or information gain), which is a measure of the distance of any given estimated model, with distribution over the data $p_j(x|\hat{\theta}_j)$, to the true model, with distribution over the data $p(x)$. The Kullback-Leibler divergence is defined as

$$KL\left(p \| \hat{p}_j\right) := \int p(x) \log \frac{p(x)}{p_j(x|\hat{\theta}_j)} \, dx. \tag{13.23}$$

The bound in (13.22e) is called the evidence lower bound (ELBO), so instead of maximising the log-likelihood directly because it is intractable, VI maximises the ELBO. If the Kullback-Leibler divergence vanishes, then we obtain equality, and the log-likelihood equals the ELBO.

To complete the analysis, we require expressions for each term in (13.22e). First, from the conditional independence implied by the graphical model in Figure 13.5 and explicitly modeled in (13.18), we have

$$\log \mathbb{P}\left(S_{1:T}, r_{1:T} | Z_{1:T}, a_{0:T-1}\right)$$

$$= \sum_{t=1}^{T} \log \mathbb{P}\left(S_t, r_t | Z_t, a_t, a_{t-1}\right) \tag{13.24}$$

$$= -\frac{1}{2} \sum_{t=1}^{T} \left\{ d_s \, \log(2\pi) + \log \det \Sigma_s^\theta(Z_t, a_{t-1}) \right.$$

$$+ \left(S_t - \mu_s^\theta(Z_t, a_{t-1}) \right)^\top \left(\Sigma_s^\theta(Z_t, a_{t-1}) \right)^{-1} \left(S_t - \mu_s^\theta(Z_t, a_{t-1}) \right) \tag{13.25}$$

$$+ \log(2\pi) + \log \det \Sigma_r^\theta(Z_t, a_t)$$

$$\left. + \left(r_t - \mu_r^\theta(Z_t, a_t) \right)^\top \left(\Sigma_r^\theta(Z_t, a_t) \right)^{-1} \left(r_t - \mu_r^\theta(Z_t, a_t) \right) \right\}.$$

One can estimate the $\mathbb{Q}$-expectation of the above expression, which appears in the ELBO (13.22e), as follows: (i) generate $\mathbb{Q}$-samples of $Z_{1:T}$, (ii) evaluate the above expression, and then (iii) compute the sample average. The Kullback-Leibler term in the ELBO (13.22e) is

$$KL\left[\mathbb{Q}(Z_{1:T} | a_{0:T}) \| \mathbb{P}(Z_{1:T} | a_{0:T}) \right]$$

$$= \mathbb{E}^{\mathbb{Q}} \left[\log \frac{\mathbb{Q}(Z_{1:T}|a_{0:T})}{\mathbb{P}(Z_{1:T}|a_{0:T})} \right] \tag{13.26a}$$

$$= \mathbb{E}^{\mathbb{Q}} \left[\log \frac{\mathbb{Q}(Z_{2:T}|Z_1, a_{1:T-1}) \, \mathbb{Q}(Z_1|a_0)}{\mathbb{P}(Z_{2:T}|Z_1 \, a_{1:T-1}) \, \mathbb{P}(Z_1|a_0)} \right] \tag{13.26b}$$

$$= \dots$$

$$= \mathbb{E}^{\mathbb{Q}} \left[\sum_{t=2}^{T} \log \frac{\mathbb{Q}(Z_t|Z_{t-1}, a_{t-1})}{\mathbb{P}(Z_t|Z_{t-1}, a_{t-1})} + \log \frac{\mathbb{Q}(Z_1|a_0)}{\mathbb{P}(Z_1|a_0)} \right] \tag{13.26c}$$

$$= \mathbb{E}^{\mathbb{Q}} \left[\sum_{t=2}^{T} KL \big[\mathbb{Q}(Z_t \mid Z_{t-1}, a_{t-1}) \parallel \mathbb{P}(Z_t \mid Z_{t-1}, a_{t-1}) \big] \right] \tag{13.26d}$$
$$+ KL \big[\mathbb{Q}(Z_1 \mid a_0) \parallel \mathbb{P}(Z_1 \mid a_0) \big] .$$

Each term in the sum is the Kullback-Leibler divergence between the $\mathbb{Q}$ and $\mathbb{P}$ one-step transition of the latent factor, conditional on the previous latent state and action. As the conditional distributions involved are all multivariate Gaussian, the one-step Kullback-Leibler divergence terms may be written explicitly as

$$KL \big[\mathbb{Q}(Z_t \mid Z_{t-1}, a_{t-1}) \parallel \mathbb{P}(Z_t \mid Z_{t-1}, a_{t-1}) \big]$$

$$= \frac{1}{2} \left[\log \frac{\det \Sigma_z^\theta (Z_{t-1}, a_{t-1})}{\det \Sigma_z^\phi (Z_{t-1}, a_{t-1}, h_T)} + \mathrm{Tr} \left(\left(\Sigma_z^\theta (Z_{t-1}, a_{t-1}) \right)^{-1} \left(\Sigma_z^\phi (Z_{t-1}, a_{t-1}, h_T) \right) \right) \right. \tag{13.27}$$

$$\left. + \Delta \mu_z^{\theta, \phi} (Z_{t-1}, a_{t-1}, h_T) \left(\Sigma_z^\theta (Z_{t-1}, a_{t-1}) \right)^{-1} \Delta \mu_z^{\theta, \phi} (Z_{t-1}, a_{t-1}, h_T) - d_z \right],$$

where $\Delta \mu_z^{\theta, \phi}(Z_{t-1}, a_{t-1}, h_T) = \left(\mu_z^\theta(Z_{t-1}, a_{t-1}) - \mu_z^\phi(Z_{t-1}, a_{t-1}, h_T) \right)$. As before, the $\mathbb{Q}$-expectation in (13.26d) may be estimated by generating $\mathbb{Q}$-samples of $Z_{1:T}$, evaluating (13.27), and then computing the sample average.

Next, we discuss how to obtain the various network parameters in the RDMM approach. The paradigm proceeds in a batch RL manner; it alternates between (i) learn the DMM network while the policy network is held frozen, and (ii) learn the policy network while the DMM network is held frozen. Part (i) of the learning process proceeds in an actor-critic manner: freeze the decoder (generative model) network parameters θ and learn the encoding (approximate posterior) network parameters ϕ by taking SGD steps of the ELBO, then freeze ϕ and use SGD to update θ. Algorithm 13.3 shows an outline of this procedure.

In simulated environments where actions do not affect the states, or when training with historical data, one may simplify learning as follows: (i) use (simulated or historical) observations and maximise sum of rewards over action network parameters ϑ, and (ii) learn the embedded DMM (encoder ϕ and decoder θ) with the action network held fixed. This last step, however, requires to alternate between maximising over the decoder and encoder networks.

Once the networks are learned using simulated or historical data, live actions may be used to update the trained networks.

Algorithm 13.3: RDMM learning algorithm.

---

1 initialize encoder, decoder, and action networks θ, ϕ, and ϑ;

2 initialize state S;

3 **for** $i \leftarrow 1$ **to** N **do**

4   **for** $t \leftarrow 1$ **to** T **do**

5   select action $a = g_a^\vartheta(S)$;

6   observe $S \xmapsto{\;a\;} (S', R)$;

7   update $S' \leftarrow S$;

8   **end**

9   simulate $Z_{1:T}$ using decoder $\mathbb{Q}$;

10   update encoding network θ using SGD to maximise ELBO;

11   update decoder network ϕ using SGD to maximise ELBO ;

12   update action network ϑ using SGD to maximise total profit;

13 **end**

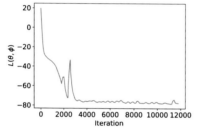 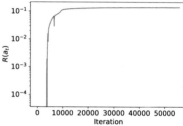

Figure 13.7 Loss function and expected reward: moving average over 100 iterations.

13.3.3 Implementation of RDMM

Next, we parameterise the ANNs of the RDMM. In a broad sense, there are three types of networks at play: (i) the GRU that encodes the observations, (ii) the ANNs that output the mean vector and the variance-covariance matrices, and (iii) the ANN that models the action.

For (i), we use a standard GRU with input size of two (corresponding to the exchange rates $x^{\in\$}$ and $x^{\pounds\$}$), five hidden layers, unidirectional, and hidden dimension equal to three. For (ii), we employ a feed-forward ANNs with two layers of 32 nodes each, these networks transform the input into two outputs, the first is the vector of mean values and the Cholesky decomposition of the variance-covariance matrix of the multivariate normal. To model an $n \times n$ variance-covariance matrix, the ANN outputs an $n(n+1)/2$-dimensional vector that characterises the lower triangular matrix in the Cholesky decomposition of the variance-covariance matrix. We reshape the $n(n+1)/2$-dimensional vector into a lower triangular matrix and ensure that the values in the diagonal are positive which is a sufficient condition for the lower-triangular matrix to be positive-definite – (Dorta et al., 2018).

Finally, for (iii) we use a feed-forward ANN with two layers of 32 nodes each, input dimension being equal to five (t, $X^{\in\$}$, $X^{\pounds\$}$, $q^{\in}$, $q^{\pounds}$), and output dimension equal to three ($a^{\in\$}$, $a^{\pounds\$}$, $a^{\in\pounds}$). To make the actions lie between a maximum and minimum range, we scale the output of the last layer, for which we use a $\tanh(x)$

activation. In the three cases: (i), (ii), (iii), we employ ReLU activations in the intermediate layers, and perform update rules according to the Adam optimizer (Kingma and Ba, 2015).

We run over 100,000 learning steps on each network in batches of 64 sample paths at a time, and 64 simulations to compute the $\mathbb{Q}$-expectations. Figure 13.7 shows the loss function (left panel), and the value of the rewards (right panel) as a function of the learning steps. Recall that the loss function in (13.22e) is the result of two calculations, the $\mathbb{Q}$-expectation of (13.25) and the Kullback–Leibler term in (13.26d), for which each term is explicitly given in (13.27).

13.4 Optimal trading in triplet

The three-dimensional optimal strategy has various features – we discuss them below. We show plots only for the RDMM model because the results for DQN are similar. The walk-the-book cost parameters are $\phi^{€\$} = \phi^{£\$} = \phi^{€£} = 0.001$, and the terminal penalty parameters are $\alpha^{€} = \alpha^{£} = 1$ – see (13.2).

First, we study the optimal action through time as a function of the inventory. The state space is five-dimensional $(t, X_t^{€\$}, X_t^{£\$}, q_t^{€}, q_t^{£})$ and the output actions is three-dimensional $(a_t^{€\$}, a_t^{£\$}, a_t^{€£})$, thus, for each plot in this section we fix several inputs and we depict the optimal strategy. Figure 13.8 displays the optimal strategy $a^{€\$}$ learnt by the RDMM when $t = 5$ weeks (left panel), $t = 7$ weeks (middle panel), $t = 9$ weeks (right panel), and the terminal date is $T = 10$ weeks. In the three plots: $q_t^{£} = 0$, $X_t^{£\$} = 1.3$, the x-axis is the inventory $q_t^{€}$, and the y-axis is the exchange rate $X_t^{€\$}$. We observe that the optimal strategy learns that as the trader approaches T, the inventory should be close to zero to avoid the terminal penalty. Note that the value of the terminal penalty parameter α^k is greater than that of the cost parameter ϕ^l. The plots for the learnt actions $a^{£\$}$ and $a^{£€}$ are similar.

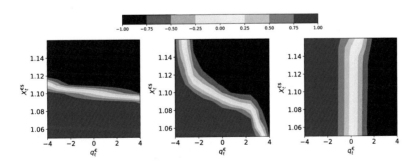

Figure 13.8 Optimal action $a_t^{€\$}$ as a function of $q_t^{€}$ and $X_t^{€\$}$. Left panel $t = 5$, middle panel $t = 7$, and right panel $t = 9$. Trading horizon $T = 10$ weeks.

In Figure 13.9 we set $t = 5$ weeks and let $X_t^{£\$} = 1.33$ in the top row (above its mean-reverting level), $X_t^{£\$} = 1.30$ in the middle row (equal to its mean-reverting level), and $X_t^{£\$} = 1.27$ in the bottom row (below its mean-reverting level). In all plots, the x-axis represents the inventory in Euros, $q_t^{€}$, and the y-axis represents the exchange rate for the pair EUR/USD, $X_t^{€\$}$. The colours represent the optimal actions learnt by the RDMM: left, middle, right columns show the actions $a_t^{€\$}$, $a_t^{£\$}$, $a_t^{€£}$, respectively. When the rate $X_t^{£\$}$ is below its mean-reverting level, the optimal action $a^{£\$} = 1$ (i.e., buy GBP and sell USD) is almost independent from the level of the rate $X_t^{€\$}$ and from the inventory $q_t^{€}$ (see bottom middle panel). Similarly, when the rate $X_t^{£\$}$ is above its mean-reverting level, the optimal action $a^{£\$} = -1$ (i.e., sell GBP and receive USD) is almost independent from the rate $X_t^{€\$}$ and from the inventory $q_t^{€}$. Thus, the optimal strategy learnt by the RDMM provides the trader with a "buy low, sell high" strategy.

The middle panel shows that when the rate $X_t^{€\$}$ is below (above) its mean-reverting value, the value of EUR in units of USD and, also, in units of GBP, are both, under-priced (over-priced). Thus, the actions $a_t^{€\$}$ and $a_t^{€£}$ are greater (smaller) than zero, i.e., the strategy buys (sells) EUR low (high).

The plot in the centre requires a more careful analysis. The equations in (13.5) show the effect that $\bar{x}^{€\$} - X_t^{€\$}$ has on $X_{t+1}^{£\$}$. If $\bar{x}^{€\$} - X_t^{€\$}$ is greater (smaller) than zero and $\bar{x}^{£\$} - X_t^{£\$} = 0$, then, on average, $X_{t+1}^{£\$}$ is greater (smaller) than $X_t^{£\$}$. Thus, when $X_t^{£\$} = \bar{x}^{£\$}$, the strategy learns that the optimal action $a_t^{£\$}$ is to buy $X_t^{£\$}$ if $X_t^{€\$} > \bar{x}^{€\$}$ and sell $X_t^{£\$}$ if $X_t^{€\$} < \bar{x}^{€\$}$, because, on average, the trader makes a profit.

Finally, in the top row (bottom row), when the exchange rate $X_t^{£\$}$ is above (below) its mean-reverting value, the optimal strategy raises (lowers) the threshold for when to buy/sell when compared with the threshold in the middle row. The RDMM learns the dynamics of the environment, so when $X_t^{£\$} = 1.33$ or $X_t^{£\$} = 1.27$, the regions where $X_{t+1}^{€\$}$ is, on average, above/below $X_t^{€\$}$, changes to those shown in the figure. For example, from equation (13.5) we see that, on average, $X_{t+1}^{€\$} > X_t^{€\$}$, if and only if $X_t^{€\$} < \bar{x}^{€\$} + \eta_0/\kappa_0 (\bar{x}^{£\$} - X_t^{£\$})$, so the threshold in the top left panel is around 1.118 ($X_t^{£\$} = 1.33$), and in the bottom left panel, the threshold is around 1.082 ($X_t^{£\$} = 1.27$) – recall that $\eta_0 = -0.3$, $\kappa_0 = 0.5$, $\bar{x}^{€\$} = 1.1$.

Next, we discuss the performance of the optimal actions in the triplet. We perform 10,000 simulations of the exchange rate process and execute the optimal actions learnt by the RDMM. In all simulations, the rate processes follow (13.5), and the trader starts with zero inventory in all currencies, i.e., $q_0^{\$} = q_0^{€} = q_0^{£} = 0$. Figure 13.10 shows a histogram of the profit and loss in USD. It is not surprising to see that in such a controlled environment, the optimal actions lose money in less than 1% of the runs because the strategies fully exploit the learnt dynamics of the environment. The mean value of the P&L is 0.14 USD and the standard deviation is 0.06. Here, 0.14 USD is the gain when the maximum quantity we buy/sell is one. Recall that we think of each trade of being of size 200,000 of the base currency, thus, the P&L of the strategy is around 28,000 USD over the course of ten weeks.

Figure 13.11 displays the paths for $q_{0:T}^{€}$ and $q_{0:T}^{£}$. The shaded region encompasses the data between the 10% and 90% quantiles, and the red line is one of the inventory paths. We observe that the strategy performs statistical arbitrages until approximately time $t = 7$ and then liquidates any outstanding inventory in

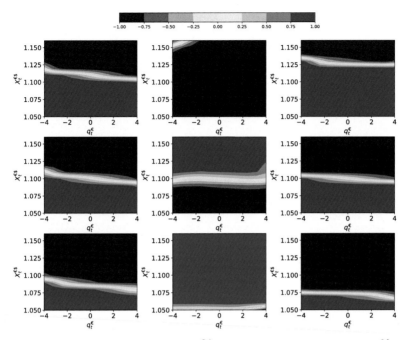

Figure 13.9 Left column: optimal action $a_t^{\in\$}$; middle column: optimal action $a_t^{\pounds\$}$; right column: optimal action $a_t^{\in\pounds}$. Top row: $X_t^{\pounds\$} = 1.33$; middle row: $X_t^{\pounds\$} = 1.30$; bottom row: $X_t^{\pounds\$} = 1.27$. Mean-reverting levels in EUR/USD and GBP/USD are $\bar{x}^{\in\$} = 1.1$ and $\bar{x}^{\pounds\$} = 1.3$.

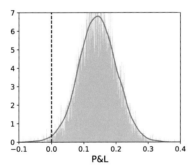

Figure 13.10 Profit and loss in USD.

EUR and GBP to finish with a flat inventory in those two currencies at time T. Had we taken the values of the impact parameters to be equal to the values of terminal inventory parameters, then we would not observe the concentration at zero of $q_T^{\in}$ and $q_T^{\pounds}$ because the strategy would in many runs find it optimal to liquidate inventory at the end of the trading horizon.

The top row in Figure 13.12 shows stylised features of the optimal actions $a_{0:T}^{\in\$}$, $a_{0:T}^{\pounds\$}$, $a_{0:T}^{\in\pounds}$ in the currency pairs $X_{0:T}^{\in\$}$, $X_{0:T}^{\pounds\$}$, $X_{0:T}^{\in\pounds}$, respectively, for one simulation. The green up-arrows are buys and the red down-arrows are sells. The bottom

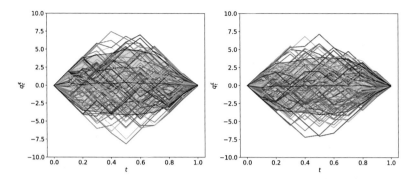

Figure 13.11 Inventory paths for $q_T^{\euro}$ and $q_T^{\pounds}$. The shaded region shows the 10%, 50%, and 90% quantiles through time.

row shows the inventories $q_{0:T}^{\euro}$ and $q_{0:T}^{\pounds}$, and the cash process $q_{0:T}^{\$}$. This figure highlights the interplay between buy-low-sell-high and inventory control. As shown in the bottom panels, the trader aims to finish with $q_T^{\euro} \approx 0$ and $q^{\pounds} \approx 0$ while taking advantage of her learnt knowledge about the exchange rate dynamics – see that most of the green arrows are below the dotted lines (mean-reverting level of the currency pairs), and, most of the red arrows are above the dotted lines.

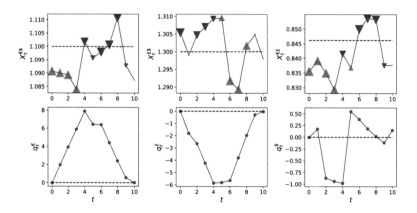

Figure 13.12 RDMM optimal strategy sample path showing FX rates, actions (size and direction of arrows), and currency inventories.

13.4.1 Remarks on RDMM versus DDQN

Next, we briefly look at the DDQN results. The DDQN and RDMM approaches produce similar optimal actions, however, as we see from comparing Figures 13.13 and 13.8, the actions from DDQN tend to be noisier and generate some unusual behaviour compared with the actions from RDMM. Moreover, the action space in RDMM can be discrete or continuous, while DDQN seeks optimal strategies over a discrete action space and tends to work only when the action space is small. Also, learning in RDMM from historical data benefits from augmenting the historical data set with simulations from the learned DMM portion of the RDMM. On the other hand, DDQN is limited to historical data; however, one may use a replay buffer to help stabilise the results. Thus, RDMM produces more stable action networks than those from the DDQN.

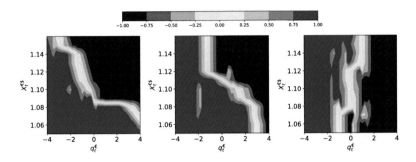

Figure 13.13 Optimal action $a_t^{\text{€\$}}$ as a function of $q_t^{\text{€}}$ and $X_t^{\text{€\$}}$. Left panel $t = 5$, middle panel $t = 7$, and right panel $t = 9$. Trading horizon $T = 10$ weeks.

13.5 Conclusions and future work

We showed how to devise a statistical arbitrage strategy that takes advantage of co-movements and mean-reversion in the exchange rates of an FX triplet. Through simulations, we illustrated the financial performance of the trading strategy based on the RDMM approach. Our approach can be applied in all asset classes where instruments are traded in electronic markets.

To benefit from all the features of RDMM, one needs to expand our study with market data. RDMM can learn a generative model for the data, which is used to simulate more data (with the same statistical properties of the original data). This can play a crucial role in the learning of optimal strategies when data are limited.

One can implement the strategies derived throughout this subchapter in live trading. For this, one proceeds as in stochastic control frameworks: learn the network parameters offline (this is equivalent to computing the value function in stochastic control); employ the learnt network parameters live to compute the optimal action. In principle, one can also update the network while new information arrives; however, this approach is computationally expensive, and instead it may be preferable to incorporate new data to update the networks in advance, e.g., overnight or a few hours before executing the strategy.

References

Cartea, Álvaro, Jaimungal, Sebastian, and Ricci, Jason. 2014. Buy low, sell high: A high frequency trading perspective. *SIAM Journal on Financial Mathematics*, **5**(1), 415–444.

Cartea, Álvaro, Jaimungal, Sebastian, and Penalva, Jose. 2015. *Algorithmic and High-Frequency Trading*. Cambridge University Press.

Cartea, Álvaro, Donnelly, Ryan, and Jaimungal, Sebastian. 2017. Algorithmic trading with model uncertainty. *SIAM Journal on Financial Mathematics*, **8**(1), 635–671.

Cartea, Álvaro, Perez Arribas, Imanol, and Sánchez-Betancourt, Leandro. 2020a. Optimal execution of foreign securities: A double-execution problem with signatures and machine learning. Available at SSRN 3562251.

Cartea, Álvaro, Jaimungal, Sebastian, and Jia, Tianyi. 2020b. Trading foreign exchange triplets. *SIAM Journal on Financial Mathematics*, **11**(3), 690–719.

Casgrain, Philippe, Ning, Brian, and Jaimungal, Sebastian. 2019. Deep Q-learning for Nash equilibria: Nash–DQN. ArXiv:1904.10554.

Cho, Kyunghyun, van Merriënboer, Bart, Bahdanau, Dzmitry, and Bengio, Yoshua. 2014. On the properties of neural machine translation: encoder–decoder approaches. Pages 103–111 in: *Proc. SSST-8, Eighth Workshop on Syntax, Semantics and Structure in Statistical Translation*. https://aclanthology.org/W14-4012.

Dorta, Garoe, Vicente, Sara, Agapito, Lourdes, Campbell, Neill D.F., and Simpson, Ivor. 2018. Structured uncertainty prediction networks. Pages 5477–5485 of: *Proc. IEEE Conference on Computer Vision and Pattern Recognition*.

Ferreira, Tadeu A. 2020. Reinforced deep Markov models with applications in automatic trading. ArXiv:2011.04391.

Guéant, Olivier. 2016. *The Financial Mathematics of Market Liquidity: From Optimal Execution to Market Making*. CRC Press.

Guéant, Olivier, and Manziuk, Iuliia. 2019. Deep reinforcement learning for market making in corporate bonds: beating the curse of dimensionality. *Applied Mathematical Finance*, **26**(5), 387–452.

Guéant, Olivier, Lehalle, Charles-Albert, and Fernandez-Tapia, Joaquin. 2012. Optimal portfolio liquidation with limit orders. *SIAM Journal on Financial Mathematics*, **3**(1), 740–764.

Guilbaud, Fabien, and Pham, Huyen. 2013. Optimal high-frequency trading with limit and market orders. *Quantitative Finance*, **13**(1), 79–94.

Hochreiter, Sepp, and Schmidhuber, Jürgen. 1997. Long short-term memory. *Neural Computation*, **9**(8), 1735–1780.

Huang, Weibing, Lehalle, Charles-Albert, and Rosenbaum, Mathieu. 2015. Simulating and analyzing order book data: The queue–reactive model. *Journal of the American Statistical Association*, **110**(509), 107–122.

Kingma, Diederik P., and Ba, Jimmy. 2015. Adam: A Method for Stochastic Optimization. In: *Proc. 3rd International Conference on Learning Representations, ICLR 2015*, Bengio, Yoshua, and LeCun, Yann (eds).

Krishnan, Rahul G., Shalit, Uri, and Sontag, David. 2015. Deep Kalman filters. ArXiv:1511.05121.

Leal, Laura, Laurière, Mathieu, and Lehalle, Charles-Albert. 2020. Learning a functional control for high-frequency finance. ArXiv:2006.09611.

Mnih, Volodymyr, Kavukcuoglu, Koray, Silver, David, Rusu, Andrei A., Veness, Joel, Bellemare, Marc G., Graves, Alex, Riedmiller, Martin, Fidjeland, Andreas K., Ostrovski, Georg, et al. 2015. Human-level control through deep reinforcement learning. *Nature*, **518**(7540), 529–533.

Ning, Brian, Lin, Franco Ho Ting, and Jaimungal, Sebastian. 2018. Double deep Q-learning for optimal execution. ArXiv:1812.06600.

Ormerod, John T., and Wand, Matt P. 2010. Explaining variational approximations. *The American Statistician*, **64**(2), 140–153.

Watkins, Christopher J.C.H., and Dayan, Peter. 1992. Q-learning. *Machine Learning*, **8**(3–4), 279–292.

Part IV

Advanced Optimization Techniques

14

Introduction to Part IV
Advanced Optimization Techniques for Banks and Asset Managers

Paul Bilokon[a], Matthew F. Dixon[b], and Igor Halperin[c]

14.1 Introduction

Innovation in computing has long served a critical role in the advancement of asset management methodology. In 1952, Harry Markowitz joined the RAND Corporation, where he met George Dantzig. With Dantzig's help, Markowitz developed the critical line algorithm for the identification of the optimal mean–variance portfolios, relying on what was later named the "Markowitz frontier" (Markowitz, 1956, 1959).

The classical mean-variance portfolio optimization framework has proliferated with the advancement of computing resources, leading to modern day optimization tools for dynamic portfolio allocation with transaction costs, taxes, and solve many other practical considerations such as long-only constraints and the asymmetry of risk. Indeed, optimization techniques help solve a wide range of problems across banking and asset management: from asset allocation to risk management, from option pricing to model calibration.

Deep Learning:

As each breakthrough innovation in computing captivates the public's attention, so too does it inspire new directions in quantitative finance. For example, deep learning models have proven remarkably successful in a wide field of applications (DeepMind, 2016; Kubota, 2017; Esteva et al., 2017) including image processing (Simonyan and Zisserman, 2014), learning in games (DeepMind, 2017), neuroscience (Poggio, 2016), energy conservation (DeepMind, 2016), skin cancer diagnostics (Kubota, 2017; Esteva et al., 2017). There are also many examples of where deep learning has been successfully applied to predictive tasks in finance.

This chapter addresses one of the most important applications of machine learning methods in quantitative finance, namely applications to asset pricing and banking. Broadly speaking, asset pricing amounts to explaining differences in asset returns in terms of other variables and parameters that can be either

[a] Imperial College, London
[b] Illinois Institute of Technology, Chicago
[c] Fidelity Investments, Boston, MA
    Published in *Machine Learning And Data Sciences For Financial Markets*, Agostino Capponi and Charles-Albert Lehalle © 2023 Cambridge University Press.

observed or inferred from data. Details of how exactly this is attained vary in a very substantial manner, depending on how such variables and their laws are chosen by the model. This approach is in stark contrast to recent developments in the field of machine learning which seek to bypass the domain expert in the model building process. A key question for quantitative finance research is how to reconcile these seemingly diametrically opposite modeling approaches? *There is a sobering realization in asset management that a "plug-and-chug" approach to modeling for asset management is far from adequate.*

Big Data:

Aside from computing resources, the advancement of machine learning would not be feasible without the proliferation of market data. Big data is now widely available across the finance industry and modeling practices have quickly moved to exploit it. For example, there is a draw towards more empirically driven modeling in asset pricing research – using ever richer sets of firm characteristics and "factors" to describe and understand differences in expected returns across assets and model the dynamics of the aggregate market equity risk premium (Gu et al., 2018).

Despite the unprecedented availability of historical data, there is no assurance that such data should be representative of the future. Equally, data coverage is far from uniform across asset space and there are often substantial challenges with ensuring accuracy in the data collection process. It stands to reason, therefore, that any attempt to put more emphasis on empirically driven research is more susceptible to the idiosyncrasies and limitations of the data. Big data does not equate to high informational content and machine learning cannot therefore be a panacea.

There are many sociological factors which have a sublime effect on how machine learning is collectively perceived among practitioners. Most notably, the investment industry is propelled by folklore around the success of machine learning. At the center of this is Renaissance Technologies' secretive Medallion Fund, which is closed to outsiders, and has earned an estimated annualized return of 35% since 1982, purportedly from early deployment of machine learning.

Another key point of contention is that the technology workforce entering the finance industry operate on the premise that iconic image classification problems, such as Kaggle's Cats vs. Dogs challenge, should be somewhat representative of prediction tasks in finance. After all, if you have a hammer, why wouldn't every problem be a nail? To illustrate why prediction in finance can be a fool's errand, we need only turn to the example of predicting daily asset returns.

As Israel et al. (2020) point out, if market participants were merely spectators of financial time series driven by micro and macro economic events, then indeed the challenge of asset return prediction would be tractable but not necessarily easy. However, traders with information that reliably predicts, say, a future rise in prices, don't sit passively on that information. Instead they start trading and this creates a feedback effect which fundamentally shifts the target. The very act of exploiting their predictive information pushes up prices, and thereby diminishes some of the predictability out of the market. And they don't stop after prices have risen just a little. They continue buying until they have exhausted their information, until prices adjust to the level that their information predicts. At such a point, there is little predictive information left to exploit. This idea, that competition in markets wipes out predictability, is the underpinning of the Nobel prize-winning work on the efficient markets hypothesis. The problem can be likened to the Cats vs. Dogs image classification problem where "cats morph into dogs once the algorithm becomes good at cat recognition" (Israel et al., 2020).

Of course, this is much too narrow an application to categorically suggest that the very notion of applying machine learning to financial markets data is flawed. There are many predictive variables beyond daily returns, e.g. price impact from limit order imbalances, liquidity, market regime change, credit events etc. Moreover, prediction is just one of many applications – nowcasting, surrogate modeling, dynamic programming, exploratory data analysis are just some of the many alternative ways in which machine learning is useful. Thus we've broadly illustrated the polarizing perspectives of machine learning, and the reality is that machine learning in asset management sits somewhere in between. To bring clarity to the utility of machine learning in finance, we must approach the intricate discussion in a more structured way. We hence distill some of the salient aspects which play a crucial role in this chapter.

Interpretability versus Auditability:

Are we asking too much or too little of machine learning? A controversial and long standing debate permeating all areas of machine learning is the question of model interpretability, i.e. can the parameters of the model be "meaningfully" interpreted to aid the understanding of the fitted machine learning model. As a baseline for interpretability, we could certainly look to human traders only to be reminded that snap decisions are often made without a rationale, perhaps overreacting to market conditions based on market sentiment or otherwise. Interpretability is a loaded term and has different meanings in different contexts. For example, statisticians would use the term to refer to a model which identifies which of its parameters are statistically significant. Interpretability is of little value if a trader can't vet the decisions, hence auditability is arguably equally or even more important than interpretability.

Is overfitting a data problem or a model problem?

As overfitting is the bain of trading strategy performance, there is a sophisticated and diverse range of perspectives on how to address overfitting. Some follow the more traditional and model based view of overfitting, others view the problem from the perspective of incorporating human knowledge and judgement, and others view overfitting as a data (imputation) problem.

All views, however, essentially agree on the notion that if we had a representative set of training data pertaining to the future, then there would be little concern about overfitting the model to data. The problem arises when the historical data is not representative of the future.

One approach which is quickly gaining traction in asset management and other areas of quantitative finance is how to incorporate other models into the estimating procedure. This is known as "model based" machine learning and in many ways shares similar motivations as choosing priors in Bayesian inference. The question then becomes one of how to embed such a model into a machine learning algorithm. A different approach is to learn the data and then simulate it under changing parameter values. This has led to the emergence of "market data generators".

The innovations presented in this chapter not only provide insight into these central questions but demonstrate the depth and breadth of contributions of machine learning to asset management. Before we introduce the first contribution to this chapter, we revisit some of the most pertinent classical financial theory as a pretext to a more technical discussion.

Classical Financial Theory:

Modern financial theory offers a number of approaches to asset pricing, which in the parlance of machine learning could be characterized as "model-based" approaches. They range from empirical regression models such as the Fama–French model, to latent factor models such as the Arbitrage Pricing Theory (APT) of Ross, to models of asset returns that are derived from a more fundamental analysis based on optimization of consumption of a representative investor in a market, see e.g. Cochrane (2001).

In particular, Cochrane showed how the one-factor CAPM model of Sharpe could be obtained as a result of consumption utility maximization with the Markowitz quadratic utility function, and under the assumption that the market is at equilibrium and all investors are identical (and the same as the representative investor) and hold the market portfolio. Furthermore, Cochrane also showed how different asset pricing models can be equivalently interpreted as particular models for the stochastic discount factor (SDF) (an "index of bad times") $M(\mathbf{F}_t, t)$ where $\mathbf{F}_t$ is a set of observed or inferred dynamic variables such as market and sector returns, macroeconomic variables etc. The SDF $M(\mathbf{F}_t, t)$ does *not* depend on characteristics of individual assets such as stocks, because with this theory the SDF is *universal* for all assets including stocks, bonds, derivatives etc. (Cochrane, 2001). All differences in returns of individual assets are explained in terms of covariances of assets' returns with the SDF $M(\mathbf{F}_t, t)$ – which in general will be different for different assets, and driven by each asset's individual characteristics.

It should be emphasized here that many of the key assumptions of the classical financial theory are made out of convenience of mathematical treatment rather than being rooted in data or empirical science. In particular, common investors in the market are highly unlikely to be fully rational, and a bounded-rational agent model of Simon (1956) may be a more accurate approach to modeling market agents. Furthermore, market dynamics are non-stationary, with regime changes that can be viewed as transitions between metastable states of the market similar to metastable systems in quantum and statistical physics (Halperin, 2020c). Market dynamics correspond to dynamics of an open, rather than a closed system, due to new money entering the market, with contributions to pension funds being the main channel of injecting the new money. Non-linearities arising from market frictions such as price impact and transaction costs are vital for producing dynamics with a reasonable long-term behavior (Halperin and Dixon, 2020; Halperin, 2020b). For more on physics-motivated approaches to analysis of markets as open and non-linear complex systems and a potential relevance of other physics-inspired methods, see Chapter 12 in Dixon et al. (2020).

While these questions address the foundations of modern financial theory, contemporaneous applications of machine learning to asset pricing take a largely "model-independent" approach. They primarily focus on relaxing restrictive linearity assumptions of classical financial models in modeling dependencies of asset returns on observables such as an individual firm's characteristics, e.g. the price-to-book ratio, and on market observables such as market or sector returns. While early work in this direction focused on predicting individual stock returns using machine learning methods involving either "shallow" methods such as various decision trees or deep learning methods, this does *not* amount to ML models for asset pricing. This is because asset pricing models seek to explain cross-section variations and longitudinal behavior of *all* assets in a given investment universe, e.g. all traded liquid stocks in the Russell 3000 universe.

14.1.1 *Pelger's asset pricing model*

Instead of taking a purely data-driven and "model-independent" machine learning approach, our first contribution, written by Marcus Pelger, offers a novel approach that combines the benefits of both classical financial theory and modern machine learning. Here we would like to briefly discuss the main technical innovations of Pelger's approach.

The first new element of Pelger's approach is using no-arbitrage constraints to regularize ML models. Unlike problems in image recognition where a signal-to-noise ratio is typically high, for financial applications such a signal is normally very low – most of financial data is *just* noise. Clearly, such data should be filtered to extract signals. Pelger's approach is to use no-arbitrage to regularize a machine learning model for asset pricing. Indeed, no-arbitrage is the most fundamental assumption of many classical financial theories, and is known to approximately hold, depending on the market, trading horizon etc., in real financial markets. Therefore, using no-arbitrage as a guiding principle to regularize a ML model appears on theoretical grounds as a better approach than using off-the-shelf regularization methods such as L2 or L1 regularization, or relying on filtering methods such as Fast Fourier Transform (FFT) to extract signals from noisy data.

Pelger's solution of enforcing no-arbitrage is rooted in the SDF approach mentioned above as a principled approach to asset pricing (Cochrane, 2001). As the mere existence of a SDF $M(\mathbf{F}_t, t)$ itself is critically dependent on the absence of arbitrage, once this assumption is accepted, any model building amounts to designing a parameterized model $M_\theta(\mathbf{F}_t, t)$ where θ is a set of trainable parameters, and training the model on available data.

Various models for the SDF $M_\theta(\mathbf{F}_t, t)$ have been considered in the financial literature. A pioneering paper on this topic written by Constantinides (1992) proposed an exponential–quadratic specification $M(x) = \exp\left(at + \sum_{i=1}^{N}(X_i(t) - \alpha_i)^2\right)$, where a, α_i are some parameters and $X_i(t)$ are Markov driver processes. Such a specification preserves positivity of the SDF for any values of $X_i(t)$. On the other hand, the CAPM model with a linear dependence of asset returns on market returns can also be interpreted as a SDF, though this time its positivity for arbitrary arguments is not guaranteed (Cochrane, 2001). Pelger's approach is focused on a linear specification of the SDF by projecting onto a set of returns in the investment universe, where non-linearities are introduced at the level of dependence of coefficients on the characteristics of individual firms. While this may not be the most general approach to the modeling of the SDF, it enables going far beyond traditional linear financial models in terms of model flexibility and predictive power.

Generative Adversarial Networks (GANs).

The second innovation of Pelger's approach is using Generative Adversarial Networks (GANs) for model training. The GAN approach is used here as a way to provide a more focused model training within the Generalized Method of Moments (GMM). The GMM method in its general form involves an infinite number of moments conditions of the type

$$\mathbb{E}\left[M\left(\mathbf{F}_t, t\right) R_{ti}^{\text{e}} g\left(\mathbf{F}_t, C_{ti}\right)\right] = 0,$$

where R_{ti}^{e} is the excess return of the ith asset at time t, and $g\left(\mathbf{F}_t, C_{ti}\right)$ is an arbitrary function of drivers $\mathbf{F}_t$ and firm-specific characteristics C_{ti}. While traditional financial approaches usually fix a set of moments used for calibration, e.g. by using 25 moments with double-sorted Fama–French portfolios, Chen et al. (2020) suggests instead using an adversarial approach with GAN into select moment conditions that produce the largest mispricing across assets.

Long-Short Term Memory Networks.

The third technical novelty of the deep learning asset pricing approach of Chen et al. (2020) is using Long-Short Term Memory (LSTM) neural networks to construct economic regimes from multidimensional time-series data containing macroeconomic variables. LSTMs capture both short- and long-term dependencies between macroeconomic variables that are needed to represent business cycles. This provides an approach to capturing the non-stationarity of macroeconomic dynamics, and appears to be a meaningful alternative to a "naïve" use of macroeconomic variables or their first differences as predictors of next-period asset returns. The GAN architecture of Chen et al. (2020) shows substantial improvements over previous results in terms of explained means and variances of asset returns, and Sharpe ratios of test portfolios. Furthermore, the chapter discusses alternative approaches of using decision trees for building more interpretable models of the SDF and asset returns.

14.2 Data wins the center stage

Our next contribution comes from Horvath, Gonzalez, and Pakkanen, who address applications of machine learning for solving traditional problems of quantitative finance such as option pricing and hedging. The unifying theme of all approaches considered in their chapter is the reliance on synthetic data rather than on real market data and is born from the perspective outlined above, namely that successful application of machine learning to quantitative finance requires a tradeoff between the plug-and-chug approach and over-used hand-crafted financial models, with their arsenal of modeling assumptions. As it takes another model (e.g. an autoencoder) to produce such synthetic data, approaches considered in this part should qualify as *surrogate model-based* approaches, which is different from more conventional machine learning approaches that operate with real data. For avoidance of doubt, the data referenced in the title of their chapter is thus not the actual data but rather synthetic data generated by a model. In this sense, approaches considered in this Part of the book are conceptually similar to traditional financial approaches, such as Monte-Carlo-based option pricing, while offering new tools from machine learning, such as neural networks, for better efficiency.

In this vein, the authors begin by introducing arguably their most celebrated and prize winning work, surrogate deep learning models for calibrating rough Bergomi models for option pricing. Prior to the advent of their work, deep learning had gained little traction in the finance industry and was regarded as a tool for prediction. At the same time, the investment banking sector was reluctant to adopt more robust and realistic option pricing models after a decade of battling with high performance computing solutions to far simpler models. Horvath et al.'s work was a match to tinder, replacing a cumbersome and intractable calibration procedure involving genetic algorithms, with a deep learning surrogate model which could eliminate one layer of computation in the calibration, speeding up calibration by a three-digit factor, and effectively rendering the rough Bergomi model calibration tractable and hence usable in practice. Their work remains one of the most exemplary motivations for deep learning to be included in the modern financial engineering toolbox.

14.2.1 Deep hedging vs. reinforcement learning for option pricing

The surrogate model outlined above deals with option pricing but does not address the problem of option hedging. While the former problem enables a formal solution that operates under the risk-neutral (pricing) probability measure $\mathbb{Q}$, hedging needs to be performed under the physical measure $\mathbb{P}$. While in the classical models such as Black–Scholes (BS) one first computes the option price and then its delta (option hedge), this is no longer the right sequence in a discrete-time setting, even if one retains the same lognormal assumptions for the underlying stock price process as in the BS model. In the discrete-time setting, it is the hedging strategy that should be chosen first according to some optimization criterion, and the option price is only determined once that strategy is specified. This was very clearly shown in discrete-time models for option pricing by Föllmer and Schweizer (1989), Schweizer (1995), and Potters et al. (2001) which demonstrated how to price options using local risk minimization.

The first data-driven and model-independent way for consistent option pricing and hedging was proposed by Halperin (2019, 2020a). The QLBS model of Halperin first restated the local risk minimization approach of Föllmer and Schweizer (1989); Schweizer (1995); Potters et al. (2001) as a Markov decision process (MDP) and showed how it could be solved in a model-based Monte Carlo setting. It then showed how the same MDP problem can be solved in a model-independent way by applying Q-learning. While the QLBS model mostly focused on quadratic utility without transaction costs, the original paper also introduced an alternative pricing and hedging scheme based on indifference pricing with the exponential utility.

The original deep hedging paper by Buehler et al. (2018) followed instead the traditional model- and simulation-based methods based on a utility-based indifference pricing approach, where the new element is introduced on the technical side, and amounts to using neural networks for function approximation. The main idea was to translate the problem of maximizing the utility function of the P&L over adapted strategy processes under a set of risk preferences, to learning the functional representation of the strategy with deep networks. Such a translation is justified on theoretical grounds by both the Doob–Dynkin lemma and the universal representation theorem.

It is instructive to further compare and contrast the two approaches. In the Q-learning based version of the QLBS model, the model learns in an offline setting from demonstrated sequences of stock prices and option hedges, treated as states and actions for reinforcement learning of *optimal* option prices and hedges. As Q-learning is an *off-policy* algorithm, with the QLBS model, the optimal hedging and pricing can be learned from data even when the latter is obtained using a sub-optimal hedging strategy.

In contrast, the deep hedging method of Buehler et al. (2018) is a simulation-based dynamic programming approach. When trajectories of the underlying are taken from data rather than being simulated, it can be viewed as *on-policy* model-based reinforcement learning. While it can also be formally viewed as an "unsupervised" approach in the sense that it only depends on prices of the underlying, a "teacher" is still implicit in this approach due to the assumption that the demonstrated trajectories correspond to an optimal policy. This assumption is hard to validate when paths of the underlying are taken from actual data in the presence of price impact, rather than being simulated from a known model.

Another point to bring to the attention of the reader is the use of the real-world measure $\mathbb{P}$ rather than the pricing measure $\mathbb{Q}$. While utility-based approaches employ the real-world measure $\mathbb{P}$, one has to rely on the Girsanov theory to use risk-neutral drifts for stock prices in this framework, which essentially amounts to finding the price of risk. The latter, by itself, is a challenging problem and while the chapter continues to serve as a starting point for understanding the importance of designing deep learning methods from theoretical results, more fundamental work in this area is needed to integrate deep learning into a probabilistic computational framework suitable for financial mathematics.

14.2.2 Market simulators

One challenge with using deep learning for option pricing and other problems in quantitative finance is data requirements. While use cases for deep learning in applications to image recognition usually employs large datasets incorporating tens or hundreds of thousands, and sometimes even millions, of examples, with option pricing the amount of available data is typically a few orders of magnitude smaller, unless one deals with intraday data.

One implication of this fact is that deep learning cannot be used for option pricing in a model-independent and purely data-driven way, without the introduction of a no-arbitrage model (see e.g. Chataigner et al., 2020). Short of dismissing deep learning on these grounds as an unsuitable approach for finance outside of intraday or high-frequency trading, an alternative approach is to rely on another machine learning model to produce artificial (synthetic) data that mimics some important characteristics of real data. In this regard, we can consider the approach as an extension of surrogate modeling in which real pricing data is compressed to a data model, and a machine-learning-based surrogate model learns the compressed representation. Such ideas have been unfashionable in the machine learning community on account of their lack of end-to-end architecture, but the finance industry is in favor of modularity on account of model risk and regulatory compliance among other reasons.

In particular, recent works on using conditional variational autoencoders (CVAE) using path signatures outlined in this chapter offer an interesting direction of research. This approach was shown to learn from small data (a few hundred or thousands paths) and generate artificial data which appear indistinguishable from real data based on statistical tests such as the Kolmogorov–Smirnov (KS) test. This is very encouraging: however one should remember that performance metrics such as the KS test or tail metrics such as VAR, CVAR or higher moments are themselves noisy, and exhibit strong fluctuations for small data. With market simulators, model risk does not disappear, of course, but is rather pushed to the tails of generated distributions. Therefore, caution should be exercised when applying deep-learning-based methods in combination with market simulators to enrich a dataset – if the problem is such that it is highly sensitive to tails of the distribution, the net effect is that we simply compound the amount of model risk, as we end up with two models, one for data generation, and another which uses this data for pricing. In our view, while these questions about potential model dependence in the tails of distribution are hard to escape for pricing or risk management, another potentially powerful and complementary direction is the application of market simulators to stressed scenario generation. In such applications, we need not *match* actual tail distributions, which is a hard problem as just explained, but rather *generate* a range of tail distributions for stress testing with scenario generation.

Perspectives on what is actually needed to constitute a reliable back-test is a vigorously debated area for practitioners, but the emerging consensus is that a multi-pronged approach is needed, and one that encompasses market data generation is likely to become increasingly prominent in the future, not least because it solves the problem of data licensing which has plagued the industry from being able to offer standardized machine learning benchmarks and collaborate with academia. Horvath et al.'s approach serves to illustrate one of the most important questions of how best to integrate machine learning with financial modeling. Whereas Pelger treats the asset price data as gospel – a reasonable assertion given strong data coverage – Horvath et al. contend with the relatively poor data coverage in option quotes. Pelger also works in the historical measure whereas Horvath et al. work in the pricing measure. These two factors are important in considering whether to simulate data or not.

14.3 Stratified models for portfolio construction

Our narrative on simulated data versus model based machine learning is further enriched by studying the final chapter in this part. Tuck et al. present the topic of a stratified modeling approach to portfolio construction and provide an innovative solution to the problem of overfitting to historical asset returns data with a model-based approach. Again, as with Pelger's work, their approach is predicated on the notion that the data is gospel and it's therefore necessary to regularize machine learning by introducing a model. However, in this case, rather than introduce a financial model, they introduce a data model as a form of regularization.

More specifically, the authors condition the asset return on a stratified variable, representing market conditions or some trading signal, with the goal of constructing mean–variance portfolio strategies conditioned on the state of the market. On the surface, such an approach is analogous to some of the early attempts at hidden Markov modeling in financial markets, introducing a latent variable to represent the well-established notion of regimes in markets – such as when the market performs historically well or poorly (a "bull" or "bear" market, respectively), or when there is historically high or low volatility. Indeed, there is a long line of literature approaching regimes from the perspective of Gaussian Mixture models and Markov switching. However, in these cases, the regime is a latent state whose probability is estimated by popular algorithms such as the EM algorithm (see Hamilton, 2010 and the references therein).

Tuck et al.'s approach is fundamentally different. Instead they treat the market regime as observable and do not approach stratified asset returns from a multi-modal inferential process, but rather directly condition the returns on the stratified variable. The idea of using discrete market observables is not new of course, but the intrinsic challenges has always been how to stratify returns on values of the market observable which do not exist in the dataset. This question is central to the robustness of such an approach as future values of the market observables, not in the training set, could very well exist in the test set.

Their approach uses a Laplacian regularization term that encourages nearby market conditions to have similar means and covariances. Crucially, this technique thus allows models for market conditions which have *not* occurred in the training data, by borrowing strength from nearby market conditions for which the data is available. At the core they solve a data imputation problem and their solution can be viewed as a graph-based approach to regularization.

An elaborate single-period portfolio construction approach is demonstrated which represents the real-world needs of practitioners. In addition to a Markowitz-style utility function, they add shorting costs, transactions costs, and risk, position, and leverage limits. The method is tested on a small universe of 18 ETFs and is found to perform well out of sample. The extension to the multi-period setting and further important methodological developments are also discussed.

These early results suggest a very different approach to performance generalization from that of Horvath et al. – the former relying on data imputation through a graph, rather than fitting a neural-network-based market data simulator to historical data. Both involve computational graphs, but the former is using the proximity of nearby market conditions to regularize the data, whereas the latter graph is being used as a network to approximate the forward map and its structure carries little to no meaning.

The regularized stratified models are not only interpretable but "auditable" – the authors are refreshingly measured as to make such a distinction. The latter property simply means that it's possible to check whether the model output is reasonable. The distinction is important in the ongoing, more general debate, about machine learning efficacy: clearly having both properties is highly desirable but, for example, having interpretability without auditability is not.

14.4 Summary

Machine learning has found numerous applications and has contributed to, rather than detracted from, the rigor of these financial disciplines. One of the advantages of machine-learning techniques is their inherent focus on data and the attention to detail when extracting information from those data. This focus enables data-driven innovation. At the same time, the emphasis on out of sample performance, by its very nature, contributes to financial stability. We very much hope that the ideas contained in these chapters will mature to generate further interest in this exciting new field and invite more researchers to build on the main concepts introduced in these seminal works.

References

Buehler, Hans, Gonon, Lukas, Teichmann Josef, and Wood, Ben. 2018. Deep hedging. Available at SSRN 3120710.

Chataigner, Marc, Crépey, Stéphane, and Dixon, Matthew. 2020. Deep local volatility. *Risks*, **8**(3), 82.

Chen, Luyang, Pelger, Markus, and Zhu, Jason. 2020. Deep learning in asset pricing. ArXiv:1904.00745.

Cochrane, John Howland. 2001. *Asset Pricing.* Princeton University Press.

Constantinides, George M. 1992. A theory of the nominal term structure of interest rates. *Review of Financial Studies,* **5**(4), 531–552.

DeepMind. 2016. DeepMind AI reduces Google data centre cooling bill by 40%. `https://deepmind.com/blog/deepmind-ai-reduces-google-data-centre-cooling-bill-40/`.

DeepMind. 2017. The story of AlphaGo so far. `https://deepmind.com/research/alphago/`.

Dixon, M. F., Halperin, I., and Bilokon, P. 2020. Machine learning in finance: from theory to practice. Springer.

Esteva, Andre, Kuprel, Brett, Novoa, Roberto A., Ko, Justin, Swetter, Susan M., Blau, Helen M., and Thrun, Sebastian. 2017. Dermatologist-level classification of skin cancer with deep neural networks. *Nature,* **542**(7639), 115–118.

Föllmer, H., and Schweizer, M. 1989. Hedging by sequential regression: an introduction to the mathematics of option trading. ASTIN Bulletin, **18**, 147–160.

Gu, Shihao, Kelly, Bryan T., and Xiu, Dacheng. 2018. Empirical asset pricing via machine learning. Chicago Booth Research Paper 18–04.

Halperin, I. 2019. The QLBS Q-learner goes nuQLear: Fitted Q iteration, inverse RL, and option portfolios. *Quantitative Finance,* **19**(9), 1469–7688.

Halperin, I. 2020a. QLBS: Q-learner in the Black–Scholes(–Merton) worlds. *Journal of Derivatives,* **28**(1), 99–122.

Halperin, Igor. 2020b. The inverted parabola world of classical quantitative finance: non-equilibrium and non-perturbative finance perspective. Available at SSRN 3669972.

Halperin, Igor. 2020c. Non-equilibrium skewness, market crises, and option pricing: Non-linear Langevin model of markets with supersymmetry. Available at SSRN 3724000

Halperin, Igor, and Dixon, Matthew. 2020. Quantum equilibrium-disequilibrium: Asset price dynamics, symmetry breaking, and defaults as dissipative instantons. *Physica A,* **537**, 122187.

Hamilton, James D. 2010. Regime switching models. Pages 202–209 of: *Macroeconometrics and Time Series Analysis,* S.N. Durlauf and L.E. Blume (eds). Palgrave Macmillan.

Israel, Ronen, Kelly, Bryan, and Moskowitz, Tobias. 2020. Can machines "learn" finance? *Journal Of Investment Management,* **18**(2).

Kubota, Taylor. 2017. Artificial intelligence used to identify skin cancer. `https://news.stanford.edu/2017/01/25/artificial-intelligence-used-identify-skin-cancer/`.

Markowitz, H. 1959. *Portfolio Selection: Efficient Diversification of Investments.* Wiley.

Markowitz, Harry. 1956. The optimization of a quadratic function subject to linear constraints. *Naval Research Logistics Quarterly,* **3**(1–2), 111–133.

Poggio, T. 2016. Deep learning: Mathematics and neuroscience. *A sponsored supplement to* Science, *Brain-Inspired Intelligent Robotics: The Intersection of Robotics and Neuroscience,* 9–12.

Potters, M., Bouchaud, J.P., and Sestovic, D. 2001. Hedged Monte Carlo: Low variance derivative pricing with objective probabilities. *Physica A,* **289**, 517–525.

Schweizer, M. 1995. Variance-optimal hedging in discrete time. *Mathematics of Operations Research,* **20**(1), 1–32.

Simon, H. 1956. Rational choice and the structure of the environment. *Psychological Review,* **63**(2). 129–138.

Simonyan, Karen, and Zisserman, Andrew. 2014. Very deep convolutional networks for large-scale image recognition. In: *International Conference on Learning Representations.*

15

Harnessing Quantitative Finance by Data-Centric Methods

Blanka Horvath[a], Aitor Muguruza Gonzalez[b]
and Mikko S. Pakkanen[c]

Abstract

Data-centric methodology, machine learning and deep learning in particular, can greatly facilitate various computational and modelling tasks in quantitative finance. In this chapter, we first demonstrate how supervised learning can help us implement and calibrate option pricing models that have previously been hard to deploy due to their analytical intractability. Secondly, we illustrate how we can discover optimal hedging strategies and arbitrage-free prices in a model-free fashion via the recent unsupervised deep hedging approach. As the availability of high-quality training samples underpins these data-centric methods, we finally outline recent work in the nascent field of market data generators, which are used to generate realistic, yet synthetic, market data for the training of financial machine learning algorithms.

15.1 Data-centric methods in quantitative finance

Deep learning has had a major impact on the possibilities of mathematical modelling in finance in recent years. The momentum in research and innovation that new computational technologies are creating in quantitative finance is observable across the financial sector and quantitative finance communities (De Spiegeleer et al., 2018; Dixon et al., 2020; Gnoatto et al., 2020; Huge and Savine, 2020; Itkin, 2014; Koshiyama et al., 2019; Liu et al., 2019; Sabate-Vidales et al., 2018). This impact is still accumulating and shifting the focus of quantitative finance research & innovation towards more data driven technologies.

The most significant transformation to our modelling practice is perhaps the realisation that – whether it is real, historical or synthetically generated – *data is increasingly becoming an integral part of the models and algorithms*, while restrictions around data have the very real potential to restrain dialogue beyond

[a] Technical University of Munich, Munich Data Science Institute, King's College London and The Alan Turing Institute
[b] Imperial College London and Kaiju Capital Management
[c] Imperial College London
Published in *Machine Learning And Data Sciences For Financial Markets*, Agostino Capponi and Charles-Albert Lehalle © 2023 Cambridge University Press.

individual firms. At the same time, this very dialogue across the sector (and beyond), with research institutions as well as regulatory bodies is essential to develop reliable risk management, model governance and efficient regulatory frameworks for the emerging technologies. This calls for our communities to direct an increased amount of research focus to terrains which (so far) have received relatively little attention by the mainstream of our investigative efforts in mathematical finance. Namely, research that is concerned with:

(A) The data itself: in particular its "representativeness" for the applications it will be used for, and at the same time, its "protectiveness" of the individuals it describes: A balance that is more challenging to strike than it seems at first site, which would most definitely benefit from a new momentum in research that is particularly focussed on quantitative finance applications.

(B) The symbiosis of the data and the models it is applied to: monitoring (with risk management in mind) the suitability of the algorithms and data with one another. That is, investigating, as data evolves, whether the models and algorithms need to be "re-trained" or need to undergo more significant "architecture updates". And, finally:

(C) The synergies with classical modelling: in other words, the lessons that can be brought into this new momentum of research from classical quantitative finance modelling practices in order to make this transformation (which seems unavoidable, whether we like it or not) as smooth and free of major disruptions as possible.

We start with a quick overview that summarises *three distinct areas of progress* in innovation (alongside with a gradually emerging fourth one), which we briefly describe below and extend upon in later sections. We see these areas as *distinct* in the sense, that they had considerably different effects on the development of the aforementioned momentum: they roughly translate to improvements in the:

(1) Speed and generality of algorithms connected to our models;
(2) Automation of processes;
(3) Ability to numerically generate data that can be succinctly described as *synthetic twins* of real datasets which raises further modelling challenges.

The literature is rapidly growing in each of the areas, as the excellent survey by Ruf and Wang (2020) illustrates, which gave a comprehensive snapshot of related research contributions as of mid-2020. Therefore, instead of extending the survey with the newest contributions up to today, we discuss here – without the claim of completeness – three areas of progress from the perspective of their transformative effect on further developments and for each of these areas we showcase some applications that exemplify the developments and highlight some further milestones. The application of new technologies in the field of quantitative finance (machine learning and deep neural networks in particular), has helped push the boundaries of the achievable further in some of the typical areas of quantitative finance activity:

(1) Speed and generality of the solutions has increased tremendously over the past years by providing numerical alternatives to established means of pricing (and hedging) of derivatives, optimal stopping, portfolio optimisation and forecasting. This in turn is changing the nature of modelling: deep learning-based algorithms facilitate speeding up calculations beyond what was possible with traditional numerical methods, and solving far higher dimensional problems than before. These numerical alternatives help us overcome the curse of dimensionality, as well as lifts the limitations imposed by the pressing need for tractability in modelling, thereby paving the way for new, more realistic model classes in finance. What is in common in these solutions is that they address problems that in principle could be solved (albeit less efficiently) by traditional numerical means, which can be used as benchmarks to monitor the correctness of outcomes. A similar speedup effect would be expected from a widespread availability of quantum computing that would facilitate further increases in computing speed by five-digit factors.

(2) A different scenario from the above is the area of developments where carefully chosen network topologies were used to design a vastly different sort of approximation tools which allow to derive (approximately) optimal solutions to problems in an automated way beyond the regimes where theoretical handcrafted (exact) solutions existed. For example hedging strategies could be derived under consideration of market frictions, in setups that are more general than the ones previously analysed and understood by classical methods.

(3) Significant progress was made in terms of generalisations of our means to simulate the dynamic evolution of financial assets, and generalisations of the very concept of a financial market model to what we now call *market generators* (or *market data simulators*). This development exemplifies a shift in the *culture* of modelling that traditionally favoured tractability over accuracy for models, towards an increased demand on the quality of simulated data and towards more elaborate means of measuring the quality of the latter.

In this chapter we showcase examples of optimisation problems in (1) and (2) and explain in simple settings how these solutions can lead to advantages in speed as described above. Furthermore, we demonstrate how the deep learning applications in (1, 2) (for example deep hedging) drives the interest for more flexible market models or model-free, data-driven market generators described in (3), and to developments that are leading to (4):

(4) Finally, but perhaps most significantly, data-driven modelling and synthetic generation of market data opens the door to further – more data focussed – avenues of quantitative modelling in finance that were so far not a core part of the traditional quantitative finance toolkit. Such avenues are for example – the increasingly important – data anonymisation techniques and related applications.

The effects of (1) and (2) are not to be underestimated. Primary effects are observable in (3) and leading to (4), while secondary effects can lead to better synergies among traditionally segregated quant disciplines, which were (so far) restrained by the mantra of tractability to their respective "corners" of modelling realities. As we become more adept at incorporating signals from different markets (Koshiyama et al., 2019) into our models and strategies, distinct quant disciplines move closer together, (Lee, 2021). While rough volatility models (Gatheral et al., 2018; Bayer et al., 2016) have already set this trend in motion, machine learning methods take this a step further: in an ideal world, we could directly create a model that captures *all* independent variables driving the evolution of observed quantities (for example all inputs relevant for price formation, or all risk factors, relevant for our risk assessment).

In reality however, the tractability of pricing and hedging methods was arguably the far more critical factor than their accuracy[1], and the limits of computation determined what methods were used to address financial problems. Recent developments summarised in (1) and (2) have effectively enabled us to speed up calculations by several orders of magnitude (see for example Bayer and Stemper, 2018; Horvath et al., 2021a) and thereby to lift the restrictive need for tractability. This in turn lightens the previous dichotomy between tractability and accuracy of modelling. One of the consequence (among many) of the previously prevalent mantra for tractability, was that several quantitative finance disciplines remained fairly segregated.

We used models that are tractable enough for the application at hand, and settled to predict (parts of) reality, that can be described by these tractable models under suitable (often idealised) conditions. In other words, we created a number of specialised tools (hammers) that were not always exactly appropriate for the problem (nail) they intended to solve.

Ironically, it is commonplace to see it the other way around: machine learning has been frequently pointed at as the popular *hammer* to which we keep presenting more or less suitable *nails*. We argue that the truth lies somewhere in between, and the challenge lies in creating a carefully crafted blend of traditional and modern methods. The recipe for this blend calls for collaborative efforts across the sector and across quant disciplines.

In order to facilitate this collaborative effort, there is an urgent need to facilitate means of sharing, measuring and processing data and thus to direct the centre of our attention at the considerations and challenges around data, to propose rich, high quality, privacy-preserving benchmark datasets for academic research, regulatory purposes and a transparent multilateral dialogue across the financial sector.

In Section 15.4 we discuss how data driven modelling enters financial applications. Training the neural network is a first step before the algorithm is used, and the training of the networks (and the training data used) shapes the performance of algorithms.

---

[1] Furthermore, statements about whether or not there are further relevant factors involved and that the model ignores (and if so, how many) were rarely known.

In Section 15.2 we highlight an example where deep learning – as an alternative to standard numerical methods – facilitated a speedup that had substantial consequences on the practical applicability of the model. For the rough Bergomi model introduced in Bayer et al. (2016) (see (15.5) below), which has a plethora of modelling advantages it was a known problem that before the availability of deep pricing the model was (prohibitively) slow to price and calibrate by classical numerical means.

Section 15.3 roughly corresponds to the area (2) of developments which we highlight in two unsupervised learning examples.

In Section 15.4 we discuss how solutions and algorithms described in Section 15.3 drive the interest for more flexible financial market models. In this section we therefore introduce the concept of *market generators*. We briefly discuss their synergies with classical models as well as how the use of market generator models necessitates a rethinking of risk management techniques. The latter creates a wealth of new questions for research (described in Section 15.5), in quantitative finance communities within finance among practitioners and academics.

15.2 Pricing and calibration by supervised learning

One of the quintessential problems in finance is the calibration of models to market data. This in plain words, comes down to choosing the parameter combination in a given stochastic financial model that renders options prices in that stochastic model, which are closest to option prices observed in the market. In practice, for a close calibration it may be necessary to evaluate option prices corresponding to a considerable number of parameter choices, until the optimal parameter combination is found. If each evaluation of a parameter choice is computationally expensive (i.e. the model is not tractable), this can become a computational bottleneck to the calibration process.

Traditionally, the tractability of financial models was arguably more critical than their accuracy. Over the past decades, considerable efforts were made to derive efficient asymptotic and numerical approximations to stochastic financial models in order to make the calibration routine feasible in reasonable computational time. One such famous example is the SABR implied volatility approximation formula of Hagan et al. (2002), which was largely responsible for the remarkable popularity of SABR model.

With this preparation in mind it is natural to consider supervised learning as an alternative numerical tool for constructing approximations to derivative prices.

15.2.1 Model calibration framework

To recall the mathematical setting of model calibration we follow closely the framework introduced in Bayer et al. (2019) and Horvath et al. (2021a). Suppose that a model is parametrised by a set of parameters Θ, i.e., by $\theta \in \Theta$. Furthermore, we consider a financial contract parametrised by a parameter $\zeta \in Z$. For example, for put and call options we generally have $\zeta = (T, K)$, the option's maturity and strike. There might be further parameters which are needed to compute prices but can be observed on the market and, hence, do not need to be calibrated. For instance, the spot price of the underlying, the interest rate, or the forward variance curve in Bergomi-type models (Bergomi, 2016) falls under this type. For this quick overview, we ignore this category. We introduce the *pricing map*

$$(\theta, \zeta) \mapsto P(\theta, \zeta), \tag{15.1}$$

the price of a financial derivative with parameters ζ in the model with parameters θ. It is this map (15.1) that we will aim to learn by supervised learning techniques either directly, for some (simple or exotic) payoff functions ζ, or indirectly, by learning the implied volatility map

$$\theta \mapsto \sigma(\theta, K, T), \tag{15.2}$$

of vanilla payoffs $\zeta(\cdot, K, T) \equiv (\cdot \mid_T - K)_+$ for some $T, K > 0$. Financial practice often prefers to work with implied volatilities rather than option prices, and we will also do so in the numerical parts of this chapter containing vanilla contracts. For the purpose of this introduction, any mention of a *price* may be, *mutatis mutandis*, replaced by the corresponding implied volatility.

Observations of market prices $\mathcal{P}(\zeta)$ for options are parametrised by ζ for a (finite) subset $\zeta \in Z' \subset Z$ of all possible option parameters.

When the model is *calibrated*, a model parameter θ is identified which minimizes a distance δ between model prices $(P(\theta, \zeta))_{\zeta \in Z'}$ and observed market prices $(\mathcal{P}(\zeta))_{\zeta \in Z'}$, i.e.,

$$\widehat{\theta} = \arg\min_{\theta \in \Theta} \delta \left((P(\theta, \zeta))_{\zeta \in Z'}, (\mathcal{P}(\zeta))_{\zeta \in Z'} \right). \tag{15.3}$$

Hence, the faster each model price $(P(\theta, \zeta))$ can be computed, the faster the calibration routine.

The most common choice of a distance function δ is a suitably weighted least squares function, i.e.,

$$\widehat{\theta} = \arg\min_{\theta \in \Theta} \sum_{\zeta \in Z'} w_\zeta \left(P(\theta, \zeta) - \mathcal{P}(\zeta) \right)^2.$$

15.2.2 *Pricing and calibration aided by deep neural networks*

It is tempting to train a neural network to perform the process of model calibration described above. These efforts of arriving at optimal model parameters, (directly) from observations in the data were pioneered by Hernandez (2017) and also followed by Dimitroff et al. (2018), Stone (2020) and others. A main characteristic of this branch of supervised learning approaches is that parameter calibration is done directly through supervised learning using data from previous, already calibrated datasets. Indeed, the idea is to directly learn the whole calibration problem, i.e., to learn the model parameters as a function of the market prices (parametrised as implied volatilities in the vanilla options case). In the formulation of (15.3), this means that we learn the mapping

$$\Pi^{-1} : (\mathcal{P}(\zeta))_{\zeta \in Z'} \mapsto \widehat{\theta}.$$

A more conservative path, taken by many authors including Ferguson and Green (2018); McGhee (2018); De Spiegeleer et al. (2018); Bayer et al. (2019); Horvath et al. (2021a), follows more closely the classical routine of calibration outlined in the previous section, consisting of evaluating an array of option prices for a selected number of parameter combinations[2]. In contrast to the classical routine however, now the calculation of each option price (which was traditionally calculated numerically) is now approximated by a neural network. Since one separates the calibration procedure from pricing, we refer to it as a two-step approach: we first learn the pricing map by a supervised learning technique that maps parameters of a financial models to prices before calibration is performed. More precisely, in step **(i)** we learn the pricing map P presented in (15.1) (off-line) and in a second step, **(ii)**, we calibrate (on-line) the model – as approximated in step **(i)** – to market data using a standard calibration routine. To formalise the two-step approach, for an option payoff ζ and a model $\mathcal{M}$ with parameters $\theta \in \Theta$ we write $\widetilde{P}(\theta, \zeta) \approx P(\theta, \zeta)$ for the approximation $\widetilde{P}$ of the true pricing map P based on a supervised learning technique. Then, in the second step, for a properly chosen distance function dist (and a properly chosen optimization algorithm) we calibrate the model by computing

$$\widehat{\theta} = \arg\min_{\theta \in \Theta} \operatorname{dist}\left(\left(\widetilde{P}(\theta, \zeta) \right)_{\zeta \in Z'}, (\mathcal{P}(\zeta))_{\zeta \in Z'} \right). \tag{15.4}$$

In summary, this method is similar to traditional routines, as the true option price has to be numerically approximated (except in very simple models like Black–Scholes) in a first step before it is calibrated to market data. The main difference is that here the approximation of the option price is now replaced by a deep neural network approximation.

- The deep neural network is only trained to approximate the option prices in the chosen model. Therefore, in this approximation, synthetic data from stochastic models is used for training. Hence, we can easily produce as many data samples for training as needed, and the training data are completely unpolluted by market imperfections.
- Decomposing the calibration problem into a two-step approach also induces a natural decomposition of the overall calibration error into pricing error (from the price approximation by the neural network) and model mis-specification. Hence, the performance of the approach is generally independent of changing market regimes – which might, of course, affect the validity of the model dynamically, as the market evolves.

---

[2] The sequence of which is determined by the chosen calibration algorithm

15.2.3 An example where deep pricing makes a difference: the rough Bergomi model

We illustrate here the advantages of this two-step approach in terms of out-of-sample performance with unseen data. We shall work with the rough Bergomi model (Bayer et al., 2016) as an example. In the abstract model framework, the rough Bergomi model is represented by $\mathcal{M}^{\text{rBergomi}}(\Theta^{\text{rBergomi}})$, with parameters $\theta = (\xi_0, \eta, \rho, H) \in \Theta^{\text{rBergomi}}$. For instance, we may choose

$$\Theta^{\text{rBergomi}} = \mathbb{R}_+ \times \mathbb{R}_+ \times [-1,1] \times]0, 1/2[,$$

to stay in a truly rough setting. The model corresponds to the following system for the log price X and the instantaneous variance V:

$$dX_t = -\frac{1}{2}V_t dt + \sqrt{V_t} dW_t, \quad \text{for } t > 0, \quad X_0 = 0, \tag{15.5a}$$

$$V_t = \xi_0 \mathcal{E}\left(\sqrt{2H}\eta \int_0^t (t-s)^{H-1/2} dZ_s\right), \quad \text{for } t > 0, \quad V_0 = v_0 > 0, \tag{15.5b}$$

where H denotes the Hurst parameter, $\eta > 0$, $\mathcal{E}(\cdot)$ the Wick exponential, and $\xi_0 > 0$ denotes the initial forward variance curve, and W and Z are correlated standard Brownian motions with correlation parameter $\rho \in [-1,1]$.

This model is particularly appealing for the supervised learning approach as Monte Carlo pricing techniques are available to perform pricing, and for small Hurst parameters, this techniques becomes computationally even slower than in the standard case. This clearly creates a bottleneck for calibration since (in particular for small values of the parameter H) the model is not computationally tractable enough for some practical purposes. For this specific example we will use deep neural networks and supervised learning to create a numerical alternative for pricing and compare the performance of these. Once Neural Networks are trained, we expose the data to an unseen market scenario e.g. data generated by a completely different model whose specific dynamics are not relevant.

We note that what we aim at achieving with this approach, is to maintain the precision of the traditional numerical method (Monte Carlo in this case) while speeding up the process. We show that the Neural Network approximation in terms of pricing (and calibration) *precision* is as good as the Monte Carlo technique, while clearly the pricing (and calibration) *speed* is by several orders of magnitude faster in the neural network case than in the traditional (Monte Carlo) case. By this we demonstrate that the speedup is not achieved at the cost of precision.

Numerical experiments presented in Bayer and Stemper (2018), Bayer et al. (2019), and Horvath et al. (2021a) demonstrate that such supervised learning algorithms can be devised (whether pointwise or grid-based) to speed up the calibration process by a factor of up to 30,000 of the original speed. Below, different learning methods and feature extraction rules are discussed to set up this supervised learning task. For the advantages and drawbacks of each, we refer the reader to the original articles.

15.2.4 Choosing the feature set

Deep learning has been incredibly successful in supervised learning problems for pricing as shown in an array of recent advances (Bayer et al., 2019; Horvath et al., 2021a). There is an inherent structure in neighbouring derivatives contracts that can be exploited by learning as well: Option prices are characterised by maturities and strikes, which are in turn governed by certain arbitrage-free relations. By sampling the implied volatility on a grid, these relations can be exploited. On the other hand, a more straightforward *pointwise* learning approach (sampling only one point on a surface for each parameter combination) permits the sampling of points on the implied volatility surface more flexibly: Bayer et al. (2019) compare both approaches in the example of the rough Bergomi model and the classical Heston model and Horvath et al. (2021a) take the analysis further by applying the *grid-based* learning approach to more complex models and contracts. We briefly present here both approaches for the case of vanilla options. We refer to Bayer et al. (2019) for an in-depth discussion of the advantages and drawbacks of each.

Pointwise learning

Step (i): Learn the map $\widetilde{P}(\theta, T, k) = \widetilde{\sigma}^{M(\theta)}(T, K)$ – that is, in equation (15.4) above we have $\zeta = (T, K)$. In the case of vanilla options ($\zeta = (T, K)$) one can rephrase this learning objective as an implied volatility problem: In the implied volatility problem the more informative implied volatility map $\widetilde{\sigma}^{M(\theta)}(T, K)$ is learned, rather than call- or put option prices $\widetilde{P}(\theta, T, K)$. We denote the artificial neural network by $\widetilde{F}(w; \theta, \zeta)$ as a function of the weights w of the neural network, the model parameters θ and the option parameters ζ. The optimisation problem to solve is the following:

$$\widehat{\omega} := \arg\min_{w \in \mathbb{R}^n} \sum_{i=1}^{N_{\text{Train}}} \eta_i (\widetilde{F}(w; \theta_i, T_i, K_i) - \widetilde{\sigma}^M(\theta_i, T_i, K_i))^2. \quad (15.6)$$

where $\eta_i \in \mathbb{R}_{>0}$ is a weight vector.

Step (ii): Solve the classical model calibration problem for the market quotes $\{\sigma_{\text{BS}}^{\text{MKT}}(K_j, T_j)\}_{j=1}^m$ to obtain

$$\widehat{\theta} := \arg\min_{\theta \in \Theta} \sum_{j=1}^m \beta_j (\widetilde{F}(\widehat{w}; \theta_i, T_i, K_i) - \sigma_{\text{BS}}^{\text{MKT}}(K_j, T_j))^2.$$

Grid-based learning

We take this idea further and design an implicit form of the pricing map that is based on storing the implied volatility surface as an image given by a grid of "pixels". Let us denote by $\Delta := \{K_i, T_j\}_{i=1, j=1}^{n, \ m}$ a fixed grid of strikes and maturities; then we propose the following two-step approach:

Step (i): Learn the map $\widetilde{F}(w, \theta) = \{\sigma^{M(\theta)}(T_i, K_j)\}_{i=1, j=1}^{n, \ m}$ via neural network where the input is a parameter combination $\theta \in \Theta$ of the stochastic model $\mathcal{M}(\theta)$ and the output is a $n \times m$ grid on the implied volatility surface approximated by a representative grid $\{\sigma^{\mathcal{M}(\theta)}(T_i, K_j)\}_{i=1,\ j=1}^{n,\ m}$ where $n, m \in \mathbb{N}$ are chosen appropriately on a hand-crafted grid of maturities and strikes. $\widetilde{F}$ takes values in $\mathbb{R}^L$ where $L = $ strikes $\times$ maturities $= nm$. The optimisation problem in the image-based implicit learning approach is:

$$\widehat{\omega} := \arg\min_{w \in \mathbb{R}^n} \sum_{i=1}^{N_{\mathrm{Train}}^{\mathrm{reduced}}} \sum_{j=1}^{L} \eta_j (\widetilde{F}(w, \theta_i)_j - \widetilde{\sigma}^{\mathcal{M}}(\theta_i, T_j, K_j))^2, \quad (15.7)$$

where $N_{\mathrm{Train}} = N_{\mathrm{Train}}^{\mathrm{reduced}} \times L$ and $\eta_i \in \mathbb{R}_{>0}$ is a weight vector.

Step (ii): Solve the minimisation problem

$$\widehat{\theta} := \arg\min_{\theta \in \Theta} \sum_{i=1}^{L} \beta_j (\widetilde{F}(\widehat{\omega}, \theta)_i - \sigma_{\mathrm{BS}}^{\mathrm{MKT}}(T_i, K_i))^2,$$

for some user-specified weights $\beta_j \in \mathbb{R}_{>0}$. We note here that $\widetilde{F}(\widehat{\omega}, \theta)$ being a Neural Network, all gradients with respect to θ are available in closed-form and are fast to evaluate.

The data generation stage for the image-based approach works as in the pointwise approach, except that the option parameters $\zeta = (T, K)$ are, fixed and are no longer part of the sampling process. This is why they appear in the general objective function of pointwise learning (15.6) but no longer appear in the objective function (15.7) of the grid-based learning above. Clearly, in this case, each output of the neural network more nuanced (as it contains an entire grid vs. a single point on the surface) but depends on the choice of the hand-crafted grid Δ. Hence, we may need to interpolate between gridpoints in order to be able to calibrate (in the calibration Step **(ii)**) also to such options, whose maturity and strike do not exactly lie on the grid, Δ, in an arbitrage-free manner, see Itkin (2014); Cohen et al. (2020); Bergeron et al. (2021).

Intermediate and related approaches include ones where the surface is learned slice-by-slice (McGhee, 2018), or via more adaptive approaches (Liu et al., 2019). We refer to Bayer et al. (2019) for more details about the different approaches.

15.2.5 *Supervised learning approaches, and their ability to loosen limitations of the tractability mantra*

Supervised learning approaches for various pricing and calibration tasks have proved incredibly successful, and this has been displayed in a wide range of impressive results (Cuchiero et al., 2020; De Spiegeleer et al., 2018; Dixon et al., 2020; Gnoatto et al., 2020; Huge and Savine, 2020; Itkin, 2014; Liu et al., 2019; Sabate-Vidales et al., 2018) that tackle computation heavy tasks. We refer the interested reader to the survey article Ruf and Wang (2020) for a full comprehensive list.

What most supervised learning approaches have in common is that they speed up pricing (and hence calibration) by at least a 3-digit factor which lifts many of the limitations imposed by the requirement of tractability and eases the use of more realistic models (Bayer et al., 2019), of higher-dimensional modelling settings (Ferguson and Green, 2018) or the possibility of more flexibly mixing them, if desired, which can be treated in the calibration stage as a model with more parameters: see Figure 15.1.

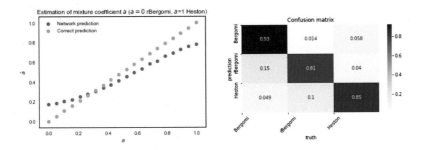

Figure 15.1 Calibration of the mixture parameter between two models (left) and three models (right).

15.3 Pricing and hedging by unsupervised deep learning

15.3.1 Deep hedging

While the supervised approach to derivatives pricing and calibration can be rather convenient when we have samples of prices given by the model we work with, it does not answer the question of how we should *hedge* the derivatives. Also, the approach is not applicable in situations where there is no underlying parametric pricing model – for example when we would like to price and hedge under real market data or data synthesised by a market generator. Fortunately, it is possible to develop an *unsupervised* deep learning approach, unsupervised in the sense that it does not require advance knowledge of either prices or hedges, that can accomplish both pricing and hedging in a model-free fashion. While the history of the application neural networks to pricing and hedging is extensive, as surveyed by Ruf and Wang (2020), here we follow the recent *deep hedging* approach introduced by Buehler et al. (2019). In a nutshell, in deep hedging we represent the hedging strategy, seen as a function of current and past market data, by a neural network and train it by optimising the hedger's profit and loss (P&L) according to our risk preferences using a set of historical price paths.[3]

To elucidate deep hedging, we consider a discrete-time market over $T \in \mathbb{N}$ periods with $d \in \mathbb{N}$ risky assets. The prices of these assets are denoted by $S_t = (S_{t,1}, \ldots, S_{t,d})$, $t = 0, \ldots, T$, and we assume here that they form an adapted,

---

[3] While we regard deep hedging here as a form of unsupervised learning, it could also be viewed as an application of *reinforcement learning*. We refer to Buehler et al. (2020) for a formulation of deep hedging using the language of reinforcement learning.

non-negative stochastic process on a filtered probability space $(\Omega, \mathcal{F}, (\mathcal{F}_t)_{t=0}^T, \mathbb{P})$, although deep hedging can ultimately be applied independently of any specific probabilistic structure. For ease of exposition, we also assume here that interest rates are zero, but this assumption is easily relaxed. However, diverting from the conventional setting of frictionless markets, we assume that any trade of size $x \in \mathbb{R}$ in asset i at time t incurs a proportional *transaction cost* $k_i |x| S_t^i$ for some constant $k_i \geq 0$. Then the liquidation value of an adapted, self-financing trading strategy $\pi_t = (\pi_{t,1}, \ldots, \pi_{t,d})$, $t = 0, \ldots, T - 1$, with zero initial wealth is

$$V(\pi; S) = \sum_{t=1}^T \pi_{t-1} \cdot (S_t - S_{t-1}) + C(\pi; S),$$

where

$$C(\pi; S) = \sum_{i=1}^d k_i \left(|\pi_{0,i}| S_0^i + \sum_{t=2}^T |\pi_{t-1,i} - \pi_{t-2,i}| S_{t-1,i} + |\pi_{T-1,i}| S_T^i \right).$$

Suppose now that our aim is to hedge a contingent claim Z which we have sold at price $p \in \mathbb{R}$ at time 0, with payoff at expiry T given by

$$Z = g(S_0, \ldots, S_T),$$

where $g : \mathbb{R}^{T+1} \to \mathbb{R}$ is a measurable function. For the moment, we treat p as a given quantity, e.g., a market quote, but we will return to pricing in Section 15.3.2 below. If we try to offset the risk from Z by trading the underlying assets, pursuing strategy π, then our final P&L is

$$\text{P\&L}_Z(p, \pi; S) = p + V(\pi; S) - Z.$$

We should then seek to tune π so that $\text{P\&L}_Z(p, \pi; S)$ becomes *optimal* according to our *risk preferences*. More specifically, given an objective function $L : \mathbb{R} \to \mathbb{R}$ that encodes our preferences, we can try to minimise risk

$$\mathbb{E}\left[L\left(\text{P\&L}_Z(p, \pi; S) \right) \right]$$

over all adapted processes π for which the expectation remains finite. Key examples of the objective function L include the quadratic loss $L(x) := x^2$ (which lends itself to quadratic hedging) and specifications via utility functions; that is, $L(x) := -U(x)$ with some utility function U.

However, optimising over all adapted processes can be hard, if not impossible, in practice, but this is where deep learning can step in. Suppose for simplicity that our trading strategies are based on prices alone; that is, our filtration $(\mathcal{F}_t)_{t=0}^T$ is generated by S via $\mathcal{F}_t = \sigma\{S_0, \ldots, S_t\}$ for any $t = 0, \ldots, T - 1$. Then, by the Doob–Dynkin lemma, for any $t = 0, \ldots, T$ there is a measurable function $g_t : \mathbb{R}^{t+1} \to \mathbb{R}$ such that

$$\pi_t = g_t(S_0, \ldots, S_t).$$

Given this functional representation of adapted strategies and the *universal approximation property* of feedforward neural networks with sufficiently many units

(Leshno et al., 1993) or layers (Kidger and Lyons, 2020), we are inspired to work with – and optimise over – the class of strategies

$$\pi_t^{(\theta_t)} = \phi_t(S_0, \ldots, S_t; \theta_t),$$

where $\phi_t(\,\cdot\,; \theta_t)$ is a feedforward neural network parameterised by vector θ_t for any $t = 0, \ldots, T - 1$. Such strategies are able to approximate the (theoretically) optimal strategies, whenever they exist, and give rise to a very flexible class of processes in any case. While for this formulation we assumed that our filtration is generated by the past and current prices alone, in practice we can also augment the *features* $S_0, \ldots, S_t$ by other relevant risk factors.

Remark 15.1 In practice, having a separate network $\phi_t(\,\cdot\,; \theta_t)$ for every $t = 0, \ldots, T - 1$ may result in too many parameters if T is large and may also be inefficient as this neglects the fact that many hedging strategies enjoy some form of *continuity* in time. One possible solution is to represent $\phi_0, \ldots, \phi_{T-1}$ by a single network that takes time and current and lagged prices as inputs; that is,

$$\pi_t^{(\theta)} = \phi(t, S_t, \ldots, S_{t-\ell}; \theta), \quad t = 0, \ldots, T - 1,$$

where $\phi(\,\cdot\,; \theta) \colon \mathbb{R}^{d(\ell+1)+1} \to \mathbb{R}^d$ is a feedforward neural network and $\ell = 0, 1, \ldots$ determines the size of the lookback window. (If the prices follow a Markov process and the payoff Z is not path-dependent, we can choose $\ell = 0$.) Another solution is to employ a *recurrent* neural network, which is able to utilise the temporal dependence, possibly non-Markovian, in the features and will, in *unfolded* form (see Goodfellow et al., 2016, Sect. 10.1), provide the functions $\phi_0, \ldots, \phi_{T-1}$. While there is no comprehensive universal approximation theory for recurrent networks, they can nevertheless work well in this task, which we demonstrate in Section 15.3.3.

Employing strategy $\pi^{(\theta)}$ represented by a neural network, the problem of finding an optimal hedging strategy then transforms to the problem of minimising $\mathbb{E}[L(\text{P\&L}_Z(p, \pi^{(\theta)}; S))]$ with respect to the parameters in $\theta = (\theta_0, \ldots, \theta_{T-1})$ – or in other words, training the neural network. In practice, we need to resort to *empirical* risk minimisation, whereby we train the network via θ to minimise

$$\frac{1}{N} \sum_{i=1}^{N} L(\text{P\&L}_Z(p, \pi^{(\theta)}; S^{(i)}))$$

for a set of price paths $S^{(1)}, \ldots, S^{(N)}$. These price paths could be IID copies of S given by a model or, more generally, samples drawn from a market generator. In principle, *real* historical price paths could also be used, but typically we need N to be at least a few thousand, which makes it difficult to find enough relevant, historical price paths with lengths matching the typical expiries of claims ranging from months to a few years. The actual training of the network is carried out by stochastic gradient descent, which iteratively updates the parameter values, starting from an initial value $\theta^{(0)}$, via

$$\theta^{(n+1)} = \theta^{(n)} - \delta \nabla_\theta \mathcal{L}_{I_n}(\theta^{(n)}),$$

where $\delta > 0$ is the learning rate and

$$\mathcal{L}_I(\theta) = \frac{1}{\#I} \sum_{i \in I} L\big(\text{P\&L}_Z\big(p, \pi^{(\theta)}; S^{(i)}\big)\big)$$

is empirical risk computed using a *mini-batch* $I \subset \{1, \ldots, N\}$. As usual, the mini-batches I_n, $n \geq 0$, are of fixed size and drawn from $\{1, \ldots, N\}$ without replacement for each pass through the data; that is, an epoch. Since the learning rate δ needs to be gradually adjusted as the training progresses, it is preferable to use an adaptive version of stochastic gradient descent such as Adam (Kingma and Ba, 2015), which also performs smoothing of the gradient updates $\nabla_\theta \mathcal{L}_{I_n}\big(\theta^{(n)}\big)$, $n \geq 0$.

15.3.2 Utility indifference pricing

To apply the deep hedging methodology to derivatives that are traded *over-the-counter*, we need to be also able to determine the price p of the claim as part of the problem. As Buehler et al. (2019) have shown, *utility indifference pricing* can be used in conjunction with deep hedging, providing a convenient solution.

Utility indifference pricing has been reviewed comprehensively in the literature (e.g., Henderson and Hobson, 2009), so we merely sketch the basic principles here. Suppose our objective function L is given by $L(x) = -U(x)$ for some utility function $U: \mathbb{R} \to \mathbb{R}$, which is assumed to be both strictly increasing and strictly concave – the key example being the exponential utility function

$$U_\lambda(x) = -\frac{1}{\lambda} \exp(-\lambda x) \tag{15.8}$$

with risk aversion parameter $\lambda > 0$. Suppose further that our existing cash balance at time 0 is $x \in \mathbb{R}$. Then $p \in \mathbb{R}$ is the *utility indifference price* of claim Z under U if it solves

$$\sup_\pi \mathbb{E}[U(x + \underbrace{V(\pi; S)}_{=\text{P\&L}_0(0, \pi; S)})] = \sup_\pi \mathbb{E}[U(x + \underbrace{p + V(\pi; S) - Z}_{=\text{P\&L}_Z(p, \pi; S)})]. \tag{15.9}$$

Intuitively, p is the compensation we ask for taking on the risk arising from Z, in order to remain at the same level of expected utility. When the underlying price process S is arbitrage-free, utility indifference pricing is arbitrage-free and if the market is complete it coincides with pricing by replication (see Henderson and Hobson, 2009).

While general utility functions can be used in indifference pricing based on deep hedging (Buehler et al., 2019, Sect. 4.5), we focus here on the special case of exponential utility (15.8), where indifference pricing becomes rather straightforward. Then equation (15.9) simplifies to

$$\sup_{\pi} \mathbb{E}[U_\lambda(\text{P\&L}_0(0,\pi;S))] = \sup_{\pi} \mathbb{E}[U_\lambda(\text{P\&L}_Z(p,\pi;S))]$$
$$= \sup_{\pi} \mathbb{E}[U_\lambda(p + \text{P\&L}_Z(0,\pi;S))]$$
$$= \exp(-\lambda p) \sup_{\pi} \mathbb{E}[U_\lambda(\text{P\&L}_Z(0,\pi;S))],$$

from which we can solve

$$p = -\frac{1}{\lambda} \log \left(\frac{\sup_\pi \mathbb{E}[U_\lambda(\text{P\&L}_0(0,\pi;S))]}{\sup_\pi \mathbb{E}[U_\lambda(\text{P\&L}_Z(0,\pi;S))]} \right).$$

So effectively we just need to solve the hedging problem *with* and *without* the claim Z, respectively, when we receive no payment initially, and compare the attained utility levels. With deep hedging, we thus first train a neural-network-based strategy $\widehat{\pi}^{(0)}$, as described above, so that empirical risk corresponding to $\text{P\&L}_0(0,\widehat{\pi}^{(0)};S)$ (i.e., without Z) under $L(x) = -U_\lambda(x)$ is optimal and another strategy $\widehat{\pi}^{(Z)}$ in the same vein optimising $\text{P\&L}_Z(0,\widehat{\pi}^{(Z)};S)$ (i.e., with Z). Finally, using these two strategies, we can estimate p via

$$\widehat{p} = -\frac{1}{\lambda} \log \left(\frac{\frac{1}{N}\sum_{i=1}^N U_\lambda(\text{P\&L}_0(0,\widehat{\pi}^{(0)};S^{(i)}))}{\frac{1}{N}\sum_{i=1}^N U_\lambda(\text{P\&L}_Z(0,\widehat{\pi}^{(Z)};S^{(i)}))} \right). \tag{15.10}$$

15.3.3 Numerical illustration

We illustrate now the use of the deep hedging methodology to price and hedge a European call option and an up-and-out call option in a discrete-time version of the Black–Scholes model under transaction costs. The price process S, now with dimensionality $d = 1$, follows

$$S_t = S_0 \exp\left(\frac{\mu}{T}t + \frac{\sigma}{\sqrt{T}} \sum_{s=1}^t \xi_s\right), \quad t = 0,\ldots,T,$$

where $S_0 > 0$, $\mu \in \mathbb{R}$ and $\sigma > 0$ are parameters and $\xi_1,\ldots,\xi_T$ are mutually independent standard normal random variables. While this model is in discrete time, for large T it approximates the complete, continuous-time Black–Scholes model. Therefore, it makes sense to compare the results of deep hedging and indifference pricing to their analytical Black–Scholes counterparts.

We specify the architecture for the functions $\phi_0,\ldots,\phi_{T-1}$ that determine our hedging strategy π using a recurrent layer based on the *long short-term memory* (LSTM) architecture of Hochreiter and Schmidhuber (1997). The LSTM architecture does not impose any low-order Markovian constraint on dependence and is thus able to encode complicated path-dependence and long-range dependence (at least in approximate sense). It has achieved state-of-the-art performance in various tasks involving sequential data, including speech and handwriting recognition (Goodfellow et al., 2016, Sect. 10.10.1). A stylised description of an LSTM layer can be given via the recursive equation

$$(y_{t+1}, c_{t+1}) = f(x_t, y_t, c_t; \theta_{\text{LSTM}}),   \tag{15.11}$$

where $x_t \in \mathbb{R}^{d'}$ is the input to the layer, $c_t \in \mathbb{R}^c$ is the state vector of the LSTM *memory cell* with $c \in \mathbb{N}$ units and $y_t \in \mathbb{R}^c$ is the layer output at time t, while θ_{LSTM} collects the parameters of the layer. The function f that determines the exact mechanics of the layer is somewhat elaborate, and we refer to Hochreiter and Schmidhuber (1997) and Goodfellow et al. (2016, Sect. 10.10.1) for details on its structure. Once unfolded, the layer can be represented graphically as in Figure 15.2. Given initial states c_0 and y_0, which are typically set to zero in practice, we can compute y_t in a *causal* manner via (15.11) as a function of $x_0, \dots, x_t$ for any $t \geq 1$. This shows that the LSTM layer is a suitable building block for an adapted trading strategy.

To specify our strategy, we employ a single LSTM layer with one-dimensional input S_t, so that $d' = d = 1$. To determine the the hedging position $\pi_t^{(\theta)} = \phi_t(S_0, \dots, S_t; \theta)$ at time t, we then map the output $y_t \in \mathbb{R}^c$ of the LSTM layer using a standard, single-unit dense layer

$$h(y) = \rho(w^T y + b), \quad y \in \mathbb{R}^c,$$

with weight vector $w \in \mathbb{R}^c$, bias $b \in \mathbb{R}$ and activation function $\rho : \mathbb{R} \to \mathbb{R}$. To summarise, for any $t = 0, \dots, T - 1$,

$$(S_0, \dots, S_t) =: (x_0, \dots, x_t) \overset{\text{LSTM}}{\longmapsto} \underbrace{y_t}_{\in \mathbb{R}^c} \overset{h}{\longmapsto} h(y_t) =: \phi_t(S_0, \dots, S_t; \theta) =: \pi_t^{(\theta)} \in \mathbb{R},$$

whereas θ is composed of the parameters θ_{LSTM}, W and b. We train θ using $N \in \mathbb{N}$ independent samples $S^{(1)}, \dots, S^{(N)}$ of S and using the exponential utility function U_λ given in (15.8). As parameter values, we use

$$S_0 := 1, \quad \mu := -\frac{\sigma^2}{2}, \quad \sigma := 0.5, \quad T := 100, \quad N := 100\,000.$$

We also generated additional $N = 100\,000$ samples of S to have an "unseen" test data set. The networks were implemented and trained in TensorFlow 2.4.1 running on Google Colaboratory[4] with a GPU accelerator.

In our first experiment, we seek to hedge a vanilla, European call option

$$Z = (S_T - K)^+$$

struck at $K = 1 = S_0$. We specify three values (*low*, *medium* and *high*, respectively) for the transaction cost level: $k := k_1 \in \{0.05\%, 0.5\%, 5\%\}$ and two values (*medium* and *high*, respectively) for the risk aversion level: $\lambda \in \{1, 10\}$. Following the financial intuition that the call option as a contract that depends positively on the underlying would only be hedged with long positions not exceeding its notional, we choose the $(0, 1)$-valued sigmoid function $x \mapsto 1/(1 + e^{-x})$ as the output layer activation function ρ.

---

[4] https://colab.research.google.com/

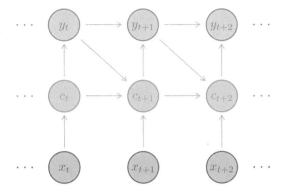

Figure 15.2 A segment of an unfolded LSTM layer.

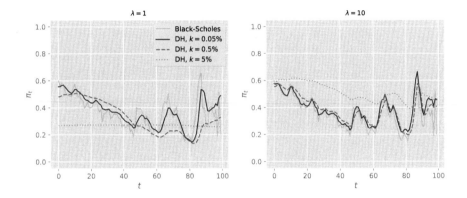

Figure 15.3 Realised paths of the deep hedging strategy π for a European call option struck at $K = 1 = S_0$ vis-à-vis those of the corresponding Black–Scholes delta hedge.

We train the strategy for each parameter combination, employing $c = 10$ units in the LSTM cell, running Adam for 40 epochs with mini-batch size 2 000.

To visualise the results and assess their financial soundness, we apply the trained strategies to samples in the test dataset. In Figure 15.3 we first plot the realised paths of the deep hedging strategy π for a particular sample in the test set and compare then to the corresponding Black–Scholes delta hedge. We observe that the deep hedging paths look like smoothed versions of the delta hedge. Indeed, the higher the transaction cost level k and the lower the risk aversion level λ are the smoother the path is, as we would expect. Under low risk aversion, $\lambda = 1$, and high transaction costs, $k = 5\%$, the realised strategy becomes effectively a static partial hedge for this sample. At the same cost level but with higher risk aversion, $\lambda = 10$, we observe that the path is still relatively smooth but biased upwards, relative to the delta hedge, essentially to compensate for the slower tracking of the variation in the price of the underlying. To gain broader understanding of the statistical properties of the realised hedges, we additionally plot them in Figure

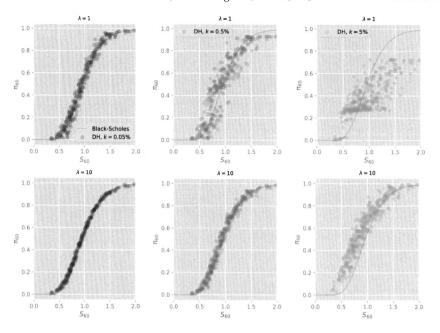

Figure 15.4 Realisations at time $t = 60$ of the deep hedging strategy π for a European call option struck at $K = 1 = S_0$ plotted against the corresponding price of the underlying, and compared to the Black–Scholes delta hedge.

15.4 for 300 samples in the test set at time $t = 60$ against the corresponding price of the underlying, i.e., S_{60}. At $k = 0.05\%$ the realised hedges are close to delta hedges, and even more so when $\lambda = 10$, since being more risk-averse we will prefer to hedge closer to replication. As k is increased, these strategies accumulate more costs, making them less viable, so the realised hedges rather expectedly then deviate more from the delta hedges. In the case $k = 5\%$ and $\lambda = 10$ we again observe the upward bias that compensates for the inviability at higher cost level to pursue a strategy that closely tracks the underlying.

We estimate the utility indifference price of the call option for each parameter combination by applying the estimator $\widehat{p}$, given in (15.10), to the test dataset in its entirety. As pointed out by Buehler et al. (2019), we can compare these utility indifference prices to asymptotic theory for small transaction costs which suggests that the indifference price p_k as a function of cost level k behaves like $p_k - p_0 \propto k^{2/3}$ for small k (e.g., Kallsen and Muhle-Karbe, 2015). In Figure 15.5 we use the Black–Scholes price p_{BS} as a proxy for the frictionless price (ignoring the discreteness of time) and plot $\log(\widehat{p}_k - p_{BS})$ against $\log(k)$ for both $\lambda = 1$ and $\lambda = 10$. The slopes of least squares fits, 0.786 and 0.715 for $\lambda = 1$ and $\lambda = 10$, respectively, are within tolerance from the scaling exponent $2/3$ predicted by the asymptotic theory.

To test the ability of the LSTM network to hedge a *path-dependent* claim, in

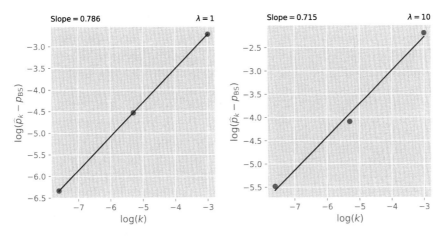

Figure 15.5 Scaling of the utility indifference price of a European call option struck at $K = 1 = S_0$, derived from the deep hedging strategy π, as a function of cost level k.

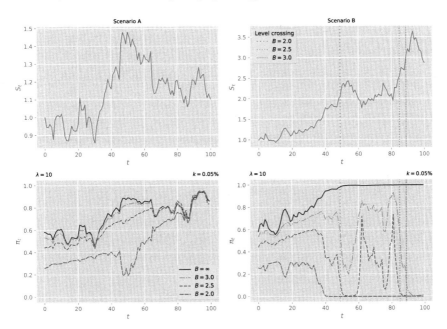

Figure 15.6 Realised paths of deep hedging strategies π for up-and-out call options struck at $K = 1 = S_0$ with barriers $B \in \{2, 2.5, 3, \infty\}$ in two different scenarios.

our second experiment we hedge a barrier option, an *up-and-out* call option

$$Z = (S_T - K)^+ \mathbf{1}_{\{\sup_{t\in[0,T]} S_t \leq B\}}$$

with knock-out barrier $B > S_0$. (For $B = \infty$ it reduces to a vanilla European call.) Such an option provides an interesting test case for deep hedging. When the price of the underlying increases we should increase the hedge position, but if the price further approaches the barrier we may in fact be inclined to decrease the position, since if the option knocks out we can unwind the hedge completely. We do not give the network the running maximum or any indicator of level crossing as a feature, instead we let the network learn to determine knock-out from prices alone. We fix $K := 1 = S_0$ and choose three barriers for the experiment, including the vanilla call for comparison purposes, specifically, $B \in \{2.0, 2.5, 3.0, \infty\}$. We adopt low level of transaction costs, $k = 0.05\%$ and high level of risk aversion, $\lambda = 10$. We retain model parameters from the earlier experiment. We only modify the network by increasing the number of units in the LSTM cell to $c = 20$, to increase the capacity of the network to capture more complex behaviour, but we keep the architecture otherwise unchanged. We train it using Adam, now for 500 epochs with increased mini-batch size 5 000.

In the context of barrier options under transaction costs, we are not aware of any published analytical results on hedging, even for the Black–Scholes model, that could be used as benchmarks for this experiment. Therefore, instead of any systematic assessment, we simply analyse the financial soundness of the realised hedges in a small case study. To this end we draw two samples from the test data set, one representing scenario A where the all of the barrier options expire in the money and another representing scenario B where all of them knock out. The realised hedges in these scenarios are presented in Figure 15.6. In scenario A, we observe that the while hedge for the lowest barrier $B = 2.0$ initially evolves largely detached from the others, all of them, including the one for the vanilla call, i.e., $B = \infty$, effectively coalesce into one after $t = 80$ once it has become overwhelmingly likely that the price of the underlying will remain above the strike but below the barriers. In scenario B, we note that the hedge for $B = 2.0$ is liquidated somewhat ahead of time before the actual knock-out happens just before $t = 50$. Amusingly, the hedge for $B = 2.5$ is also unwound at that time, but it is quickly rebuilt as the price of the underlying just barely misses the barrier and reverts down. When finally the level $B = 2.5$ and $B = 3.0$ are breached, both hedges for both barrier options are promptly liquidated, while the vanilla call remains fully hedged. The utility indifference prices $\widehat{p}$ for $B = 2, 2.5, 3.0$ are 0.131, 0.177 and 0.192, respectively, and while there are no numerical benchmarks to compare them to, we note that they are of reasonable magnitude, in correct order and all below the price 0.201 of the vanilla call, as we would expect.

These two experiments highlight the remarkable flexibility of the deep hedging approach, especially using the LSTM architecture, to adapt from hedging a simple claim to a complex one, while including market frictions. Such results remain out of the reach of even the best analytical tools available today.

15.4 The increasing symbiosis of models with the data and the generation of synthetic market datasets: *"Market Generators"*

From a conceptual point of view perhaps even more fundamental than the "direct" impacts in terms of speed and generality discussed in point (**B**) of the introduction is an emerging symbiosis of algorithms and the data. To exemplify this, let us observe a deep neural network designed for some financial application given by the triplet

$$\underbrace{(\text{Architecture, Objective function;}}_{Network} \quad \text{Training data}) \Rightarrow \text{Algorithm} \quad (15.12)$$

Once the learning phase is concluded, the trained network (henceforth simply "algorithm") can then in turn be applied to test data (typically real market data):

$$(\text{Algorithm;} \quad \underbrace{\text{Test data}}_{\text{Market data}}) \Rightarrow \text{Output} \quad (15.13)$$

to produce some output (say an option price or an investment strategy etc.).

The influence and impacts of this symbiosis summarised in (15.12) go both ways: Data used in training phase influences the algorithms, and similarly, the chosen network architecture imposes restrictions on the structure of data to which it can be applied.

The former comes as no surprise: Deep learning models being so flexible, the data used for training influences the type of output of the application. We elaborate on this in section 15.4.1 below. The latter (i.e. the case when network architectures may need updating due to substantial structural changes in the data) calls for rethinking risk management routines profoundly, in order to determine whether a simple re-training of the network is sufficient, or a more substantial update of the network architecture is necessary, see Horvath et al. (2021b) for examples. Effects of the influence of an ill-suited architecture to the data at hand are visible for example in Horvath et al. (2021b), which highlights how a suitable network architecture has decisive implications on network performance. Risk management should therefore include ongoing monitoring on the structure of the data and the suitability of the network architecture to that data as well as regular updates of the training data that algorithms are exposed to.

15.4.1 The case for more flexible data-driven market models

The technological advances and tools developed in recent years create the desire for market models that are directly data-driven and closely reflect evolving market reality. The implications of the influence of data on algorithms becomes quite visible in Buehler et al. (2019): When training data is numerically generated by the Black–Scholes model,

$$\underbrace{(\text{Architecture, Objective function;}}_{\text{Deep heding engine}} \quad \underbrace{\text{Training data}}_{\text{B-S, Heston}}) \Rightarrow \text{Algorithm}_{\text{B-S,H}}$$

the trained network (the hedging engine) approximates the corresponding delta as an optimal hedging strategy when applied to (Black–Scholes generated) test data. Analogous effects can be seen when the training data is generated by Heston paths.

$$(\text{Algorithm}_{\text{BS,H}};\ \underbrace{\text{Test data}}_{\text{B-S, Heston}}) \Rightarrow \text{investment strategies}_{\text{B-S,H}}$$

In fact, the approaches presented in the previous section and in Buehler et al. (2019) are inherently model agnostic. For a given collection of sample paths provided to the network in the training phase, the trained network outputs strategies that are (approximately) optimal in market regimes where path distributions resemble the ones presented during the training phase.

$$(\underbrace{\text{Architecture, Objective function};}_{\text{Deep hedging engine}}\ \underbrace{\text{Training data}}_{?}) \Rightarrow \text{Algorithm}_?$$

It is then natural to ask how we can obtain training datasets that can generate optimal hedging strategies when applied to real market data, i.e that reflect distributions observed in market data as closely as possible.

$$(\text{Algorithm}_?;\ \underbrace{\text{Test data}}_{\text{real market data}}) \Rightarrow \text{investment strategies}$$

The more realistic the market paths presented to the network during the training phase, the better performance can be expected on real data. The performance is – to exaggerate slightly – as good as the data that you provide during the training phase which gives a first incentive to look more closely at point **(A)**, the properties of the data itself. This also drives the interest for more flexibility in market models, to be able to closely follow changes in distribution of the data.

Classical models are known to suffer from a relative inflexibility when major shifts in the market occur. Even the more realistic modern examples such as (15.5) have a fixed number of parameters which naturally makes them – despite a plethora of advantages – somewhat limited in their flexibility. The deep hedging example and similar model-agnostic neural network based financial applications highlighted use cases for even more flexible means of creating synthetic market data: The inflexibility imposed by fixed number of parameters in classical models can be lifted by applying neural networks (and in particular deep generative models) as powerful approximators of functions and distributions to build generative models for financial markets: *market scenario generators* (or simply *market generators*). These models are capable of closely reflecting real market dynamics, in a model-free and directly data driven way (Bühler et al., 2020; Kondratyev and Schwarz, 2019; Snow, 2020).

Generative models – such as Restricted Boltzmann Machines (Kondratyev and Schwarz, 2019), Generative Adverserial Networks (Xu et al., 2020; Snow, 2020; Wiese et al., 2019; Ni et al., 2020) and Variational Autoencoders (Bühler et al., 2020) – are based on the idea of transforming random samples of latent variables to distributions observed in data samples via differentiable functions, which are approximated by a neural network by backpropagation.

$$\underbrace{(\text{Architecture, Objective function;}}_{\text{Deep generative model: RBM, GAN, VAE}} \underbrace{\text{Training data }}_{\text{real or synthetic dataset}}) \Rightarrow \text{synthetic data}$$

The application of such frameworks to financial settings is what we refer to as *market generators*. Market generators – in contrast to classical models – are so flexible that they are capable of generating data samples that are statistically *indistinguishable* from a given original dataset. A key question here is how to quantify the similarity of financial time series with one another or evaluate the quality of market generators? Or in other words: what are good objective function for market generators and good metrics to measure the similarity of stochastic processes? This question can either be answered from an application focussed standpoint, for example evaluating the performance of the hedging network on real market data (see also Antonov et al., 2018) as a means of quality of the synthetic data used for training; or from a more universal standpoint (see Chevyrev and Oberhauser, 2018; Oberhauser and Király, 2019; Bühler et al., 2020) by formulating a suitable metric measuring the *similarity* or the distance between stochastic processes as indicated in Figure 15.7.

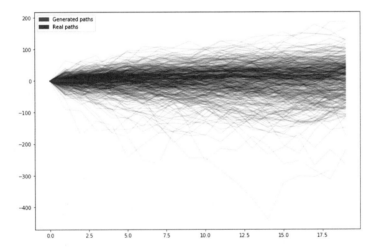

Figure 15.7 The image demonstrates two sets of sample paths. The set of paths in red is to be replicated by a generative model, i.e. it is the input data for a market generator. The set of paths in blue is synthetically generated by a market generator. To evaluate the the quality of the generated paths, one needs to formulate a *similarity metric* that measures how close the two sets of (random) paths are to one another. Suitable metrics can be formulated using the signatures of the paths, as demonstrated in Chevyrev and Oberhauser (2018); Oberhauser and Király (2019); Bühler et al. (2020). Such metrics are central to market generators.

15.4.2 The case for classical models

Market generation is currently still in its infancy. The flexibility of these models, the great variety in which they appear and their inherent synergy with the data are all factors that make them difficult to risk manage. Classical models on the other hand are well understood and a wealth of research and analysis of their properties is available, and so is decades worth of experience in their risk management. Furthermore, classical models have also been developed with the aim to reflect (despite their relative inflexibility) a selection of stylized facts of financial markets (Cont, 2001). Some model families (Gatheral et al., 2018) have the impressive ability to exhibit these with a remarkable precision and provide an overarching theoretical basis for their behaviour across markets and applications. Most importantly, the distribution and dynamical properties of classical models can be carefully controlled by a handful set of parameters at each point in time. They can therefore conveniently provide training sets for deep learning algorithms (see Horvath et al., 2021b) where certain properties of the training set are present or absent to test the robustness of a deep learning algorithm under those market conditions.

15.4.3 Synergies between the classical and modern approaches, and further risk management considerations

Classical asymptotic methods and the type of (supervised) deep learning methods described in Section 15.2 inherently complement each other: Asymptotic expansions address extreme, limiting scenarios, typically where one of the observed quantities (strike, time to maturity or a combination of these) is very small or very large, while the neural network in 15.3 a priori does not address these regimes. Furthermore, risk management will greatly benefit from sensitivity results (Bartl et al., 2020), and the available analytical and numerical solutions classical models also provide much-needed benchmarks – in the cases where these are applicable – and control variates for deep learning applications to facilitate risk management of the latter, as showcased in Antonov et al. (2020).

On the other hand, approximate solutions obtained by neural networks can facilitate finding analytical solutions for classical models under market conditions where these hadn't been available.

Finally, flexible DNN-based models can be applied to smoothly interpolate between classical models as shifts in the data occur in a fully data driven way (Kidger et al., 2020; Bühler et al., 2020; Ni et al., 2020).

15.5 Outlook and challenges with data at centre stage

Applications of Market Generators*; and new waters of modelling*

The symbiosis of data and modern algorithms (15.12) discussed above highlights the need for research on data privacy considerations and for providing high-quality representative benchmark datasets to enable a means of sharing training and test data across institutions.

It should also be noted that market generators can provide solutions to these issues: The flexibility of market generators opens the door to applications that were, till now, not typically a core part of financial modelling in the classical setting.

(i) Data anonymisation: When the available data is confidential, it is desirable to generate anonymised datasets that are representative of the true underlying distribution of the data but cannot be traced back to their origins. Financial data and medical data are often proprietary, or confidential. When testing investment strategies or the effectiveness of a treatment it is imperative not to be able to trace back the datasets to the individual client or patient.

Evaluating whether the produced data is representative of the distribution that a dataset stems from, depends on the distributional properties (evaluation metrics) that we control for. Thus the question of adequate performance evaluation metrics is a central matter for research on market generators, see Bühler et al. (2020). The choice of evaluation metric can also influence the level of anonymity achieved by such generative procedures, where there is typically a trade-off between the representativeness of a dataset and the level of anonymity it can guarantee. The study presented in Kondratyev et al. (2020) is, in part, devoted to understanding the latter question in more detail. Similar considerations are in place regarding our ability to trace back models used by market competitors.

(ii) Small original training datasets: Though big data as a concept is ubiquitous today, in more situations than not, the amount of data available for training a neural network is small rather than large. When there are natural restrictions on the number of available original samples (constraints on the number of experiments, restrictions on the access to data), the available data may not be sufficient to train the neural network application at hand (e.g. hedging engine). Clearly, the more complex the application, the more data samples are needed to train it.

Generative models for sparse data environments therefore need to be relatively parsimonious and trainable on a low number of data samples (here we tacitly assume that the samples are representative of the distribution). Once such a generative network is available, more complex neural network applications can also be trained, using the market generator that produces the necessary amount of training samples for the latter. Further practical applications of market generators include (but are not limited to) the following use cases:

(iii) Outlier detection: Once the distribution of a dataset can be statistically identified and reproduced (even if the data does not follow any known parametric distribution) it becomes possible to identify *typical regimes* – that is, data points that are typical for the distribution – as well as outliers or *atypical events.* Detecting outliers and atypical events enables us to identify occurrences of regime switching in a market, gives the basis for fraud detection and for the identification of human-machine interfaces in automation; that is, such events that alert an automated machine to hand over the handling of an atypical process to a human with appropriate responsibilities.

(iv) Backtesting: When developing a trading strategy, carrying out a *backtest* to measure how the strategy would perform in a realistic environment is of crucial importance. However, using historical data may result in overfitting of the trading strategy. Having a market simulator capable of generating realistic, independent samples of market paths would allow a more robust backtest, less prone to overfitting.

 (v) Risk management of portfolios – be it of financial derivatives or trading strategies – is of utmost importance. A realistic market simulator can be used to generate synthetic paths to estimate various risk metrics, such as Value at Risk (VaR).

These applications are to date fairly unexplored, however they are gaining more and more relevance in a landscape where data increasingly assumes a central role in quantitative applications. And so, market generators have the very real potential of creating a whole new era of financial modelling.

References

Antonov, A., Baldeaux, J. F., and Sesodia, R. (2018). Quantifying model performance. Available at SSRN 3299615.

Antonov, A., Konikov, M., and Piterbarg, V. (2020). Neural networks with asymptotics control. Available at SSRN 3544698.

Bartl, D., Drapeau, S., Obłój, J., and Wiesel, J. (2020). Data driven robustness and uncertainty sensitivity analysis. ArXiv 2006.12022.

Bayer, C., Friz, P., and Gatheral, J. (2016). Pricing under rough volatility. *Quantitative Finance*, 16(6):887–904.

Bayer, C., Horvath, B., Muguruza, A., Stemper, B., and Tomas, M. (2019). On deep calibration of (rough) stochastic volatility models. ArXiv 1908.08806.

Bayer, C. and Stemper, B. (2018). Deep calibration of rough stochastic volatility models. ArXiv 1810.03399.

Bergeron, M., Fung, N., Hull, J., and Poulos, Z. (2021). Variational autoencoders: A hands-off approach to volatility. ArXiv 2102.03945.

Bergomi, L. (2016). *Stochastic Volatility Modeling*. CRC Press.

Buehler, H., Gonon, L., Teichmann, J., and Wood, B. (2019). Deep hedging. *Quantitative Finance*, **19**(8), 1271–1291.

Buehler, H., Gonon, L., Teichmann, J., Wood, B., Mohan, B., and Kochems, J. (2020). Deep hedging: Hedging derivatives under generic market frictions using reinforcement learning. Available at SSRN 3355706.

Bühler, H., Horvath, B., Lyons, T., Arribaz, I. P., and Wood, B. (2020). A data-driven market simulator for small data environments. Available at SSRN 3632431.

Chevyrev, I. and Oberhauser, H. (2018). Signature moments to characterize laws of stochastic processes. ArXiv 1810.10971.

Cohen, S. N., Reisinger, C., and Wang, S. (2020). Detecting and repairing arbitrage in traded option prices. *Applied Mathematical Finance*, **27**(5), 345–373.

Cont, R. (2001). Empirical properties of asset returns: stylized facts and statistical issues. *Quantitative Finance*, **1**(2), 223–236.

Cuchiero, C., Khosrawi, W., and Teichmann, J. (2020). A generative adversarial network approach to calibration of local stochastic volatility models. ArXiv 2005.02505.

De Spiegeleer, J., Madan, D. B., Reyners, S., and Schoutens, W. (2018). Machine learning for quantitative finance: fast derivative pricing, hedging and fitting. *Quantitative Finance*, **18**(10), 1635–1643.

Dimitroff, G., Röder, D., and Fries, C. P. (2018). Volatility model calibration with convolutional neural networks. Available at SSRN 3252432.

Dixon, M., Crépey, S., and Chataigner, M. (2020). Deep local volatility. ArXiv 2007.10462.

Ferguson, R. and Green, A. (2018). Deeply learning derivatives. ArXiv 1809.02233.

Gatheral, J., Jaisson, T., and Rosenbaum, M. (2018). Volatility is rough. *Quantitative Finance*, **18**(6). 933–949.

Gnoatto, A., Picarelli, A., and Reisinger, C. (2020). Deep xVA solver – a neural network based counterparty credit risk management framework. ArXiv 2005.02633.

Goodfellow, I., Bengio, Y., and Courville, A. (2016). *Deep Learning*. MIT Press.

Hagan, P. S., Kumar, D., Lesniewski, A. S., and Woodward, D. E. (2002). Managing smile risk. *Wilmott Magazine*, July 2002.

Henderson, V. and Hobson, D. (2009). Utility indifference pricing: An overview. Pages 44–74 of: *Indifference Pricing: Theory and Applications*, R. Carmona (ed). Princeton University Press.

Hernandez, A. (2017). Model calibration with neural networks. *Risk.net*.

Hochreiter, S. and Schmidhuber, J. (1997). Long short-term memory. *Neural Computation*, **9**(8), 1735–1780.

Horvath, B., Muguruza, A., and Tomas, M. (2021a). Deep learning volatility: a deep neural network perspective on pricing and calibration in (rough) volatility models. *Quantitative Finance*, **21**(1), 11–27.

Horvath, B., Teichmann, J., and Zurič, Z. (2021b). Deep hedging under rough volatility. Available at SSRN 3778043.

Huge, B. and Savine, A. (2020). Differential machine learning. Available at SSRN 3591734.

Itkin, A. (2014). To sigmoid-based functional description of the volatility smile. ArXiv 1407.0256.

Kallsen, J. and Muhle-Karbe, J. (2015). Option pricing and hedging with small transaction costs. *Mathematical Finance*, **25**(4), 702–723.

Kidger, P. and Lyons, T. (2020). Universal approximation with deep narrow networks. Pages 2306–2327 of: *Proc 33rd Conference on Learning Theory*, J. Abernethy and S. Agarwal (eds).

Kidger, P., Morrill, J., Foster, J., and Lyons, T. (2020). Neural controlled differential equations for irregular time series. ArXiv 2005.08926.

Kingma, D. P. and Ba, J. L. (2015). Adam: a method for stochastic optimization. In *Proc. Third International Conference on Learning Representations*.

Kondratyev, A. and Schwarz, C. (2019). The market generator. Available at SSRN 3384948.

Kondratyev, A., Schwarz, C., and Horvath, B. (2020). Data anonymisation, outlier detection and fighting overfitting with restricted Boltzmann machines. Available at SSRN 3526436.

Koshiyama, A. S., Firoozye, N., and Treleaven, P. C. (2019). Generative adversarial networks for financial trading strategies fine-tuning and combination. ArXiv 1901.01751.

Lee, G. (2021). Union beckons for the three quant tribes. *Risk.net*.

Leshno, M., Lin, V. Y., Pinkus, A., and Schocken, S. (1993). Multilayer feedforward networks with a nonpolynomial activation function can approximate any function. *Neural Networks*, **6**(6), 861–867.

Liu, S., Borovykh, A., Grzelak, L. A., and Oosterlee, C. W. (2019). A neural network-based framework for financial model calibration. *Journal of Mathematics in Industry*, **9**(9), 28 pages.

McGhee, W. A. (2018). An artificial neural network representation of the SABR stochastic volatility model. Available at SSRN 3288882.

Ni, H., Liao, S., Szpruch, L., Wiese, M., and Xiao, B. (2020). Conditional Sig-Wasserstein GANs for time series generation. ArXiv 2006.05421.

Oberhauser, H. and Király, F. (2019). Kernels for sequentially ordered data. *Journal of Machine Learning Research*, **20**(31), 1–45.

Ruf, J. and Wang, W. (2020). Neural networks for option pricing and hedging: a literature review. *Journal of Computational Finance*, **24**(1), 1–46.

Sabate-Vidales, M., Siska, D., and Szpruch, L. (2018). Unbiased deep solvers for parametric PDEs. ArXiv 1810.05094.

Snow, D. (2020). MTSS-GAN: Multivariate time series simulation generative adversarial networks. Available at SSRN 3616557.

Stone, H. (2020). Calibrating rough volatility models: a convolutional neural network approach. *Quantitative Finance*, **20**(3), 379–392.

Wiese, M., Knobloch, R., Korn, R., and Kretschmer, P. (2019). Quant GANs: Deep generation of financial time series. ArXiV 1907.06673.

Xu, T., Wenliang, L. K., Munn, M., and Acciaio, B. (2020). COT-GAN: Generating sequential data via causal optimal transport. ArXiv 2006.08571.

16

Asset Pricing and Investment with Big Data

Markus Pelger[a]

Abstract

We survey the most recent advances of using machine learning methods to explain differences in expected asset returns and form profitable portfolios. We discuss how to build better machine learning estimators by incorporating economic structure in the form of a no-arbitrage model. A no-arbitrage constraint in the objective function helps estimating asset pricing models in spite of the low signal-to-noise ratio in financial return data. We show how to include this economic constraint in large dimensional factor models, deep neural networks and decision trees. The resulting models strongly outperform conventional machine learning models in terms of Sharpe ratios, explained variation and pricing errors.

16.1 Overview

We survey the most recent advances of using machine learning methods to explain differences in asset returns and form profitable portfolios. Asset prices depend on a large set of economic variables and the functional form of this dependency is unknown and likely complex. Machine learning methods offer a promising solution as they can fit flexible functional forms while regularization provides robust fits allowing for many potential explanatory variables. However, machine-learning tools are designed to work well for prediction tasks in a high signal-to-noise environment. As asset returns in efficient markets seem to be dominated by unforecastable news, it is hard to predict their risk premia with off-the-shelf methods. We discuss how to build better machine learning estimators by incorporating economic structure in the form of a no-arbitrage model. Our empirical analysis focuses on the set of all US equities.

First, we illustrate the intuition with high-dimensional factor models in Section 16.3. Then, we compare the ability of deep neural networks to explain asset prices when used for simple return prediction and for estimating a structural no-arbitrage model in Section 16.4. We show that there is a substantial improvement among

[a] Stanford University
Published in *Machine Learning And Data Sciences For Financial Markets*, Agostino Capponi and Charles-Albert Lehalle © 2023 Cambridge University Press.

all asset pricing metrics by using a no-arbitrage condition as criterion function, constructing the most informative test assets with an adversarial approach (GAN) and extracting the states of the economy from many macroeconomic time series with a time-series network (LSTM). Third, we discuss in Section 16.5 that this insight extends to decision trees which offer an interpretable alternative to model complex functional relationships. Last but not least, we discuss the investment implications in 16.6.

Asset pricing and investment are two sides of the same coin. So far, most papers have separated the construction of profitable investment strategies with machine learning into two steps. First, advanced methods extract signals for predicting future returns. Second, these signals are used to form profitable portfolios, which are typically long-short investments based on total predicted returns. However, we argue that these two steps should be merged together, that is machine learning techniques should extract the signals that are the most relevant for the overall portfolio design, which is exactly what is done in a no-arbitrage model.

16.2 No-arbitrage pricing and investment

We start with a brief review of asset pricing models. Our goal is to explain the differences in the cross-section of returns R for individual stocks. Let $R_{t+1,i}$ denote the return of asset i at time $t + 1$. The fundamental no-arbitrage assumption is equivalent to the existence of a stochastic discount factor (SDF) M_{t+1} such that for any return in excess of the risk-free rate $R^e_{t+1,i} = R_{t+1,i} - R^f_{t+1}$, it holds

$$\mathbb{E}_t\left[M_{t+1}R^e_{t+1,i}\right] = 0 \quad \Leftrightarrow \quad \mathbb{E}_t[R^e_{t+1,i}] = \underbrace{\left(-\frac{\mathrm{Cov}_t(R^e_{t+1,i}, M_{t+1})}{\mathrm{Var}_t(M_{t+1})}\right)}_{\beta^{\mathrm{SDF}}_{t,i}} \cdot \underbrace{\frac{\mathrm{Var}_t(M_{t+1})}{\mathbb{E}_t[M_{t+1}]}}_{\lambda_t},$$

where $\beta^{\mathrm{SDF}}_{t,i}$ is the exposure to systematic risk and λ_t is the price of risk. $\mathbb{E}_t[.]$ denotes the expectation conditional on the information at time t. The SDF is an affine transformation of the tangency portfolio. Without loss of generality we consider the SDF formulation

$$M_{t+1} = 1 - \sum_{i=1}^{N} \omega_{t,i} R^e_{t+1,i} = 1 - \omega_t^\top R^e_{t+1}.$$

The fundamental pricing equation $\mathbb{E}_t[R^e_{t+1}M_{t+1}] = 0$ implies the SDF weights

$$\omega_t = \mathbb{E}_t[R^e_{t+1}R^{e\top}_{t+1}]^{-1}\mathbb{E}_t[R^e_{t+1}], \tag{16.1}$$

which are the portfolio weights of the conditional mean-variance efficient portfolio.[1] We define the tangency portfolio as $F_{t+1} = \omega_t^\top R^e_{t+1}$ and will refer to this

---

[1] Any portfolio on the globally efficient frontier achieves the maximum Sharpe ratio. These portfolio weights represent one possible efficient portfolio. An alternative formulation would be $M_{t+1} = 1 - \sum_{i=1}^{N} \omega_{t,i}(R^e_{t+1,i} - \mathbb{E}_t[R^e_{t+1,i}])$ which results in the conventional conditional tangency portfolio weights $\omega_t = \mathrm{Cov}_t(R^e_{t+1})^{-1}\mathbb{E}_t[R^e_{t+1}]$.

traded factor as the SDF. The asset pricing equation can now be formulated as

$$\mathbb{E}_t[R^e_{t+1,i}] = \frac{\text{Cov}_t(R^e_{t+1,i}, F_{t+1})}{\text{Var}_t(F_{t+1})} \cdot \mathbb{E}_t[F_{t+1}] = \beta^{\text{SDF}}_{t,i} \mathbb{E}_t[F_{t+1}].$$

Hence, no-arbitrage implies a 1-factor model

$$R^e_{t+1,i} = \beta^{\text{SDF}}_{t,i} F_{t+1} + \epsilon_{t+1,i}$$

with $\mathbb{E}_t[\epsilon_{t+1,i}] = 0$ and $\text{Cov}_t(F_{t+1}, \epsilon_{t+1,i}) = 0$. Conversely, the factor model formulation implies the stochastic discount factor formulation above.

Different asset pricing model impose different structures on the SDF weights ω and SDF loadings β_{SDF}. The estimation challenge arises from modeling the conditional expectation $\mathbb{E}_t[.]$ which can depend in a complex way on a large number of asset-specific and macroeconomic variables. This is where machine learning tools are essential to deal in a flexible way with the large dimensionality.

The most common way is to translate the problem into an unconditional asset pricing model on sorted portfolios. Under additional assumptions one could obtain a valid SDF M_{t+1} conditional on a set of asset-specific characteristics C_t by its projection on the return space:

$$M_{t+1} = 1 - \omega_t^\top R^e_t \qquad \text{with } \omega_{t,i} = f\left(C_{t,i}\right),$$

where C_t is a $N \times L$ matrix of K characteristics observed for N stocks and $f(\cdot)$ is a general, potentially nonlinear and non-separable function. Most of the reduced-form asset pricing models approximate this function by a (potentially infinite) set of simple managed portfolios $f_j(\cdot)$, such that $f\left(C_{t,i}\right) \approx \sum_{j=1}^J f_j\left(C_{t,i}\right) \tilde{w}_j$. The SDF then becomes a linear combination of (potentially infinitely many) managed portfolios with constant weights $\tilde{\omega}_j$:

$$M_{t+1} = 1 - \sum_{j=1}^J \tilde{w}_j \tilde{R}^e_{t+1,j} \qquad \text{with } \tilde{R}^e_{t+1,j} = \sum_{i=1}^N f_j\left(C_{t,i}\right) R^e_{t+1,i}, \qquad (16.2)$$

where $\tilde{R}_{t+1}$ are the returns of managed portfolios that correspond to different basis functions in the characteristic space. The number of basis portfolios increases by the complexity of the basis functions and the number of characteristics. The most common managed portfolios are sorted on characteristic quantiles, that is, they use indicator functions based on characteristic quantiles to approximate $f(C_{t,i})$. Popular sorts are the size and value double-sorted portfolios of Fama and French (1992), that are also used to construct their long–short factors. The linear factor model literature imposes the additional assumption that a small number of risk factors based on characteristic managed portfolios should span the SDF.

The estimation of an asset pricing model yields investment opportunities with an attractive risk return trade-off. Estimating the SDF weights ω yields a tradeable portfolio, which in a correct model, should have the highest possible conditional Sharpe ratio. The SDF loadings β^{SDF} predict future asset returns up to a proportionality constant and hence identify the future relative performance of assets. Last but not least, an asset pricing model identifies mispriced assets which represent "alpha" investment opportunities.

16.3 Factor models

The workhorse models in equity asset pricing are based on linear factor models exemplified by Fama and French (1993, 2015). Finding the "right" factors has become the central question of asset pricing. Harvey and Zhu (2016) document that more than 300 published candidate factors have predictive power for the cross-section of expected returns. As argued by Cochrane (2011), this "factor zoo" leads to the question of which risk factors are important and which factors are subsumed by others. Recently, new methods have been developed to study the cross-section of returns in the linear framework but accounting for the large amount of conditioning information. This chapter focuses on the work of Lettau and Pelger (2020b,a) who extend principal component analysis (PCA) to account for no-arbitrage. They show that a no-arbitrage penalty term makes it possible to overcome the low signal-to-noise ratio problem in financial data and find the information that is relevant for the pricing kernel. Kelly et al. (2019) apply PCA to stock returns projected on characteristics to obtain a conditional multi-factor model where the loadings are linear in the characteristics. Pelger (2020) combines high-frequency data with PCA to capture non-parametrically the time-variation in factor risk. Pelger and Xiong (2020) show that macroeconomic states are relevant to capture time-variation in PCA-based factors. Kozak et al. (2020) estimate the SDF based on characteristic sorted factors with a modified elastic net regression. Giglio and Xiu (2021) complement the SDF with PCA-based factors to conduct asset pricing tests. Bryzgalova et al. (2020a) propose a Bayesian solution for the factor zoo. Kozak et al. (2020) and Lettau and Pelger (2020a) show that there is a small number of linear combinations of risk factors that can explain returns, but that a small number of conventional risk factors might not be sufficient: in other words, the sparsity seems to be in the rotated factor space, and hence conventional selection methods like lasso applied directly to a large number of conventional factors seem to perform worse.

We assume that excess returns follow a standard approximate factor model and the assumptions of the arbitrage pricing theory are satisfied. This means that excess returns of an asset i, $R_{t,i}^e$, have a systematic component captured by K factors and a nonsystematic, idiosyncratic component capturing asset-specific risk. The approximate factor structure allows the nonsystematic risk to be weakly dependent. We observe the excess return of N assets over T time periods:

$$R_{t,i}^e = F_t \beta_i^\top + e_{t,i} \qquad i = 1,\dots,N, \ t = 1,\dots,T \qquad (16.3)$$

$$\Longleftrightarrow \underbrace{R^e}_{T\times N} = \underbrace{F}_{T\times K} \underbrace{\beta^\top}_{K\times N} + \underbrace{e}_{T\times N}, \qquad (16.4)$$

where the loadings (or betas) β, and in a latent factor model also the unknown factors, have to be estimated. Σ_R and Σ_F are the variance-covariance matrices of returns and factors, respectively, and Σ_e is the variance-covariance matrix of e. Given factors F, we can compute the maximal Sharpe ratio from the tangency portfolio of the mean-variance frontier that is spanned by F as

$$\omega_{\mathrm{F}} = \Sigma_F^{-1} \mu_F,$$

where μ_F and Σ_F are the mean and variance-covariance matrix of F. The implied SDF is given by $M_t = 1 - \omega_{\mathrm{F}}^\top (F_t - \mathbb{E}[F_t])$.

In this subsection we review the method of Lettau and Pelger (2020a) to find the most important factors for explaining asset returns and bringing order to the chaos of factors. Their methodology uses large financial data sets to identify factors that simultaneously explain the time series and cross-section of stock returns. The estimation approach is a generalization of the widely utilized principal component analysis (PCA), e.g. in Connor and Korajczyk (1986, 1988). Statistical factor analysis based on PCA extracts factors that capture comovement but does not incorporate any information contained in the means of the data. Therefore, it is not surprising that PCA factors do not capture the differences in mean risk premia of assets. Lettau and Pelger (2020a) propose an alternative estimator, risk-premium PCA (RP-PCA), that incorporates information in the first and second moments of the data yielding a more efficient estimator than standard PCA. The risk-premium PCA estimator can be interpreted as a generalized PCA with an additional penalty term that accounts for cross-sectional pricing errors, thus combining PCA factor analysis with the arbitrage pricing theory (APT) of Ross (1976). The objective of finding factors that can explain comovement and the cross-section of expected returns simultaneously is based on fundamental insights of APT: Systematic time-series factors also determine cross-sectional risk premia. The RP-PCA exploits this connection explicitly.

We will work in a large-dimensional panel, that is, both N and T are large. Under the assumption that the factors and residuals are uncorrelated, the covariance matrix of the returns consists of a systematic and idiosyncratic part:

$$\mathrm{Var}(R^e) = \beta \mathrm{Var}(F) \beta^\top + \mathrm{Var}(e).$$

Since the largest eigenvalues of $\mathrm{Var}(R^e)$ are driven by the factors, PCA can be used to estimate the loadings and factors. Note that standard PCA estimators of latent factors use the information contained in the second moments but ignore information that is contained in the first moment.

If R^e contains only excess returns, the role of means is explicitly given by Ross' arbitrage pricing theory (APT), which implies that expected excess returns are explained by the exposure to the risk factors multiplied by the risk premium of the factors. If the factors are excess returns, the APT implies:

$$\mathbb{E}[R_{t+1,i}^e] = \beta_i \mathbb{E}[F].$$

Factors identified by standard PCA explain as much time variation as possible. Conventional statistical factor analysis applies PCA to the sample covariance matrix $\frac{1}{T} R^{e\top} R^e - \overline{R^e}\,\overline{R^e}^\top$ where $\overline{R^e}$ denotes the sample mean of excess returns. Hence, $\widehat{\beta}_{\mathrm{PCA}}$ are estimated as the eigenvectors of the K largest eigenvalues of the sample covariance matrix. The factors are estimated as $\widehat{F}_{\mathrm{PCA}} = R^e \widehat{\beta}_{\mathrm{PCA}} \left(\widehat{\beta}_{\mathrm{PCA}}^\top \widehat{\beta}_{\mathrm{PCA}} \right)^{-1}$.

It is straightforward to express the PCA loadings and factors as solutions to the minimization of the following objective function:

$$\widehat{F}_{\text{PCA}}, \widehat{\beta}_{\text{PCA}} = \arg\min_{\beta,F} \frac{1}{NT} \sum_{n=1}^{N} \sum_{t=1}^{T} \left((R_{t,i}^e - \overline{R^e}_n) - (F_t - \overline{F}) \beta_i^\top \right)^2. \quad (16.5)$$

The risk-premium-PCA (RP-PCA) estimator modifies the objective function so that cross-sectional pricing errors are taken into account. The RP-PCA objective function minimizes a weighted average of the unexplained variation and cross-sectional pricing errors:

$$\widehat{F}_{\text{RP}}, \widehat{\beta}_{\text{RP}} = \arg\min_{\beta,F} \underbrace{\frac{1}{NT} \sum_{n=1}^{N} \sum_{t=1}^{T} (R_{t,i}^e - F_t \beta_i^\top)^2}_{\text{unexplained TS variation}} + \gamma \underbrace{\frac{1}{N} \sum_{n=1}^{N} \left(\overline{R_{t+1,i}^e} - \overline{F} \beta_i^\top \right)^2}_{\text{XS pricing error}},$$

$$(16.6)$$

where $\gamma \geq -1$ is the weight of the average cross-sectional pricing error relative to the times-series error in standard PCA. It is straightforward to show that minimizing Equation (16.6) is equivalent to applying PCA to the matrix

$$\Sigma_{\text{RP}} = \frac{1}{T} R^{e\top} R^e + \gamma \overline{R^e}\,\overline{R^e}^\top. \quad (16.7)$$

Note that Σ_{RP} is equal to the variance-covariance matrix of R^e if $\gamma = -1$. Thus, standard PCA using the variance-covariance matrix is a special case of RP-PCA. RP-PCA with $\gamma > -1$ can be understood as PCA applied to a matrix that "overweights" the means. As in standard PCA, the eigenvectors of the K largest eigenvalues of Σ_{RP} are proportional to the loadings $\widehat{\beta}_{\text{RP}}$. In PCA, the eigenvalues are equal to factor variances, while eigenvalues in RP-PCA are equal to a more generalized notion of "signal strength" of a factor that includes the information in the mean. RP-PCA factors are estimated by a regression of the returns on the estimated loadings, that is, $\widehat{F}_{\text{RP}} = R^e \widehat{\beta}_{\text{RP}} \left(\widehat{\beta}_{\text{RP}}^\top \widehat{\beta}_{\text{RP}} \right)^{-1}$.[2]

Lettau and Pelger (2020b) develop the asymptotic inferential theory for the RP-PCA estimator under a general approximate factor model and show that it dominates conventional estimation based on PCA if there is information in the mean. They distinguish between strong and weak factors.[3] Strong factors essentially affect all underlying assets. The market-wide return is an example of a strong factor in asset pricing applications. RP-PCA can estimate these factors more efficiently than PCA as it efficiently combines information in first and second moments of the data. Weak factors affect only a subset of the underlying assets and are harder to detect. Many asset-pricing factors fall into this category.

---

[2] In latent factor models only the product $F\beta^\top$ is identified. For any full rank $K \times K$ matrix H the factors FH^{-1} and loadings $\beta H^\top$ yield the same factor model. We use the standard convention to normalize the loadings $\beta^\top \beta / N = I_K$ and assume that the factors are uncorrelated.

[3] Lettau and Pelger (2020b) generalize the spiked covariance models from random matrix theory and properties in Onatski (2012) to analyze the asymptotic behavior of the RP-PCA estimator for weak factors.

RP-PCA can find weak factors with high Sharpe ratios, which cannot be detected with PCA, even if an infinite amount of data is available.

The empirical analysis of Lettau and Pelger (2020a) uses excess returns of a large cross-section of single-sorted decile portfolios constructed from 37 anomaly characteristics. The empirical findings can be summarized as follows. First, PCA is not a reliable method to estimate latent asset pricing factors and is dominated by RP-PCA. They show that even for 25 double-sorted portfolios that follow a clear factor structure, PCA can fail to detect the underlying factor structure, while RP-PCA reliably finds all relevant asset-pricing factors. Second, they show that a small number of factors is sufficient to fit the first and second moments of the 370 anomaly portfolios. The RP-PCA method extracts five significant factors that together yield a high Sharpe ratio (SR), small pricing errors, and capture most of the time-series variation in the data.

Table 16.1 shows the out-of-sample asset pricing results for five RP-PCA factors compared with five PCA and the Fama–French 5-factor model. RP-PCA essentially explains the same amount of variation as PCA, while also explaining average returns and achieving substantially higher mean-variance efficiency. Figure 16.1 illustrates the superior out-of-sample profitability of the RP-PCA tangency portfolio relative to other factor models. Figure 16.2 shows that five RP-PCA factors are sufficient to capture the pricing information in the 370 portfolios. The first factor is long-only in all portfolios and is highly correlated with the market return. Two additional factors capture time-series variation but play no role in the cross-section of returns or the Sharpe ratio of the implied SDF. The remaining two factors are relevant for the cross-section and Sharpe ratio but are less critical for the time-series variation. All results hold in-sample as well as out-of-sample, suggesting that RP-PCA is stable and robust. These results show the importance of using a no-arbitrage structure in the estimation of asset pricing factors from large data sets that have only a weak signal. The RP-PCA estimator achieves this parsimoniously and efficiently without adding any computational burden.

Table 16.1 This table shows out-of-sample asset pricing results with different factor models. We report the out-of-sample maximal Sharpe ratios, root-mean-squared pricing errors and unexplained idiosyncratic variation for $K = 5$ factors and RP-weight $\gamma = 10$. The data are monthly returns for $N = 370$ decile portfolios for $T = 650$ time observations from November 1963 to December 2017. The out-of-sample results use a rolling window of 240 months. Additional details about the data and implementation are in Lettau and Pelger (2020a).

	Sharpe ratio	Pricing error	Unexplained variation
RP-PCA	0.45	0.12	12.70%
PCA	0.17	0.14	12.56%
Fama–French 5	0.31	0.21	13.66%

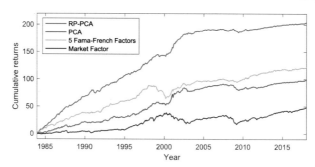

Figure 16.1 The figure shows the out-of-sample cumulative excess returns for the SDFs based on different factor models. The data are monthly returns for $N = 370$ decile portfolios for $T = 650$ time observations from November 1963 to December 2017. The out-of-sample results use a rolling window of 240 months for the SDF estimation. RP-PCA and PCA use five factors and the RP-weight $\gamma = 10$. Additional details about the data and implementation are in Lettau and Pelger (2020a).

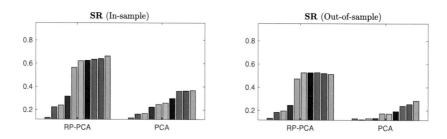

Figure 16.2 This figure shows the maximum Sharpe ratios for different numbers of factors estimated with RP-PCA and PCA. RP-PCA uses $\gamma = 10$. The data are monthly returns for $N = 370$ decile portfolios for $T = 650$ time observations from November 1963 to December 2017. The out-of-sample results use a rolling window of 240 months. Additional details about the data and implementation are in Lettau and Pelger (2020a).

16.4 Deep learning in asset pricing

16.4.1 Forecasting

Expected returns can depend in a complex way on large amount of firm-specific and macroeconomic information. It is a natural idea to use machine learning techniques like deep neural networks to deal with the high dimensionality and complex functional dependencies of the problem. This section focuses on the model of Chen et al. (2020). They show that including the no-arbitrage constraint in the learning algorithm significantly improves the risk premium signal and makes it possible to better explain individual stock returns.

A lot of pathbreaking contributions have recently been made in studying the impact of characteristics on returns with flexible machine learning model, but without imposing an underlying risk model or a no-arbitrage condition. In their pioneering work Gu et al. (2020) conduct a comparison of machine learning methods for predicting the panel of individual US stock returns and demonstrate the benefits of flexible methods. Messmer (2017) and Feng et al. (2020b) follow a similar approach as Gu et al. (2020) to predict stock returns with neural networks. Bianchi et al. (2021) provide a comparison of machine learning methods for predicting bond returns in the spirit of Gu et al. (2020). Freyberger et al. (2020) use Lasso selection methods to estimate the risk premia of stock returns as a non-linear additive function of characteristics. Gu et al. (2021) extend the linear conditional factor model of Kelly et al. (2019) to a non-linear factor model using an autoencoder neural network. Sadhwani et al. (2021) predict mortgage prepayments, delinquencies, and foreclosures with deep neural networks. Sirignano and Cont (2019) use deep neural networks to predict the direction of high-frequency price moves with order flow history.

Predicting asset returns yields an estimate of conditional expected returns $\mu_{t,i}$ and maps into a cross-sectional asset pricing model. Conditional expected returns $\mu_{t,i}$ are proportional to the loadings in the 1-factor formulation:

$$\mu_{t,i} := \mathbb{E}_t[R^e_{t+1,i}] = \beta_{t,i}\mathbb{E}_t[F_{t+1}].$$

Hence, up to a time-varying proportionality constant the SDF weights and loadings are equal to $\mu_{t,i}$. In this subsection we will consider the best performing forecasting approach of Gu et al. (2020), which uses deep neural networks, for asset pricing. They use a feedforward network (FFN), which estimates the conditional mean function $\mu(\cdot): \mathbb{R}^p \times \mathbb{R}^q \to \mathbb{R}$ by minimizing the average sum of squared prediction errors:

$$\hat{\mu} = \min_{\mu} \frac{1}{T} \sum_{t=1}^{T} \frac{1}{N_t} \sum_{i=1}^{N_t} \left(R^e_{t+1,i} - \mu(I_t, C_{t,i})\right)^2,$$

where $I_t \times C_{t,i} \in \mathbb{R}^p \times \mathbb{R}^q$ denotes all the variables in the information set at time t. We denote by I_t all p macroeconomic conditioning variables that are not asset specific, e.g. inflation rates or the market return, while $C_{t,i}$ are q firm-specific characteristics, e.g. the size or book-to-market ratio of firm i at time t.

16.4.2 No-arbitrage model

Chen et al. (2020) propose a non-parametric adversarial estimation approach and show that it can be interpreted as a data-driven way to construct informative test assets. Finding the SDF weights is equivalent to solving a method of moment problem. The conditional no-arbitrage moment condition implies infinitely many unconditional moment conditions

$$\mathbb{E}[M_{t+1}R^e_{t+1,i}g(I_t, C_{t,i})] = 0 \qquad (16.8)$$

for any function $g(.) : \mathbb{R}^p \times \mathbb{R}^q \to \mathbb{R}^D$, where $I_t \times C_{t,i} \in \mathbb{R}^p \times \mathbb{R}^q$ denotes all the variables in the information set at time t and D is the number of moment conditions. The unconditional moment conditions can be interpreted as the pricing errors for a choice of portfolios and times determined by $g(.)$. The challenge lies in finding the relevant moment conditions to identify the SDF.

A well-known formulation includes 25 moments that corresponds to pricing the 25 size and value double-sorted portfolios of Fama and French (1993). For this special case each g corresponds to an indicator function if the size and book-to-market values of a company are in a specific quantile. Another special case is to consider only unconditional moments, i.e. setting g to a constant. This corresponds to minimizing the unconditional pricing error of each stock.

The SDF portfolio weights $\omega_{t,i} = \omega(I_t, C_{t,i})$ and risk loadings $\beta_{t,i} = \beta(I_t, C_{t,i})$ are general functions of the information set, that is, $\omega : \mathbb{R}^p \times \mathbb{R}^q \to \mathbb{R}$ and $\beta : \mathbb{R}^p \times \mathbb{R}^q \to \mathbb{R}$. For example, the SDF weights and loadings in the Fama–French 3-factor model are a special case, where both functions are approximated by a two-dimensional kernel function that depends on the size and book-to-market ratio of firms. The Fama–French 3-factor model only uses firm-specific information but no macroeconomic information, e.g. the loadings cannot vary based on the state of the business cycle.

Chen et al. (2020) use an adversarial approach to select the moment conditions that lead to the largest mispricing:

$$\min_{\omega} \max_{g} \frac{1}{N} \sum_{j=1}^{N} \left\| \mathbb{E}\left[\left(1 - \sum_{i=1}^{N} \omega(I_t, I_{t,i}) R^e_{t+1,i}\right) R^e_{t+1,j} g(I_t, C_{t,j}) \right] \right\|^2, \quad (16.9)$$

where the function ω and g are normalized functions chosen from a specified functional class. This is a minimax optimization problem. These types of problems can be modeled as a zero-sum game, where one player, the asset pricing modeler, wants to choose an asset pricing model, while the adversary wants to choose conditions under which the asset pricing model performs badly. This can be interpreted as first finding portfolios or times that are the most mispriced and then correcting the asset pricing model to also price these assets. The process is repeated until all pricing information is taking into account, that is the adversary cannot find portfolios with large pricing errors. Note that this is a data-driven generalization for the research protocol conducted in asset pricing in the last decades. Assume that the asset pricing modeler uses the Fama–French 5-factor model, that is M is spanned by those five factors. The adversary might propose momentum sorted test assets, that is g is a vector of indicator functions for different quantiles of past returns. As these test assets have significant pricing errors with respect to the Fama–French five factors, the asset pricing modeler needs to revise her candidate SDF, for example, by adding a momentum factor to M. Next, the adversary searches for other mispriced anomalies or states of the economy, which the asset pricing modeler will exploit in her SDF model.

A special case assumes a linear structure in the factor portfolio weights $\omega_{t,i} = \theta^\top C_{t,i}$ and linear conditioning in the test assets:

$$\frac{1}{N} \sum_{j=1}^{N} \mathbb{E}\left[\left(1 - \frac{1}{N} \sum_{i=1}^{N} \theta^\top C_{t,i} R^e_{t+1,i}\right) R^e_{t+1,j} C_{t,j} \right] = 0 \Leftrightarrow \mathbb{E}\left[(1 - \theta^\top \tilde{F}_{t+1}) \tilde{F}_{t+1}^\top \right] = 0,$$

where $\tilde{F}_{t+1} = \frac{1}{N} \sum_{i=1}^{N} C_{t,i} R^e_{t+1,i}$ are q characteristic managed factors. Such characteristic managed factors based on linearly projecting onto quantiles of characteristics are exactly the input to PCA in Kelly et al. (2019) or the elastic net mean-variance optimization in Kozak et al. (2020). The solution to minimizing the sum of squared errors in these moment conditions is a simple mean-variance optimization for the q characteristic managed factors that is, $\theta = \left(\mathbb{E}\left[\tilde{F}_{t+1}\tilde{F}_{t+1}^{\top}\right]\right)^{-1}\mathbb{E}\left[\tilde{F}_{t+1}\right]$ are the weights of the tangency portfolio based on these factors. Chen et al. (2020) choose this specific linear version of the model as it maps directly into the linear approaches that have already been successfully used in the previous subsection. This linear framework essentially captures the class of linear factor models.

16.4.3 Economic dynamics

Chen et al. (2020) introduce a novel way to use neural networks to extract economic conditions from complex time series. They propose Long-Short-Term-Memory (LSTM) networks to summarize the dynamics of a large number of macroeconomic time series in a small number of economic states. More specifically, their LSTM approach aggregates a large dimensional panel cross-sectionally into a small number of time-series and extracts from those a non-linear time-series model. The key element is that it can capture short and long-term dependencies which are necessary for detecting business cycles. A Recurrent Neural Network (RNN) with LSTM estimates the hidden macroeconomic state variables. Instead of directly passing macroeconomic variables I_t as covariates to the feedforward network, Chen et al. (2020) extract their dynamic patterns with an LSTM and only pass on a small number of hidden states capturing these dynamics.

Many macroeconomic variables themselves are not stationary. Hence, researchers need to first perform transformations, which typically take the form of some difference of the time-series. There is no reason to assume that the pricing kernel has a Markovian structure with respect to the macroeconomic information, in particular after transforming them into stationary increments. For example, business cycles can affect pricing but the GDP growth of the last period is insufficient to learn if the model is in a boom or a recession. Hence, we need to include lagged values of the macroeconomic variables and find a way to extract the relevant information from a potentially large number of lagged values.

As an illustration, Figure 16.3 shows the time-series of the S&P 500 price together with its log difference to remove the obvious non-stationarity. Using only the last observation of the differenced data obviously results in a loss of information and cannot identify the cyclical dynamic patterns. On the other hand simply including all lagged values of the increments of a large number of macroeconomic time-series as additional covariates in a non-parametric model blows up the parameter space, and neglects the intrinsic time-series structure in those explanatory variables. Chen et al. (2020) show how to extract only the relevant dynamic information, which is then passed on as an input to another model.

Formally, we have a sequence of stationary vector-valued processes $\{x_0, \ldots, x_t\}$ where we set x_t to the stationary transformation of I_t at time t, i.e. typically an

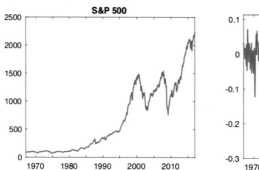
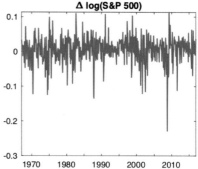

Figure 16.3 This figure shows the illustrative macroeconomic time-series of S&P 500 prices and log returns.

increment. Our goal is to estimate a functional mapping h that transforms the time-series x_t into "state processes" $h_t = h(x_0, \ldots, x_t)$ for $t = 1, \ldots, T$. The simplest transformation is to simply take the last increment, that is $h_t^\Delta = h^\Delta(x_0, \ldots, x_t) = x_t$. This approach is used in most papers including Gu et al. (2020) and neglects the serial dependency structure in x_t.

Macroeconomic time-series variables are strongly cross-sectionally dependent, that is, there is redundant information which could be captured by some form of factor model. A cross-sectional dimension reduction is necessary as the number of time-series observations in our macroeconomic panel is of a similar magnitude as the number of cross-sectional observations. Ludvigson and Ng (2007) advocate the use of PCA to extract a small number K_h of factors which is a special case of the function $h^{\text{PCA}}(x_0, \ldots, x_t) = W_x x_t$ for $W_x \in \mathbb{R}^{p \times K_h}$. This aggregates the time series to a small number of latent factors that explain the correlation in the innovations in the time series, but PCA cannot identify the current state of the economic system which depends on the dynamics.

RNNs are a family of neural networks for processing sequences of data. They estimate non-linear time-series dependencies for vector valued sequences in a recursive form. A vanilla RNN model takes the current input variable x_t and the previous hidden state h_{t-1}^{RNN} and performs a non-linear transformation to get the current state h_t^{RNN}.

$$h_t^{\text{RNN}} = h^{\text{RNN}}(x_0, \ldots, x_t) = \sigma(W_h h_{t-1}^{\text{RNN}} + W_x x_t + w_0),$$

where σ is the non-linear activation function. Intuitively, a vanilla RNN combines two steps: First, it summarize cross-sectional information by linearly combining a large vector x_t into a lower-dimensional vector. Second, it is a non-linear generalization of an autoregressive process where the lagged variables are transformations of the lagged observed variables. This type of structure is powerful if only the immediate past is relevant, but it is not suitable if the time series dynamics are driven by events that are further back in the past. Conventional RNNs can encounter problems with exploding and vanishing gradients when considering longer time lags. This is why Chen et al. (2020) use the more complex Long-Short-Term-Memory cells. The LSTM is designed to deal with lags of unknown and potentially long duration in the time series, which makes it well-suited to detect business cycles.

The LSTM approach can deal with both the large dimensionality of the system and a very general functional form of the states while allowing for long-term dependencies. Intuitively, an LSTM uses different RNN structures to model short-term and long-term dependencies and combines them with a non-linear function. We can think of an LSTM as a flexible hidden state space model for a large-dimensional system. On the one hand it provides a cross-sectional aggregation similar to a latent factor model. On the other hand, it extracts dynamics similar in spirit to state space models, like for example the simple linear Gaussian state space model estimated by a Kalman filter. The strength of the LSTM is that it combines both elements in a general non-linear model. Chen et al. (2020) show that an LSTM can successfully extract a business cycle pattern which essentially captures deviations of a local mean from a long-term mean.

The output of the LSTM is the function $h^{\text{LSTM}}(\cdot)$, which yields a small number of state processes $h_t = h^{\text{LSTM}}(x_0, \ldots, x_t)$ which Chen et al. (2020) use instead of the macroeconomic variables I_t as an input to their SDF network. Note, that each state h_t depends only on current and past macroeconomic increments and has no look-ahead bias.

16.4.4 Model architecture

The model architecture is summarized in Figure 16.4. For a given conditioning function $g(\cdot)$, Chen et al. (2020) estimate $\hat{\omega}$ by minimizing the weighted sample moments, which can be interpreted as weighted sample mean pricing errors. They construct the conditioning function $\hat{g}$ via a conditional network with a similar neural network architecture. Both, the SDF network and the conditional network each use an FFN network combined with an LSTM that estimates the macroeconomic hidden state variables, i.e. instead of directly using I_t as an input each network summarizes the whole macroeconomic time series information in the state process h_t (respectively h_t^g for the conditional network). The two LSTMs are based on the criteria function of the two networks, that is h_t are the hidden states that can minimize the pricing errors, while h_t^g generate the test assets with the largest mispricing. The loss function $L(\omega|g, h_t^g, h_t, C_{t,i})$ is the average of sample moments in Equation 16.9, yielding the following estimation problem:

$$\{\hat{\omega}, \hat{h}_t, \hat{g}, \hat{h}_t^g\} = \min_{\omega, h_t} \max_{g, h_t^g} L(\omega|g, h_t^g, h_t, C_{t,i}).$$

16.4.5 Empirical results

The empirical analysis in Chen et al. (2020) is based on a data set of all available US stocks from CRSP with monthly returns from 1967 to 2016 combined with 46 time-varying firm-specific characteristics and 178 macroeconomic time series.

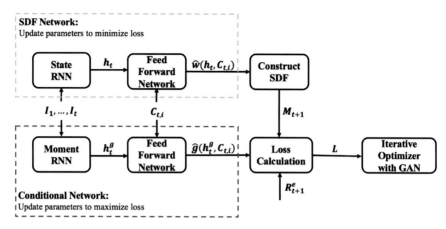

Figure 16.4 This figure shows the model architecture of GAN (Generative Adversarial Network) with RNN (Recurrent Neural Network) with LSTM cells. The SDF network has two parts: (1) An LSTM estimates a small number of macroeconomic states. (2) These states, together with the firm-characteristics, are used in an FFN to construct a candidate SDF for a given set of test assets. The conditioning network also has two networks: the first it creates its own set of macroeconomic states; the second, it combines with the firm-characteristics in an FFN to find mispriced test assets for a given SDF, M. These two networks compete until convergence: that is, until neither the SDF nor the test assets can be improved.

It includes the most relevant pricing anomalies and forecasting variables for the equity risk premium. The models are estimated on the first 20 years of data, tuning parameters are selected on the five-year validation data and the remaining 25 years are the out-of-sample test data. Their approach outperforms out-of-sample all other benchmark approaches, which include linear models and deep neural networks that forecast risk premia instead of solving a GMM type problem. The linear model applies mean-variance optimization with elastic net penalty on long–short factors, while the forecast model (Forecast) uses a feed forward neural network. Table 16.2 compares the models out-of-sample with respect to the Sharpe ratio implied by the pricing kernel, the explained variation and explained average returns of individual stocks. Their GAN model has an annual out-of-

Table 16.2 This table compares the performance of different SDF models. It shows the out-of-sample monthly Sharpe ratio (SR) of the SDF, explained time-series variation (EV) and cross-sectional mean R^2 for the GAN, Forecast and Linear model. Additional details about the data and implementation are in Chen et al. (2020).

Model	Sharpe Ratio	Explained Variation	Explained Mean
GAN	0.75	0.08	0.23
Forecast	0.44	0.04	0.15
Linear	0.50	0.04	0.19

sample Sharpe ratio of 2.6 compared to 1.7 for the linear special case of their model and 1.5 for the deep learning forecasting approach. At the same time they can explain 8% of the variation of individual stock returns and explain 23% of the expected returns of individual stocks, which is substantially higher than the other benchmark models. On standard test assets based on single- and double-sorted anomaly portfolios their asset pricing model reveals an unprecedented pricing performance. In fact, on all 46 anomaly sorted decile portfolios they achieve a cross-sectional R^2 higher than 90%. Figure 16.5 illustrates how the SDF portfolios translate into attractive investment opportunities. Note that all standard risk measures including maximum losses or drawdown are low for the GAN SDF.

Figure 16.6 summarizes the effect of conditioning on the hidden macroeconomic state variables. First, they add the 178 macroeconomic variables as predictors to all networks without reducing them to the hidden state variables. The performance for the out-of-sample Sharpe ratio of the Linear, Forecast and GAN model completely collapses. Conditioning only on the last normalized macroe-

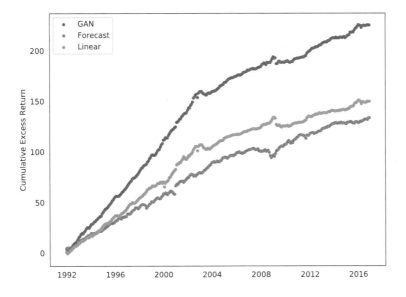

Figure 16.5 The figure shows the out-of-sample cumulative excess returns for the SDF for GAN, Forecast and Linear. Each factor is normalized by its standard deviation for the time interval under consideration.

conomic observation, which is usually an increment, does not allow the detection of a dynamic structure, e.g. a business cycle. Even worse, including the large number of irrelevant variables actually lowers the performance compared to a model without macroeconomic information. Although the models use a form of regularization, a too large number of irrelevant variables makes it harder to select

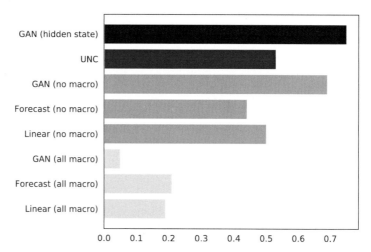

Figure 16.6 This figure shows the out-of-sample Sharpe ratio of SDFs for different inclusions of the macroeconomic information. The GAN (hidden states) is our reference model. UNC is a special version of our model that uses only unconditional moments (but includes LSTM macroeconomic states in the FFN network for the SDF weights). GAN (no macro), Forecast (no macro) and Linear (no macro) use only firm-specific information as conditioning variables but no macroeconomic variables. GAN (all macro), Forecast (all macro), Linear (all macro) include all 178 macro variables as predictors (respectively conditioning variables) without using an LSTM to transform them into macroeconomic states.

those that are actually relevant. GAN without the macroeconomic but only firm-specific variables has an out-of-sample Sharpe ratio that is around 10% lower than with the macroeconomic hidden states. This is another indication that it is relevant to include the dynamics of the time series. The UNC model uses only unconditional moments as the objective function; that is, they use a constant conditioning function g, but include the LSTM hidden states in the factor weights. The Sharpe ratio is around 20% lower than the GAN with hidden states. Hence, it is not only important to include all characteristics and the hidden states in the weights and loadings of SDF but also in the conditioning function g in order to identify the assets and times that matter for pricing.

16.5 Decision trees in asset pricing

Decision trees are an appealing alternative to neural networks as they are easier to interpret. Bryzgalova et al. (2020b) show how to build a cross-section of asset returns, that is, a small set of basis assets that capture complex information contained in a given set of stock characteristics and span the SDF. They use decision trees to generalize the concept of conventional sorting and introduce a new approach to the robust recovery of the SDF, which endogenously yields optimal portfolio splits. They propose to use their small set of informative and interpretable basis assets as test assets for asset pricing models, as pricing them is equivalent to spanning the SDF. Moritz and Zimmerman (2016), Gu et al. (2020), and Rossi (2018) also rely on decision trees in estimating conditional moments of stock returns, but do not use an asset pricing objective in the estimation. Moritz and Zimmerman (2016) apply tree-based models to studying momentum, while Gu et al. (2020) use random forest to model expected returns on stocks as a function of characteristics. Rossi (2018) uses Boosted Regression Trees to form conditional mean-variance efficient portfolios based on the market portfolio and the risk-free asset. Since Bryzgalova et al. (2020b) use decision trees not for a direct prediction of returns but for constructing a set of basis assets that span the efficient frontier, none of the standard pruning algorithms available in the literature is applicable in their setting because of its global optimization nature. Their novel method prunes the trees based on an asset pricing criterion.

Bryzgalova et al. (2020b) use decision trees as basis portfolios in Equation 16.2. Their Asset Pricing Trees (AP-Trees) have two key elements: (1) the construction of conditional tree portfolios and (2) the *pruning* of the overall portfolio set based on the SDF spanning requirement. Figure 16.7 shows a simple decision tree based on a sequence of *conditional* consecutive splits. For example, one could start by dividing the universe of stocks into two groups based on the individual stock's market cap, then within each group – by their value, then by size again, and so on. The nodes of such a tree also correspond to managed portfolios and reflect the conditional impact of characteristics in a simple and transparent way:

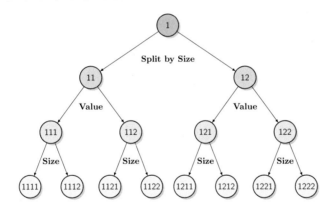

Figure 16.7 The figure presents an example of an AP-Tree of depth 3 based on size and book-to market. The first 50/50 split is done by size, the second by value, and the last one by size again. The portfolio label corresponds to the path along the tree that identifies it, with "1" standing for going left, while "2" stands for going right.

Relying on a different list of the variables employed, the order and depth of the splits, it produces a very diverse and rich set of portfolios. As a result, all the decision trees' final and intermediate nodes represent a high-dimensional set of possible investment strategies, easily ensuring that each portfolio is well-diversified. Importantly, while any individual portfolio (tree node) has a clear economic interpretation, together the collection of such trees is extremely flexible, and can easily span the SDF.

They start with the whole set of potential managed portfolios offered by AP-Trees and develop a new approach to reduce them to a small number of interpretable test assets, the process they refer to as *pruning*. The decision on where to make a split along the tree follows an intuitive criteria: Assets are combined together in higher-level nodes, making the original portfolios redundant, only if their combination spans the SDF as well as the original, granular trading strategies. This requirement naturally maps the pruning process into robust SDF recovery within a mean-variance framework. For example, in Figure 16.7 they split further a portfolio of the smallest 50% of stocks, sorted by value, only if including the small-value and small-growth portfolios (in addition to other assets) in the SDF results in a higher total Sharpe ratio than including the combined small cap portfolio. This objective function, global Sharpe ratio, is completely different from the one used in standard decision trees, because the decision of doing the split is not local, and depends not only on the features of the parent and children nodes, but also how they co-move with the other potential basis assets.

16.5.1 SDF recovery as a mean-variance optimization problem

As outlined in Equation 16.2 the SDF becomes in general a linear combination of a large number managed portfolios: $M_t = 1 - \sum_{j=1}^{J} w_j \tilde{R}_{t,j}$. Given the managed portfolios, finding the SDF weights w is generally equivalent to finding the tangency portfolio with the highest Sharpe ratio in the mean–variance space. AP-Trees form the set of basis assets that reflect the relevant information conditional on characteristics and could be used to build the SDF. However, using all the potential portfolios is often not feasible due to the curse of dimensionality: For example, using trees with depth three for three characteristics results in 216 nodes, and with 10 characteristics the number of basis portfolios explodes to 8,000. Hence, Bryzgalova et al. (2020b) introduce a technique to shrink the dimension of the basis assets, with the key goal of retaining *both* the relevant information contained in characteristics and portfolio interpretability.

Bryzgalova et al. (2020b) find SDF weights by solving a mean-variance optimization problem with elastic net shrinkage applied to all final and intermediate nodes of AP-Trees. This approach combines three crucial features: (1) It shrinks the contribution of the assets that do not help in explaining variation. (2) It shrinks the sample mean of tree portfolios towards their average return, which is crucial, since estimated means with large absolute values are likely to be very noisy, introducing a bias. (3) It includes a lasso-type shrinkage to obtain a sparse representation of the SDF, selecting a small number of AP-Tree basis assets.

The search of a robust tangency portfolio can effectively be decomposed into two separate steps. First, they construct a robust mean-variance efficient frontier using the standard optimization with shrinkage terms. Then, they select the optimal portfolio located on the robust frontier on the validation data. Denote by $\hat{\mu}$ and $\hat{\Sigma}$ the sample mean and covariance matrix of all AP-Tree portfolios.

1. *Mean-variance portfolio construction with elastic net*:
   For a given set of values of tuning parameters μ_0, λ_1 and λ_2, use the training dataset to solve

$$\text{minimize} \quad \frac{1}{2}w^\top \hat{\Sigma} w + \lambda_1 ||w||_1 + \frac{1}{2}\lambda_2 ||w||_2^2$$
$$\text{subject to} \quad w^\top \mathbf{1} = 1$$
$$w^\top \hat{\mu} \geq \mu_0,$$

where $\mathbf{1}$ denotes a vector of ones, $||\omega||_2^2 = \sum_{i=1}^N \omega_i^2$ and $||\omega||_1 = \sum_{i=1}^N |w_i|$, and N is the number of assets.

2. *Tracing out the efficient frontier*: Select tuning parameters μ_0, λ_1 and λ_2 to maximize the Sharpe ratio on a validation sample of the data.

Without imposing any shrinkage on the portfolio weights for the SDF, the problem has an explicit solution, $\hat{\omega}_{\text{naive}} = \hat{\Sigma}^{-1}\hat{\mu}$. Their estimator is a shrinkage version. Tracing out the efficient frontier (without an elastic net penalty, $\lambda_1 = \lambda_2 = 0$) out-of-sample to select the tangency problem is equivalent to applying conventional in-sample mean-variance optimization but with a sample mean shrunk toward the cross-sectional average. It results in the weights

$$\hat{\omega}_{\text{robust}} = \hat{\Sigma}^{-1}\left(\hat{\mu} + \lambda_0 \mathbf{1}\right),$$

with a one-to-one mapping between the target mean μ_0 and mean shrinkage λ_0. The robust portfolio is equivalent to a weighted average of the naive tangency portfolio and the minimum-variance portfolio. Tracing out the robust efficient frontier out-of-sample that it includes a ridge penalty (i.e., no lasso penalty, $\lambda_1 = 0$, but general λ_2) is equivalent to conventional in-sample mean-variance optimization but with a shrunk sample mean and a sample covariance matrix shrunk toward a diagonal matrix. It has the weights

$$\hat{\omega}_{\text{robust}} = \left(\hat{\Sigma} + \lambda_2 I_N\right)^{-1}\left(\hat{\mu} + \lambda_0 \mathbf{1}\right).$$

Their approach generalizes the SDF estimation approach of Kozak et al. (2020) by including a mean shrinkage to the variance shrinkage and sparsity. Bryzgalova et al. (2020b) show that tracing out the whole efficient frontier is generally equivalent to different levels of shrinkage on the mean return, and generally does not have to be zero, which is imposed in Kozak et al. (2020). In fact, using cross-validation to find the optimal value of this shrinkage, they find that it in most cases it is not equal to zero. Intuitively, since the estimation of expected returns is severely contaminated with measurement error, it is likely that extremely high or low rates of return (relative to their peers) are actually overestimated/underestimated simply due to chance, and, hence, if left unchanged, would bias the SDF recovery.

Their estimator can also be interpreted as a robust approach to the mean-variance optimization problem. The robust mean-variance optimization is equivalent to finding the mean-variance efficient solution under a worst case outcome for estimation uncertainty. Given uncertainty sets for the achievable Sharpe ratio

S_{SR}, estimated mean S_μ and estimated variance S_Σ, the robust estimation solves

$$\min_{w} \max_{\mu, \Sigma \in S_{SR} \cap S_\mu \cap S_\Sigma} w^\top \Sigma w \quad \text{such that } w^\top \mathbf{1} = 1, \quad w^\top \hat{\mu} = \mu_0.$$

Each shrinkage has a direct correspondence to an uncertainty set: Mean shrinkage provides robustness against Sharpe ratio estimation uncertainty, variance shrinkage governs robustness against variance estimation uncertainty, and lasso controls robustness against mean estimation uncertainty. A higher mean shrinkage can also be interpreted as a higher degree of risk aversion of a mean-variance optimizer.

Many trading restrictions can easily be incorporated by simply removing undesirable nodes. Bryzgalova et al. (2020b) provide an example of using minimum market capitalization as one such restriction, but the same procedure can be applied to other cases as well.

16.5.2 *Empirical results*

Bryzgalova et al. (2020b) obtain monthly equity return data for all US stocks from January 1964 to December 2016, yielding 53 years total. They use the same 10 firm-specific characteristics as defined on the Kenneth French Data Library. The SDF weights and portfolio components are estimated on the training data (first 20 years). The shrinkage parameters are chosen on the validation data (10 years). All performance metrics are calculated out-of-sample on the test sample (23 years).

Historically, there has been only one way to build a set of basis assets that reflects more than two or three characteristics at the same time: Bundling several separate cross-sections together, usually either as a combination of several double or single sorts. By construction these portfolios exclude or at least drastically limit any interaction effects between the characteristics. Bryzgalova et al. (2020b) consider the most important conventional sorts as benchmarks: (a) Sets of 10 quintile portfolios, uniformly sorted by characteristics (50 assets altogether), (b) Sets of 10 decile portfolios (100 assets), (c) A combination of six double-sorted portfolios, with each based on size and some other characteristic (54 assets), and (d) A combination of 25 double-sorted portfolios, with each based on size and some other characteristic (225 assets). Note that conventional long–short factors are usually based on single- or double-sorted quantile portfolios and hence are also included in the span of these basis assets.

As a second benchmark Bryzgalova et al. (2020b) also consider machine learning predictions methods to map the characteristic information into a small number of portfolios. Decile-sorted portfolios based on predicted expected returns, which often yields a large spread in realized returns as well, are a popular choice in many recent studies of stock characteristics. While prediction-based portfolios constructed from multiple characteristics often have high Sharpe ratios, there is nothing per se in their construction that suggests they should be spanning the SDF, projected on the space of characteristics. Bryzgalova et al.

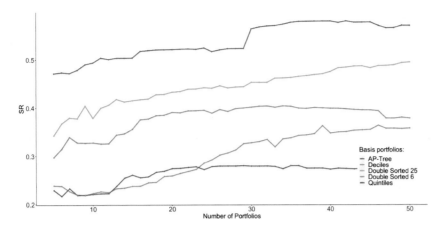

Figure 16.8 This figure shows out-of-sample monthly Sharpe ratios of SDFs based on 10 characteristics as a function of the number of basis assets constructed with AP-Trees, 10 quintile sorts, 10 decile sorts, combination of double sorts based on size and the other characteristic (either 6 or 25 double sorted assets per specific portfolio). We apply robust shrinkage with lasso to all basis assets and choose the optimal validation mean and variance shrinkage. Additional details about the data and implementation are in Bryzgalova et al. (2020b).

(2020b) use the best performing deep neural network from Gu et al. (2020) to predict the next period's returns based on the current period's characteristics and sort the stocks into quantile portfolios based on the prediction. In addition, they consider random forest, that exploits a collection of decision trees based on characteristics to predict future returns. Both deep neural networks and random forests, are the two best methods to predict future stock returns according to Gu et al. (2020) and hence serve as an appropriate benchmark that subsumes other prediction methods. They calculate the robust mean-variance efficient portfolio from the quantile prediction portfolios for both methods (labeled DL-MV for deep learning and RF-MV for random forest forecasting). In addition, they also consider a simple long–short factor from buying the highest prediction quantile and selling the lowest prediction quantile, labeled DL-LS and RF-LS.

Figures 16.8 and 16.9 compare the out-of-sample Sharpe ratios of the SDFs spanned by a small number of portfolios for AP-Trees and conventionally sorted portfolios or return-prediction portfolios. First, AP-Trees clearly stand out in terms of Sharpe ratios that are always two to three times larger than those of alternative basis assets. Second, conventional sorting methods, that correspond to coarse kernel approximations, do not reflect most of the investment opportunities as they neglect interaction effects. Third, it is also clear that leading return-prediction portfolios display a subpar performance. This clearly shows that while a flexible off-the-shelf machine-learning forecasting method does a good job at predicting returns per se, it is not necessarily the right tool to build portfolios spanning the SDF.

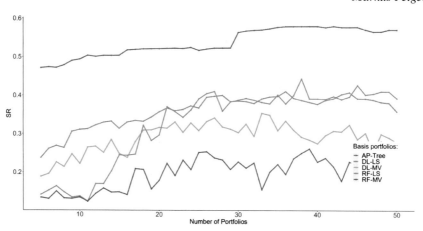

Figure 16.9 This figure shows out-of-sample monthly Sharpe ratios of SDFs based on 10 characteristics as a function of the number of basis assets constructed with AP-Trees and forecasted sorted portfolios based on deep learning DL-MV and random forest RF-MV. We apply robust shrinkage with lasso to all basis assets and choose the optimal validation mean and variance shrinkage. We also include long-short portfolios denoted as DL-LS and RF-LS based on a highest and lowest prediction quantile. The number of portfolios correspond the number of AP-Tree portfolios and number of quantiles for the prediction. Additional details about the data and implementation are in Bryzgalova et al. (2020b).

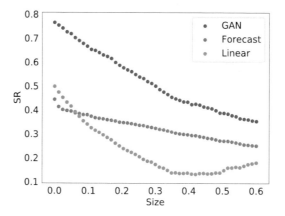

Figure 16.10 This figure shows the out-of-sample Sharpe ratios of the SDF for GAN, Forecast, and Linear after we have set the portfolio weights ω to zero if the market capitalization at the time of investment are below a specified cross-sectional quantile. Additional details about the data and implementation are in Chen et al. (2020).

16.6 Directions for future research

Avramov et al. (2021) raise the concern that the performance of machine learning portfolios could deteriorate in the presence of trading costs due to high turnover or extreme positions. This important insight can be taken into account when constructing machine learning investment portfolios. Figure 16.10 shows the out-of-sample Sharpe ratios of the SDF portfolios of Chen et al. (2020) after we have set the SDF weights ω to zero for stocks with market capitalization below a specified cross-sectional quantile at the time of portfolio construction. The idea is to remove stocks that are more prone to trading frictions. There is a clear trade-off between trading-frictions and achievable Sharpe ratios. However, this indicates that a machine learning portfolio can be estimated to optimally trade-off the trading frictions and a high risk-adjusted return. For example, GAN without 40% of the smallest stocks still has an annual SR of 1.73. Note that these are all lower bounds as GAN has not been re-estimated without these stocks, but we have just set the portfolio weights of the stocks below the cutoffs to zero.

So far, most papers have separated the construction of profitable machine learning portfolios into two steps. In the first step, machine learning methods extract signals for predicting future returns. In a second step, these signals are used to form profitable portfolios, which are typically long-short investments based on prediction. However, we argue that these two steps should be merged together, that is machine learning techniques should extract the signals that are the most relevant for the overall portfolio design. This is exactly what we achieve when the objective is the estimation of the SDF, which is the conditionally mean-variance efficient portfolio. A step further is to include trading-frictions directly in this estimation, that is, machine learning techniques should extract the signals that are the most relevant for portfolio design under constraints. A promising step into this direction is presented in Bryzgalova et al. (2020b) who estimate mean-variance efficient portfolios with decision trees that can easily incorporate constraints and Cong et al. (2020) who use a reinforcement learning approach that could also include constraints.

Another promising direction for future research is to use machine learning methods to include alternative data in the information set. An example is Ke et al. (2020), who use natural language processing to extract sentiment information from news articles. The machine learning method can bring alternative data into a format which can be used as an input to the approaches discussed in this chapter.

References

Avramov, D., Cheng, S., and Metzker, L. 2021. Machine learning versus economic restrictions: evidence from stock return predictability. Available at SSRN 3450322.

Bianchi, D., Büchner, M., and Tamoni, A. 2021. Bond risk premia with machine learning. *Review of Financial Studies*, **34**(2), 1046–1089.

Bryzgalova, S., Julliard, C., and Huang, J. 2020a. Bayesian solutions for the factor zoo: we just ran two quadrillion models. Available at SSRN 3481736.

Bryzgalova, S., Pelger, M., and Zhu, J. 2020b. Forest through the trees: building cross-sections of stock returns. Available at SSRN 3493458.

Chen, L., Pelger, M., and Zhu, J. 2020. Deep learning in asset pricing. Available at SSRN 3350138.

Cochrane, J. H. 2011. Presidential address: Discount rates. *Journal of Finance*, **66**(4), 1047–1108.

Cong, L. Will, Tang, Ke, Wang, Jingyuan, and Zhang, Yang. 2020. AlphaPortfolio for investment and economically interpretable AI. Available at SSRN 3554486.

Connor, G., and Korajczyk, R. 1988. Risk and return in an equilibrium APT: Application to a new test methodology. *Journal of Financial Economics*, **21**, 255–289.

Connor, Gregory, and Korajczyk, Robert A. 1986. Performance measurement with the arbitrage pricing theory: A new framework for analysis. *Journal of Financial Economics*, **15**(3), 373–394.

Fama, E. F., and French, K. R. 1993. Common risk factors in the returns on stocks and bonds. *Journal of Financial Economics*, **33**(1), 3–56.

Fama, E. F., and French, K. R. 2015. A five-factor asset pricing model. *Journal of Financial Economics*, **116**(1), 1–22.

Fama, Eugene F., and French, Kenneth R. 1992. The cross-section of expected stock returns. *Journal of Finance*, **47**(2), 427–465.

Feng, Guanhao, He, Jingyu, and Polson, Nicholas G. 2020b. Deep learning for predicting asset returns. ArXiv:1804.09314.

Freyberger, Joachim, Neuhierl, Andreas, and Weber, Michael. 2020. Dissecting characteristics nonparametrically. *Review of Financial Studies*, **33**(5), 2326–2377.

Giglio, S., and Xiu, D. 2021. Asset pricing with omitted factors. *Journal of Political Economy*, **129**(7), 1947–1990.

Gu, S., Kelly, B., and Xiu, D. 2020. Empirical asset pricing via machine learning. *Review of Financial Studies*, **33**(5), 2223–2273.

Gu, S., Kelly, B., and Xiu, D. 2021. Autoencoder asset pricing models. *Journal of Econometrics*, **222**(1B), 429–450.

Harvey, C. R., Y. Liu, and Zhu, H. 2016. ... and the cross-section of expected returns. *Review of Financial Studies*, **29**(1), 5–68.

Ke, Zheng T., Kelly, Bryan, and Xiu, Dacheng. (2020). Predicting returns with text data. Available at SSRN 3389884.

Kelly, B., Pruitt, S., and Su, Y. 2019. Characteristics are covariances: A unified model of risk and return. *Journal of Financial Economics*, **134**(3), 501–524.

Kozak, Serhiy, Nagel, Stefan, and Santosh, Shrihari. 2020. Shrinking the cross section. *Journal of Financial Economics*, **135**(2), 271–292.

Lettau, M., and Pelger, M. 2020a. Factors that fit the time-series and cross-section of stock returns. *Review of Financial Studies*, **33**(5), 2274–2325.

Lettau, Martin, and Pelger, Markus. 2020b. Estimating latent asset pricing factors. *Journal of Econometrics*, **218**(1), 1–31.

Ludvigson, S., and Ng, S. 2007. The empirical risk return relation: A factor analysis approach. *Journal of Financial Economics*, **83**(1), 171–222.

Messmer, Marcial. 2017. Deep learning and the cross-section of expected returns. Available at SSRN 3081555.

Moritz, B., and Zimmerman, T. 2016. Tree-based conditional portfolio sorts: The relation between past and future stock returns. Available at SSRN 2740751.

Onatski, A. 2012. Asymptotics of the principal components estimator of large factor models with weakly influential factors. *Journal of Econometrics*, 244–258.

Pelger, M. 2020. Understanding systematic risk: A high-frequency approach. *Journal of Finance*, **75**(4), 2179–2220.

Pelger, M., and Xiong, R. 2020. State-varying factor models of large dimensions. *Journal of Business & Economic Statistics*. DOI: 10.1080/07350015.2021.1927744.

Ross, S. A. 1976. The arbitrage theory of capital asset pricing. *Journal of Economic Theory*, **13**, 341–360.

Rossi, A. G. 2018. Predicting stock market returns with machine learning. Working paper, University of Maryland, Smith School of Business.

Sadhwani, Apaar, Giesecke, Kay, and Sirignano, Justin. 2021. Deep learning for mortgage risk. *Journal of Financial Econometrics* **19**(2), 313–368.

Sirignano, Justin, and Cont, Rama. 2019. Universal features of price formation in financial markets: perspectives from deep learning. *Quantitative Finance*, **19**(9), 1449–1459.

Portfolio Construction Using Stratified Models

Jonathan Tuck[a], Shane Barratt[a] and Stephen Boyd[a]

Abstract

In this chapter we develop models of asset return mean and covariance that depend on some observable market conditions, and use these to construct a trading policy that depends on these conditions, and the current portfolio holdings. After discretizing the market conditions, we fit Laplacian regularized stratified models for the return mean and covariance. These models have a different mean and covariance for each market condition, but are regularized so that nearby market conditions have similar models. This technique allows us to fit models for market conditions that have not occurred in the training data, by borrowing strength from nearby market conditions for which we do have data. These models are combined with a Markowitz-inspired optimization method to yield a trading policy that is based on market conditions. We illustrate our method on a small universe of 18 ETFs, using three well known and publicly available market variables to construct 1000 market conditions, and show that it performs well out of sample. The method, however, is general, and scales to much larger problems, that presumably would use proprietary data sources and forecasts along with publicly available data.

17.1 Introduction

Trading policy.

We consider the problem of constructing a trading policy that depends on some observable market conditions, as well as the current portfolio holdings. We denote the asset daily returns as $y_t \in \mathbb{R}^n$, for $t = 1, \ldots, T$. The observable market conditions are denoted as z_t. We assume these are discrete or categorical, so we have $z_t \in \{1, \ldots, K\}$. We denote the portfolio asset weights as $w_t \in \mathbb{R}^n$, with $\mathbf{1}^T w_t = 1$, where $\mathbf{1}$ is the vector with all entries one. The trading policy has the form

$$\mathcal{T} : \{1, \ldots, K\} \times \mathbb{R}^n \to \mathbb{R}^n,$$

[a] Stanford University
Published in *Machine Learning And Data Sciences For Financial Markets*, Agostino Capponi and Charles-Albert Lehalle© 2023 Cambridge University Press.

where $w_t = \mathcal{T}(z_t, w_{t-1})$, i.e., it maps the current market condition and previous portfolio weights to the current portfolio weights. In this chapter we refer to z_t as the market conditions, since in our example it is derived from market conditions, but in fact it could be anything known before the portfolio weights are chosen, including proprietary forecasts or other data. Our policy $\mathcal{T}$ is a simple Markowitz-inspired policy, based on a Laplacian regularized stratified model of the asset return mean and covariance; see, e.g., Markowitz (1952); Grinold and Kahn (1999); Boyd et al. (2017).

Laplacian regularized stratified model.

We model the asset returns, conditioned on market conditions, as Gaussian,

$$y \mid z \sim \mathcal{N}(\mu_z, \Sigma_z),$$

with $\mu_z \in \mathbb{R}^n$ and $\Sigma_z \in \mathbf{S}^n_{++}$ (the set of symmetric positive definite $n \times n$ matrices), $z = 1, \ldots, K$. This is a stratified model, with stratification feature z. We fit this stratified model, i.e., determine the means $\mu_1, \ldots, \mu_K$ and covariances $\Sigma_1, \ldots, \Sigma_K$, by minimizing the negative log-likelihood of historical training data, plus a regularization term that encourages nearby market conditions to have similar means and covariances. This technique allows us to fit models for market conditions which have not occurred in the training data, by borrowing strength from nearby market conditions for which we do have data. Laplacian regularized stratified models are discussed in, e.g., Danaher et al. (2014); Saegusa and Shojaie (2016); Tuck et al. (2019); Tuck et al. (2021); Tuck and Boyd (2022b,a). One advantage of Laplacian regularized stratified models is they are interpretable. They are also auditable: we can easily check if the results are reasonable.

This chapter.

In this chapter we present a single example of developing a trading policy as described above. Our example is small, with a universe of 18 ETFs, and we use market conditions that are publicly available and well known. Given the small universe and our use of widely available market conditions, we cannot expect much in terms of performance, but we will see that the trading algorithm performs well out of sample. Our example is meant only as a simple illustration of the ideas; the techniques we decribe can easily scale to a universe of thousands of assets, and use proprietary forecasts in the market conditions. We have made the code for this chapter available online at `https://github.com/cvxgrp/lrsm_portfolio`.

Outline.

We start by reviewing Laplacian regularized models in §17.2. In §17.3 we describe the data records and dataset we use. In §17.4 we describe the economic conditions with which we will stratify our return and risk models. In §17.5 and §17.6 we describe, fit, and analyze the stratified return and risk models, respectively. In §17.7 we give the details of how our stratified return and risk models are used to create the trading policy $\mathcal{T}$. We mention a few extensions and variations of the methods in §17.8.

17.1.1 Related work

A number of studies show that the underlying covariances of equities change during different market conditions, such as when the market performs historically well or poorly (a "bull" or "bear" market, respectively), or when there is historically high or low volatility (Erb et al., 1994; Longin and Solnik., 2001; Ang and Bekaert, 2003, 2004; Borland, 2012). Modeling the dynamics of underlying statistical properties of assets is an area of ongoing research. Many model these statistical properties as occurring in hard regimes (i.e., where the statistical properties are the same within a given regime), and utilize methods such as hidden Markov models (Ryden et al., 1998; Hastie et al., 2009; Nystrup et al., 2018) or greedy Gaussian segmentation (Hallac et al., 2019) to model the transitions and breakpoints between the regimes. In contrast, this chapter assumes a hard regime model of our statistical parameters, but our chief assumption is, informally speaking, that similar regimes have similar statistical parameters.

Asset allocation based on changing market conditions is a sensible method for active portfolio management (Ang and Bekaert, 2002; Ang and Timmermann, 2011; Nystrup et al., 2015; Petre, 2015). A popular method is to utilize convex optimization control policies to dynamically allocate assets in a portfolio, where the time-varying statistical properties are modeled as a hidden Markov model (Nystrup et al., 2019).

17.2 Laplacian regularized stratified models

In this section we review Laplacian regularized stratified models, focusing on the specific models we will use; for more detail see Tuck et al. (2021); Tuck and Boyd (2022b). We are given data records of the form $(z, y) \in \{1, \ldots, K\} \times \mathbb{R}^n$, where z is the feature over which we stratify, and y is the outcome. We let $\theta \in \Theta$ denote the parameter values in our model. The stratified model consists of a choice of parameter $\theta_z \in \Theta$ for each value of z. In this chapter we will construct two stratified models. One is for return, where $\theta_z \in \Theta = \mathbb{R}^n$ is an estimate or forecast of return in market condition z. The other is for return covariance, where $\theta_z \in \Theta = \mathbf{S}_{++}^n$ is the inverse covariance or precision matrix, and $\mathbf{S}_{++}^n$ denotes the set of symmetric positive definite $n \times n$ matrices. (We use the precision matrix since it is the natural parameter in the exponential family representation of a Gaussian, and renders the fitting problems convex.)

To choose the parameters $\theta_1, \ldots, \theta_K$, we minimize

$$\sum_{k=1}^{K} (\ell_k(\theta_k) + r(\theta_k)) + \mathcal{L}(\theta_1, \ldots, \theta_K). \tag{17.1}$$

Here ℓ_k is the loss function, that depends on the training data y_i, for $z_i = k$, typically a negative log-likelihood under our model for the data. The function r is the local regularizer, chosen to improve out of sample performance of the model.

The last term in (17.1) is the Laplacian regularization, which encourages neighboring values of z, under some weighted graph, to have similar parameters.

It is characterized by $W \in \mathbf{S}^K$, a symmetric weight matrix with zero diagonal entries and nonnegative off-diagonal entries. The Laplacian regularization has the form

$$\mathcal{L}(\theta_1, \ldots, \theta_K) = \frac{1}{2} \sum_{i,j=1}^{K} W_{ij} \|\theta_i - \theta_j\|^2,$$

where the norm is the Euclidean or ℓ_2 norm when θ_z is a vector, and the Frobenius norm when θ_z is a matrix. We think of W as defining a weighted graph, with edges associated with positive entries of W, with edge weight W_{ij}. The larger W_{ij} is, the more encouragement we give for θ_i and θ_j to be close.

When the loss and regularizer are convex, the problem (17.1) is convex, and so in principle is tractable (Boyd and Vandenberghe, 2004). The distributed method introduced in Tuck et al. (2021), which exploits the properties that the first two terms in the objective are separable across k, while the last term is separable across the entries of the parameters, can easily solve very large instances of the problem.

A Laplacian regularized stratified model typically includes several hyper-parameters, for example that scale the local regularization, or scale some of the entries in W. We adjust these hyper-parameters by choosing some values, fitting the Laplacian regularized stratified model for each choice of the hyper-parameters, and evaluating the true loss function on a (held-out) validation set. (The true loss function is often but not always the same as the loss function used in the fitting objective (17.1).) We choose hyper-parameters that give the least, or nearly least, true loss on the validation data, biasing our choice toward larger values, i.e., more regularization.

We make a few observations about Laplacian regularized stratified models. First, they are interpretable, and we can check them for reasonableness by examining the values θ_z, and how they vary with z. At the very least, we can examine the largest and smallest values of each entry (or some function) of θ_z over $z \in \{1, \ldots, K\}$.

Second, we note that a Laplacian regularized stratified model can be created even when we have no training data for some, or even many, values of z. The parameter values for those values of z are obtained by borrowing strength from their neighbors for which we do have data. In fact, the parameter values for values of z for which we have no data are weighted averages of their neighbors. This implies a number of interesting properties, such as a maximum principle: Any such value lies between the minimum and maximum values of the parameter over those values of z for which we have data.

17.3 Dataset

Our example considers $n = 18$ ETFs as the universe of assets, listed in Table 17.1. These ETFs were chosen because they broadly represent the market. Each data record has the form (y_t, z_t), where $y_t \in \mathbb{R}^{18}$ is the daily *active* return of each asset with respect to VTI, an ETF which broadly tracks the total stock market, from

Table 17.1 Universe of 18 ETFs.

Asset	Description
AGG	iShares Core US Aggregate Bond ETF
DBC	PowerShares DB Commodity Index Tracking Fund
GLD	SPDR Gold Shares
IBB	iShares Nasdaq Biotechnology ETF
ITA	iShares US Aerospace & Defense ETF
PBJ	Invesco Dynamic Food & Beverage ETF
TLT	iShares 20 Plus Year Treasury Bond ETF
VNQ	Vanguard Real Estate Index Fund ETF
VTI	Vanguard Total Stock Market Index Fund ETF
XLB	Materials Select Sector SPDR Fund
XLE	Energy Select Sector SPDR Fund
XLF	Financial Select Sector SPDR Fund
XLI	Industrial Select Sector SPDR Fund
XLK	Technology Select Sector SPDR Fund
XLP	Consumer Staples Select Sector SPDR Fund
XLU	Utilities Select Sector SPDR Fund
XLV	Health Care Select Sector SPDR Fund
XLY	Consumer Discretionary Select Sector SPDR Fund

market close on day $t-1$ until market close on day t, and z_t represents the market condition known by the market close on day $t-1$, described later in §17.4. (The daily active return of each asset with respect to VTI is the daily return of that asset minus the daily return of VTI.) Henceforth, when we refer to return or risk we mean active return or active risk, with respect to our benchmark VTI. The benchmark VTI has zero active return and risk.

Our dataset spans March 2006 to December 2019, for a total of 3461 data points. We first partition it into two subsets. The first, using data from 2006–2014, is used to fit the return and risk models as well as to choose the hyper-parameters in the return and risk models and the trading policy. The second subset, with data in 2015–2019, is used to test the trading policy. We then randomly partition the first subset into two parts: a training set consisting of 80% of the data records, and a validation set consisting of 20% of the data records. Thus we have three datasets: a training data set with 1779 data points in the date range 2006–2014, a validation set with 445 data points also in the date range 2006–2014, and a test dataset with 1237 data points in the date range 2015–2019. We use 9 years of data to fit our models and choose hyper-parameters, and 5 years of later data to test the trading policy. In order to minimize the influence of outliers in the models, return data in the training and validation datasets were winsorized (clipped) at their 1st and 99th percentiles. The return data in the test dataset was not winsorized.

Table 17.2 Correlation of the market indicators over the training and validation period, 2006–2014.

	Volatility	Inflation	Mortgage
Volatility	1	−0.13	−0.28
Inflation	–	1	0.28
Mortgage	–	–	1

17.4 Stratified market conditions

Each data record also includes the market condition z known on the previous day's market close. To construct the market condition z, we start with three (real-valued) market indicators.

Market implied volatility.

The volatility of the market is a commonly used economic indicator, with extreme values associated with market turbulence (French et al., 1987; Schwert, 1989; Aggarwal et al., 1999; Chun et al., 2020). Here, volatility is measured by the 15-day moving average of the CBOE volatility index (VIX) on the S&P 500 (CBOE, 2020), lagged by an additional day.

Inflation rate.

The inflation rate measures the percentage change of purchasing power in the economy (Wynne and Sigalla, 1994; Boyd et al., 1996, 2001; Boyd and Champ, 2003; Hung, 2003; Mahyar, 2017). The inflation rate is published by the United States Bureau of Labor Statistics (USBLS, 2020) as the percent change of the consumer price index (CPI), which measures changes in the price level of a representative basket of consumer goods and services, and is updated monthly.

30-year US mortgage rates.

This metric is the interest rate charged by a mortgage lender on 30-year mortgages, and the change of this rate is an economic indicator correlated with economic spending (Cava, 2016; Sutton et al., 2017). The 30-year US mortgage rate are published by the Federal Home Loan Mortgage Corporation, a public government-sponsored enterprise, and is generally updated weekly (FRED, 2020). Here, this market condition is the 8-week rolling percent change of the 30-year US mortgage rate.

These three economic indicators are not particularly correlated over the training and validation period, as can be seen in Table 17.2.

Discretization.

Each of these market indicators is binned into deciles, labeled $1, \ldots, 10$. (The decile boundaries are computed using the data up to 2015.) The total number of stratification feature values is then $K = 10 \times 10 \times 10 = 1000$. We can think of z as a 3-tuple of deciles, in $\{1, \ldots, 10\}^3$, or encoded as a single value $z \in \{1, \ldots, 1000\}$.

The market conditions over the entire dataset are shown in Figure 17.1, with the

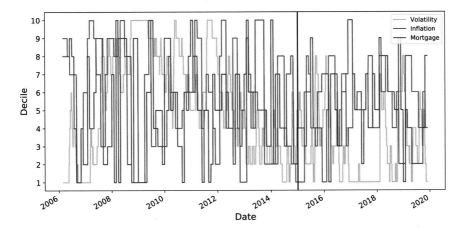

Figure 17.1 Stratification feature values over time. The vertical line at 2015 separates the training and validation period (2006–2014) from the test period (2015–2019).

vertical line at 2015 indicating the boundary between the training and validation period (2006–2014) and the test period (2015–2019). The average value of $\|z_{t+1} - z_t\|_1$ (interpreting them as vectors in $\{1, \ldots, 10\}^3$) is around 0.35, meaning that on each day, the market conditions change by around 0.35 deciles on average.

Data scarcity.

The market conditions can take on $K = 1000$ possible values. In the training/validation datasets, only 346 of 1000 market conditions appear, so there are 654 market conditions for which there are no data points. The most populated market condition, which corresponds to market conditions $(9, 0, 0)$, contains 42 data points. The average number of data points per market condition in the training/validation data is 2.22.

For about 65% of the market conditions, we have *no* training data. This scarcity of data means that the Laplacian regularization is critical in constructing models of the return and risk that depend on the market conditions.

In the test dataset, only 188 of the economic conditions appear. The average number of data points per market condition in the test dataset is 1.24. Only 71 economic conditions appear in both the training/validation and test datasets. In the test data, there are only 442 days (about 36% of the 1237 test data days) in which the market conditions for that day were observed in the training/validation datasets.

Regularization graph.

Laplacian regularization requires a weighted graph that tells us which market conditions are 'close'. Our graph is the Cartesian product of three chain graphs (Tuck et al., 2021), which link each decile of each indicator to its successor (and predecessor). This graph on the 1000 values of z has 2700 edges. Each edge connects two adjacent deciles of one of our three economic indicators. We assign three different positive weights to the edges, depending on which indicator they link. We denote these as

$$\gamma_{\text{vol}}, \quad \gamma_{\text{inf}}, \quad \gamma_{\text{mort}}. \tag{17.2}$$

These are hyper-parameters in our Laplacian regularization. Each of the nonzero entries in the weight matrix W is one of these values. For example, the edge between $(3,1,4)$ and $(3,2,4)$, which connects two values of z that differ by one decile in Inflation, has weight γ_{inf}.

17.5 Stratified return model

In this section we describe the stratified return model. The model consists of a return vector $\theta_z = \mu_z \in \mathbb{R}^{18}$ for each of $K = 1000$ different market conditions, for a total of $Kn = 18000$ parameters.

The loss in (17.1) is a Huber penalty,

$$\ell_k(\mu_k) = \sum_{t:z_t=k} \mathbf{1}^T H(\mu_k - y_t),$$

where H is the Huber penalty (applied entrywise above),

$$H(z) = \begin{cases} z^2, & |z| \le M \\ 2M|z| - M^2, & |z| > M, \end{cases}$$

where $M > 0$ is the half-width, which we fix at the reasonable value $M = 0.01$. (This corresponds to the 79th percentile of absolute return on the training dataset.) The Huber loss is utilized because it is robust (or less sensitive) to outliers. We use quadratic or ℓ_2 squared local regularization in (17.1),

$$r(\mu_k) = \gamma_{\text{ret,loc}} \|\mu_k\|_2^2,$$

where the positive regularization weight $\gamma_{\text{ret,loc}}$ is another hyper-parameter.

The Laplacian regularization contains the three hyper-parameters (17.2), so overall our stratified return model has four hyper-parameters.

17.5.1 Hyper-parameter search

To choose the hyper-parameters for the stratified return model, we start with a coarse grid search, which evaluates combinations of hyper-parameters over a large range. We evaluate all combinations of

$$\gamma_{\text{ret,loc}} = 0.001, 0.01, 0.1,$$
$$\gamma_{\text{vol}} = 1, 10, 100, 1000, 10000, 100000$$
$$\gamma_{\text{inf}} = 1, 10, 100, 1000, 10000, 100000$$
$$\gamma_{\text{mort}} = 1, 10, 100, 1000, 10000, 100000$$

a total of 648 combinations, and select the hyper-parameter combination that yields the largest correlation between the return estimates and the returns over

Table 17.3 Correlations to the true returns over the training set and the held-out validation set for the return models.

Model	Train correlation	Validation correlation
Stratified return model	0.093	0.054
Common return model	0.018	0.001

the validation set. (Thus, our true loss is negative correlation of forecast and realized returns.) The hyper-parameters

$$(\gamma_{\text{ret,loc}}, \gamma_{\text{vol}}, \gamma_{\text{inf}}, \gamma_{\text{mort}}) = (0.01, 10, 100, 10000)$$

gave the best results over this coarse hyper-parameter grid search.

We then perform a second hyper-parameter grid search on a finer grid of values centered around the best values from the coarse search. We test all combinations of

$$\gamma_{\text{ret,loc}} = 0.0075, 0.01, 0.0125,$$
$$\gamma_{\text{vol}} = 2, 5, 10, 20, 50,$$
$$\gamma_{\text{inf}} = 20, 50, 100, 200, 500,$$
$$\gamma_{\text{mort}} = 2000, 5000, 10000, 20000, 50000,$$

a total of 375 combinations. The final hyper-parameter values are

$$(\gamma_{\text{ret,loc}}, \gamma_{\text{vol}}, \gamma_{\text{inf}}, \gamma_{\text{mort}}) = (0.01, 20, 50, 5000). \tag{17.3}$$

These can be roughly interpreted as follows. The large value for γ_{mort} tells us that our return model should not vary much with mortgage rate, and the smaller values for γ_{vol} and γ_{inf} tells us that our return model can vary more with volatility and inflation.

17.5.2 Final stratified return model

Table 17.3 shows the correlation coefficient of the return estimates to the true returns over the training and validation sets, for the stratified return model and the common return model, i.e., the empirical mean over the training set. The stratified return model estimates have a larger correlation with the realized returns in both the training set and the validation set.

Table 17.4 summarizes some of the statistics of our stratified return model over the 1000 market conditions, along with the common model value. We can see that each forecast varies considerably across the market conditions. Note that the common model values are the averages over the training data; the median, minimum, and maximum are over the 1000 market conditions.

Table 17.4 Return predictions, in percent daily return. The first column gives the common return model; the second, third, and fourth columns give median, minimum, and maximum return predictions over the 1000 market conditions for the Laplacian regularized stratified model. All returns are relative to VTI, which has zero return.

Asset	Common	Median	Min	Max
AGG	−0.015	−0.064	−0.109	0.045
DBC	−0.049	−0.050	−0.131	0.076
GLD	−0.007	−0.017	−0.111	0.130
IBB	0.040	0.045	−0.053	0.132
ITA	0.022	0.029	−0.062	0.059
PBJ	0.009	0.007	−0.038	0.096
TLT	0.011	−0.053	−0.162	0.092
VNQ	0.015	0.008	−0.229	0.064
VTI	0	0	0	0
XLB	0.003	0.014	−0.033	0.066
XLE	−0.001	0.020	−0.081	0.113
XLF	−0.023	−0.047	−0.341	0.039
XLI	0.008	0.015	−0.053	0.052
XLK	0.001	0.003	−0.045	0.081
XLP	0.006	−0.001	−0.040	0.062
XLU	−0.009	−0.017	−0.067	0.072
XLV	0.012	0.011	−0.029	0.055
XLY	0.014	0.007	−0.048	0.049

17.6 Stratified risk model

In this section we describe the stratified risk model, i.e., a return covariance that depends on z. For determining the risk model, we can safely ignore the (small) mean return, and assume that y_t has zero mean. (The return is small, so the squared return is negligible.) The model consists of $K = 1000$ inverse covariance matrices $\Sigma_k^{-1} = \theta_k \in \mathbf{S}_{++}^{18}$, indexed by the market conditions. Our stratified risk model has $Kn(n + 1)/2 = 171000$ parameters.

The loss in (17.1) is the negative log-likelihood on the training set (scaled, with constant terms ignored),

$$\ell_k(\theta_k) = \mathrm{Tr}(S_k \Sigma_k^{-1}) - \log \det(\Sigma_k^{-1})$$

where $S_k = \frac{1}{n_k} \sum_{t:z_t = k} y_t y_t^T$ is the empirical covariance matrix of the data y for which $z = k$, and n_k is the number of data samples with $z = k$. (When $n_k = 0$, we take $S_k = 0$.) We found that local regularization did not improve the model performance, so we take local regularization $r = 0$. All together our stratified risk model has the three Laplacian hyper-parameters (17.2).

Table 17.5 Average negative log-likelihood (scaled, with constant terms ignored) over the training and validation sets for the stratified and common risk models.

Model	Train loss	Validation loss
Stratified risk model	−6.69	−1.45
Common risk model	3.47	4.99

17.6.1 Hyper-parameter search

We start with a coarse grid search over all 216 combinations of

$$\gamma_{\text{vol}} = 0.01, 0.1, 1, 10, 100, 1000,$$

$$\gamma_{\text{inf}} = 0.01, 0.1, 1, 10, 100, 1000,$$

$$\gamma_{\text{mort}} = 0.01, 0.1, 1, 10, 100, 1000,$$

selecting the hyper-parameter combination with the smallest negative log-likelihood (our true loss) on the validation set. The hyper-parameters

$$(\gamma_{\text{vol}}, \gamma_{\text{inf}}, \gamma_{\text{mort}}) = (0.1, 10, 100)$$

gave the best results.

We then perform a second search on a finer grid, focusing on hyper-parameter value near the best values from the coarse search. We evaluate all 125 combinations of

$$\gamma_{\text{vol}} = 0.02, 0.05, 0.1, 0.2, 0.5,$$

$$\gamma_{\text{inf}} = 2, 5, 10, 20, 50,$$

$$\gamma_{\text{mort}} = 20, 50, 100, 200, 500.$$

For the stratified risk model, the final hyper-parameter values chosen are

$$(\gamma_{\text{vol}}, \gamma_{\text{inf}}, \gamma_{\text{mort}}) = (0.2, 20, 50).$$

It is interesting to compare these to the hyper-parameter values chosen for the stratified return model, given in (17.3). Since the losses for return and risk models are different, we can scale the hyper-parameters in the return and risk to compare them. We can see that they are not the same, but not too different, either; both choose γ_{inf} larger than γ_{vol}, and γ_{mort} quite a bit larger than γ_{vol}.

17.6.2 Final stratified risk model

Table 17.5 shows the average negative log likelihood (scaled, with constant terms ignored) over the training and held-out validation sets, for both the stratified risk model and the common risk model, i.e., the empirical covariance. We can see that the stratified risk model has substantially better loss on the training and validation sets.

Table 17.6 summarizes some of the statistics of our stratified return model asset volatilities, i.e., $((\Sigma_z)_{ii})^{1/2}$, expressed as daily percentages, over the 1000

Table 17.6 Forecasts of volatility, expressed in percent daily return. The first column gives the common model; the second, third, and fourth columns give median, minimum, and maximum volatility predictions over the 1000 market conditions for the Laplacian regularized stratified model. Volatilities are of return relative to VTI, so VTI has zero volatility.

Asset	Common	Median	Min	Max
AGG	1.314	0.906	0.586	4.135
DBC	1.285	1.070	0.778	3.870
GLD	1.671	1.269	0.982	5.201
IBB	0.905	0.823	0.694	2.120
ITA	0.618	0.557	0.492	1.428
PBJ	0.650	0.513	0.437	1.915
TLT	1.816	1.334	0.809	5.828
VNQ	1.328	0.786	0.666	4.409
VTI	0	0	0	0
XLB	0.771	0.641	0.507	1.703
XLE	1.019	0.857	0.686	2.401
XLF	1.190	0.617	0.389	4.401
XLI	0.500	0.440	0.370	1.045
XLK	0.515	0.465	0.387	1.057
XLP	0.759	0.576	0.455	2.425
XLU	0.882	0.749	0.639	2.186
XLV	0.701	0.509	0.428	2.108
XLY	0.535	0.442	0.355	1.154

market conditions, along with the common model asset volatilities. We can see that the predictions vary considerably across the market conditions, with a few varying by a factor almost up to ten. Table 17.7 summarizes the same statistics for the correlation of each asset with AGG, an aggregate bond market ETF. Here we see dramatic variation, for example, the correlation between XLI (an industrials ETF) and AGG varies from -79% to +82% over the market conditions.

17.7 Trading policy and backtest

17.7.1 Trading policy

In this section we give the details of how we use our stratified return and risk models to construct the trading policy $\mathcal{T}$.

At the beginning of each day t, we use the previous day's market conditions z_t to allocate our current portfolio according to the weights w_t, computed as the solution of the Markowitz-inspired problem (Boyd et al., 2017)

$$\begin{aligned}
\text{maximize} \quad & \mu_{z_t}^T w - \gamma_{sc}\kappa^T(w)_- - \gamma_{tc}\tau_t^T|w - w_{t-1}| \\
\text{subject to} \quad & w^T\Sigma_{z_t}w \le \sigma^2, \quad \mathbf{1}^T w = 1, \\
& \|w\|_1 \le L_{max}, \quad w_{min} \le w \le w_{max},
\end{aligned} \quad (17.4)$$

with optimization variable $w \in \mathbb{R}^{18}$, where $w_- = \max\{0, -w\}$ (elementwise), and the absolute value is elementwise. We describe each term and constraint below.

Table 17.7 Forecasts of correlations with the aggregate bond index AGG. The first column gives the common model; the second, third, and fourth columns give median, minimum, and maximum correlation predictions over the 1000 market conditions for the Laplacian regularized stratified model.

Asset	Common	Median	Min	Max
AGG	1	1	1	1
DBC	0.492	0.416	−0.384	0.952
GLD	0.684	0.524	0.093	0.971
IBB	0.250	0.063	−0.585	0.917
ITA	0.024	−0.051	−0.807	0.875
PBJ	0.565	0.384	0.006	0.946
TLT	0.935	0.897	0.803	0.994
VNQ	−0.345	0.021	−0.932	0.652
XLB	−0.214	−0.232	−0.749	0.808
XLE	−0.205	−0.185	−0.935	0.619
XLF	−0.520	−0.289	−0.970	0.042
XLI	−0.107	−0.108	−0.789	0.816
XLK	0.154	0.075	−0.705	0.846
XLP	0.714	0.579	0.344	0.973
XLU	0.555	0.458	0.142	0.939
XLV	0.607	0.429	−0.106	0.962
XLY	−0.061	−0.026	−0.701	0.844

- *Return forecast.* The first term in the objective, $\mu_{z_t}^T w$, is the expected return under our forecast mean, which depends on the current market conditions.

- *Shorting cost.* The second term $\gamma_{sc} \kappa^T (w)_-$ is a shorting cost, with $\kappa \in \mathbb{R}_+^{18}$ the vector of shorting cost rates. (For simplicity we take the shorting cost rates as constant.) The positive hyper-parameter γ_{sc} scales the shorting cost term, and is used to control our shorting aversion.

- *Transaction cost.* The third term $\gamma_{tc} \tau_t^T |w - w_{t-1}|$ is a transaction cost, with $\tau_t \in \mathbb{R}_+^{18}$ the vector of transaction cost rates used on day t. We take τ_t as one-half the average bid-ask spread of each asset for the previous 15 trading days (excluding the current day). We summarize the bid-ask spreads of each asset over the training and holdout periods in Table 17.8. The positive hyper-parameter γ_{tc} scales the transaction cost term, and is used to control the turnover.

- *Risk limit.* The constraint $w^T \Sigma_z w \leq \sigma^2$ limits the (daily) risk (under our risk model, which depends on market conditions) to σ, which corresponds to an annualized risk of $\sqrt{250}\sigma$.

- *Leverage limit.* The constraint $\|w\|_1 \leq L_{max}$ limits the portfolio leverage, or equivalently, it limits the total short position $\mathbf{1}^T (w)_-$ to no more than $(L_{max} - 1)/2$.

- *Position limits.* The constraint $w_{min} \leq w \leq w_{max}$ (interpeted elementwise) limits the individual weights.

Table 17.8 One-half the mean bid-ask spread of each asset, over the training and validation periods and the holdout period.

Asset	Training/validation period	Holdout period
AGG	0.000298	0.000051
DBC	0.000653	0.000324
GLD	0.000112	0.000048
IBB	0.000418	0.000181
ITA	0.000562	0.000175
PBJ	0.000966	0.000637
TLT	0.000157	0.000048
VNQ	0.000394	0.000066
VTI	0.000204	0.000048
XLB	0.000310	0.000098
XLE	0.000181	0.000077
XLF	0.000359	0.000200
XLI	0.000295	0.000079
XLK	0.000324	0.000093
XLP	0.000298	0.000095
XLU	0.000276	0.000099
XLV	0.000271	0.000070
XLY	0.000334	0.000059

Parameters.

Some of the constants in the trading policy (17.4) we simply fix to reasonable values. We fix the shorting cost rate vector to $(0.0005)\mathbf{1}$, i.e., 5 basis points for each asset. We take $\sigma = 0.0045$, which corresponds to an annualized volatility (defined as $\sqrt{250}\sigma$) of around 7.1%. We take $L_{\max} = 2$, which means the total short position cannot exceed one half of the portfolio value. (A portfolio with a leverage of 2 is commonly referred to as a *150/50 portfolio*.) We fix the position limits as $w_{\min} = -0.25\mathbf{1}$ and $w_{\max} = 0.4\mathbf{1}$, meaning we cannot short any asset by more than 0.25 times the portfolio value, and we cannot hold more than 0.4 times the portfolio value of any asset.

Hyper-parameters.

Our trading policy has two hyper-parameters, γ_{sc} and γ_{tc}, which control our aversion to shorting and trading, respectively.

17.7.2 Backtests

Backtests are carried out starting from a portfolio of all VTI and a starting portfolio value of $v = 1$ dollars. On day t, after computing w_t as the solution to (17.4), we compute the value of our portfolio v_t by

$$r_{t,\mathrm{net}} = r_t^T w_t - \kappa^T (w_t)_- - (\tau_t^{\mathrm{sim}})^T |w_t - w_{t-1}|, \qquad v_t = v_{t-1}(1 + r_{t,\mathrm{net}}),$$

Here $r_t \in \mathbb{R}^{18}$ is the vector of asset returns on day t, $r_t^T w_t$ is the gross return of the portfolio for day t, τ_t^{sim} is one-half the realized bid-ask spread on day t, and $r_{t,\mathrm{net}}$

Table 17.9 Annualized return and risk for the stratified model policy over the train and validation periods.

	Return	Risk
Train	11.9%	6.25%
Validation	10.2%	6.88%

is the net return of the portfolio for day t including shorting and transaction costs. In particular, *our backtests take shorting and transaction costs into account.* Note also that in the backtests, we use the actual realized bid-ask spread on that day (which is not known at the beginning of the day) to determine the true transaction cost, whereas in the policy, we use the trailing 15 day average (which is known at the beginning of the day).

Our backtest is a bit simplified. Our simulation assumes dividend reinvestment. We account for the shorting and transaction costs by adjusting the portfolio return, which is equivalent to splitting these costs across the whole portfolio; a more careful treatment might include a small cash account. For portfolios of very high value, we would add an additional nonlinear transaction cost term, for example proportional to the 3/2-power or the square of $|w_t - w_{t-1}|$ (Almgren and Chriss, 2000; Boyd et al., 2017).

17.7.3 Hyper-parameter selection

To choose values of the two hyper-parameters in the trading policy, we carry out multiple backtest simulations over the training set. We evaluate these backtest simulations by their realized return (net, including costs) over the validation set.

We perform a grid search, testing all 625 pairs of 25 values of each hyper-parameter logarithmically spaced from 0.1 to 10. The annualized return on the validation set, as a function of the hyper-parameters, are shown in Figure 17.2. We choose the final values

$$\gamma_{sc} = 8.25, \quad \gamma_{tc} = 1.47,$$

shown on Figure 17.2 as a star.

These values are themselves interesting. Roughly speaking, we should plan our trades as if the shorting cost were more than 8.25 times the actual cost, and the transaction cost is about 1.5 times the true transaction cost. The blue and purple region at the bottom of the heat map indicates poor validation performance when the transaction cost parameter is too low, i.e., the policy trades too much.

Table 17.9 gives the annualized return and risk for the policies over the train and validation periods.

Common model trading policy.

We will compare our stratified model trading policy to a common model trading policy, which uses the constant return and risk models, along with the same

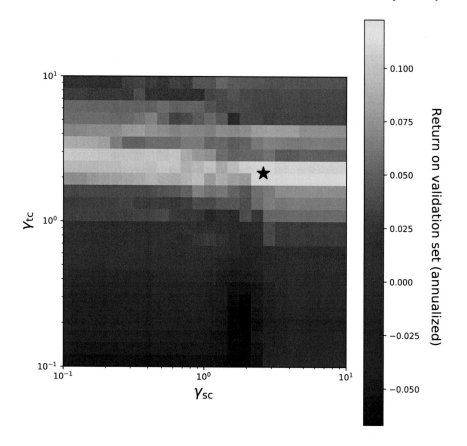

Figure 17.2 Heatmap of the annualized return on the validation set as a function of the two hyper-parameters γ_{sc} and γ_{tc}. The star shows the hyper-parameter combination used in our trading policy.

Markowitz policy (17.4). In this case, none of the parameters in the optimization problem change with market conditions, and the only parameter that changes in different days is w_{t-1}, the previous day's asset weights, which enters into the transaction cost.

We also perform a grid search for this trading policy, over the same 625 pairs of the hyper-parameters. For the common model trading policy, we choose the final values

$$\gamma_{sc} = 1, \quad \gamma_{tc} = 0.38.$$

17.7.4 *Final trading policy results*

We backtest our trading policy on the test dataset, which includes data from 2015–2019. We remind the reader that no data from this date range was used to create, tune, or validate any of the models, or to choose any hyper-parameters.

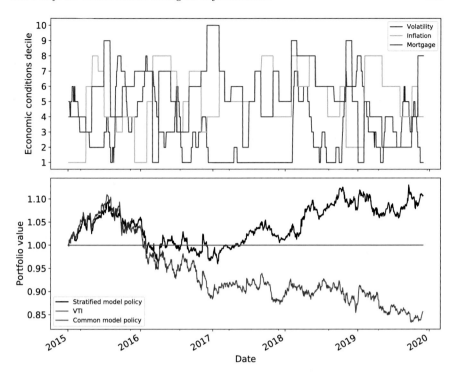

Figure 17.3 Plot of economic conditions (top) and cumulative portfolio value for the stratified model and the common model (bottom) over the test period. The horizontal blue line is the cumulative portfolio value for buying and holding the benchmark VTI.

For comparison, we also give results of a backtest using the constant return and risk models.

Figure 17.3 plots the economic conditions over the test period (top) as well as the active portfolio value (i.e., value above the benchmark VTI) for our stratified model and common model. Buying and holding the benchmark VTI gives zero active return, and a constant active portfolio value of 1. The superior performance of the stratified model policy, e.g., higher Sharpe ratio, is evident in this plot.

Table 17.10 shows the annualized active return, annualized active risk, annualized active Sharpe ratio (return divided by risk), and maximum drawdown of the active portfolio value for the policies over the test period. We remind the reader that we are fully accounting for the shorting and transaction cost, so the turnover of the policy is accounted for in these backtest metrics.

The results are impressive when viewed in the following light. First, we are using a very small universe of only 18 ETFs. Second, our trading policy uses only three widely available market conditions, and indeed, only their deciles. Third, the policy was entirely developed using data prior to 2015, with no adjustments made for the next five years. (In actual use, one would likely re-train the model periodically, perhaps every quarter or year.)

Table 17.10 Annualized active return, active risk, active Sharpe ratios, and maximum drawdown of the active portfolio value for the three policies over the test period (2015–2019).

	Return	Risk	Sharpe ratio	Maximum drawdown
Stratified model policy	2.55%	8.42%	0.302	13.4%
Common model policy	0.003%	7.47%	0.038	16.3%

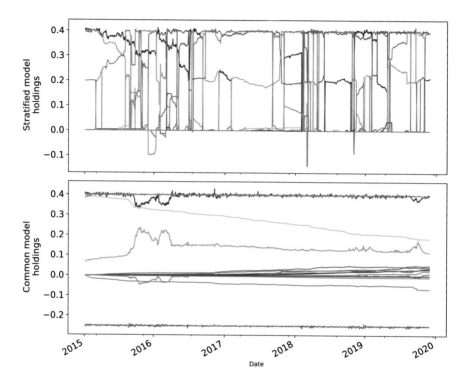

Figure 17.4 Asset weights of the stratified model policy (top) and of the common model policy (bottom), over the test period. The first time period asset weights, which are all VTI, are not shown.

Comparison of stratified and constant policies.

In Figure 17.4, we plot the asset weights of the stratified model policy (top) and of the common model policy (bottom), over the test period. (The variations in the common model policy holdings come from a combination of a daily rebalancing of the assets and the transaction cost model.) The top plot shows that the weights in the stratified policy change considerably with market conditions. The common model policy is mainly concentrated in just seven assets, GLD (gold), IBB (biotech), ITA (aerospace & defense), XLE (energy), XLV (health care), and XLY (consumer discretionary) (which is effectively cash when considering active returns and risks). Notably, both portfolios are long-only.

Table 17.11 The top four rows give the regression model coefficients of the active portfolio returns on the Fama–French factors; the fifth row gives the intercept or alpha value.

Factor	Stratified model policy	Common model policy
MKTRF	–0.001362	0.139547
SMB	0.279307	0.235330
HML	–0.361305	–0.448571
UMD	–0.174945	–0.108064
Alpha	0.000085	–0.000215

Factor analysis.

We fit a linear regression model of the active returns of the two policies over the test set to four of the Fama–French factors (Fama and French, 1992, 1993; French, 2021):

- *MKTRF*, the value-weighted return of United States equities, minus the risk free rate,
- *SMB*, the return on a portfolio of small size stocks minus a portfolio of big size stocks,
- *HML*, the return on a portfolio of value stocks minus a portfolio of growth stocks, and
- *UMD*, the return on a portfolio of high momentum stocks minus a portfolio of low or negative momentum stocks.

We also include an intercept term, commonly referred to as alpha. Table 17.11 gives the results of these fits. Relative to the common model policy, the stratified model policy active returns are much less positively correlated to the market, shorter the size factor, longer the value factor, and shorter the momentum factor. Its active alpha is around 2.13% annualized. (The common model policy's active alpha is around –5.38% annualized.) While not very impressive on its own, this alpha seems good considering it was accomplished with just 18 ETFs, and using only three widely available quantities in the policy.

17.8 Extensions and variations

We have presented a simple (but realistic) example only to illustrate the ideas, which can easily be applied in more complex settings, with a far larger universe, a more complex trading policy, and using proprietary forecasts of returns and quantities used to judge market conditions. We describe some extensions and variations on our method below.

Multi-period optimization.

For simplicity we use a policy that is based on solving a single-period Markowitz problem. The entire method immediately extends to policies based on multi-period optimization. For example, we would fit separate stratified models of return and risk for the next 1-day, 5-day, 20-day, and 60-day periods (roughly daily, weekly, monthly, quarterly), all based on the same current market conditions. These data are fed into a multi-period optimizer as described in Boyd et al. (2017).

Joint modeling of return and risk.

In this chapter we have created separate Laplacian regularized stratified models for return and risk. The advantage of this approach is that we can judge each model separately (and with different true objectives), and use different hyper-parameter values. It is also possible to fit the return mean and covariance *jointly*, in one stratified model, using the natural parameters in the exponential family for a Gaussian, Σ^{-1} and $\Sigma^{-1}\mu$. The resulting log-likelihood is jointly concave, and a Laplacian regularized model can be directly fit.

Low-dimensional economic factors.

When just a handful (such as in our example, three) base quantities are used to construct the stratified market conditions, we can bin and grid the values as we do in this chapter. This simple stratification of market conditions preserves interpretability. If we wish to include more raw data in our stratification of market conditions, simple binning and enumeration is not practical. Instead we can use several techniques to handle such situations. The simplest is to perform dimensionality-reduction on the (higher-dimensional) economic conditions, such as principal component analysis (Pearson, 1901) or low-rank forecasting (Barratt et al., 2020), and appropriately bin these low-dimensional economic conditions. These economic conditions may then be related on a graph with edge weights decided by an appropriate method, such as nearest neighbor weights.

Structured covariance estimation.

It is quite common to model the covariance matrix of returns as having structure, e.g., as the sum of a diagonal matrix plus a low-rank matrix (Richard et al., 2012; Fan et al., 2016). This structure can be added by a combination of introducing new variables to the model and encoding constraints in the local regularization. In many cases, this structure constraint turns the stratified risk model fitting problem into a non-convex problem, which may be solved approximately.

Multi-linear interpolation.

In the approach presented above, the economic conditions are categorical, i.e., take on one of $K = 1000$ possible values at each time t, based on the deciles of three quantities. A simple extension is to use multi-linear interpolation (Weiser and Zarantonello, 1988; Davies, 1997) to determine the return and risk to use in the Markowitz optimizer. Thus we would use the actual quantile of the three market quantiities, and not just their deciles. In the case of risk, we would apply the interpolation to the precision matrix Σ_t^{-1}, the natural parameter in the exponential family description of a Gaussian.

End-to-end hyper-parameter optimization.

In the example presented in this chapter there are a total of nine hyper-parameters to select. We keep things simple by separately optimizing the hyper-parameters for the stratified return model, the stratified risk model, and the trading policy. This approach allows each step to be checked independently. It is also possible to simultaneously optimize all of the hyper-parameters with respect to a single backtest, using, for example, CVXPYlayers (Agrawal et al., 2019, 2020) to differentiate through the trading policy.

Stratified ensembling.

The methods described in this chapter can be used to combine or emsemble a collection of different return forecasts or signals, whone performance varies with market (or other) conditions. We start with a collection of return predictions, and combine these (ensemble them) using weights that are a function of the market conditions. We develop a stratified selection of the combining weights.

17.9 Conclusions

We argue that stratified models are interesting and useful in portfolio construction and finance. They can contain a large number of parameters, but unlike, say, neural networks, they are fully interpretable and auditable. They allow arbitrary variation across market conditions, with Laplacian regularization there to help us come up with reasonable models even for market conditions for which we have no training data. The maximum principle mentioned on page 320 tells us that a Laplacian regularized stratified model will never do anything crazy when it encounters values of z that never appeared in the training data. Instead it will use a weighted sum of other values for which we do have training data. These weights are not just any weights, but ones carefully chosen by validation.

The small but realistic example we have presented is only meant to illustrate the ideas. The very same ideas and method can be applied in far more complex and sophisticated settings, with a larger universe of assets, a more complex trading policy, and incorporating proprietary data and forecasts.

Acknowledgements

The authors gratefully acknowledge discussions with and suggestions from Ronald Kahn, Raffaele Savi, and Andrew Ang. We thank an anonymous reviewer for making several suggestions on an early draft, as well as catching some inconsistencies between the data and algorithm as described in the chapter and as implemented in the code. Jonathan Tuck is supported by a Stanford Graduate Fellowship in Science and Engineering.

References

Aggarwal, R., Inclan, C., and Leal, R. 1999. Volatility in emerging stock markets. *Journal of Financial and Quantitative Analysis*, **34**(1), 33–55.

Agrawal, A., Amos, B., Barratt, S., Boyd, S., Diamond, S., and Kolter, Z. 2019. Differentiable Convex optimization layers. In: *Advances in Neural Information Processing Systems.*

Agrawal, A., Barratt, S., Boyd, S., and Stellato, B. 2020 (10–11 Jun). Learning convex optimization control policies. Pages 361–373 of: *Proceedings of the 2nd Conference on Learning for Dynamics and Control.*

Almgren, R., and Chriss, N. 2000. Optimal execution of portfolio transactions. *Journal of Risk*, **3**(2), 5–39.

Ang, A., and Bekaert, G. 2002. International asset allocation with regime shifts. *Review of Financial Studies*, **15**(4), 1137–1187.

Ang, A., and Bekaert, G. 2003 (Nov.). How do regimes affect asset allocation? Tech. rept. 10080. National Bureau of Economic Research.

Ang, A., and Bekaert, G. 2004. How regimes affect asset allocation. *Financial Analysts Journal*, **60**(2), 86–99.

Ang, A., and Timmermann, A. 2011 (June). Regime changes and financial markets. Tech. rept. 17182. National Bureau of Economic Research.

Barratt, S., Dong, Y., and Boyd, S. 2020. *Low rank forecasting.* ArXiv 2101.12414.

Borland, L. 2012. Statistical signatures in times of panic: markets as a self-organizing system. *Quantitative Finance*, **12**(9), 1367–1379.

Boyd, J., and Champ, B. 2003. Inflation and financial market performance: what have we learned in the last ten years? Tech. rept. 0317. Federal Reserve Bank of Cleveland.

Boyd, J., Levine, R., and Smith, B. 1996 (Oct.). Inflation and financial market performance. Tech. rept. Federal Reserve Bank of Minneapolis.

Boyd, J., Levine, R., and Smith, B. 2001. The impact of inflation on financial sector performance. *Journal of Monetary Economics*, **47**(2), 221–248.

Boyd, S., and Vandenberghe, L. 2004. *Convex Optimization.* Cambridge University Press.

Boyd, S., Busseti, E., Diamond, S., Kahn, R., Koh, K., Nystrup, P., and Speth, J. 2017. Multi-period trading via convex optimization. *Foundations and Trends in Optimization*, **3**(1), 1–76.

Cava, G. La. 2016 (July). Housing prices, mortgage interest rates and the rising share of capital income in the United States. BIS Working Papers 572. Bank for International Settlements.

CBOE (Chicago Board Options Exchange). 2020. CBOE volatility index. `http://www.cboe.com/vix`.

Chun, D., Cho, H., and Ryu, D. 2020. Economic indicators and stock market volatility in an emerging economy. *Economic Systems*, **44**(2), 100788.

Danaher, P., Wang, P., and Witten, D. 2014. The joint graphical lasso for inverse covariance estimation across multiple classes. *Journal of the Royal Statistical Society*, **76**(2), 373–397.

Davies, S. 1997. Multidimensional triangulation and interpolation for reinforcement learning. Pages 1005–1011 of: *Advances in Neural Information Processing Systems 9*, Mozer, M. C., Jordan, M., and Petsche, T. (eds). MIT Press.

Erb, C., Harvey, C., and Viskanta, T. 1994. Forecasting international equity correlations. *Financial Analysts Journal*, **50**(6), 32–45.

Fama, E., and French, K. 1992. The cross-section of expected stock returns. *Journal of Finance*, **47**(2), 427–465.

Fama, E., and French, K. 1993. Common risk factors in the returns on stocks and bonds. *Journal of Financial Economics*, **33**(1), 3–56.

Fan, J., Liao, Y., and Liu, H. 2016. An overview of the estimation of large covariance and precision matrices. *Econometrics Journal*, **19**(1), C1–C32.

FRED (Federal Reserve Economic Data, Federal Reserve Bank of St. Louis). 2020. 30-Year Fixed Rate Mortgage Average in the United States (MORTGAGE30US). `https://fred.stlouisfed.org/series/MORTGAGE30US`.

French, K. 2021. Description of Fama/French Factors. `https://mba.tuck.dartmouth.edu/pages/faculty/ken.french/data_library.html#Research`.

French, K., Schwert, W., and Stambaugh, R. 1987. Expected stock returns and volatility. *Journal of Financial Economics*, **19**(1), 3.

Grinold, R., and Kahn, R. 1999. *Active Portfolio Management: A Quantitative Approach for Producing Superior Returns and Controlling Risk*. McGraw-Hill.

Hallac, D., Nystrup, P., and Boyd, S. 2019. Greedy Gaussian segmentation of multivariate time series. *Advances in Data Analysis and Classification*, **13**(3), 727–751.

Hastie, T., Tibshirani, R., and Friedman, J. 2009. *The Elements of Statistical Learning: Data Mining, Inference, and Prediction*. Springer.

Hung, F.-S. 2003. Inflation, financial development, and economic growth. *International Review of Economics & Finance*, **12**(1), 45–67.

Longin, F., and Solnik., B. 2001. Correlation structure of international equity markets during extremely volatile periods. *Journal of Finance*, **56**(2), 649–676.

Mahyar, H. 2017. The effect of inflation on financial development indicators in Iran (2000–2015). *Studies in Business and Economics*, **12**(2), 53–62.

Markowitz, H. 1952. Portfolio selection. *Journal of Finance*, **7**(1), 77–91.

Nystrup, P., Hansen, B., Madsen, H., and Lindström, E. 2015. Regime-Based versus static asset allocation: Letting the data speak. *Journal of Portfolio Management*, **42**(1), 103–109.

Nystrup, P., Madsen, H., and Lindström, E. 2018. Dynamic portfolio optimization across hidden market regimes. *Quantitative Finance*, **18**(1), 83–95.

Nystrup, P., Boyd, S., Lindström, E., and Madsen, H. 2019. Multi-period portfolio selection with drawdown control. *Annals of Operations Research*, **282**(1), 245–271.

Pearson, K. 1901. On lines and planes of closest fit to systems of points in space. *The London, Edinburgh, and Dublin Philosophical Magazine and Journal of Science*, **2**(11), 559–572.

Petre, G. 2015. A case for dynamic asset allocation for long term investors. *Procedia Economics and Finance*, **29**, 41–55.

Richard, E., Savalle, P.-A., and Vayatis, N. 2012. Estimation of simultaneously sparse and low rank matrices. Pages 51–58 of: *Proc. 29th International Conference on Machine Learning*. Madison, WI, USA: Omnipress.

Ryden, T., Terasvirta, T., and Asbrink, S. 1998. Stylized facts of daily return series and the hidden Markov model. *Journal of Applied Econometrics*, **13**(3), 217–244.

Saegusa, T., and Shojaie, A. 2016. Joint estimation of precision matrices in heterogeneous populations. *Electronic Journal of Statistics*, **10**(1), 1341–1392.

Schwert, W. 1989. Why does stock market volatility change over time? *Journal of Finance*, **44**(5), 1115–1153.

Sutton, G., Mihaljek, D., and Subelytė, A. 2017 (Oct.). Interest rates and house prices in the United States and around the world. BIS Working Papers 665. Bank for International Settlements.

Tuck, J., and Boyd, S. 2022a. Eigen-stratified models. *Optimization and Engineering*, **23**, 397–419. https://doi.org/10.1007/s11081-020-09592-x.

Tuck, J., and Boyd, S. 2022b. Fitting Laplacian regularized stratified Gaussian models. *Optimization and Engineering*, **23**, 895–915. https://doi.org/10.1007/s11081-021-09611-5.

Tuck, J., Hallac, D., and Boyd, S. 2019. Distributed majorization–minimization for Laplacian regularized problems. *IEEE/CAA Journal of Automatica Sinica* **6**(1), 45–52.

Tuck, J., Barratt, S., and Boyd, S. 2021. A distributed method for fitting Laplacian regularized stratified models. *Journal of Machine Learning Research*, **22**(1), 2795–2831.

USBLS (United States Bureau of Labor Statistics). 2020. Consumer price index. https://www.bls.gov/cpi/.

Weiser, A., and Zarantonello, S. 1988. A note on piecewise linear and multilinear table interpolation in many dimensions. *Mathematics of Computation*, **50**(181), 189–196.

Wynne, M., and Sigalla, F. 1994. The consumer price index. *Economic and Financial Policy Review*, **2**(Feb), 1–22.

Part V

New Frontiers for Stochastic Control in Finance

18

Introduction to Part V
Machine Learning and Applied Mathematics: a Game of Hide-and-Seek?

Gilles Pagès[a]

Stochastic control is a mature theory that has been for quite a while at the core of applied probability and the main field of its industrial applications. The liberalization of financial markets in the 1970s in the United States, in particular the emergence of options and derivatives markets, brought to the fore other fields of probability theory such as stochastic calculus and Monte Carlo simulation (i.e. numerical probability). But stochastic control is never far away when dealing with portfolio management, optimal asset allocation, long term contracts on energy markets or any other form of "monitoring" in finance, especially in incomplete markets. However, in this field, more than anywhere else, "time is money"; continuous-time stochastic control mainly relied on the numerical solution of the Hamilton–Jacobi–Bellman equations, something highly dependent on optimization methods and hence vulnerable to the curse of dimensionality. From a more probabilistic viewpoint, Markovian discrete time control, even if making it possible to progress in, say, moderately high dimensions via the dynamic programming principle (see Bellman, 1957), also remains globally limited in high dimension because of the two-fold difficulties associated with the determination of the value function and the backward step-by-step search of the optimal control.

Surprisingly enough, two recent events have almost simultaneously shaken up this observation. On the one hand, the development of mean-field games (see Lasry and Lions, 2018, and Carmona and Delarue, 2018, for a probabilistic approach) which enhanced the importance of sophisticated stochastic models (McKean–Vlasov type differential equations, see McKean, 1967, and particle methods). Such games, which aim at modeling and analyzing the behavior of a homogeneous population of a very large number of agents, crucially bring into play stochastic control problems in very high dimensions.

The second event is the (third!) revival – almost the "rebirth" – of "deep" connectionist learning methods thanks to the work of Hinton, Le Cun and Bengio (see LeCun et al., 2015) after their crushing victory at the 2012 *ImageNet* challenge (www.image-net.com), which happened following several ups and downs

[a] LPSM, Sorbonne-Université
Published in *Machine Learning And Data Sciences For Financial Markets*, Agostino Capponi and Charles-Albert Lehalle© 2023 Cambridge University Press.

since the 1950s. We will come back to this incredible story in the next section[1]. As for the *Imagenet* challenge, it is a question of training a parametrized function which maps/classifies pictures of a database to what they represent (man/woman, car/truck, dog/cat, etc.). This wide parametrized family of functions is known as a *neural network*. For a given input, the network computes an output that is compared to the (known) expected answer. The learning or training phase consists in correcting in an adaptive way the error of prediction by a reinforcement rule, in practice a backpropagated gradient descent (GD). Hence the terminology of *supervised learning*. The input dimension is often enormous, but that of the parameter network to be trained while learning the mapping (e.g. a picture to its output "he" or "she") may be even more so. This GD procedure therefore lives in a space of several tens of thousands of dimensions: the GPT3 network reached 175 billion parameters (see `https://openai.com/blog/openai-api/`)!

The whole community of applied sciences has been impacted by such striking events. Dimension 10 could be no longer be regarded as "high"... And the idea of using these neural networks, especially their mysterious training techniques, quickly took root, especially in the stochastic control community, often, but not always, in connection with finance. For their part, investment banks, always on the lookout for technical advances that would give them a competitive advantage, accompanied and sometimes preceded the craze.

Part V, *New Frontiers in Stochastic Control in Finance*, is an illustration of this convergence of interests – in every sense – since all the contributions involve neural networks or at least stochastic optimization methods to solve control problems, and several of them (well, at least two) use these tools to solve problems arising from mean-field games. Under the assumption that readers are *a priori* more familiar with stochastic control, or even mean-field games, than with neural networks, machine learning (ML) and artificial intelligence (AI), it may be helpful in this introduction to explore briefly the history of those connectionist scientists whose work has often developed alongside mathematicians but without much interaction between them, either because there was no demand, or because it was not always welcome.

A brief history of artificial intelligence

Artificial intelligence has always fascinated human societies since the beginning of the first industrial revolution. In the form of the revisited Promethean myth of Prometheus, how can we breathe life and give a soul to a recreated physical frame? Think of Mary Shelley (1797–1851) who imagined Dr. Frankenstein's creature. Or, almost 100 years later, among many others, consider the French writer Gaston Leroux in his popular novel *The Bloody Doll*[2]. Even more recently,

---

[1] Since 2012 the exploits have multiplied: among others, *AlphaGo* beating the world champion of Go in 2015; *AlphaFold* and *AlphaFold2* more than doubling the efficiency of the virtual folding of proteins since 2014, both devised by the Deepmind company (now a subsidiary of Google); and, more recently, automatic text generation by the GPT3 network.

[2] *La poupée sanglante*, Gaston Leroux, 1923, Tallandier, Paris.

think of Isaac Asimov who published *I, Robot* in 1950. But such imaginings were undoubtedly a little too ambitious at the time to go beyond fantasy novels. Yet it was also in the 19th century that Ada Lovelace (1815–1852), during a long-term collaboration with Charles Babbage (1791–1871), wrote in 1843 what has been since considered as the first computer program, with the aim of calculating the Bernoulli numbers on the (never built) *Analytical Engine*, a computing device designed by Babbage. A century later, Alan Turing (1912–1954), in a first attempt to unify mathematics, logic and algorithmics, imagined his eponymous machine as the ultimate judge of any algorithm, thus building the foundations of a theory halfway between AI and what is more prosaically called *computer science* or *data processing*. A major obstacle was that the computer still had to be invented. . .

Turing, after (co-)breaking the Enigma code of the Third Reich during World War II, tackled the design of an effective computing machine, supported by the British government, but his attempts remained unsuccessful for various reasons too complex to explain here. It was across the Atlantic, within the Manhattan project devoted to building an atomic bomb, that in 1943 the first "Turing-complete" computer (*ENIAC*, for *Electronic Numerical Integrator And Computer*) was designed (and "inaugurated" in 1946). Even if the original A-bombs were created without the use of the computing power of ENIAC, the latter played a crucial and recognized role in later developments. Indeed the Monte Carlo method was imagined by Stanislas Ulam one evening when, annoyed by constantly losing his "Solitaire" card games, he tried unsuccessfully to calculate his probability of success, only finally to surrender and resolve to simulate repeated games on ENIAC and to estimate it instead. Seduced by the idea, von Neumann urged replacing cards with particles to simulate neutron equations and solve certain Boltzmann-type PDEs. Enrico Fermi (1901–1954) conceived its "brother" *FERMIAC* for the same purpose. Monte Carlo was simply the code name of the method. Such simulations were extensively used during the development of the H-bomb to cut through various controversies between Ulam and his fellow physicist Edward Teller (1908–2003). The "Teller–Ulam" H-bomb was patented by the two men and the article describing the Monte Carlo method, co-authored with Nicholas Metropolis (1915–1999), was published later (Metropolis and Ulam, 1949).

It was also during World War II that the first "artificial neuron" emerged, in its most basic form, in the minds of Warren McCulloch and Walter Pitts as a system aggregate of vector-valued data. More formally, the principle was to aggregate a vector $x = (x^1, \ldots, x^d) \in \mathbb{R}^d$ into a scalar using a vector of weights $w = (w^1, \ldots, w^d)$ via an inner product $w \cdot x = \sum_{i=1}^{d} w^i x^i$ (see McCulloch and Pitts, 1943). The reason for doing this was to realize Hebb's reinforcement rule. But what was it good for? It was too simple to be a credible model of neurons for biologists. In 1958, Frank Rosenblatt, while at the Cornell Aeronautical Laboratory, added a threshold function $\mathbf{1}_{\{w \cdot x \geq \alpha\}}$ and proposed a *supervised learning algorithm* allowing it to "teach" the system, by training in a finite number of steps, the weights and threshold values. The aim was to classify data sharing a bi-modal *feature* (in today's language, see Rosenblatt, 1958). By *supervised* we mean here that the output of the system is compared to the exact answer and its weights w^i are then recursively updated "accordingly". Behind the word "accordingly" is hidden a kind of stochastic gradient descent (SGD), minimizing the classifying error – interpreted as success/failure reinforcement process. This raised some public enthusiasm for the cupboard-sized machine, enough for it to be included as an attraction on television shows. AI was revealed to the public. Soon "sorrowful minds" (sic) – Marvin Minsky and Seymour Papert – pointed out (see Minsky and Papert, 1969) that, in fact, this first perceptron was simply a linear classifier; this was AI's first winter. However, the concept of AI was growing and taking shape, notably under the leadership of Norbert Wiener[3]. In 1960, Bernard Widrow and his PhD student Marcian Hoff took McCullogh & Pitts' neuron, got rid of the thresholding function, and developed a supervised learning algorithm that produced linear regression of data, even in a multi-dimensional input–output framework (see Widrow and Hoff, 1960). They named it ADALINE for ADAptive LINEar neuron. It is a perceptron without a hidden layer. The resulting formal learning algorithm was in fact already known in numerical analysis as the recursive inversion of a positive-definite matrix. But rediscovering from a radically different starting point than numerical analysis was already a strong signal of the originality and the power of this "neural guided intuition".

We have to wait until 1986 to see Paul Werbos (1947– .) introduce the (feedforward) perceptron with a hidden layer and initiate its calibration by backpropagation of the gradient. It was then developed in a somewhat resounding way by David Rumelhart, Geoffrey Hinton and Ronald Williams (Rumelhart et al., 1986), the first two of whom were experimental psychologists and defined their work as mathematical psychology.

Hinton (1947– .), a graduate in experimental psychology from King's College Cambridge, undertook and defended in 1978 his PhD thesis at the University of Edinburgh on neural networks in computer science, even though AI was then still in "hibernaton". At the University of San Diego, in the early 1980s, he met Rumelhart (1942–2011), another pioneer of artificial neural networks, with whom he systematically developed the backpropagation form of the (stochastic) gradient method. It consisted in calibrating the weights of a network by recursive and adaptive error corrections. He worked then on Werbos' feedforward perceptron with one hidden layer. In terms of optimization, adaptivity is akin to an SGD attached to the empirical measurement of the database, but with a huge level of complexity never attained since the seminal contribution of Robbins and Monro (1951). This adaptive approach is an alternative to the so-called "batch" approach where each update of the weights requires scrolling through the whole database. This optimization phase by SGD also experienced its winter at the end of the 20th century, eclipsed by simulated annealing and genetic algorithms to be reborn, ubiquitous, with deep learning.

In fact, as so often in science, the story is not as simple because, in the

---

[3] Norbert Wiener (1894–1964) was a child prodigy, the inventor of cybernetics – an ancestor of robotics – and is better known to probabilists for defining Brownian motion (or the Wiener process!) in rigorous mathematical terms.

1970s, IBM computer scientists had already developed a sophisticated automatic differentiation process (AAD for Adjoint Automatic Differentiation) based on the formula for differentiation of compound functions and in almost all points similar to an iteration of this backpropagation algorithm. So, if we consider n differentiable vector fields $f_k : \mathbb{R}^d \to \mathbb{R}^d$, $k = 1, \ldots, n$,

$$J_x(f_n \circ \cdots \circ f_1)^* = J_{y_1}(f_1)^* \circ \cdots \circ J_{y_n}(f_n)^*, \; y_{k+1} = f_k(y_k), \; k = 1, \ldots, n-1, \; y_1 = x,$$

where $*$ means transpose, and $J_x(f) = \left[\frac{\partial f_i}{\partial x^j}(x) \right]$ denotes the Jacobian of f at $x = (x^1, \ldots, x^d)$. One first performs a forward computation of the auxiliary variables y_k, then one goes backward to compute the successive Jacobians.

Unfortunately, the antennae of the two communities seemingly did not fruitfully cross at the time (except in the case of Werbos).

Meanwhile, Yann Le Cun[4] completed and defended his (PhD) thesis *Modèles connexionnistes de l'apprentissage*[5] in June 1987 at Univerité Pierre et Marie Curie in Paris (now an eponymous campus of Sorbonne University). The most significant parts of his thesis (on a back-propagation method for training a multi-layered perceptron) had already been published in articles (in French) from 1986. Geoffrey Hinton who had immediately noticed the first of these, came to Paris as a distinguished member of the jury for the defense and brought Le Cun back to Canada as a post-doc at the University of Toronto where he had a position. Yann Le Cun would not return to France (to work), moving from Toronto to ATT and NYU. He would confide later that his work was triggered by reading a debate from 1975, during the "Entretiens de Royaumont", between the linguist Noam Chomsky and the psychologist Jean Piaget on innate or acquired language learning.

This period marked a reappearance of neural networks in scientific news and, with it, of prophecies concerning AI that are worthy of science fiction novels. This induced, just as in the first breakthrough of AI in the 1950s, the release of science-fiction movies, like *Terminator*, announcing the advent of the reign of machines. More prosaically, at the beginning of the 1990s, the French bank *Crédit Mutuel de Bretagne* adopted the automatic reading of checks, addresses and postal codes using neural networks. Was this a tribute to the MNIST database[6] made up of thousands of handwritten digital images (see yann.lecun.com/exdb/mnist/)?

Everything seemed to be going well but nobody really knew "why". Moreover, the limited performance of computers, difficulties in accessing CRAY super-computers (which ruled the world of high performance computing in the 1980s) hampered both the development and the ambitions of technologies such as pattern recognition, automatic translation, classification, ... Just as in the 1950s, too much hope and hype worked against the pursued objective. In particular, the lack of explanation as to why it worked led to questioning the reliability and the scalability with regard to applications such as those we know today. The time was still not right.

---

[4] Known professionally as Yann LeCun when writing papers in English.
[5] Connectionist learning models.
[6] Modified or Mixed National Institute of Standards and Technology.

Meanwhile, from a theoretical point of view, things were moving, first in the USSR, and then in the USA, thanks to the efforts of Russian mathematicians Vladimir Vapnik (1936– .) and Alexey Chervonenkis (1938–2014). At the dawn of the 1970s, Vapnik and Chervonenkis laid the foundations of the statistical learning theory by defining what is now called, in their honor the *VC*-dimension, for assessing the complexity of a classification problem (see Vapnik, 1989). This quantity is involved in a probabilistic inequality that relates the *learning error* rate of a classifier on a given dataset to the *generalization error*, i.e. the mis-classification rate observed when "feeding" the same (trained) classifier with a new dataset, different but statistically similar to the original one. This inequality is the first measure for the phenomenon of over-parameterization or overfitting: by learning too well the original database (which is always possible by increasing the number of parameters), the classifier becomes unable to efficiently classify anything else. A compromise must be found. And thanks to their high *VC*-dimension, as evaluated by Vapnik, neural networks can hardly be considered as suitable architectures for performing efficient automatic classifications. In any case, they are much less suitable than the support vector machines (SVM) imagined by the two authors few years later (see Boucheron et al., 2005 for a mathematical account).

These SVMs provide a classification method linked to Mercer's theorem involving kernels and based on the embedding of the data in higher-dimensional spaces in which a linear partition becomes possible. A shift occurred at the end of the 1990s and the prospects of neural networks (NN) became much less bright, at least from a theoretical point of view – in particular under the impetus of an academic statistical community fascinated by the theory of learning, which is incidentally well suited to careful mathematical analysis. If these SVMs have enjoyed significant success, especially in their ability to learn fast, it is appears today that the most striking and persistent contribution of Vapnik and Chervonenkis is that they pointed out and established a measure for an irreducible conflict between learning and generalization errors. Applicable to all types of data, whatever their size, nature and origin, the Vapnik–Chervonenkis inequality appears as a sort of Heisenberg's uncertainty principle for data-science, sometimes even able to replace models.

But Hinton and Le Cun were obstinate. They were soon joined in this challenge by Yoshua Bengio (1964– .), a young researcher noticed by Le Cun right after his PhD defense (at McGill) on speech recognition by neural networks. Le Cun invited him to come to the ATT–Bell Labs where they were in daily and profitable contact with Vapnik.

To tell the truth, at the dawn of the 2000s, they were still a little isolated in their belief in the potential of connectionist methods and in the innovations they were developing: recurrent, convolutional neurons, an understanding of overfitting, etc. Scientific journals were still rejecting their articles.

In 2003, they joined forces, got funding from the Canadian Institute for Advanced Research and threw themselves headlong into a scaling up of connectionist methods that led to what is known today as *Deep Learning*, though still without convincing those around them: the largest worldwide conference of neural networks (Neural Information Processing Systems, NIPS, now NeurIPS) declined to let them organise a section in 2007. They hurriedly set up a satellite conference attended by more than 300 people. At the time they were about 600 delegates attending NIPS (compared with more than 10 000 today). The next few years saw many changes: first, two years later, on voice recognition (NLP); then image processing (classification, detection,. . .) around 2011. Building on this progress, they (in fact Hinton and collaborators) entered their *Supervision* project into ILSVRC2012 (Imagenet Large Scale Visual Recognition Challenge 2012), the most competitive image recognition competition based on *Imagenet*, the largest image base at that time. The project included many recent advances, such as convolutional neurons inspired by image processing and initiated by Le Cun. And *Supervision* won. Not only won, but crushed the competition like never before. Google recruited Hinton in 2012; and Facebook (now Meta), Le Cun in 2013. In 2016, Deep Mind and its program Alpha Go crushed the world champion of Go. And in two participations, in 2016 and 2018, in a contest about deployment of proteins, the Alpha Fold and Alpha Fold 2 programs doubled in two stages (from 40% up to 80%) the reconstruction rates of protein folding previously obtained by the best bioinformaticians. On March 27, 2019, Bengio, Hinton and Le Cun received the prestigious 1 million dollar Turing Prize from the ACM (Association of Computing Machinery) for their contribution "of major and lasting technical importance to the Information Theory field". The tide had turned: it was the stuff of legend.

Perhaps out of prudence or for the sake of precision, the renaissance in the area has spread and is popularly known as "Machine Learning" rather than "Artificial Intelligence" which maybe had the painful connotations of the "winters" in the late 1960s and 1990s.

From the side of applied mathematics, rumor has been building for some time: deep networks can "learn" functions from samples of inputs–outputs without suffering (too much) from the curse of dimensionality. All the prejudices of the past fade and soon vanish. A new generation of applied mathematicians is at the helm and high performance computation has became routine with the availability since 2007 of GPUs (Graphic Processing Units) for massive parallel computation from NVIDIA and ATI.

Will machine learning conquer applied mathematics?

In mathematical finance, attention is focused on the Monte Carlo method: but it is too slow. Why not train a deep network once and for all to learn on simulated data pricing and hedging formulas of derivatives? Once trained, it will compute incomparably faster. And why not ultimately only rely on historical data, although quants and traders have always been reluctant to? The training (viewed as a kind of warm-up) will take a lot time because, if deep networks may learn better than SVM, they learn more slowly. The article "Deep Hedging" (see Buehler et al., 2019) was the first to explore this vein through a collaboration between the JPMorgan bank and academics from ETH Zürich. Taking advantage of their financial expertise they proposed many possible loss functions for the training based on various risk measures (quadratic risk, expected shortfall, etc.) or pricing methods (e.g. indifference price). But, contrary to the dreams of many traders after the emergence of deep learning, Buehler et al. were not able to get rid of the diffusion models driving the underlying traded asset dynamics. Many others rushed into the subject in connection with finance: see Becker et al. (2019) with deep optimal stopping; or applications of reservoir computations; or the attention paid to the "deep pricing" of *callable* derivatives by practitioners.

We can therefore try to emulate or approximate a function which is represented by the expectation of a sophisticated stochastic process (hard to simulate) by the output function of a deep neural network. We could then minimize it without (too much) trouble, thereby opening up unexpected perspectives in the most computationally intensive field of applied probability: stochastic control. As a by-product, one has access to efficient stochastic optimization procedures for exploring high-dimensional spaces. Combining the two raises hopes for overcoming the curse of dimension in stochastic control at the heart of all numerical methods. As expected, this is a major theme of this handbook which runs through all the contributions.

All that would not have been possible, at least not so quickly, without open source software libraries such as *TensorFlow* (from Google), *PyTorch* (from Facebook), or *Keras* which combines both and more. Such libraries provide robust procedures for stochastic optimization (e.g. SGD), not just those tailored for training deep neural networks. It has seen a revolution for quants working in investment banks or hedge funds, used to developing their own propriety codes. For academics, it also provides free access to huge libraries. Thus, pre-processing of the dataset yields an optimal efficiency for stochastic optimization algorithm. Overall, these libraries have turned out to be tremendous accelerators for both research and testing in the worlds of academics and of practitioners.

Yet mathematicians are mathematicians and they are still reluctant to use a method without understanding "how" it works. After the first universal approximation results by Cybenko (1989) or Hornik et al. (1989) in the late 1980s, even the best specialists of functional analysis failed to shed light on breaking the curse of dimensionality in the 1990s. Thus, for a feedforward perceptron with n units on its (single) hidden layer, the rate of approximation of a C^r function f on (a compact subset) of $\mathbb{R}^d$ is of order $O(n^{-\frac{r}{d}})$: see Attali and Pagès (1997) among many others. Some improvements may yield $O(n^{\frac{r}{d-1}})$: see Maiorov (1999). Barron (1993) established a $O(n^{-\frac{1}{2}})$ rate under a seemingly dimension-free Fourier condition on C^1-functions f. Unfortunately, such functions become sparse as d increases.

With the recent revival of connectionist methods, new approaches have been proposed to explain their efficiency. Let us briefly mention a few ideas arising from the probability and mathematical finance communities. What can be learned from interpreting a neural network as a controlled ordinary differential equation driven by the combination of a small number of vector fields? Can the signature of semi-martingale or rough paths be used to to analyze the dependence of neural networks upon dimension? Meanwhile investigations to speed up stochastic gradient descents gave birth to so many variants and avatars that trying to make an inventory of them would be in vain. This explosion resulted from the joint efforts of both the optimization and the computer science communities. From the "probabilistic" side, let us mention connections with the Langevin equation either in its standard or McKean–Vlasov versions, combined with entropy regularization in order to improve the efficiency of SGD (see Hu et al., 2019).

Inside new frontiers

The chapters that make up this Part tackle the core of the interplay between stochastic control, neural networks and stochastic optimization. All authors have made efforts to present in a highly pedagogical way the state of the art of the problem under consideration. Their aim is to support non-specialist readers in their journey through the land of high-dimensional control (before reaching its frontiers). For each topic a selected bibliography is provided from the genesis of the problem to the more recent developments. This explains why few references are given in my brief presentation.

In Chapter 19 Zhou proposes a new approach to solving a stochastic control problem of a Brownian diffusion process with a path-dependent and terminal cost function that combines path-dependent and terminal costs. He describes a new and original resolution method whose novel feature introduces an exploration of the state space using a Langevin algorithm i.e. a gradient descent associated to the cost/loss function with an exogenous additional noise to improve the exploration. The temperature is tuned as the solution of a control problem. To avoid bang-bang extreme solutions, an entropic regularization is introduced that helps to overcome critical difficulties such as the curse of dimensionality.

Chapter 22 starts with the probabilistic representation of semi-linear and fully non-linear PDEs by Backward Stochastic Equations (BSDE) of first and second order and their use in solving them numerically. The authors discuss the curse of dimensionality beyond dimension 3 in the case of the classical numerical analysis approach (finite differences); or 7 in the case of regression. A brief review of neural-based deterministic and probabilistic methods is presented and the authors propose a new method that takes advantage of a Markovian dynamic programming principle. The authors propose a scheme combining neural networks and AAD.

Chapters 21 and 23 are devoted to numerical aspects of mean-field games which are by nature huge-dimensional stochastic control problems modeling a large number of interacting homogeneous agents (driven by a controlled diffusion depending on the state and a flow of distributions). The two (already classical) problems to be solved are the search for a Nash equilibrium of the system (MFG) equilibrium on the one hand and the MFC problem in which all the agents co-operate to minimize the cost function depending on a McKean–Vlasov equation. It models competitive and cooperative games and recently met with great success in economics and finance. The PDE related to the MFG Nash equilibrium is an HJB equation coupled with a Kolmogorov–Fokker–Planck equation (or Forward–Backward SDE in probabilistic language) whereas MFC appears as a control problem of the McKean–Vlasov equation.

In Chapter 20 the method proposed for solving these control problems is the introduction of a particle system (to make simulating the Vlasov feature possible) and search optimal controls as a family of neural networks optimized by stochastic gradient. In Chapter 21, the authors focus on the setting where the agents do not know the model, leading to an approach by reinforcement learning. They propose for both MFG and MFC problems a two-time-scale approach solved by a Q-learning algorithm. In each of these chapters various examples inspired by stylized models are treated with numerical implementations to illustrate the numerical methods.

Chapter 23 is focused on a recent model of neural network, the Generative Adversorial Network introduced in 2014 by Ian Goodfellow (a former student of Bengio) and others. The system is made up of two networks: a generator G and a discriminator D. The training of the network consists in solving a min–max problem based on a mutual Jensen–Shannon entropy criterion of two parametrized families of probability distributions. The numerical instability that appears can be overcome by substituting Wasserstein distance for the entropy. Several optimization methods are proposed to attain this Nash equilibrium by an SGD (with inverted signs). The convergence is studied through a diffusion approximation of the procedure, either with a finite horizon or on the long-range (invariant distribution). Simulations are presented including applications to asset pricing and simulation of time series of financial data.

Acknowledgement.

I thank V. Lemaire for helpful discussions and references.

References

Attali, J.-G., and Pagès, G. 1997. Approximations of functions by a multilayer perceptron: a new approach. *Neural Networks*, **10**(6), 1069–1081.

Barron, A. R. 1993. Universal approximation bounds for superpositions of a sigmoidal function. *IEEE Trans. on Information Theory*, **39**, 930–945.

Becker, S., Cheridito, P., and Jentzen, A. 2019. Deep optimal stopping. *Journal of Machine Learning Research*, **20**, 1–25.

Bellman, R. 1957. *Dynamic Programming*. Reprinted with a new introduction by Stuart Dreyfus (2010). Princeton University Press.

Boucheron, S., Bousquet, O., and Lugosi, G. 2005. Theory of classification: a survey of some recent advances, *ESAIM: Probability and Statistics*. **9**, 323–375.

Buehler, H., Gonon, L., Teichmann, J., and Wood, B. 2019. Deep hedging. *Quantitative Finance*, **19**(8), 1271–1291.

Carmona, R., and Delarue, F. 2018. *Probabilistic Theory of Mean Field Games with Applications*. Volume I. *Mean Field FBSDEs, Control, and Games*. Volume II. *Mean Field Games with Common Noise and Master Equations*. Springer.

LeCun, Y., Bengio, Y., and Hinton G. 2015. Deep learning. *Nature*, **521**(7553), 436–44.

Cybenko, G. 1989. Approximation by superpositions of a sigmoidal function. *Math. Control Signals Systems*, **2**(4), 303–314.

Hornik, K., Stinchcombe, M., and White, H. 1989. Multilayer feedforward networks are universal approximators. *Neural Networks*, **2**(5), 359–366.

Hu, K., Ren, Z., Siska, D., and Szpruch, L. 2019. *Mean-field Langevin dynamics and energy landscape of neural networks.* 1905.07769.

Lasry, J.-M., and Lions, P.-L. 2018. Mean-field games with a major player. *C. R. Math. Acad. Sci. Paris*, **356**(8), 886–890.

Maiorov, V. E. 1999. On best approximation by ridge functions. *Journal of Approximation Theory*, **99**(1), 68–94.

McCulloch, W. S., and Pitts, W. H. 1943. A logical calculus of the ideas immanent in nervous activity. *Bulletin of Mathematical Biophysics*, **5**, 115–133.

McKean, H. P. 1967. Propagation of chaos for a class of nonlinear parabolic equations. *Lecture Series in Differential Equations*, **7**, 41–57. Catholic Univ., Washington, DC.

Metropolis, N., and Ulam, S. 1949. The Monte Carlo method. *J. Amer. Statist. Assoc*, **44**, 335–341.

Minsky, M. L., and Papert, S. 1969. *Perceptrons: An Introduction to Computational Geometry.* New augmented edition (1988). MIT Press.

Robbins, H., and Monro, S. 1951. A stochastic approximation method. *Annals of Mathematical Statistics*, **22**(3), 400–407.

Rosenblatt, F. 1958. The perceptron: A probabilistic model for information storage and organization in the brain. *Psychological Review*, **65**(6), 386–408.

Rumelhart, D. E., Hinton, G. E., and Williams, R. J. 1986. Learning representations by back-propagating errors. *Nature*, **323**(6088), 533–536.

Vapnik, V. N. 1989. *Statistical Learning Theory.* Wiley.

Widrow, B., and Hoff, M. E. 1960. Adaptive switching circuits. Pages 96–104 of: *IRE WESCON Convention Record*. Reprinted in *Neurocomputing*, MIT Press.

19

The Curse of Optimality, and How to Break it?

Xun Yu Zhou[a]

Abstract

We strive to seek optimality, but often find ourselves trapped in bad "optimal" solutions that are either local optimizers, or are too rigid to leave any room for errors, or are simply based on wrong models or erroneously estimated parameters. A way to break this "curse of optimality" is to engage exploration through randomization. Exploration broadens search space, provides flexibility, and facilitates learning via trial and error. We review some of the latest development in this exploratory approach in the stochastic control setting with continuous time and spaces.

19.1 Introduction

Optimal solutions derived from various optimization theories are often bad traps that hinder practical use. An example that immediately comes to mind is a local optimizer out of the first-order condition. Another example is the bang–bang control in optimal control theory: an optimal control takes only extreme values when the control variable appears linearly in the Hamiltonian. Such a control is too sensitive to estimation errors and thus tends to be very unstable and hardly usable.

Classical theories also often take the "separation principle" between estimation and optimization; see Wonham (1968) for example. One typically assumes a model, estimates model parameters based on past data, and then optimizes as if the underlying model was correct. Think of a gambler at an array of slot machines ("one-armed bandits") that have different but unknown probabilities of winning. He has to decide how many times to play each machine and in what order so as to maximize the expected gains. The classical estimation-and-optimization approach will tackle the problem in the following way: playing each machine for n rounds, where n is sufficiently and judiciously large, and observing the outcomes. If, say, Machine 1 has returned the most gains, then the gambler will

[a] Columbia University

Published in *Machine Learning And Data Sciences For Financial Markets*, Agostino Capponi and Charles-Albert Lehalle © 2023 Cambridge University Press.

believe it is indeed the best machine and he will henceforth play this machine *only*.

The flaw of this approach is evident: Machine 1 may well be a sub-optimal machine and sticking to it subsequently may result in falling into a bad trap. This example is a precursor of what is now widely known as a *reinforcement learning* (RL) problem. The RL approach would take the bandits problem in a different way and formulate it as one that trades off near-term and long-term gains. Specifically, the gambler carefully balances between greedily exploiting what has been learned so far to choose the machine that yields near-term higher rewards, and continuously exploring the rest of the machines to acquire more information to potentially achieve long-term benefits. The so-called *ε-greedy strategy* (Sutton and Barto, 2018) exemplifies this idea: at the nth play the gambler tosses a biased coin with heads occurring with a probability $1 - \varepsilon_n$ and tails with a probability ε_n. He then plays the *current* best machine if heads appears and the other machines at random (with uniform probability) if tails appears. Here $\varepsilon_n > 0$ is a small number and ought to get smaller as n becomes larger.

The ε-greedy strategy is a *randomized* strategy: at each play, instead of deterministically and definitely playing a particular machine, the gambler lets a coin flip decide which machine to play. The problem now becomes one of designing the scheme for $\{\varepsilon_n\}_{n \in \mathbb{N}}$ to achieve a good balance between exploration (learning) and exploitation (optimizing). A notable feature, and indeed one that is essentially different from the classical approach, is that the gambler is no longer interested in estimating the winning probabilities of the machines; rather he is focusing on learning the best sequence $\{\varepsilon_n\}_{n \in \mathbb{N}}$. In other words, *he learns his best strategies instead of learning a model*. This underpins the basic tenet in RL: An agent does not pre-assume a structural model nor attempt to estimate an existing model's parameters but, instead, gradually learns the best (or near-best) strategies based on trial and error, through interactions with the black-box environment and incorporation of the responses of these interactions.[1] This learning approach addresses to large extent the problem of "curse of optimality" that is due to engaging a wrong model.

The *exploration through randomization* approach may also be employed to break the curse of optimality even in problems where learning is not necessary. Take for example a non-convex optimization problem where the function to be minimized is completely known. Still, the first-order condition and the associated algorithms such as the gradient descent (GD) give only local minima. *Simulated annealing*, independently proposed by Kirkpatrick et al. (1983) and Cerny (1985), performs randomization at each iteration of the GD algorithm to get the iterates out of any possible trap of a local minimum. Specifically, at each iteration, the algorithm randomly samples a solution close to the current one and moves to it

---

[1] This sounds strikingly different from the model-based approach; but a careful reflection would reveal that it is exactly how people, especially babies and young children, learn things. Take learning a new language for example. Adults usually start with learning the grammar (the model) before actually speaking, whereas babies directly learn to speak (strategies) through interactions and trial-and-error. It is widely held that the latter learn a language much faster and more effectively than the former.

according to a probability distribution. This scheme facilitates a broader search or exploration for the global optimum with the risk of moving to worse solutions at some iterations. The risk is however controlled by slowly cooling down over time the "temperature" which is used to characterize the level of exploration. Another example is to use randomization to smooth out an overly sensitive (and hence unstable) optimal bang-bang control that takes only extreme actions.

Randomization uses a probability distribution (measure) to replace a deterministic action. However, the latter can be embedded into the former as a Dirac measure. To avoid the situation in which the optimal distribution turns out to be a Dirac measure, one can force a minimal level of exploration. In the RL literature, entropy has been used to measure the level of exploration and the *entropy-regularized* (also termed as "softmax") exploratory formulation has been proposed, mostly in the discrete-time and discrete-space Markov Decision Processes (MDPs) setting. In this formulation, exploration enters explicitly into the optimization objective as a regularization term, with a trade-off weight (the *temperature* parameter) imposed on the entropy of the exploration strategy; see Ziebart et al. (2008), Nachum et al. (2017), and Neu et al. (2017) and the references therein. Wang et al. (2020) was the first to extend this formulation to the setting of stochastic control with continuous time and continuous state and action (control) spaces. They derived a stochastic relaxed control formulation to model the repetitive learning in RL, and used the differential entropy to regularize the exploration. They showed that the optimal distribution for exploration is a *Gibbs measure* or a *Boltzmann distribution* of the form $\pi(u) \propto e^{\frac{1}{\lambda}H(u)}$ where λ is the temperature and H is the Hamiltonian. When the state depends on the action u linearly and the reward is quadratic in u, the Hamiltonian is quadratic in u and hence the Gibbs measure specializes to a Gaussian distribution (under some technical assumptions), which in turn justifies the widely used *Gaussian exploration* (Haarnoja et al., 2017). Wang and Zhou (2020) further applied this result to a continuous-time Markowitz mean–variance portfolio selection problem, and devised an RL algorithm to learn the efficient investment strategies without any knowledge about the key parameters such as the stocks' mean returns and volatility rates.

Motivated by considerations other than RL, Gao et al. (2022) applied the general framework and results of Wang et al. (2020) to the temperature control problem for Langevin diffusions. A Langevin diffusion is a continuous-time version of a simulated annealing algorithm – the Langevin algorithm – to find the global minimum of a non-convex function. The temperature process controls the level of random noises injected into the algorithm. The selection of this process can be formulated as a classical stochastic control problem, whose optimal solution is nevertheless bang–bang and hence extremely prone to mis-specifications in the model. Gao et al. (2022) took the entropy-regularized framework of Wang et al. (2020) by randomizing this temperature process, and concluded that a truncated exponential distribution is optimal for sampling temperatures and in turn sampling the noises to be injected into the Langevin algorithm.

This chapter reviews the approaches and main results in Wang et al. (2020),

Wang and Zhou (2020), and Gao et al. (2022), albeit in a finite-time horizon instead of the infinite one, argues that exploration through randomization can effectively address the curse of optimality in settings including but not limited to RL, and suggests some open research questions.

The remainder of this chapter proceeds as follows. In Section 19.2 we present the entropy-regularized exploratory stochastic control problem based on the notion of exploration through randomization. Section 19.3 derives the optimal distributions for sampling actions to control the dynamics. Section 19.4 gives a concrete application of the general theory to the sampling problem of the Langevin algorithm. In Section 19.5 we discuss the algorithmic aspects of the general theory in the RL context. Finally, Section 19.6 concludes.

19.2 Entropy-regularized exploratory formulation

19.2.1 Classical stochastic control

Let $T > 0$, $b : [0,T] \times \mathbb{R}^d \times U \mapsto \mathbb{R}^d$ and $\sigma : [0,T] \times \mathbb{R}^d \times U \mapsto \mathbb{R}^{d \times n}$ be given. The classical stochastic control problem is to control the *state* (or *feature*) dynamics, a stochastic differential equation (SDE):

$$dx_s^u = b(s, x_s^u, u_s)ds + \sigma(s, x_s^u, u_s)dW_s, \ s \in [0,T]. \tag{19.1}$$

The process $u = \{u_s, 0 \leq s \leq T\}$, defined on a filtered probability space $(\Omega, \mathcal{F}, \mathbb{P}; \{\mathcal{F}_s\}_{s \geq 0})$ along with a standard $\{\mathcal{F}_s\}_{s \geq 0}$-adapted, n-dimensional Brownian motion $W = \{W_s, s \geq 0\}$, is an admissible (*open-loop*) control, denoted by $u \in \mathcal{A}^{\mathrm{cl}}$, if

(i) it is an $\{\mathcal{F}_s^W\}_{s \geq 0}$-adapted measurable process taking values in U, where $\{\mathcal{F}_s^W\}_{s \geq 0} \subset \{\mathcal{F}_s\}_{s \geq 0}$ is the natural filtration generated by the Brownian motion, and $U \subset \mathbb{R}^m$ is the *action space* representing the constraints on an agent's decisions (*controls* or *actions*); and

(ii) for any given initial condition $x_0^u = x_0 \in \mathbb{R}^d$, the SDE (19.1) admits solutions $x^u = \{x_s^u, 0 \leq s \leq T\}$ on the same filtered probability space, whose distributions are all identical.[2]

Given $x_0^u = x_0 \in \mathbb{R}^d$ at time $t = 0$, the objective of the control problem is to find $u \in \mathcal{A}^{\mathrm{cl}}$ so that the total reward

$$J(u) := \mathbb{E}\left[\int_0^T r\left(s, x_s^u, u_s\right) ds + h(x_T^u)\right] \to \max \tag{19.2}$$

where $r : [0,T] \times \mathbb{R}^d \times U \mapsto \mathbb{R}$ and $h : \mathbb{R}^d \mapsto \mathbb{R}$ are the running and terminal reward functions respectively.

In the classical setting where the model is fully known (namely, when the

---

[2] Throughout this chapter, admissible controls are defined in the *weak* sense, namely, the filtered probability space and the Brownian motion are also *part* of the control. This is to ensure, among other things, that dynamic programming works; see Yong and Zhou (1999, Chapter 4). For simplicity, however, we will refer to, for example, only the process u as a control.

functions b, σ, r and h are fully specified), one can solve this problem by Bellman's dynamic programming in the following manner; see e.g. Yong and Zhou (1999) for a systematic account of the method. Define the *optimal value function*

$$V^{\mathrm{cl}}(t,x) := \sup_{u \in \mathcal{A}^{\mathrm{cl}}} \mathbb{E}\left[\int_t^T r\left(s, x_s^u, u_s\right) ds + h(x_T^u)\Big| x_t^u = x\right], \quad (t,x) \in [0,T] \times \mathbb{R}^d,$$

(19.3)

where (and throughout this chapter) t and x are generic variables representing respectively the current time and state of the system dynamics.[3]

If $V^{\mathrm{cl}} \in C^{1,2}([0,T] \times \mathbb{R}^d)$, then it satisfies the *Hamilton–Jacobi–Bellman (HJB) equation*

$$\begin{cases} v_t(t,x) + \sup_{u \in U} H(t,x,u,v_x(t,x),v_{xx}(t,x)) = 0, & (t,x) \in [0,T) \times \mathbb{R}^d; \\ v(T,x) = h(x) \end{cases}$$

(19.4)

where H is the (generalized) *Hamiltonian* (Yong and Zhou, 1999, Chapters 3 & 4)

$$H(t,x,u,p,P) = \tfrac{1}{2}\mathrm{tr}\left[\sigma(t,x,u)'P\sigma(t,x,u)\right] + p \cdot b(t,x,u) + f(t,x,u),$$
$$(t,x,u,p,P) \in [0,T] \times \mathbb{R}^d \times U \times \mathbb{R}^d \times \mathbb{R}^{d \times d},$$

(19.5)

where $\mathrm{tr}(A)$ denotes the trace of a square matrix A.

Let

$$u^*(t,x) := \mathrm{argmax}_{u \in U} H(t,x,u,v_x(t,x),v_{xx}(t,x)), \quad (t,x) \in [0,T) \times \mathbb{R}^d. \quad (19.6)$$

This is a *deterministic* mapping from the current time and state to the action space U, which is an instance of a *feedback policy* (or *feedback law*). It is important to understand the differences and relationship between an open-loop control and a feedback policy. The former is a stochastic process – so it is a function of the time t and the state of nature ω; and the latter is a deterministic function of the time t and the state of the system x. Throughout this chapter we call the former a *control* and the latter a *policy*. A policy u can *generate* a control by substituting u into the system dynamics (19.1) starting from any present time–state pair $(t,x) \in [0,T) \times \mathbb{R}^d$.

The verification theorem dictates that u^* is an optimal policy in the sense that it generates an optimal control for the problem (19.3) with *any* $(t,x) \in [0,T) \times \mathbb{R}^d$ via $u_s^* = u^*(s,x_s^*)$ where x^* is the solution to (19.1) upon substituting u_s with $u^*(s,x_s^*)$.

Equation (19.6) stipulates that at any give time and state, the optimal action is guided by the Hamiltonian, *deterministically* and *rigidly*. Moreover, this action policy is derived off-line at $t = 0$ and *will* be carried out throughout, *assuming*, that is, the model is completely specified.

---

[3] In the classical control theory literature, V is termed simply the "value function". However, in what follows, as is customary in the RL literature, we will also use the term *value function* for any given feedback policy. Hence, to avoid confusion, we call V the "*optimal* value function".

19.2.2 Exploratory formulation

As we have discussed in the introduction, there are various reasons why the agent may be unable or unwilling to execute the "optimal" policy (19.6), and will instead need to explore through randomization. For example, in the case when the underlying model is not known, the agent is not able to maximize the unknown Hamiltonian in (19.6), and hence employs exploration to interact with and learn the best strategies through trial and error. The exploration is modelled by a *distribution* of controls $\pi = \{\pi_s(\cdot), s \geq 0\}$ over the control space U from which each trial is sampled. Here π is a density-function-valued stochastic process; i.e. $\pi_s(\cdot, \omega)$ is a probability density function on U for any $(s, \omega) \in [0, T] \times \Omega$. We therefore extend the notion of controls to distributions when exploration is called for. A classical control $u = \{u_s, s \geq 0\}$ can be regarded as a Dirac distribution $\pi_s(\cdot) = \delta_{u_s}(\cdot)$.

This subsection and the next one largely follow the formulation and analysis in Wang et al. (2020), except that we are in the setting of a finite-time horizon while Wang et al. (2020) is for the infinite-time horizon. However, all the results in the current setting can be derived analogously.

Given a distributional control π, the agent repeatedly sample *classical* controls from π for N rounds over the same time horizon to control the dynamics and observe the corresponding values of the total reward. As explained in Wang et al. (2020), when $N \to \infty$, by the law of large numbers the limiting system dynamics under π becomes

$$dX_s^\pi = \tilde{b}(s, X_s^\pi, \pi_s)ds + \tilde{\sigma}(s, X_s^\pi, \pi_s)dW_s, \quad s \in [0, T], \tag{19.7}$$

where the coefficients $\tilde{b}$ and $\tilde{\sigma}$ are defined as

$$\tilde{b}(s, y, \pi) := \int_U b(s, y, u)\pi(u)du, \quad y \in \mathbb{R}^d, \ \pi \in \mathcal{P}(U), \tag{19.8}$$

and

$$\tilde{\sigma}(s, y, \pi) := \sqrt{\int_U \sigma^2(s, y, u)\pi(u)du}, \quad y \in \mathbb{R}^d, \ \pi \in \mathcal{P}(U), \tag{19.9}$$

with $\mathcal{P}(U)$ being the set of density functions of probability measures on U that are absolutely continuous with respect to the Lebesgue measure.

We call (19.7) the *exploratory formulation* of the controlled state dynamics, and $\tilde{b}(\cdot, \cdot)$ and $\tilde{\sigma}(\cdot, \cdot)$ in (19.8) and (19.9), respectively, the *exploratory drift* and the *exploratory volatility*.

Similarly, the reward function r in (19.2) is modified to the *exploratory reward*

$$\tilde{r}(s, y, \pi) := \int_U r(s, y, u)\pi(u)du, \quad y \in \mathbb{R}^d, \ \pi \in \mathcal{P}(U). \tag{19.10}$$

19.2.3 Entropy regularization

Given the exploratory formulation, it seems natural to set the objective to maximize

$$\mathbb{E}\left[\int_0^T \tilde{r}\left(s, X_s^\pi, \pi_s\right) ds + h(X_T^\pi)\right] \tag{19.11}$$

subject to (19.7) under $X_0^\pi = x_0$. However, it is entirely possible that the optimal distributional control for this problem is just Dirac, and hence we would then be in the realm of classical stochastic control. Indeed this happens when the so-called Roxin condition is satisfied; see Yong and Zhou (1999, Chapter 2). Thus, in order to encourage a *genuine* exploration, we need to regulate its level. We use Shannon's *differential entropy* to measure the level of exploration:

$$\mathcal{H}(\pi) := -\int_U \pi(u) \ln \pi(u) du, \quad \pi \in \mathcal{P}(U),$$

and require the total expected entropy to maintain a minimum level

$$-\mathbb{E}\int_0^T \int_U \pi_s(u) \ln \pi_s(u) \, du \, ds \geq a \tag{19.12}$$

where $a > 0$ is given. Taking the Lagrange multiplier of this exploration constraint we arrive at the following new objective:

$$\mathbb{E}\left[\int_0^T \left(\tilde{r}\left(s, X_s^\pi, \pi_s\right) - \lambda \int_U \pi_s(u) \ln \pi_s(u) du\right) ds + h(X_T^\pi)\right] \to \max, \tag{19.13}$$

where $\lambda > 0$ is the Lagrange multiplier, which can also be regarded as an exogenous exploration weighting parameter capturing the trade-off between exploitation (the original reward function) and exploration (the entropy). This constant is also known as the *temperature* parameter.

Denote by $\mathcal{B}(U)$ the Borel algebra on U. A density-function-valued process $\pi = \{\pi_s(\cdot), 0 \leq s \leq T\}$, defined on a filtered probability space $(\Omega, \mathcal{F}, \mathbb{P}; \{\mathcal{F}_s\}_{s \geq 0})$ along with a standard $\{\mathcal{F}_s\}_{s \geq 0}$-adapted, n-dimensional Brownian motion $W = \{W_s, s \geq 0\}$, is an admissible distributional control, denoted by $\pi \in \mathcal{A}$, if

(i) for each $0 \leq s \leq T$, $\pi_s(\cdot) \in \mathcal{P}(U)$ a.s.;

(ii) for each $A \in \mathcal{B}(U)$, $\{\int_A \pi_s(u) du, 0 \leq s \leq T\}$ is $\{\mathcal{F}_s^W\}_{s \geq 0}$-adapted measurable process;

(iii) the SDE (19.7) with $X_0^\pi = x_0$ admits solutions $x^\pi = \{x_s^\pi, 0 \leq s \leq T\}$ on the same filtered probability space, whose distributions are all identical.

19.3 Optimal distributional policies

To solve the entropy-regularized exploratory control problem (19.13), we again apply dynamic programming. Introduce the optimal value function

$$V(t,x) :=$$

$$\sup_{\pi \in \mathcal{A}} \mathbb{E}\left[\int_0^T \left(\int_U r\left(s, X_s^\pi, u\right) \pi_s(u)\, du - \lambda \int_U \pi_s(u) \ln \pi_s(u) du \right) ds + h(X_T^\pi) \,\middle|\, X_t^\pi = x \right].$$

(19.14)

Using standard arguments, we deduce that V satisfies the HJB equation

$$v_t(t,x) + \sup_{\pi \in \mathcal{P}(U)} \int_U \left[H(t, x, u, v_x(t,x), v_{xx}(t,x)) - \lambda \ln \pi(u) \right] \pi(u) du = 0$$

(19.15)

$$(t,x) \in [0,T) \times \mathbb{R}^d,$$

with the terminal condition $v(T,x) = h(x)$.

Noting that $\pi \in \mathcal{P}(U)$ if and only if

$$\int_U \pi(u) du = 1 \quad \text{and} \quad \pi(u) \ge 0 \text{ a.e.} \quad \text{on } U,$$

(19.16)

we can solve the (constrained) maximization problem on the left hand side of (19.15) to get a *feedback* policy:

$$\pi^*(u; t, x) = \frac{1}{Z(\lambda, t, x, v_x(t,x), v_{xx}(t,x))} \exp\left(\frac{1}{\lambda} H(t, x, u, v_x(t,x), v_{xx}(t,x)) \right),$$

(19.17)

where $u \in U$, $(t,x) \in [0,T] \times \mathbb{R}^d$, and

$$Z(\lambda, t, x, v_x(t,x), v_{xx}(t,x)) := \int_U \exp\left(\frac{1}{\lambda} H(t, x, u, v_x(t,x), v_{xx}(t,x)) \right) du$$

(19.18)

is the normalizing factor that makes $\pi^*(\cdot; t, x)$ a density function.

The optimal policy (19.17) is a deterministic function of the variables u, t and x. For each given time–state pair (t,x), $\pi^*(\cdot; t, x)$ is the density function of a Gibbs measure. When the temperature λ is very high, all the actions are chosen in largely equal probabilities. When the temperature cools down, i.e., $\lambda \to 0$, the distribution increasingly concentrates around the (global) maximizers of the Hamiltonian, giving rise to something resembling the ε-greedy policies in multi-armed bandit problems. When $\lambda = 0$, the distribution degenerates into the Dirac measure on the maximizers of the Hamiltonian which is the classical optimal control.

In the linear–quadratic (LQ) case when b, σ are linear in x and u and r, h quadratic in x and u, the Hamiltonian is quadratic in u. In the infinite horizon case, Wang et al. (2020) proved that the Gibbs measure specializes to the Gaussian distribution under some technical assumptions. We expect the same to be true for the current case of a finite time horizon, although there may be some technical subtleties. Moreover, Wang and Zhou (2020) applied the LQ results to a continuous-time mean–variance portfolio selection problem and devised an algorithm for solving it without needing to know the parameters of the underlying stocks.

In RL there is a widely used *heuristic* exploration strategy called the *Boltzmann exploration*, which assigns the following probability to an action a when in state s_t at time t:

$$p(s_t, a) = \frac{e^{Q_t(s_t,a)/\lambda}}{\sum_{a=1}^m e^{Q_t(s_t,a)/\lambda}}, \quad a = 1, 2, \ldots, m, \tag{19.19}$$

where $Q_t(s, a)$ is the *Q-function* value of a state–action pair (s, a), and $\lambda > 0$ is again a temperature parameter that controls the level of exploration; see e.g. Bridle (1990), Cesa-Bianchi et al. (2017), and Sutton and Barto (2018). There is a clear resemblance between (19.17) and (19.19). This in turn suggests that the continuous counterpart of the Q-function is the Hamiltonian, given that the former is not well defined and cannot be used to rank and select actions in the continuous setting (Tallec et al., 2019). The importance of this observation is twofold: the fact that we are able to derive a result that reconciles with an eminent heuristic strategy in the discrete setting, verifies and justifies the entropy-regularized exploratory formulation for the continuous setting, and, more importantly, the formulation lays a *theoretical underpinning* of the Boltzmann exploration, thereby providing an explaination of a largely heuristic approach.[4]

Putting (19.17) back into (19.15), we obtain the following (elegant) form of the HJB equation

$$v_t(t, x) + \lambda \ln Z(\lambda, t, x, v_x(t, x), v_{xx}(t, x)) = 0, \quad (t, x) \in [0, T] \times \mathbb{R}^d; \quad v(T, x) = h(x). \tag{19.20}$$

This equation, called the *exploratory HJB equation*, appears to be a new type of parabolic partial differential equation (PDE), which would provide a whole wealth of new research problems. For example, what do we know about its well-posedness (existence and uniqueness) in both the classical and viscosity senses? How does its solution, along with its first- and second-order derivatives, depend on the temperature $\lambda > 0$? As a result, how does the optimal policy (19.17), along with its mean, variance and entropy, depend on λ? Does the solution converge when $\lambda \to 0$ and, if yes, what is the convergence rate? Some of these questions have been answered in Tang et al. (2022).

Another significant direction for research is in the choice of the temperature λ. In this section, as in Wang et al. (2020), λ is set to be an *exogenous* constant. However, the agent is supposed to learn more, and hence need less, exploration as time goes by. So it seems plausible that λ should depend on time and indeed decay over time. On the other hand, it seems also reasonable that λ should depend on the system state to optimize its use. In other words, λ ought to be *endogenous*. How can we then formulate the problem to optimize the temperature process?

---

[4] A formula of the type (19.17) was first derived in Wang et al. (2020, eq. (17)), but the connection with Boltzmann exploration and Gibbs measure was not noted there.

19.4 Non-convex optimization and Langevin diffusions

While the entropy-regularized exploratory formulation was originally motivated by RL in Wang et al. (2020), its use may go beyond RL, which this section will demonstrate. The presentation follows Gao et al. (2022), although we take a finite-horizon setup as opposed to that of the infinite-horizon in Gao et al. (2022).

Consider a finite-dimensional optimization problem:

$$\min_{x \in \mathbb{R}^d} f(x), \qquad (19.21)$$

where $f : \mathbb{R}^d \to [0, \infty)$ is a *non-convex* function. The traditional gradient descent (GD) algorithm may be trapped in a local optimum. The Langevin algorithm injects noise into GD in order to get out of the trap:

$$X_{k+1} = X_k - \eta f_x(X_k) + \sqrt{2\eta\beta_k}\xi_k, \quad k = 0, 1, 2, \ldots, \qquad (19.22)$$

where f_x is the gradient of f, $\eta > 0$ is the step size, $\{\xi_k\}$ is i.i.d Gaussian noise and $\{\beta_k\}$ is a sequence of the temperature parameters that typically decays over time to zero. The continuous-time version of this algorithm is the so-called *overdamped Langevin diffusion*:

$$dX_s = -f_x(X_s)dt + \sqrt{2\beta_s}dW_s, \quad X_0 = x_0, \qquad (19.23)$$

where $x_0 \in \mathbb{R}^d$ is an initialization, $W = \{W_s : s \geq 0\}$ is a standard d-dimensional Brownian motion with $W_0 = 0$, and $\beta = \{\beta_s : s \geq 0\}$ is an adapted, nonnegative stochastic process, which is also called the *temperature process* of the Langevin diffusion.

When $\beta_s \equiv \beta > 0$, under some mild assumptions on f, the solution of (19.23) admits a unique stationary distribution which is the Gibbs measure with density $\pi(x) = \frac{1}{Z(\beta)}e^{-\frac{1}{\beta}f(x)}$ (Chiang et al., 1987). When β becomes small, this measure increasingly concentrates on the *global* minimum of f. This provides a theoretical justification of using Langevin diffusion (19.23) to sample noises for the Langevin algorithm (19.22).

A natural problem is to control the temperature process $\{\beta_t : t \geq 0\}$ so that the performance of the continuous-time version of the Langevin algorithm (19.23) is optimized. Specifically, given an arbitrary initialization $X_0 = x_0 \in \mathbb{R}^d$, a computing budget $T > 0$, and the range of the temperature $U = [a, b]$ where $0 \leq a < b < \infty$, we aim to solve the following stochastic control problem where the temperature process is taken as the control:

$$\text{Minimize} \quad \mathbb{E}[f(X_T)],$$

$$\text{subject to} \quad \begin{cases} \text{equation (19.23)}, \\ \{\beta_s : 0 \leq s \leq T\} \text{ is adapted}, \\ \beta_s \in U \text{ a.e.} s \in [0, T], \text{ a.s.} \end{cases} \qquad (19.24)$$

This is a classical control problem. Its HJB equation is:

$$v_t(t, x) + \min_{\beta \in [a, 1]} [\beta \text{tr}(v_{xx}(t, x)) - f_x(x) \cdot v_x(t, x)] = 0, \quad x \in \mathbb{R}^d; \quad v(T, x) = f(x). \qquad (19.25)$$

Then the verification theorem yields that an optimal feedback policy is "bang–bang": $\beta^*(t,x) = b$ if $\mathrm{tr}(v_{xx}(x)) < 0$, and $\beta^*(t,x) = a$ otherwise. This policy stipulates that one should, in some time–state pairs, heat at the highest possible temperature, while in others cool down completely, depending on the sign of $\mathrm{tr}(v_{xx}(t,x))$. This policy, while *theoretically* optimal, is clearly too *rigid* to achieve good performance in practice as it concentrates on two extreme actions only, and a computational error of $v_{xx}(t,x)$ may cause drastic change from one extreme to the other. This motivates us to use the exploratory formulation and entropy regularization in order to *smooth out* the temperature processes. Note that here the motivation is no longer from "learning" per se as we can perfectly well assume that the functional form f is given and known.

We now present our entropy-regularized exploratory formulation of the problem. Instead of a classical control $\{\beta_s : 0 \le s \le T\}$ where $\beta_s \in U = [a,b]$ for $s \in [0,T]$, we consider a distributional control $\pi = \{\pi_s(\cdot) : 0 \le s \le T\}$, which represents a randomization of classical controls over the control space U where a temperature $\beta_s \in U$ can be sampled from this distribution whose probability density function is $\pi_s(\cdot)$ at time s. The optimal value function of the exploratory problem is

$$V(t,x) := \inf_{\pi \in \mathcal{A}} \mathbb{E}\left[-\lambda \int_0^T \int_U \pi_s(u) \ln \pi_s(u)\, du\, ds + f(X_T^\pi)\Big| X_t^\pi = x\right], \quad (19.26)$$

where the system dynamic is

$$dX_s^\pi = -f_x(X_s^\pi)dt + \tilde{\sigma}(\pi_s)dW_s, \quad (19.27)$$

with

$$\tilde{\sigma}(\pi) := \sqrt{\int_U 2u\pi(u)du}. \quad (19.28)$$

This problem is a special case of the general problem formulated in the previous section (except that we now have a minimization problem instead of a maximization one). Applying the general results there, we obtain the following optimal feedback policy:

$$\pi^*(u; t,x) = \frac{1}{Z(\lambda, v_{xx}(t,x))} \exp\left(-\frac{1}{\lambda}[\mathrm{tr}(v_{xx}(t,x))u]\right), \quad (19.29)$$

where $u \in U$, $(t,x) \in [0,T] \times \mathbb{R}^d$, and

$$Z(\lambda, v_{xx}(t,x)) := \int_U \exp\left(-\frac{1}{\lambda}[\mathrm{tr}(v_{xx}(t,x))u]\right) du > 0.$$

This is a *truncated* (in U) *exponential distribution* with the (state-dependent) parameter $c(t,x) := \mathrm{tr}(v_{xx}(t,x))/\lambda$, and we do not require either $\mathrm{tr}(v_{xx}(x)) > 0$ (i.e. v is in general non-convex) or $c(t,x) > 0$ here.

The HJB equation is

$$v_t(t,x) - f_x(x) \cdot v_x(x) - \lambda \ln(Z(\lambda, v_{xx}(t,x))) = 0, \quad (t,x) \in [0,T) \times \mathbb{R}^d, \quad (19.30)$$

with $v(T,x) = f(x)$.

To apply the obtained results to sample the Langevin algorithm (19.22), we can take the following steps. First, we solve the HJB equation (19.30) to get v. Second, with the initialization $X_0 = x_0$, and for each $k = 0, 1, 2, \ldots$, we sample β_k from $\pi^*(\cdot; \eta_k, X_k)$ where π^* is determined by (19.29), X_k is the current iterate, and η_k is the cumulative step size. Finally we apply (19.22) to move to the next iterate where ξ_k is independently sampled from a standard Gaussian distribution. For a numerical experiment comparing the performance of this method (albeit based on the infinite-horizon model) with other benchmarks, see Gao et al. (2022).

19.5 Algorithmic considerations for RL

The previous sections are mainly about the *theory* of an entropy-regularized exploratory formulation. We now discuss some aspects of the algorithm design in the RL context. Specifically, we need to design RL algorithms to *learn* the optimal solutions of the entropy-regularized problems and to output implementable policies, without assuming any knowledge about the underlying parameters or attempting to estimate these parameters.

First thing to note is that some of the theoretical results presented earlier already have algorithmic implications. For example, if we know Gaussian is optimal, then we will need to learn only two parameters (mean and variance). If an exponential distribution is optimal, then there is only one parameter to learn. Making use of this information could dramatically simplify the corresponding algorithms and speed up their convergence.

The following discussion, however, is more general without targeting for a particular distribution. It is a generalization of the algorithm developed in Wang and Zhou (2020) for the mean–variance portfolio selection problem. The two key steps involved in our algorithm are *policy evaluation* and *policy improvement*, as standard in RL for MDPs (Sutton and Barto, 2018).

First we define the *value function* of a given distributional policy π. Note that π generates an open-loop distributional control through the exploratory dynamics (19.7) in the same way as in classical control. Specifically, for each given current time–state pair $(t,x) \in [0,T) \times \mathbb{R}^d$, π generates an open-loop control

$$\pi_s(u) := \pi(u; s, X_s^\pi) \tag{19.31}$$

where $\{X_s^\pi, t \le s \le T\}$ solves (19.7) with $X_t^\pi = x$ when the policy π is applied and assuming that $\{\pi_s(\cdot), t \le s \le T\} \in \mathcal{A}$. Now define the value function of π:

$$V^\pi(t,x) := \mathbb{E}\left[\int_t^T \left(\int_U r(s, X_s^\pi, u)\pi_s(u)du - \lambda \int_U \pi_s(u)\ln\pi_s(u)du \right)ds \right.$$
$$\left. + h(X_T^\pi)\Big| X_t^\pi = x \right]. \tag{19.32}$$

In an RL algorithm, one starts with an initial policy π_0.[5] For each given π_k,

---

[5] The choice of this initialization can also be guided by the theory. For instance, if the theory stipulates

$k = 0, 1, 2, \ldots$, policy evaluation is carried out to obtain its value function V^{π_k}. Then, a policy improvement theorem specifies the next policy π_{k+1}, and the iterations go on. We now describe these steps.

For the policy evaluation, we follow Doya (2000) for learning the value function V^π under any arbitrarily given admissible policy π. By Bellman consistency, we have

$$V^\pi(t, x) = \mathbb{E}\left[\int_t^{t'} \left(\int_U r\left(s, X_s^\pi, u\right) \pi_s(u)\, du - \lambda \int_U \pi_s(u) \ln \pi_s(u) du \right) ds \right.$$
$$\left. + V^\pi(t', X_{t'}^\pi) \middle| X_t^\pi = x \right], \tag{19.33}$$

for any $(t, x) \in [0, T) \times \mathbb{R}^d$ and $t' \in (t, T]$. This is actually analogous to the Bellman principle of optimality for the *optimal* value function. Rearranging this equation and dividing both sides by $t' - t$, we obtain

$$\mathbb{E}\left[\frac{V^\pi(t', X_{t'}^\pi) - V^\pi(t, X_t^\pi)}{t' - t} \right.$$
$$\left. + \frac{1}{t' - t} \int_t^{t'} \left(\int_U r\left(s, X_s^\pi, u\right) \pi_s(u)\, du - \lambda \int_U \pi_s(u) \ln \pi_s(u) du \right) ds \middle| X_t^\pi = x \right] = 0.$$

Letting $t' \to t$ in the left hand side motivates the definition of the *temporal difference* (TD) error

$$\delta_t := \dot{V}_t^\pi + \int_U r\left(t, X_t^\pi, u\right) \pi_t(u)\, du - \lambda \int_U \pi_t(u) \ln \pi_t(u) du, \tag{19.34}$$

where $\dot{V}_t^\pi := \frac{d}{dt} V^\pi(t, X_t^\pi)$ is the sample-wise total derivative of V^π along (t, X_t^π).

The objective of the policy evaluation procedure is to minimize the expected total squared TD error in order to find the value function V^π. In general, this can be done as follows. Denote by V^θ and π^ϕ respectively the parametrized value function and policy (upon using regressions or neural networks, or taking advantage of any known parametric forms of them), with θ, ϕ being the vectors of suitable dimensions to be learned. We then minimize

$$C(\theta, \phi) := \tfrac{1}{2}\mathbb{E}\left[\int_0^T |\delta_t|^2 dt \right]$$
$$= \tfrac{1}{2}\mathbb{E}\left[\int_0^T \left| \dot{V}_t^\theta + \int_U r(t, X_t^\phi, u) \pi_t^\phi(u)\, du - \lambda \int_U \pi_t^\phi(u) \ln \pi_t^\phi(u) du \right|^2 dt \right],$$

where $\pi^\phi = \{\pi_t^\phi(\cdot),\ 0 \leq t \leq T\}$ is generated from π^ϕ with respect to a given initial state $X_0 = x_0$ at time 0. To approximate $C(\theta, \phi)$, we first discretize $[0, T]$ into small intervals $[t_i, t_{i+1}]$, $i = 0, 1, \ldots, l$, with an equal length Δt, where $t_0 = 0$ and $t_{l+1} = T$. Then we collect a set of samples $\mathcal{D} = \{(t_i, x_i),\ i = 0, 1, \ldots, l+1\}$ in the following way. The initial sample is $(0, x_0)$ for $i = 0$. Now, at each $t_i, i = 0, 1, \ldots, l$, we sample $\pi_{t_i}^\phi$ to obtain $u_i \in U$ and then use the *constant* control $u_t \equiv u_i$ to control the (classical) system dynamics (19.1) during $[t_i, t_{i+1})$. We observe the state x_{i+1}

---

that Gaussian is optimal, then we can choose π_0 as Gaussian with some initial values of the mean and variance.

at the next time instant t_{i+1} along with the reward r_i collected over $[t_i, t_{i+1}]$. We then approximate $\dot{V}_t^\theta$ by

$$\dot{V}^\theta(t_i, x_i) := \frac{V^\theta(t_{i+1}, x_{i+1}) - V^\theta(t_i, x_i)}{\Delta t},$$

and approximate $C(\theta, \phi)$ by

$$C(\theta, \phi) = \frac{1}{2} \sum_{(t_i, x_i) \in \mathcal{D}} \left(\dot{V}^\theta(t_i, x_i) + r_i + \lambda \int_U \pi_{t_i}^\phi(u) \ln \pi_{t_i}^\phi(u) du \right)^2 \Delta t. \quad (19.35)$$

Finally, we seek a $(\theta^*, \phi^*)'$ that minimizes $C(\theta, \phi)$ using stochastic gradient descent algorithms; see, for example, Goodfellow et al. (2016, Chapter 8). This in turn leads to the value function V^{θ^*}, concluding the policy evaluation step.[6]

The policy improvement step is to update the next policy based on the current policy π along with the corresponding value function V^π, the latter having been found by the policy evaluation. Assuming $V^\pi \in C^{1,2}([0,T] \times \mathbb{R}^d) \cap C^0([0,T] \times \mathbb{R}^d)$, and that the policy $\tilde{\pi}$ defined by

$$\tilde{\pi}(u; t, x) = \frac{1}{Z(\lambda, t, x, V_x^\pi(t, x), V_{xx}^\pi(t, x))} \exp\left(\frac{1}{\lambda} H(t, x, u, V_x^\pi(t, x), V_{xx}^\pi(t, x)) \right)$$
$$(19.36)$$

generates admissible (open-loop) distributional controls for the exploratory dynamics (19.7). Then we can prove that $\tilde{\pi}$ is better than π in that

$$V^{\tilde{\pi}}(t, x) \geq V^\pi(t, x), \quad (t, x) \in [0, T] \times \mathbb{R}^d. \quad (19.37)$$

There is an obvious resemblance between the updating rule (19.36) and the optimal policy (19.17). Their proofs are also similar: $\tilde{\pi}$ achieves the supremum in (19.15) where v is replaced with V^π. For a proof in the mean–variance setting, see Wang and Zhou (2020).

19.6 Conclusion

In this chapter, we have put forth the notion of "curse of optimality" to capture the theoretical and empirical observations that traditional approaches to optimization often end with unfavorable solutions that are not globally optimal, or too extreme to be useful, or outright irrelevant in practice. We find that an entropy-regularized exploratory reformulation of the problem, originally motivated by balancing exploration and exploitation for reinforcement learning, may provide viable solutions to *all* these setbacks. This is because the randomization involved in such a formulation helps escape from local traps, broadens search space and reduces the desire to be "perfect" (extreme) by allowing more flexibility and

---

[6] In a recent paper, Jia and Zhou (2022) consider a general policy evaluation problem with continuous time and space. Applying a martingale approach, the authors find that the mean-square TD error method introduced here actually minimizes temporal variations rather than achieving accurate evaluation. They derive alternative policy evaluation methods based on martingality, some of which correspond to well-studied TD algorithms such as TD(0) and TD(λ) for disctete-time MDPs.

Xun Yu Zhou

accommodation. In the realm of continuous time and state/action spaces, this is still a largely uncharted research area where open problems abound.

References

Bridle, John S. 1990. Training stochastic model recognition algorithms as networks can lead to maximum mutual information estimation of parameters. Pages 211–217 of: *Advances in Neural Information Processing Systems*.

Cerny, V. 1985. Thermodynamical approach to the traveling salesman problem: an efficient simulation algorithm. *Journal of Optimization Theory and Applications*, **45**(1), 41–51.

Cesa-Bianchi, Nicolò, Gentile, Claudio, Lugosi, Gábor, and Neu, Gergely. 2017. Boltzmann exploration done right. Pages 6284–6293 of: *Advances in Neural Information Processing Systems*.

Chiang, Tzuu-Shuh, Hwang, Chii-Ruey, and Sheu, Shuenn Jyi. 1987. Diffusion for global optimization in $\mathbb{R}^n$. *SIAM Journal on Control and Optimization*, **25**(3), 737–753.

Doya, Kenji. 2000. Reinforcement Learning In Continuous Time and Space. *Neural Computation*, **12**(1), 219–245.

Gao, Xuefeng, Xu, Zuo Quan, and Zhou, Xun Yu. 2022. State-dependent temperature control for Langevin diffusions. *SIAM Journal on Control and Optimization*, **60**(3), 1250–1268.

Goodfellow, Ian, Bengio, Yoshua, and Courville, Aaron. 2016. *Deep Learning*. MIT Press.

Haarnoja, Tuomas, Tang, Haoran, Abbeel, Pieter, and Levine, Sergey. 2017. Reinforcement learning with deep energy-based policies. Pages 1352–1361 of: *Proceedings of the 34th International Conference on Machine Learning*.

Jia, Yanwei, and Zhou, Xun Yu. 2022. Policy evaluation and temporal-difference learning in continuous time and space: a martingale approach. *Journal of Machine Learning Research*, **23**, 1–55.

Kirkpatrick, S., Gelatt, J., and Vecchi, M. 1983. Optimization by simulated annealing. *Science*, **220**(4598), 671–680.

Nachum, Ofir, Norouzi, Mohammad, Xu, Kelvin, and Schuurmans, Dale. 2017. Bridging the gap between value and policy based reinforcement learning. Pages 2775–2785 of: *Advances in Neural Information Processing Systems*.

Neu, Gergely, Jonsson, Anders, and Gómez, Vicenç. 2017. A unified view of entropy-regularized markov decision processes. ArXiv:1705.07798.

Sutton, Richard S., and Barto, Andrew G. 2018. *Reinforcement Learning: An Introduction*. MIT Press.

Tallec, Corentin, Blier, Léonard, and Ollivier, Yann. 2019. Making Deep Q-learning methods robust to time discretization. ArXiv:1901.09732.

Tang, Wenpin, Zhang, Yuming Paul, and Zhou, Xun Yu. 2022. Exploratory HJB equations and their convergence. *SIAM Journal on Control and Optimization*, to appear.

Wang, Haoran, and Zhou, Xun Yu. 2020. Continuous-time mean–variance portfolio selection: A reinforcement learning framework. *Mathematical Finance*, **30**, 1273–1308.

Wang, Haoran, Zariphopoulou, Thaleia, and Zhou, Xun Yu. 2020. Reinforcement learning in continuous time and space: A stochastic control approach. *Journal of Machine Learning Research*, **21**, 1–34.

Wonham, Murray. 1968. On the separation theorem of stochastic control. *SIAM Journal on Control*, **6**(2), 312–326.

Yong, Jiongmin, and Zhou, Xun Yu. 1999. *Stochastic Controls: Hamiltonian Systems and HJB Equations*. Springer .

Ziebart, Brian D., Maas, Andrew L., Bagnell, J. Andrew, and Dey, Anind K. 2008. Maximum Entropy Inverse Reinforcement Learning. Pages 1433–1438 of: *AAAI*, vol. 8. Chicago, IL, USA.

Deep Learning for Mean Field Games and Mean Field Control with Applications to Finance

René Carmona[a], and Mathieu Laurière[a]

Abstract

Financial markets and more generally macro-economic models involve a large number of individuals interacting through variables such as prices resulting from the aggregate behavior of all the agents. Mean field games have been introduced to study Nash equilibria for such problems in the limit when the number of players is infinite. The theory has been extensively developed in the past decade, using both analytical and probabilistic tools, and a wide range of applications have been discovered, from economics to crowd motion. More recently the interaction with machine learning has attracted a growing interest. This aspect is particularly relevant to solve very large games with complex structures, in high dimension or with common sources of randomness. In this chapter, we review the literature on the interplay between mean field games and deep learning, with a focus on three families of methods. A special emphasis is given to financial applications.

20.1 Introduction

Most applications in financial engineering rely on numerical implementations which suffer from the curse of dimensionality. Recent developments in machine learning and the availability of powerful and readily available public domain packages have triggered a renewal of interest in financial numerics: deep learning technology has pushed the limits of computational finance, and one can now tackle problems which seemed out of reach a few years ago.

On a different part of the spectrum of scientific research, several sub-fields of economics experienced significant transitions: the emergence of continuous time stochastic models in macro-economics, and the development of general equilibrium theory in finance created a commonality between the two fields. This convergence provided a fertile ground for mean field game and mean field control theories, which naturally appeared as the tools of choice for theoreticians and applied mathematicians and economists.

[a] ORFE, Princeton University. Work supported by NSF grant DMS-1716673 and ARO grant W911NF-17-1-0578.

Fashions come and go, and there is no point in trying to catch a train which already left the station. However, machine learning and mean field theories are here to stay. Gaining a deep understanding of the inner workings of both paradigms is the first step in the recognition of their immense potential, and their limitations. Harnessing their synergy will lead to breakthroughs and spectacular progress. This chapter is a modest attempt to lay some ground for this symbiosis.

20.1.1 Literature review

Economics and finance are two of the fields of choice to which the early contributors to the theory of Mean Field Games (MFGs) and Mean Field Control (MFC) paid special attention. To wit, more than 30 pages of the introductory chapter *Learning by Examples* of Carmona and Delarue (2018a) were devoted to applications to these fields. Additionally, a more recent review of the literature which appeared on this very subject since 2018 can be found in Carmona (2020). Here we shall mention some of the most striking applications, and emphasize those which triggered progress in numerical methods and especially applications of Machine Learning (ML) techniques.

Historically, macro-economic models have been cast as general equilibrium problems and solved as such. However, many of these models, see for example Krusell and Smith (1998); Aiyagari (1994); Bewley (1986); Huggett (1993), carry all the elements of MFGs, and have since been revisited in light of our understanding of MFGs. See for example Nuño (2017); Nuño and Moll (2018) and Sargent et al. (2021), and Achdou et al. (2014, 2017) for a numerical point of view.

But as emphasized in Carmona (2020), the crusade of Brunnermeier and Sannikov arguing for the merging of macro-economics and finance models through the common use of continuous time helped the convergence of economists, financial engineers and applied mathematicians toward the use of a common language and a common set of models. MFG models play a crucial role in this evolution.

The analysis of systemic risk inherent in large banking networks was a natural ground for mean field models of interactions. The early work (Garnier et al., 2013) and the more recent model (Nadtochiy and Shkolnikov, 2019) lead to challenging Partial Differential Equation (PDE) problems, while Carmona et al. (2015) offers a simple model which can be solved explicitly, both in its finitely many player version and in its infinite player form, and for which the master equation can be derived and solved explicitly. Subsequent and more realistic models involving delays like Carmona et al. (2018) or interactions through default times like Elie et al. (2020) are unfortunately more difficult to solve, even numerically.

Some of the applications of mean field games lie at the intersection of macro-economics and financial engineering. As an example, we mention the MFG models for Bertrand and Cournot equilibria of oil production introduced in Guéant (2019); Guéant et al. (2011), and revisited later on in Chan and Sircar (2015, 2017) with an interesting emphasis on exhaustibility. A more mathematical PDE analysis of the model has been carried out in Graber and Bensoussan (2018).

Like the macro-economic general equilibrium models mentioned earlier, models of bank run such as Rochet and Vives (2004) are *screaming* for a MFG reformulation and this was first done in Nutz (2018) and Carmona et al. (2017), and later on in Bertucci (2018) by analytic methods, introducing a new class of MFGs of timing. However, like in macro-economics realistic models require the introduction of a common noise, making the theoretical solution much more involved, and numerical implementations even more difficult.

The high frequency markets offer, without the shadow of a doubt, one of the most computer intensive financial engineering application one can think of. While most existing papers on the subject revolve around the properties of price impact, see e.g. Carmona and Lacker (2015); Cardaliaguet and Lehalle (2018); Cartea and Jaimungal (2016), modeling the interaction between a large number of market makers and trading programs is certainly a exciting challenge that the methodology and the numerical tools developed for the analysis of MFGs should make it possible to meet.

The introduction of the MFG paradigm opened the door to the search for solutions of large population equilibrium models which could not be imagined to be solvable before. Still, the actual solutions of practical applications had to depend on the development of efficient numerical algorithms implementing the MFG principles. This was done early in the development of the theory: see, for example, Achdou and Capuzzo-Dolcetta (2010); Achdou and Laurière (2020); Laurière (2020) and the references therein. More recently, the release in the public domain of powerful software packages such as `TensorFlow` has made it possible to test at a very low cost the possible impact of machine learning (ML) tools in the solution of challenging problems for MFGs and MFC, whether these problems were formulated in the probabilistic approach (Fouque and Zhang, 2020; Carmona and Laurière, 2019; Germain et al., 2022a) or the analytical approach (Al-Aradi et al., 2018; Carmona and Laurière, 2021; Ruthotto et al., 2020; Cao et al., 2020; Lin et al., 2020; Laurière, 2020). These new methods combine neural network approximations and stochastic optimization techniques to solve McKean–Vlasov control problems, mean field forward-backward stochastic differential equations (FBSDE) or mean field PDE systems.

While the present chapter concentrates on ML applications of MFGs and MFC in finance, the reader should not be surprised if they recognize a strong commonality of ideas and threads with the chapter (Angiuli et al., 2021) dealing with reinforcement learning for MFGs with a special focus on a two-timescale procedure, and the chapter (Germain et al., 2022b) offering a review of neural-network-based algorithms for stochastic control and PDE applications in finance.

The rest of this chapter is organized as follows. In the rest of this section, we define the MFG and MFC problems. In §20.2, we present a direct ML method for MKV control. We then turn our attention to method related to the optimality conditions of MFGs and MFC. A neural-network based shooting method for generic MKV FBSDE systems is discussed in §20.3. A deep learning method for mean field PDE systems is presented in §20.4. We conclude in §20.5.

20.1.2 Definition of the problems

The parameters of our models are a time horizon $T > 0$, integers d and k for the dimensions of the state space $Q \subseteq \mathbb{R}^d$ and the action space $\mathbb{R}^k$. Typically, the state space Q will be the whole space $\mathbb{R}^d$. We shall use the notation $Q_T = [0,T] \times Q$, and $\langle \cdot, \cdot \rangle$ for the inner product of two vectors of compatible dimensions. We denote by $\mathcal{P}_2(Q)$ the space of probability measures on Q which integrate the square of the norm when Q is unbounded.

Now let $f : Q \times \mathcal{P}_2(Q) \times \mathbb{R}^k \to \mathbb{R}, (x,m,\alpha) \mapsto f(x,m,\alpha)$ and $g : Q \times \mathcal{P}_2(Q) \to \mathbb{R}, (x,m) \mapsto g(x,m)$ be functions giving respectively the instantaneous running cost and the terminal cost, let $b : Q \times \mathcal{P}_2(Q) \times \mathbb{R}^k \to \mathbb{R}^d, (x,m,\alpha) \mapsto b(x,m,\alpha)$ be its drift function, and let $\sigma > 0$ be the volatility of the state's evolution (for simplicity we focus on the case of a non-degenerate diffusion although some of the methods presented below can also be applied when the diffusion is degenerate). These functions could be allowed to also depend on time at the expense of heavier notation. Here, x, m and α play respectively the role of the state of the agent, the mean field term (i.e. the population's distribution), and the control used by the agent. In general, the mean field term is a probability measure. However, in some cases, we will assume that this probability measure has a density which is in $L^2(Q)$.

Definition 20.1 (MFG equilibrium)   When considering the mean field game problem for a given initial distribution $m_0 \in \mathcal{P}_2(Q)$, we call a Nash equilibrium a flow of probability measures $\hat{m} = (\hat{m}(t,\cdot))_{0 \le t \le T}$ in $\mathcal{P}_2(Q)$ and a feedback control $\hat{\alpha} : Q_T \to \mathbb{R}^k$ satisfying the following two conditions:

(1) $\hat{\alpha}$ minimizes $J_{\hat{m}}^{\mathrm{MFG}}$ where, for $m = (m(t,\cdot))_{0 \le t \le T}$,

$$J_m^{\mathrm{MFG}} : \alpha \mapsto \mathbb{E}\left[\int_0^T f(X_t^{m,\alpha}, m(t,\cdot), \alpha(t, X_t^{m,\alpha}))dt + g(X_T^{m,\alpha}, m(T,\cdot)) \right]$$
(20.1)

under the constraint that the process $X^{m,\alpha} = (X_t^{m,\alpha})_{t \ge 0}$ solves the stochastic differential equation (SDE)

$$dX_t^{m,\alpha} = b(X_t^{m,\alpha}, m(t,\cdot), \alpha(t, X_t^{m,\alpha}))dt + \sigma dW_t, \qquad t \ge 0, \quad (20.2)$$

where W is a standard d-dimensional Brownian motion, and $X_0^{m,\alpha}$ has distribution m_0;
(2) For all $t \in [0,T]$, $\hat{m}(t,\cdot)$ is the probability distribution of $X_t^{\hat{m},\hat{\alpha}}$.

In the definition, (20.1), of the cost function, the subscript m is used to emphasize the dependence on the mean field flow, which is fixed when an infinitesimal agent performs their optimization. The second condition ensures that if all the players use the control $\hat{\alpha}$ identified in the first bullet point, the law of their individual states is indeed $\hat{m}$.

Using the same drift, running and terminal cost functions and volatility, we can also consider the corresponding McKean–Vlasov (MKV for short) control, or mean field control (MFC for short) problem. This optimization problem corresponds to a social optimum and is phrased as an optimal control problem. It can be interpreted as a situation in which all the agents cooperate to minimize the average cost.

Definition 20.2 (MFC optimum) A feedback control $\alpha^* : Q_T \to \mathbb{R}^k$ is an optimal control for the MKV control (or MFC) problem for a given initial distribution $m_0 \in \mathcal{P}_2(Q)$ if it minimizes

$$J^{\mathrm{MFC}} : \alpha \mapsto \mathbb{E}\left[\int_0^T f(X_t^\alpha, m^\alpha(t, \cdot), \alpha(t, X_t^\alpha))dt + g(X_T^\alpha, m^\alpha(T, \cdot))\right] \quad (20.3)$$

where $m^\alpha(t, \cdot)$ is the probability distribution of the law of X_t^α, under the constraint that the process $X^\alpha = (X_t^\alpha)_{t \geq 0}$ solves the stochastic differential equation of McKean–VLasov type:

$$dX_t^\alpha = b(X_t^\alpha, m^\alpha(t, \cdot), \alpha(t, X_t^\alpha))dt + \sigma dW_t, \qquad t \geq 0, \quad (20.4)$$

X_0^α having distribution m_0.

The applications of MFC are not limited to social optima in very large games. These problems also arise for example in risk management (Andersson and Djehiche, 2011) or in optimal control with a cost involving a conditional expectation (Achdou et al., 2020; Nutz and Zhang, 2020). If $m^* = m^{\alpha^*}$ is the flow of state distribution for an optimal control α^*, then:

$$J^{\mathrm{MFC}}(\alpha^*) = J_{m^*}^{\mathrm{MFG}}(\alpha^*) \leq J_{\hat{m}}^{\mathrm{MFG}}(\hat{\alpha}).$$

In general the inequality is strict, which leads to the notion of price of anarchy (Carmona et al., 2019).

To simplify the presentation, we have introduced MFG and MFC in a basic formulation where the interactions occur through the distribution of states. However, in many applications, the interactions occur through the distribution of controls or through the joint distribution of states and controls. This aspect will be illustrated in some of the examples discussed below.

20.2 Direct method for MKV control

In this section, we present a direct approach to the numerical solution of McKean–Vlasov control problems. It hinges on an approach developed for standard control problems, in which the control feedback function is restricted to a parametric family of functions, especially a class of neural networks whose parameters are learned by stochastic optimization (Gobet and Munos, 2005; Han and E, 2016). This method was extended to the mean field setting in Fouque and Zhang (2020); Carmona and Laurière (2019). We illustrate this approach with the solution of a price impact model.

20.2.1 Description of the method

Since MFC is an optimization problem, it is natural to leverage stochastic optimization tools from machine learning directly applied to the definition (20.3)–(20.4). We introduce three approximations leading to a formulation more amenable to numerical treatment.

Approximation steps.

First, we restrict the set of (feedback) controls to be the set of neural networks with a given architecture. We introduce new notation to define this class of controls. We denote by:

$$\mathbf{L}_{d_1,d_2}^{\psi} = \left\{ \phi : \mathbb{R}^{d_1} \to \mathbb{R}^{d_2} \mid \text{there exists } (\beta, w) \in \mathbb{R}^{d_2} \times \mathbb{R}^{d_2 \times d_1}, \right.$$

$$\left. \text{for all } i \in \{1, \ldots, d_2\}, \ \phi(x)_i = \psi \left(\beta_i + \sum_{j=1}^{d_1} w_{i,j} x_j \right) \right\}$$

the set of layer functions with input dimension d_1, output dimension d_2, and activation function $\psi : \mathbb{R} \to \mathbb{R}$. Typical choices for ψ are the ReLU function $\psi(x) = x^+$ or the sigmoid function $\psi(x) = 1/(1 + e^{-x})$. Building on this notation and denoting by $\circ$ the composition of functions, we define:

$$\mathbf{N}_{d_0,\ldots,d_{\ell+1}}^{\psi} = \left\{ \varphi : \mathbb{R}\mathbb{R}^{d_0} \to \mathbb{R}^{d_{\ell+1}} \mid \text{there exists } (\phi_i)_{i=0,\ldots,\ell-1} \in \times_{i=0}^{i=\ell-1} \mathbf{L}_{d_i,d_{i+1}}^{\psi}, \right.$$

$$\left. \text{there exists } \phi_\ell \in \mathbf{L}_{d_\ell,d_{\ell+1}}^{\mathrm{id}}, \varphi = \phi_\ell \circ \phi_{\ell-1} \circ \cdots \circ \phi_0 \right\}$$

the set of regression neural networks with ℓ hidden layers and one output layer, the activation function of the output layer being the identity $\psi = \mathrm{id}$. The number ℓ of hidden layers, the numbers $d_0, d_1, \ldots, d_{\ell+1}$ of units per layer, and the activation functions, is what is called the architecture of the network. Once it is fixed, the actual network function $\varphi \in \mathbf{N}_{d_0,\ldots,d_{\ell+1}}^{\psi}$ is determined by the remaining real-valued parameters:

$$\theta = (\beta^{(0)}, w^{(0)}, \beta^{(1)}, w^{(1)}, \ldots\ldots, \beta^{(\ell-1)}, w^{(\ell-1)}, \beta^{(\ell)}, w^{(\ell)})$$

defining the functions $\phi_0, \phi_1, \ldots, \phi_{\ell-1}$ and ϕ_ℓ respectively. The set of such parameters is denoted by Θ. For each $\theta \in \Theta$, the function φ computed by the network will be denoted by $\alpha_\theta \in \mathbf{N}_{d_0,\ldots,d_{\ell+1}}^{\psi}$. As it should be clear from the discussion of the previous section, here, we are interested in the case where $d_0 = d + 1$ (since the inputs are time and state) and $d_{\ell+1} = k$ (i.e., the control dimension).

Our first approximation is to minimize J^{MFC} defined by (20.3)–(20.4) over $\alpha \in \mathbf{N}_{d+1,d_1,\ldots,d_\ell,k}^{\psi}$, or equivalently, to minimize over $\theta \in \Theta$ the function:

$$\mathbf{J} : \theta \mapsto \mathbb{E} \left[\int_0^T f(X_t^{\alpha_\theta}, m^{\alpha_\theta}(t, \cdot), \alpha_\theta(t, X_t^{\alpha_\theta})) dt + g(X_T^{\alpha_\theta}, m^{\alpha_\theta}(T, \cdot)) \right]$$

where $m^{\alpha_\theta}(\cdot, t)$ is the law of $X_t^{\alpha_\theta}$, under the constraint that the process X^{α_θ} solves the SDE (20.4) with feedback control α_θ.

Next, we approximate the probability distribution of the state. A (computationally) simple option is to replace it by the empirical distribution of a system of N interacting particles. Given a feedback control α, we denote by $(\underline{X}_t^\alpha)_t = (X_t^{1,\alpha}, \ldots, X_t^{N,\alpha})_t$ the solution of the system:

$$dX_t^{i,\alpha} = b(X_t^{i,\alpha}, m_t^{N,\alpha}, \alpha(t, X_t^{i,\alpha}))dt + \sigma dW_t^i, \qquad t \geq 0, \quad i = 1, \ldots, N \quad (20.5)$$

where

$$m_t^{N,\alpha} = \frac{1}{N} \sum_{j=1}^{N} \delta_{X_t^{j,\alpha}},$$

is the empirical measure of the N particles, $(W^i)_{i=1,\ldots,N}$ is a family of N independent d-dimensional Brownian motions, and the initial positions $(X_0^{i,\alpha})_{i=1,\ldots,N}$ are i.i.d. with distributions m_0. The N stochastic differential equations in (20.5) are coupled via their drifts through the empirical measure $m_t^{N,\alpha}$. The controls are distributed in the sense that the control used in the equation for $X^{i,\alpha}$ is a function of t and $X_t^{i,\alpha}$ itself, and not of the states of the other particles. Despite their dependence due to the coupling, it is expected that the empirical measures converge when $N \to \infty$ to the solution of the SDE (20.4). Not only does this convergence holds, but in this limit, the individual particle processes $(X_t^{i,\alpha})_{0 \leq t \leq T}$ become independent in this limit. This fundamental result is known under the name of *propagation of chaos*: see Carmona and Delarue (2018b) for details and the role this result plays in the theory of MFGs and MFC. As per this second approximation, the new problem is to minimize over $\theta \in \Theta$ the function

$$\mathbf{J}^N : \theta \mapsto \frac{1}{N} \sum_{i=1}^{N} \mathbb{E}\left[\int_0^T f(X_t^{i,\alpha_\theta}, m_t^{N,\alpha_\theta}, \alpha_\theta(t, X_t^{i,\alpha_\theta}))dt + g(X_T^{i,\alpha_\theta}, m_T^{N,\alpha_\theta})\right],$$

under the dynamics (20.5) with control α_θ.

Our third approximation is to discretize time. Let N_T be a positive integer, let $\Delta t = T/N_T$ and $t_n = n\Delta t$, $n = 0, \ldots, N_T$. We now minimize over $\theta \in \Theta$ the function

$$\mathbf{J}^{N,\Delta t} : \theta \mapsto \frac{1}{N} \sum_{i=1}^{N} \mathbb{E}\left[\sum_{n=0}^{N_T-1} f(\check{X}_{t_n}^{i,\alpha_\theta}, \check{m}_{t_n}^{N,\alpha_\theta}, \alpha_\theta(t_n, \check{X}_{t_n}^{i,\alpha_\theta}))\Delta t + g(\check{X}_T^{i,\alpha_\theta}, \check{m}_T^{N,\alpha_\theta})\right],$$

$$(20.6)$$

under the dynamic constraint:

$$\check{X}_{t_{n+1}}^{i,\alpha_\theta} = \check{X}_{t_n}^{i,\alpha_\theta} + b(\check{X}_{t_n}^{i,\alpha_\theta}, \check{m}_{t_n}^{N,\alpha_\theta}, \alpha_\theta(t, \check{X}_{t_n}^{i,\alpha_\theta}))\Delta t + \sigma \Delta \check{W}_n^i, \qquad n = 0, \ldots, N_T - 1,$$

$$(20.7)$$

and the initial positions $(\check{X}_0^{i,\alpha_\theta})_{i=1,\ldots,N}$ are i.i.d. with distribution given by the density m_0, where

$$\check{m}_{t_n}^{N,\alpha_\theta} = \frac{1}{N} \sum_{j=1}^{N} \delta_{\check{X}_{t_n}^{j,\alpha_\theta}},$$

and the $(\Delta \check{W}_n^i)_{i=1,\ldots,N; \ n=0,\ldots,N_T-1}$ are i.i.d. random variables with Gaussian distribution $\mathcal{N}(0, \Delta t)$.

Under suitable assumptions on the model and the neural network architecture, the difference between $\inf_\theta \mathbf{J}^{N,\Delta t}(\theta)$ and $\inf_\alpha J^{\mathrm{MFC}(\alpha)}$ goes to 0 as N_t, N and the number of parameters in the neural network go to infinity. See Carmona and Laurière (2019) for details.

Optimization procedure.

Two obvious difficulties have to be overcome. First the fact that the cost function (20.6) is in general non-convex. Second, the parameter θ is typically high dimensional. But the cost (20.6) being written as an expectation, it is reasonable to rely on a form of stochastic gradient descent (SGD) algorithm. The randomness in our problem comes from the initial positions $\check{\underline{X}}_0 = (\check{X}_0^{i,\alpha_\theta})_i$ and the random shock innovations $(\Delta\check{\underline{W}}_n)_{n=0,...,N_T} = (\Delta\check{W}_n^i)_{i,n}$. Hence $S = (\check{\underline{X}}_0, (\Delta\check{\underline{W}}_n)_n)$ is going to play the role of a random sample in SGD. Given a realization of S and a choice of parameter θ, we can construct the trajectory $(\check{X}_{t_n}^{i,\alpha_\theta,S})_{i=1,...,N,n=0,...,N_T}$ by following (20.7), and compute the induced cost:

$$\mathbf{J}_S^{N,\Delta t}(\theta) =$$
$$\frac{1}{N}\sum_{i=1}^N\left[\sum_{n=0}^{N_T-1} f(\check{X}_{t_n}^{i,\alpha_\theta,S},\check{m}_{t_n}^{N,\alpha_\theta,S},\alpha_\theta(t_n,\check{X}_{t_n}^{i,\alpha_\theta,S}))\Delta t + g(\check{X}_T^{i,\alpha_\theta,S},\check{m}_T^{N,\alpha_\theta,S})\right]. \quad (20.8)$$

We give our SGD procedure in Algorithm 20.1. The most involved step is the computation of the gradient $\nabla_\theta\mathbf{J}_S^{N,\Delta t}(\theta^{(k)})$ with respect to θ. However, modern programming libraries (such as `TensorFlow` or `PyTorch`) perform this computation automatically using backpropagation, simplifying dramatically the code. The present method is thus extremely straightforward to implement: in contrast to the methods based on optimality conditions, there is no need to derive by hand any PDE, any FBSDE, or compute gradients. We work directly with the definition of the MFC.

Besides this aspect, the main reasons behind the success of this method are the expressive power of neural networks and the fact that there is a priori no limitation on the number K of iterations because the samples S come from Monte-Carlo simulations and not from a training set of data. In the implementation of this method, using mini-batches and ADAM (Kingma and Ba, 2014) can help improving convergence.

20.2.2 Numerical illustration: a price impact model

We consider a financial application originally solved as a mean field game in the weak formulation by Carmona and Lacker (2015), and revisited in the book of Carmona and Delarue (2018a, Sections 1.3.2 and 4.7.1) in the strong formulation. This is a problem of optimal execution in the presence of price impact resulting from a large group of traders affecting the price of a single asset through their aggregate trades: a large number of buy orders will push the price up while a large number of sell orders will deflate the price. This aggregate impact on the price is the source of the *mean field* in the model. As a consequence, this model

---

Algorithm 20.1 SGD for MFC

Data: An initial parameter $\theta^{(0)} \in \Theta$; a number of steps K; a sequence $(\beta^{(k)})_{k=0,\ldots,K-1}$
    of learning rates.
Result: A parameter θ such that α_θ approximately minimizes J^{MFC}
begin
    for k = 0, 1, 2, ..., K − 1 **do**
        Pick $S = (\check{\underline{X}}_0, (\Delta \check{\underline{W}}_n)_n)$
        Compute the gradient $\nabla_\theta J_S^{N,\Delta t}(\theta^{(k)})$, see (20.8)
        Set $\theta^{(k+1)} = \theta^{(k)} - \beta^{(k)} \nabla_\theta J_S^{N,\Delta t}(\theta^{(k)})$
    return $\theta^{(K)}$

---

is an instance of mean field problems with interactions through the distribution of controls, introduced by Gomes et al. (2014); Gomes and Voskanyan (2016) who coined the term *extended MFG*.

Here we shall consider the MFC counterpart. The method described above can readily be adapted to solve MFC with interactions through the control's distribution by computing the empirical distribution of controls for an interacting system of N particles. For the sake of completeness, we recall the model and the derivation of a succinct MFC formulation. See the aforementioned references for the N-agent problem and more details (in the MFG setting). A typical and infinitesimal trader's inventory at time t is denoted by X_t. We assume that it evolves according to the SDE:

$$dX_t = \alpha_t \, dt + \sigma \, dW_t,$$

where α_t represents the rate of trading, and W is a standard Brownian motion.

Denoting by $v_t^\alpha = \mathcal{L}(\alpha_t)$ the law of the control at time $t \in [0,T]$, we assume that the evolution of the price is given by:

$$dS_t = \gamma \left(\int_{\mathbb{R}} a \, dv_t^\alpha(a) \right) dt + \sigma_0 \, dW_t^0,$$

where γ and σ_0 are positive constants, and the Brownian motion W^0 is independent from W. The *price impact* effect is taken into account by the fact that the price $(S_t)_{0 \le t \le T}$ of the asset is influenced by the average trading rate of all the traders.

If we denote by K_t the amount of cash held by the trader at time t, the dynamics of the process $(K_t)_{0 \le t \le T}$ are given by:

$$dK_t = -\left(\alpha_t S_t + c_\alpha(\alpha_t) \right) dt,$$

where the function $a \mapsto c_\alpha(a)$ is a non-negative convex function satisfying $c_\alpha(0) = 0$. It corresponds to the cost for trading at rate a. At time t, the trader's total wealth, denoted by V_t, is the sum of the cash holding and the value of the inventory marked at the price S_t, i.e., $V_t = K_t + X_t S_t$. Using the self-financing condition of Black–Scholes theory, the evolution of the trader's wealth is:

$$dV_t = dK_t + X_t\,dS_t + S_t\,dX_t$$
$$= \left[-c_\alpha(\alpha_t) + \gamma X_t \int_{\mathbb{R}} a\,dv_t^\alpha(a) \right] dt + \sigma S_t\,dW_t + \sigma_0 X_t\,dW_t^0. \tag{20.9}$$

We assume that the trader is subject to a running cost for holding an inventory, modeled by a function c_X of their inventory, and to a terminal liquidation constraint at maturity T represented by a scalar function g. Thus, the trader's cost function, to be minimized, is defined by:

$$J(\alpha) = \mathbb{E}\left[\int_0^T c_X(X_t)\,dt + g(X_T) - V_T \right].$$

Taking into account (20.9), this cost can be rewritten in terms of X only as:

$$J(\alpha) = \mathbb{E}\left[\int_0^T \left(c_\alpha(\alpha_t) + c_X(X_t) - \gamma X_t \int_{\mathbb{R}} a\,dv_t^\alpha(a) \right) dt + g(X_T) \right].$$

Following the Almgren–Chriss linear price impact model, we assume that the functions c_X, c_α and g are quadratic. Thus, the cost is of the form:

$$J(\alpha) = \mathbb{E}\left[\int_0^T \left(\frac{c_\alpha}{2}\alpha_t^2 + \frac{c_X}{2}X_t^2 - \gamma X_t \int_{\mathbb{R}} a\,dv_t^\alpha(a) \right) dt + \frac{c_g}{2}X_T^2 \right].$$

Let us stress that this problem is an extended MFC: the population distribution is not frozen during the optimization over α, and the interactions occur through the distribution of controls v^α.

This model is solved by reinforcement learning techniques (for both MFC and the corresponding MFG) in Angiuli et al. (2021). Here, we present results obtained with the deep learning method described above, see Algorithm 20.1. The results displayed in Figure 20.3 show that the control is linear, as expected from the theory, and the distribution moves towards 0 while becoming more concentrated. In other words, at the beginning the traders have a relatively large inventory with a large variance across the population, and they liquidate to end up with smaller inventories and less variance. One can see that towards the end of the time interval, the control learnt is not exactly linear, probably because a regions has been less explored than the rest leading to a less accurate training. For these results, we used the parameters: $T = 1$, $c_X = 2$, $c_\alpha = 1$, $c_g = 0.3$, $\sigma = 0.5$ and the value of γ indicated in the captions. Moreover, in the algorithm we took $N = 2000$ particles and $N_T = 50$ time steps. We see in Figure 20.1 that when $\gamma = 0.2$, the optimal control is to constantly liquidate. However, as shown in Figure 20.2, when $\gamma = 1$, the traders start by liquidating by towards the end of the time interval, they buy. This can be explained by the fact that with a higher γ, the price impact effect is stronger and the traders can use phenomenon to increase their wealth by collectively buying and hence increasing the price. In each case, the neural network manages to learn a control which approximately matches the semi-explicit one obtained by reducing the problem to an system of ordinary differential equations (ODE) as explained in Carmona and Delarue (2018a, Sections 1.3.2 and 4.7.1).

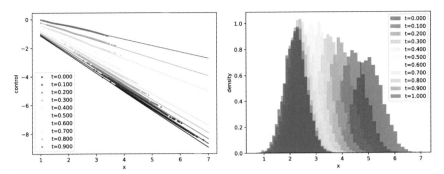

Figure 20.1 Price impact MFC example solved by Algorithm 20.1. Left: Control learnt (dots) and exact solution (lines). Right: associated empirical state distribution. Here, $\gamma = 0.2$.

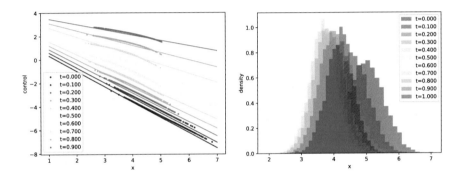

Figure 20.2 Price impact MFC example solved by Algorithm 20.1. Left: Control learnt (dots) and exact solution (lines). Right: associated empirical state distribution. Here, $\gamma = 1$.

20.3 Deep BSDE method for MKV FBSDEs

In this section, we present an extension to the mean field regime of the DeepBSDE method introduced in E et al. (2017) and analyzed in Han and Long (2020). The latter uses neural networks to learn the solution of BSDEs. It relies on a (stochastic) shooting method, where one tries to find a suitable starting point in order to match a given terminal condition. These ideas have been extended to the mean field setting to solve forward-backward systems of McKean–Vlasov SDEs in Fouque and Zhang (2020); Carmona and Laurière (2019); Germain et al. (2022a). After presenting the main ideas, we illustrate the performance of the method on a systemic risk MFG model introduced in Carmona et al. (2015) for which explicit solutions exist.

20.3.1 Description of the method

As explained in Carmona and Delarue (2018a), an MFG equilibrium can be reduced to the solution of a forward-backward system of SDEs (FBSDE for short) which reads:

$$\begin{cases} dX_t = b(t, X_t, \mathcal{L}(X_t), \hat{\alpha}(t, X_t, \mathcal{L}(X_t), Y_t)) \, dt \, + \sigma \, dW_t \\ dY_t = - \, \partial_x H(t, X_t, \mathcal{L}(X_t), Y_t, Z_t, \hat{\alpha}(t, X_t, \mathcal{L}(X_t), Y_t)) \, dt + Z_t dW_t, \end{cases} \tag{20.10}$$

with initial condition X_0 having distribution m_0 and terminal condition $Y_T = \partial_x g(X_T, \mathcal{L}(X_T))$. H is the Hamiltonian:

$$H(t, x, \mu, y, z, \alpha) = b(t, x, \mu, \alpha) \cdot y + \sigma \cdot z + f(t, x, \mu, \alpha),$$

and $\hat{\alpha}$ denotes its minimizer. Solutions of MFC can also be characterized through a similar FBSDE system, but in the latter case, the backward equation involves a partial derivative of the Hamiltonian with respect to the measure argument. See Carmona and Delarue (2018a, Section 6.4.2) for more details. Moreover, such FBSDE systems can also be obtained using dynamic programming, in which case Y represents the value function instead of its gradient.

All these systems are particular cases of the following general system of forward-backward SDEs of McKean–Vlasov type (MKV FBSDE for short), the system (20.10) being derived from the application of the Pontryagin stochastic maximum principle applied to our original MKV control problem:

$$\begin{cases} dX_t = B\left(t, X_t, \mathcal{L}(X_t), Y_t\right) \, dt + \sigma \, dW_t, \\ dY_t = - \, F\left(t, X_t, \mathcal{L}(X_t), Y_t, \sigma^\dagger Z_t\right) \, dt + Z_t \, dW_t, \end{cases} \tag{20.11}$$

with initial condition X_0 having distribution m_0 and terminal condition $Y_T = G(X_T, \mathcal{L}(X_T))$.

The solution strategy is to replace the backward equation forced on us due to the optimization, by a forward equation and to treat its initial condition, which is what we are looking for, as a control for a new optimization problem. This strategy has been successfully applied to problems in economic contract theory where it is known as Sannikov's trick. See for example Kohlmann and Zhou (2000); Cvitanić and Zhang (2013); Cvitanić et al. (2018). Put it plainly, the strategy is a form of shooting method: the controller chooses the initial point and the volatility of the Y process, and penalizes them proportionally to how far they are from matching the terminal condition. Specifically, we minimize over $y_0 : \mathbb{R}^d \to \mathbb{R}^d$ and $z : \mathbb{R}_+ \times \mathbb{R}^d \to \mathbb{R}^{d \times d}$ the cost functional

$$J_{\text{FBSDE}}(y_0, z) = \mathbb{E}\left[\left| Y_T^{y_0, z} - G(X_T^{y_0, z}, \mathcal{L}(X_T^{y_0, z})) \right|^2 \right] \tag{20.12}$$

where $(X^{y_0, z}, Y^{y_0, z})$ solves

$$\begin{cases} dX_t^{y_0, z} = B\left(t, X_t^{y_0, z}, \mathcal{L}(X_t^{y_0, z}), Y_t^{y_0, z}\right) \, dt + \sigma \, dW_t, \\ dY_t^{y_0, z} = - \, F\left(t, X_t^{y_0, z}, \mathcal{L}(X_t^{y_0, z}), Y_t^{y_0, z}, \sigma^\dagger z(t, X_t^{y_0, z})\right) \, dt + z(t, X_t^{y_0, z}) \, dW_t, \end{cases} \tag{20.13}$$

with *initial* condition $X_0^{y_0, z} = \xi_0 \in L^2(\Omega, \mathcal{F}_0, \mathbb{P}; \mathbb{R}^d)$ and $Y_0^{y_0, z} = y_0(X_0)$. In some sense, the above problem is an optimal control problem of MKV dynamics if we view $(X_t^{y_0, z}, Y_t^{y_0, z})$ as state and (y_0, z) as control. It is rather special because the control is initial value and the volatility of the second component of the state, and looked for among feedback functions of the first component of the state. Under suitable conditions, the optimally controlled process $(X_t, Y_t)_t$ solves the FBSDE system (20.11) and vice versa.

In the same spirit as the method presented in §20.2, we consider a finite-size population with N particles and replace the controls y_0 and z by neural networks, say $y_{0,\theta}$ and z_ω with parameters θ and ω respectively. We then discretize time with steps of size Δt. Let us denote by $\mathbf{J}_{\mathrm{FBSDE,S}}^{N,\Delta t}(\theta, \omega)$ the analog of (20.8) for the cost function (20.12) stemming from the MKV FBSDE. Finally, we use SGD to perform the optimization. The method is summarized in Algorithm 20.2. It is similar to Algorithm 20.1, so we only stress the main differences. The two neural networks could be taken with different architectures and their parameters optimized with different learning rates. In §20.3.2 below, we illustrate the performance of this method on MKV FBSDEs coming from an MFG model of systemic risk.

Algorithm 20.2 SGD for MKV FBSDE

Data: An initial parameter $\theta^{(0)}, \omega^{(0)} \in \Theta$; a number of steps K; a sequence $(\beta^{(k)})_{k=0,\dots,K-1}$ of learning rates.
Result: Parameters (θ, ω) such that $(y_{0,\theta}, z_\omega)$ approximately minimizes J_{FBSDE}
begin
    for k = 0, 1, 2, ..., K − 1 **do**
        Pick $S = (\check{\underline{X}}_0, (\Delta \check{\underline{W}}_n)_n)$
        Simulate N trajectories for (20.13) with $y_0 = y_{0,\theta^{(k)}}$ and $z = z_{\theta^{(k)}}$
        Compute the gradient $\nabla_{(\theta,\omega)} \mathbf{J}_{\mathrm{FBSDE,S}}^{N,\Delta t}(\theta^{(k)}, \omega^{(k)})$
        Set $(\theta^{(k+1)}, \omega^{(k+1)}) = (\theta^{(k)}, \omega^{(k)}) - \beta^{(k)} \nabla_{(\theta,\omega)} \mathbf{J}_{\mathrm{FBSDE,S}}^{N,\Delta t}(\theta^{(k)})$
    return $(\theta^{(K)}, \omega^{(K)})$

20.3.2 Numerical illustration: a toy model of systemic risk

The following MFG model was introduced in Carmona et al. (2015) as an example which can be solved explicitly with a common noise, for finitely many players as well as in the mean field limit, in the open loop case as well as the closed loop set-up, and for which the master equation can be derived and solved explicitly. Individual players are financial institutions, and their states are the logarithms of their cash reserves. We assume that their evolutions are given by one dimensional diffusion processes involving a common noise $\mathbf{W}^0$ and an independent idiosyncratic noise $\mathbf{W}$. The costs take into account the rates of lending and borrowing and penalize departure from the aggregate state of the other institutions.

Because of the presence of the common noise the best response needs to be computed when the conditional flow of distributions is fixed. Due to the linear–quadratic nature of the model, the optimization is performed given the flow of the conditional mean log-monetary reserves $\bar{\mathbf{m}} = (\bar{m}_t)_{t \in [0,T]}$ which is adapted to the filtration generated by $\mathbf{W}^0$. Assuming that the log-monetary reserve of a bank satisfies the SDE:

$$dX_t = [a(\bar{m}_t - X_t) + \alpha_t] \, dt + \sigma \left(\rho \, dW_t^0 + \sqrt{1 - \rho^2} \, dW_t \right).$$

where $a > 0$ and $\rho \in [0, 1]$ builds dependence between the random shocks. The first term comes from the fact that the bank is assumed to borrow or lend to each other bank at a rate proportional to the difference between their log-monetary reserves. The term in α_t represents the rate at which the bank borrows or lends to a central bank. Each institution tries to minimize its expected cost given by:

$$J^{\mathrm{MFG}}(\bar{m}, \alpha)$$
$$= \mathbb{E}\left[\int_0^T \left(\frac{1}{2}\alpha_t^2 - q\alpha_t(\bar{m}_t - X_t) + \frac{\epsilon}{2}(\bar{m}_t - X_t)^2 \right) dt + \frac{c}{2}(\bar{m}_T - X_T)^2 \right],$$

where q, ϵ, c and σ are positive constants satisfying $q \leq \epsilon^2$ so that the running cost is jointly convex in the state and the control variables. Here, q can be interpreted as a parameter chosen by a regulator to incentivize borrowing or lending: if the log-monetary reserve X_t of the bank is smaller than the average $\bar{m}_t$, then the bank has an incentive to choose a positive control α_t, meaning that it borrows; similarly, if $X_t > \bar{m}_t$, then the banks has an incentive to choose $\alpha_t > 0$. For more details on the interpretation of this model in terms of systemic risk, the reader is referred to Carmona et al. (2015). The model is of linear-quadratic type and hence has an explicit solution through a Riccati equation, which we use as a benchmark for comparison with our numerical results.

Note that since this example is a MFG, we can not use Algorithm 20.1 to compute directly the equilibrium. Instead, we use Algorithm 20.2 to solve the appropriate FBSDE system (we omit this system here for brevity; see Carmona et al. (2015)). In order to deal with the additional randomness induced by the common noise, we add $\bar{m}_t$ as an argument of the neural networks playing the roles $y_0(\cdot)$ and $z(\cdot)$ introduced above.

Figure 20.3 displays sample trajectories of X^i and Y^i for three different values of i. We can see that the approximation is better for X^i than for Y^i, particularly towards the end of the time interval. This is probably due to the fact that the latter is supposed to solved a BSDE but we replaced it by a forward equation so errors accumulate along time. However the error decreases with the number of time steps, particles and units in the neural network. See Carmona and Laurière (2019) for more details. For the numerical tests presented here, we used $\sigma = 0.5$, $\rho = 0.5$, $q = 0.5$, $\epsilon = q^2 + 0.5 = 0.75$, $a = 1$, $c = 1.0$ and $T = 0.5$.

20.4 DGM method for mean field PDEs

In this section we present an adaptation of the Deep Galerkin Method (DGM) introduced in Sirignano and Spiliopoulos (2018) for MFGs and MFC. In this context, it can be used to solve the forward-backward PDE system (Al-Aradi et al., 2018; Carmona and Laurière, 2021; Ruthotto et al., 2020; Cao et al., 2020; Lin et al., 2020; Laurière, 2020) or some forms of the Master equation (Laurière,

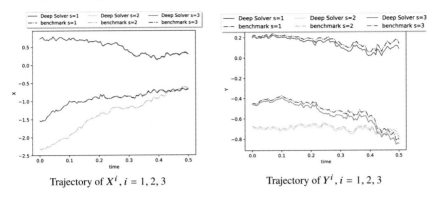

Trajectory of X^i, $i = 1, 2, 3$ Trajectory of Y^i, $i = 1, 2, 3$

Figure 20.3 Systemic risk MFG example solved by Algorithm 20.2. Three sample trajectories: solution computed by deep solver (full lines, in cyan, blue, and green) and by analytical formula (dashed lines, in orange, red and purple).

2020). The key idea is to replace the unknown function by a neural network and to tune the parameters to minimize a loss function based on the residual of the PDE. After presenting the main ideas, we illustrate this method on a model of optimal execution.

20.4.1 Description of the method

We present the method with the example of a finite horizon MFG model on a compact domain. We rewrite the MFG PDE system as a minimization problem where the control is the pair of density and value function, and the loss function is the sum of the PDE residuals and terms taking into account the boundary conditions. The same approach can be adapted to the ergodic setting, where initial and terminal conditions are replaced by normalization conditions see Carmona and Laurière (2021). In a finite horizon MFG as defined in Definition 20.1, the optimal control can be characterized (under suitable conditions, see Lasry and Lions (2007)), as:

$$\hat{\alpha}(x, m(t), \nabla u(t,x)) = \operatorname{argmin}_{a \in \mathbb{R}^k} \Big(f(x, m(t), a) + \nabla u(t,x) \cdot b(x, m(t), a) \Big),$$

where (m, u) solve the forward-backward PDE system on Q_T:

$$0 = \partial_t m(t,x) - \nu \Delta m(t,x) + \operatorname{div}\Big(m(t,x) \partial_q H^*(x, m(t), \nabla u(t,x)) \Big) \qquad (20.14)$$

$$0 = \partial_t u(t,x) + \nu \Delta u(t,x) + H^*(x, m(t), \nabla u(t,x)) \qquad (20.15)$$

with the initial and terminal conditions:

$$m(0,x) = m_0(x), \quad u(T,x) = g(x, m(T)), \qquad x \in Q,$$

where $\nu = \frac{\sigma^2}{2}$ and H^* is the optimized Hamiltonian, defined as:

$$H^*(x, m, q) := \min_{a \in \mathbb{R}^k} \Big(f(x, m, a) + q \cdot b(x, m, a) \Big).$$

The Kolmogorov–Fokker–Planck (KFP) equation, (20.14), describes the evolution of the population distribution, while the Hamilton–Jacobi–Bellman (HJB) equation, (20.15), describes the evolution of the value function. The latter is obtained for instance by a dynamic programming argument for the infinitesimal player's optimization problem given the flow of population density. These PDEs are coupled, hence we can not solve one before the other one.

For simplicity, we replace the domain Q by a compact subset $\tilde{Q}$. We denote $\tilde{Q}_T = [0,T] \times \tilde{Q}$. We introduce the following loss function:

$$L(m,u) = L^{(\mathrm{KFP})}(m,u) + L^{(\mathrm{HJB})}(m,u) \tag{20.16}$$

where

$$L^{(\mathrm{KFP})}(m,u) = C^{(\mathrm{KFP})} \left\| \partial_t m - \nu\Delta m + \mathrm{div}\left(m \partial_q H^*(x,m(t),\nabla u) \right) \right\|_{L^2(\tilde{Q})} \\ + C_0^{(\mathrm{KFP})} \left\| m(0) - m_0 \right\|_{L^2(\tilde{Q})} \tag{20.17}$$

and

$$L^{(\mathrm{HJB})}(m,u) = C^{(\mathrm{HJB})} \left\| \partial_t u + \nu\Delta u + H^*(\cdot,m(t),\nabla u) \right\|_{L^2(\tilde{Q}_T)} \\ + C_T^{(\mathrm{HJB})} \left\| u(T) - g(\cdot,m(T)) \right\|_{L^2(\tilde{Q})}. \tag{20.18}$$

Each component of the loss L in (20.16) encodes one of the two PDEs of the optimality system (20.14)–(20.15) with one term for the PDE residual and one term for the initial or terminal condition. The positive constants $C^{(\mathrm{KFP})}$, $C_0^{(\mathrm{KFP})}$, $C^{(\mathrm{HJB})}$, and $C_T^{(\mathrm{HJB})}$ give more or less importance to each component. On a bounded domain with boundary condition, more penalty terms could be included. Note that $L(m,u) = 0$ if (m,u) is a smooth enough solution to the PDE system (20.14)–(20.15). A similar system can be derived to characterize the optimal control of a MFC and the method can be adapted to this setting. See Carmona and Laurière (2021).

Replacing m and u by neural networks is the lynchpin of the algorithm. We denote by m_{θ_1} and u_{θ_2} these neural nets parameterized by θ_1 and θ_2 respectively. As in the method discussed in the previous sections, the integrals on $\tilde{Q}_T$ (resp. $\tilde{Q}$) are interpreted as expectations with respect to a uniform random variable over $\tilde{Q}_T$ (resp. $\tilde{Q}$), and we use SGD to minimize the total loss function L. More precisely, for a given $\mathbf{S} = (S, S_0, S_T)$, where S is a finite set of points in $\tilde{Q}_T$ and S_0 and S_T are finite sets of points in Q, we define the empirical loss function as

$$L_{\mathbf{S}}(\theta) = L_{\mathbf{S}}^{(\mathrm{KFP})}(\theta) + L_{\mathbf{S}}^{(\mathrm{HJB})}(\theta), \qquad \theta = (\theta_1,\theta_2) \tag{20.19}$$

where

$$
L_{\mathbf{S}}^{(\mathrm{KFP})}(\theta) = C^{(\mathrm{KFP})} \left(\frac{1}{|S|} \sum_{(t,x) \in S} \left| \partial_t m_{\theta_1}(t,x) - \nu \Delta m_{\theta_1}(t,x) \right. \right.
$$

$$
\left. \left. + \operatorname{div}\left(m_{\theta_1}(t,x) \partial_q H^*(x, m_{\theta_1}(t), \nabla u_{\theta_2}(t,x)) \right) \right|^2 \right)^{1/2}
$$

$$
+ C_0^{(\mathrm{KFP})} \left(\frac{1}{|S_0|} \sum_{x \in S_0} |m(0,x) - m_0(x)|^2 \right)^{1/2}
$$

and

$$
L_{\mathbf{S}}^{(\mathrm{HJB})}(\theta)
$$
$$
= C^{(\mathrm{HJB})} \left(\frac{1}{|S|} \sum_{x \in S} \left| \partial_t u_{\theta_2}(t,x) + \nu \Delta u_{\theta_2}(t,x) + H^*(x, m_{\theta_1}(t), \nabla u_{\theta_2}(t,x)) \right|^2 \right)^{1/2}
$$
$$
+ C_T^{(\mathrm{HJB})} \left(\frac{1}{|S_T|} \sum_{x \in S_T} \left| u_{\theta_2}(T,x) - g(x, m_{\theta_1}(T)) \right|^2 \right)^{1/2}.
$$

The method is summarized in Algorithm 20.3. The two neural networks could be taken with different architectures and their parameters optimized with different learning rates. The convergence of the neural network approximation was discussed in Sirignano and Spiliopoulos (2018) in the context of a single PDE using a standard universal approximation theorem. Unfortunately, this does not shed any light on the rate of convergence. A rate of convergence can be obtained by using more constructive approximation results with neural networks. See Carmona and Laurière (2021) and the references therein. In turn, this property leads to bounds on both the loss function of the algorithm and the error on the value function of the control problem. However, to the best of our knowledge, the convergence of the algorithm towards approximately optimal parameters remains to be proved, as for the other methods presented in this chapter.

Algorithm 20.3 DGM for MFG PDE system

Data: Initial parameters $\theta^{(0)} = (\theta_1^{(0)}, \theta_2^{(0)}) \in \Theta$; a number of steps K; a sequence $(\beta^{(k)})_{k=0,\ldots,K-1}$ of learning rates.

Result: Parameters $\theta = (\theta_1, \theta_2)$ such that $(m_{\theta_1}, u_{\theta_2})$ approximately minimize L defined in (20.16)

begin
    for $k = 0, 1, 2, \ldots, K-1$ **do**
        Pick $\mathbf{S} = (S, S_0, S_T)$
        Compute the gradient $\nabla_\theta L_{\mathbf{S}}(\theta^{(k)})$, see (20.19)
        Set $\theta^{(k+1)} = \theta^{(k)} - \beta^{(k)} \nabla_\theta L_{\mathbf{S}}(\theta^{(k)})$
    return $\theta^{(K)}$

An important advantage of the DGM method is its flexibility and its generality:

386   *René Carmona and Mathieu Laurière*

in principle, it can be applied to almost any PDE since it is agnostic to the structure of the PDE in question, or in the extension described above, of the PDE system. In tailoring the strategy to the specifics of our system, our main challenge was the choice of the relative weights to be assigned to the various terms in the aggregate loss function. If they are not chosen appropriately, SGD can easily be stuck in local minima. For example, if the weights $C_0^{\text{(KFP)}}$ and $C_T^{\text{(HJB)}}$ are not large enough, the neural networks might find trivial solutions minimizing the residuals while ignoring the initial and terminal conditions. However, if these weights are too large, the neural networks might satisfy very well these conditions without solving very precisely each PDE on the interior of the domain. See Carmona and Laurière (2021) for a more detailed discussion on this aspect.

20.4.2 Numerical illustration: a crowded trade model

This is a model of optimal execution similar to the one studied in §20.2.2. Here, we follow Cardaliaguet and Lehalle (2018) and we assume that a broker is instructed by a client to liquidate Q_0 shares of a specific stock by a fixed time horizon T, and that this broker is representative of a large population of brokers trying to do exactly the same thing. Such a situation occurs when a large number of index trackers suddenly need to rebalance their portfolios because the composition of the market index they track is changed. So our typical broker tries to maximize the quantity:

$$\mathbb{E}\left[X_T + Q_T(S_T - AQ_T) - \phi \int_0^T |Q_t|^2\, dt\right]$$

where at time $t \in [0,T]$, S_t is the price of the stock, Q_t is the inventory (i.e. number of shares) held by the broker, and X_t is their wealth. The constant $\phi > 0$ weigh a penalty for holding inventory through time while $A > 0$ plays a similar role at the terminal time. The dynamics of these three state variables are given by:

$$\begin{cases} dS_t = \gamma \bar{\mu}_t\, dt + \sigma\, dW_t \\ dQ_t = \alpha_t\, dt \\ dX_t = -\alpha_t(S_t + \kappa \alpha_t)\, dt. \end{cases}$$

The time evolution of the price S_t is subject to random shocks with standard deviation σ where the innovation dW_t is given by the increments of a standard Brownian motion, and a drift accounting for a permanent price impact $\gamma \bar{\mu}_t$ resulting from the aggregate trading rate $\bar{\mu}_t$ of all the brokers multiplied by a constant $\gamma > 0$. The rate of trading α_t is the control of the broker. Finally, the constant $\kappa > 0$ account for a quadratic transaction cost. Except for the fact that $\bar{\mu}_t$ is here endogenous, this is the model considered in Cartea and Jaimungal (2016), to which a deep learning method has been applied in Leal et al. (2020) to approximate the optimal control on real data.

Remark 20.3 The current model has two major differences with the model considered earlier in §20.2.2. It does not belong to the class of linear-quadratic models because the transaction costs entering the dynamics are quadratic in the control. But most importantly, the broker's inventory does not have a Brownian component. The presence of a quadratic variations term in the dynamics of the inventory was demonstrated in Carmona and Webster (2019) running econometric tests on high frequency market data. This was one of the reasons for the choice of the model used in §20.2.2. Surprisingly, it is shown in Carmona and Leal (2021) that the inclusion of a Brownian motion component in the dynamics of the inventory process Q_t does not require significant changes to the proof, including the form of the ansatz for the value function.

In any case, when the flow $(\bar{\mu}_t)_{0 \leq t \leq T}$ is fixed, the optimization problem involved in the computation of the best response reduces to an HJB equation whose solution $V(t, x, s, q)$ can be found as in Cartea and Jaimungal (2016) by formulating the ansatz $V(t, x, s, q) = x + qs + v(t, q)$ for some function v. Rewriting the HJB equation one sees that v must solve the equation:

$$-\gamma \bar{\mu} q = \partial_t v - \phi q^2 + \sup_\alpha \{\alpha \partial_q v - \kappa \alpha^2\}$$

with terminal condition $v(T, q) = -Aq^2$, the optimal control being $\alpha_t^*(q) = \frac{\partial_q v(t,q)}{2\kappa}$. Accordingly, if we denote by $m(t, \cdot)$ the distribution of inventories at time t, the aggregate trading rate is given by:

$$\bar{\mu}_t = \int \alpha_t^*(q) m(t, dq) = \int \frac{\partial_q v(t,q)}{2\kappa} m(t, dq),$$

in equilibrium since we use the optimal control. Since the evolution of the inventory distribution can be captured by the Kolmogorov–Fokker–Planck partial differential equation:

$$\partial_t m + \partial_q \left(m \frac{\partial_q v(t,q)}{2\kappa} \right) = 0,$$

with a given initial condition $m(0, \cdot) = m_0$, the solution of the MFG can be characterized by the PDE system is (see Cardaliaguet and Lehalle, 2018 for more details):

$$
\begin{cases}
-\gamma \bar{\mu} q = \partial_t v - \phi q^2 + \dfrac{|\partial_q v(t,q)|^2}{4\kappa} \\[2ex]
\partial_t m + \partial_q \left(m \dfrac{\partial_q v(t,q)}{2\kappa} \right) = 0 \\[2ex]
\bar{\mu}_t = \int \dfrac{\partial_q v(t,q)}{2\kappa} m(t, dq) \\[2ex]
m(0, \cdot) = m_0, v(T, q) = -Aq^2.
\end{cases}
\qquad (20.20)
$$

Note that the mean field interactions are through $\bar{\mu}_t$, which is a non-local (in space) term involving the derivative of the solution to the HJB equation.

This PDE system has been solved with the DGM method in Al-Aradi et al.

(2018) after a change of variable for the distribution. Here, for the sake of numerical illustration, we present results based on directly solving this system by following the methodology discussed above, suitably modified for the time-dependent PDE system (20.20). The initial and terminal conditions are imposed by penalization. The non-local term is estimated with Monte Carlo samples. For the results presented here, we used the following values for the parameters: $T = 1$, $\sigma = 0.3$, $A = 1$, $\phi = 1$, $\kappa = 1$, $\gamma = 1$, and a Gaussian initial distribution with mean 4 and variance 0.3.

The evolution of the distribution m is displayed in Figure 20.4 while the value function v and the optimal control α^* are displayed in Figure 20.5. As expected from the theory, the distribution concentrates close to zero and we recover a linear control, which matches the optimal one obtained with semi-explicit formula (see Cardaliaguet and Lehalle, 2018 for more details). For the neural network approximating the density, on the last layer, we used an exponential activation function. This ensures that the density is always non-negative.

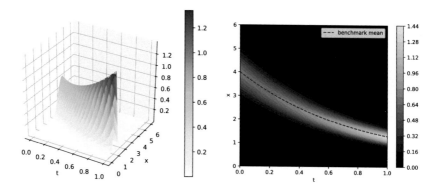

Figure 20.4 Trade crowding MFG example solved by Algorithm 20.3. Evolution of the distribution m: surface (left) and contour (right). The dashed red line corresponds to the mean obtained by the semi-explicit formula.

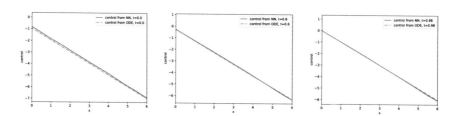

Figure 20.5 Trade crowding MFG example solved by Algorithm 20.3. Optimal control α^* (dashed line) and learnt control (full line) at three different time steps.

20.5 Conclusion

In this chapter we have presented three families of strategies to solve MFC and MFG. The first one is designed to solve MFC problems by directly trying to minimize the cost functional after replacing the control function by a neural network. The second approach tackles generic MKV FBSDE systems and uses a (stochastic) shooting method where the unknown starting point Y_0 of the backward variable and the Z-component are learnt as neural network functions of the state. The last approach solves mean field PDE systems by minimizing the residuals when the unknown functions are replaced by neural networks. We have illustrated these methods on stylized models arising in finance. The expressive power of neural networks let us expect that these methods will allow researchers and practitioners to solve much more complex and realistic models. The development of sample efficient methods able to learn solutions on real data while taking into account mean field interactions seem particularly relevant for future applications.

References

Achdou, Yves, and Capuzzo-Dolcetta, Italo. 2010. Mean field games: numerical methods. *SIAM J. Numer. Anal.*, **48**(3), 1136–1162.

Achdou, Yves, and Laurière, Mathieu. 2020. Mean Field Games and Applications: Numerical Aspects. In: *Mean Field Games*. C.I.M.E. Foundation Subseries, vol. 2281. Springer International Publishing.

Achdou, Yves, Buera, Francisco J., Lasry, Jean-Michel, Lions, Pierre-Louis, and Moll, Benjamin. 2014. Partial differential equation models in macroeconomics. *Philos. Trans. R. Soc. Lond. Ser. A Math. Phys. Eng. Sci.*, **372**(2028), 20130397, 19.

Achdou, Yves, Han, Jiequn, Lasry, Jean-Michel, Lions, Pierre-Louis, and Moll, Benjamin. 2017. Income and wealth distribution in macroeconomics: A continuous-time approach. Tech. Rept. National Bureau of Economic Research.

Achdou, Yves, Lauriere, Mathieu, and Lions, Pierre-Louis. 2020. Optimal control of conditioned processes with feedback controls. *J. Math. Pures Appl.*, **148**, 308–341.

Aiyagari, S. Rao. 1994. Uninsured idiosyncratic risk and aggregate saving. *Quarterly Journal of Economics*, **109**(3), 659–84.

Al-Aradi, Ali, Correia, Adolfo, Naiff, Danilo, Jardim, Gabriel, and Saporito, Yuri. 2018. Solving nonlinear and high-dimensional partial differential equations via deep learning. ArXiv:1811.08782.

Andersson, Daniel, and Djehiche, Boualem. 2011. A maximum principle for SDEs of mean-field type. *Appl. Math. Optim.*, **63**(3), 341–356.

Angiuli, Andrea, Fouque, Jean-Pierre, and Laurière, Mathieu. 2021. Reinforcement learning for mean field games, with applications to economics. In: *Handbook on Machine Learning in Financial Markets: A Guide to Contemporary Practices*, A. Capponi and C.-A. Lehalle (eds), Cambridge University Press.

Bertucci, Charles. 2018. Optimal stopping in mean field games, an obstacle problem approach. *J. Math. Pures Appl. (9)*, **120**, 165–194.

Bewley, Truman. 1986. Stationary monetary equilibrium with a continuum of independently fluctuating consumers. In: *Contributions to Mathematical Economics in Honor of Gerard Debreu*, Werner Hildenbrand, and Andreu Mas-Collel (eds), North-Holland.

Cao, Haoyang, Guo, Xin, and Laurière, Mathieu. 2020. Connecting GANs and MFGs. ArXiv:2002.04112.

Cardaliaguet, Pierre, and Lehalle, Charles-Albert. 2018. Mean field game of controls and an application to trade crowding. *Math. Financ. Econ.*, **12**(3), 335–363.

Carmona, René. 2020. Applications of mean field games to economic theory. (AMS Short Course.) ArXiv:2012.05237.

Carmona, René, and Delarue, François. 2018a. *Probabilistic Theory of Mean Field Games with Applications. I, Mean Field FBSDEs, Control, and Games.* Springer.

Carmona, René, and Delarue, François. 2018b. *Probabilistic Theory of Mean Field Games with Applications. II, Mean Field Games with Common Noise and Master Equations.* Springer.

Carmona, René, and Lacker, Daniel. 2015. A probabilistic weak formulation of mean field games and applications. *Ann. Appl. Probab.*, **25**(3), 1189–1231.

Carmona, René, and Laurière, Mathieu. 2019. Convergence analysis of machine learning algorithms for the numerical solution of mean field control and games: II – the finite horizon case. *Annals of Applied Probability*, to appear. ArXiv:1908.01613.

Carmona, René, and Laurière, Mathieu. 2021. Convergence analysis of machine learning algorithms for the numerical solution of mean field control and games I: The ergodic case. *SIAM Journal on Numerical Analysis*, **59**(3), 1455–1485.

Carmona, René, and Leal, Laura. 2021. Optimal execution with quadratic variation inventories. Tech. Rept., Princeton University.

Carmona, René, and Webster, Kevin. 2019. The self-financing equation in high-frequency markets. *Finance & Stochastics*, **23**, 729–759.

Carmona, René, Fouque, Jean-Pierre, and Sun, Li-Hsien. 2015. Mean field games and systemic risk. *Commun. Math. Sci.*, **13**(4), 911–933.

Carmona, René, Delarue, François, and Lacker, Daniel. 2017. Mean field games of timing and models for bank runs. *Appl. Math. Optim.*, **76**(1), 217–260.

Carmona, René, Fouque, Jean-Pierre, Mousavi, Seyyed Mostafa, and Sun, Li-Hsien. 2018. Systemic risk and stochastic games with delay. *J. Optim. Theory Appl.*, **179**(2), 366–399.

Carmona, René, Graves, Christy V., and Tan, Zongjun. 2019. Price of anarchy for mean field games. Pages 349–383 of: *CEMRACS 2017 – Numerical Methods for Stochastic Models: Control, Uncertainty Quantification, Mean-Field*. ESAIM Proc. Surveys, vol. 65. EDP Sci.

Cartea, Álvaro, and Jaimungal, Sebastian. 2016. Incorporating order-flow into optimal execution. *Math. Financ. Econ.*, **10**(3), 339–364.

Chan, Patrick, and Sircar, Ronnie. 2015. Bertrand and Cournot mean field games. *Appl. Math. Optim.*, **71**(3), 533–569.

Chan, Patrick, and Sircar, Ronnie. 2017. Fracking, renewables, and mean field games. *SIAM Rev.*, **59**(3), 588–615.

Cvitanić, Jakša, and Zhang, Jianfeng. 2013. *Contract Theory in Continuous-Time Models*. Springer.

Cvitanić, Jakša, Possamaï, Dylan, and Touzi, Nizar. 2018. Dynamic programming approach to principal-agent problems. *Finance Stoch.*, **22**(1), 1–37.

E, Weinan, Han, Jiequn, and Jentzen, Arnulf. 2017. Deep learning-based numerical methods for high-dimensional parabolic partial differential equations and backward stochastic differential equations. *Commun. Math. Stat.*, **5**(4), 349–380.

Elie, Romuald, Ichiba, Tomoyuki, and Laurière, Mathieu. 2020. Large banking systems with default and recovery: a mean field game model. ArXiv:2001.10206.

Fouque, Jean-Pierre, and Zhang, Zhaoyu. 2020. Deep learning methods for mean field control problems with delay. *Frontiers in Applied Mathematics and Statistics*, **6(11)**.

Garnier, Josselin, Papanicolaou, George, and Yang, Tzu-Wei. 2013. Large deviations for a mean field model of systemic risk. *SIAM J. Financial Math.*, **4**(1), 151–184.

Germain, Maximilien, Mikael, Joseph, and Warin, Xavier. 2022a. Numerical resolution of McKean–Vlasov FBSDEs using neural networks. *Methodology and Computing in Applied Probability*, https://doi.org/10.1007/s11009-022-09946-1.

Germain, Maximilien, Pham, Huyên, and Warin, Xavier. 2022b. Neural networks-based algorithms for stochastic control and PDEs in finance. In: *Handbook on Machine Learning in Financial Markets: A Guide to Contemporary Practices*, A. Capponi and C.-A. Lehalle (eds), Cambridge University Press.

Gobet, Emmanuel, and Munos, Rémi. 2005. Sensitivity analysis using Itô–Malliavin calculus and martingales, and application to stochastic optimal control. *SIAM J. Control Optim.*, **43**(5), 1676–1713.

Gomes, Diogo A., and Voskanyan, Vardan K. 2016. Extended deterministic mean-field games. *SIAM J. Control Optim.*, **54**(2), 1030–1055.

Gomes, Diogo A., Patrizi, Stefania, and Voskanyan, Vardan. 2014. On the existence of classical solutions for stationary extended mean field games. *Nonlinear Anal.*, **99**, 49–79.

Graber, P. Jameson, and Bensoussan, Alain. 2018. Existence and uniqueness of solutions for Bertrand and Cournot mean field games. *Appl. Math. Optim.*, **77**(1), 47–71.

Guéant, Olivier. 2019. *Mean Field Games and Applications to Economics*. PhD. thesis, Université Paris Dauphine.

Guéant, Olivier, Lasry, Jean-Michel, and Lions, Pierre-Louis. 2011. Mean field games and applications. Pages 205–266 of: *Paris–Princeton Lectures on Mathematical Finance 2010*. Lecture Notes in Math., vol. 2003. Springer.

Han, Jiequn, and E, Weinan. 2016. Deep learning approximation for stochastic control problems. *Deep Reinforcement Learning Workshop, NeurIPS*. ArXiv:1611.07422.

Han, Jiequn, and Long, Jihao. 2020. Convergence of the deep BSDE method for coupled FBSDEs. *Probability, Uncertainty and Quantitative Risk*, **5**(1), 1–33.

Huggett, Mark. 1993. The risk-free rate in heterogeneous-agent incomplete-insurance economies. *Journal of Economic Dynamics and Control*, **17**(5-6), 953–969.

Kingma, Diederik P., and Ba, Jimmy. 2014. Adam: A method for stochastic optimization. ArXiv:1412.6980.

Kohlmann, Michael, and Zhou, Xun Yu. 2000. Relationship between backward stochastic differential equations and stochastic controls: a linear-quadratic approach. *SIAM J. Control Optim.*, **38**(5), 1392–1407.

Krusell, Per, and Smith, Anthony A. 1998. Income and wealth heterogeneity in the macroeconomy. *Journal of Political Economy*, **106**(5), 867–896.

Lasry, Jean-Michel, and Lions, Pierre-Louis. 2007. Mean field games. *Jpn. J. Math.*, **2**(1), 229–260.

Laurière, Mathieu. 2020. Numerical methods for mean field games and mean field type control. (AMS Short Course.) ArXiv:2106.06231.

Leal, Laura, Laurière, Mathieu, and Lehalle, Charles-Albert. 2020. Learning a functional control for high-frequency finance. ArXiv:2006.09611.

Lin, Alex Tong, Fung, Samy Wu, Li, Wuchen, Nurbekyan, Levon, and Osher, Stanley J. 2020. APAC-Net: Alternating the population and agent control via two neural networks to solve high-dimensional stochastic mean field games. ArXiv:2002.10113.

Nadtochiy, Sergey, and Shkolnikov, Mykhaylo. 2019. Particle systems with singular interaction through hitting times: application in systemic risk modeling. *Ann. Appl. Probab.*, **29**(1), 89–129.

Nuño, Galo. 2017. Optimal social policies in mean field games. *Appl. Math. Optim.*, **76**(1), 29–57.

Nuño, Galo, and Moll, Benjamin. 2018. Social optima in economies with heterogeneous agents. *Review of Economic Dynamics*, **28**, 150–180.

Nutz, Marcel. 2018. A mean field game of optimal stopping. *SIAM J. Control Optim.*, **56**(2), 1206–1221.

Nutz, Marcel, and Zhang, Yuchong. 2020. Conditional optimal stopping: a time-inconsistent optimization. *Ann. Appl. Probab.*, **30**(4), 1669–1692.

Rochet, Jean-Charles, and Vives, Xavier. 2004. Coordination failures and the lender of last resort: was Bagehot right after all? *Journal of the European Economic Association*, **2**(6), 1116–1147.

Ruthotto, Lars, Osher, Stanley J., Li, Wuchen, Nurbekyan, Levon, and Fung, Samy Wu. 2020. A machine learning framework for solving high-dimensional mean field game and mean field control problems. *Proc. National Academy of Sciences*, **117**(17), 9183–9193.

Sargent, Thomas J., Wang, Neng, and Yang, Jinqiang. 2021. Earnings growth and the wealth distribution. *Proc. National Academy of Sciences*, `https://www.pnas.org/doi/full/10.1073/pnas.2025368118`.

Sirignano, Justin, and Spiliopoulos, Konstantinos. 2018. DGM: a deep learning algorithm for solving partial differential equations. *J. Comput. Phys.*, **375**, 1339–1364.

Reinforcement Learning for Mean Field Games, with Applications to Economics

Andrea Angiuli[a], Jean-Pierre Fouque[a] and Mathieu Laurière[b]

Abstract

Mean field games (MFG) and mean field control problems (MFC) are frameworks to study Nash equilibria or social optima in games with a continuum of agents. These problems can be used to approximate competitive or cooperative games with a large finite number of agents and have found a broad range of applications, in particular in economics. In recent years, the question of learning in MFG and MFC has garnered interest, both as a way to compute solutions and as a way to model how large populations of learners converge to an equilibrium. Of particular interest is the setting where the agents do not know the model, which leads to the development of reinforcement learning (RL) methods. After reviewing the literature on this topic, we present a two-timescale approach with RL for MFG and MFC, which relies on a unified Q-learning algorithm. The main novelty of this method is to simultaneously update an action-value function and a distribution but with different rates, and in a model-free fashion. Depending on the ratio of the two learning rates, the algorithm learns either the MFG or the MFC solution. To illustrate this method, we apply it to two finite-horizon problems: a mean field problem of accumulated consumption with HARA utility function, and a trader's optimal liquidation problem.

21.1 Introduction

Dynamic games with many players are pervasive in today's highly connected world. In many models, the agents are indistinguishable since they have the same dynamics and cost functions. Moreover, the interactions are often anonymous since each player is influenced only by the empirical distribution of all the agents. However, such games are intractable when the number of agents is very large. Mean field games were introduced by Lasry and Lions (2007) and Huang et al. (2006) to tackle such situations by passing to the limit and considering games

[a] Department of Statistics and Applied Probability, UCSB. Work supported by NSF grant DMS-1953035.
[b] Operations Research and Financial Engineering Department, Princeton University. Work supported by NSF grant DMS-1716673 and ARO grant W911NF-17-1-0578
Published in *Machine Learning And Data Sciences For Financial Markets*, Agostino Capponi and Charles-Albert Lehalle© 2023 Cambridge University Press.

with an infinite number of players interacting through the population distribution. Although the standard formulation of MFG focuses on finding Nash equilibria, social optima arising in a cooperative setting have also been studied under the term mean field control (Bensoussan et al., 2013) or control of McKean–Vlasov dynamics (Lasry and Lions, 2007). Equilibria or social optima in such games can be characterized in a tractable way through forward-backward systems of partial differential equations (PDE) or stochastic differential equations (SDE) (Carmona and Delarue, 2018a; Lasry and Lions, 2007).

Mean field games with interactions through the controls, sometimes called "extended", occur when the dynamics or the cost function of a typical player explicitly depends on the distribution of the *controls* of the other players, and not just on the distribution of their states. Such games were first introduced by Gomes et al. (2014) and Gomes and Voskanyan (2016) and their investigation quickly garnered interest.

Interactions through the controls' distribution are particularly relevant in economics and finance; see, e.g., Huang (2013), Gomes and Saúde (2014), Carmona and Lacker (2015), Chan and Sircar (2015), Graber and Bensoussan (2018) and Cardaliaguet and Lehalle (2018), and see Carmona (2020) for a recent survey. Some aspects of the PDE approach and the probabilistic approach to such games have been treated respectively by Bertucci et al. (2019), Bonnans et al. (2019), Kobeissi (2019), and by Carmona and Lacker (2015). As in many fields, linear-quadratic models are particularly appealing due to their tractability; see, e.g., Alasseur et al. (2019) and Graber (2016) for applications to energy production.

The approach we propose is based on ideas from reinforcement learning (RL). Applications of RL in economics and finance have recently attracted a lot of interest; see, e.g., Charpentier et al. (2020). However, since our problems involve mean-field interactions, the population distribution requires a special treatment. In our setup, the agent is feeding an action to the environment which produces the next state and the reward (or cost). The environment also updates in an automatic way (without decision) the joint distribution of states and controls. Then, the agent update their Q-matrix and proceeds (see the diagram in Figure 21.1). The environment can be viewed as a "black box" or as a "simulator" depending on the problem, but, in any case, it generates the new state if the dynamics is unknown and the reward if not computable by the agent. It is also interesting to note that even in cases where the dynamics and the reward structure are known but complicated, our algorithm can be viewed as a numerical method for computing the optimal strategy for the corresponding MFG or MFC problems.

Since the introduction of MFG theory, several numerical methods have been proposed; see, e.g., Achdou and Laurière (2020), Laurière (2020) and the references therein. Recently, several methods to solve MFGs based on machine learning tools have been proposed relying either on the probabilistic approach (Fouque and Zhang, 2020; Carmona and Laurière, 2019; Germain et al., 2019; Min and Hu, 2021) or the analytical approach (Al-Aradi et al., 2018; Carmona and Laurière, 2021; Ruthotto et al., 2020; Cao et al., 2020; Lin et al., 2020; Laurière, 2020). They combine neural network approximations and stochastic

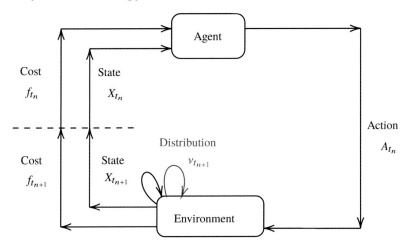

Figure 21.1 At time t_n of an experiment, the agent feeds the action A_{t_n} to the environment, which outputs the associated cost $f_{t_{n+1}} = f(X_{t_n}, A_{t_n}, \nu_{t_n})$ for $n < N_T$ (and $f_{t_{n+1}} = g(X_{t_n}, \nu_{t_n})$ for $n = N_T$) and the new state $X_{t_{n+1}} \sim p(X_{t_n}, A_{t_n}, \nu_{t_n})$. Besides these outputs provided to the agent, the environment keeps track of $X_{t_{n+1}}$ and $\nu_{t_{n+1}}$ for the next iteration.

optimization techniques to solve McKean–Vlasov control problems, mean field FBSDE or mean field PDE systems; see Carmona et al. (2022) for a recent survey and applications to finance. These methods are based on the knowledge of the full model, but the question of learning solutions to MFG and MFC without full knowledge of the model have also attracted a surge of interest.

As far as learning methods for mean field problems are concerned, most works focus either on MFG or on MFC. Yang et al. (2018b) use a mean field approximation in the context of multi-agent reinforcement learning (MARL) to reduce the computational cost. Yang et al. (2018a) use inverse reinforcement learning to learn the dynamics of a mean field game on a graph. To approximate stationary MFG solutions, Guo et al. (2019) use fixed point iterations on the distribution combined with tabular Q-learning to learn the best response at each iteration. Anahtarci et al. (2020) combine this kind of learning scheme together with an entropic regularization. Convergence of an actor-critic method for linear-quadratic MFG has been studied by Fu et al. (2019). Model-free learning for finite horizon MFG has been studied by Mishra et al. (2020) using a backward scheme. Fictitious play without or with reinforcement learning has been studied respectively by Cardaliaguet and Hadikhanloo (2017) and Hadikhanloo and Silva (2019), and byElie et al. (2020), Perrin et al. (2020), and Xie et al. (2020), or online mirror descent (Hadikhanloo, 2017; Pérolat et al., 2021). These iterative methods have been proved to converge under a monotonicity condition which is weaker than the strict contraction property used to ensure convergence of fixed point iterations. They can be extended to continuous space problems using deep reinforcement learning; see, e.g., Perrin et al. (2021). A two-timescale approach to solve MFG with finite state and action spaces has been proposed by Mguni et al. (2018) and Subramanian and Mahajan (2019).

To learn MFC optima, Subramanian and Mahajan (2019) designed a gradient based algorithm. Model-free policy gradient method has been proved to converge for linear-quadratic problems in Carmona et al. (2019a) and Wang et al. (2020), whereas Q-learning for a "lifted" Markov decision process on the space of distributions were studied by Carmona et al. (2019b) and Gu et al. (2019, 2020). Optimality conditions and propagation of chaos type result for mean field Markov decision processes were studied by Motte and Pham (2019).

We proposed (Angiuli et al., 2022) a unified two-timescale Q-learning algorithm to solve both MFG and MFC problems in an infinite horizon stationary regime. The key idea is to iteratively update estimates of the distribution and the Q-function with different learning rates. Suitably choosing these learning rates enables the algorithm to learn the solution of the MFG or the one of the MFC. A slow updating of the distribution of the state leads to the Nash equilibrium of the competitive MFG and the algorithm learns the corresponding optimal strategy. A rapid updating of the distribution leads to learning of the optimal control of the corresponding cooperative MFC. Moreover, in contrast with other approaches, our algorithm does not require the environment to output the population distribution which means that a single agent can learn the solution of mean field problems.

In the present chapter, we extend this algorithm in two directions: finite horizon setting, and "extended" mean field problems which involve the distribution of controls as well. That demonstrates the flexibility of our two-timescale algorithm and broadens the range of applications.

The rest of the chapter is organized as follows. In §21.2, we introduce the framework of finite horizon mean field games and mean field control problems. In §21.3, we present the main ideas behind the two-timescale approach in this context. Based on this perspective, we introduce in §21.4 a reinforcement learning algorithm to solve MFC and MFG problems. We then illustrate this method on two examples: a mean field accumulation problem in §21.5 and an optimal execution problem for a mean field of traders in §21.6. We then conclude in §21.7.

Notation.

For a random variable X, $\mathcal{L}(X)$ denotes its law. d and k are two positive integers corresponding respectively to the state and the action dimensions. Unless otherwise specified, ν will be used to denote a state-action distribution, and its first and second marginals will respectively be denoted by μ and θ.

21.2 Finite horizon mean field problems

In this section, we introduce the framework of mean field games and mean field control problems in finite horizon. For the sake of consistency with the MFG literature, we use a continuous time formalism. For the link with finite player games; see, e.g., Carmona and Delarue (2018a).

21.2.1 Mean field games

Let $(\Omega, \mathcal{F}, \mathbb{F} = (\mathcal{F}_t)_{0 \le t \le T}, \mathbb{P})$ be a filtered probability space, where the filtration supports an m-dimensional Brownian motion $W = (W_t)_{0 \le t \le T}$ and an initial condition $\xi \in L^2(\Omega, \mathcal{F}_0, \mathbb{P}; \mathbb{R}^d)$. Let $f : [0, T] \times \mathbb{R}^d \times \mathcal{P}_2(\mathbb{R}^{d+k}) \times \mathbb{R}^k \to \mathbb{R}$ and $g : \mathbb{R}^d \times \mathcal{P}_2(\mathbb{R}^d) \to \mathbb{R}$ be respectively a running cost function and a terminal cost function. Let $b : [0, T] \times \mathbb{R}^d \times \mathcal{P}_2(\mathbb{R}^{d+k}) \times \mathbb{R}^k \to \mathbb{R}^d$ be a drift function and let $\sigma : [0, T] \times \mathbb{R}^d \times \mathcal{P}_2(\mathbb{R}^{d+k}) \times \mathbb{R}^k \to \mathbb{R}^{d \times m}$ be a volatility function.

A mean field game equilibrium is defined as a pair

$$(\hat{\alpha}, \hat{v}) = ((\hat{\alpha}_t)_{t \in [0,T]}, (\hat{v}_t)_{t \in [0,T]}) \in \mathbb{A} \times C([0, T], \mathcal{P}_2(\mathbb{R}^{d+k})),$$

where $\mathbb{A}$ is the set of admissible controls, namely progressively measurable processes that are square integrable, such that

1. $\hat{\alpha}$ solve the standard stochastic control problem when $v = \hat{v}$:

$$\inf_{\alpha \in \mathbb{A}} J_v(\alpha) = \inf_{\alpha \in \mathbb{A}} \mathbb{E}\left[\int_0^T f(t, X_t^{\alpha,v}, v_t, \alpha_t)dt + g(X_T^{\alpha,v}, \hat{\mu}_T)\right],$$

where $\hat{\mu}_T$ is the first marginal of v_T corresponding to the terminal state distribution subject to

$$dX_t^{\alpha,v} = b(t, X_t^{\alpha,v}, v_t, \alpha_t)dt + \sigma(t, X_t^{\alpha,v}, v_t, \alpha_t)dW_t, \quad X_0^{\alpha,v} = \xi.$$

2. $\hat{v}_t = \mathcal{L}(X_t^{\hat{\alpha}}, \hat{\alpha}_t)$ for all $0 \le t \le T$.

The solution can be characterized from either the PDE viewpoint (leading to a coupled Hamilton–Jacobi–Bellman and Kolmogorov–Fokker–Planck equations) (Lasry and Lions, 2007; Huang et al., 2006) or from a probabilistic viewpoint (Carmona and Delarue, 2018b). Within the probabilistic viewpoint, there are two approaches, both of which are formulated with FBSDEs. The backward variable can represent either the value function of a typical player or the derivative of this value function. See Carmona and Delarue (2018b, Volume 1, Chapters 3 and 4) for more details. These analytical and probabilistic approaches also lead to computational methods, as long as the model is known. However, when the model is not known, one needs to develop other tools, as we will discuss in the next sections.

21.2.2 Mean field control

In contrast with the MFG problem – which corresponds to a Nash equilibrium, the mean field control (MFC) problem is an optimization problem. It can be interpreted as the problem posed to a social planner trying to find the optimal behavior of a population so as to minimize a social cost (i.e., a cost averaged over the whole population). It is an optimal control problem for a McKean–Vlasov dynamics: Find α^* which satisfies:

$$\inf_{\alpha \in \mathbb{A}} J(\alpha) = \inf_{\alpha \in \mathbb{A}} \mathbb{E}\left[\int_0^T f(t, X_t^\alpha, v_t^\alpha, \alpha_t)dt + g(X_T^\alpha, \mu_T^\alpha)\right],$$

subject to

$$dX_t^\alpha = b(t, X_t^\alpha, \nu_t^\alpha, \alpha_t)dt + \sigma(t, X_t^\alpha, \nu_t^\alpha, \alpha_t)dW_t, \quad X_0^\alpha = \xi.$$

where ν_t^α is a shorthand notation for $\mathcal{L}(X_t^\alpha, \alpha_t)$ and μ_T^α is its first marginal at terminal time T. The dynamics of X involves the law of this process, hence the terminology McKean–Vlasov dynamics (McKean Jr, 1966). To alleviate notation we will sometimes write $\nu^* = \nu^{\alpha^*}$ for the law of the optimally controlled process.

Remark 21.1   Although the two problems look similar, in general they have different solutions, i.e., $\hat\alpha \neq \alpha^*$ and $\hat\nu \neq \nu^*$, even when the functions in the cost and the dynamics are the same; see, e.g., Carmona et al. (2019c).

Remark 21.2   Although the mean field paradigm is the same, the special case where the interactions are only through the state distribution (i.e., the first marginal of ν) has attracted more interest in the literature than the present general setup. However interactions through the distribution of controls appears in many applications, particularly in economics and finance as already mentioned in the introduction. See next sections for some examples.

Remark 21.3   Although the reinforcement learning literature typically focuses on infinite horizon discounted problems, we focus here on finite horizon problems. This will cause some numerical difficulties but is crucial for many applications.

21.3 Two-timescale approach

21.3.1 Discrete formulation

To simplify the presentation and to be closer to the traditional reinforcement learning setup, we consider a discrete time model with a finite number of states and actions. Let $\mathcal{X}$ and $\mathcal{A}$ be finite sets corresponding to spaces of states and actions respectively, which can correspond to discretized version of the continuous spaces (possibly after a truncation) used in the previous section. We denote by $\Delta^{|\mathcal{X}|}$ the simplex in dimension $|\mathcal{X}|$, which we identify with the space of probability measures on $\mathcal{X}$. $\Delta^{|\mathcal{X}| \times |\mathcal{A}|}$ is defined similarly on the product space $\mathcal{X} \times \mathcal{A}$. Moreover, we consider a discrete time setting, which here again could come from a suitable discretization of the continuous time evolution used in the previous section. We take a uniform grid, say $t_n = n \times \Delta t$, $n = 0, 1, 2, \ldots, N_T$, where $\Delta t = T/N_T > 0$ is the time step. Hence X_{t_n} and α_{t_n} in this section can be interpreted as approximation of the state and the action at time t_n in the previous section. The state follows a random evolution in which $X_{t_{n+1}}$ is determined as a function of the current state X_{t_n}, the action α_{t_n}, the state-action population distribution ν_{t_n} at time t_n, and some noise. We introduce the transition probability function:

$$p(x'|x, a, \nu), \qquad (x, x', a, \nu) \in \mathcal{X} \times \mathcal{X} \times \mathcal{A} \times \Delta^{|\mathcal{X} \times \mathcal{A}|},$$

which gives the probability to jump to state x' when being at state x and using action a and when the population distribution is v. For simplicity, we consider the homogeneous case where this function does not depend on time, which corresponds, in the continuous formulation, to the case where both b and σ are time-independent. Restoring this time-dependence if needed is a straightforward procedure.

We now consider the MFG cost function given by: for $v = (v_{t_n})_{n=0,...,N_T}$

$$\widetilde{J}_v(\alpha) = \mathbb{E}\left[\sum_{n=0}^{N_T-1} f(X_{t_n}^{\alpha,v}, \alpha_{t_n}, v_{t_n}) + g(X_{t_{N_T}}^{\alpha,v}, \mu_{t_{N_T}})\right],$$

where $\mu_{t_{N_T}}$ is the first marginal of $v_{t_{N_T}}$. Again, for simplicity, we assume that f does not depend on time. The process $X^{\alpha,v}$ has a given initial distribution $\mu_0 \in \Delta^{|X|}$ and follows the dynamics

$$\mathbb{P}(X_{t_{n+1}}^{\alpha,v} = x' | X_{t_n}^{\alpha,v} = x, \alpha_{t_n} = a, v_{t_n} = v) = p(x'|x, a, v).$$

Given a population distribution sequence v, the value function of an infinitesimal player is

$$V_v(x) = \inf_\alpha V_v^\alpha(x),$$

where

$$V_v^\alpha(x) = \mathbb{E}\left[\sum_{n=0}^{N_T-1} f(X_{t_n}^{\alpha,v}, \alpha_{t_n}, v_{t_n}) + g(X_{t_{N_T}}^{\alpha,v}, \mu_{t_{N_T}}) \,\Big|\, X_0^{\alpha,v} = x\right].$$

Note that the $\widetilde{J}$ and the V functions are related by:

$$\widetilde{J}_v(\alpha) = \mathbb{E}_{X_0 \sim \mu_0}[V_v^\alpha(X_0)].$$

In contrast, we also consider the MFC cost function

$$\widetilde{J}(\alpha) = \mathbb{E}\left[\sum_{n=0}^{N_T-1} f(X_n^\alpha, \alpha_{t_n}, v_n^\alpha) + g(X_{t_{N_T}}^\alpha, \mu_{t_{N_T}}^\alpha)\right],$$

where $v_{t_n}^\alpha = \mathcal{L}(X_{t_n}^\alpha, \alpha_{t_n})$ is the state-action distribution at time t_n of X^α controlled by α. The process X^α has initial distribution μ_0 and dynamics

$$\mathbb{P}(X_{t_{n+1}}^\alpha = x' | X_{t_n}^\alpha = x, \alpha_{t_n} = a, v_{t_n} = v_{t_n}^\alpha) = p(x'|x, a, v_{t_n}^\alpha).$$

Since the dynamics is of MKV type, in general, the value function in an MFC problem is the value function of the social planner and it takes the distribution v as input; see, e.g., Laurière and Pironneau (2014), Pham and Wei (2016), Carmona et al. (2019b), Motte and Pham (2019), Gu et al. (2019) andDjete et al. (2019). However, when the population is already evolving according to the sequence of distributions v^α generated by a control α, the cost-to-go of an infinitesimal agent starting at position x and using control α too is simply a function of its position and is given by

$$V^\alpha(x) = \mathbb{E}\left[\sum_{n=0}^{N_T-1} f(X_n^\alpha, \alpha_{t_n}, v_{t_n}^\alpha) + g(X_{t_{N_T}}^\alpha, \mu_{t_{N_T}}^\alpha) \,\Big|\, X_0^\alpha = x\right].$$

21.3.2 Action-value function

The state value function is useful as far as the value of the game or control problem is concerned. However, it does not provide any information about the equilibrium or optimal control, namely, $\hat{\alpha}$ or α^* respectively. For this reason, one can introduce the state-action value function, also called Q-function, which takes as inputs not only a state x but also an action a. For a standard Markov Decision Process (MDP) without mean field interactions, the Q-function characterizes the optimal cost-to-go if one starts at state x and uses action a before starting using the optimal control. To approximate this function, one of the most popular methods in RL is the so-called Q-learning (Watkins, 1989). See e.g. Sutton and Barto (2018, Chapter 3) for more details.

Before moving on to the mean-field setup, let us recall that in the traditional setup, the definition of the optimal Q-function, denoted by Q^*, is given by:

$$\begin{cases} Q^*_{N_T}(x,a) = g(x), \qquad (x,a) \in X \times \mathcal{A}, \\ Q^*_n(x,a) = \min_\alpha \mathbb{E}\left[\sum_{n'=n}^{N_T-1} f(X_{t_{n'}}, \alpha_{n'}(X_{t_{n'}})) + g(X_{t_{N_T}}) \,\Big|\, X_{t_n} = x, A_{t_n} = a \right], \\ \qquad n < N_T, (x,a) \in X \times \mathcal{A}, \end{cases}$$

where $\alpha_{n'}(\cdot) = \alpha(t_{n'}, \cdot)$. Using dynamic programming, it can be shown that $(Q^*_n)_n$ is the solution of the Bellman equation:

$$\begin{cases} Q^*_{N_T}(x,a) = g(x), \qquad (x,a) \in X \times \mathcal{A}, \\ Q^*_n(x,a) = f(x,a) + \sum_{x' \in X} p(x'|x,a) \min_{a'} Q^*_{n+1}(x',a'), \quad n < N_T, (x,a) \in X \times \mathcal{A}. \end{cases}$$

The corresponding optimal value function $(V^*_n)_n$ is given by:

$$V^*_n(x) = \min_a Q^*_n(x,a), \qquad n \le N_T, x \in X.$$

As mentioned above, one of the main advantages of computing the action-value function instead of the value function is that from the former, one can directly recover the optimal control at time n by computing $\arg\min_{a \in \mathcal{A}} Q^*_n(x,a)$. This is particularly important in order to design model-free methods, as we will see in the next section.

The above approach can be adapted to solve MFG by noticing that, when the population behavior is given, the problem posed to a single representative agent is a standard MDP. It can thus be tackled using a Q-function which implicitly depends on the population distribution: given $\nu = (\nu_{t_n})_{n=0,\dots,N_T}$

$$\begin{cases} Q^*_{N_T,\nu}(x,a) = g(x,\mu_{t_{N_T}}), \qquad (x,a) \in X \times \mathcal{A}, \\ Q^*_{n,\nu}(x,a) = f(x,a,\nu_{t_n}) \\ \qquad + \sum_{x' \in X} p(x'|x,a,\nu_{t_n}) \min_{a'} Q^*_{n+1,\nu}(x',a'), \qquad n < N_T, (x,a) \in X \times \mathcal{A}. \end{cases}$$

This function characterizes, at each time step t_n, the optimal cost-to-go for an agent starting at time t_n at state x, using action a for the first step, and then acting optimally for the rest of the time steps, while the population evolution is given by $v = (v_{t_n})_n$. However, to find the Nash equilibrium, it is not sufficient to compute the Q-function for an arbitrary sequence of distributions v: we want to find $Q^*_{v^*}$ where v^* is the population evolution generated by the optimal control computed from $Q^*_{v^*}$. In what follows, we will directly aim at the Q-function $Q^*_{v^*}$ via a two-timescale approach.

In the MFC problem the population distribution is not fixed while each player optimizes because all the agents cooperate to choose a distribution which is optimal from the point of view of the whole society. As a consequence, the optimization problem can not be recast as a standard MDP. However we will show below that it is still possible to compute the social optimum using a modified Q-function that does not involve explicitly the population distribution. This major difficulty is treated in detail in the context of infinite horizon in our previous work (Angiuli et al., 2022).

21.3.3 Unification through a two-timescale approach

A simple approach to compute the MFG solution is to iteratively update the state-action value function, Q, and the population distribution, v: Starting with an initial guess $v^{(0)}$, repeat for $k = 0, 1, \ldots$,

1. Solve the backward equation for $Q^{(k+1)} = Q^*_{v^{(k)}}$, which characterizes the optimal state-action value function of a typical player if the population behavior is given by $v^{(k)}$:

$$
\begin{cases}
Q^{(k+1)}_{N_T}(x,a) = g(x, \mu^{(k)}_{t_{N_T}}), & (x,a) \in X \times \mathcal{A}, \\
Q^{(k+1)}_n(x,a) = f(x, a, v^{(k)}_{t_n}) \\
\quad + \sum_{x' \in X} p(x'|x, a, v^{(k)}_{t_n}) \min_{a'} Q^{(k+1)}_{n+1}(x', a'), & n < N_T, (x,a) \in X \times \mathcal{A}.
\end{cases}
$$
(21.1)

2. Solve the forward equation for $\mu^{(k+1)}$ (resp. $v^{(k+1)}$), which characterizes the evolution of the population state distribution (resp. state-action distribution) if everyone uses the optimal control $\alpha^{(k+1)}_{t_n}(x) = \arg\min_a Q^{(k+1)}_n(x,a)$ coming from the above Q-function (assuming this control is uniquely defined for simplicity):

$$
\begin{cases}
\mu^{(k+1)}_{t_0}(x) = \mu_{t_0}(x), & x \in X, \\
v^{(k+1)}_{t_0}(x,a) = \mu_{t_0}(x)\mathbf{1}_{a=\alpha^{(k+1)}_{t_n}(x)}, & (x,a) \in X \times \mathcal{A}, \\
\mu^{(k+1)}_{t_{n+1}}(x) = \sum_{x' \in X} \mu^{(k+1)}_{t_n}(x')p(x|x', \alpha^{(k+1)}_{t_n}(x'), v^{(k+1)}_{t_n}), & 0 \le n < N_T, x \in X, \\
v^{(k+1)}_{t_{n+1}}(x,a) = \mu^{(k+1)}_{t_{n+1}}(x)\mathbf{1}_{a=\alpha^{(k+1)}_{t_{n+1}}(x)}, & 0 \le n < N_T, (x,a) \in X \times \mathcal{A}.
\end{cases}
$$
(21.2)

Here the evolution of the joint state-action population distribution is simply the product of the state distribution and a Dirac mass:

$$v_{t_n}^{(k+1)} = \mu_{t_n}^{(k+1)} \otimes \delta_{\alpha_{t_n}^{(k+1)}}.$$

This is because we assumed that the optimal control is given by a deterministic function from $\mathcal{X}$ to $\mathcal{A}$. If we were using randomized control, we would need to replace the Dirac mass by the distribution of controls.

To alleviate notation, let us introduce the operators

$$\widetilde{\mathcal{T}} \colon (\Delta^{|\mathcal{X} \times \mathcal{A}|})^{N_T+1} \to (\mathbb{R}^{|\mathcal{X} \times \mathcal{A}|})^{N_T+1} \quad \text{and} \quad \widetilde{\mathcal{P}} \quad (\mathbb{R}^{|\mathcal{X} \times \mathcal{A}|})^{N_T+1} \to (\Delta^{|\mathcal{X} \times \mathcal{A}|})^{N_T+1}$$

such that (21.1) and (21.2) become

$$Q^{(k+1)} = \widetilde{\mathcal{T}}(v^{(k)}), \qquad v^{(k+1)} = \widetilde{\mathcal{P}}(Q^{(k+1)}).$$

If this iteration procedure converges, we have $Q^{(k+1)} \to Q^{(\infty)}, v^{(k+1)} \to v^{(\infty)}$ as $k \to +\infty$ for some $Q^{(\infty)}, v^{(\infty)}$ satisfying

$$Q^{(\infty)} = \widetilde{\mathcal{T}}(v^{(\infty)}), \qquad v^{(\infty)} = \widetilde{\mathcal{P}}(Q^{(\infty)}),$$

which implies that $v^{(\infty)}$ is the state-action equilibrium distribution of the MFG solution, and the associated equilibrium control is given by

$$\alpha_{t_n}^{(\infty)}(x) = \arg\min_a Q_n^{(\infty)}(x, a)$$

for each n.

However, this procedure fails to converge in many MFG by lack of strict contraction property. To remedy this issue, a simple twist is to introduce some kind of damping. Building on this idea, we introduce the following iterative procedure, where $(\rho_Q^{(k)})_{k \geq 0}$ and $(\rho_v^{(k)})_{k \geq 0}$ are two sequences of learning rates:

$$Q^{(k+1)} = (1 - \rho_Q^{(k)})Q^{(k)} + \rho_Q^{(k)}\widetilde{\mathcal{T}}(v^{(k)}), \qquad v^{(k+1)} = (1 - \rho_v^{(k)})v^{(k)} + \rho_v^{(k)}\widetilde{\mathcal{P}}(Q^{(k+1)}).$$

For the sake of brevity, let us introduce the operators

$$\mathcal{T} \colon (\mathbb{R}^{|\mathcal{X} \times \mathcal{A}|})^{N_T+1} \times (\Delta^{|\mathcal{X} \times \mathcal{A}|})^{N_T+1} \to (\mathbb{R}^{|\mathcal{X} \times \mathcal{A}|})^{N_T+1}$$

and

$$\mathcal{P} \colon (\mathbb{R}^{|\mathcal{X} \times \mathcal{A}|})^{N_T+1} \times (\Delta^{|\mathcal{X} \times \mathcal{A}|})^{N_T+1} \to (\Delta^{|\mathcal{X} \times \mathcal{A}|})^{N_T+1}$$

$$\mathcal{T}(Q, v) = \widetilde{\mathcal{T}}(v) - Q, \qquad \mathcal{P}(Q, v) = \widetilde{\mathcal{P}}(Q) - v.$$

Then the above iterations can be written as

$$Q^{(k+1)} = Q^{(k)} + \rho_Q^{(k)}\mathcal{T}(Q^{(k)}, v^{(k)}), \qquad v^{(k+1)} = v^{(k)} + \rho_v^{(k)}\mathcal{P}(Q^{(k+1)}, v^{(k)}). \quad (21.3)$$

If $\rho_v^{(k)} < \rho_Q^{(k)}$, the Q-function is updated at a faster rate, while it is the converse if $\rho_v^{(k)} > \rho_Q^{(k)}$. We can thus intuitively guess that these two regimes should converge to different limits. Similar ideas were studied by Borkar (1997, 2008) in the so-called two-timescales approach. The key insight comes from rewriting the (discrete time) iterations in continuous time as a pair of ODEs. From Borkar (2008, Chapter 6, Theorem 2), we expect to have the following two situations:

- If $\rho_v^{(k)} < \rho_Q^{(k)}$, the system (21.3) tracks the ODE system

$$\dot{Q}^{(t)} = \frac{1}{\epsilon}\mathcal{T}(Q^{(t)}, v^{(t)}),$$
$$\dot{v}^{(t)} = \mathcal{P}(Q^{(t)}, v^{(t)}),$$

where $\rho_v^{(k)}/\rho_Q^{(k)}$ is thought of being of order $\epsilon \ll 1$. Hence, for any fixed $\tilde{v}$, the solution of

$$\dot{Q}^{(t)} = \frac{1}{\epsilon}\mathcal{T}(Q^{(t)}, \tilde{v}),$$

is expected to converge as $\epsilon \to 0$ to a $Q^{\tilde{v}}$ such that $\mathcal{T}(Q^{\tilde{v}}, \tilde{v}) = 0$. This condition can be interpreted as the fact that $Q^{\tilde{v}} = (Q_{t_n}^{\tilde{v}})_{n=0,\dots,N_T}$ is the state-action value function of an infinitesimal agent facing the crowd distribution sequence $\tilde{v} = (\tilde{v}_{t_n})_{n=0,\dots,N_T}$. Then the second ODE becomes

$$\dot{v}^{(t)} = \mathcal{P}(Q^{v^{(t)}}, v^{(t)}),$$

which is expected to converge as $t \to +\infty$ to a $v^{(\infty)}$ satisfying

$$\mathcal{P}(Q^{v^{(\infty)}}, v^{(\infty)}) = 0.$$

This condition means that $v^{(\infty)}$ and the associated control given by $\hat{\alpha}_n(x) = \arg\min_a Q_n^{v^{(\infty)}}(x, a)$ form a Nash equilibrium.
- If $\rho_v^{(k)} > \rho_Q^{(k)}$, the system (21.3) tracks the ODE system

$$\dot{Q}^{(t)} = \mathcal{T}(Q^{(t)}, v^{(t)}),$$
$$\dot{v}^{(t)} = \frac{1}{\epsilon}\mathcal{P}(Q^{(t)}, v^{(t)}),$$

where $\rho_Q^{(k)}/\rho_v^{(k)}$ is thought of being of order $\epsilon \ll 1$. Here, for any fixed $\tilde{Q}$, the solution of

$$\dot{v}^{(t)} = \frac{1}{\epsilon}\mathcal{P}(\tilde{Q}, v^{(t)}),$$

is expected to converge as $\epsilon \to 0$ to a $v^{\tilde{Q}}$ such that $\mathcal{P}(\tilde{Q}, v^{\tilde{Q}}) = 0$, meaning that $v^{\tilde{Q}} = (v_{t_n}^{\tilde{Q}})_{n=0,\dots,N_T}$ is the distribution evolution of a population in which every agent uses control $\tilde{\alpha}_{t_n}(x) = \arg\min_a \tilde{Q}_n(x, a)$ at time t_n. In fact, the definitions of $\tilde{\alpha}_{t_n}$ and $v^{\tilde{Q}}$ need to be *modified* to take into account the first action (x, a). For the details of this crucial step for handling MFC, we refer to Angiuli et al. (2022).

Then the first ODE becomes

$$\dot{Q}^{(t)} = \frac{1}{\epsilon}\mathcal{T}(Q^{(t)}, v^{Q^{(t)}}),$$

which is expected to converge as $t \to +\infty$ to a $Q^{(\infty)}$ such that

$$\mathcal{T}(Q^{(\infty)}, v^{Q^{(\infty)}}) = 0.$$

This condition (in the *modified* MFC setup) means that the control $\hat{\alpha}_{t_n}(x) = \arg\min_a Q_n^{(\infty)}(x, a)$ is an MFC optimum and the induced optimal distribution is $v^{Q^{(\infty)}}$.

The above iterative procedure is purely deterministic and allows us to understand the rationale behind the two-timescale approach. However, in practice we rarely have access to the operators $\mathcal{T}$ and $\mathcal{P}$. Instead, we will consider that we only have access to noisy versions and we use intuition from stochastic approximation to design an algorithm. Instead of assuming that we know the dynamics or the reward functions, we will simply assume that the learning agent can interact with an environment from which she can sample stochastic transitions.

21.4 Reinforcement learning algorithm

21.4.1 Reinforcement learning

Reinforcement learning is a branch of machine learning which studies algorithms to solve an MDP based on trials and errors. An MDP describes the sequential interaction of an agent with an environment. Let $\mathcal{X}$ and $\mathcal{A}$ be the state and action space respectively. At each time t_n, the agent observes its current state $X_{t_n} \in \mathcal{X}$ and chooses an action $A_{t_n} \in \mathcal{A}$. Due to the agent's action, the environment provides the new state of the agent $X_{t_{n+1}}$ and incurs a cost $f_{t_{n+1}}$. The goal of the agent is to find an optimal strategy (or policy) π^* which assigns to each state an action in order to minimize the aggregated discounted costs. The aim of RL is to design methods which allow the agent to learn (an approximation of) π^* by making repeated use of the environment's outputs but without knowing how the environment produces the new state and the associated cost. A detailed overview of this field can be found in Sutton and Barto (2018) (although RL methods are often presented with reward maximization objectives, we consider cost minimization problems for the sake of consistency with the MFG literature).

Here and in what follows, we use *policy π* instead of control α as the algorithm uses in fact ε-greedy policies. In the limit $\varepsilon \to 0$ the optimal policy is in fact a deterministic control.

21.4.2 Algorithm

In this section we propose an extension of the Unified Two-Timescales Mean Field Q-learning (U2-MF-QL) algorithm discussed in our previous work (Angiuli et al., 2022).

Q-learning is one of the most popular procedure in RL introduced in the seminal work of Watkins (1989). It is designed to solve problems with finite and discrete state and action spaces, $\mathcal{X}$ and $\mathcal{A}$. It is based on the evaluation of the optimal action-value function $Q^*(x, a)$, defined in the case of an infinite horizon minimization problem as

$$Q^*(x,a) = \min_{\pi} \mathbb{E}\left[\sum_{n=0}^{\infty} \rho^n f_{t_{n+1}}(X_{t_n}, \pi(X_{t_n})) \,\middle|\, X_{t_0} = x, A_{t_0} = a\right],$$

which represents the optimal expected aggregated discounted cost when starting in the state x and choosing the first action to be a. The optimal action at state x is provided by the argmin of $Q^*(x, \cdot)$, i.e., $\pi^*(x) = \arg\min_{\mathcal{A}} Q^*(x, \cdot)$. However Q^* is a priori unknown. In order to learn Q^* by trials and errors, an approximate version

Q of the table Q^* is constructed through a stochastic approximation procedure based on the Bellman equation given by

$$Q^*(x, a) = \mathbb{E}\left[f_{t_0}(X_{t_0}, A_{t_0}) + \rho Q^*(X_{t_1}, \pi^*(X_{t_1})) \,\middle|\, X_{t_0} = x, A_{t_0} = a \right]. \quad (21.4)$$

At each step, an action is taken, which leads to a cost and to a new state. On the one hand, it is interesting to act efficiently in order to avoid high costs, and on the other hand it is important to improve the quality of the table Q by trying actions and states which have not been visited many times so far. This is the so-called exploitation–exploration trade-off. The trade-off between exploration of the unknown environment and exploitation of the currently available information is taken care of by an ε-greedy policy based on Q. The algorithm chooses the action that minimizes the immediate cost with probability $1 - \varepsilon$, and a random action otherwise.

The U2-MF-QL algorithm represents a unified approach to solve asymptotic Mean Field Games and Mean Field Control problems based on the relationship between two learning rates relative to the update rules of the Q table and the distribution of the population μ respectively. Based on the intuition presented in §21.3, a choice of learning rates (ρ^Q, ρ^μ) such that $\rho^Q > \rho^\mu$ allows the algorithm to solve an MFG problem. The estimation of Q is updated at a faster pace with respect to the distribution which behaves as quasi-static mimicking the freezing of the flow of measures characteristic of the solving scheme discussed in §21.2.1. On the other hand, learning rates satisfying $\rho^Q < \rho^\mu$ allow the algorithm to update instantaneously the control function (Q table) at any change of the distribution reproducing the MFC framework. Under suitable assumptions, one may expect the asymptotic problems to be characterized by controls that are independent of time. In this case, the learning goals reduce to a control function valid for every time point and the asymptotic distribution of the states of the population.

The finite horizon framework presented in §§21.2.1 and 21.2.2 differs from the asymptotic case discussed in Angiuli et al. (2022) in several ways other than the restriction on the finite time interval $[0, T]$. First, the mean field interaction is through the joint distribution of states and actions of the population rather than the marginal distribution of the states. Further, both the control rule and the mean field distribution are generally time dependent. Due to these differences, the 2-dimensional matrix Q in U2-MF-QL is replaced by a 3-dimensional matrix $\mathbf{Q} := (Q_n(\cdot, \cdot))_{n=0,\dots,N_T} = (Q(\cdot, \cdot, t_n))_{n=0,\dots,N_T}$ in the finite horizon version of the algorithm (U2-MF-QL-FH). The extra dimension is introduced to learn a time dependent control function.

The Unified Two-Timescales Mean Field Q-learning for Finite Horizon problems (U2-MF-QL-FH) is designed to solve problems with finite state and action spaces in finite and discrete time. The same algorithm can be applied to MFG and MFC problems where the interaction with the population is through the marginal distribution of the states $\mu \in \mathcal{P}(\mathcal{X})$ or the law of the controls $\theta \in \mathcal{P}(\mathcal{A})$. In these cases the estimation of the flow of marginal distributions is obtained through the vectors $(\mu_{t_n})_{n=0,\dots,N_T}$ (resp. $(\theta_{t_n})_{n=0,\dots,N_T}$) defined on

---

Algorithm 21.1 Unified Two-Timescales Mean Field Q-learning – Finite Horizon

---

Require: $\tau = \{t_0 = 0, \ldots, t_{N_T} = T\}$ with $t_0 = 0 < \cdots < t_{N_T} = T$: time steps,

     $X = \{x_0, \ldots, x_{|X|-1}\}$: finite state space,

     $\mathcal{A} = \{a_0, \ldots, a_{|\mathcal{A}|-1}\}$: finite action space,

     μ_0 : initial distribution of the representative player,

     ε : factor related to the ε−greedy policy,

     tol_v, tol_Q : break rule tolerances.

1: **Initialization:** $Q_n(\cdot, \cdot) := Q(\cdot, \cdot, t_n) = 0$ for all $(x, a) \in X \times \mathcal{A}$, for all $t_n \in \tau$,

     $v_{t_n}^0 = \frac{1}{|X \times \mathcal{A}|} J_{|X| \times |\mathcal{A}|}$ for $n = 0, \ldots, N_T$ where $J_{d \times m}$ is an $d \times m$ unit matrix

2: **for** each episode $k = 1, 2, \ldots$ **do**

3:      **Observe** $X_{t_0}^k \sim \mu_0$

4:      **for** $n \leftarrow 0$ to N_T **do**

5:          **Choose action** $A_{t_n}^k$ using the ε-greedy policy derived from $Q_n^{k-1}(X_{t_n}^k, \cdot)$

6:          **Update** v:

         $v_{t_n}^k = v_{t_n}^{k-1} + \rho_k^v(\delta(X_{t_n}^k, A_{t_n}^k) - v_{t_n}^{k-1})$

         where $\delta(X_{t_n}^k, A_{t_n}^k) = \left(\mathbf{1}_{x,a}(X_{t_n}^k, A_{t_n}^k)\right)_{x \in X, a \in \mathcal{A}}$

         Observe cost $f_{t_{n+1}} = f(X_{t_n}^k, A_{t_n}^k, v_{t_n}^k)$ and state $X_{t_{n+1}}^k$ provided by the environment

7:          **Update** Q_n:

         $Q_n^k(x, a) :=$

$$:= \begin{cases} Q_n^{k-1}(x, a) + \rho_{x,a,k}^{Q_n}[\mathcal{B} - Q_n^{k-1}(x, a)] & \text{if } (X_{t_n}^k, A_{t_n}^k) = (x, a) \\ Q_n^{k-1}(x, a) & \text{o.w.} \end{cases}$$

         where

$$\mathcal{B} := \begin{cases} f_{t_{n+1}} + \rho \min_{a' \in \mathcal{A}} Q_{n+1}^{k-1}(X_{t_{n+1}}^k, a'), & \text{if } t_n < T \\ f_{t_{n+1}}, & \text{o.w.} \end{cases}$$

8:      **end for**

9:      **if** $\|v_{t_n}^k - v_{t_n}^{k-1}\|_1 \leq tol_v$ and $\|Q_n^k - Q_n^{k-1}\|_{1,1} < tol_Q$ for all $n = 0, \ldots, N_T$ **then**

10:          break

11:      **end if**

12: **end for**

---

the space X (resp. $\mathcal{A}$). The initialization is given by $\mu_{t_n}^0 = [\frac{1}{|X|}, \ldots, \frac{1}{|X|}]$ (resp. $\theta_{t_n}^0 = [\frac{1}{|\mathcal{A}|}, \ldots, \frac{1}{|\mathcal{A}|}]$), for $n = 0, \ldots, N_T$. The update rule at episode k is given by $\mu_{t_n}^k = \mu_{t_n}^{k-1} + \rho_k^\mu(\delta(X_{t_n}) - \mu_{t_n}^{k-1})$ (resp. $\theta_{t_n}^k = \theta_{t_n}^{k-1} + \rho_k^\theta(\delta(A_{t_n}) - \theta_{t_n}^{k-1})$, where $\delta(X_{t_n}) = [\mathbf{1}_{x_0}(X_{t_n}), \ldots, \mathbf{1}_{x_{|X|-1}}(X_{t_n})]$ (resp. $\delta(A_{t_n}) = [\mathbf{1}_{a_0}(A_{t_n}), \ldots, \mathbf{1}_{a_{|\mathcal{A}|-1}}(A_{t_n})]$), for $n = 0, \ldots, N_T$.

21.4.3 Learning rates

Algorithm 21.1 is based on two stochastic approximation rules for the distribution v and the 3D matrix Q. The design of the learning is discussed widely in the literature, in a general context by Borkar (1997) and Borkar (2008), and with focus in reinforcement learning by Borkar and Konda (1997) and Even-Dar and Mansour (2003). Based on experimental evidences, we define the learning rates appearing in Algorithm 21.1 as follows:

$$\rho^{Q_n}_{x,a,k} = \frac{1}{(1 + N_T \#|(x,a,t_n,k)|)^{\omega^Q}}, \qquad \rho^\nu_k = \frac{1}{(1 + k)^{\omega^\nu}}, \qquad (21.5)$$

where $\#|(x,a,t_n,k)|$ is counting the number of visits of the pair (x,a) at time t_n until episode k. Differently from the asymptotic version of the algorithm presented in Angiuli et al. (2022) for which each pair (x,a) has a unique counter for all time points, in the finite horizon formulation a distinct counter $\#|(x,a,t_n,k)|$ is defined for each time point t_n. This choice of learning rates allows to update each matrix Q_n in an asynchronous way. The exponent ω^Q can take values in $(\frac{1}{2}, 1]$. As presented in §21.3.3, the pair (ω^Q, ω^ν) is chosen depending on the particular problem to solve. In a competitive framework (MFG), these parameters have to be searched in the set of values for which the condition $\rho^Q > \rho^\nu$ is satisfied at each iteration. On the other hand, a good choice for the cooperative case (MFC) should satisfy the condition $\rho^Q < \rho^\nu$.

21.4.4 Application to continuous problems

Although it is presented in a setting with finite state and action spaces, the application of the algorithm U2-MF-QL-FH can be extended to continuous problems. Such adaptation requires truncation and discretization procedures to time, state and action spaces which should be calibrated based on the specific problem.

In practice, a continuous time interval $[0, T]$ would be replaced by a uniform discretization $\tau = \{t_n\}_{n \in \{0,\dots,N_T\}}$. The environment would provide the new state and reward at these discrete times. The continuous state would be projected on a finite set $\mathcal{X} = \{x_0, \dots, x_{|\mathcal{X}|-1}\} \subset \mathbb{R}^d$. Likewise, actions will be provided to the environment in a finite set $\mathcal{A} = \{a_0, \dots, a_{|\mathcal{A}|-1}\} \subset \mathbb{R}^k$, where the projected distribution ν would be estimated. Then Algorithm 21.1 is run on those spaces.

In the problems presented in §21.6, we will use the benchmark linear-quadratic models given in continuous time and space for which we present explicit formulas. In that case, we use an Euler discretization of the dynamics followed by a projection on $\mathcal{X}$. We do not address here the error of approximation since the purpose of this comparison with a benchmark is mainly for illustration.

21.5 A mean field accumulation problem

21.5.1 Description of the problem

A further application of mean field theory to economics was given by the mean field capital accumulation problem in Huang (2013). In that paper, the author studied an extension of the the classical one-agent modeling of optimal stochastic growth to an infinite population of symmetric agents. We introduce the model following the author's presentation.

At discrete time $t \in \mathbb{Z}_+$, the wealth of the representative agent is represented by a process $X_t^{\alpha,\theta}$ characterized by the dynamics

$$X_{t+1}^{\alpha,\theta} = G\left(\int a d\theta_t(a), W_t\right) \alpha_t \tag{21.6}$$

where $\alpha = (\alpha_t)_{0 \le t \le T}$ is the controlled variable denoting the agent's investment for production, $G\left(\int a d\theta_t(a), W_t\right)$ is the production function, $\theta = (\theta_t)_{0 \le t \le T}$ is the mean field term represented by the law of the investment level of the population, $\int a d\theta_t(a)$ is its mean, and $W = (W_t)_{0 \le t \le T}$ is a random disturbance. At each time t, the control α_t can only take values in $[0, X_t^{\alpha,\theta}]$ so that $\text{Supp}(\theta_t) \subseteq [0, X_t^{\alpha,\theta}]$, implying that borrowing is not allowed. The wealth remaining after investment is all consumed, i.e. the consumption variable c_t is equal to $c_t = X_t^{\alpha,\theta} - \alpha_t$. The model is based on the following assumptions:

(A1) W is a random noise source with support D_W. The initial state X_{t_0} is a positive random variable independent of W with mean m_0;

(A2) The function $G : [0, \infty) \times D_W \mapsto [0, \infty)$ is continuous. If $w \in D_W$ is fixed, $G(z, w)$ is a decreasing function of z;

(A3) $\mathbb{E}G(0, W) < \infty$ and $\mathbb{E}G(z, W) > 0$ for each $z \in [0, \infty)$.

The multiplicative factor G in the dynamics of the wealth process $X_t^{\alpha,\theta}$ shows the direct dependence of the wealth on both the individual investment and the population aggregated investment. Further, assumption (A2) relates to the negative mean field impact explained as the loss in production efficiency when the aggregated investment increases. An example for the function G is given by

$$G(z, w) = \frac{\beta w}{1 + \delta z^\eta},$$

where β, δ, η are non negative parameters. Let W be a positive random noise with mean equal to 1. Then $D_W \subset [0, \infty)$ and (A2)–(A3) are satisfied.

The goal of the agent is to optimize the expected aggregated discounted utility of consumption given by

$$J(\alpha, \theta) = \mathbb{E}\sum_{t=0}^{T} \rho^t v(c_t) = \mathbb{E}\sum_{t=0}^{T} \rho^t v(X_t^{\alpha,\theta} - \alpha_t), \tag{21.7}$$

where $\rho \in (0, 1]$ is the discount factor. In particular, Huang (2013) analysed the case of a Hyperbolic Absolute Risk Aversion (HARA) utility function defined as

$$v(c_t) = v(X_t^{\alpha,\theta} - \alpha_t) := \frac{1}{\gamma}(X_t^{\alpha,\theta} - \alpha_t)^\gamma, \tag{21.8}$$

where $\gamma \in (0, 1)$.

21.5.2 Solution of the MFG

In a competitive game setting, the resulting mean field game problem has solution given by Theorem 3 of Section 3.2 and Theorem 6 of Section 4 in Huang (2013). Let denote the functions $\Phi(z)$, $\phi(z)$ and $\Psi(z)$ as follows

$$\Phi(z) = \rho\mathbb{E}G^\gamma(z,W), \quad \phi(z) = \Phi(z)^{\frac{1}{\gamma-1}}, \quad \Psi(z) = \mathbb{E}G(z,W).$$

Suppose that the mean field interaction is through $(z_t)_{t=0,\dots,T}$ the first moment of the flow of measures $\boldsymbol{\theta} = (\theta_t)_{t=0,\dots,T}$. The relative value function is defined as

$$V^{\boldsymbol{\theta}}(t,x) = \sup_\alpha \mathbb{E}\left[\sum_{s=t}^T \rho^s v(X_s^{\alpha,\theta} - \alpha_s)|X_t^{\alpha,\theta} = x\right].$$

The value function is equal to $V^{\boldsymbol{\theta}}(t,x) = \frac{1}{\gamma}D_t^{\gamma-1}x^\gamma$, where D_t can be obtained using the recursive formula

$$D_t = \frac{\phi(z_t)D_{t+1}}{1 + \phi(z_t)D_{t+1}}, \qquad D_T = 1.$$

The optimal control w.r.t. $\boldsymbol{\theta}$ is given by

$$\hat{\alpha}_t(x) = \frac{x}{1 + \phi(z_t)D_{t+1}}, \quad t \leq T-1, \qquad \hat{\alpha}_T = 0.$$

The equivalent of the Nash equilibrium in the mean field limit is obtained by solving the fixed point equation

$$(\Lambda_0, \dots, \Lambda_{T-1})(z_0, \dots, z_{T-1}) = (z_0, \dots, z_{T-1}),$$

where

$$
\begin{cases}
\Lambda_0(z_0, \dots, z_{T-1}) := \frac{1+\phi(z_{T-1})+\cdots+\phi(z_{T-1})\cdots\phi(z_1)}{1+\phi(z_{T-1})+\cdots+\phi(z_{T-1})\cdots\phi(z_0)}m_0, \\[2mm]
\Lambda_k(z_0, \dots, z_{T-1}) := \\
\quad := \frac{1+\phi(z_{T-1})+\cdots+\phi(z_{T-1})\cdots\phi(z_{k+1})}{1+\phi(z_{T-1})+\cdots+\phi(z_{T-1})\cdots\phi(z_0)}\Psi(z_{k-1})\cdots\Psi(z_0)m_0, \quad \text{for } 1 \leq k \leq T-2, \\[2mm]
\Lambda_{T-1}(z_0, \dots, z_{T-1}) := \\
\quad := \frac{1}{1+\phi(z_{T-1})+\cdots+\phi(z_{T-1})\cdots\phi(z_0)}\Psi(z_{T-2})\cdots\Psi(z_0)m_0, \quad \text{for } k = T-1.
\end{cases}
$$

Example 21.4 A simple example was proposed in Section 3.3 of Huang (2013). Let T be equal to 2 and (z_0, z_1) be given. The solution is defined by

$$D_0 = \frac{\phi(z_1)\phi(z_0)}{1 + \phi(z_1) + \phi(z_1)\phi(z_0)}, \quad D_1 = \frac{\phi(z_1)}{1 + \phi(z_1)}, \quad D_2 = 1,$$

with controls

$$\hat{\alpha}_0(x) = \frac{(1 + \phi(z_1))x}{1 + \phi(z_1) + \phi(z_1)\phi(z_0)}, \quad \hat{\alpha}_1(x) = \frac{x}{1 + \phi(z_1) + \phi(z_1)\phi(z_0)}, \quad \hat{\alpha}_2(x) = 0.$$

21.5.3 Solution of the MFC

We now turn our attention to the cooperative setting. For this problem, we are not aware of any explicit solution for the social optimum. Instead, we employ the numerical method proposed by Carmona and Laurière (2019) and use the result as a benchmark. We recall how this method works in our context. The initial problem is to minimize over α:

$$J(\alpha) = \mathbb{E} \sum_{t=0}^{T} \rho^t v(c_t) = \mathbb{E} \sum_{t=0}^{T} \rho^t v(X_t^\alpha - \alpha_t),$$

subject to: X_0^α has a fixed distribution and

$$X_{t+1}^\alpha = G(\mathbb{E}[\alpha_t], W_t)\alpha_t, \quad t > 0.$$

This problem is approximated by the following one. We fix an architecture of neural network with input in $\mathbb{R}^2$ and output in $\mathbb{R}$. Such neural networks are going to play the role of the control function, in a Markovian feedback form. The inputs are the time and space variables, and the output is the value of the control. Then the goal is to minimize over parameters ω of neural networks with this architecture the following function:

$$\widetilde{J}^N(\omega) = \mathbb{E}\left[\frac{1}{N}\sum_{i=1}^{N}\sum_{t=0}^{T}\rho^t v(c_t^i)\right] = \mathbb{E}\left[\frac{1}{N}\sum_{i=1}^{N}\sum_{t=0}^{T}\rho^t v(X_t^{i,\varphi_\omega} - \varphi_\omega(t, X_t^{i,\varphi_\omega}))\right],$$

subject to: $X_0^{i,\varphi_\omega}, i = 1, \ldots, N$ are i.i.d. with fixed distribution and

$$X_t^{i,\varphi_\omega} = G\left(\frac{1}{N}\sum_{j=1}^{N}\varphi_\omega(t, X_t^{j,\varphi_\omega}), W_t^i\right)\varphi_\omega(t, X_t^{i,\varphi_\omega}), \quad t > 0, i = 1, \ldots, N.$$

Notice that the parameters ω are used to compute the X_t^{i,φ_ω} for every i and every t. The mean of the control $\mathbb{E}[\alpha_t]$ is replaced by an empirical average over N samples. For this problem, an approximate minimizer is computed by running stochastic gradient descent (SGD for short) or one of its variants. At iteration k, we have a candidate ω_k for the parameters of the neural network. We randomly pick initial positions $\underline{X}_0 := (X_0^{i,\varphi_{\omega_k}})_{i=1,\ldots,N}$ and noises $\underline{W} := (W_t^i)_{t=1,\ldots,T,i=1,\ldots,N}$. Based on this, we simulate trajectories $(X_t^{i,\varphi_{\omega_k}})_{t,i}$ and compute the associated cost, namely the term inside the expectation in the definition of $\widetilde{J}^N(\omega)$:

$$L(\omega_k; \underline{X}_0, \underline{W}) := \frac{1}{N}\sum_{i=1}^{N}\sum_{t=0}^{T}\rho^t v(X_t^{i,\varphi_{\omega_k}} - \varphi_{\omega_k}(t, X_t^{i,\varphi_{\omega_k}})).$$

Using backpropagation, the gradient $\nabla_\omega L(\omega_k; \underline{X}_0, \underline{W})$ of this cost with respect to ω is computed, and it is used to update the parameters. We thus obtain ω_{k+1} defined by:

$$\omega_{k+1} = \omega_k - \eta_k \nabla_\omega L(\omega_k; \underline{X}_0, \underline{W}),$$

where $\eta_k > 0$ is the learning rate used at iteration k. In our implementation for the numerical results presented below, instead of the plain SGD algorithm we used Adam optimizer (Kingma and Ba, 2014).

21.5.4 Numerical results

In this section, numerical results of the application of the U2-MF-QL-FH algorithm to the mean field capital accumulation problem are presented. The interaction with the population is through the law of the controls. Algorithm 21.1 was adapted to this case as discussed in §21.4.2.

The problem analyzed is a specific case of the Example 21.4. For more details we refer to Huang (2013, Sections 6.3 and 7, Example 18).

The production function is defined as follows:

$$G(z, W) = g(z)W, \qquad g(z) = \frac{1}{\rho \mathbb{E}[W^\gamma]} \frac{C}{1 + (C-1)z^3}; \qquad (21.9)$$

here W has support $D_W = \{0.9, 1.3\}$ with corresponding probabilities $[0.75, 0.25]$, C is equal to 3, the discount factor ρ is equal to 0.95 and the parameter γ of the utility function defined in equation (21.8) is equal to 0.2. The distribution of $X_0^{\alpha,\theta}$ is uniform in $[0, 1]$.

This problem is characterized by discrete time and continuous state and action spaces. In order to apply the U2-MF-QL-FH algorithm, these spaces are truncated and discretized as discussed in §21.4.4. They have been chosen large enough to make sure that the state is within the boundary most of the time. In practice, this would have to be calibrated in a model-free way through experiments. In this example, for the numerical experiments, we used the knowledge of the model.

The action space is given by $\mathcal{A} = \{a_0 = 0, \ldots, a_{|\mathcal{A}|-1} = 4\}$ and the state space by $\mathcal{X} = \{x_0 = 0, \ldots, x_{|\mathcal{X}|-1} = 4\}$. The step size for the discretization of the state and action spaces is given by 0.05.

Algorithm 21.1 is adapted to this particular example. Since borrowing is not allowed, the set of admissible action at state x is given by $\mathcal{A}(x) = \{a \in \mathcal{A} \text{ if } a \leq x\} \subseteq \mathcal{A}$. The exploitation-exploration trade off is tackled on each episode using an ε−greedy policy. Supposed that the agent is in state x, the algorithm chooses a random action in $\mathcal{A}(x)$ with probability ε and the action in $\mathcal{A}(x)$ which results optimal based on the current estimation with probability $1 - \varepsilon$. In our example, the value of epsilon is fixed to 0.15.

The following numerical results show how the U2-MF-QL-FH algorithm is able to learn an approximation of the control function and the mean field term in the MFG and MFC cases depending on the choice of the parameters $(\omega^Q, \omega^\theta)$.

Learning of the controls

Figures 21.2a,b, 21.3a,b, 21.4a,b: controls learned by the algorithm.
The controls learned by the U2-MF-QL-FH algorithm are compared with the benchmark solutions. Each plot corresponds to a different time point $t \in \{0, 1, 2\}$. The x−axis represents the state variable x. The y−axis relates to the action $\alpha_t(x)$.

The blue (resp. green) markers show the benchmark control function for the MFG (resp. MFC) problem. The red markers are the controls learned by the algorithm. The plots show how the algorithm converges to different solutions based on the choice of the pair $(\omega^Q, \omega^\theta)$. On the left, the choice $(\omega^Q, \omega^\theta) = (0.55, 0.85)$ produces the approximation of the solution of the MFG. On the right, the set of parameters $(\omega^Q, \omega^\theta) = (0.7, 0.05)$ lets the algorithm learn the solution of the MFC problem. The results presented in the figures are averaged over 10 runs.

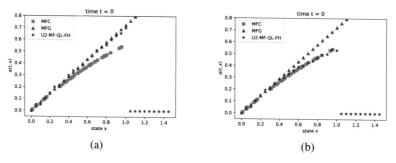

(a)   (b)

Figure 21.2 (a): Learned controls for MFG at time 0. (b): Learned controls for MFC at time 0.

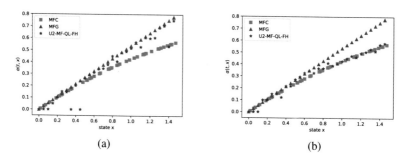

(a)   (b)

Figure 21.3 (a): Learned controls for MFG at time 1. (b) Learned controls for MFC at time 1.

Learning of the mean field

Figures 21.5a,b; 21.6a,b; 21.7a,b: $\mathbb{E}[\alpha_t]$ learned by the algorithm.

The estimation of the first moment of the distribution of the controls evolves with respect the number of learning episodes. The estimated quantity is compared with the benchmarks presented in §§21.5.2 and 21.5.3. Each plot corresponds to a different time point $t \in \{0, 1, 2\}$. The x-axis represents the learning episode k. The y-axis relates to the estimate of the first moment of the mean field $\mathbb{E}[\alpha_t^k]$ obtained by episode k. The blue (resp. green) line shows the benchmark solution for the MFG (resp. MFC) problem. The red dots are the estimates learned by the algorithm. On the left, the algorithm reaches the solution of the MFG based on the parameters $(\omega^Q, \omega^\theta) = (0.55, 0.85)$. On the right, the values $(\omega^Q, \omega^\theta) =$

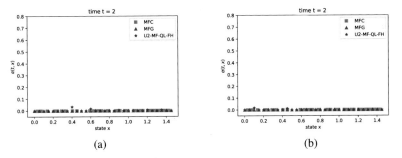

Figure 21.4 (a): Learned controls for MFG at time 2. (b): Learned controls for MFC at time 2.

$(0.7, 0.05)$ allows the algorithm to converge to the solution of the MFC problem. The results presented in the figures are averaged over 10 runs.

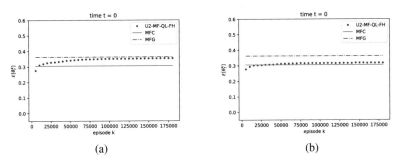

Figure 21.5 (a): Learned control's mean for MFG at time 0. (b): Learned control's mean for MFC at time 0.

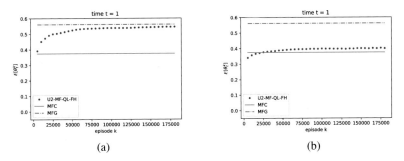

Figure 21.6 (a): Learned control's mean for MFG at time 1. (b): Learned control's mean for MFC at time 1.

21.6 A mean field execution problem

We now consider the *Price Impact Model* as an example of the application to finance originally studied by Carmona and Lacker (2015), and presented in

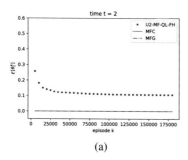 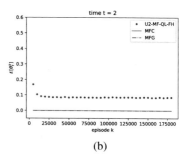

(a)   (b)

Figure 21.7 (a): Learned control's mean for MFG at time 2. (b): Learned control's mean for MFC at time 2.

Carmona and Delarue (2018a, Sections 1.3.2 and 4.7.1). This model addresses the question of optimal execution in the context of high frequency trading when a large group of traders want to buy or sell shares before a given time horizon T (e.g., one day). The price of the stock is influenced by the actions of the traders: if they buy, the price goes up, whereas if they sell, the price goes down. This effect is stronger if a significant proportion of traders buy or sell at the same time. Incorporating such a price impact naturally leads to a problem with mean field interactions through the traders' actions.

Approaching this problem as a mean field game, the inventory of the representative trader is modeled by a stochastic process $(X_t)_{0 \leq t \leq T}$ such that

$$dX_t = \alpha_t dt + \sigma dW_t, \quad t \in [0, T],$$

where α_t corresponds to the trading rate and W is a standard Brownian motion. The price of the asset $(S_t)_{0 \leq t \leq T}$ is influenced by the trading strategies of all the traders through the mean of the law of the controls $(\theta_t = \mathcal{L}(\alpha_t))_{0 \leq t \leq T}$ as follows:

$$dS_t = \gamma \left(\int_{\mathbb{R}} a d\theta_t(a) \right) dt + \sigma_0 dW_t^0, \quad t \in [0, T],$$

where γ and σ_0 are constants and the Brownian motion W^0 is independent from W. The amount of cash held by the trader at time t is denoted by the process $(K_t)_{0 \leq t \leq T}$. The dynamic of K is modeled by

$$dK_t = -[\alpha_t S_t + c_\alpha(\alpha_t)]dt,$$

where the function $\alpha \mapsto c_\alpha(\alpha)$ is a non-negative convex function satisfying $c_\alpha(0) = 0$, representing the cost for trading at rate α. The wealth V_t of the trader at time t is defined as the sum of the cash held by the trader and the value of the inventory with respect to the price S_t:

$$V_t = K_t + X_t S_t.$$

Applying the self-financing condition from Black–Scholes theory, the changes over time of the wealth V are given by the equation:

$$dV_t = dK_t + X_t dS_t + S_t dX_t$$

$$= \left[-c_\alpha(\alpha_t) + \gamma X_t \int_{\mathbb{R}} a d\theta_t(a) \right] dt + \sigma S_t dW_t + \sigma_0 X_t dW_t^0. \tag{21.10}$$

We assume that the trader is subject to a running liquidation constraint modeled by a function c_X of the shares they hold, and to a terminal liquidation constraint at maturity T represented by a scalar function g. Thus, the cost function is defined by:

$$J(\alpha) = \mathbb{E}\left[\int_0^T c_X(X_t) dt + g(X_T) - V_T \right],$$

where the terminal wealth V_T is taken into account with a negative sign as the cost function is to be minimized. From equation (21.10), it follows that

$$J(\alpha) = \mathbb{E}\left[\int_0^T f(t, X_t, \theta_t, \alpha_t) dt + g(X_T) \right],$$

where the running cost is defined by

$$f(t, x, \theta, \alpha) = c_\alpha(\alpha) + c_X(x) - \gamma x \int_{\mathbb{R}} a d\theta(a),$$

for $0 \le t \le T$, $x \in \mathbb{R}^d$, $\theta \in \mathcal{P}(\mathbb{A})$ and $\alpha \in \mathbb{A} = \mathbb{R}$. We assume that the functions c_X and g are quadratic and that the function c_α is strongly convex in the sense that its second derivative is bounded away from 0. Such a particular case is known as the Almgren-Chriss linear price impact model. Thus, the control is chosen to minimize:

$$J(\alpha) = \mathbb{E}\left[\int_0^T \left(\frac{c_\alpha}{2} \alpha_t^2 + \frac{c_X}{2} X_t^2 - \gamma X_t \int_{\mathbb{R}} a d\theta_t(a) \right) dt + \frac{c_g}{2} X_T^2 \right],$$

over $\alpha \in \mathbb{A}$. To summarize, the running cost consists of three components. The first term represents the cost for trading at rate α. The second term takes into consideration the running liquidation constraint in order to penalize unwanted inventories. The third term defines the actual price impact. Finally, the terminal cost represents the terminal liquidation constraint.

21.6.1 The MFG trader problem

Referring to §21.2.1, the MFG problem is solved by first solving a standard stochastic control problem where the flow of distribution of control is given and then, solving a fixed point problem ensuring that this flow of distribution is identical to the flow of distributions of the optimal control. We adopt here the FBSDE approach where the backward variable represents the derivative of the value function. In other words, the optimal control is obtained by minimizing the Hamiltonian

$$H(x, \alpha, \theta, y) = \left(\frac{c_\alpha}{2} \alpha^2 + \frac{c_X}{2} x^2 - \gamma x \int_{\mathbb{R}} a d\theta(a) \right) + \alpha y, \tag{21.11}$$

to obtain

$$\hat{\alpha}_t = -\frac{1}{c_\alpha} Y_t, \tag{21.12}$$

where (X,Y) solves the FBSDE system obtained via the Pontryagin approach:

$$\begin{cases} dX_t = -\dfrac{1}{c_\alpha} Y_t \, dt + \sigma \, dW_t, & X_0 \sim \mu_0 \\[2mm] dY_t = -\left(c_X X_t + \dfrac{\gamma}{c_\alpha} \mathbb{E}[Y_t] \right) dt + Z_t \, dW_t, & Y_T = c_g X_T. \end{cases} \tag{21.13}$$

21.6.2 Solution of the MFG problem

The solution of the mean field game case is discussed in detail in Carmona and Delarue (2018a, Sections 1.3.2 and 4.7.1). In a nutshell, one takes expectation in (21.13) to obtain a system of forward-backward ODEs for the mean of X_t denoted by $\bar{x}_t$ and the mean of Y_t denoted by $\bar{y}_t$. This system is solved using the ansatz $\bar{y}_t = \bar{\eta}_t \bar{x}_t + \bar{\chi}_t$. The coefficient function $\bar{\eta}_t$ satisfies a Riccati equation which admits the solution:

$$\bar{\eta}_t = \frac{-C(e^{(\delta^+ - \delta^-)(T-t)} - 1) - c_g(\delta^+ e^{(\delta^+ - \delta^-)(T-t)} - \delta^-)}{(\delta^- e^{(\delta^+ - \delta^-)(T-t)} - \delta^+) - c_g B(e^{(\delta^+ - \delta^-)(T-t)} - 1)},$$

for $t \in [0,T]$, where $B = 1/c_\alpha$, $C = c_X$, $\delta^\pm = -D \pm \sqrt{R}$, with $D = -\gamma/(2c_\alpha)$, $R = D^2 + BC$ and $\bar{x}_0 = \mathbb{E}[X_0]$. Additionally, we found $\bar{\chi}_t = 0$, and

$$\bar{x}_t = \bar{x}_0 e^{-\int_0^t \frac{\bar{\eta}_s}{c_\alpha} ds}.$$

The FBSDE system (21.13) is solved by replacing $\mathbb{E}[Y_t]$ with the explicit expression for $\bar{y}_t = \bar{\eta}_t \bar{x}_t + \bar{\chi}_t$, and using the ansatz $Y_t = \eta_t X_t + \chi_t$. One finds the following explicit formulas for the coefficient functions η_t and χ_t:

$$\eta_t = -c_\alpha \sqrt{c_X/c_\alpha} \frac{c_\alpha \sqrt{c_X/c_\alpha} - c_g - (c_\alpha \sqrt{c_X/c_\alpha} + c_g) e^{2\sqrt{c_X/c_\alpha}(T-t)}}{c_\alpha \sqrt{c_X/c_\alpha} - c_g + (c_\alpha \sqrt{c_X/c_\alpha} + c_g) e^{2\sqrt{c_X/c_\alpha}(T-t)}},$$

$$\chi_t = (\bar{\eta}_t - \eta_t)\bar{x}_t.$$

Finally, the optimal control (21.12) is given by $\hat{\alpha}_t = \hat{\alpha}(t, X_t)$ where

$$\hat{\alpha}(t,x) = -\frac{1}{c_\alpha}\left(\eta_t x + (\bar{\eta}_t - \eta_t)\bar{x}_t \right). \tag{21.14}$$

21.6.3 The MFC trader problem

In the case of mean field control (i.e., control of McKean–Vlasov dynamics), following Acciaio et al. (2018, Theorem 3.2) and Laurière and Tangpi (2020, Section 5.3.2), we find that the optimal control is given by

$$\alpha_t^* = -\frac{1}{c_\alpha}(Y_t - \gamma \mathbb{E}[X_t]), \tag{21.15}$$

which differs from the equilibrium control (21.12) from the MFG solution because the optimality condition in the MFC case involves the derivative of the Hamiltonian (21.11) with respect to the distribution of controls. More precisely, we have

$$0 = \partial_\alpha H(X_t, \alpha_t, \theta_t, Y_t) + \tilde{\mathbb{E}}\left[\partial_\theta H(\tilde{X}_t, \tilde{\alpha}_t, \tilde{\theta}_t, \tilde{Y}_t)(\alpha_t)\right] = c_\alpha \alpha_t + Y_t - \gamma \mathbb{E}[X_t].$$

Then, the corresponding FBSDE system becomes

$$\begin{cases} dX_t = -\dfrac{1}{c_\alpha}(Y_t - \gamma\mathbb{E}[X_t])\,dt + \sigma dW_t, \quad X_0 \sim \mu_0 \\ dY_t = -\left(c_X X_t + \dfrac{\gamma}{c_\alpha}\mathbb{E}[Y_t] - \dfrac{\gamma^2}{c_\alpha}\mathbb{E}[X_t]\right)dt + Z_t dW_t, \quad Y_T = c_g X_T. \end{cases} \quad (21.16)$$

As a consequence, the two FBSDE systems, (21.13) and (21.16) respectively, for MFG and MFC differ.

21.6.4 Solution of the MFC problem

The approach used to obtain the solution of the MFC problem is similar to what was presented in §21.6.2 for the MFG, but taking into consideration the extra terms due to the derivative of the Hamiltonian with respect to the distribution of controls.

First, taking the expectation in (21.16), one obtains the following system of forward-backward ODEs:

$$\begin{cases} \dot{\bar{x}}_t = -\dfrac{1}{c_\alpha}(\bar{y}_t - \gamma\bar{x}_t), \quad \bar{x}_0 = x_0, \\ \dot{\bar{y}}_t = -\left(c_X \bar{x}_t + \dfrac{\gamma}{c_\alpha}\bar{y}_t - \dfrac{\gamma^2}{c_\alpha}\bar{x}_t\right), \quad \bar{y}_T = c_g\bar{x}_T. \end{cases} \quad (21.17)$$

Using the ansatz $\bar{y}_t = \bar{\phi}_t \bar{x}_t + \bar{\psi}_t$, we deduce that the coefficient functions $\bar{\phi}_t$ and $\bar{\psi}_t$ must satisfy

$$\begin{cases} \dot{\bar{\phi}}_t + 2\dfrac{\gamma}{c_\alpha}\bar{\phi}_t - \dfrac{1}{c_\alpha}\bar{\phi}_t^2 + c_X - \dfrac{\gamma^2}{c_\alpha}, \quad \bar{\phi}_T = c_g, \\ \dot{\bar{\psi}}_t + \dfrac{1}{c_\alpha}(\gamma - \bar{\phi}_t)\bar{\psi}_t = 0, \quad \bar{\psi}_T = 0. \end{cases} \quad (21.18)$$

From the second equation, we get $\bar{\psi}_t = 0$ for all $t \in [0, T]$, and solving the Riccati equation for $\bar{\phi}_t$, we obtain:

$$\bar{\phi}_t = -\frac{1}{R}\frac{(c_2 + Rc_g)c_1 e^{(T-t)(c_2-c_1)} - c_2(c_1 + Rc_g)}{(c_2 + Rc_g)e^{(T-t)(c_2-c_1)} - (c_1 + Rc_g)}, \quad (21.19)$$

where $c_{1/2} = \frac{-a \pm \sqrt{a^2 - 4b}}{2}$ are the roots of $c^2 + ac + b = 0$, with $a = 2\gamma R$, $b = R(\gamma^2 R - c_X)$, and $R = 1/c_\alpha$.

Using $\bar{y}_t = \bar{\phi}_t \bar{x}_t$ in the first equation of (21.17), we obtain a first-order linear equation for $\bar{x}_t$ which admits the solution

$$\bar{x}_t = \bar{x}_0 e^{-\frac{1}{c_\alpha}\left(\int_0^t \bar{\phi}_s ds - \gamma t\right)}. \quad (21.20)$$

The solution of the McKean–Vlasov FBSDE system (21.16) is obtained using the ansatz $Y_t = \phi_t X_t + \psi_t$. Observe that the drift terms in the equations for Y_t in the systems (21.13) and (21.16) have the same linear component $-c_X X_t$. Due to this similarity, the slope coefficient functions η_t and ϕ_t are identical;

$$\eta_t = \phi_t, \quad \text{for all} \quad t \in [0, T].$$

However, the function $\psi_t = (\bar{\phi}_t - \phi_t)\bar{x}_t$ differs from χ_t in the MFG case due to the new formulations of $\bar{\phi}_t$ and $\bar{x}_t$ given in (21.19) and (21.20). Finally, the optimal control (21.15) is given by $\alpha_t^* = \alpha^*(t, X_t)$ where

$$\alpha^*(t, x) = -\frac{1}{c_\alpha}\left(\phi_t x + (\bar{\phi}_t - \phi_t - \gamma)\bar{x}_t\right). \tag{21.21}$$

21.6.5 Numerical results

In this section, numerical results of the application of the U2-MF-QL-FH algorithm to the trader problem are discussed. As in the case of the mean field capital accumulation problem, the interaction with the population is through the law of the controls. Algorithm 21.1 is adapted to this case as discussed in §21.4.2.

We consider the problem defined by the choice of parameters: $c_\alpha = 1$, $c_x = 2$, $\gamma = 1.75$, and $c_g = 0.3$. The time horizon is equal to $T = 1$. The distribution of the inventory process at initial time X_{t_0} is Gaussian with mean 0.5 and standard deviation 0.3. The volatility of the process X_t is given by $\sigma = 0.5$.

This problem is characterized by continuous time and continuous state and action spaces. In order to solve this problem using the U2-MF-QL-FH algorithm, truncation and discretization techniques together with a projector operator are applied. The time interval $[0, T]$ is uniformly discretized as $\tau = \{t_0, \ldots, t_{N_T} = T\}$ with $\Delta t = 1/16$. The state and action spaces are truncated and discretized as discussed in §21.4.4. The truncation parameters are chosen large enough to make sure that the state is within the boundary most of the time.

In the MFG (resp. MFC), the action space is given by $\mathcal{A} = \{a_0 = -2.5, \ldots, a_{|\mathcal{A}|-1} = 1\}$ (resp. $\mathcal{A} = \{a_0 = -0.25, \ldots, a_{|\mathcal{A}|-1} = 5\}$) and the state space by $\mathcal{X} = \{x_0 = -1.5, \ldots, x_{|\mathcal{X}|-1} = 1.75\}$ (resp. $\mathcal{X} = \{x_0 = -0.75, \ldots, x_{|\mathcal{X}|-1} = 4\}$). The step size for the discretization of the spaces $\mathcal{A}$, and $\mathcal{X}$ is given by $\Delta_a = \Delta_x = \sqrt{\Delta t} = 1/4$. The exploitation-exploration trade off is tackled on each episode using an ε-greedy policy. Supposed the agent is in state x, the algorithm picks the action that is optimal based on the current estimates with probability $1 - \epsilon$ and a random action in $\mathcal{A}$ with probability ε. In particular, the value of epsilon is fixed to 0.1.

The following numerical results show how the U2-MF-QL-FH algorithm is able to learn an approximation of the control function and the mean field term in the MFG and MFC cases depending on the choice of the parameters $(\omega^Q, \omega^\theta)$.

Learning of controls

Figures 21.8a,b; 21.9a,b; 21.10a,b: controls learned by the algorithm.

The controls learned by the U2-MF-QL-FH algorithm are compared with the theoretical solutions. Each plot corresponds to a different time point $t \in \{0, 0.5, 1\}$. The layout is the same applied for the mean field capital accumulation problem in §21.5.4. On the left, the choice $(\omega^Q, \omega^\theta) = (0.55, 0.85)$ produces the approximation of the solution of the MFG. On the right, the values of the parameters $(\omega^Q, \omega^\theta) = (0.65, 0.15)$ lets the algorithm to approach the solution of the MFC problem. The accuracy of the approximation is better at initial times and degrades towards the final horizon showing an higher complexity of the tuning of the algorithm to this problem. The results presented in the figures are averaged over 10 runs.

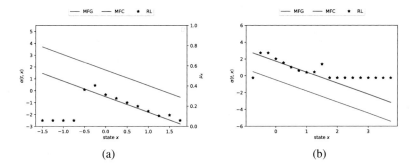

(a)   (b)

Figure 21.8 (a): Learned controls for MFG at time 0. (b): Learned controls for MFC at time 0.

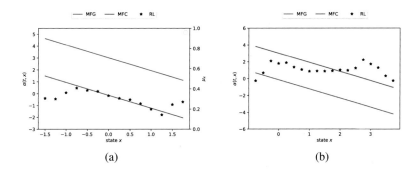

(a)   (b)

Figure 21.9 (a): Learned controls for MFG at time 7/16. (b): Learned controls for MFC at time 7/16.

Learning of the mean field

Figures 21.11a,b; 21.12a,b; 21.13a,b: $\mathbb{E}[\theta_t]$ learned by the algorithm. The estimation of the first moment of the distribution of the controls evolves with respect to the number of learning episodes. Each plot corresponds to a different time point $t \in \{0, 0.5, 1\}$. The layout is the same described in §21.5.4. On the

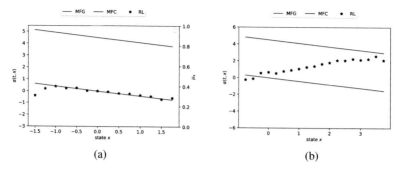

(a)   (b)

Figure 21.10 (a): Learned controls for MFG at time 15/16. (b): Learned controls for MFC at time 15/16.

left, the solution of the MFG is obtained choosing $(\omega^Q, \omega^\theta) = (0.55, 0.85)$. On the right, the MFC solution is approached by the set of parameters $(\omega^Q, \omega^\theta) = (0.65, 0.15)$. The results presented in the figures are averaged over 10 runs.

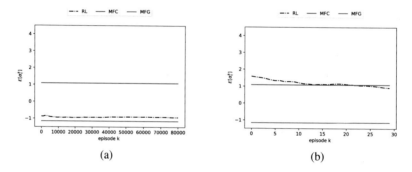

(a)   (b)

Figure 21.11 (a): Learned control's mean for MFG at time 0. (b): Learned control's mean for MFC at time 0.

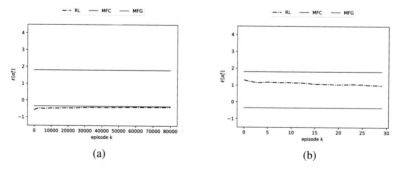

(a)   (b)

Figure 21.12 (a): Learned control's mean for MFG at time 7/16. (b): Learned control's mean for MFC at time 7/16.

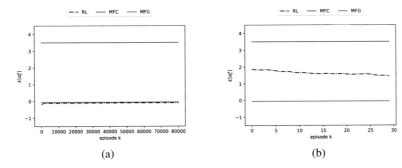

Figure 21.13 (a): Learned control's mean for MFG at time 15/16. (b): Learned control's mean for MFC at time 15/16.

21.7 Conclusion

In this chapter, we have presented a reinforcement learning algorithm which can be used to approximate solutions of mean field games or mean field control problems in the case of interaction through the distribution of controls and in finite horizon. The method unifies the two problems through a two-timescale perspective. We have illustrated the algorithm on two examples: an optimal investment problem with HARA utility function, and an optimal liquidation problem.

The main ingredients of the algorithm are the learning rates for the Q-matrix and for the distribution of controls. Their relative decay with respect to the number of episodes is the key quantity to stir the algorithm towards learning the optimal controls for MFG or MFC problems. Roughly speaking, updating the Q-matrix faster (resp. slower) than the distribution of controls leads to the MFG (resp. MFC) solution. Convergence follows by applying Borkar's results as shown by Angiuli et al. (2022) in the case of infinite horizon problems. Choosing these rates in an optimal way remains the main challenge in specific applications. In particular, we expect that allowing these rates to depend on the time steps could lead to improved results. This aspect is left for future investigations.

The algorithm presented here is the context of finite space via the Q-matrix even though the proposed examples are originally in continuous space and then discretized. Dealing directly with a continuous space is the topic of the ongoing work on deep reinforcement learning for mean field problems (Angiuli and Hu, 2021).

The area of reinforcement learning for mean field problems is extremely rich with a huge potential for applications in various disciplines. It is in its infancy, and we hope that the results and explanations presented here will be helpful to newcomers interested in this direction of research.

References

Acciaio, Beatrice, Backhoff-Veraguas, Julio, and Carmona, Rene. 2018. Extended mean field control problems: stochastic maximum principle and transport perspective. *SIAM J. Control Optim.*, to appear.

Achdou, Yves, and Laurière, Mathieu. 2020. Mean Field Games and Applications: Numerical Aspects. In: *Mean Field Games*. C.I.M.E. Foundation Subseries, vol. 2281. Springer.

Al-Aradi, Ali, Correia, Adolfo, Naiff, Danilo, Jardim, Gabriel, and Saporito, Yuri. 2018. Solving nonlinear and high-dimensional partial differential equations via deep learning. ArXiv:1811.08782.

Alasseur, Clemence, Ben Tahar, Imen, and Matoussi, Anis. 2019. An Extended Mean Field Game for Storage in Smart Grids. *J. Optimization Theory and Applications*, to appear.

Anahtarci, Berkay, Kariksiz, Can Deha, and Saldi, Naci. 2020. Q-learning in regularized mean-field games. ArXiv:2003.12151.

Angiuli, Andrea, and Hu, Ruimeng. 2021. Deep reinforcement learning for mean field games and mean field control problems in continuous spaces. In preparation.

Angiuli, Andrea, Fouque, Jean-Pierre, and Laurière, Mathieu. 2022. Unified reinforcement Q-learning for mean field game and control problems. *Mathematics of Control, Signals, and Systems*, **34**, 217–271. ArXiv:2006.13912.

Bensoussan, Alain, Frehse, Jens, Yam, Phillip, et al. 2013. *Mean Field Games and Mean Field Type Control Theory*. Springer.

Bertucci, Charles, Lasry, Jean-Michel, and Lions, Pierre-Louis. 2019. Some remarks on mean field games. *Comm. Partial Differential Equations*, **44**(3), 205–227.

Bonnans, Frédéric J., Hadikhanloo, Saeed, and Pfeiffer, Laurent. 2019. Schauder Estimates for a Class of Potential Mean Field Games of Controls. ArXiv:1902.05461.

Borkar, Vivek S. 1997. Stochastic approximation with two time scales. *Systems & Control Letters*, **29**(5), 291–294.

Borkar, Vivek S. 2008. *Stochastic Approximation: A Dynamical Systems Viewpoint*. Cambridge University Press; Hindustan Book Agency.

Borkar, Vivek S, and Konda, Vijaymohan R. 1997. The actor-critic algorithm as multi-time-scale stochastic approximation. *Sadhana*, **22**(4), 525–543.

Cao, Haoyang, Guo, Xin, and Laurière, Mathieu. 2020. Connecting GANs and MFGs. ArXiv:2002.04112.

Cardaliaguet, Pierre, and Hadikhanloo, Saeed. 2017. Learning in mean field games: the fictitious play. *ESAIM: Control, Optimisation and Calculus of Variations*, **23**(2), 569–591.

Cardaliaguet, Pierre, and Lehalle, Charles-Albert. 2018. Mean field game of controls and an application to trade crowding. *Math. Financ. Econ.*, **12**(3), 335–363.

Carmona, René. 2020. Applications of mean Field Games to economic theory. (AMS Short Course) ArXiv:2012.05237.

Carmona, René, and Delarue, François. 2018a. *Probabilistic Theory of Mean Field Games with Applications. I: Mean Field FBSDEs, Control, and Games*. Probability Theory and Stochastic Modelling, vol. 83. Springer.

Carmona, René, and Delarue, François. 2018b. *Probabilistic Theory of Mean Field Games with Applications. I–II*. Springer.

Carmona, René, and Lacker, Daniel. 2015. A probabilistic weak formulation of mean field games and applications. *Ann. Appl. Probab.*, **25**(3), 1189–1231.

Carmona, René, and Laurière, Mathieu. 2019. Convergence analysis of machine learning algorithms for the numerical solution of mean field control and games. II: the finite horizon case. *Annals of Applied Probability*, to appear. ArXiv:1908.01613.

Carmona, René, and Laurière, Mathieu. 2021. Convergence analysis of machine learning algorithms for the numerical solution of mean field control and games. I: the ergodic case. *SIAM Journal on Numerical Analysis*, **59**(3), 1455–1485.

Carmona, René, Laurière, Mathieu, and Tan, Zongjun. 2019a. Linear–quadratic mean-field reinforcement learning: convergence of policy gradient methods. Preprint.

Carmona, René, Laurière, Mathieu, and Tan, Zongjun. 2019b. Model-free mean-field reinforcement learning: mean-field MDP and mean-field Q-learning. Preprint.

Carmona, René, Graves, Christy V., and Tan, Zongjun. 2019c. Price of anarchy for mean field games. Pages 349–383 of: *CEMRACS 2017 – Numerical Methods for Stochastic Models: Control, Uncertainty Quantification, Mean-Field*. ESAIM Proc. Surveys, vol. 65. EDP Sci., Les Ulis.

Carmona, René, , and Laurière, Mathieu. 2022. Deep learning for mean field games, with applications to finance. In *Machine Learning in Financial Markets: A Guide to Contemporary Practices*, A. Capponi and C.-A. Lehalle (eds), Cambridge University Press.

Chan, Patrick, and Sircar, Ronnie. 2015. Bertrand and Cournot mean field games. *Appl. Math. Optim.*, **71**(3), 533–569.

Charpentier, Arthur, Elie, Romuald, and Remlinger, Carl. 2020. Reinforcement learning in economics and finance. ArXiv:2003.10014.

Djete, Mao Fabrice, Possamaï, Dylan, and Tan, Xiaolu. 2019. McKean–Vlasov optimal control: the dynamic programming principle. ArXiv:1907.08860.

Elie, Romuald, Perolat, Julien, Laurière, Mathieu, Geist, Matthieu, and Pietquin, Olivier. 2020. On the convergence of model-free learning in mean field games. In: *Proc. of AAAI*.

Even-Dar, Eyal, and Mansour, Yishay. 2003. Learning rates for Q-learning. *Journal of Machine Learning Research*, **5**(Dec), 1–25.

Fouque, Jean-Pierre, and Zhang, Zhaoyu. 2020. Deep learning methods for mean field control problems with delay. *Frontiers in Applied Mathematics and Statistics*, **6**(11).

Fu, Zuyue, Yang, Zhuoran, Chen, Yongxin, and Wang, Zhaoran. 2019. Actor–critic provably finds Nash equilibria of linear–quadratic mean-field games. ArXiv:1910.07498.

Germain, Maximilien, Mikael, Joseph, and Warin, Xavier. 2019. Numerical resolution of McKean–Vlasov FBSDEs using neural networks. ArXiv:1909.12678.

Gomes, Diogo A., and Saúde, João. 2014. Mean field games models – a brief survey. *Dyn. Games Appl.*, **4**(2), 110–154.

Gomes, Diogo A., and Voskanyan, Vardan K. 2016. Extended deterministic mean-field games. *SIAM J. Control Optim.*, **54**(2), 1030–1055.

Gomes, Diogo A., Patrizi, Stefania, and Voskanyan, Vardan. 2014. On the existence of classical solutions for stationary extended mean field games. *Nonlinear Anal.*, **99**, 49–79.

Graber, P. Jameson. 2016. Linear quadratic mean field type control and mean field games with common noise, with application to production of an exhaustible resource. *Appl. Math. Optim.*, **74**(3), 459–486.

Graber, P. Jameson, and Bensoussan, Alain. 2018. Existence and uniqueness of solutions for Bertrand and Cournot mean field games. *Appl. Math. Optim.*, **77**(1), 47–71.

Gu, Haotian, Guo, Xin, Wei, Xiaoli, and Xu, Renyuan. 2019. Dynamic programming principles for learning MFCs. ArXiv:1911.07314.

Gu, Haotian, Guo, Xin, Wei, Xiaoli, and Xu, Renyuan. 2020. Q-learning for mean-field controls. ArXiv:2002.04131.

Guo, Xin, Hu, Anran, Xu, Renyuan, and Zhang, Junzi. 2019. Learning mean-field games. Pages 4966–4976 of: *Advances in Neural Information Processing Systems*.

Hadikhanloo, Saeed. 2017. Learning in anonymous nonatomic games with applications to first-order mean field games. ArXiv:1704.00378.

Hadikhanloo, Saeed, and Silva, Francisco J. 2019. Finite mean field games: fictitious play and convergence to a first order continuous mean field game. *J. Mathématiques Pures et Appliquées*, **132**, 362–297.

Huang, Minyi. 2013. A mean field capital accumulation game with HARA utility. *Dynamic Games and Applications*, **3**(4), 446–472.

Huang, Minyi, Malhamé, Roland P., and Caines, Peter E. 2006. Large population stochastic dynamic games: closed-loop McKean–Vlasov systems and the Nash certainty equivalence principle. *Commun. Inf. Syst.*, **6**(3), 221–251.

Kingma, Diederik P., and Ba, Jimmy. 2014. Adam: A method for stochastic optimization. ArXiv:1412.6980.

Kobeissi, Ziad. 2019. On classical solutions to the mean field game system of controls. ArXiv:1904.11292.

Lasry, Jean-Michel, and Lions, Pierre-Louis. 2007. Mean field games. *Jpn. J. Math.*, **2**(1), 229–260.

Laurière, Mathieu. 2020. Numerical methods for mean field games and mean field type control. (AMS Short Course.) ArXiv:2106.06231.

Laurière, Mathieu, and Pironneau, Olivier. 2014. Dynamic programming for mean-field type control. *C. R. Math. Acad. Sci. Paris*, **352**(9), 707–713.

Laurière, Mathieu, and Tangpi, Ludovic. 2020. Convergence of large population games to mean field games with interaction through the controls. ArXiv:2004.08351.

Lin, Alex Tong, Fung, Samy Wu, Li, Wuchen, Nurbekyan, Levon, and Osher, Stanley J. 2020. APAC-Net: Alternating the population and agent control via two neural networks to solve high-dimensional stochastic mean field games. ArXiv:2002.10113.

McKean Jr, Henry P. 1966. A class of Markov processes associated with nonlinear parabolic equations. *PNAS*, **56**(6), 1907–1911.

Mguni, David, Jennings, Joel, and de Cote, Enrique Munoz. 2018. Decentralised learning in systems with many, many strategic agents. In: *32nd AAAI Conference on Artificial Intelligence.*

Min, Ming, and Hu, Ruimeng. 2021. Signatured deep fictitious play for mean field games with common noise. ArXiv:2106.03272.

Mishra, Rajesh K., Vasal, Deepanshu, and Vishwanath, Sriram. 2020. Model-free reinforcement learning for non-stationary mean field games. Pages 1032–1037 of: *59th IEEE Conference on Decision and Control.* IEEE.

Motte, Médéric, and Pham, Huyên. 2019. Mean-field Markov decision processes with common noise and open-loop controls. ArXiv:1912.07883.

Pérolat, Julien, Perrin, Sarah, Elie, Romuald, Laurière, Mathieu, Piliouras, Georgios, Geist, Matthieu, Tuyls, Karl, and Pietquin, Olivier. 2021. Scaling up mean field games with online mirror descent. ArXiv:2103.00623

Perrin, Sarah, Pérolat, Julien, Laurière, Mathieu, Geist, Matthieu, Elie, Romuald, and Pietquin, Olivier. 2020. Fictitious play for mean field games: continuous time analysis and applications. In preparation.

Perrin, Sarah, Laurière, Mathieu, Pérolat, Julien, Geist, Matthieu, Élie, Romuald, and Pietquin, Olivier. 2021. Mean field games flock! The reinforcement learning way. Accepted to IJCAI'21. ArXiv:2105.07933.

Pham, Huyên, and Wei, Xiaoli. 2016. Discrete time McKean–Vlasov control problem: a dynamic programming approach. *Applied Mathematics & Optimization*, **74**(3), 487–506.

Ruthotto, Lars, Osher, Stanley J, Li, Wuchen, Nurbekyan, Levon, and Fung, Samy Wu. 2020. A machine learning framework for solving high-dimensional mean field game and mean field control problems. *PNAS*, **117**(17), 9183–9193.

Subramanian, Jayakumar, and Mahajan, Aditya. 2019. Reinforcement learning in stationary mean-field games. In: *18th International Conference on Autonomous Agents and Multiagent Systems.*

Sutton, Richard S, and Barto, Andrew G. 2018. *Reinforcement Learning: An Introduction.* MIT Press.

Wang, Weichen, Han, Jiequn, Yang, Zhuoran, and Wang, Zhaoran. 2020. Global convergence of policy gradient for linear-quadratic mean-field control/game in continuous time. ArXiv:2008.06845.

Watkins, Christopher John Cornish Hellaby. 1989. *Learning from Delayed Rewards.* PhD. thesis, King's College, Cambridge.

Xie, Qiaomin, Yang, Zhuoran, Wang, Zhaoran, and Minca, Andreea. 2020. Provable fictitious play for general mean-field games. ArXiv:2010.04211.

Yang, Jiachen, Ye, Xiaojing, Trivedi, Rakshit, Xu, Huan, and Zha, Hongyuan. 2018a. Deep mean field games for learning optimal behavior policy of large populations. In: *International Conference on Learning Representations*.

Yang, Yaodong, Luo, Rui, Li, Minne, Zhou, Ming, Zhang, Weinan, and Wang, Jun. 2018b. Mean field multi-agent reinforcement learning. Pages 5567–5576 of: *International Conference on Machine Learning*.

Neural Networks-Based Algorithms for Stochastic Control and PDEs in Finance

Maximilien Germain[a], Huyên Pham[b] and Xavier Warin[c]

Abstract

This chapter presents machine learning techniques and deep reinforcement learning-based algorithms for the efficient resolution of nonlinear partial differential equations and dynamic optimization problems arising in investment decisions and derivative pricing in financial engineering. We survey recent results in the literature, present new developments, notably in the fully nonlinear case, and compare the different schemes illustrated by numerical tests on various financial applications. We conclude by highlighting some future research directions.

22.1 Breakthrough in the resolution of high-dimensional nonlinear problems

The numerical resolution of control problems and nonlinear PDEs – arising in several financial applications such as portfolio selection, hedging, or derivatives pricing – is subject to the so-called "curse of dimensionality", making impractical the discretization of the state space in dimension greater than three by using classical PDE resolution methods such as finite differences schemes. Probabilistic regression Monte Carlo methods based on a Backward Stochastic Differential Equation (BSDE) representation of semilinear PDEs have been developed in Zhang (2004), Bouchard and Touzi (2004), Gobet et al. (2005) to overcome this obstacle. These mesh-free techniques are successfully applied upon dimension six or seven, nevertheless, their use of regression methods requires a number of basis functions growing fastly with the dimension. What can be done to further increase the dimension of numerically solvable problems?

A breakthrough with deep learning based-algorithms has been made in the last five years towards this computational challenge, and we mention the recent survey by Beck et al. (2020). The main interest in the use of machine learning techniques for control and PDEs is the ability of deep neural networks to efficiently represent

[a] EDF R&D and Université de Paris
[b] Université de Paris, and Laboratoire de Finance des Marchés de l'Energie
[c] EDF R&D, and Laboratoire de Finance des Marchés de l'Energie
     Published in *Machine Learning And Data Sciences For Financial Markets*, Agostino Capponi and Charles-Albert Lehalle© 2023 Cambridge University Press.

high-dimensional functions without using spatial grids, and with no curse of dimensionality (Grohs et al., 2018; Hutzenthaler et al., 2020). Although the use of neural networks for solving PDEs is not new, see e.g. Dissanayake and Phan-Thien (1994), the approach has been successfully revived with new ideas and directions. Neural networks have experienced increasing popularity since the works on Reinforcement Learning for solving the game of Go by Google DeepMind teams. These empirical successes and the introduced methods permit the solution of control problems in moderate or large dimension. Moreover, recently developed open source libraries like Tensorflow and Pytorch also offer an accessible framework for implementing these algorithms.

A first natural use of neural networks for stochastic control concerns the discrete time setting, with the study of Markov Decision Processes, either by brute force or by using dynamic programming approaches. In the continuous-time setting, and in the context of PDE resolution, we present various methods. A first kind of scheme is rather generic and can be applied to a variety of PDEs coming from a wide range of applications. Other schemes rely on BSDE representations, strongly linked to stochastic control problems. In both cases, numerical evidence seems to indicate that the methods can be used in high dimension, greater than 10 and up to 1000 in certain studies. Some theoretical results also illustrate the convergence of specific algorithms. These advances pave the way for new methods dedicated to the study of large population games, examined in the context of mean-field games and mean-field control problems.

The outline of this chapter is as follows. We first focus on some schemes for discrete time control in §22.2 before presenting generic machine learning schemes for PDEs in §22.3.1. Then we review BSDE-based machine learning methods for semilinear equations in §22.3.2. Existing algorithms for fully nonlinear PDEs are detailed in §22.3.3 before presenting new BSDE schemes designed to treat this more difficult case. Numerical tests on CVA pricing and portfolio selection are conducted in §22.4 to compare the different approaches. Finally, we highlight in §22.5 further directions and perspectives including recent advances for the resolution of mean field games and mean field control problems with or without a model.

22.2 Deep learning approach to stochastic control

We present in this section some recent breakthroughs in the numerical resolution of stochastic control in high dimension by means of machine learning techniques. We consider a model-based setting in discrete-time, i.e., a Markov decision process.

Let us fix a probability space $(\Omega, \mathcal{F}, \mathbb{P})$ equipped with a filtration $\mathbb{F} = (\mathcal{F}_t)_t$ representing the available information at any time $t \in \mathbb{N}$ (note that $\mathcal{F}_0$ is the trivial σ-algebra). The evolution of the system is described by a model dynamics for the state process $(X_t)_{t \in \mathbb{N}}$ valued in $\mathcal{X} \subset \mathbb{R}^d$:

$$X_{t+1} = F(X_t, \alpha_t, \varepsilon_{t+1}), \quad t \in \mathbb{N}, \tag{22.1}$$

where $(\varepsilon_t)_t$ is a sequence of i.i.d. random variables valued in E, with ε_{t+1} $\mathcal{F}_{t+1}$-

measurable containing all the noisy information arriving between t and $t + 1$, and where $\alpha = (\alpha_t)_t$ is the control process valued in $A \subset \mathbb{R}^q$. The dynamics function F is a measurable function from $\mathbb{R}^d \times \mathbb{R}^q \times E$ into $\mathbb{R}^d$, and assumed to be known. Given a running cost function f, a finite horizon $T \in \mathbb{N}^*$, and a terminal cost function, the problem is to minimize over control process α a functional cost

$$J(\alpha) = \mathbb{E}\Big[\sum_{t=0}^{T-1} f(X_t, \alpha_t) + g(X_T) \Big]. \tag{22.2}$$

In some relevant applications, we may require constraints on the state and control in the form: $(X_t, \alpha_t) \in S, t \in \mathbb{N}$, for some subset S of $\mathbb{R}^d \times \mathbb{R}^q$. This can be handled by relaxing the state/constraint and introducing into the costs a penalty function $L(x, a)$: $f(x, a) \leftarrow f(x, a) + L(x, a)$, and $g(x) \leftarrow g(x) + L(x, a)$. For example, if the constraint set is in the form: $S = \{(x, a) \in \mathbb{R}^d \times \mathbb{R}^q : h_k(x, a) = 0, k = 1, \ldots, p, h_k(x, a) \geq 0, k = p+1, \ldots, m\}$, then one can take as penalty functions:

$$L(x, a) = \sum_{k=1}^{p} \mu_k |h_k(x, a)|^2 + \sum_{k=p+1}^{m} \mu_k \max(0, -h_k(x, a)), \tag{22.3}$$

where μ_k are penalization parameters (large in practice), see e.g. Han and E (2016).

22.2.1 Global approach

The method consists simply in approximating at any time t, the feedback control, i.e. a function of the state process, by a neural network (NN):

$$\alpha_t \simeq \pi^{\theta_t}(X_t), \quad t = 0, \ldots, T - 1,$$

where π^θ is a feedforward neural network on $\mathbb{R}^d$ with parameters θ; and then to minimize over the global set of parameters $\theta = (\theta_0, \ldots, \theta_{T-1})$ the quantity (playing the role of loss function)

$$\tilde{J}(\theta) = \mathbb{E}\Big[\sum_{t=0}^{T-1} f(X_t^\theta, \pi^{\theta_t}(X_t^\theta)) + g(X_T^\theta) \Big],$$

where X^θ is the state process associated with the NN feedback controls:

$$X_{t+1}^\theta = F(X_t^\theta, \pi^{\theta_t}(X_t^\theta), \varepsilon_{t+1}), \quad t = 0, \ldots, T - 1.$$

This basic idea of approximating control by parametric function of the state was proposed in Gobet and Munos (2005), and updated with the use of (deep) neural networks by Han and E (2016). This method met success thanks to its simplicity and the easy accessibility of common libraries like TensorFlow for optimizing the parameters of the neural networks. Some recent extensions of this approach dealt with stochastic control problems with delay, see Han and Hu (2021). However, global optimization over such a huge set of parameters $\theta = (\theta_0, \ldots, \theta_{T-1})$ may suffer from being stuck in suboptimal traps and thus does not converge, especially for large horizon T. An alternative is to consider controls $\alpha_t \simeq \pi^\theta(t, X_t)$, for

$t = 0, \ldots, T - 1$, with a single neural network π^θ giving more stabilized results as studied by Fécamp et al. (2019). We focus here on feedback controls, which is not a restriction as we are in a Markov setting. For path-dependent control problems, we may consider recurrent neural networks to take into consideration the past of state trajectory as input of the policy.

22.2.2 Backward dynamic programming approach

Bachouch et al. (2022) proposed methods that combine ideas from numerical probability and deep reinforcement learning. Their algorithms are based on the classical dynamic programming (DP), (deep) neural networks for the approximation/learning of the optimal policy and value function, and Monte Carlo regressions with performance and value iterations.

The first algorithm, called NNContPI, is a combination of dynamic programming and the approach in Han and E (2016). It learns sequentially the control by NN $\pi^\theta(\cdot)$ and performance iterations, and is designed as follows:

Algorithm 22.1: NNContPI

---

1   **Input:** the training distributions $(\mu_t)_{t=0}^{T-1}$;
2   **Output:** estimates of the optimal strategy $(\hat{\pi}_t)_{t=0}^{T-1}$;
3   **for** $t = T - 1, \ldots, 0$ **do**
4   Compute $\hat{\pi}_t := \pi^{\hat{\theta}_t}$ with

$$\hat{\theta}_t \in \arg\min_\theta \mathbb{E}\left[f\left(X_t, \pi^\theta(X_t)\right) + \sum_{s=t+1}^{T-1} f\left(X_s^\theta, \hat{\pi}_s\left(X_s^\theta\right)\right) + g\left(X_T^\theta\right) \right]$$

(22.4)

(22.5)

where $X_t \sim \mu_t$ and where $\left(X_s^\theta\right)_{s=t+1}^{T}$ is defined by induction as:

$$\begin{cases} X_{t+1}^\theta &= F\left(X_t, \pi^\theta(X_t), \varepsilon_{t+1}\right), \\ X_{s+1}^\theta &= F\left(X_s^\theta, \hat{\pi}_s(X_s^\theta), \varepsilon_{s+1}\right), \quad \text{for } s = t+1, \ldots, T-1. \end{cases}$$

The second algorithm, refered to as Hybrid-Now, combines optimal policy estimation by neural networks and dynamic programming principle, and relies on an hybrid procedure between value and performance iteration to approximate the value function by neural network $\Phi^\eta(\cdot)$ on $\mathbb{R}^d$ with parameters η.

The convergence analysis of Algorithms NNContPI and Hybrid-Now are studied in Huré et al. (2021), and various applications in finance are implemented in Bachouch et al. (2022). These algorithms are well-designed for control problems with continuous control space $A = \mathbb{R}^q$ or a ball in $\mathbb{R}^q$. In the case where the control

---

Algorithm 22.2: Hybrid-Now

---

1 **Input:** the training distributions $(\mu_t)_{t=0}^{T-1}$;
2 **Output:**
3 – estimate of the optimal strategy $(\hat{\pi}_t)_{t=0}^{T-1}$;
4 – estimate of the value function $(\hat{V}_t)_{t=0}^{T-1}$;
5 Set $\hat{V}_T = g$;
6 **for** $t = T - 1, \ldots, 0$ **do**
7   Compute:

$$\hat{\theta}_t \in \arg\min_\theta \mathbb{E}\left[f\left(X_t, \pi^\theta(X_t)\right) + \hat{V}_{t+1}(X_{t+1}^\theta) \right] \qquad (22.6)$$

    where $X_t \sim \mu_t$, and $X_{t+1}^\theta = F(X_t, \pi^\theta(X_t), \varepsilon_{t+1})$;
8   Set $\hat{\pi}_t := \pi^{\hat{\theta}_t}$;
9   Compute

$$\hat{\eta}_t \in \arg\min_\eta \mathbb{E}\left| f(X_t, \hat{\pi}_t(X_t)) + \hat{V}_{t+1}(X_{t+1}^{\hat{\theta}_t}) - \Phi^\eta(X_t) \right|^2. \qquad (22.7)$$

   Set $\hat{V}_t = \Phi^{\hat{\eta}_t}$;

---

space A is finite, it is relevant to randomize controls, and then use classification methods by approximating the distribution of controls with neural networks and Softmax activation functions.

22.3 Machine learning algorithms for nonlinear PDEs

By a change of time scale, the Markov decision process (22.1)–(22.2) can be obtained from the time discretization of a continuous-time stochastic control problem with controlled diffusion dynamics on $\mathbb{R}^d$,

$$dX_t = b(X_t, \alpha_t)\, dt + \sigma(X_t, \alpha_t)\, dW_t, \qquad (22.8)$$

and cost functional to be minimized over control process α valued in A:

$$J(\alpha) = \mathbb{E}\left[\int_0^T f(X_t, \alpha_t)\, dt + g(X_T) \right]. \qquad (22.9)$$

In this case, it is well known, see e.g. Pham (2009), that the dynamic programming Bellman equation leads to a PDE of the form

$$\begin{cases} \partial_t u + H(x, D_x u, D_x^2 u) = 0, & \text{on } [0, T) \times \mathbb{R}^d \\ u(T, .) = g & \text{on } \mathbb{R}^d, \end{cases} \qquad (22.10)$$

where $H(x, z, \gamma) = \inf_{a \in A} \left[b(x, a) \cdot z + \frac{1}{2}\mathrm{tr}(\sigma\sigma^\top(x, a)\gamma) + f(x, a) \right]$ is the so-called Hamiltonian function. The numerical resolution of such a class of second-order parabolic PDEs will be addressed in this section.

22.3.1 Deterministic approach by neural networks

In the schemes below, differential operators are evaluated by automatic differentiation of the network function approximating the solution of the PDE. Machine learning libraries such as Tensorflow or Pytorch allow to efficiently compute these derivatives. The PDE problem studied is

$$
\begin{cases}
\partial_t u + \mathcal{F}u = 0 & \text{on } [0,T) \times \Lambda \\
u(T,\cdot) = g & \text{on } \Lambda \\
u(t,x) = h(t,x) & \text{on } [0,T) \times \partial\Lambda,
\end{cases}
\tag{22.11}
$$

where $\mathcal{F}$ is a space differential operator, and Λ a subset of $\mathbb{R}^d$.

Deep Galerkin Method (Sirignano and Spiliopoulos, 2017).

The Deep Galerkin Method is a meshfree machine learning algorithm for solving PDEs on a domain, eventually with boundary conditions. The principle is to sample time and space points according to a training measure, e.g. uniform on a bounded domain, and minimize a performance measure quantifying how well a neural network satisfies the differential operator and boundary conditions. The method consists in minimizing over neural network $\mathcal{U}\colon \mathbb{R} \times \mathbb{R}^d \to \mathbb{R}^d$, the L^2 loss

$$
\mathbb{E}|\partial_t \mathcal{U}(\tau,\kappa) + \mathcal{F}\mathcal{U}(\tau,\kappa)|^2 + \mathbb{E}|\mathcal{U}(0,\xi) - g(\xi)|^2 + \mathbb{E}|\mathcal{U}(\tau,\kappa) - h(\tau,\kappa)|^2 \tag{22.12}
$$

with κ, τ, ξ independent random variables in $\Lambda \times [0,T] \times \partial\Lambda$. Sirignano and Spiliopoulos (2017) also prove a convergence result (without rate) for the Deep Galerkin method. This method has been tested on financial problems by Al-Aradi et al. (2018). A major advantage of this method is its adaptability to a large range of PDEs with or without boundary conditions. Indeed the loss function is straightforwardly modified according to changes in the constraints one wishes to enforce on the PDE solution. A related approach is the deep parametric PDE method, see Khoo et al. (2020), and Glau and Wunderlich (2020) applied to option pricing.

Other approximation methods.

(i) *Physics-informed neural networks* (Raissi et al., 2019). Physics-informed neural networks use both data (obtained for a limited amount of samples from a PDE solution), and theoretical dynamics to reconstruct solutions from PDEs. The convergence of this method in the second-order linear parabolic (or elliptic) case was proved in Shin et al. (2020), see also Gräser and Srinivasan (2020).

(ii) *Deep Ritz method* (E and Yu, 2018). The Deep Ritz method focuses on the resolution of the variational formulation from elliptic problems where the integral is evaluated by randomly sampling time and space points, as in the

Deep Galerkin method (Sirignano and Spiliopoulos, 2017) and the minimization is performed over the parameters of a neural network. This scheme is tested on Poisson equation with different types of boundary conditions. In Müller and Zeinhofer (2019) the convergence of the Deep Ritz algorithm is studied.

22.3.2 Probabilistic approach by neural networks

Semi-linear case

In this section, we consider semilinear PDEs of the form

$$
\begin{cases}
\partial_t u + \mu \cdot D_x u + \frac{1}{2}\mathrm{Tr}(\sigma\sigma^\intercal D_x^2 u) = f(\cdot,\cdot,u,\sigma^\intercal D_x u) & \text{on } [0,T) \times \mathbb{R}^d \\
u(T,\cdot) = g & \text{on } \mathbb{R}^d
\end{cases}
\tag{22.13}
$$

for which we have the forward backward SDE representation

$$
\begin{cases}
Y_t = g(X_T) - \int_t^T f(s,X_s,Y_s,Z_s)\mathrm{d}s - \int_t^T Z_s \cdot \mathrm{d}W_s, & 0 \le t \le T, \\
X_t = X_0 + \int_0^t \mu(s,X_s)\mathrm{d}s + \int_0^t \sigma(s,X_s)\mathrm{d}W_s,
\end{cases}
\tag{22.14}
$$

via the (nonlinear) Feynman–Kac formula: $Y_t = v(t,X_t)$, $Z_t = \sigma^\intercal(t,X_t)D_x v(t,X_t)$, $0 \le t \le T$, see Pardoux and Peng (1990).

Let π be a subdivision $\{t_0 = 0 < t_1 < \cdots < t_N = T\}$ with modulus $|\pi| := \sup_i \Delta t_i$, $\Delta t_i := t_{i+1} - t_i$, satisfying $|\pi| = O\left(\frac{1}{N}\right)$, and consider the Euler–Maruyama discretization $(X_i)_{i=0,\dots,N}$ defined by

$$
X_i = X_0 + \sum_{j=0}^{i-1} \mu(t_j,X_j)\Delta t_j + \sum_{j=0}^{i-1} \sigma(t_j,X_j)\Delta W_j,
\tag{22.15}
$$

where $\Delta W_j := W_{t_{j+1}} - W_{t_j}$, $j = 0,\dots,N$. Sample paths of $(X_i)_i$ act as training data in the machine learning setting. Thus our training set can be chosen as large as desired, which is relevant for training purposes as sit does not lead to overfitting.

The time discretization of the BSDE (22.14) can be written in backward induction as

$$
Y_i^\pi = Y_{i+1}^\pi - f(t_i,X_i,Y_i^\pi,Z_i^\pi)\Delta t_i - Z_i^\pi.\Delta W_i, \quad i = 0,\dots,N-1,
\tag{22.16}
$$

which can be described as conditional expectation formulae

$$
\begin{cases}
Y_i^\pi = \mathbb{E}_i\left[Y_{i+1}^\pi - f(t_i,X_i,Y_i^\pi,Z_i^\pi)\Delta t_i\right] \\
Z_i^\pi = \mathbb{E}_i\left[\frac{\Delta W_i}{\Delta t_i}Y_{i+1}^\pi\right], \qquad i = 0,\dots,N-1,
\end{cases}
\tag{22.17}
$$

where $\mathbb{E}_i$ is a notation for the conditional expectation with respect to $\mathcal{F}_{t_i}$.

Deep BSDE scheme (E et al., 2017; Han et al., 2017).

The essence of this method is to write down the backward equation (22.16) as a forward equation. One approximates the initial condition Y_0 and the Z-component at each time by network functions taking the forward process X as input. The objective function to optimize is the error between the reconstructed dynamics and the true terminal condition. More precisely, the problem is to minimize over network functions $\mathcal{U}_0 \colon \mathbb{R}^d \to \mathbb{R}$, and sequences of network functions $\mathcal{Z} = (\mathcal{Z}_i)_i$, $\mathcal{Z}_i \colon \mathbb{R}^d \to \mathbb{R}^d$, $i = 0, \ldots, N-1$, the global quadratic loss function

$$J_G(\mathcal{U}_0, \mathcal{Z}) = \mathbb{E} \left| Y_N^{\mathcal{U}_0, \mathcal{Z}} - g(X_N) \right|^2, \tag{22.18}$$

where $(Y_i^{\mathcal{U}_0, \mathcal{Z}})_i$ is defined by forward induction as

$$Y_{i+1}^{\mathcal{U}_0, \mathcal{Z}} = Y_i^{\mathcal{U}_0, \mathcal{Z}} + f(t_i, X_i, Y_i^{\mathcal{U}_0, \mathcal{Z}}, \mathcal{Z}_i(X_i))\Delta t_i + \mathcal{Z}_i(X_i) \cdot \Delta W_i, \quad i = 0, \ldots, N-1,$$

starting from $Y_0^{\mathcal{U}_0, \mathcal{Z}} = \mathcal{U}_0(X_0)$. The output of this scheme, for the solution $(\widehat{\mathcal{U}}_0, \widehat{\mathcal{Z}})$ to this global minimization problem, supplies an approximation $\widehat{\mathcal{U}}_0$ of the solution $u(0,.)$ to the PDE at time 0, and approximations $Y_i^{\widehat{\mathcal{U}}_0, \widehat{\mathcal{Z}}}$ of the solution to the PDE (22.13) at times t_i evaluated at X_{t_i}, i.e., of $Y_{t_i} = u(t_i, X_{t_i})$, $i = 0, \ldots, N$. The convergence of this algorithm through *a posteriori* error was studied by Han and Long (2020), see also Jiang and Li (2021). A variant has been proposed by Chan-Wai-Nam et al. (2019) which introduces a single neural network $\mathcal{Z}(t, x) \colon [0, T] \times \mathbb{R}^d \mapsto \mathbb{R}^d$ instead of N independent neural networks. This simplifies the optimization problem and leads to more stable solutions. A close method introduced by Raissi (2018) uses also a single neural network $\mathcal{U}(t, x) \colon [0, T] \times \mathbb{R}^d \mapsto \mathbb{R}$ and estimates Z as the automatic derivative in the space of $\mathcal{U}$. We also refer to Jacquier and Oumgari (2019) for a variation of this deep BSDE scheme to curve-dependent PDEs arising in the pricing under a rough volatility model, to Nüskens and Richter (2019) for approximation methods for Hamilton–Jacobi–Bellman PDEs, and to Kremsner et al. (2020) for an extension of the deep BSDE scheme to elliptic PDEs with applications in insurance.

Deep Backward Dynamic Programming (DBDP) (Huré et al., 2020).

The method builds upon the backward dynamic programming relation (22.16) stemming from the time discretization of the BSDE, and approximates simultaneously at each time step t_i the processes (Y_{t_i}, Z_{t_i}) with neural networks trained with the forward diffusion process X_i as input. The scheme can be implemented in two similar versions:

(1) *DBDP1.* Starting from $\widehat{\mathcal{U}}_N^{(1)} = g$, proceed by backward induction for $i = N-1, \ldots, 0$, by minimizing over network functions $\mathcal{U}_i \colon \mathbb{R}^d \to \mathbb{R}$, and $\mathcal{Z}_i \colon \mathbb{R}^d \to \mathbb{R}^d$ the quadratic loss function

$$J_i^{(B1)}(\mathcal{U}_i, \mathcal{Z}_i)$$
$$= \mathbb{E} \left| \widehat{\mathcal{U}}_{i+1}^{(1)}(X_{i+1}) - \mathcal{U}_i(X_i) - f(t_i, X_i, \mathcal{U}_i(X_i), \mathcal{Z}_i(X_i))\Delta t_i - \mathcal{Z}_i(X_i) \cdot \Delta W_i \right|^2,$$

and update $(\widehat{\mathcal{U}}_i^{(1)}, \widehat{\mathcal{Z}}_i^{(1)})$ as the solution to this local minimization problem.

(2) *DBDP2.* Starting from $\widehat{\mathcal{U}}_N^{(2)} = g$, proceed by backward induction for $i = $

$N-1,\dots,0$, by minimizing over C^1 network functions $\mathcal{U}_i \colon \mathbb{R}^d \to \mathbb{R}$ the quadratic loss function

$$J_i^{(B2)}(\mathcal{U}_i) \tag{22.19}$$

$$= \mathbb{E}\Big|\widehat{\mathcal{U}}_{i+1}^{(2)}(X_{i+1}) - \mathcal{U}_i(X_i) - f(t_i, X_i, \mathcal{U}_i(X_i), \sigma(t_i, X_i)^\top D_x \mathcal{U}_i(X_i))\Delta t_i \tag{22.20}$$

$$- D_x \mathcal{U}_i(X_i)^\top \sigma(t_i, X_i)\Delta W_i\Big|^2, \tag{22.21}$$

where $D_x \mathcal{U}_i$ is the automatic differentiation of the network function $\mathcal{U}_i$. Update $\widehat{\mathcal{U}}_i^{(2)}$ as the solution to this local minimization problem, and set $\widehat{\mathcal{Z}}_i^{(2)} = \sigma^\top(t_i, \cdot)D_x \mathcal{U}_i^{(2)}$.

The output of DBDP supplies an approximation $(\widehat{\mathcal{U}}_i, \widehat{\mathcal{Z}}_i)$ of the solution $u(t_i, \cdot)$ and its gradient $\sigma^\top(t_i, \cdot)D_x u(t_i, \cdot)$ to the PDE (22.13) on the time grid t_i, for $i = 0, \dots, N-1$. The study of the approximation error due to the time discretization and the choice of the loss function is accomplished in Huré et al. (2020).

Variants and extensions of DBDP schemes

(i) A regression-based machine learning scheme inspired by regression Monte Carlo methods for numerically computing condition expectations in the time discretization (22.17) of the BSDE, is given as follow: starting from $\widehat{\mathcal{U}}_N = g$, proceed by backward induction for $i = N-1,\dots,0$, in two regression problems:

    (a) Minimize over network functions $\mathcal{Z}_i \colon \mathbb{R}^d \to \mathbb{R}^d$

$$J_i^{r,Z}(\mathcal{Z}_i) = \mathbb{E}\Big|\frac{\Delta W_i}{\Delta t_i}\widehat{\mathcal{U}}_{i+1}(X_{i+1}) - \mathcal{Z}_i(X_i)\Big|^2 \tag{22.22}$$

    and update $\widehat{\mathcal{Z}}_i$ as the solution to this minimization problem.

    (b) Minimize over network functions $\mathcal{U}_i \colon \mathbb{R}^d \to \mathbb{R}$

$$J_i^{r,Y}(\mathcal{U}_i) = \mathbb{E}\Big|\widehat{\mathcal{U}}_{i+1}(X_{i+1}) - \mathcal{U}_i(X_i) - f(t_i, X_i, \mathcal{U}_i(X_i), \widehat{\mathcal{Z}}_i(X_i))\Delta t_i\Big|^2 \tag{22.23}$$

    and update $\widehat{\mathcal{U}}_i$ as the solution to this minimization problem.

Compared to these regression-based schemes, the DBDP scheme simultaneously estimates the pair component (Y, Z) through the minimization of the loss functions $J_i^{(B1)}(\mathcal{U}_i, \mathcal{Z}_i)$ (or $J_i^{(B2)}(\mathcal{U}_i)$ for the second version), $i = N-1, \dots, 0$. Interestingly, the convergence of the DBDP scheme can be confirmed by computing at each time step the infimum of loss function, which should vanish for the exact solution (up to the time discretization). In contrast, the infimum of the loss functions in usual regression-based schemes is unknown for the true solution as it is supposed to match the residual of L^2-projection. Therefore the scheme accuracy cannot be directly verified.

(ii) The DBDP scheme is based on local resolution, and was first used to solve linear PDEs, see Sabate Vidales et al. (2018). It is also suitable for solving variational inequalities and can be used to value American options as shown in Huré et al. (2020). Alternative methods consists in using the Deep Optimal Stopping scheme (Becker et al., 2019a) or the method from Becker et al. (2019b). Some tests on Bermudan options were also performed by Liang et al. (2019) and Fujii et al. (2019) with some refinements of the Deep BSDE scheme.

(iii) The **Deep Splitting (DS) scheme** in Beck et al. (2019a) combines ideas from the DBDP2 and regression-based schemes. Indeed the current regression-approximation on Z is estimated by the automatic differentiation of the neural network computed at the previous optimization step. The current approximation of Y is then computed by a regression-type optimization problem. It can be seen as a local version of the global algorithm from Raissi (2018) or as a step by step Feynman–Kac approach. As the scheme is a local one, it can be used to value American options. The convergence of this method was studied by Germain et al. (2020).

(iv) Local resolution permits the addition of other constraints such as those on a replication portfolio using facelifting techniques as in Kharroubi et al. (2020).

(v) The **Deep Backward Multistep (MDBDP) scheme** (Germain et al., 2020) is described as follows: for $i = N-1, \ldots, 0$, minimize over network functions $\mathcal{U}_i : \mathbb{R}^d \to \mathbb{R}$, and $\mathcal{Z}_i : \mathbb{R}^d \to \mathbb{R}^d$ the loss function

$$J_i^{MB}(\mathcal{U}_i, \mathcal{Z}_i) \tag{22.24}$$

$$= \mathbb{E}\Big|g(X_N) - \sum_{j=i+1}^{N-1} f(t_j, X_j, \widehat{\mathcal{U}}_j(X_j), \widehat{\mathcal{Z}}_j(X_j))\Delta t_j - \sum_{j=i+1}^{N-1} \widehat{\mathcal{Z}}_j(X_j) \cdot \Delta W_j$$

$$- \mathcal{U}_i(X_i) - f(t_i, X_i, \mathcal{U}_i(X_i), \mathcal{Z}_i(X_i))\Delta t_i - \mathcal{Z}_i(X_i) \cdot \Delta W_i\Big|^2 \tag{22.25}$$

and update $(\widehat{\mathcal{U}}_i, \widehat{\mathcal{Z}}_i)$ as the solution to this minimization problem. This output provides an approximation $(\widehat{\mathcal{U}}_i, \widehat{\mathcal{Z}}_i)$ of the solution $u(t_i, \cdot)$ to the PDE (22.13) at times t_i, $i = 0, \ldots, N-1$.

MDBDP is a machine learning version of the Multi-step Forward Dynamic Programming method studied by Bender and Denk (2007) and Gobet and Turkedjiev (2014). Instead of solving at each time step two regression problems, the multi-step approach lets us consider only a single minimization as in the DBDP scheme. Compared to the latter, the multi-step consideration is expected to provide better accuracy by reducing the propagation of errors in the backward induction as it can be shown by comparing the error estimated in Germain et al. (2020) and Huré et al. (2020) at both theoretical and numerical levels.

22.3.3 Case of fully nonlinear PDEs

In this subsection, we consider fully nonlinear PDEs in the form

$$\begin{cases} \partial_t u + \mu \cdot D_x u + \frac{1}{2}\mathrm{Tr}(\sigma\sigma^\intercal D_x^2 u) = F(\cdot,\cdot,u,D_x u,D_x^2 u) & \text{on } [0,T)\times\mathbb{R}^d \\ u(T,\cdot) = g & \text{on } \mathbb{R}^d. \end{cases}$$
(22.26)

For this purpose, we introduce a forward diffusion process X in $\mathbb{R}^d$ as in (22.14), and associated to the linear part $\mathcal{L}$ of the differential operator on the left-hand side of the PDE (22.26). Since the function F contains the dependence both on the gradient $D_x u$ and the Hessian $D_x^2 u$, we can shift the linear differential operator (left-hand side) of the PDE (22.26) into the function F. However, in practice, this linear differential operator associated to a diffusion process X is used for training simulations in SGD of machine learning schemes. We refer to Section 3.1 in Pham et al. (2021) for a discussion about the choice of the parameters μ, σ. In what follows, we assume for simplicity that $\mu = 0$, and σ is a constant invertible matrix.

Let us derive formally a BSDE representation for the nonlinear PDE (22.26) on which we shall rely for designing our machine learning algorithm. Assuming that the solution u to this PDE is smooth C^2, and denoting by (Y,Z,Γ) the triple of $\mathbb{F}$-adapted processes valued in $\mathbb{R}\times\mathbb{R}^d\times\mathbb{S}^d$, defined by

$$Y_t = u(t,X_t), \quad Z_t = D_x u(t,X_t), \quad \Gamma_t = D_x^2 u(t,X_t), \quad 0 \le t \le T, \tag{22.27}$$

a direct application of Itô's formula to $u(t,X_t)$, yields that (Y,Z,Γ) satisfies the backward equation

$$Y_t = g(X_T) - \int_t^T F(s,X_s,Y_s,Z_s,\Gamma_s)ds - \int_t^T Z_s^\intercal \sigma dW_s, \quad 0 \le t \le T. \tag{22.28}$$

Compared to the case of a semi-linear PDE of the form (22.13), the key point is the approximation/learning of the Hessian matrix $D_x^2 u$, hence of the Γ-component of the BSDE (22.28). We present below different approaches to the approximation of the Γ-component. To the best of our knowledge, no theoretical convergence result is available for machine learning schemes in the fully nonlinear case but several methods show good empirical performance.

Deep 2BSDE scheme (Beck et al., 2019b).

This scheme relies on the 2BSDE representation of Cheridito et al. (2007):

$$\begin{cases} Y_t = g(X_T) - \int_t^T F(s,X_s,Y_s,Z_s,\Gamma_s)ds - \int_t^T Z_s^\intercal \sigma dW_s, \\ Z_t = D_x g(X_T) - \int_t^T A_s ds - \int_t^T \Gamma_s \sigma dW_s, \quad 0 \le t \le T, \end{cases} \tag{22.29}$$

with $A_t = \mathcal{L}D_x u(t,X_t)$. The idea is to adapt the Deep BSDE algorithm to the fully nonlinear case. Again, we treat the backward system (22.29) as a forward equation by approximating the initial conditions Y_0, Z_0 and the A- and Γ-components of the 2BSDE at each time by network functions taking the forward process X as input, and aiming to match the terminal condition.

Second-order DBDP (2DBDP) (Pham et al., 2021).

The basic idea is to adapt the DBDP scheme by approximating the solution u and its gradient $D_x u$ by network functions $\mathcal{U}$ and $\mathcal{Z}$, and then the Hessian $D_x^2 u$ by the automatic differentiation $D_x \mathcal{Z}$ of the network function $\mathcal{Z}$ (or double automatic differentiation $D_x^2 \mathcal{U}$ of the network function $\mathcal{U}$), via a learning approach relying on the time discretization of the BSDE (22.28). It turns out that such a method approximates poorly Γ inducing instability of the scheme: indeed, while the unique pair solution (Y, Z) to classical BSDEs (22.14) completely characterizes the solution to the related semilinear PDE and its gradient, the relation (22.28) does not let us characterize directly the triple (Y, Z, Γ). This approach was proposed and tested in Pham et al. (2021) where the automatic differentiation is performed on the previous value of $\mathcal{Z}$ with a truncation $\mathcal{T}$ which enables reduction of instabilities.

Second-order multistep schemes.

To overcome the instability in the approximation of the Γ-component in the second-order DBDP scheme, we propose a finer approach based on a suitable probabilistic representation of the Γ-component for learning accurately the Hessian function $D_x^2 u$ by using also Malliavin weights. We start from the training simulations of the forward process $(X_i)_i$ on the uniform grid $\pi = \{t_i = i|\pi|, i = 0, \ldots, N\}$, $|\pi| = T/N$, and notice that $X_i = \mathcal{X}_{t_i}$, $i = 0, \ldots, N$ when μ and σ are constants. The approximation of the value function u and its gradient $D_x u$ is learnt simultaneously on the grid π but requires in addition a preliminary approximation of the Hessian $D_x^2 u$ in the fully nonlinear case. This will be performed by a regression-based machine learning scheme on a subgrid $\hat{\pi} \subset \pi$, which allows to reduce the computational time of the algorithm.

We propose three versions of second-order MDBDP based on different representations of the Hessian function. For the second and the third one, we need to introduce a subgrid $\hat{\pi} = \{t_{\hat{\kappa}\ell}, \ell = 0, \ldots, \hat{N}\} \subset \pi$, of modulus $|\hat{\pi}| = \hat{\kappa}|\pi|$, for some $\hat{\kappa} \in \mathbb{N}^*$, with $N = \hat{\kappa}\hat{N}$.

- *Version 1:* Extending the methodology introduced in Pham et al. (2021), the current Γ-component at step i can be estimated by automatic differentiation of the Z-component at the previous step while the other Γ-components are estimated by automatic differentiation of their associated Z-components:

$$\Gamma_i \simeq D_x Z_{i+1}, \quad \Gamma_j \simeq D_x Z_j, \quad j > i. \tag{22.30}$$

- *Version 2:* The time discretization of (22.28) on the time grid $\hat{\pi}$, where $(Y_\ell^{\hat{\pi}}, Z_\ell^{\hat{\pi}}, \Gamma_\ell^{\hat{\pi}})$ denotes an approximation of the triple

$$\left(u(t_{\hat{\kappa}\ell}, X_{\hat{\kappa}\ell}), D_x u(t_{\hat{\kappa}\ell}, X_{\hat{\kappa}\ell}), D_x^2 u(t_{\hat{\kappa}\ell}, X_{\hat{\kappa}\ell}) \right), \quad \ell = 0, \ldots, \hat{N},$$

leads to the standard representation formula for the Z component:

$$Z_\ell^{\hat\pi} = \mathbb{E}_{\hat\kappa\ell}\left[Y_{\ell+1}^{\hat\pi}\hat{H}_\ell^1\right], \quad \ell = 0,\ldots,\hat{N}-1, \tag{22.31}$$

(recall that $\mathbb{E}_{\hat\kappa\ell}$ denotes the conditional expectation with respect to $\mathcal{F}_{t_{\hat\kappa\ell}}$), with the Malliavin weight of order one:

$$\hat{H}_\ell^1 = (\sigma^\intercal)^{-1}\frac{\hat{\Delta}W_\ell}{|\hat\pi|}, \quad \hat{\Delta}W_\ell := W_{t_{\hat\kappa(\ell+1)}} - W_{t_{\hat\kappa\ell}}. \tag{22.32}$$

By direct differentiation, we then obtain an approximation of the Γ component as

$$\Gamma_\ell^{\hat\pi} \simeq \mathbb{E}_{\hat\kappa\ell}\left[D_x u(t_{\hat\kappa(\ell+1)}, X_{\hat\kappa(\ell+1)})\hat{H}_\ell^1\right]. \tag{22.33}$$

Moreover, by introducing the antithetic variable

$$\hat{X}_{\hat\kappa(\ell+1)} = X_{\hat\kappa\ell} - \sigma\hat{\Delta}W_\ell, \tag{22.34}$$

we then propose the following regression estimator of $D_x^2 u$ on the grid $\hat\pi$ for $\ell = 0,\ldots,\hat{N}-1$ with

$$\begin{cases} \hat{\Gamma}^{(1)}(t_{\hat\kappa\hat{N}}, X_{\hat\kappa\hat{N}}) &= D^2 g(X_{\hat\kappa\hat{N}}) \\ \hat{\Gamma}^{(1)}(t_{\hat\kappa\ell}, X_{\hat\kappa\ell}) &= \mathbb{E}_{\hat\kappa\ell}\left[\frac{D_x u(t_{\hat\kappa(\ell+1)}, X_{\hat\kappa(\ell+1)}) - D_x u(t_{\hat\kappa(\ell+1)}, \hat{X}_{\hat\kappa(\ell+1)})}{2}\hat{H}_\ell^1\right]. \end{cases} \tag{22.35}$$

- *Version 3:* Alternatively, the time discretization of (22.28) on $\hat\pi$ yields the iterated conditional expectation relation:

$$Y_\ell^{\hat\pi} = \mathbb{E}_{\hat\kappa\ell}\left[g(X_{\hat\kappa\hat{N}}) - |\hat\pi|\sum_{m=\ell}^{\hat{N}-1} F(t_{\hat\kappa m}, X_{\hat\kappa m}, Y_m^{\hat\pi}, Z_m^{\hat\pi}, \Gamma_m^{\hat\pi})\right], \quad \ell = 0,\ldots,\hat{N}, \tag{22.36}$$

By (double) integration by parts, and using Malliavin weights on the Gaussian vector X, we obtain a multistep approximation of the Γ-component:

$$\Gamma_\ell^{\hat\pi} \simeq \mathbb{E}_{\hat\kappa\ell}\left[g(X_{\hat\kappa\hat{N}})\hat{H}_{\ell,\hat{N}}^2 - |\hat\pi|\sum_{m=\ell+1}^{\hat{N}-1} F(t_{\hat\kappa m}, X_{\hat\kappa m}, Y_m^{\hat\pi}, Z_m^{\hat\pi}, \Gamma_m^{\hat\pi})\hat{H}_{\ell,m}^2\right], \tag{22.37}$$

for $\ell = 0,\ldots,\hat{N}$, where

$$\hat{H}_{\ell,m}^2 = (\sigma^\intercal)^{-1}\frac{\hat{\Delta}W_\ell^m(\hat{\Delta}W_\ell^m)^\intercal - (m-\ell)|\hat\pi|I_d}{(m-\ell)^2|\hat\pi|^2}\sigma^{-1}, \quad \hat{\Delta}W_\ell^m := W_{t_{\hat\kappa m}} - W_{t_{\hat\kappa\ell}}. \tag{22.38}$$

By introducing again the antithetic variables

$$\hat{X}_{\hat\kappa m} = X_{\hat\kappa\ell} - \sigma\hat{\Delta}W_\ell^m, \quad m = \ell+1,\ldots,\hat{N}, \tag{22.39}$$

we then propose another regression estimator of $D_x^2 u$ on the grid $\hat{\pi}$ with

$$\hat{\Gamma}^{(2)}(t_{\hat{k}\ell}, X_{\hat{k}\ell}) \tag{22.40}$$

$$= \mathbb{E}_{\hat{k}\ell} \left[\frac{g(X_{\hat{k}\hat{N}}) + g(\hat{X}_{\hat{k}\hat{N}})}{2} \hat{H}^2_{\ell,\hat{N}} \right. \tag{22.41}$$

$$- \frac{|\hat{\pi}|}{2} \sum_{m=\ell+1}^{\hat{N}-1} \Big(F\big(t_{\hat{k}m}, X_{\hat{k}m}, u(t_{\hat{k}m}, X_{\hat{k}m}), D_x u(t_{\hat{k}m}, X_{\hat{k}m}), \hat{\Gamma}^{(2)}(t_{\hat{k}m}, X_{\hat{k}m})\big) \tag{22.42}$$

$$+ F\big(t_{\hat{k}m}, \hat{X}_{\hat{k}m}, u(t_{\hat{k}m}, \hat{X}_{\hat{k}m}), D_x u(t_{\hat{k}m}, \hat{X}_{\hat{k}m}), \hat{\Gamma}^{(2)}(t_{\hat{k}m}, \hat{X}_{\hat{k}m})\big) \tag{22.43}$$

$$\left. - 2F\big(t_{\hat{k}\ell}, X_{\hat{k}\ell}, u(t_{\hat{k}\ell}, X_{\hat{k}\ell}), D_x u(t_{\hat{k}\ell}, X_{\hat{k}\ell}), \hat{\Gamma}^{(2)}(t_{\hat{k}\ell}, X_{\hat{k}\ell})\big)\Big) \hat{H}^2_{\ell,m} \right], \tag{22.44}$$

for $\ell = 0, \ldots, N-1$, and $\hat{\Gamma}^{(2)}(t_{\hat{k}\hat{N}}, X_{\hat{k}\hat{N}}) = D^2 g(X_{\hat{k}\hat{N}})$. The correction term $-2F$ evaluated at time $t_{\hat{k}\ell}$ in $\hat{\Gamma}^{(2)}(t_{\hat{k}\ell}, X_{\hat{k}\ell})$ does not add bias since

$$\mathbb{E}_{\hat{k}\ell} \left[F\big(t_{\hat{k}\ell}, X_{\hat{k}\ell}, u(t_{\hat{k}\ell}, X_{\hat{k}\ell}), D_x u(t_{\hat{k}\ell}, X_{\hat{k}\ell}), \hat{\Gamma}^{(2)}(t_{\hat{k}\ell}, X_{\hat{k}\ell})\big) \hat{H}^2_{\ell,m} \right] = 0, \tag{22.45}$$

for all $m = \ell+1, \ldots, \hat{N}-1$, and by Taylor expansion of F at second order, we see that, together with the antithetic variable, it lets us control the variance when the time step goes to zero.

Remark 22.1 In the case where the function g has some regularity property, one can avoid the integration by parts at the terminal data component in the above expression of $\hat{\Gamma}^{(2)}$. For example, when g is C^1, $\frac{g(X_{\hat{k}\hat{N}})+g(\hat{X}_{\hat{k}\hat{N}})}{2} \hat{H}^2_{\ell,\hat{N}}$ is alternatively replaced in $\hat{\Gamma}^{(2)}$ expression by $(Dg(X_{\hat{k}\hat{N}}) - Dg(\hat{X}_{\hat{k}\hat{N}}))\hat{H}^1_{\ell,\hat{N}}$, while when it is C^2 it is replaced by $D^2 g(X_{\hat{k}\hat{N}})$.

Remark 22.2 We point out that in our machine learning setting for the versions 2 and 3 of the scheme, we only solve two optimization problems by time step instead of three as in Fahim et al. (2011). One optimization is dedicated to the computation of the Γ-component but the $\mathcal{U}$- and $\mathcal{Z}$-components are simultaneously learned by the algorithm.

We can now describe the three versions of second-order MDBDP schemes for the numerical resolution of the fully nonlinear PDE (22.26). We emphasize that these schemes do not require *a priori* that the solution to the PDE be smooth.

The proposed algorithms 22.3, 22.4, 22.5 are in backward iteration, and involve one optimization at each step. Moreover, as the computation of Γ requires a further derivation for Algorithms 22.4 and 22.5, we may expect that the additional propagation error varies according to $\frac{|\pi|}{|\hat{\pi}|} = \frac{1}{\hat{k}}$, and thus the convergence of the scheme when $\hat{k}$ is large. In the numerical implementation, the expectation in the loss functions are replaced by empirical average and the minimization over network functions is performed by stochastic gradient descent.

Algorithm 22.3: Second-order Explicit Multistep DBDP (2EMDBDP)

---

1 **for** $i = N - 1, \ldots, 0$ **do**

2   If $i = N - 1$, update $\widehat{\Gamma}_i = D^2 g$, otherwise $\widehat{\Gamma}_i = D_x \widehat{Z}_{i+1}$, $\widehat{\Gamma}_j = D_x \widehat{Z}_j$,
    $j \in [\![i + 1, N - 1]\!]$,   /* Update Hessian */

3   Minimize over network functions $\mathcal{U} \colon \mathbb{R}^d \to \mathbb{R}$, and $\mathcal{Z} \colon \mathbb{R}^d \to \mathbb{R}^d$ the
    loss function at time t_i:

$$J_i^{MB}(\mathcal{U}, \mathcal{Z}) \tag{22.46}$$

$$= \mathbb{E} \left| g(X_N) - |\pi| \sum_{j=i+1}^{N-1} F(t_j, X_j, \widehat{\mathcal{U}}_j(X_j), \widehat{\mathcal{Z}}_j(X_j), \widehat{\Gamma}_j(X_j)) \right.$$

$$- \sum_{j=i+1}^{N-1} \widehat{\mathcal{Z}}_j(X_j)^{\mathsf{T}} \sigma \Delta W_j - \mathcal{U}(X_i)$$

$$\left. - |\pi| F(t_i, X_i, \mathcal{U}(X_i), \mathcal{Z}(X_i), \widehat{\Gamma}_i(X_{i+1})) - \mathcal{Z}(X_i) \cdot \sigma \Delta W_i \right|^2. \tag{22.47}$$

    Update $(\widehat{\mathcal{U}}_i, \widehat{\mathcal{Z}}_i)$ as the solution to this minimization problem
    /* Update the function and its derivative */

4 **end**

22.3.4 *Limitations of the machine learning approach*

Let us point out some important features for a successful application, and thus limitations, of deep learning methods in the resolution of PDEs.

- A key point is the ability to represent the solution by the network. In particular, PDEs arising from complex mechanics structure could be hardly approximated by neural networks.

- A second point is to avoid local minima far away from the solution.

22.4 Numerical applications

We test our different algorithms on various examples and by varying the state space dimension. If not stated otherwise, we choose the maturity $T = 1$. In each example we use an architecture composed of two hidden layers with $d + 10$ neurons. We apply Adam gradient descent (Kingma and Ba, 2014) with a decreasing learning rate, using the Tensorflow library (Abadi et al., 2016). Each numerical experiment is conducted using a node composed of two Intel ® Xeon® Gold 5122 Processors, 192 Go of RAM, and 2 GPU nVidia® Tesla® V100 16Go. We use a batch size of 1000.

Algorithm 22.4: Second-order Multistep DBDP (2MDBDP)

---

1 **for** $\ell = \hat{N}, \ldots, 0$ **do**

2   If $\ell = \hat{N}$, update $\widehat{\Gamma}_\ell = D^2 g$, otherwise minimize over network functions $\Gamma \colon \mathbb{R}^d \to \mathbb{S}^d$ the loss function

$$\mathcal{J}_\ell^{1,M}(\Gamma) = \mathbb{E}\left|\Gamma(X_{\hat{k}\ell}) - \frac{\widehat{\mathcal{Z}}_{\hat{k}(\ell+1)}(X_{\hat{k}(\ell+1)}) - \widehat{\mathcal{Z}}_{\hat{k}(\ell+1)}(\hat{X}_{\hat{k}(\ell+1)})}{2}\hat{H}_\ell^1\right|^2.$$

                                                                                    (22.48)

     Update $\widehat{\Gamma}_\ell$ the solution to this minimization problem   `/* Update Hessian */`

3   **for** $k = \hat{k} - 1, \ldots, 0$ **do**

4   Minimize over network functions $\mathcal{U} \colon \mathbb{R}^d \to \mathbb{R}$, and $\mathcal{Z} \colon \mathbb{R}^d \to \mathbb{R}^d$ the loss function at time t_i, $i = (\ell - 1)\hat{k} + k$:

$$J_i^{MB}(\mathcal{U}, \mathcal{Z}) \hspace{6cm} (22.49)$$

$$= \mathbb{E}\Bigg| g(X_N) - |\pi| \sum_{j=i+1}^{N-1} F(t_j, X_j, \widehat{\mathcal{U}}_j(X_j), \widehat{\mathcal{Z}}_j(X_j), \widehat{\Gamma}_\ell(X_j))$$

$$- \sum_{j=i+1}^{N-1} \widehat{\mathcal{Z}}_j(X_j)^\intercal \sigma \Delta W_j - \mathcal{U}(X_i)$$

$$- |\pi| F(t_i, X_i, \mathcal{U}(X_i), \mathcal{Z}(X_i), \widehat{\Gamma}_\ell(X_i)) - \mathcal{Z}(X_i) \cdot \sigma \Delta W_i\Bigg|^2.$$

                                                                                   (22.50)

         Update $(\widehat{\mathcal{U}}_i, \widehat{\mathcal{Z}}_i)$ as the solution to this minimization problem `/* Update the function and its derivative */`

5   **end**

6 **end**

22.4.1 Numerical tests on credit valuation adjustment pricing

We consider an example of a model from Henry-Labordere (2017) for the pricing of CVA in a d-dimensional Black–Scholes model

$$\mathrm{d}X_t = \sigma X_t \, \mathrm{d}W_t, \; X_0 = 1_d \hspace{3cm} (22.58)$$

with $\sigma > 0$, given by the nonlinear PDE

$$\begin{cases} \partial_t u + \frac{\sigma^2}{2}\mathrm{Tr}(x^\top D_x^2 u \; x) + \beta(u_+ - u) = 0 & \text{on } [0, T] \times \mathbb{R}^d \\ u(T, x) = |\sum_{i=1}^d x_i - d| - 0.1 & \text{on } \mathbb{R}^d \end{cases} \hspace{1cm} (22.59)$$

with a straddle-type payoff. We compare our results with the DBDP scheme (Huré et al., 2020) with the ones from the Deep BSDE solver (Han et al., 2017). The results in Table 22.1 are averaged over 10 runs and the standard deviation is written in parentheses. We use ReLu activation functions.

Algorithm 22.5: Second-order Multistep Malliavin DBDP (2M²DBDP)

1 **for** $\ell = \hat{N}, \ldots, 0$ **do**

2 $\quad$ If $\ell = \hat{N}$, update $\widehat{\Gamma}_\ell = D^2 g$, otherwise minimize over network functions $\Gamma \colon \mathbb{R}^d \to \mathbb{S}^d$ the loss function

$$\mathcal{J}_\ell^{2,M}(\Gamma) \tag{22.51}$$

$$= \mathbb{E}\bigg|\Gamma(X_{\hat{k}\ell}) - \frac{D^2 g(X_{\hat{k}\hat{N}}) + D^2 g(\hat{X}_{\hat{k}\hat{N}})}{2} \tag{22.52}$$

$$+ \frac{|\hat{\pi}|}{2} \sum_{m=\ell+1}^{\hat{N}-1} \Big(F\big(t_{\hat{k}m}, X_{\hat{k}m}, \widehat{\mathcal{U}}_{\hat{k}m}(X_{\hat{k}m}), \widehat{\mathcal{Z}}_{\hat{k}m}(X_{\hat{k}m}), \widehat{\Gamma}_m(X_{\hat{k}m})\big) \tag{22.53}$$

$$+ F\big(t_{\hat{k}m}, \hat{X}_{\hat{k}m}, \widehat{\mathcal{U}}_{\hat{k}m}(\hat{X}_{\hat{k}m}), \widehat{\mathcal{Z}}_{\hat{k}m}(\hat{X}_{\hat{k}m}), \widehat{\Gamma}_m(\hat{X}_{\hat{k}m})\big) \tag{22.54}$$

$$- 2F\big(t_{\hat{k}\ell}, \hat{X}_{\hat{k}\ell}, \widehat{\mathcal{U}}_{\hat{k}\ell}(\hat{X}_{\hat{k}\ell}), \widehat{\mathcal{Z}}_{\hat{k}\ell}(\hat{X}_{\hat{k}\ell}), \widehat{\Gamma}_\ell(\hat{X}_{\hat{k}\ell})\big) \Big) \hat{H}_{\ell,m}^2 \bigg|^2. \tag{22.55}$$

$\quad$ Update $\widehat{\Gamma}_\ell$ the solution to this minimization problem $\quad$ /* Update Hessian */

3 $\quad$ **for** $k = \hat{k} - 1, \ldots, 0$ **do**

4 $\quad\quad$ Minimize over network functions $\mathcal{U} \colon \mathbb{R}^d \to \mathbb{R}$, and $\mathcal{Z} \colon \mathbb{R}^d \to \mathbb{R}^d$ the loss function at time t_i, $i = (\ell - 1)\hat{k} + k$:

$$J_i^{MB}(\mathcal{U}, \mathcal{Z}) \tag{22.56}$$

$$= \mathbb{E}\bigg|g(X_N) - |\pi| \sum_{j=i+1}^{N-1} F(t_j, X_j, \widehat{\mathcal{U}}_j(X_j), \widehat{\mathcal{Z}}_j(X_j), \widehat{\Gamma}_\ell(X_j))$$

$$- \sum_{j=i+1}^{N-1} \widehat{\mathcal{Z}}_j(X_j)^\intercal \sigma \Delta W_j - \mathcal{U}(X_i)$$

$$- |\pi| F(t_i, X_i, \mathcal{U}(X_i), \mathcal{Z}(X_i), \widehat{\Gamma}_\ell(X_i)) - \mathcal{Z}(X_i) \cdot \sigma \Delta W_i\bigg|^2. \tag{22.57}$$

$\quad\quad$ Update $(\widehat{\mathcal{U}}_i, \widehat{\mathcal{Z}}_i)$ as the solution to this minimization problem /* Update the function and its derivative */

5 $\quad$ **end**

6 **end**

We observe in Table 22.1 that both algorithms give very close results and are able to solve the nonlinear pricing problem in high dimension d. The variance of the results is quite small and similar from one to another but increases with the dimension. The same conclusions arise when solving the PDE for the larger maturity $T = 2$.

Table 22.1 CVA value with $X_0 = 1$, $T = 1$, $\beta = 0.03$, $\sigma = 0.2$ and 50 time steps.

Dimension d	DBDP (Huré et al., 2020)	DBSDE (Han et al., 2017)
1	0.05950 (0.000257)	0.05949 (0.000264)
3	0.17797 (0.000421)	0.17807 (0.000288)
5	0.25956 (0.000467)	0.25984 (0.000331)
10	0.40930 (0.000623)	0.40886 (0.000196)
15	0.52353 (0.000591)	0.52389 (0.000551)
30	0.78239 (0.000832)	0.78231 (0.001266)

22.4.2 Portfolio allocation in stochastic volatility models

We consider several examples from Pham et al. (2021) that we solve with Algorithms 22.3 (2EMDBDP), 22.4 (2MDBDP), and 22.5 (2M²DBDP) above. Notice that some comparison tests with the 2DBSDE scheme (Beck et al., 2019b) have been already made in Pham et al. (2021). For a resolution with $N = 120$, $\hat{N} = 30$, the execution of our multitep algorithms takes between 10000 s and 30000 s (depending on the dimension) with a number of gradient descent iterations fixed at 4000 at each time step except 80000 at the first one. We use tanh as an activation function.

We consider a portfolio selection problem formulated as follows. There are n risky assets of uncorrelated price process $P = (P^1, \ldots, P^n)$ with dynamics governed by

$$dP_t^i = P_t^i \sigma(V_t^i)\left[\lambda_i(V_t^i)dt + dW_t^i\right], \quad i = 1, \ldots, n, \tag{22.60}$$

where $W = (W^1, \ldots, W^n)$ is an n-dimensional Brownian motion, $\lambda = (\lambda^1, \ldots, \lambda^n)$ is the market price of risk of the assets, σ is a positive function (e.g. $\sigma(v) = e^v$, corresponding to the Scott model), and $V = (V^1, \ldots, V^n)$ is the volatility factor modeled by an Ornstein–Uhlenbeck (OU) process

$$dV_t^i = \kappa_i[\theta_i - V_t^i]dt + \nu_i dB_t^i, \quad i = 1, \ldots, n, \tag{22.61}$$

with $\kappa_i, \theta_i, \nu_i > 0$, and $B = (B^1, \ldots, B^n)$ a n-dimensional Brownian motion, such that $d\langle W^i, B^j \rangle = \delta_{ij}\rho_{ij}dt$, with $\rho_i := \rho_{ii} \in (-1, 1)$. An agent can invest at any time an amount $\alpha_t = (\alpha_t^1, \ldots, \alpha_t^n)$ in the stocks, which generates a wealth process $X = X^\alpha$ governed by

$$dX_t = \sum_{i=1}^{n} \alpha_t^i \sigma(V_t^i)\left[\lambda_i(V_t^i)dt + dW_t^i\right].$$

The objective of the agent is to maximize her expected utility from terminal wealth:

$$\mathbb{E}\left[U(X_T^\alpha)\right] \quad \leftarrow \quad \text{maximize over } \alpha \tag{22.62}$$

It is well known that the solution to this problem can be characterized by the dynamic programming method (see e.g. Pham, 2009), which leads to the Hamilton–Jacobi–Bellman equation for the value function on $[0, T) \times \mathbb{R} \times \mathbb{R}^n$:

$$
\begin{cases}
\partial_t u + \sum_{i=1}^{n} \left[\kappa_i(\theta_i - v_i)\partial_{v_i} u + \frac{1}{2} v_i^2 \partial_{v_i}^2 u \right] \\
= \frac{1}{2} R(v)\frac{(\partial_x u)^2}{\partial_{xx}^2 u} + \sum_{i=1}^{n} \left[\rho_i \lambda_i(v_i)v_i \frac{\partial_x u \partial_{xv_i}^2 u}{\partial_{xx}^2 u} + \frac{1}{2}\rho_i^2 v_i^2 \frac{(\partial_{xv_i}^2 u)^2}{\partial_{xx}^2 u} \right] \\
u(T, x, v) = U(x), \qquad x \in \mathbb{R}, \ v \in \mathbb{R}^n,
\end{cases}
$$

with a Sharpe ratio $R(v) := |\lambda(v)|^2$, for $v = (v_1, \ldots, v_n) \in (0, \infty)^n$. The optimal portfolio strategy is then given in feedback form by $\alpha_t^* = \hat{a}(t, X_t^*, V_t)$, where $\hat{a} = (\hat{a}_1, \ldots, \hat{a}_n)$ is given by

$$
\hat{a}_i(t, x, v) \tag{22.63}
$$

$$
= -\frac{1}{\sigma(v_i)}\left(\lambda_i(v_i)\frac{\partial_x u}{\partial_{xx}^2 u} + \rho_i v_i \frac{\partial_{xv_i}^2 u}{\partial_{xx}^2 u} \right), \ (t, x, v = (v_1, \ldots, v_n)) \in [0, T) \times \mathbb{R} \times \mathbb{R}^n, \tag{22.64}
$$

for $i = 1, \ldots, n$.

We shall test this example when the utility function U is of exponential form: $U(x) = -\exp(-\eta x)$, with $\eta > 0$, and under different cases for which explicit solutions are available. We refer to Pham et al. (2021) where these solutions are described.

(1) *Merton problem.* This corresponds to a degenerate case where the factor V, hence the volatility σ and the risk premium λ are constant ($v_i = \theta_i$, $v_i = 0$). We train our algorithms with the forward process

$$
X_{k+1} = X_k + |\lambda|\Delta t_k + \Delta W_k, \qquad k = 0, \ldots, N, \quad X_0 = x_0. \tag{22.65}
$$

(2) *One risky asset: $n = 1$.* We train our algorithms with the forward process

$$
X_{k+1} = X_k + \lambda(\theta)\Delta t_k + \Delta W_k, \quad k = 0, \ldots, N-1, \qquad X_0 = x_0 \tag{22.66}
$$

$$
V_{k+1} = V_k + v\Delta B_k, \qquad k = 0, \ldots, N-1, \quad V_0 = \theta. \tag{22.67}
$$

We test our algorithm with $\lambda(v) = \lambda v$, $\lambda > 0$, for which we have an explicit solution.

(3) *No leverage effect, i.e., $\rho_i = 0$, $i = 1, \ldots, n$.* We train with the forward process

$$
X_{k+1} = X_k + \sum_{i=1}^{n} \lambda_i(\theta_i)\Delta t_k + \Delta W_k, \qquad k = 0, \ldots, N-1, \quad X_0 = x_0 \tag{22.68}
$$

$$
V_{k+1}^i = V_k^i + v_i \Delta B_k^i, \qquad k = 0, \ldots, N-1, \quad V_0^i = \theta_i. \tag{22.69}
$$

We test our algorithm with $\lambda_i(v) = \lambda_i v_i$, $\lambda_i > 0$, $i = 1, \ldots, n$, $v = (v_1, \ldots, v_n)$, for which we have an explicit solution.

Table 22.2 Estimate of $u(0,1)$ in the Merton problem with $N = 120$, $\hat{N} = 30$. Average and standard deviation observed over 10 independent runs are reported. The theoretical solution is -0.50662.

	Average	Standard deviation	Relative error (%)
Pham et al. (2021)	−0.50561	0.00029	0.20
2EMDBDP	**−0.50673**	**0.00019**	**0.022**
2MDBDP	−0.50647	0.00033	0.030
2M²DBDP	−0.50644	0.00022	0.035

Table 22.3 Estimate of $u(0,1,\theta)$ on the one-asset problem with stochastic volatility ($d = 2$) and $N = 120$, $\hat{N} = 30$. Average and standard deviation observed over 10 independent runs are reported. The exact solution is -0.53609477.

	Average	Standard deviation	Relative error (%)
Pham et al. (2021)	−0.53431	0.00070	0.34
2EMDBDP	**−0.53613**	**0.00045**	**0.007**
2MDBDP	−0.53772	0.00046	0.304
2M²DBDP	−0.53205	0.00050	0.755

Merton Problem.

We take $\eta = 0.5$, $\lambda = 0.6$, $N = 120$, $\hat{N} = 30$, $T = 1$, $x_0 = 1$. We plot in Figure 22.1 the neural networks approximation of u, $D_x u$, $D_x^2 u$, and the feedback control $\hat{a}$ (for one asset) computed from our different algorithms, together with their analytic values (in orange). As also reported in the estimates of Table 22.2, the multistep algorithms improve significantly the results obtained in Pham et al. (2021), where the estimation of the Hessian is not really accurate (see the blue curve in Figure 22.1).

One asset $n = 1$ in Scott volatility model.

We take $\eta = 0.5$, $\lambda = 1.5$, $\theta = 0.4$, $\nu = 0.4$, $\kappa = 1$, $\rho = -0.7$, $T = 1$, $x_0 = 1$. For all tests we choose $N = 120$, $\hat{N} = 30$ and $\sigma(v) = e^v$. We report in Table 22.3 the relative error between the neural networks approximation of $u, D_x u, D_x^2 u$ computed from our different algorithms and their analytic values. It turns out that the multistep extension of Pham et al. (2021), namely the 2EMDBDP scheme, yields a very accurate approximation, much better than the other algorithms, and also a reduction of the standard deviation.

No leverage in the Scott model.

In the case with one asset we take $\eta = 0.5$, $\lambda = 1.5$, $\theta = 0.4$, $\nu = 0.2$, $\kappa = 1$, $T = 1$, $x_0 = 1$. For all tests we choose $N = 120$, $\hat{N} = 30$ and $\sigma(v) = e^v$. We report in Table 22.4 the relative error between the neural network approximation of $u, D_x u, D_x^2 u$ computed from our different algorithms and their analytic values.

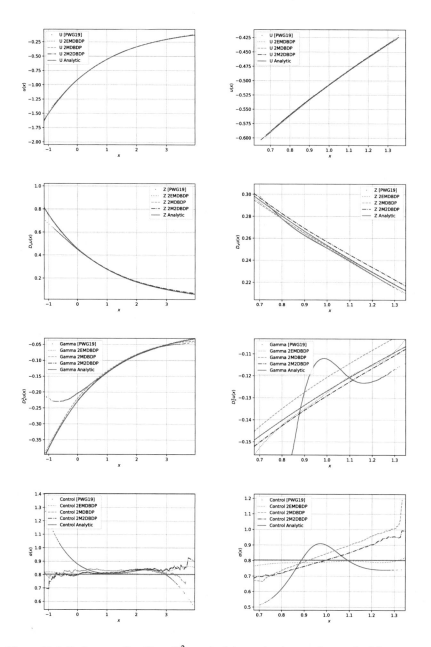

Figure 22.1 Estimates of u, $D_x u$, $D_x^2 u$ and of the optimal control α on the Merton problem with $N = 120$, $\hat{N} = 30$. We take $x_0 = 1$, at the left $t = 0.5042$, and at the right $t = 0.0084$.

All the algorithms yield quite accurate results, but compared to the case with correlation in Table 22.3, it appears here that the best performance in terms of precision is achieved by Algorithm $2M^2$DBDP.

Table 22.4 Estimate of $u(0, 1, \theta)$, with 120 time steps on the no-leverage problem with one asset ($d = 2$) and $N = 120$, $\hat{N} = 30$. Average and standard deviation observed over 10 independent runs are reported. The exact solution is -0.501566.

	Average	Standard deviation	Relative error (%)
Pham et al. (2021)	−0.49980	0.00073	0.35
2EMDBDP	−0.50400	0.00229	0.485
2MDBDP	−0.50149	**0.00024**	0.015
2M²DBDP	**−0.50157**	0.00036	**0.001**

Table 22.5 Estimate of $u(0, 1, \theta)$, with 120 time steps on the no-leverage problem with four assets ($d = 5$) and $N = 120$, $\hat{N} = 30$. Average and standard deviation observed over 10 independent runs are reported. The theoretical solution is -0.44176462.

	Average	Standard deviation	Relative error (%)
Pham et al. (2021)	−0.43768	0.00137	0.92
2EMDBDP	**−0.4401**	**0.00051**	**0.239**
2MDBDP	−0.43796	0.00098	0.861
2M²DBDP	−0.44831	0.00566	1.481

In the case with four assets ($n = 4$, $d = 5$), we take

$$\eta = 0.5, \quad \lambda = \begin{pmatrix} 1.5 & 1.1 & 2. & 0.8 \end{pmatrix}, \quad \theta = \begin{pmatrix} 0.1 & 0.2 & 0.3 & 0.4 \end{pmatrix},$$
$$v = \begin{pmatrix} 0.2 & 0.15 & 0.25 & 0.31 \end{pmatrix}, \quad \kappa = \begin{pmatrix} 1. & 0.8 & 1.1 & 1.3 \end{pmatrix}.$$

The results are reported in Table 22.5. We observe that the algorithm in Pham et al. (2021) provides a not so accurate outcome, while its multistep version (2EMDBDP scheme) reduces the relative error and the standard deviation ten-fold.

In the case with nine assets ($n = 9$, $d = 10$), we take

$$\eta = 0.5, \quad \lambda = \begin{pmatrix} 1.5 & 1.1 & 2 & 0.8 & 0.5 & 1.7 & 0.9 & 1 & 0.9 \end{pmatrix},$$
$$\theta = \begin{pmatrix} 0.1 & 0.2 & 0.3 & 0.4 & 0.25 & 0.15 & 0.18 & 0.08 & 0.91 \end{pmatrix},$$
$$v = \begin{pmatrix} 0.2 & 0.15 & 0.25 & 0.31 & 0.4 & 0.35 & 0.22 & 0.4 & 0.15 \end{pmatrix},$$
$$\kappa = \begin{pmatrix} 1 & 0.8 & 1.1 & 1.3 & 0.95 & 0.99 & 1.02 & 1.06 & 1.6 \end{pmatrix}.$$

The results are reported in Table 22.6. The approximation is less accurate than in lower dimensions, but we observe again that compared to the one-step scheme in Pham et al. (2021), the multistep versions improve significantly the standard deviation of the result. However the best performance precision-wise is obtained here by the Pham et al. (2021) scheme.

Table 22.6 Estimate of $u(0, 1, \theta)$, with 120 time steps on the no-leverage problem with nine assets ($d = 10$) and $N = 120$. Average and standard deviation (SD) observed over ten independent runs are reported. The theoretical solution is –0.27509173.

	$\hat{N}$	Average	SD	Relative error (%)
Pham et al. (2021)		**–0.27920**	0.05734	**1.49**
2EMDBDP		–0.26631	**0.00283**	3.19
2MDBDP	30	–0.28979	0.00559	5.34
2MDBDP	60	–0.28549	0.00948	3.78
2MDBDP	120	**–0.28300**	0.01129	**2.87**
2M^{2}DBDP	30	NC	NC	NC

22.5 Extensions and perspectives

Solving mean-field control and mean-field games through McKean–Vlasov FBSDEs.

These methods solve the optimality conditions for mean-field problems through the stochastic Pontryagin principle from Carmona and Delarue (2018). The law of the solution influences the coupled FBSDEs dynamics so they are of McKean–Vlasov type.

Variations around the Deep BSDE method (Han et al., 2017) were used to solve such a system by Carmona and Laurière (2019), Fouque and Zhang (2020). Germain et al. (2019) used the merged method from Chan-Wai-Nam et al. (2019) and solved several numerical examples in dimension 10 by introducing an efficient law estimation technique. Carmona and Laurière (2019) also proposed another method dedicated to mean-field control to directly tackle the optimization problem with a neural network as the control in the stochastic dynamics. The N-player games, before going to the mean-field limit of an infinite number of players, were solved by Hu (2019); Han et al. (2020).

Solving mean-field control through the master Bellman equation and symmetric neural networks.

Germain et al. (2021b) solved the master Bellman equation arising from the dynamic programming principle applied to mean-field control problems (see Pham and Wei, 2017). They approximated the value function evaluated on the empirical measure stemming from particle simulation of a training forward process. The symmetry between i.i.d. particles is enforced by optimizing over exchangeable high-dimensional neural networks, invariant by permutation of their inputs. In a companion paper (Germain et al., 2021a) they provided a rate for the particle method convergence.

Reinforcement Learning for mean-field control and mean-field games (Carmona et al., 2019; Anahtarcı et al., 2019; Angiuli et al., 2020; Gu et al., 2020; Guo et al., 2020).

Some works focus on similar problems but with unknown dynamics. Thus they rely on trajectories sampled from a simulator and reinforcement learning – especially Q-learning – to estimate the state action value function and optimal control without a model. The idea is to optimize a neural network by relying on a memory of past state action transitions used to train the network in order for it to satisfy the Bellman equation on samples from memory replay.

Machine learning framework for solving high-dimensional mean-field game and mean-field control problems (Ruthotto et al., 2020).

This chapter has focussed on potential mean-field games in which the cost functions depending on the law can be written as the linear functional derivative of a function with respect to a measure. A Lagrangian method with deep Galerkin-type penalization is used. In this case the potential is approached by a neural network and solving mean-field games amounts to solving an unconstrained optimization problem.

Deep quantum neural networks (Sakuma, 2020).

We briefly mention some work studying the use of deep quantum neural networks which exploit the quantum superposition properties by replacing bits with "qubits". Promising results are obtained when using these networks for regression in financial contexts such as implied volatility estimation. Future work may study the application of such neural networks to control problems and PDEs.

Path signature for path-dependent PDE (Sabate Vidales et al., 2020).

This work extends previously developed methods for solving state-dependent PDEs to the linear path-dependent setting, coming for instance from the pricing and hedging of path-dependent options. A path-dependent Feynman–Kac representation is numerically computed through a global minimization over neural networks. The authors show that using LSTM networks taking the forward process's path signatures (coming from the rough paths literature) as input yields better results than taking the discretized path as input of a feedforward network.

References

Abadi, M., Barham, P., Chen, J., Chen, Z., Davis, A., Dean, J., Devin, M., Ghemawat, S., Irving, G., Isard, M., Kudlur, M., Levenberg, J., Monga, R., Moore, S., Murray, D.G., Steiner, B., Tucker, P., Vasudevan, Warden, V., Wicke, M., Yu, Y., and Zheng, X. 2016. TensorFlow: A system for large-scale machine learning. Pages 265–283 of: *12th USENIX Symposium on Operating Systems Design and Implementation (OSDI 16)*.

Al-Aradi, A., Correia, A., Naiff, D., Jardim, G., and Saporito, Y. 2018. Solving nonlinear and high-dimensional partial differential equations via deep earning. ArXiv:1811.08782.

Anahtarcı, B., Karıksız, C. Deha, and Saldi, N. 2019. Fitted Q-Learning in mean-field games. ArXiv:1912.13309.

Angiuli, A., Fouque, J.-P., and Laurière, M. 2020. Unified reinforcement Q-learning for mean field game and control problems. ArXiv:2006.13912.

Bachouch, A., Huré, C., Pham, H., and Langrené, N. 2022. Deep neural networks algorithms for stochastic control problems on finite horizon: numerical computations. *Methodol. Comput. Appl. Probab.*, **24**(1), 143–178.

Beck, C., Becker, S., Cheridito, P., Jentzen, A., and Neufeld, A. 2019a. Deep splitting method for parabolic PDEs. ArXiv:1907.03452, 07.

Beck, C., E, W., and Jentzen, A. 2019b. Machine learning approximation algorithms for high-dimensional fully nonlinear partial differential equations and second-order backward stochastic differential equations. *J. Nonlinear Sci.*, **29**(4), 1563–1619.

Beck, C., Hutzenthaler, M., Jentzen, A., and Kuckuck, B. 2020. An overview on deep learning-based approximation methods for partial differential equations. ArXiv:2012.12348.

Becker, S., Cheridito, P., and Jentzen, A. 2019a. Deep optimal stopping. *J. Mach. Learn. Res.*, **20**, 1–25.

Becker, S., Cheridito, P., Jentzen, A., and Welti, T. 2019b. Solving high-dimensional optimal stopping problems using deep learning. ArXiv:1908.01602.

Bender, C., and Denk, R. 2007. A forward scheme for backward SDEs. *Stochastic Process. Appl.*, **117**(12), 1793–1812.

Bouchard, B., and Touzi, N. 2004. Discrete-time approximation and Monte-Carlo simulation of backward stochastic differential equations. *Stochastic Process. Appl.*, **111**(2), 175–206.

Carmona, R., and Delarue, F. 2018. *Probabilistic Theory of Mean Field Games with Applications. Volumes I and II.* Springer.

Carmona, R., and Laurière, M. 2019. Convergence analysis of machine learning algorithms for the numerical solution of mean-field control and games: II The finite horizon case. ArXiv:1908.01613.

Carmona, R., Laurière, M., and Tan, Z. 2019. Model-free mean-field reinforcement learning: Mean-field MDP and mean-gield Q-learning. ArXiv:1910.12802.

Chan-Wai-Nam, Quentin, Mikael, Joseph, and Warin, Xavier. 2019. Machine learning for semi-linear PDEs. *J. Sci. Comput.*, **79**(02), 1667–1712.

Cheridito, P., Soner, H.M., Touzi, N., and Victoir, N. 2007. Second-order backward stochastic differential equations and fully nonlinear parabolic PDEs. *Comm. Pure Appl. Math.*, **60**(7), 1081–1110.

Dissanayake, M.W.M., and Phan-Thien, N. 1994. Neural network-based approximations for solving partial differential equations. *Commun. Numer. Methods Eng.*, **10**(3), 195–201.

E, W., and Yu, B. 2018. The deep Ritz method: A deep learning-based numerical algorithm for solving variational problems. *Commun. Math. Stat.*, **6**, 1–12.

E, W., Han, J., and Jentzen, A. 2017. Deep learning-based numerical methods for high dimensional parabolic partial differential equations and backward stochastic differential equations. *Commun. Math. Stat.*, **5**(4), 349–380.

Fahim, A., Touzi, N., and Warin, X. 2011. A probabilistic numerical method for fully nonlinear parabolic PDEs. *Ann. Appl. Probab.*, **21**(4), 1322–1364.

Fécamp, S., Mikael, J., and Warin, X. 2019. Risk management with machine-learning-based algorithms. ArXiv:1902.05287.

Fouque, J-.P., and Zhang, Z. 2020. Deep learning methods for mean field control problems with delay. *Frontiers in Applied Mathematics and Statistics*, **6**:11.

Fujii, M., Takahashi, A., and Takahashi, M. 2019. Asymptotic expansion as prior knowledge in deep learning method for high dimensional BSDEs. *Asia Pacific Financial Markets*, **26**(3), 391–408.

Germain, M., Mikael, J., and Warin, X. 2019. Numerical resolution of McKean–Vlasov FBSDEs using neural networks. ArXiv:1909.12678.

Germain, M., Pham, H., and Warin, X. 2020. Deep backward multistep schemes for nonlinear PDEs and approximation error analysis. ArXiv:2006.01496v1.

Germain, M., Pham, H., and Warin, X. 2021a. Rate of convergence for particles approximation of PDEs in Wasserstein space. ArXiv:2103.00837.

Germain, M., Pham, H., Warin, X., and Laurière, M. 2021b. Solving mean-field PDEs with symmetric neural networks. *In preparation.*

Glau, K., and Wunderlich, L. 2020. The deep parametric PDE method: application to option pricing. ArXiv:2012.06211.

Gobet, E., and Munos, R. 2005. Sensitivity analysis using Itô-Malliavin calculus and martingales, and application to stochastic optimal control. *SIAM J. Control Optim.*, **43**(5), 1676–1713.

Gobet, E., and Turkedjiev, P. 2014. Linear regression MDP scheme for discrete backward stochastic differential equations under general conditions. *Math. Comp.*, **85**(03).

Gobet, E., Lemor, J-P., and Warin, X. 2005. A regression-based Monte Carlo method to solve backward stochastic differential equations. *Ann. Appl. Probab.*, **15**(3), 2172–2202.

Gräser, C. and Srinivasan, P. A. A. 2020. Error bounds for PDE-regularized learning. ArXiv:2003.06524.

Grohs, P., Hornung, F., Jentzen, A., and von Wurstemberger, P. 2018. A proof that rectified deep neural networks overcome the curse of dimensionality in the numerical approximation of Black–Scholes partial differential equation. *Memoirs of the American Mathematical Society*, to appear.

Gu, H., Guo, X., Wei, X., and Xu, R. 2020. Q-learning algorithm for mean-field controls, with convergence and complexity analysis. ArXiv:2002.04131.

Guo, X., Hu, A., Xu, R., and Zhang, J. 2020. A general framework for learning mean-field games. ArXiv:2003.06069.

Han, J., and E, W. 2016. Deep learning approximation for stochastic control problems. *Deep Reinforcement Learning Workshop.*

Han, J., and Hu, R. 2021. Recurrent Neural Networks for Stochastic Control Problems with Delay. ArXiv:2101.01385.

Han, J., and Long, J. 2020. Convergence of the deep BSDE method for coupled FBSDEs. *Probab. Uncertain. Quant. Risk*, **5**(1), 1–33.

Han, J., Hu, R. and Long, J. 2020. Convergence of deep fictitious play for stochastic differential games. ArXiv:2008.05519.

Han, J., Jentzen, A., and E, W. 2017. Solving high-dimensional partial differential equations using deep learning. *Proc. Natl. Acad. Sci. USA*, **115**(07).

Henry-Labordere, P. 2017. Deep primal–dual algorithm for BSDEs: Applications of machine learning to CVA and IM. Available at SSRN 3071506.

Hu, R. 2019. Deep fictitious play for stochastic differential games. ArXiv:1903.09376.

Huré, C., Pham, H., Bachouch, A., and Langrené, N. 2021. Deep neural networks algorithms for stochastic control problems on finite horizon: convergence analysis. *SIAM J. Numer. Anal.*, **59**(1), 525–557.

Huré, C., Pham, H., and Warin, X. 2020. Deep backward schemes for high-dimensional nonlinear PDEs. *Math. Comp.*, **89**(324), 1547–1580.

Hutzenthaler, M., Jentzen, A., Kruse, T., and Nguyen, T.A. 2020. A proof that rectified deep neural networks overcome the curse of dimensionality in the numerical approximation of semilinear heat equation. *SN Partial Differential Equations and Applications*, **1**(10), 1–34.

Jacquier, A., and Oumgari, M. 2019. Deep curve-dependent PDEs for affine rough volatility. ArXiv:1906.02551.

Jiang, Y., and Li, J. 2021. Convergence of the deep BSDE method for FBSDEs with non-Lipschitz coefficients. ArXiv:2101.01869.

Kharroubi, I., Lim, T., and Warin, X. 2020. Discretization and Machine Learning Approximation of BSDEs with a Constraint on the Gains-Process. ArXiv:2002.02675.

Khoo, Y., Lu, J., and Ying, L. 2020. Solving parametric PDE problems with artificial neural networks. *European Journal of Applied Mathematics*, 1–15.

Kingma, D. P., and Ba, J. 2014. Adam: A method for stochastic optimization. In: Proc. 3rd International Conference for Learning Representations, San Diego.

Kremsner, S., Steinicke, A., and Szölgyenyi, M. 2020. A deep neural network algorithm for semilinear elliptic PDEs with applications in insurance mathematics. ArXiv:2010.15757.

Liang, J., Xu, Z., and Li, P. 2019. Deep learning-based least square forward-backward stochastic differential equation solver for high-dimensional derivative pricing. ArXiv:1907.10578.

Müller, J. and Zeinhofer, M. 2019. Deep Ritz revisited. ArXiv:1912.03937.

Nüskens, N., and Richter, L. 2019. Solving high-dimensional Hamilton–Jacobi–Bellman PDEs using neural networks: perspective from the theory of controlled diffusions and measures on path space. ArXiv:2005.05409.

Pardoux, E., and Peng, S. 1990. Adapted solution of a backward stochastic differential equation. *Systems & Control Letters*, **14**(1), 55–61.

Pham, H. 2009. *Continuous-time Stochastic Control and Optimization with Financial Applications*. Springer.

Pham, H., and Wei, X. 2017. Dynamic programming for optimal control of stochastic McKean–Vlasov dynamics. *SIAM J. Control Optim.*, **55**(2), 1069–1101.

Pham, H., Warin, X., and Germain, M. 2021. Neural networks-based backward scheme for fully nonlinear PDEs. *SN Partial Differential Equations and Applications*, **2**(16).

Raissi, M. 2018. Forward–backward stochastic neural networks: Deep learning of high-dimensional partial differential equations. ArXiv:1804.07010.

Raissi, M., Perdikaris, P., and Karniadakis, G.E. 2019. Physics-informed neural networks: A deep learning framework for solving forward and inverse problems involving nonlinear partial differential equations. *J. Comput. Phys.*, **378**, 686 – 707.

Ruthotto, L., Osher, S. J., Li, W., Nurbekyan, L., and Fung, S. W. 2020. A machine learning framework for solving high-dimensional mean field game and mean field control problems. *Proc. Natl. Acad. Sci. USA*, **117**(17), 9183–9193.

Sabate Vidales, M., Siska, D., and Szpruch, L. 2018. Unbiased deep solvers for parametric PDEs. ArXiv:1810.05094v2.

Sabate Vidales, M., Siska, D., and Szpruch, L. 2020. Solving path dependent PDEs with LSTM networks and path signatures. ArXiv:2011.10630v1.

Sakuma, T. 2020. Application of deep quantum neural networks to finance. ArXiv:2011.07319v1.

Shin, Y., Darbon, J., and Em Karniadakis, G. 2020. On the convergence of physics informed neural networks for linear second-order elliptic and parabolic type PDEs. ArXiv:2004.01806.

Sirignano, J., and Spiliopoulos, K. 2017. DGM: A deep learning algorithm for solving partial differential equations. *J. Comput. Phys.*, **375**(08).

Zhang, J. 2004. A numerical scheme for BSDEs. *Ann. Appl. Probab.*, **14**(1), 459–488.

23

Generative Adversarial Networks: Some Analytical Perspectives

Haoyang Cao[a] and Xin Guo[b]

Abstract

Ever since their debut, generative adversarial networks (GANs) have attracted a tremendous amount of attention. Over recent years, varieties of GAN models have been developed and tailored to different practical applications. Meanwhile, some issues regarding the performance and the training of GANs have been recognised and investigated from various theoretical perspectives. This chapter will start from an introduction to GANs from an analytical perspective, then move onto the training of GANs via stochastic-differential-equation approximation and finally discuss some applications of GANs in computing high-dimensional mean field games as well as tackling questions in mathematical finance.

23.1 Introduction

Generative adversarial networks (GANs), were introduced in 2014 to the machine learning community by Goodfellow et al. (2014). The key idea behind GANs is to interpret the process of generative modeling as a competing game between two neural networks: a generator G and a discriminator D. The generator attempts to fool the discriminator by converting random noise into sample data, while the discriminator tries to identify whether the input sample is fake or true.

Since their introduction, GANs have enjoyed great empirical success, with a wide range of applications especially in image generation and natural language processing, including high resolution image generation (Denton et al., 2015; Radford et al., 2015), image inpainting (Yeh et al., 2016), image super-resolution (Ledig et al., 2017), visual manipulation (Zhu et al., 2016), text-to-image synthesis (Reed et al., 2016), video generation (Vondrick et al., 2016), semantic segmentation (Luc et al., 2016), and abstract reasoning diagram generation (Kulharia et al., 2017).

Despite the empirical success of GANs, there are well-recognized issues in

[a] The Alan Turing Institute
[b] University of California, Berkeley
Published in *Machine Learning And Data Sciences For Financial Markets*, Agostino Capponi and Charles-Albert Lehalle © 2023 Cambridge University Press.

GANs training, such as the vanishing gradient when the discriminator significantly outperforms the generator (Arjovsky and Bottou, 2017), mode collapse – which is believed to be linked to gradient exploding (Salimans et al., 2016), and the challenge of GANs convergence (Barnett, 2018). To improve the performance of GANs training, various approaches have been proposed to ameliorate these issues, including different choices of network architectures, loss functions, and regularization. See for instance, the comprehensive survey on these techniques in Wiatrak et al. (2019) and the references therein. Meanwhile, there has been a growing research interest in the theoretical understanding of GANs training. Berard et al. (2020) proposed a novel visualization method for the GANs training process through the gradient vector field of loss functions. In a deterministic GANs training framework, Mescheder et al. (2018) demonstrated that regularization improved the convergence performance of GANs. Conforti et al. (2020) and Domingo-Enrich et al. (2020) analyzed a generic zero-sum minimax game including that of GANs, and connected the mixed Nash equilibrium of the game with the invariant measure of Langevin dynamics.

Recently, GANs have attracted attention in the mathematical finance community, largely due to the clear analogy between simulation of financial time series data and image generation, see for instance Wiese et al. (2019) and Wiese et al. (2020). In response to the growing interests in GANs and their computational potential for high-dimensional control problems, stochastic games and backward stochastic differential equations, this chapter provides a gentle introduction to GANs from an analytical perspective, and highlights some of the latest developments in GANs training in the framework of stochastic differential equations, and reviews several representatives GANs applications in asset pricing and simulations of financial time series data.

Throughout this chapter, the following notations will be adopted, unless otherwise specified.

- The set of k continuously differentiable functions over some domain $\mathcal{X} \subset \mathbb{R}^d$ is denoted by $C^k(\mathcal{X})$ for $k = 0, 1, 2, \ldots$; in particular when $k = 0$, $C^0(\mathcal{X}) = C(\mathcal{X})$ denotes the set of continuous functions.

- Let $p \geq 1$. $L^p_{\text{loc}}(\mathbb{R}^d)$ denotes the set of functions f defined on $\mathbb{R}^d$ such that for any compact subset $\mathcal{X}$, $\int_{\mathcal{X}} \|f(x)\|_p^p \, dx < \infty$.

- Let $J = (J_1, \ldots, J_d)$ be a d-tuple multi-index of order $|J| = \sum_{i=1}^{d} J_i$. For a function $f \in L^1_{\text{loc}}(\mathbb{R}^d)$, its Jth-weak derivative $D^J f \in L^1_{\text{loc}}(\mathbb{R}^d)$ is a function such that for any smooth and compactly supported test function g,

$$\int_{\mathbb{R}^d} D^J f(x) g(x) \, dx = (-1)^{|J|} \int_{\mathbb{R}^d} f(x) \nabla^J g(x) \, dx.$$

- The Sobolev space $W^{k,p}_{\text{loc}}(\mathbb{R}^d)$ is a set of functions f on $\mathbb{R}^d$ such that for any d-tuple multi-index J with $|J| \leq k$, $D^J f \in L^p_{\text{loc}}(\mathbb{R}^d)$.

23.2 Basics of GANs: an analytical view

GANs as generative models.

GANs fall into the category of generative models. The process of generative modeling is to approximate an unknown probability distribution $\mathbb{P}_r$ by constructing a class of suitable parametrized probability distributions $\mathbb{P}_\theta$. That is, given a latent space $\mathcal{Z}$ and a sample space $\mathcal{X}$, define a latent variable $Z \in \mathcal{Z}$ with a fixed probability distribution $\mathbb{P}_z$ and a family of functions $G_\theta \colon \mathcal{Z} \to \mathcal{X}$ parametrized by θ. Then $\mathbb{P}_\theta$ is defined as the probability distribution of $G_\theta(Z)$, i.e., Law$(G_\theta(Z))$.

In contrast to other generative models, GANs consist of two competing components: a generator G and a discriminator D. In particular, the generator G is implemented using a neural network (NN), i.e., function approximators via specific graph structures and network architectures, which we denote by $G = G_\theta$ as a parametrized function. Meanwhile, another neural network for the discriminator D assigns a score between 0 to 1 to an input sample, either from the true distribution $\mathbb{P}_r$ or the approximated distribution $\mathbb{P}_\theta = $ Law$(G_\theta(Z))$; we denote the parametrized D as D_ω. A higher score from the discriminator D would indicate that the sample is more likely to be from the true distribution. GANs are trained by optimizing G and D iteratively until D can no longer distinguish between samples from $\mathbb{P}_r$ and those from $\mathbb{P}_\theta$.

GANs as minimax games.

Mathematically, GANs are minimax games as

$$\min_G \max_D \left\{ \mathbb{E}_{X \sim \mathbb{P}_r}[\log D(X)] + \mathbb{E}_{Z \sim \mathbb{P}_z}[\log(1 - D(G(Z)))] \right\}. \tag{23.1}$$

In particular, fixing G and optimizing for D in (23.1), the optimal discriminator would be

$$D_G^*(x) = \frac{p_r(x)}{p_r(x) + p_\theta(x)},$$

where p_r and p_θ are density functions of $\mathbb{P}_r$ and $\mathbb{P}_\theta$ respectively. Plugging the above D_G^* back to (23.1), the following equation holds:

$$\min_G \left\{ \mathbb{E}_{X \sim \mathbb{P}_r} \left[\log \frac{p_r(X)}{p_r(X) + p_\theta(X)} \right] + \mathbb{E}_{Y \sim \mathbb{P}_\theta} \left[\log \frac{p_\theta(Y)}{p_r(Y) + p_\theta(Y)} \right] \right\}$$

$$= -\log 4 + 2 \left\{ \frac{1}{2} \mathbb{E}_{X \sim \mathbb{P}_r} \left[\log \frac{p_r(X)}{\frac{p_r(X) + p_\theta(X)}{2}} \right] + \frac{1}{2} \mathbb{E}_{Y \sim \mathbb{P}_\theta} \left[\log \frac{p_\theta(Y)}{\frac{p_r(Y) + p_\theta(Y)}{2}} \right] \right\}$$

$$= -\log 4 + 2\mathrm{JS}(\mathbb{P}_r, \mathbb{P}_\theta).$$

That is to say, training of GANs with (23.1) being the objective is equivalent to minimizing the Jensen–Shannon (JS) divergence between $\mathbb{P}_r$ and $\mathbb{P}_\theta$. In other words, through optimization over discriminators, GANs are essentially minimizing proper divergences between the true distribution and the generated distribution over some sample space $\mathcal{X}$.

GANs and optimal transport.

This view of GANs as an optimization problem with an appropriate divergence function has been instrumental for addressing the instability of GANs training. Variants of GANs with different divergences have been proposed to improve the performance of GANs. For instance, Nowozin et al. (2016) and Nock et al. (2017) extended the JS divergence in Goodfellow et al. (2014) to the broader class of f-divergence. This extension provides the flexibility of choosing various f functions for the loss function in GANs training. Srivastava et al. (2019) explored scaled Bregman divergence to resolve the issue of support mismatch bewteen $\mathbb{P}$ and $\mathbb{Q}$ in the use of f-divergence and Bregman divergence. This is achieved through introducing a noisy base measure μ such that μ is a mixture of $\mathbb{P}$ and $\mathbb{Q}$ convolved with some Gaussian distributions. Arjovsky et al. (2017) adopted Wasserstein-1 distance that enjoys higher smoothness with respect to the model parameters and consequently leads to a much more stable training of GANs. Guo et al. (2021) proposed relaxed Wasserstein divergence by generalizing Wasserstein-1 distance with Bregman cost functions to first bypass the restriction on data information geometry in WGAN and achieve faster training. Salimans et al. (2018) and Sanjabi et al. (2018) utilized the Sinkhorn loss instead of the optimal transport type of loss by interpolating with energy distance and adding entropy regularization. This can significantly reduce the computational burden of optimal transport costs and increase the stability of training.

The flexibility of choosing appropriate divergence functions, especially the development of Wasserstein GANs (WGANs), leads to the natural connection between GANs and optimal transport problems, established in Cao et al. (2020b), which identifies sufficient conditions to recast GANs in the framework of optimal transport.

The idea behind this link is intuitive: GANs as generative models are minimax games with the goal of minimizing the "error" of the generated sample data against the true sample data; this error is measured under appropriate divergence functions between the true distribution and the generated distribution. Now if this error is viewed as a cost of transporting/fitting the generated distribution into the true distribution, GANs become optimal transport problems.

Indeed, this connection is explicit in the case of WGANs, via the Kantonovich duality:

Theorem 23.1 *Suppose that $\mathbb{P}_r \in L^1(\mathcal{X})$ and $G \in L^1(\mathbb{P}_z)$ where*

$$L^1(\mathbb{P}_z) = \left\{ f : \mathcal{Z} \to \mathbb{R} : \int_{\mathcal{Z}} |f(z)| \mathbb{P}_z(dz) < \infty \right\}.$$

WGAN is an optimal transport problem between Law$(G(Z))$ *and* $\mathbb{P}_r$.

As seen in Cao et al. (2020b), this connection goes beyond the framework of WGANs. Indeed, take any Polish space $\mathcal{X}$ with metric l, then $\mathcal{X} \times \mathcal{X}$ is also a Polish space with metric l'. Denote by $\mathcal{P}(\mathcal{X})$ the set of all probability distributions over the sample space $\mathcal{X}$, define a generic divergence function

$$W : \mathcal{P}(\mathcal{X}) \times \mathcal{P}(\mathcal{X}) \mapsto \mathbb{R}^+,$$

and take a class of GANs with this divergence W. If W can be written as an appropriate optimal cost W_c and if such an optimal transport problem has a duality representation, then the GANs model is a transport problem: the discriminator locates the best coupling among Π_G under a given G, and the generator refines the set of possible couplings Π_G to minimize the divergence.

There are earlier studies connecting GANs and optimal transport problems, by different approaches and from different perspectives. For instance, Salimans et al. (2018) defines a novel divergence called the minibatch energy distance, based on solutions of three associated optimal transport problems. This new divergence is then used to replace the JS divergence for the vanilla GANs. Note that this minibatch energy distance itself is not an optimal transport cost. In Lei et al. (2019), a geometric interpretation of WGANs from the perspective of optimal transport is provided: the latent random variable from the latent space is mapped to the sample space via an optimal mass transport so that the resulted distribution can minimize its Wasserstein distance against the true distribution.

GANs and MFGs.

In addition to this relation between GANs and optimal transport, Cao et al. (2020b) further associated GANs with mean field games (MFGs), and designed a new algorithm for computing MFGs. This connection between MFGs and GANs can be seen conceptually in Table 23.1.

Table 23.1 A first link between GANs and MFGs

	GANs	MFGs
Generator G	NN for approximating the map $G : \mathcal{Z} \mapsto \mathcal{X}$	NN for solving the Hamilton–Jacobi–Bellman equation
Characterization of $\mathbb{P}_r$	Sample data	Fokker–Planck equation for consistency
Discriminator D	NN measuring divergence between $\mathbb{P}_\theta$ and $\mathbb{P}_r$	NN for measuring differential residual from the FP equation

Evidently, there is more than one way to establish this connection between MFGs and GANs. Alternatively, one can switch the roles of the generator and discriminator and view the mean field term as a generator and the value function as a discriminator.

Example.

For certain classes of MFGs, such an interpretation of MFGs as GANs may be explicit. For instance, take the class of periodic MFGs from Cirant and Nurbekyan (2018) on the flat torus $\mathbb{T}^d$ and a finite time horizon $[0, T]$. Such an MFG minimizes the following cost:

$$J_m(t,\alpha) = \mathbb{E}\left[\int_t^T L(X_t^\alpha,\alpha(X_t^\alpha)) + f(X_t^\alpha,m(X_t^\alpha))\,dt\right], \quad t \in [0,T] \quad (23.2)$$

where $X^\alpha = (X_t^\alpha)_t$ is a d-dimensional process with dynamics

$$dX_t^\alpha = \alpha(X_t^\alpha)dt + \sqrt{2\epsilon}dW_t.$$

Here α is a control policy, L and f constitute the running cost and $m(t,\cdot)$, for $t \in [0,T]$, denotes the probability density of X_t^α at time t.

Now, consider the convex conjugate of the running cost L, namely,

$$H_0(x,p) = \sup_{\alpha\in\mathbb{R}^d}\{\alpha\cdot p - L(x,\alpha)\},$$

and write $F(x,m) = \int^m f(x,z)\,dz$. Then this class of MFGs can be characterized by the following coupled PDE system as illustrated in Cirant and Nurbekyan (2018),

$$\begin{cases} -\partial_s u - \epsilon\Delta_x u + H_0(x,\nabla_x u) = f(x,m), \\ \partial_s m - \epsilon\Delta_x m - \mathrm{div}\left(m\nabla_p H_0(x,\nabla u)\right) = 0, \\ m > 0,\ m(0,\cdot) = m^0(\cdot),\ u(T,\cdot) = u^T(\cdot). \end{cases} \quad (23.3)$$

Here the first equation is a Hamilton–Jacobi–Bellman (HJB) equation governing the value function and the second is a Fokker–Planck (FP) equation governing the evolution of the optimally controlled state process, with m^0 and u^T the initial functions for $m(t,\cdot)$ and $u(t,\cdot)$, respectively.

Note that this system of equations (23.3) is equivalent to the following minimax game:

$$\inf_{u\in C^2([0,T]\times\mathbb{T}^d)}\sup_{m\in C^2([0,T]\times\mathbb{T}^d)}\Phi(m,u), \quad (23.4)$$

where

$$\Phi(m,u) = \int_0^T\int_{\mathbb{T}^d}[m(-\partial_t u - \epsilon\Delta_x u) + mH_0(x,\nabla_x u) - F(x,m)]\,dx\,dt$$
$$+ \int_{\mathbb{T}^d}\left[m(T,x)u(T,x) - m^0(x)u(0,x) - m(x,T)u^T(x)\right]\,dx.$$

Therefore, by (23.4), the connection between GANs and MFGs is transparent.

Having established the interpretation of MFGs as GANs, the immediate question to ask is whether GANs can be understood as MFGs. Cao et al. (2020b) further showed that GANs can also be seen as MFGs, under the Pareto optimality criterion.

Theorem 23.2 *GANs in* Goodfellow et al. (2014) *are MFGs under the Pareto optimality criterion, assuming that the latent variables Z and true data X are both i.i.d. sampled, respectively, with $\mathbb{E}[|\log(D(X))|]$, $\mathbb{E}[|\log(1-D(G(Z)))|] < \infty$ for all possible D and G.*

The above theorem shows that the theoretical framework of GANs in Goodfellow et al. (2014) can be seen as MFGs under the Pareto optimality criterion, where the generator network is a representative player of infinitely many identical players working in collaboration to defeat the discriminator. In practical training of GANs, however, only finitely many data points, i.e., N latent variables $\{Z_i\}_{i=1}^{N}$ and M samples from the unknown true distribution $\{X_j\}_{j=1}^{M}$, are available and therefore GANs in practice can be interpreted as N-player cooperative games with players being interchangeable and hence adopting the same strategy. Here, $Z_i \overset{\text{i.i.d.}}{\sim} \mathbb{P}_z$, $X_j \overset{\text{i.i.d.}}{\sim} \mathbb{P}_r$ and $\{X_j\}_{j=1}^{M} \perp \{Z_i\}_{i=1}^{N}$. The state process for player i is given by the feedforward process within its generator network G_i, with the initial layer being Z_i and the final layer being the generated sample $G_i(Z_i)$. Since the players are interchangeable and collaborating, a common generator network G is adopted by all N players to form a symmetric strategy profile $\mathbf{S}^G$ for the N-player game. These players face with a discriminator $D^{N,M} \in \mathcal{D} = \{D | D : X \to [0,1]\}$ that favors the true samples X_j's. In particular, the collective cost for the N players of choosing a common generator G is given by

$$J^{N,M}\left(\mathbf{S}^G; \{Z_i\}_{i=1}^{N}, \{X_j\}_{j=1}^{M}\right) = \max_{D \in \mathcal{D}} \frac{\sum_{i=1}^{N} \sum_{j=1}^{M} \log\left[D(X_j)\right] + \log\left[1 - D(G(Z_i))\right]}{N \cdot M}$$

and $D^{N,M}$ is given by

$$D^{N,M} = D^{N,M}(\cdot; G, \{Z_i\}_{i=1}^{N}, \{X_j\}_{j=1}^{M})$$
$$= \arg\max_{D \in \mathcal{D}} \frac{\sum_{i=1}^{N} \sum_{j=1}^{M} \log\left[D(X_j)\right] + \log\left[1 - D(G(Z_i))\right]}{N \cdot M}.$$

Definition 23.3 (Pareto optimality)   A strategy profile $\mathbf{S}^{G^{N,*}}$ among all possible symmetric strategy profiles is said to be Pareto optimal if for any symmetric strategy profile $\mathbf{S}^G$,

$$J^{N,M}\left(\mathbf{S}^{G^{N,*}}; \{Z_i\}_{i=1}^{N}, \{X_j\}_{j=1}^{M}\right) \leq J^{N,M}\left(\mathbf{S}^G; \{Z_i\}_{i=1}^{N}, \{X_j\}_{j=1}^{M}\right).$$

Before characterizing $G^{N,*}$ and $D^{N,M}$, the cost $J^{N,M}$ using the empirical measures $\delta_r^M = \frac{1}{M}\sum_{j=1}^{M} \delta_{X_j}$ and $\delta_G^N = \frac{1}{N}\sum_{i=1}^{N} \delta_{G(Z_i)}$ can be rewritten as follows:

$$J^{N,M}\left(\mathbf{S}^G; \{Z_i\}_{i=1}^{N}, \{X_j\}_{j=1}^{M}\right) = \max_{D \in \mathcal{D}} \int_X \log D(x)\delta_r^M(x) + \log[1 - D(x)]\delta_G^N(x) dx.$$

Then $D^{N,M}$ and $G^{N,*}$ are naturally characterized by the two empirical distributions.

Proposition 23.4   *Under a given G, a particular $D^{N,M}$ is given by*

$$D^{N,M}(x) = \begin{cases} \frac{\delta_r^M(x)}{\delta_r^M(x) + \delta_G^N(x)}, & x \in \{G(Z_1), \ldots, G(Z_N), X_1, \ldots, X_M\}; \\ \frac{1}{2}, & \text{otherwise}; \end{cases}$$

in fact, for $x \notin \{G(Z_1), \ldots, G(Z_N), X_1, \ldots, X_M\}$, $D^{N,M}(x)$ can take any value in $[0,1]$.

Theorem 23.5 *The set of possible $G^{N,*}$'s is given by*

$$\mathcal{G}^{N,*} = \left\{ G \in \mathcal{G} : \delta_G^N = \delta_r^M \right\},$$

provided that $\mathcal{G}^{N,*} \neq \emptyset$.

The above results show that in practice training of GANs over finitely many samples, the generator can recover the empirical distribution of true samples at best. Moreover, the non-emptiness of $\mathcal{G}^{N,*}$ highly depends on the design of G network architecture. This will be discussed in detail in Section 23.3. Theoretically, however, N and M can be taken to infinity, leading the N-player cooperative games into MFGs with Pareto optimality criterion as stated in Theorem 23.2. Here the mean field information is given by $G\#\mathbb{P}_z = \lim_{N\to\infty} \delta_G^N$ and the convergence of N-player games to MFGs is guaranteed by the law of large numbers and the continuous mapping theorem.

23.3 GANs training

In the previous section, it was pointed out that there have been many practical methods for improving the performance of GANs training, apart from choosing a proper network architecture. This section is intended to provide mathematical explanation for these practical methods by analyzing GANs training via stochastic differential equation approximation. Before going into detail about GANs training, it is worth revisiting the objective of GANs.

Equilibrium of GANs training

GANs are trained by optimizing G and D iteratively until D can no longer distinguish between true samples and generated samples. Recall that G_θ denotes the generator parametrized by the neural network with the set of parameters $\theta \in \mathbb{R}^{d_\theta}$, and D_ω denotes the discriminator parametrized by the other neural network with the set of parameters $\omega \in \mathbb{R}^{d_\omega}$. Under a fixed network architecture, the parametrized version of GANs training is to find

$$v_U^{\text{GAN}} = \min_\theta \max_\omega L_{\text{GAN}}(\theta, \omega),$$

where $L_{\text{GAN}}(\theta, \omega) = \mathbb{E}_{X\sim\mathbb{P}_r}[\log D_\omega(X)] + \mathbb{E}_{Z\sim\mathbb{P}_z}[\log(1 - D_\omega(G_\theta(Z)))].$

(23.5)

Remark 23.6 From a game theory viewpoint, the objective in (23.5), if attained, is in fact the upper value of the two-player zero-sum game of GANs. Meanwhile, the lower value of the game is given by the following maximin problem,

$$v_L^{\text{GAN}} = \max_\omega \min_\theta L_{\text{GAN}}(\theta, \omega). \tag{23.6}$$

Clearly the following relation holds,

$$v_L^{\text{GAN}} \leq v_U^{\text{GAN}}. \tag{23.7}$$

Moreover, if there exists a pair of parameters (θ^*, ω^*) such that both (23.5) and (23.6) are attained, then (θ^*, ω^*) is a Nash equilibrium of this two-player zero-sum game. Indeed, if L_{GAN} is convex in θ and concave in ω, then there is no duality gap hence the equality in (23.7) holds by the minimax theorem (see Von Neumann, 1959, and Sion, 1958).

It is worth noting that conditions for such an equality in (23.7) is usually not satisfied in many common GANs models, as observed in Zhu et al. (2020) and analyzed in Guo and Mounjid (2020).

GANs training via SGD.

As in most deep learning models, stochastic gradient descent (SGD) (or one of its variants) is a standard approach for solving the optimization problem in GANs training. Accordingly, the evolution of parameters of θ and ω in (23.5) by SGD from the current step t to the next step $t + 1$ is

$$
\begin{aligned}
\omega_{t+1} &= \omega_t + \alpha_d \nabla_\omega L_{\text{GAN}}(\theta_t, \omega_t), \\
\theta_{t+1} &= \theta_t - \alpha_g \nabla_\theta L_{\text{GAN}}(\theta_t, \omega_{t+1}).
\end{aligned}
\tag{23.8}
$$

Here the α_d and α_g denote the step sizes of updating the discriminator and the generator, respectively. This evolution (23.8) corresponds to the alternating updating scheme of the algorithm in Goodfellow et al. (2014) where at each iteration, the discriminator is updated before the generator. One of the main challenges for GANs training is the convergence of such an alternating SGD.

GANs training and SDEs approximation.

GANs training is performed on a data set

$$
\mathcal{D} = \{(z_i, x_j)\}_{1 \leq i \leq N, 1 \leq j \leq M},
$$

where $\{z_i\}_{i=1}^N$ are sampled from $\mathbb{P}_z$ and $\{x_j\}_{j=1}^M$ are real image data following the unknown distribution $\mathbb{P}_r$. The objective of GANs is to solve the following minimax problem:

$$
\min_\theta \max_\omega \Phi(\theta, \omega),
\tag{23.9}
$$

for some cost function Φ, with Φ of a separable form

$$
\Phi(\theta, \omega) = \frac{\sum_{i=1}^N \sum_{j=1}^M J(D_\omega(x_j), D_\omega(G_\theta(z_i)))}{N \cdot M}.
\tag{23.10}
$$

When the stochastic gradient algorithm (SGA) is performed to solve the minimax problem (23.9), the full gradients of Φ with respect to θ and ω, denoted by g_θ and g_ω respectively, are estimated over a mini-batch $\mathcal{B}$ of batch size B, denoted by $g_\theta^{\mathcal{B}}$ and $g_\omega^{\mathcal{B}}$.

Let $\eta_t^\theta > 0$ and $\eta_t^\omega > 0$ be the learning rates at iteration $t = 0, 1, 2, \ldots$, for θ and ω respectively; then solving the minimax problem (23.9) with SGA and *alternating parameter update* implies descent of θ along g_θ and ascent of ω along g_ω at each iteration: within each iteration, the minibatch gradient for θ and ω are calculated on different batches. In order to emphasize this difference, $\bar{\mathcal{B}}$ represents the minibatch for θ and $\mathcal{B}$ for that of ω, with $\bar{\mathcal{B}} \overset{\text{i.i.d.}}{\sim} \mathcal{B}$. The one-step update can be written as follows:

$$\begin{cases} \omega_{t+1} = \omega_t + \eta_t^\omega g_\omega^{\mathcal{B}}(\theta_t, \omega_t), \\ \theta_{t+1} = \theta_t - \eta_t^\theta g_\theta^{\bar{\mathcal{B}}}(\theta_t, \omega_{t+1}). \end{cases} \tag{23.11}$$

Some practical training of GANs uses *simultaneous parameter update* between the discriminator and the generator, corresponding to a similar yet subtly different form

$$\begin{cases} \omega_{t+1} = \omega_t + \eta_t^\omega g_\omega^{\mathcal{B}}(\theta_t, \omega_t), \\ \theta_{t+1} = \theta_t - \eta_t^\theta g_\theta^{\mathcal{B}}(\theta_t, \omega_t). \end{cases} \tag{23.12}$$

For the ease of exposition, the learning rates are assumed to be constant $\eta_t^\theta = \eta_t^\omega = \eta$, with η viewed as the time interval between two consecutive parameter updates. In Guo and Mounjid (2020), the optimal (variable) learning rate for GANs training is studied under a stochastic control framework.

Write $g_\theta^{i,j}$ and $g_\omega^{i,j}$ for $\nabla_\theta J(D_\omega(x_j), D_\omega(G_\theta(z_i)))$ and $\nabla_\omega J(D_\omega(x_j), D_\omega(G_\theta(z_i)))$, respectively, and define the following covariance matrices

$$\Sigma_\theta(\theta, \omega) = \frac{\sum_i \sum_j [g_\theta^{i,j}(\theta, \omega) - g_\theta(\theta, \omega)][g_\theta^{i,j}(\theta, \omega) - g_\theta(\theta, \omega)]^T}{N \cdot M},$$

$$\Sigma_\omega(\theta, \omega) = \frac{\sum_i \sum_j [g_\omega^{i,j}(\theta, \omega) - g_\omega(\theta, \omega)][g_\omega^{i,j}(\theta, \omega) - g_\omega(\theta, \omega)]^T}{N \cdot M};$$

then as the batch size B gets sufficiently large, the classical central limit theorem leads to the following approximation of (23.11),

$$\begin{cases} \omega_{t+1} = \omega_t + \eta g_\omega^{\mathcal{B}}(\theta_t, \omega_t) \approx \omega_t + \eta g_\omega(\theta_t, \omega_t) + \frac{\eta}{\sqrt{B}} \Sigma_\omega^{\frac{1}{2}}(\theta_t, \omega_t) Z_t^1, \\ \theta_{t+1} = \theta_t - \eta g_\theta^{\mathcal{B}}(\theta_t, \omega_{t+1}) \approx \theta_t - \eta g_\theta(\theta_t, \omega_{t+1}) + \frac{\eta}{\sqrt{B}} \Sigma_\theta^{\frac{1}{2}}(\theta_t, \omega_{t+1}) Z_t^2, \end{cases} \tag{23.13}$$

with independent random variables $Z_t^1 \sim N(0, I_{d_\omega})$ and $Z_t^2 \sim N(0, I_{d_\theta})$, $t = 0, 1, \dots$.

If we could ignore the difference between t and $t + 1$, then the approximation could be written in the following form

$$d\begin{pmatrix} \Theta_t \\ \mathcal{W}_t \end{pmatrix} = \begin{pmatrix} -g_\theta(\Theta_t, \mathcal{W}_t) \\ g_\omega(\Theta_t, \mathcal{W}_t) \end{pmatrix} dt + \sqrt{2\beta^{-1}} \begin{pmatrix} \Sigma_\theta(\Theta_t, \mathcal{W}_t)^{\frac{1}{2}} & 0 \\ 0 & \Sigma_\omega(\Theta_t, \mathcal{W}_t)^{\frac{1}{2}} \end{pmatrix} dW_t, \tag{23.14}$$

where $\beta = \frac{2B}{\eta}$ and $\{W_t\}_{t \geq 0}$ are standard $(d_\theta + d_\omega)$-dimensional Brownian motions. This would be the approximation for GANs training of (23.12).

Taking the subtle difference between t and $t + 1$ into consideration and thus the interaction between the generator and the discriminator, the approximation for the GANs training process of (23.11) should be

$$
d \begin{pmatrix} \Theta_t \\ \mathcal{W}_t \end{pmatrix}
$$

$$
= \left[\begin{pmatrix} -g_\theta(\Theta_t, \mathcal{W}_t) \\ g_\omega(\Theta_t, \mathcal{W}_t) \end{pmatrix} + \frac{\eta}{2} \begin{pmatrix} \nabla_\theta g_\theta(\Theta_t, \mathcal{W}_t) & -\nabla_\omega g_\theta(\Theta_t, \mathcal{W}_t) \\ -\nabla_\theta g_\omega(\Theta_t, \mathcal{W}_t) & -\nabla_\omega g_\omega(\Theta_t, \mathcal{W}_t) \end{pmatrix} \begin{pmatrix} -g_\theta(\Theta_t, \mathcal{W}_t) \\ g_\omega(\Theta_t, \mathcal{W}_t) \end{pmatrix} \right] dt
$$

$$
+ \sqrt{2\beta^{-1}} \begin{pmatrix} \Sigma_\theta(\Theta_t, \mathcal{W}_t)^{\frac{1}{2}} & 0 \\ 0 & \Sigma_\omega(\Theta_t, \mathcal{W}_t)^{\frac{1}{2}} \end{pmatrix} d\mathcal{W}_t. \tag{23.15}
$$

Equations (23.14) and (23.15) can be written in more compact forms as

$$
d \begin{pmatrix} \Theta_t \\ \mathcal{W}_t \end{pmatrix} = b_0(\Theta_t, \mathcal{W}_t) dt + \sigma(\Theta_t, \mathcal{W}_t) d\mathcal{W}_t, \tag{23.16}
$$

$$
d \begin{pmatrix} \Theta_t \\ \mathcal{W}_t \end{pmatrix} = b(\Theta_t, \mathcal{W}_t) dt + \sigma(\Theta_t, \mathcal{W}_t) d\mathcal{W}_t. \tag{23.17}
$$

where $b(\theta, \omega) = b_0(\theta, \omega) + \eta b_1(\theta, \omega)$, with

$$
b_0(\theta, \omega) = \begin{pmatrix} -g_\theta(\theta, \omega) \\ g_\omega(\theta, \omega) \end{pmatrix}, \tag{23.18}
$$

$$
b_1(\theta, \omega) = \frac{1}{2} \begin{pmatrix} \nabla_\theta g_\theta(\theta, \omega) & -\nabla_\omega g_\theta(\theta, \omega) \\ -\nabla_\theta g_\omega(\theta, \omega) & -\nabla_\omega g_\omega(\theta, \omega) \end{pmatrix} \begin{pmatrix} -g_\theta(\theta, \omega) \\ g_\omega(\theta, \omega) \end{pmatrix}
$$

$$
= -\frac{1}{2} \nabla b_0(\theta, \omega) b_0(\theta, \omega) - \begin{pmatrix} \nabla_\omega g_\theta(\theta, \omega) g_\omega(\theta, \omega) \\ 0 \end{pmatrix}, \tag{23.19}
$$

and $\sigma(\theta, \omega) = \sqrt{2\beta^{-1}} \begin{pmatrix} \Sigma_\theta(\Theta_t, \mathcal{W}_t)^{\frac{1}{2}} & 0 \\ 0 & \Sigma_\omega(\Theta_t, \mathcal{W}_t)^{\frac{1}{2}} \end{pmatrix}. \tag{23.20}$

Note the term $-\frac{\eta}{2} \begin{pmatrix} \nabla_\omega g_\theta(\theta, \omega) g_\omega(\theta, \omega) \\ 0 \end{pmatrix}$ for (23.17), which highlights the interaction between the generator and the discriminator in GANs training process.

In Cao and Guo (2020), it is shown that these coupled SDEs are indeed the continuous-time approximations of GANs training processes, with precise error bound analysis, where the approximations are under the notion of weak approximation as in Li et al. (2019).

Theorem 23.7 *Fix an arbitrary time horizon $\mathcal{T} > 0$ and take the learning rate $\eta \in (0, 1 \wedge \mathcal{T})$ and the number of iterations $N = \left\lfloor \frac{\mathcal{T}}{\eta} \right\rfloor$. Suppose that*

(1a) *$g_\omega^{i,j}$ is twice continuously differentiable, and $g_\theta^{i,j}$ and $g_\omega^{i,j}$ are Lipschitz, for any $i = 1, \ldots, N$ and $j = 1, \ldots, M$;*

(1b) *Φ is of $C^3(\mathbb{R}^{d_\theta + d_\omega})$, $\Phi \in W_{\text{loc}}^{4,1}(\mathbb{R}^{d_\theta + d_\omega})$, and for any multi-index $J = (J_1, \ldots, J_{d_\theta + d_\omega})$ with $|J| = \sum_{i=1}^{d_\theta + d\omega} J_i \leq 4$, there exist $k_1, k_2 \in \mathbb{N}$ such that*

$$
|D^J \Phi(\theta, \omega)| \leq k_1 \left(1 + \left\| \begin{pmatrix} \theta \\ \omega \end{pmatrix} \right\|_2^{2k_2} \right)
$$

for $\theta \in \mathbb{R}^{d_\theta}$, $\omega \in \mathbb{R}^{d_\omega}$ almost everywhere;

(1c) $(\nabla_\theta g_\theta)g_\theta$, $(\nabla_\omega g_\theta)g_\omega$, $(\nabla_\theta g_\omega)g_\theta$ and $(\nabla_\omega g_\omega)g_\omega$ are all Lipschitz.

Then, given any initialization $\theta_0 = \theta$ and $\omega_0 = \omega$, for any test function $f \in C^3(\mathbb{R}^{d_\theta+d_\omega})$ such that for any multi-index J with $|J| \leq 3$ there exist $k_1, k_2 \in \mathbb{N}$ satisfying

$$|\nabla^J f(\theta,\omega)| \leq k_1\left(1 + \left\|\begin{pmatrix}\theta\\\omega\end{pmatrix}\right\|_2^{2k_2}\right),$$

the following weak approximation holds

$$\max_{t=1,\dots,N}\left|\mathbb{E}f(\theta_t,\omega_t) - \mathbb{E}f(\Theta_{t\eta},\mathcal{W}_{t\eta})\right| \leq C\eta^2 \qquad (23.21)$$

for constant $C \geq 0$, where (θ_t,ω_t) and $(\Theta_{t\eta},\mathcal{W}_{t\eta})$ are given by (23.11) and (23.17), respectively.

Theorem 23.8 *Fix an arbitrary time horizon $\mathcal{T} > 0$, take the learning rate $\eta \in (0, 1 \wedge \mathcal{T})$ and the number of iterations $N = \left\lfloor \frac{\mathcal{T}}{\eta}\right\rfloor$. Suppose*

(2a) *$\Phi(\theta,\omega)$ is continuously differentiable, $\Phi \in W^{3,1}_{\mathrm{loc}}(\mathbb{R}^{d_\theta+d_\omega})$ and for any multi-index $J = (J_1,\dots,J_{d_\theta+d_\omega})$ with $|J| = \sum_{i=1}^{d_\theta+d\omega} J_i \leq 3$, there exist $k_1, k_2 \in \mathbb{N}$ such that $D^J\Phi$ satisfies*

$$|D^J\Phi(\theta,\omega)| \leq k_1\left(1 + \left\|\begin{pmatrix}\theta\\\omega\end{pmatrix}\right\|_2^{2k_2}\right)$$

for $\theta \in \mathbb{R}^{d_\theta}$, $\omega \in \mathbb{R}^{d_\omega}$ almost everywhere;
(2b) *$g_\theta^{i,j}$ and $g_\omega^{i,j}$ are Lipschitz for any $i = 1,\dots,N$ and $j = 1,\dots,M$.*

Then, given any initialization $\theta_0 = \theta$ and $\omega_0 = \omega$, for any test function $f \in C^2(\mathbb{R}^{d_\theta+d_\omega})$ such that for any multi-index J with $|J| \leq 2$ there exist $k_1, k_2 \in \mathbb{N}$ satisfying

$$|\nabla^J f(\theta,\omega)| \leq k_1\left(1 + \left\|\begin{pmatrix}\theta\\\omega\end{pmatrix}\right\|_2^{2k_2}\right),$$

then the following weak approximation holds

$$\max_{t=1,\dots,N}\left|\mathbb{E}f(\theta_t,\omega_t) - \mathbb{E}f(\Theta_{t\eta},\mathcal{W}_{t\eta})\right| \leq C\eta \qquad (23.22)$$

for constant $C \geq 0$, where (θ_t,ω_t) and $(\Theta_{t\eta},\mathcal{W}_{t\eta})$ are given by (23.12) and (23.16), respectively.

The above theorems from Cao and Guo (2020) make it possible to analyze the long-run behavior of GANs training via the invariant measure of the SDEs.

Long-run behavior of GANs training via the invariant measure of SDEs.

The invariant measure here in the context of GANs training can be interpreted in the following sense. First of all, the invariant measure μ^* describes the joint probability distribution of the generator and discriminator parameters $(\Theta^*, \mathcal{W}^*)$ in equilibrium. For instance, if the training process converges to the unique minimax point (θ^*, ω^*) for $\min_\theta \max_\omega \Phi(\theta, \omega)$, the invariant measure is the Dirac mass at (θ^*, ω^*). Having the distribution of (θ^*, ω^*), especially the marginal distribution of Θ^*, helps to characterize the probability distribution of the generated samples, $\text{Law}(G_{\Theta^*}(Z))$, and this distribution is in particular useful in the evaluation of GANs performance via metrics such as inception score and Frèchet inception distance. (See Salimans et al. (2016) and Heusel et al. (2017) for more details on these metrics). Besides, from a game perspective, the pair of conditional laws $(\text{Law}(\Theta^* | \mathcal{W}^*), \text{Law}(\mathcal{W}^* | \Theta^*))$ can be seen as the mixed strategies adopted by the generator and discriminator in equilibrium, respectively.

Theorem 23.9 *Assume the following conditions hold for* (23.17)*:*

(3a) *both b and σ are bounded and smooth and have bounded derivatives of any order;*

(3b) *there exist some positive real numbers r and M_0 such that for any $\begin{pmatrix} \theta & \omega \end{pmatrix}^T \in \mathbb{R}^{d_\theta + d_\omega}$,*

$$\begin{pmatrix} \theta & \omega \end{pmatrix} b(\theta, \omega) \leq -r \left\| \begin{pmatrix} \theta \\ \omega \end{pmatrix} \right\|_2, \quad \text{if } \left\| \begin{pmatrix} \theta \\ \omega \end{pmatrix} \right\|_2 \geq M_0;$$

(3c) *$\mathcal{A}$ is uniformly elliptic, i.e., there exists $l > 0$ such that for any $\begin{pmatrix} \theta \\ \omega \end{pmatrix}, \begin{pmatrix} \theta' \\ \omega' \end{pmatrix} \in \mathbb{R}^{d_\theta + d_\omega}$,*

$$\begin{pmatrix} \theta' & \omega' \end{pmatrix}^T \sigma(\theta, \omega) \sigma(\theta, \omega)^T \begin{pmatrix} \theta' \\ \omega' \end{pmatrix} \geq l \left\| \begin{pmatrix} \theta' \\ \omega' \end{pmatrix} \right\|_2^2,$$

then (23.17) *admits a unique invariant measure μ^* with an exponential convergence rate.*

Similar results hold for the invariant measure of (23.16) *with b replaced by b_0.*

The assumptions (1a)–(1c), (2a)–(2b) and (3a) for the regularity conditions of the drift, the volatility, and the derivatives of loss function Φ, are more than mathematical convenience. They are essential constraints on the growth of the loss function with respect to the model parameters, necessary for avoiding the explosive gradient encountered in the training of GANs. Moreover, these conditions put restrictions on the gradients of the objective functions with respect to the parameters. By the chain rule, it requires both careful choices of network structures as well as particular forms of the loss function Φ.

Dynamics of training loss and FDR.

To have a more quantifiable characteristic of the long-run behavior of GANs training, the analysis of the training loss dynamics reveals a fluctuation–dissipation relation (FDR) the GANs training.

Theorem 23.10 *Assume the existence of an invariant measure μ^* for (23.17), then*

$$\mathbb{E}_{\mu^*}\left[\|\nabla_\theta\Phi(\Theta^*,\mathcal{W}^*)\|_2^2 - \|\nabla_\omega\Phi(\Theta^*,\mathcal{W}^*)\|_2^2\right]$$

$$= \beta^{-1}\mathbb{E}_{\mu^*}\left[\mathrm{Tr}\left(\Sigma_\theta(\Theta^*,\mathcal{W}^*)\nabla_\theta^2\Phi(\Theta^*,\mathcal{W}^*) + \Sigma_\omega(\Theta^*,\mathcal{W}^*)\nabla_\omega^2\Phi(\Theta^*,\mathcal{W}^*)\right)\right]$$

$$- \frac{\eta}{2}\mathbb{E}_{\mu^*}\left[\nabla_\theta\Phi(\Theta^*,\mathcal{W}^*)^T\nabla_\theta^2\Phi(\Theta^*,\mathcal{W}^*)\nabla_\theta\Phi(\Theta^*,\mathcal{W}^*)\right.$$

$$\left. + \nabla_\omega\Phi(\Theta^*,\mathcal{W}^*)^T\nabla_\omega^2\Phi(\Theta^*,\mathcal{W}^*)\nabla_\omega\Phi(\Theta^*,\mathcal{W}^*)\right]. \qquad (23.23)$$

The corresponding FDR for the simultaneous update case of (23.16) is

$$\mathbb{E}_{\mu^*}\left[\|\nabla_\theta\Phi(\Theta^*,\mathcal{W}^*)\|_2^2 - \|\nabla_\omega\Phi(\Theta^*,\mathcal{W}^*)\|_2^2\right] =$$

$$\beta^{-1}\mathbb{E}_{\mu^*}\left[\mathrm{Tr}\left(\Sigma_\theta(\Theta^*,\mathcal{W}^*)\nabla_\theta^2\Phi(\Theta^*,\mathcal{W}^*) + \Sigma_\omega(\Theta^*,\mathcal{W}^*)\nabla_\omega^2\Phi(\Theta^*,\mathcal{W}^*)\right)\right].$$

Note that this FDR relation for GANs training is analogous to that for stochastic gradient descent algorithm on a pure minimization problem in Yaida (2019) and Liu and Theodorou (2019). This FDR relation in GANs reveals the crucial difference between GANs training of discriminator and generator networks versus training of two independent neural networks. It connects the microscopic fluctuation from the noise of SGA with the macroscopic dissipation phenomena related to the loss function. In particular, the quantity $\mathrm{Tr}(\Sigma_\theta\nabla_\theta^2\Phi + \Sigma_\omega\nabla_\omega^2\Phi)$ links the covariance matrices Σ_θ and Σ_ω from SGAs with the loss landscape of Φ, and reveals the trade-off of the loss landscape between the generator and the discriminator.

Alternatively, the evolution of the squared norm of the parameters leads to a different type of FDR that will be practically useful for learning rate scheduling.

Theorem 23.11 *Assume the existence of an invariant measure μ^* for (23.16), then*

$$\mathbb{E}_{\mu^*}\left[\Theta^{*,T}\nabla_\theta\Phi(\Theta^*,\mathcal{W}^*) - \mathcal{W}^{*,T}\nabla_\omega\Phi(\Theta^*,\mathcal{W}^*)\right] =$$

$$\beta^{-1}\mathbb{E}_{\mu^*}\left[\mathrm{Tr}(\Sigma_\theta(\Theta^*,\mathcal{W}^*) + \Sigma_\omega(\Theta^*,\mathcal{W}^*))\right] \qquad (23.24)$$

Scheduling of learning rate.

Notice that the quantities in (23.24), including the parameters (θ,ω) and first-order derivatives of the loss function g_θ, g_ω, $g_\theta^{i,j}$ and $g_\omega^{i,j}$, are computationally inexpensive. Therefore, (23.24) enables customized scheduling of learning rate, instead of predetermined scheduling ones such as Adam or RMSprop optimizer.

For instance, recall that $g_\theta^\mathcal{B}$ and $g_\omega^\mathcal{B}$ are respectively unbiased estimators for g_θ and g_ω, and

$$\hat{\Sigma}_\theta(\theta,\omega) = \frac{\sum_{k=1}^B [g_\theta^{I_k,J_k}(\theta,\omega) - g_\theta^\mathcal{B}(\theta,\omega)][g_\theta^{I_k,J_k}(\theta,\omega) - g_\theta^\mathcal{B}(\theta,\omega)]^T}{B-1},$$

$$\hat{\Sigma}_\omega(\theta,\omega) = \frac{\sum_{k=1}^B [g_\omega^{I_k,J_k}(\theta,\omega) - g_\omega^\mathcal{B}(\theta,\omega)][g_\omega^{I_k,J_k}(\theta,\omega) - g_\omega^\mathcal{B}(\theta,\omega)]^T}{B-1}$$

are respectively unbiased estimators of $\Sigma_\theta(\theta,\omega)$ and $\Sigma_\omega(\theta,\omega)$. Now in order to improve GANs training with the simultaneous update, one can introduce two tunable parameters $\epsilon > 0$ and $\delta > 0$ to have the following scheduling:

$$\text{if } \left| \frac{\Theta^T g_\theta^\mathcal{B}(\Theta_t, \mathcal{W}_t) - \mathcal{W}_t^T g_\omega^\mathcal{B}(\Theta_t, \mathcal{W}_t)}{\beta^{-1}\text{Tr}(\hat{\Sigma}_\theta(\Theta_t, \mathcal{W}_t) + \hat{\Sigma}_\omega(\Theta_t, \mathcal{W}_t))} - 1 \right| < \epsilon, \text{ then update } \eta \text{ by } (1-\delta)\eta.$$

23.4 Applications of GANs

23.4.1 Computing MFGs via GANs

Based on the conceptual connection between GANs and MFGs, Cao et al. (2020b) proposed a new computational approach for MFGs, using two neural networks in an adversarial way, summarized in Algorithm 23.1, in which

- u_θ is the NN approximation of the unknown value function u for the HJB equation,
- m_ω is the NN approximation for the unknown mean information function m.

Note that Algorithm 23.1 can be adapted for broader classes of dynamical systems with variational structures. Such GANs structures are exploited in Yang et al. (2020) and Yang and Perdikaris (2019) to synthesize complex systems governed by physical laws.

To test the performance of Algorithm 23.1, they considered a class of ergodic MFGs with the following payoff function:

$$\hat{J}_m(\alpha) = \liminf_{T \to \infty} \frac{1}{T} \mathbb{E}\left[\int_t^T L(X_t^\alpha, \alpha(X_t^\alpha)) + f(X_t^\alpha, m(X_t^\alpha))\, dt \right], \qquad (23.25)$$

subject to $dX_t^\alpha = \alpha(X_t^\alpha)dt + dW_t$, with the cost of control and running cost given by

$$L(x,\alpha) = \frac{1}{2}|\alpha|^2 + 2\pi^2 \left[-\sum_{i=1}^d \sin(2\pi x_i) + \sum_{i=1}^d |\cos(2\pi x_i)|^2 \right] - 2\sum_{i=1}^d \sin(2\pi x_i),$$

$$f(x,m) = \ln(m).$$

Algorithm 23.1 MFGANs

At $k = 0$, initialize θ and ω. Let N_θ and N_ω be the number of training steps of the inner-loops and K be that of the outer-loop. Let $\beta_i > 0$, $i = 1, 2$.

for $k \in \{0, \ldots, K - 1\}$ **do**

Let $m = 0$, $n = 0$.

Sample $\{(s_i, x_i)\}_{i=1}^{B_d}$ on $[0, T] \times \mathbb{R}^d$ according to a predetermined distribution p_{prior}, where B_d denotes the number of training samples for updating loss related to FP residual.

Let $\hat{L}_D(\theta, \omega) = \hat{L}_{FP}(\theta, \omega) + \beta_D \hat{L}_{init}(\omega)$, with

$$\hat{L}_{FP} = \frac{1}{B_d} \left\{ \sum_{i=1}^{B_d} \left[\partial_s m_\omega(s_i, x_i) + \text{div} \left[m_\omega(s_i, x_i) b(s_i, x_i, m(s_i, x_i), \alpha_{\theta, \omega}^*(s_i, x_i)) \right] \right. \right.$$

$$\left. \left. - \frac{\sigma^2}{2} \Delta_x m_\omega(s_i, x_i) \right]^2 \right\},$$

$$\hat{L}_{init} = \frac{\sum_{i=1}^{B_d} \left[m_\omega(0, x_i) - m^0(x_i) \right]^2}{B_d},$$

where m^0 is a known density function for the initial distribution of the states and $\beta_D > 0$ is the weight for the penalty on the initial condition of m.

for $m \in \{0, \ldots, N_\omega - 1\}$ **do**

$\omega \leftarrow w - \alpha_d \nabla_\omega \hat{L}_D$ with learning rate α_d.

Increase m.

end for

Sample $\{(s_j, x_j)\}_{j=1}^{B_g}$ on $[0, T] \times \mathbb{R}$ according to a predetermined distribution p_{prior}, where B_g denotes the number of training samples for updating loss related to HJB residual.

Let $\hat{L}_G(\theta, \omega) = \hat{L}_{HJB}(\theta, \omega) + \beta_G \hat{L}_{term}(\theta)$, with

$$\hat{L}_{HJB} = \frac{1}{B_g} \left\{ \sum_{j=1}^{B_g} \left[\partial_s u_\theta(s_j, x_j) + \frac{\sigma^2}{2} \Delta_x u_\theta(s_j, x_j) + H_\omega \left(s_j, x_j, \nabla_x u_\theta(s_j, x_j) \right) \right]^2 \right\},$$

$$\hat{L}_{term} = \frac{\sum_{j=1}^{B_g} u_\theta(T, x_j)^2}{B_g},$$

where $\beta_G > 0$ is the weight for the penalty on the terminal condition of u.

for $n \in \{0, \ldots, N_\theta - 1\}$ **do**

$\theta \leftarrow \theta - \alpha_g \nabla_\theta \hat{L}_G$ with learning rate α_g.

Increase n.

end for

Increase k.

end for

Return θ, ω

In this class of mean field games, the associated HJB and FP equations are

$$\begin{cases} -\epsilon \Delta u + H_0(x, \nabla u) = f(x,m) + \bar{H}, \\ -\epsilon \Delta m - \mathrm{div}\left(m\nabla_p H_0(x, \nabla u)\right) = 0, \\ \int_{\mathbb{T}^d} u(x)\,dx = 0; \; m > 0, \; \int_{\mathbb{T}^d} m(x)\,dx = 1, \end{cases} \tag{23.26}$$

where the convex conjugate H_0 is given by $H_0(x,p) = \sup_\alpha \{\alpha \cdot p - \frac{1}{2}|\alpha|^2\} - \tilde{f}(x)$. Here, the periodic value function u, the periodic density function m, and the unknown $\bar{H}$ can be explicitly derived. Indeed, assuming the existence of a smooth solution $(m, u, \bar{H})$, m in the second equation in (23.26) can be written as

$$m(x) = \frac{e^{2u(x)}}{\int_{\mathbb{T}^d} e^{2u(x')}\,dx'}.$$

Hence the solution to (23.26) is given by $u(x) = \sum_{i=1}^d \sin(2\pi x_i)$ and

$$\bar{H} = \ln\left(\int_{\mathbb{T}^d} e^{2\sum_{i=1}^d \sin(2\pi x_i)}\,dx\right).$$

The optimal control policy is also explicitly given by

$$\alpha^* = \arg\max_\alpha \{\nabla_x u \cdot \alpha - L(x, \alpha)\}$$

$$= \nabla_x u = 2\pi \left(\cos(2\pi x_1) \quad \cdots \quad \cos(2\pi x_d)\right) \in \mathbb{R}^d.$$

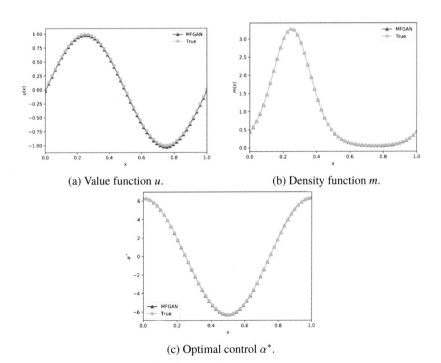

(a) Value function u.

(b) Density function m.

(c) Optimal control α^*.

Figure 23.1 One-dimensional test case.

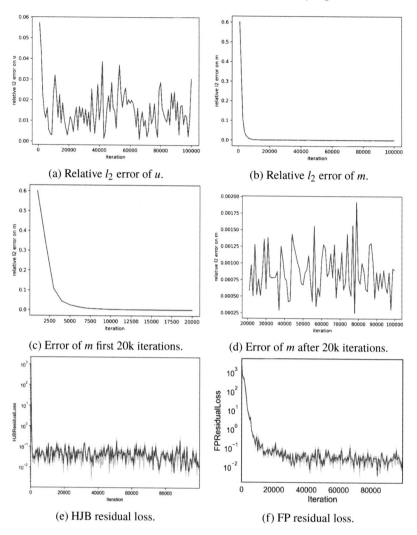

(a) Relative l_2 error of u.

(b) Relative l_2 error of m.

(c) Error of m first 20k iterations.

(d) Error of m after 20k iterations.

(e) HJB residual loss.

(f) FP residual loss.

Figure 23.2 Losses and errors in the one-dimensional test case.

Algorithm 23.1 is first tested on a one-dimensional case, with the result highlighted in Figures 23.1 and 23.2. Figures 23.1(a) and 23.1(b) show the learnt functions of u and m against the true ones, respectively, and 23.1(c) shows the optimal control, with the accuracy of the learnt functions versus the true ones. The plots of loss in Figures 23.2(a) and 23.2(b), depict the evolution of relative l_2 error as the number of outer iterations grows to K. Within 10^5 iterations, the relative l_2 error of u oscillates around 3×10^{-2}, and the relative l_2 errors of m decreases below 10^{-3}. The evolution of the HJB and FP differential residual loss is shown in Figures 23.2(e) and 23.2(f), respectively. In theses figures, the solid line is the average loss among three experiments, with standard deviation captured by the shadow around the line. Both differential residuals first rapidly descend to the magnitude of 10^{-2} and then the descent slows down accompanied by oscillation.

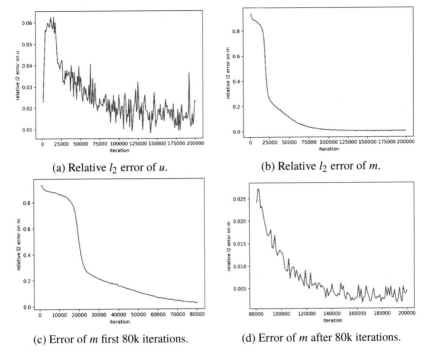

(a) Relative l_2 error of u.

(b) Relative l_2 error of m.

(c) Error of m first 80k iterations.

(d) Error of m after 80k iterations.

Figure 23.3 Input of dimension 4.

Algorithm 23.1 is then applied to a four-dimensional case, with the result shown in Figure 23.3. Within 2×10^5 iterations, the relative l_2 error of u decreases below 2×10^{-2} and that of m decreases to 4×10^{-3}.

Note the potential power of GANs training when compared with a similar experiment in Test Case 4 in Carmona and Laurière (2019) without the adversarial training for two neural networks: algorithms in Carmona and Laurière (2019) need a significantly larger number of iterations to achieve the same level of accuracy.

A concurrent work of Lin et al. (2020) used a primal-dual variational formulation associated with the coupled HJB–FP system as in Cirant and Nurbekyan (2018), to recast MFGs as GANs in a different way, seeing the density function as the generator and the value function as the discriminator. Based on this alternative interpretation, a GANs-based algorithm named APAC-Net was proposed. Through numerical experiments, this algorithm was shown to be able to solve certain classes of MFGs in dimension up to 100.

23.4.2 GANs in mathematical finance

There are essentially two different frameworks in which GANs have been adopted in the mathematical finance literature. The first is to reformulate a constrained control and optimization problem as a minimax problem so that the generator and discriminator networks can be constructed for computational purpose. The second one is to draw an analogy between simulation of financial time series data and image generation such that various statistical and distributional properties can be exploited for performance evaluations. We will review here several representative works in each category.

Asset pricing and minimax problem.

Chen et al. (2019) is one of the earliest works to identify the minimax structure in a non-linear model for asset pricing. Its primary idea is to exploit the no-arbitrage condition and recast the constrained problem into the minimax framework of GANs. Their objective is to estimate the pricing kernel or stochastic discount factor (SDF) that summarizes the information of the cross-section of returns for different stocks.

Specifically, take the return of asset $i \in \{1, \ldots, n\}$ at time $t+1$ as $R_{t+1,i}$ and the excess return as $R_{t+1,i}^e = R_{t+1,i} - R_{t,i}$. Let M_{t+1} be the SDF satisfying the no-arbitrage condition,

$$\mathbb{E}_t\left[M_{t+1}R_{t+1,i}^e\right] = 0 \iff \mathbb{E}_t\left[R_{t+1,i}^e\right] = \left(-\frac{\text{Cov}_t(R_{t+1,i}^e, M_{t+1})}{\text{Var}_t(M_{t+1})}\right) \cdot \frac{\text{Var}_t(M_{t+1})}{\mathbb{E}_T[M_{t+1}]},$$

where $\mathbb{E}_t$ stands for expectation conditional on some suitable information by time t. Then assume that

$$M_{t+1} = 1 - \omega^T R_{t+1}^e = 1 - F_{t+1}, \quad \beta_{t,i} = -\frac{\text{Cov}_t(R_{t+1,i}^e, M_{t+1})}{\text{Var}_t(M_{t+1})},$$

where $\omega = (\omega_1, \ldots, \omega_i, \ldots, \omega_n)$ denotes the SDF weights vector which is also the weights vector of the conditional mean-variance efficient portfolio, and $\beta_{t,i}$ denotes the time-varying exposure to systematic risk for asset i.

In this one-factor model setup, the main quantities to be estimated are the two vectors ω and $\beta_t = (\beta_{t,1}, \ldots, \beta_{t,i}, \ldots, \beta_{t,n})$. To handle the no-arbitrage constraint, they utilize the unconditional moment conditions: given any σ-algebra generated by some random variable Z, $\mathcal{F} = \sigma(Z)$,

$$Y = \mathbb{E}[X|\mathcal{F}] \implies \mathbb{E}[Xf(Z)] = \mathbb{E}[Yf(Z)],$$

for any measurable function f. In particular, let the choice of information be $\sigma(I_t, I_{t,i})$, where I_t represents the macroeconomic conditions at time t whereas $I_{t,i}$ denotes information at time t for the specific stock i, then

$$\omega_i = \omega_{t,i} = \omega(I_t, I_{t,i}), \quad \beta_{t,i} = \beta(I_t, I_{t,i}).$$

Consequently, the no-arbitrage condition implies

$$\mathbb{E}\left[M_{t+1}R_{t+1}^e g(I_t, I_{t,i})\right] = 0,$$

for any measurable function g; if $\hat{\omega}$ and $\hat{\beta}$ correspond to the correct SDF $\hat{M}$, this is equivalent to

$$\max_g \frac{1}{N}\sum_{j=1}^N \left\|\mathbb{E}\left[\hat{M}_{t+1}R_{t+1,j}^e g(I_t, I_{t,j})\right]\right\|^2 = 0.$$

Now, estimating SDF that satisfies the no-arbitrage condition is transformed into the minimax game

$$\min_{\omega} \max_{g} \frac{1}{N} \sum_{j=1}^{N} \left\| \mathbb{E}\left[\hat{M}_{t+1} R_{t+1,j}^{e} g(I_t, I_{t,j}) \right] \right\|^2,$$

a natural GANs structure.

This proposed GANs model is then compared with an alternative model with the no-arbitrage condition relaxed to a first moment condition given by the one-factor model, $\mathbb{E}[R_{t+1,j}^{e}] \propto \mathbb{E}[F_{t+1}]$. It is further compared with a second alternative model with both ω and g assumed to be linear. The GAN model is shown to outperform uniformly in terms of Sharp ratio, explained variation, and cross-sectional mean R^2.

GANs as financial time series data simulators

Another application of GANs is to generate financial time series data for both equity and derivatives. We consider below a selection, by no means comprehensive, of some representative works.

In Wiese et al. (2019), the main objective is to build a simulator for equity option markets. Instead of dealing with option price directly, which is subject to the no-arbitrage constraint, they work with an equivalent and less constrained form called discrete local volatility (DLV).

In this formulation, the time-varying DLV σ_t is seen as a function of strike K and maturity M. The generator takes the state variable $S_t = f(\sigma_t, \ldots, \sigma_0)$ as well as some random noise Z_{t+1} as inputs and set $X_{t+1} = \log \sigma_{t+1} = g(Z_{t+1}, S_t)$. The discriminator tries to distinguish the true (X_{t+1}, S_t) and the generated $(\tilde{X}_{t+1}, \tilde{S}_t)$. Other calibration techniques such as PCA are also incorporated.

This formulation is compared among different neural network based simulators. The performance evaluation is based on four types of criteria: the distributional metric which is the distance between the empirical probability distribution functions of the generated and historical data, the distributional scores given by skewness and kurtosis scores, the dependency score through the autocorrelation function score for the log-return process and finally the cross-correlation scores for the log-DLV and the DLV log returns. Their numerical results show that the GANs model outperforms the other benchmark models such as vector autoregressive models, TCN models, and quasi maximum likelihood estimation.

In a closely related work, Wiese et al. (2020) proposed a special GANs model called the Quant GAN. The main characteristic of Quant GANs is taking temporal convolutional networks (TCNs) as the generator. By choosing appropriate kernel size K and dilation factor D, TCNs can carry long-time dependency and avoid abnormal behavior of gradients over time. They show that with Lipschitz constraint on the choices of activation functions and weights, the generated process has as many number of moments as the input latent variable. Finally, they use the inverse Lambert W transform for the real asset log-return processes to copy with the heavy-tail property in the GAN training. In the Lambert W transform, a random variable X with mean μ, variance σ^2 and cumulative distribution function F_X is transformed into

$$Y = \frac{X - \mu}{\sigma} \exp\left(\frac{\delta(X - \mu)^2}{2\sigma^2}\right) + \mu,$$

with a proper choice of nonnegative parameter δ so that Y has heavier tail than X if $\delta > 0$. The inverse Lambert W transform is its inverse process.

They propose two different approaches to utilizing TCNs: one is to use the pure TCNs to directly generate time series, the other is to use TCNs to generate drift and volatility processes and add another network to represent the noise. They test the simple GARCH model for comparison purposes, with the evaluation of the models based on distributional metrics and dependence scores. In particular, the former includes Wasserstein distance and DY metric, i.e., a measurement of the distance between the estimated likelihoods from real and generated data, and the latter include ACF score and the leverage effect score. Their results show that the GANs model with pure TCNs perform the best for the majority of the tests, and that both GAN models dominate the GARCH model.

Other related works include Takahashi et al. (2019) and Zhang et al. (2019). The GANs model in Takahashi et al. (2019) captures statistical properties exhibited in real financial data, such as linear unpredictability, the heavy-tailed price return distribution, volatility clustering, leverage effects, the coarse-fine volatility correlation, and the gain/loss asymmetry. The GANs model in Zhang et al. (2019) is used to predict stock prices from historical stock data, where long-short-term-memory is adopted as the generator and multi-layer perceptron as the discriminator. In particular, the generator acts as a function characterizing the unknown and possibly complex relation between stock price in the future and historical data.

There are other extensions of GAN models. For instance, embracing the general idea of adversarial training in GANs, Cuchiero et al. (2020) proposes a generative adversarial approach for (robust) calibration of local stochastic volatility models; the generation of volatility surfaces follows neural SDEs, where a special deep-hedging-based variance reduction technique is applied and the adversarial training idea is embedded in evaluating the simulated volatility surfaces: the loss function may come from a family of candidate loss functions to ensure robustness. Recently, a GANs model called COT-GAN has been proposed in Xu et al. (2020) based on causal optimal transport theory. In this work, the temporal causality condition naturally leads to an adversarial framework for GANs and a mixed Sinkhorn distance is proposed to calculate the optimal transport cost with reduced bias. This new framework could be used for generating sequential data including financial time series.

23.5 Conclusion and discussion

This notes covers three major aspects of GANs, essentials of GANs in the optimization and game framework, GANs training via stochastic analysis, and recent applications of GANs in mathematical finance.

Despite its vast popularity and power in data and image generation, GANs face many challenges in implementation and training and remain largely undeveloped in theory. For instance, the well-posedness of GANs as a minimax game has not been fully understood until Guo and Mounjid (2020) in which the convexity issue is analyzed in details. The connection between mean field games and GANs via the minimax structure presents GANs' potential computing power for high dimensional control and optimization problems with variational structures. The next natural test field is forward-backward-stochastic-differential equations, where there is a natural variational structure to retrofit for the minimax game of GANs. Beyond computational power, more explorations are needed to see if convergence and computation complexity results can be obtained, especially given the SDE approximation of GANs training. A small step in this direction is Guo and Mounjid (2020), which formulates simple stochastic control problems for learning rate and batch size analysis and shows their impact on error and variance reduction. One also wonders if the empirical success of GAN in data generation can be replicated in the general area of simulation and if a robust theoretical analysis can be established.

References

Arjovsky, Martin, and Bottou, Léon. 2017. Towards principled methods for training generative adversarial networks. In: *International Conference on Learning Representations*.

Arjovsky, Martin, Chintala, Soumith, and Bottou, Léon. 2017. Wasserstein generative adversarial networks. Pages 214–223 of: *International Conference on Machine Learning*.

Barnett, Samual A. 2018. Convergence problems with generative adversarial networks (GANs). ArXiv:1806.11382.

Berard, Hugo, Gidel, Gauthier, Almahairi, Amjad, Vincent, Pascal, and Lacoste-Julien, Simon. 2020. A closer look at the optimization landscape of generative adversarial networks. In: *International Conference on Learning Representations*.

Cao, Haoyang, and Guo, Xin. 2020. SDE approximations of GANs training and its long-run behavior. ArXiv:2006.02047.

Cao, Haoyang, Guo, Xin, and Laurière, Mathieu. 2020b. Connecting GANs, MFGs and OT. ArXiv:2002.04112.

Carmona, René, and Laurière, Mathieu. 2019. Convergence analysis of machine learning algorithms for the numerical solution of mean field control and games: II – The finite horizon case. Arxiv:1908.01613.

Chen, Luyang, Pelger, Markus, and Zhu, Jason. 2019. Deep learning in asset pricing. Available at SSRN 3350138.

Cirant, Marco, and Nurbekyan, Levon. 2018. The variational structure and time-periodic solutions for mean-field games systems. ArXiv:1804.08943.

Conforti, Giovanni, Kazeykina, Anna, and Ren, Zhenjie. 2020. Game on random environment, mean-field Langevin system and neural networks. ArXiv:2004.02457.

Cuchiero, Christa, Khosrawi, Wahid, and Teichmann, Josef. 2020. A generative adversarial network approach to calibration of local stochastic volatility models. *Risks*, **8**(4), 101.

Denton, Emily L, Chintala, Soumith, Szlam, Arthur, and Fergus, Rob. 2015. Deep generative image models using a Laplacian pyramid of adversarial networks. Pages 1486–1494 of: *Advances in Neural Information Processing Systems.*

Domingo-Enrich, Carles, Jelassi, Samy, Mensch, Arthur, Rotskoff, Grant M, and Bruna, Joan. 2020. A mean-field analysis of two-player zero-sum games. ArXiv:2002.06277.

Goodfellow, Ian J., Pouget-Abadie, Jean, Mirza, Mehdi, Xu, Bing, Warde-Farley, David, Ozair, Sherjil, Courville, Aaron, and Bengio, Yoshua. 2014. Generative adversarial nets. Pages 2672–2680 of: *Advances in Neural Information Processing Systems.*

Guo, Xin, and Mounjid, Othmane. 2020. Convergence of GANs training: a game and stochastic control methodology. Arxiv:2112.00222.

Guo, Xin, Hong, Johnny, Lin, Tianyi, and Yang, Nan. 2021. Relaxed Wasserstein with applications to GANs. Pages 3325–3329 of: *Proc. IEEE International Conference on Acoustic, Speech and Signal Processing (ICASSP).*

Heusel, Martin, Ramsauer, Hubert, Unterthiner, Thomas, Nessler, Bernhard, and Hochreiter, Sepp. 2017. GANs trained by a two time-scale update rule converge to a local Nash equilibrium. Pages 6626–6637 of: *Advances in Neural Information Processing Systems.*

Kulharia, Viveka, Ghosh, Arnab, Mukerjee, Amitabha, Namboodiri, Vinay, and Bansal, Mohit. 2017. Contextual RNN-GANs for abstract reasoning diagram generation. Pages 1382–1388 of: *Proc. 31st AAAI Conference on Artificial Intelligence.*

Ledig, Christian, Theis, Lucas, Huszár, Ferenc, Caballero, Jose, Cunningham, Andrew, Acosta, Alejandro, Aitken, Andrew, Tejani, Alykhan, Totz, Johannes, Wang, Zehan, et al. 2017. Photo-realistic single image super-resolution using a generative adversarial network. Pages 4681–4690 of: *Proc. IEEE Conference on Computer Vision and Pattern Recognition.*

Lei, Na, Su, Kehua, Cui, Li, Yau, Shing-Tung, and Gu, Xianfeng David. 2019. A geometric view of optimal transportation and generative model. *Computer Aided Geometric Design,* **68**, 1–21.

Li, Qianxiao, Tai, Cheng, and E, Weinan. 2019. Stochastic modified equations and dynamics of stochastic gradient algorithms I: mathematical foundations. *Journal of Machine Learning Research,* **20**(40), 1–47.

Lin, Alex Tong, Fung, Samy Wu, Li, Wuchen, Nurbekyan, Levon, and Osher, Stanley J. 2020. APAC-Net: Alternating the population and agent control via two neural networks to solve high-dimensional stochastic mean field games. ArXiv:2002.10113.

Liu, Guan-Horng, and Theodorou, Evangelos A. 2019. Deep learning theory review: An optimal control and dynamical systems perspective. ArXiv:1908.10920.

Luc, Pauline, Couprie, Camille, Chintala, Soumith, and Verbeek, Jakob. 2016. Semantic segmentation using adversarial networks. ArXiv:1611.08408.

Mescheder, Lars, Geiger, Andreas, and Nowozin, Sebastian. 2018. Which training methods for GANs do actually converge? Pages 3481—-3490 of: *International Conference on Machine Learning.*

Nock, Richard, Cranko, Zac, Menon, Aditya K, Qu, Lizhen, and Williamson, Robert C. 2017. f-GANs in an information geometric nutshell. Pages 456–464 of: *Advances in Neural Information Processing Systems.*

Nowozin, Sebastian, Cseke, Botond, and Tomioka, Ryota. 2016. f-GAN: training generative neural samplers using variational divergence minimization. Pages 271–279 of: *Proc. 30th International Conference on Neural Information Processing Systems.*

Radford, Alec, Metz, Luke, and Chintala, Soumith. 2015. Unsupervised representation learning with deep convolutional generative adversarial networks. ArXiv:1511.06434.

Reed, Scott, Akata, Zeynep, Yan, Xinchen, Logeswaran, Lajanugen, Schiele, Bernt, and Lee, Honglak. 2016. Generative adversarial text to image synthesis. Pages 1060–1069 of: *33rd International Conference on Machine Learning.*

Salimans, Tim, Goodfellow, Ian, Zaremba, Wojciech, Cheung, Vicki, Radford, Alec, and Chen, Xi. 2016. Improved techniques for training GANs. Pages 2234–2242 of: *Advances in Neural Information Processing Systems.*

Salimans, Tim, Zhang, Han, Radford, Alec, and Metaxas, Dimitris. 2018. Improving GANs using optimal transport. In: *International Conference on Learning Representations*.

Sanjabi, Maziar, Ba, Jimmy, Razaviyayn, Meisam, and Lee, Jason D. 2018. On the convergence and robustness of training GANs with regularized optimal transport. Pages 7091–7101 of: *Advances in Neural Information Processing Systems*.

Sion, Maurice. 1958. On general minimax theorems. *Pacific Journal of Mathematics*, **8**(1), 171–176.

Srivastava, Akash, Greenewald, Kristjan, and Mirzazadeh, Farzaneh. 2019. BreGMN: scaled-Bregman generative modeling networks. ArXiv:1906.00313.

Takahashi, Shuntaro, Chen, Yu, and Tanaka-Ishii, Kumiko. 2019. Modeling financial time-series with generative adversarial networks. *Physica A: Statistical Mechanics and its Applications*, **527**(Aug), 121261.

Von Neumann, John. 1959. On the theory of games of strategy. *Contributions to the Theory of Games*, **4**, 13–42.

Vondrick, Carl, Pirsiavash, Hamed, and Torralba, Antonio. 2016. Generating videos with scene dynamics. Pages 613–621 of: *Advances in Neural Information Processing Systems*.

Wiatrak, Maciej, Albrecht, Stefano V, and Nystrom, Andrew. 2019. Stabilizing generative adversarial networks: a survey. ArXiv:1910.00927.

Wiese, Magnus, Bai, Lianjun, Wood, Ben, Morgan, J P, and Buehler, Hans. 2019. Deep hedging: learning to simulate equity option markets. ArXiv:1911.01700.

Wiese, Magnus, Knobloch, Robert, Korn, Ralf, and Kretschmer, Peter. 2020. Quant GANs: deep generation of financial time series. *Quantitative Finance*, **20**(9), 1419–1440.

Xu, Tianlin, Wenliang, Li K, Munn, Michael, and Acciaio, Beatrice. 2020. COT-GAN: Generating sequential data via causal optimal transport. ArXiv:2006.08571.

Yaida, Sho. 2019. Fluctuation-dissipation relations for stochastic gradient descent. In: *International Conference on Learning Representations*.

Yang, Liu, Zhang, Dongkun, and Karniadakis, George Em. 2020. Physics-informed generative adversarial networks for stochastic differential equations. *SIAM Journal on Scientific Computing*, **42**(1), A292–A317.

Yang, Yibo, and Perdikaris, Paris. 2019. Adversarial uncertainty quantification in physics-informed neural networks. *Journal of Computational Physics*, **394**, 136–152.

Yeh, Raymond, Chen, Chen, Lim, Teck Yian, Hasegawa-Johnson, Mark, and Do, Minh N. 2016. Semantic image inpainting with perceptual and contextual losses. ArXiv:1607.07539.

Zhang, Kang, Zhong, Guoqiang, Dong, Junyu, Wang, Shengke, and Wang, Yong. 2019. Stock market prediction based on generative adversarial network. *Procedia Computer Science*, **147**, 400–406.

Zhu, Banghua, Jiao, Jiantao, and Tse, David. 2020. Deconstructing generative adversarial networks. *IEEE Transactions on Information Theory*, **66**(11), 7155–7179.

Zhu, Jun-Yan, Krähenbühl, Philipp, Shechtman, Eli, and Efros, Alexei A. 2016. Generative visual manipulation on the natural image manifold. Pages 597–613 of: *European Conference on Computer Vision*. Springer.

CONNECTIONS WITH THE REAL ECONOMY

Part VI

Nowcasting with Alternative Data

24

Introduction to Part VI
Nowcasting is Coming

Michael Recce[a]

Not long ago twenty–twenty hindsight was the best we could hope for. The health of economies and business are analyzed retrospectively. It often takes weeks to months for the management of a company to fully understand performance. Economic statistics are released by governments with a lag and are updated multiple times over a period of months or years after they are initially published. The retrospective information on the performance of companies and economies is also insufficiently granular. Clearly better and more timely information is an advantage in business and in investing. Technology, software, algorithms and data are fueling a trend to provide more detailed and more timely information about the current performance of businesses and economies. Macro and micro economic information will soon be available in real-time, or nowcast, and this book describes some of the important areas of progress that are making this happen.

This trend is the logical extension of a technology driven sequence that has its roots several decades ago. Today individual business sectors, companies, and economies are at a wide range of stages of adoption of this sequence of improvements to their visibility into current performance. So understanding the progression, helps to clarify the trend and to project what we might expect as the trend continues. This book includes five important directions that are key parts of this trend.

24.1 Micro before macro

It is certainly easier for a company to study the structured data that it has about its own business than it is to combine data across a large number of companies within a country to understand the macro economic environment of the country. The management team of a company needs to know what is working and what isn't working to be able to make decisions. Businesses that have a better understanding of their marketplace have a competitive advantage. These companies are able to respond more quickly to customer interests, or sales performance of subsets of their products or in subsets of their markets.

Most public and private companies only understand their performance weeks

[a] CEO, AlphaROC Inc., New York
Published in *Machine Learning And Data Sciences For Financial Markets*, Agostino Capponi and Charles-Albert Lehalle © 2023 Cambridge University Press.

or months after the business transactions have occurred. Even after these delays companies usually have less information about their own customers and transactions then they would like to have. Advances in technology and software have created a competitive arms race in most business sectors, rewarding companies who most quickly and completely understand their data. This advance has given rise to a wide range of business services and products industries generating an accelerating sequence of advances during the past few decades.

Businesses are investing in services, technology, and people in order to gain more insight into their performance in the marketplace. The consulting services industry, supporting this effort, alone is estimated to be over $250 Billion in 2020. The term "digital transformation" has been a driving mantra from businesses for over two decades, as they spend to get more timely and more detailed information about what is working and what is not. Software products from databases, to business analytics, to data visualization tools, customer relationship management software, enterprise resource planning software, and price optimization software are just some examples of the investments being made by businesses to increase the detail and timeliness of information about their customers and transactions. Businesses are also hiring data scientists and data engineers, at an increasing rate, in order to get insights that are more timely, actionable and detailed.

This drive for competitive advantage through investment in technology, software and services has led to rapid, but non-uniformly distributed advances in the real-time information that management teams have to make better decisions. In recent years the focus in successful start-ups is being agile, building minimum viable products, and iterating them rapidly with detailed nowcast views of functionality that engages their customer base. These newer companies are starting in a world based on expectations that nowcast information on customers and products is available.

24.2 Advance driven by Moore's law

The quest for better information sooner has always existed, but the rate of change is largely due to exponential scaling of the compute power of integrated circuits, and the exponential decline in the cost of computing described by Moore's law. In expensive computer power, has led to faster communication, more detail in any transaction, to the development of software systems to process these data, and most recently to the development of improvements in statistical analysis and machine learning to gain insights from these data. While many describe Moore's law slowing down after 50 years of doubling transistors in integrated circuits every two years. The increasing density of transistors led to simultaneous exponential increase in computer power, and decrease in the cost of computer power. The impact of this rapid change in available technology has barely started to permeate all aspects of business processes and products. The internet of things is one of the clear growth areas of pervasive technology from smart appliances to smart homes to smart cities. It is clear that the rise of detailed, timely data on business and commerce will continue its exponential growth for decades to come.

Data is now being recorded everywhere and about everything. As the volume of this "digital residue" increases, it provides an additional path for nowcasting micro and macro economic information. These data are often unstructured, in contrast to the data that is processed within the databases of a business. It is useful to refer to the data within a firm as its internal data, and this data from sensors and devices outside a business as external data. The unstructured, external data recorded outside of a business is now often called alternative data. More about this external, alternative data later, but first it is useful to look at the utility that has been derived from processing structured, internal data.

24.3 The CEO dashboard

A decade ago, in firms that had undergone a sufficient level of digital transformation, a trend emerged to construct CEO dashboards. A CEO dashboard, is a nowcast view of the current state of the business that is built from the firm's structured, internal data. This nowcast data indicates which parts of the business are struggling and which parts are performing well. These dashboards make it possible for the management team to make more well informed operational decisions that are grounded in the current dynamics of the marketplace. Important decisions on inventory placement, pricing, staffing, product development directions, supply chains and threats from competitors. Companies that have these dashboards have a clear competitive advantage, pushing the level of detail and the timeliness of the data to continue to improve.

This increasing detail in CEO dashboards is along directions of core key performance indicators (KPIs) that are often specialized to business type or sector. For example in retail relevant KPIs include the number of new stores that were opened and the year over year growth of business in the established stores. While the mean growth rate is important, it is even more useful to understand the performance of each individual store, and the distribution across the network of stores. Perhaps stores in one region have a higher revenue per square foot than stores in another region.

The most important direction in business detail of CEO dashboards is an analysis of cohorts, or types of customers. These cohorts could be separated by their demographic, psychographic or geographic properties. Expansion of a business into a new cohort, or type of customer, is one of the biggest drivers of near-term rapid growth. This is a total addressable market (TAM) expansion.

Growth can come from a set of different sources. For example, if the revenue comes from a new customer then the company can expect to get future revenue from the customer, until at least, the average lifetime value for that customer has been reached. While if the revenue is from an existing customer then the lifetime value from that customer is being depleted. If the new customer comes from a new cohort, then the company might expect more rapid growth of new customers as it begins to engage with a new group of customers.

This detail in CEO dashboards also shows the loss of customers to competitors. When a business is losing market share it is crucial to know where the battle front is, who the competitors are, which cohorts are most affected, and which products or services are deemed to be insufficiently competitive. Businesses are investing to gain this nowcast information, because it clearly provides operational advantages.

24.4 Internet companies led progress in nowcasting

Innovations in nowcasting have not occurred at the same rate in every industry sector. Internet software based products and services have, in general, advanced the most rapidly. Starting with the new technology tools obviates the need for technological transformation. Also, delivering products and services in near real-time, requires a skill set that makes nowcasting of internal data more easily within reach.

These businesses also led the way in innovations for cohort based analysis. Products are recommended to a customer based on a demographic, psychographic, and geographic analysis of the similarity of what is known about a customer and about other customers of the same product. A good example of this was the Netflix crowdsourced competition that was run ten years ago. It gave a million dollars to the team that produced the best recommendations on films to watch. The solution involved analyzing prior films watched to group customers into cohorts and interpreting the ratings of others within a cohort.

The core transactions at many internet companies, like search engines and social media companies, are inherently unstructured. This necessitated the development of algorithms, software and systems to ingest, analyze and interpret unstructured information efficiently. Structured data is labeled and has a consistent data type. For example, most data in spreadsheets is structured, where the column heading is the label and the value in that column for each row has the same data type (i.e. date, real number, or a word). Unstructured data is everything else, essentially implying that work must be done to separate the data into its relevant parts, label each of these parts and to then interpret the value.

During the past two decades the advances in the processing of unstructured transactions have been phenomenal. People and objects can be recognized in photographs. Digital assistants interpret unstructured questions and provide answers. Also, digital advertising has become almost uncanny in how accurately a personalized ad is selected. All of these are examples of large-scale processing of unstructured data using machine learning algorithms and cloud computing methods.

24.5 CEO dashboard from Alternative data

If an internet advertising company can accurately predict the right ad to show someone then the company knows what the user wants to buy. Knowing what customers want to buy across all geographies and all sectors, is tantamount to knowing which companies are winning in the marketplace. In addition the internet advertising company knows, at least with some probability, when a new customer starts buying a product, and maybe why a choice was made to switch from the prior merchant. The internet advertising company has used machine learning methods to classify the person seeing the ad into a cohort, and therefore knows if the company is expanding its TAM into new cohorts, and the competitiveness of each product for each cohort.

The same methods that are used by the internet advertising company to decide what ad to show can be used to construct the CEO dashboard of the merchant. This dashboard is now being constructed from unstructured data that is external to the firm. It is less complete, more noisy and biased in comparison with a CEO dashboard that is constructed from internal data. It is less complete because it is a subsample of the customers and the transactions of the firm. These external transactions are not labeled and errors can be introduced by incorrectly labeling data.

However, in other ways it is more complete. It is more complete because it can interpret a broader picture of customer behavior, when the customer is not transacting with the company. It indicates why customers are churning and who is the closest competitor for each of the cohorts of customers. This dashboard can both reproduce the internal view that is available to the CEO and simultaneously see what the customers chose when they are not transacting with the firm.

Since there is bias and noise in external data, the process of constructing a CEO dashboard is improved by including data from multiple different sources. The primary data used in internet advertising is clickstream data. Payment processing transactions are also very informative, as well as job listings, satellite imaging, cell phone location information, and many other sources.

Once constructed this nowcast CEO dashboard, from external data, provides to a competitor or an investor an informational advantage with enormous insights. Partial CEO dashboards can be constructed for all of the participants in a marketplace. This can be used to select a future winner, or to provide guidance to a participant in the market for how to survive or how to thrive. Eventually a sufficient number of reconstructed CEO dashboards may provide a mechanism for very detailed and timely nowcasting of macro economic data.

24.6 Nowcasting with alternative data

The unstructured data used in the advertising example above is called clickstream data, which is the sequence of web sites that are visited, and sometimes search terms entered prior to reaching a web site. Analysis can also be performed on web sites to determine a set of keywords associated with the web site. In these data the search terms alone have been shown to be a leading indicator of interests and intent to purchase. In 2009 Choi and Varian demonstrated that search term data could be used to nowcast economic statistics months before they are published. In Chapter 25 of this book Ferrara and Simoni extend this work, and evaluate methods for preselection of the most useful search terms to nowcast economic indicators.

All of these types of alternative data can be described as transactions. A particular web site was visited or this keyword was typed at this time by this cookie. An item was purchased on this day at this time or this is the number of cars in the professional spaces at this home supply store in this location, at this time, and on this date.

Another key source of data for both micro and macroeconomic nowcasting is the rapidly growing number of satellites that capture images of the earth. Satellite data may indicate how busy a business or a factory is by counting cars in a parking lot. Crop yields can be estimated, as well as processing and stockpiling of commodities like metals and oil. In Chapter 29, de Franchis and colleagues describe the challenges and methods in using satellite data for nowcasting macroeconomic data and financial business analysis.

Job listings are highly informative for constructing a detailed view of a business. They indicate the directions where a business is growing. For example seasonal job hiring is a signal that indicates the expected level of business in the peak quarter. Similarly, high turnover of sales staff is not a positive signal on the health of a business. Job listings also show business to business relationships, and the hiring company looks for expertise in the vendor products that are being used.

The weather has been effectively used to predict commerce. For example when the weather is good in a region people are more likely to go shopping. Cell phone location data provides a correlated, complementary and more detailed indication than the weather. With cell phone location data there is a direct measurement of, for example, the fact that consumers went shopping. Credit card transaction data is an even better predictor of commerce than cell phone location, since credit card transaction data can directly measure how much was spent. Chapter 27 describes recent work by Fleder and Shaw on methods for using time series analysis to nowcast this type of information from credit card data. This provides a nowcast view of the performance of a company and the types of products that are being sold.

Almost by definition these alternative sources are incomplete. The data was created for a completely different purpose. Only a biased subset of the transaction activity is available, and the data are unstructured. This data has to be processed, using methods like natural language processing (NLP) to identify and to label parts of the transaction that convey relevant information. In Chapter 28, Mann and collaborators present a survey of NLP methods applied to alternative data for financial applications. For example a credit card transaction might have a merchant name, a merchant address and payment method and an amount. While this information is in the transaction it must be labeled and interpreted using a complex, evolving set of methods.

External data and internal data can be combined to provide a more complete picture of a business and its interactions than that which was available only using internal data. Corporations are beginning to acquire external data to supplement their understanding of their business. For example it is important to know how close the nearest competitor is for each cohort of customers. In addition to knowing more about the competitive landscape between individual companies it is very important to understand the forces at play in the larger ecosystem - or the macro picture of the environment in which the company and its competitors are operating. Nowcasting advances and methods described in this book are part of a continuing trend to increase the use of data, technology and software for competitive insights.

This progress is not without challenges. Most of the alternative data sets are so biased, noisy and difficult to process that effort spent produces limited returns. In addition access to these data will face continuing privacy pressure. In Chapter 26 Jain describes these challenges, and others, and how they are overcome to nowcast macro economic data.

It is indisputable that in the coming years there will be more data, more processing power, and machine learning algorithms will continue to improve. The people and process changes that are required to achieve results with these data require changes that take time and may never occur in some established businesses. Other companies will gain an informational competitive advantage with these methods, and will in many cases win in the marketplace. It seems clear that gradually this type of information about businesses and economies will become commonplace and nowcasting will become the norm.

25

Data Preselection in Machine Learning Methods: An Application to Macroeconomic Nowcasting with Google Search Data

Laurent Ferrara[a] and Anna Simoni[b]

Abstract

In this chapter, we present some machine learning (ML) econometric methods that allow us to conveniently exploit Google search data for macroeconomic nowcasting. In particular, we focus on the issue of preselection of variables from a large set of Google search categories before entering them into ML approaches in order to nowcast macroeconomic variables. We consider two ML approaches allowing us to estimate linear regression models starting from large information sets: factor extraction and ridge regularisation. As an application we consider euro area GDP growth nowcasting using weekly Google search data. Empirical results tend to suggest that estimating a ridge regression associated with an *ex ante* preselection procedure seems a pertinent strategy in terms of nowcasting accuracy.

25.1 Introduction

Nowcasting macroeconomic activity has proved extremely helpful to policy-makers or investors in order to track, in real-time, fluctuations of broad macroeconomic aggregates such as GDP, inflation, consumption or employment. Indeed, it turns out that statistical institutes tend to publish official figures of macroeconomic aggregates with a delay ranging from one to two months, depending on the country. As a standard example, to be aware of the official aggregate economic activity of the first quarter of the year, as measured by GDP, we need sometimes to wait until mid-May. Yet, it turns out that economists receive plenty of information during this period of time and are thus tempted to come up with their own assessment of current economic activity.

In this respect, nowcasting tools have been widely developed in order to be able to inform decision-makers and economic agents about the current state of the economic activity, at any date of the year. Giannone et al. (2008) were the first to put forward time series econometric models, namely dynamic factor models,

[a] Skema Business School, University Côte d'Azur & QuantCube Technology
[b] CREST, CNRS, ENSAE, École Polytechnique, Institut Polytechnique de Paris
Published in *Machine Learning And Data Sciences For Financial Markets*, Agostino Capponi and Charles-Albert Lehalle © 2023 Cambridge University Press.

in order to compute high-frequency nowcasts of US GDP growth. Beyond the US economy, many nowcasting tools have been proposed to monitor macroeconomic aggregates for advanced economies (see, among others, Angelini et al., 2011 for the euro area, or Bragoli, 2017, for Japan), for emerging economies (see, among others, Bragoli et al., 2015 for Brazil or Modugno et al., 2016 for Turkey) or for the world economy as a whole (Ferrara and Marsilli, 2018).

As regards the US economy, both Atlanta Fed and New York Fed propose regular updates of current and future quarter-over-quarter GDP growth rates, expressed in annualized terms. If we focus on the recent Covid-19 crisis, surprisingly, as March 25, 2020, the nowcast of Atlanta Fed for 2020q1 was still very high, equal to 3.1%, while it was clear at that time that the Covid-19 crisis was about to have a major negative impact on the economic activity in March, and thus on first quarter of the year. Indeed, it turns out that official GDP growth was about −5% for the first quarter of 2020. Also, the GDP nowcast from New York Fed was a bit lower at 1.49% as March 20, but still high.

How is that? In fact, all those macroeconomic nowcasting tools only integrate official information, such as production, sales, surveys, that are released by official sources with a lag. Some price variables, such as stock prices that are reacting more rapidly to news, are also included in the nowcasting tools, but they do not strongly contribute to the indicator. So how can we improve macroeconomic nowcasting tools to reflect high-frequency evolutions of economic activity, especially in times of crises? A solution is to investigate *alternative data*, that are available on a high-frequency basis, by opposition to *official data*. Indeed, over the years, we saw a surge in the number of accessible alternative data (multi-lingual social media, satellite imagery, localization data, textual databases, . . .), that can be efficiently used for macroeconomic nowcasting (see a review in Buono et al., 2018). In this chapter, we focus on a specific source of information, namely Google data, but the methodologies we put forward can be adapted to many other types of big datasets containing alternative data.

Google trend data have been put forward by Choi and Varian (2009) and Choi and Varian (2012) and have been then widely and successfully used in the empirical literature to forecast and/or to nowcast various macroeconomic aggregates. For example, applications include household consumption (Choi and Varian, 2012), unemployment rate (D'Amuri and Marcucci, 2017), building permits (Coble and Pincheira, 2017), car sales (Nymand-Andersen and Pantelidis, 2018) or GDP growth by using Google search data (Ferrara and Simoni, 2020; Goetz and Knetsch, 2019). Forecasting prices with Google data has been also considered, for example by Seabold and Coppola (2015) who focus on a set of Latin American countries for which publication delays are quite large. Though there is still a debate as regards the marginal gain of Google data when correctly controlling for usual macroeconomic information (see Li, 2016), overall, it seems that Google data can be extremely useful when economists do not have access to information or when information is fragmented, as for example when dealing with emerging economies (see Carriere-Swallow and Labbe, 2013) or low-income developing countries (Narita and Yin, 2018). This chapter contributes to this debate and focuses on the usefulness of Google search data (GSD in the remaining of the chapter) to nowcast macroeconomic aggregates. In particular, we stress that, due to the large dimension of GSD and to their particular structure, specific econometric methodologies have to be used in order to make a meaningful forecasting.

Related to this background, a crucial issue that practitioners face when dealing with alternative sources of data relates to the optimal use of large datasets for nowcasting/forecasting purposes. From a theoretical point of view, we could think that the more the data, the better the forecasting accuracy, as the information set is larger. From a practical point of view, however, this is less clear and it turns out that preselecting data from the original large dataset is most of the time leading to fruitful results. For example, against the background of bridge equations augmented with dynamic factors, Barhoumi et al. (2010) empirically show that factors estimated on a small database lead to competitive results for nowcasting French GDP compared with the most disaggregated data. From a theoretical point of view, Boivin and Ng (2006) suggest that larger databases lead to poor forecast when idiosyncratic errors are cross-correlated or when the forecasting power comes from a factor that is dominant in a small database but is dominated in a larger dataset. An empirical way to circumvent this issue is to target more accurately the variable to be nowcast. For example, Bai and Ng (2008) show that forming targeted predictors enables to improve the accuracy of inflation forecasts while Schumacher (2010) shows that targeting German GDP within a dynamic factor model is a performing strategy.

In this chapter, our objective is to assess the gain in GDP growth nowcasting accuracy obtained by preselecting a small core dataset from the initial large database of Google information. Preselection means that only that subset of the Google series that is more correlated with the GDP growth series is used in the nowcasting model. The other available series are discarded. The rationale is that the GDP growth to be nowcast is highly predictable by a subset of the Google search variables. We carry out a preselection of the data by looking at their correlation with the target variable, namely GDP growth, through the thresholding approach suggested by Bai and Ng (2008) and Fan and Lv (2008). We consider two types of machine learning approaches that deal with large database: factor models and ridge regression. We first propose here a formal presentation of those approaches, which are applied after preselection. For estimation of Factor models we present the method based on principal component analysis (PCA) and its sparsified version (Sparse PCA). Then, for both the PCA and the ridge regression approaches we empirically compare the gain that we get by preselecting data when trying to asses quarterly GDP growth in the euro area using a large database of weekly Google search variables. As a nowcasting tool, we focus on a simple linear bridge regression that relates GDP growth with standard economic explanatory variables and Google search variables. This tool enables us to easily account for the frequency mismatch between GDP growth and explanatory variables.

The salient results are the following. First, we get that in general GSD are useful for nowcasting GDP growth. Indeed, at the beginning of the quarter, when there is no available economic information about the current quarter, nowcasts that only integrate Google data deliver reasonable root mean squared forecasting errors (RMSFEs), in the sense that they are similar in size (though obviously higher) to those obtained at the end of the quarter when economic information is fully accounted for. Second, preselecting the GSD leads to much better results in terms of nowcasting accuracy, especially in the case of the ridge regression. Results are more mixed as regards factor models for which we do not come up with systematic gains. Last, we assess the results throughout the business cycle by comparing nowcasts estimated during a relatively calm period (2014–16) and those during a marked slowdown in economic activity (2017–18). We observe that there are large gains in nowcasting during the low phase of the cycle when we preselect data, in particular when using a ridge regression. Overall, empirical results tend to show that using a ridge regression associated with an *ex ante* preselection procedure appears as a good strategy for macroeconomic nowcasting using Google search data. Obviously, those empirical results have to be extended to other large databases than GSD, other macroeconomic variables beyond GDP growth and also to other ML approaches.

25.2 Structure of Google search database

The Google data that we use in this study are weekly Google search data that have been available to us by the European Central Bank.[1] They are updated every Tuesday and they relate to queries done with Google search machines. The queries are assigned by Google to particular categories using natural language processing methods.

Those data differ from Google Trends data which are indices expressing the weekly average search share of a particular term in a particular region compared with the largest value over the requested period. More precisely, for a given period of time and region, Google Trends provide the ratio between the search share for a keyword of week w over the maximum search share for the same keyword over the considered period. Instead, GSD are indexes of weekly volume changes of Google queries. GSD are normalized at 1 at the first week of January 2004 which is the first week of availability of these data. Then, the following values indicate the deviation from the first value. However, there is no information about the search volume. Both GSD and Google Trends are grouped by category but the categories and subcategories differ for these two types of data. In addition GSD are grouped by country. For GSD, there is a total of 26 categories divided in 270 subcategories for each country.

Treating weekly data is particularly challenging as the number of entire weeks present in every quarter is not always the same, and a careful analysis has to be done when incorporating these data. Another difficulty is that original data are not seasonally adjusted. Thus, we take the growth rate over 52 weeks to eliminate

[1] The Google search sata are not publicly available but they can be obtained for research purposes by making a request to the European Central Bank, Directorate General for Statistics.

the seasonality within the data. This is a standard approach when dealing with weekly variables, that could possibly be improved by developing ad hoc statistical methods to seasonally adjust weekly time series.

25.3 The nowcasting approach

25.3.1 Linear Bridge equation

To nowcast macroeconomic quarterly aggregates (for instance quarterly GDP growth rate) a useful tool is the linear bridge equation model, which allows to construct nowcasts by using predictors available at different frequencies. Let t denote a given quarter of interest identified by its last month, for example the first quarter of 2020 is dated by $t = \text{March2020}$. In the case of weekly Google search data and of predictors coming from official sources, a bridge equation model to nowcast any series of interest Y_t for a specific quarter t is the following, for $t = 1, \ldots, T$:

$$Y_t = \beta_0 + \beta_1' x_{t,o} + \beta_2' x_{t,g} + \varepsilon_t, \qquad \mathbf{E}[\varepsilon_t | x_{t,o}, x_{t,g}] = 0, \qquad (25.1)$$

where $x_{t,o}$ is the N_o-vector containing *official* variables, and $x_{t,g}$ is the N_g-vector of *Google* variables and ε_t is an unobservable error term. As an example, predictors coming from official sources are *soft* variables, such as opinion surveys, and *hard* variables, such as industrial production or sales. Because variables in $x_{t,o}$ and in $x_{t,g}$ are sampled over different frequencies – for instance monthly and weekly, respectively – and are released with various reporting lags, the relevant information set for calculating the nowcast evolves within the quarter. In the remaining of this chapter we assume that the highest observed frequency in the available data is weekly, then a given quarter is made up of thirteen weeks. Thus, by denoting with $x_{t,j,i}^{(w)}$, $j \in \{o, g\}$, the ith series in the N_j-vector $x_{t,j}$ released at week $w \in \{1, \ldots, 13\}$ of quarter t, we define the relevant information set at week w of a quarter t as:

$$\Omega_t^{(w)} := \{x_{t,j,i}^{(w)}, j \in \{o, g\}, i = 1, \ldots, N_j, \text{ such that } x_{t,j,i}^{(w)} \text{ is released at week } \leq w\}.$$

We note that $\Omega_t^{(w)}$ is not a matrix but a list of variables that are in the information set. For simplicity, we keep in $\Omega_t^{(w)}$ only the observations relative to the current quarter t and do not consider past observations. The predictors are in the relevant information set only for the weeks corresponding to their release, consequently the data set is unbalanced.

To explicitly account for the different frequencies of the variables, one can replace model (25.1) by a model for each week w, denoted by $M1, \ldots, M13$ and defined as:

$$\widehat{Y}_{t|w} := \mathbf{E}[Y_t | \Omega_t^{(w)}], \qquad t = 1, \ldots, T \quad \text{and} \quad w = 1, \ldots, 13$$
$$\text{and} \quad \mathbf{E}[Y_t | \Omega_t^{(w)}] = \beta_{0,w} + \beta_{1,w}' x_{t,o}^{(w)} + \beta_{2,w}' x_{t,g}^{(w)}, \qquad (25.2)$$

where $\beta_{1,w,i} = 0$ if $x_{t,o,i}^{(w)} \notin \Omega_t^{(w)}$ and $\beta_{2,w,i} = 0$ if $x_{t,g,i}^{(w)} \notin \Omega_t^{(w)}$. For instance, as the first observation of industrial production relative to the current quarter t

is only released in week 9, then we set the corresponding $\beta_{1,w,i} = 0$ for every $w = 1, \ldots, 8$. In our analysis below, Y_t denotes the Euro area GDP growth rate (EA–GDP in the following). Thus, the bridge equations (25.2) exploit weekly information to obtain more accurate nowcasts of EA–GDP. The idea of having thirteen models is that a researcher that wants to nowcast the current-quarter values of Y_t will use the model corresponding to the current week of the quarter. For instance, to nowcast the current-quarter value of Y_t at the end of week 2, model $M2$ will be used.

25.3.2 Preselection of Google search variables

In this chapter we assume that the dimension in $x_{t,g}$ is high while the dimension of $x_{t,o}$ is low. The generalisation to the case where both vectors have high dimension is straightforward. There is a trade-off in using all the variables in $x_{t,g}$: on the one hand including all the available GSD increases the information that can be exploited to predict Y_t, but on the other hand the use of a large number of predictors increases the noise in the estimation process. Therefore, one is better off throwing away some of the Google search variables if their correlation with the variable to be forecast is small. This amounts to implementing a given selection procedure.

Let X_g be the $T \times N_g$ matrix of GSD: $X_g = (x_{1,g}, \ldots, x_{T,g})'$ and Y be the T-vector of outcome variables. An intuitive procedure for variable selection consists in targeting, that is looking at the correlation between each predictor and Y_t. Thus, if we assume that the columns of X_g have been centered to have zero mean, we compute:

$$\omega_j := \frac{x'_{g,j} Y}{\|x_{g,j}\|}, \tag{25.3}$$

where $x_{g,j}$ is the jth column of X_g. Then, we select the variables in X_g that have the absolute value of ω_j larger than a given threshold λ; that is:

$$\widehat{M}_g := \widehat{M}_g(\lambda) := \left\{ 1 \leq j \leq N_g : |\omega_j| \text{ is among the first } [\lambda T] \text{ largest of all} \right\},$$

where $[\lambda T]$ denotes the integer part of λT. This selection method has been proposed in different settings by Bair et al. (2006), Fan and Lv (2008) and Bai and Ng (2008).

While the ease of this selection procedure makes it very attractive, according to Bai and Ng (2008), given the dependent nature of our data and because other variables other than GSD could be available, the selection cannot be based just on the bivariate relation between Y_t and $x_{t,g,j}$. Indeed, a simple forecast based on lagged values of Y_t, and possibly on official variables $x_{t,o}$, is always available as an alternative nowcasting procedure. Therefore, we need to control for these predictors when considering the correlation between Y_t and each Google search variable. Similarly, Barut et al. (2016) noticed that if we know that some variables are responsible for the outcomes Y_t, this knowledge should be taken into account when selecting the remaining variables. Their procedure is called *conditional sure independent screening* while the procedure discussed in Bai and Ng (2008) is called *hard thresholding*. The two procedures work similarly and, applied to our setting, they consist in evaluating the marginal predictive power of each Google search variable $x_{t,g,j}$ after controlling for $x_{t,o}$. That is, following Bai and Ng (2008), the algorithm is the following:

1. for each $j = 1, \ldots, N_g$, regress Y_t on a constant, $x_{t,o}$ and $x_{t,g,j}$, and compute the corresponding t-statistics t_j associated with $x_{t,g,j}$.
2. select the variables in X_g that have the absolute value $|t_j|$ larger than a given threshold λ:

$$\widehat{M}_g := \widehat{M}_g(\lambda) := \left\{ 1 \leq j \leq N_g : |t_j| \text{ is among the first } [\lambda T] \text{ largest of all} \right\}.$$

We denote by $X_{g,\lambda}$ the $(T \times \widehat{s}_g)$ matrix consisting of the columns of X_g corresponding to the indices in $\widehat{M}_g$, where $\widehat{s}_g := |\widehat{M}_g|$. The following notation is useful for the next part of the chapter:

$$X_{g,\lambda} = \begin{pmatrix} x'_{1,g,\lambda} \\ \vdots \\ x'_{T,g,\lambda} \end{pmatrix} = \begin{pmatrix} x_{1,g,\lambda,1} & \cdots & x_{1,g,\lambda,\widehat{s}_g} \\ \vdots & \ddots & \vdots \\ x_{T,g,\lambda,1} & \cdots & x_{T,g,\lambda,\widehat{s}_g} \end{pmatrix},$$

where $x_{t,g,\lambda}$ is a $\widehat{s}_g$-vector.

Alternative procedures for preselection have been proposed in the literature. For instance, Bai and Ng (2008) have proposed using *soft thresholding* methods that perform subset selection and shrinkage simultaneously, like the Lasso, the elastic net and least angle regressions. An advantage of these methods is that they take into account the information in all the predictors simultaneously when they do model selection and so avoid to select predictor that are too similar. Our experience with GSD points for a preference for the *hard thresholding* procedure such as the *conditional sure independent screening*. However, it could be interesting to carry out a large empirical comparison of the various types of thresholding.

25.4 Methods based on the construction of an index: PCA and sparse PCA (SPCA)

When dealing with a large number of data, a first idea is to try to reduce the dimension of the dataset by summarizing the information in a few number of latent factors. In this respect, by using the diffusion index forecasting framework of (Stock and Watson, 2002) one can write the Bridge equation for week w as

$$Y_t^{(w)} = \beta_{0,w} + \beta'_{1,w} x_{t,o}^{(w)} + \gamma'_w F_t^{(w)} + \epsilon_t^{(w)}, \tag{25.4}$$

where the error term $\epsilon_t^{(w)}$ is orthogonal to $(x_{t,o}^{(w)}, x_{t,g}^{(w)})$ and the latent factors $F_t^{(w)}$ (for week w) are a k-dimensional stationary process with covariance matrix $\mathbf{E}[F_t^{(w)} F_t^{(w)'}] = I_k$, where k is low. The latent factors $F_t^{(w)}$ are observed through $x_{t,g}^{(w)}$ since the latter are assumed to be related to the latent factors as follows:

$$x_{t,g}^{(w)} = \Lambda F_t^{(w)} + u_t^{(w)}, \tag{25.5}$$

where $u_t^{(w)}$ is an N_g-vector of stationary processes that are orthogonal to $F_t^{(w)}$ and such that $\mathbf{E}[u_t^{(w)} u_t^{(w)\prime}] = \Psi$ of full rank for all N_g. The matrix $\Lambda := [\lambda_1', \ldots, \lambda_{N_g}']'$ of factor loadings is an $(N_g \times k)$ non-random matrix of full rank k for all N_g, and λ_j is a k-vector for every $j = 1, \ldots, N_g$. The diffusion index nowcast of $Y_t^{(w)}$ is given by $\widehat{\beta}_{0,w} + \widehat{\beta}_{1,w}' x_{t,o}^{(w)} + \widehat{\beta}_{2,w}' \widehat{F}_t^{(w)}$, where $\widehat{\beta}_{j,w}$ are the OLS estimates, for $j = 0, 1, 2$, and $\widehat{F}_t^{(w)}$ are the principal component estimates of $F_t^{(w)}$. As $\widehat{F}_t^{(w)}$ can be written as a linear combination of the elements of $x_{t,g}^{(w)}$, then the Diffusion Index Bridge equation (25.4) for week w becomes

$$Y_t^{(w)} = \beta_{0,w} + \beta_{1,w}' x_{t,o}^{(w)} + \overline{\beta}_{2,w}' x_{t,g}^{(w)} + \epsilon_t^{(w)}, \tag{25.6}$$

where $\overline{\beta}_{2,w}$ is a restricted version of $\beta_{2,w}$ in (25.2), see Bai and Ng (2008). Therefore, in our setting, we can interpret F_t as a *Google search* index.

In the following, we sometimes omit the superindex (w) to lighten the notation. Let $\widehat{M}_g$ be the set of selected indices, $\widehat{s}_g := |\widehat{M}_g|$, and $X_{g,\lambda}$ be the $(T \times \widehat{s}_g)$ matrix consisting of those columns of X_g corresponding to the indices in $\widehat{M}_g$. As in the previous section the columns of X_g are assumed to have been centered to have mean 0. The singular value decomposition (SVD) of $X_{g,\lambda}$ is

$$X_{g,\lambda} = U_{g,\lambda} D_{g,\lambda} V_{g,\lambda}', \tag{25.7}$$

where $U_{g,\lambda}$ is a $(T \times m)$ matrix whose columns are the eigenvectors of $X_{g,\lambda} X_{g,\lambda}'$ and $m := \operatorname{rank}(X_{g,\lambda})$, $V_{g,\lambda}$ is a $(\widehat{s}_g \times m)$ matrix whose columns are the eigenvectors of $X_{g,\lambda}' X_{g,\lambda}$, and $D_{g,\lambda}$ is an m-diagonal matrix containing the singular values $d_{g,j}$ assumed to be ordered: $d_{g,1} \geq d_{g,2} \geq \cdots \geq d_{g,m}$. The jth column of $U_{g,\lambda}$ is called the jth supervised principal component of X_g. Denote the jth column of $U_{g,\lambda}$ by $u_{g,\lambda,j}$.

Minimization of the criterion $\sum_{i=1}^{N} \sum_{t=1}^{T} (x_{t,g,\lambda,j} - \lambda_i' F_t)^2 / (\widehat{s}_g T)$ under the restriction that $F'F/T = I_k$, where $F := [F_1, \ldots, F_T]'$ is a $(T \times k)$ matrix, gives an estimate of the jth column of F as $\widehat{F}_j = \sqrt{T} u_{g,\lambda,j}$. Moreover, Λ is estimated by $\widehat{\Lambda} = X_{g,\lambda}' \widehat{F}/T = X_{g,\lambda}[u_{g,\lambda,1}, \ldots, u_{g,\lambda,k}]/\sqrt{T}$. Therefore, one can use the supervised PCA to estimate the factors $F_t^{(w)}$ in (25.4) and then fit model (25.4) as

$$\widehat{Y}_t^{(w)} = \widehat{\beta}_{0,w} + \widehat{\beta}_{1,w}' x_{t,o}^{(w)} + \sqrt{T}[u_{t,g,\lambda,1}^{(w)}, \ldots, u_{t,g,\lambda,k}^{(w)}] \widehat{\gamma}_w. \tag{25.8}$$

Because of (25.7) we can write

$$U_{g,\lambda} = X_{g,\lambda} V_{g,\lambda} D_{g,\lambda}^{-1}$$
$$= X_{g,\lambda} W_{g,\lambda}$$

and so

$$[u_{t,g,\lambda,1}^{(w)}, \ldots, u_{t,g,\lambda,k}^{(w)}] = X_{g,\lambda} W_{g,\lambda}[e_1, \ldots, e_k],$$

where e_j is an m-vector of zeros except for the entry in the jth position that is equal to one. Therefore, the fitted model (25.4) can be further written as

$$\widehat{Y}_t^{(w)} = \widehat{\beta}_{0,w} + \widehat{\beta}_{1,w}' x_{t,o}^{(w)} + x_{t,g,\lambda}^{(w)} W_{g,\lambda}[e_1, \ldots, e_k] \widehat{\gamma}_w \sqrt{T}$$
$$= \widehat{\beta}_{0,w} + \widehat{\beta}_{1,w}' x_{t,o}^{(w)} + x_{t,g,\lambda}^{(w)} \overline{\beta}_{2,w}, \tag{25.9}$$

that is, the fitted diffusion index Bridge equation (25.4) can be seen as the estimated model that uses all the preselected Google variables and where the estimation is restricted to satisfying $\bar{\beta}_{2,w} = W_{g,\lambda}[e_1,\ldots,e_k]\hat{\gamma}_w \sqrt{T}$. This version is useful for making forecast since, if we have new predictors $\tilde{x}_{t,o}^{(w)}$ and $\tilde{x}_{t,g,\lambda}^{(w)}$ then,

$$\tilde{Y}_t^{(w)} = \hat{\beta}_{0,w} + \hat{\beta}_{1,w}'\tilde{x}_{t,o}^{(w)} + \tilde{x}_{t,g,\lambda}^{(w)}\bar{\beta}_{2,w}.$$

Sparse PCA.

Instead of using the SVD decomposition of $X_{g,\lambda}$ to construct the supervised principal components of X_g to get an estimate of the factors, one can use a sparsified version of it, called *sparse PCA* (SPCA), as proposed by Zou et al. (2006). The idea of SPCA is to enforce sparsity in the factor loadings Λ in (25.5) so that elements of Λ that are close to zero with standard PCA are set exactly equal to zero with SPCA. This favours the interpretability of the results which is particulary useful when the dimension $\hat{s}_g$ of the preselected Google search categories is large.

SPCA modifies the minimization problem of PCA by adding two penalty terms: one that deals with possible multicollinearity of the matrix $X_{g,\lambda}^{(w)'}X_{g,\lambda}^{(w)}$ (ℓ_2-type penalty) and one that imposes sparsity (ℓ_1-type penalty). More precisely, before adding the two penalty terms, SPCA requires us to reformulate the PCA problem by imposing the structure of the PC estimator for F and uses the properties of the trace operator to write the PCA criterion as

$$\frac{1}{\hat{s}_gT}\|X_{g,\lambda} - F\Lambda'\|_F^2 = \frac{1}{\hat{s}_gT}\sum_{j=1}^{\hat{s}_g}\sum_{t=1}^{T}(x_{t,g,\lambda,j} - \lambda_j'F_t)^2$$

$$= \frac{1}{\hat{s}_gT}\sum_{j=1}^{\hat{s}_g}\sum_{t=1}^{T}(x_{t,g,\lambda,j} - \lambda_j'\sqrt{T}V_{g,\lambda,(1:k)}'x_{t,g,\lambda})^2,$$

where $V_{g,\lambda,(1:k)}$ denotes the $(\hat{s}_g \times k)$ submatrix made of the first k columns of $V_{g,\lambda}$ and $\|\cdot\|_F$ denotes the Frobenius norm, see Zou et al. (2006). Therefore, by using the general notation A and B to denote two $(\hat{s}_g \times k)$ matrices, we replace the minimization problem of the PCA by the following minimization problem:

$$\min_{A,B\in\mathbb{R}^{\hat{s}_g\times k}} \sum_{t=1}^{T}\|x_{t,g,\lambda} - AB'x_{t,g,\lambda}\|^2 + \mu_1\sum_{j=1}^{k}\|\beta_j\|^2 + \sum_{j=1}^{k}\mu_{2,j}\|\beta_j\|_1$$

$$\text{such that} \quad A'A = I_k. \tag{25.10}$$

where $\alpha_j, \beta_j \in \mathbb{R}^{\hat{s}_g}$, $A = [\alpha_1,\ldots,\alpha_k]$, $B = [\beta_1,\ldots,\beta_k]$, $\|\beta_j\|^2 = \sum_{i=1}^{N_g}\beta_{ji}^2$ and $\|\beta_j\|_1 = \sum_{i=1}^{N_g}|\beta_{ji}|$. This is the optimization problem considered in Zou et al. (2006).

This minimization can be implemented in two steps where, in the first step, the minimization is performed with respect to B, for a given A, and in the second step the minimization is performed with respect to A, for a given B. The algorithm is the following and is adapted from Zou et al. (2006).

Algorithm of Zou et al. (2006, page 272)

1. Let the initial value of A be set equal to $V_{g,\lambda,(1:k)}$, where $V_{g,\lambda,(1:k)}$ is a $(\widehat{s}_g \times k)$ matrix whose columns are the first k eigenvectors of $X'_{g,\lambda}X_{g,\lambda}$.

2. Given this value of A, solve the following Elastic Net problem for $j = 1, \ldots k$:

$$\beta_j = \arg\min_{\beta \in \mathbb{R}^{\widehat{s}_g}} (\alpha_j - \beta)' X'_{g,\lambda} X_{g,\lambda} (\alpha_j - \beta) + \mu_1 \|\beta\|^2 + \mu_{2,j} \|\beta\|.$$

3. For a fixed $B = [\beta_1, \ldots, \beta_k]$, compute the SVD of $X'_{g,\lambda} X_{g,\lambda} B = U_B D_B V'_B$, then set $A = U_B V'_B$.

4. Repeat steps 2–3 until convergence.

5. Normalization: $V_{g,\lambda,j} = \beta_j / \|\beta_j\|, j = 1, \ldots, k$.

25.5 Methods based on regularisation: ridge

The bridge equation to nowcast EA–GDP is the one given in (25.2) with $x_{t,g}^{(w)}$ the N_g-vector of variables coming from GSD. Instead of using PCA or SPCA to restrict the coefficients in (25.9) and to construct an index of GSD, we can apply a ridge regularisation procedure to the Bridge equation model with preselected regressors:

$$Y_t^{(w)} = \beta_{0,w} + \beta'_{1,w} x_{t,o}^{(w)} + \beta'_{2,w} x_{t,g,\lambda}^{(w)} + \epsilon_t. \tag{25.11}$$

A regularisation procedure is useful when the number $\widehat{s}_g$ of preselected Google search categories is large compared to the time dimension T. Ridge regularisation is also known as Tikhonov regularisation in inverse problem theory (Tikhonov, 1963).

Let $\beta := (\beta_{0,w}, \beta'_{1,w}, \beta'_{2,w})'$ and write $X_{t,\lambda} := (1, x_{t,o}^{(w)'}, x_{t,g,\lambda}^{(w)'})'$, where $x_{t,g,\lambda}^{(w)}$ is the vector containing only the preselected Google variables. Ridge regularisation procedure estimates β in equation (25.11) by minimizing a penalized residuals sum of squares where the penalty is given by the Euclidean squared norm $\| \cdot \|_2^2$. We apply the ridge regularisation to the selected predictors $X_{t,\lambda}$. By using model (25.2) for each week $w \in \{1, \ldots, 13, \}$, Ferrara and Simoni (2020) have defined the *ridge after model selection* estimator as: $\widehat{\beta}^{(w)} = \widehat{\beta}^{(w)}(\alpha)$ where

$$\widehat{\beta}^{(w)}(\alpha) := \arg\min_{\beta; \beta_{2,w,i}=0, \, i\in\overline{M}_g^c} \left\{ \frac{1}{T} \sum_{t=1}^{T} \left(Y_t - \beta_0 - \beta'_{1,w} x_{t,o}^{(w)} - \beta'_{2,w} x_{t,g}^{(w)} \right)^2 + \alpha \|\beta\|_2^2 \right\},$$

and $\alpha > 0$ is a regularization parameter that tunes the amount of shrinkage. Without loss of generality, we can assume that the selected elements of $x_{t,g}^{(w)}$ corresponding to the indices in $\widehat{M}_g$ are the first elements of the vector. Then, we can write $\widehat{\beta}^{(w)}$ as $\widehat{\beta}^{(w)} = (\widehat{\beta}_\lambda^{(w)'}, \mathbf{0}')'$ where

$$\widehat{\beta}_\lambda^{(w)} = \left(\frac{1}{T} \sum_{t=1}^{T} X_{t,\widehat{M}} X'_{t,\widehat{M}} + \alpha I \right)^{-1} \frac{1}{T} \sum_{t=1}^{T} X'_{t,\widehat{M}} Y_t,$$

here $\mathbf{0}$ is a column vector of zeros of compatible dimension, and I is the identity matrix. The tuning parameter α can be selected by using for instance Generalized cross-validation (GCV) technique (see Li, 1986, 1987) whose idea is to choose a value for α for which the MSFE is as small as possible.

25.6 Nowcasting Euro area GDP growth: Empirical results

In this section we analyse the performance of Google search data to nowcast the euro area quarterly GDP growth rate (EA–GDP in the following) which is the variable Y_t in model (25.1)–(25.2). GDP data are stemming from Eurostat. The official macroeconomic series that we use as regressors $x_{t,o}$ for our empirical analysis are (i) the growth rate of the index of industrial production for the euro area as a whole provided by Eurostat, which is a global measure of hard data, denoted by IP_t, and (ii) a composite index of opinion surveys from various sectors computed by the European Commission (the so-called *euro area Sentiment Index*) denoted by S_t. Thus, $x_{t,o} = (IP_t, S_t)'$ and both these series are monthly.

There are two main difficulties when dealing with this type of data. First, there is a frequency mismatch with the EA–GDP which is available quarterly, while IP_t and S_t are available monthly and GSD are available weekly. The second difficulty is due to the fact that official series and Google search data are released with various reporting lags, leading thus to an unbalanced information set at each point in time within the quarter. In the literature, this issue is refered to as *ragged-edge database*: see Angelini et al. (2011). As discussed in §25.3, we deal with these issues by considering a different model for every week w of the quarter, i.e. the thirteen models given by equation (25.2).

For the macroeconomic series we mimic the exact release dates as published by Eurostat, see the scheme of the release timeline in Figure 25.1. For S_t and IP_t, for instance, this means that the first survey of the quarter, referring to the first month, typically arrives in week $w = 5$. Then, the second survey of the quarter, related to the second month, is available in week $w = 9$. The IP_t for the first month of the quarter is available about 45 days after the end of the reference month, that is generally in week $w = 11$. Finally, the final survey, related to the third month of the quarter, is available in week $w = 13$.

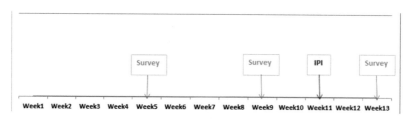

Figure 25.1 Timeline of data release in a pseudo real-time exercise within the quarter. Series considered are: Survey, S_t, and Industrial Production index, IP_t.

A stylized fact of the GSD is that they are not all extremely correlated with the EA–GDP, at least during periods of economic stability. Figure 25.2 shows that the number of Google search variable with a correlation with EA–GDP larger than

0.20 is relatively small. For this reason we can expect GSD to be informative for nowcasting only in periods for which economists do not have access to official information, as it is the case at the beginning of the quarter. As shown in Ferrara and Simoni (2020), GSD also performs very well for nowcasting in periods of recession. Indeed, they show that during the recession phase of 2008q1–2009q2, a broader information set is needed to adequately assess the state of the economy, while a core selected dataset is likely to be sufficient during expansions.

As our objective in this empirical application is to assess the gain in forecasting from preselecting GSD, we have implemented both the PCA and ridge procedures previously described, with and without any preselection step. The training sample starts on 2005q1 and ends two quarters before the start of the out-of-sample nowcasting period. We consider two different out-of-sample nowcasting periods: (i) a period of cyclical stability (2014q1–2016q1) and (ii) a period that exhibits a sharp downturn in EA–GDP (2017q1–2018q4). As underlined in Ferrara and Simoni (2020), the gain from preselection on nowcasting accuracy is likely to differ depending on business cycles. We carry out a recursive scheme method, that is, the parameters are re-estimated at each new nowcasting quarter using all the past information available until the penultimate quarter before the out-of-sample nowcasting one.

We also compare nowcasts obtained with and without GSD in a pseudo real-time exercise, that is, by using historical data but by accounting for their ragged-edge nature. The aim is to assess: (i) whether GSD are informative when there is

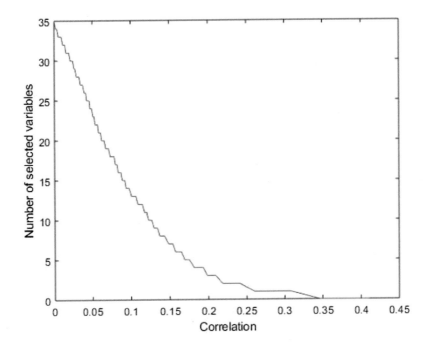

Figure 25.2 Plot of the number of Google search variables versus the correlation with current EA–GDP (computed for the first training sample).

no official data available for the forecaster, that is, $x_{t,o}$ is empty, and (ii) whether GSD remain informative when official data become available. Then, we look at the effects of preselecting GSD before estimating the model. The criterion we use for comparison is the Root Mean Squared Forecasting Error (RMSFE).

Tables 25.1 and 25.2 refer to RMSFEs obtained with PCA with three factors for the two nowcasting periods 2014q1–2016q1 and 2017q1–2018q4, respectively. The preselection method that we consider here is the targeted selection of Bai and Ng (2008) based on the absolute value of the t-statistics associated to each Google search variable. We select the value of the threshold λ that minimizes the MSFE of the nowcast of the EA–GDP of the last three quarters of the training sample.

Overall, we do not find strong evidence of positive effects linked to data preselection. Results suggest that when we use the PCA approach for the tranquil period 2014q1–2016q1 we do better when we do not preselect Google search data. On the other hand, it turns out that preselection appears to be useful for nowcasting during the period 2017q1–2018q4 when GDP is dropping, except during weeks 7–10. This results is interesting and suggests that when coupled with PCA preselection is suitable depending on the business cycle.

Other interesting facts are the following. First, the RMSFEs at the beginning of the quarter, when only GSD are used, are quite reasonable, meaning that they can be efficiently integrated into real-time macroeconomic analyses. Second, a comparison between the first and third rows of the tables shows that when official data come up, the benefit of using GSD disappears. Indeed, RMSFEs stemming from models without any Google data are similar for week $w = 5$ and are even lower later in the quarter.

Tables 25.3 and 25.4 refer to RMSFEs obtained from the ridge models, with and without preselection, for the two nowcasting periods 2014q1–2016q1 and 2017q1–2018q4. For the ridge regularisation approach, the threshold and the regularisation parameter have been chosen in order to minimise the out-of-sample RMSFEs. The minimisation procedure is not exhaustive in the sense that we have compared the RMSFE values obtained by retaining the first 1, 5, 10 and 100 Google search variables with largest absolute t-statistics. Concerning the α we have considered ten values in the grid [0, 1]. This way of choosing regularisation parameters is of course not feasible in real situations but our aim here is only to show that if the regularisation parameters are adequately chosen, Google search variables can have a high informational contents to nowcast EA–GDP.

Overall, results stemming from ridge regularisation show evidence of a large gain in nowcasting accuracy when a preselection procedure is implemented. This gain is especially strong during the 2017q1–2018q4 period for which we have a shift in EA–GDP growth. We also note that the ridge approach with preselection systematically outperforms the approach that does not integrate any GSD, meaning that even when controlling by economic variables, we get a gain by introducing Google variables at any point within the quarter.

Table 25.1 RMSFE corresponding to nowcasting based on PCA applied to the 13 models (25.2) with the variables: S_t, IP_t and Google data. *t-stat* refers to preselection based on the t-statistics with threshold chosen by looking at the forecast of the last three in-sample periods. *No presel* refers to PCA without preselection. "No Google" refers to models that do not include Google search data.

	PCA – Nowcasting period: 2014q1–2016q1 (with and without preselection)												
	M1	M2	M3	M4	M5	M6	M7	M8	M9	M10	M11	M12	M13
t-stat	0.3239	0.4657	0.2485	0.2516	0.1788	0.3129	0.3823	0.3722	0.3465	0.4406	0.2466	0.2831	0.3141
No presel.	0.2299	0.2145	0.2168	0.2334	0.1831	0.1883	0.1707	0.2139	0.2308	0.2355	0.3767	0.2578	0.2443
No Google					0.1807				0.1897		0.1928		0.2017

Table 25.2 RMSFE corresponding to nowcasting based on PCA applied to the 13 models (25.2) with the variables: S_t, IP_t and Google data. *t-stat* refers to preselection based on the t-statistics with threshold chosen by looking at the forecast of the last three in-sample periods. *No presel* refers to PCA without preselection. "No Google" refers to models that do not include Google search data.

	PCA – Nowcasting period: 2017q1–2018q4 (with and without preselection)												
	M1	M2	M3	M4	M5	M6	M7	M8	M9	M10	M11	M12	M13
t-stat	0.3599	0.2911	0.3161	0.3386	0.3534	0.2896	0.4726	0.7121	0.5362	0.5056	0.2748	0.2966	0.329
No presel.	0.3813	0.3376	0.3455	0.3363	0.4294	0.4368	0.4381	0.4346	0.457	0.4485	0.2983	0.3041	0.3219
No Google					0.4340				0.4841		0.2871		0.3177

Table 25.3 RMSFE corresponding to nowcasting based on ridge regularisation applied to the 13 models (25.2) with the variables: S_t, IP_t and Google data. *t-stat* refers to preselection based on the *t*-statistics with threshold chosen by looking at the forecast of the last three in-sample periods. *No presel* refers to ridge regularisation without preselection. "No Google" refers to models that do not include Google search data.

	M1	M2	M3	M4	M5	M6	M7	M8	M9	M10	M11	M12	M13
	Ridge – Nowcasting period: 2014q1–2016q1 (with and without preselection)												
t-stat	0.1562	0.1668	0.1813	0.1954	0.1210	0.1219	0.1232	0.0923	0.1012	0.1120	0.2550	0.1062	0.1933
No presel.	0.3613	0.5662	0.3646	0.4419	0.7513	0.2801	0.5685	0.4628	0.4845	0.3553	0.3688	0.3985	1.1055
No Google					0.1807				0.1897		0.1928		0.2017

Table 25.4 RMSFE corresponding to nowcasting based on ridge regularisation applied to the 13 models (25.2) with the variables: S_t, IP_t and Google data. *t-stat* refers to preselection based on the *t*-statistics with threshold chosen by looking at the forecast of the last three in-sample periods. *No presel* refers to ridge regularisation without preselection. "No Google" refers to models that do not include Google search data.

	M1	M2	M3	M4	M5	M6	M7	M8	M9	M10	M11	M12	M13
	Ridge – Nowcasting period: 2017q1–2018q4 (with and without preselection)												
t-stat	0.2140	0.1546	0.2991	0.2635	0.0896	0.1100	0.1778	0.3005	0.1310	0.1406	0.0393	0.0449	0.0826
No presel.	0.9650	1.0971	0.6263	0.8409	0.5469	0.8446	0.5709	0.7579	0.5636	0.5214	0.5814	0.6636	0.4725
No Google					0.4340				0.4841		0.2871		0.3177

25.7 Conclusions

In this chapter, we show how one can use ML-based econometric techniques to incorporate the information contained in Google search data in order to now-cast macroeconomic aggregates. The Google search data that we use differ from standard Google Trends as they are indexes of weekly volume changes of some specific Google queries already grouped by categories. We discuss the issue of variables preselection among a large set of Google search categories before entering the data into Machine Learning approaches in order to nowcast macroeconomic variables. We also discuss the issues of frequency mismatch and of reporting lags that differ among the various variables included into the model. All these concerns are relevant for practical implementation of macroeconomic nowcasting with Google search data.

We describe the linear Bridge equation model that is generally considered as a convenient framework for macroeconomic nowcasting in the literature. We focus on two ML approaches that can be used to account for large informational sets: an approach based on factor extraction and an approach based on ridge regularisation.

As an application we consider euro area GDP growth nowcasting using weekly Google search data. Empirical results of our analysis for the nowcasting periods 2014q1–2016q1 and 2017q1–2018q4 tend to suggest that estimating a ridge regression associated with an *ex ante* preselection procedure appears as a pertinent strategy for macroeconomic nowcasting using Google search data. Obviously, those empirical results have to be extended to other periods, other macroeconomic variables beyond EA–GDP, large databases other than Google search data, and also to other ML approaches. Other econometric methods could be also considered to handle mixed-frequency data, such as the MIDAS approach put forward by Ghysels et al. (2007).

References

Angelini, E., Camba-Mendez, G., Giannone, D., Reichlin, L., and Ruenstler, G. 2011. Short-term forecasts of euro area GDP growth. *Economic Journal*, **14**, C25–C44.

Bai, J., and Ng, S. 2008. Forecasting economic time series using targeted predictors. *Journal of Econometrics*, **146**(2), 304 – 317.

Bair, E., Hastie, T., Paul, D., and Tibshirani, R.. 2006. Prediction by supervised principal components. *Journal of the American Statistical Association*, **101**(473), 119–137.

Barhoumi, K., Darne, O., and Ferrara, L.. 2010. Are disaggregate data useful for forecasting French GDP with dynamic factor models ? *Journal of Forecasting*, **29**, 132–144.

Barut, E., Fan, J., and Verhasselt, A.. 2016. Conditional sure independence screening. *Journal of the American Statistical Association*, **111**(515), 1266–1277.

Boivin, J., and Ng, S. 2006. Are more data always better for factor analysis? *Journal of Econometrics*, **132**, 169–194.

Bragoli, D. 2017. Nowcasting the Japanese economy. *International Journal of Forecasting*, **33**(2), 390–402.

Bragoli, D., Metelli, L., and Modugno, M. 2015. The importance of updating: Evidence from a Brazilian nowcasting model. *OECD Journal: Journal of Business Cycle Measurement and Analysis*, **2015**(1), 5–22.

Buono, D., Kapetanios, G., Marcellino, M., Mazzi, G.-L., and Papailias, F. 2018. Big data econometrics: Nowcasting and early estimates. Tech. rept. 82. Working Paper Series, Universita Bocconi. Available at SSRN 3206554.

Carriere-Swallow, Y., and Labbe, F. 2013. Nowcasting with Google trends in an emerging market. _Journal of Forecasting_, **32**(4), 289–298.

Choi, H., and Varian, H. 2009. Predicting initial claims for unemployment insurance using Google trends. Google Technical Report.

Choi, H., and Varian, H. 2012. Predicting the present with Google trends. Google Technical Report.

Coble, D., and Pincheira, P. 2017. Nowcasting building permits with Google trends. MPRA Paper 76514. University Library of Munich, Germany.

D'Amuri, F., and Marcucci, J. 2017. The predictive power of Google searches in forecasting unemployment. _International Journal of Forecasting_, **33**, 801–816.

Fan, J., and Lv, J. 2008. Sure independence screening for ultrahigh dimensional feature space. _Journal of the Royal Statistical Society B_, **70**, 849–911.

Ferrara, L., and Marsilli, C. 2018. Nowcasting global economic growth: A factor-augmented mixed-frequency approach. _The World Economy_, **42**(3), 846–875.

Ferrara, L., and Simoni, A. 2020. When are Google data useful to nowcast GDP? An approach via pre-selection and shrinkage. ArXiv:2007.00273.

Ghysels, Eric, Sinko, Arthur, and Valkanov, Rossen. 2007. MIDAS regressions: Further results and new directions. _Econometric Reviews_, **26**(1), 53–90.

Giannone, Domenico, Reichlin, Lucrezia, and Small, David. 2008. Nowcasting: the real-time informational content of macroeconomic data. _Journal of Monetary Economics_, **55**(4), 665–676.

Goetz, T., and Knetsch, T. 2019. Google data in bridge equation models for German GDP. _International Journal of Forecasting_, **35**(1), 45–66.

Li, K.-C. 1986. Asymptotic optimality of C_L and generalized cross-validation in ridge regression with application to spline smoothing. _Annals of Statistics_, **14**(3), 1101–1112.

Li, K.-C. 1987. Asymptotic optimality for C_p, C_L, cross-validation and generalized cross-validation: discrete index set. _Annals of Statistics_, **15**(3), 958–975.

Li, X. 2016. Nowcasting with big data: Is Google useful in the presence of other information? Available at `https://www.dropbox.com/s/phrqn91214hiw1v/20181120LiXinyuanJMP.pdf?dl=0`

Modugno, M., Soybilgen, B., and Yazgan, E. 2016. Nowcasting Turkish GDP and news decomposition. _International Journal of Forecasting_, **32**(4), 1369–1384.

Narita, F., and Yin, R. 2018. In search for information: Use of Google Trends' data to narrow information gaps for low-income developing countries. Tech. Rept. WP/18/286. IMF Working Paper.

Nymand-Andersen, P., and Pantelidis, E. 2018. Google econometrics: Nowcasting euro area car sales and big data quality requirements. Tech. Rept. European Central Bank.

Schumacher, C. 2010. Factor forecasting using international targeted predictors: The case of German GDP. _Economics Letters_, **107**(2), 95–98.

Seabold, S., and Coppola, A. 2015 (Aug.). Nowcasting prices using Google trends: an application to Central America. Policy Research Working Paper Series 7398. The World Bank.

Stock, J., and Watson, M. 2002. Forecasting using principal components from a large number of predictors. _Journal of the American Statistical Association_, **97**, 1167–1179.

Tikhonov, A.N. 1963. Regularization of incorrectly posed problems. _Soviet Math. Dokl._, **4**, 1624–1627.

Zou, H., Hastie, T., and Tibshirani, R. 2006. Sparse principal component analysis. _Journal of Computational and Graphical Statistics_, **15**(2), 265–286.

26

Alternative Data and ML for Macro Nowcasting

Apurv Jain[a]

Abstract

Worldwide macroeconomic data suffer from three fundamental problems – high dimensionality, a staggered release schedule, and poor data quality. Nowcasts are a popular set of tools that address the first two problems, and with the advent of alternative data we now have a chance to address the issue of poor data quality. This chapter provides an overview of nowcasting techniques, discusses the need for an *ex ante* hypothesis to guide alternative data selection, and compares typical alternative and traditional datasets on several factors including timeliness and granularity. Finally, a case study is presented that establishes that search data can statistically and economically significantly improve US government employment data in terms of timeliness and accuracy – a new result. The case study nowcasts revisions to Non-Farm Payrolls (NFP) three months in advance of the government data, proves these revisions are new and not noise in the framework of Mankiw et al. (1984), controls for Wall Street analyst predictions, and finds that machine learning techniques such as random forest and elastic net provide a substantial improvement over traditional linear regression methods.

26.1 The fundamental problems of macro data

Imagine having the difficult job of a Fed chair in April 2020: COVID-19 is spreading across the world and we are not yet sure of the economic impact of the disease on the United States. We wait for the employment situation report (Non-Farm Payrolls or NFP), an important and trusted indicator for the month of March 2020 that will be released on April 3. The March estimate of $-701,000$ job losses is a big change from the $+273,000$ jobs gained in February. As if that were not bad enough, the data reported for the March 2020 employment situation worsened over time. The May employment report, released in the first week of June, reveals that the labor market was worse than expected as the actual jobs lost in March were $-1,373,000$ – considerably higher than the first estimate of $-701,000$. A similar issue plagued the unemployment rate which declined from

[a] MacroXStudio, San Francisco, CA. Parts of this chapter were written while the author was a visiting researcher at Harvard Business School.
Published in *Machine Learning And Data Sciences For Financial Markets*, Agostino Capponi and Charles-Albert Lehalle© 2023 Cambridge University Press.

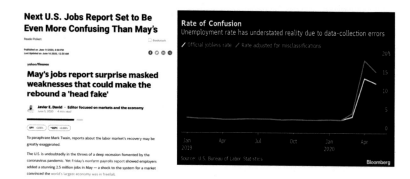

Figure 26.1 News headlines illustrate the confusion of real-time noisy economic data. Used with permission of Bloomberg Finance L.P. Source: Bloomberg L.P., June 14, 2020, and Yahoo finance, June 5 2020.

14.5% to 13.3% in the month of May, but, after adjusting for data errors, appears to have been 16.4% – an *increase*. Fulfilling the dual mandate (Chicago Fed, 2020a) of "stable prices and maximum sustainable employment" would be easier with better data[1].

These considerations give us a sense of the fundamental problems with measuring and managing such an enormously complex mechanism as the economic activity of a country. The data collected to support this difficult task are *imperfect*: they are many, staggered, slow, incomplete, subject to methodological change, and noisy (hence frequently revised)[2], coarse, and potentially biased[3,4]. We can organize these problems into are three broad categories:

1. The Big Data or high-dimensionality problem arises when the number of features (p) is significantly greater than the number of observations (T), written $p \gg T$. Thousands of statistical variables may get measured, but not all of them many be relevant for situation to hand. For example, FRED, a database maintained by the St. Louis Fed, has more than 786,000 series and the highest number of variables utilized for a nowcast is around 500.

2. Economic data releases are measured at various frequencies: some, such as real GDP, may be quarterly and others, such as Initial Jobless Claims, may be weekly. The data are also released at different times, with various lags and revisions. This gives rise to the staggered or "jagged edge" problem.

---

[1] Orphanides (2001) and Orphanides and Williams (2007) cover the issues of data revisions and their applications to central banking. Orphanides documented how the Fed overreacted to bad measures of inflation in the 1970s, causing the monetary policy to be too easy.

[2] Kliesen (2014) notes some of these fundamental problems as does Jain (2019). Croushore (2011a) provides an overview for how impactful data revisions are, especially when it comes to forecasting, policy-making, business cycle turning point predictions, and other macro research. It seems that the list should include asset allocation as a practical matter.

[3] "Real-Time Data Research" (Philadelphia Fed, 2021a) is a valuable resource of the appropriate data for better testing any forecasting methodology after accounting for revisions.

[4] For example, Aruoba (2008) documents predictable biases in US data.

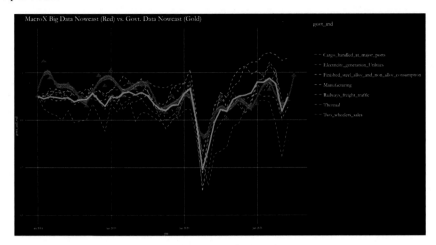

Figure 26.2 An alternative – or big – data-based nowcast can be quite useful in making real-time decisions. The thick red line with embedded triangles shows a satellite and social data based nowcast for India that is at least one month ahead of the typical government data releases and captures the big economic recovery in July 2021 ahead of the various government data (dotted lines) and a traditional nowcast (yellow line) of those government data. Nowcasts and data source Nowcasts and data courtesy MacroXStudio.

Table 26.1 Data revision for some major US economic releases. A ∗ indicates significant difference in "good" (data release > series average) and "bad" (data release < series average) economic times via 1000 bootstrapped confidence intervals. The computations *exclude* COVID-19 affected periods which would increase the revision magnitudes substantially, to focus on the long-term mean and patterns. Data source: Bloomberg.

Data	Mean	SD	%Revision in "good" times	%Revision in "bad" times	SD in "good" times	SD in "bad" times
US GDP growth (QoQ)	2.2%	2.4%	26%	80%*	1.35%	2.13%*
US NFP change (mom)	96.4	212.1	26%	46%*	77.5	214.6*
US Retail sales (mom)	0.33%	0.42%	47%	136%*	0.25%	0.26%*

3. The poor data quality problem includes lagged availability (typically ranging from 1 to 3 months), issues of imprecise measurements and sampling, a lack of detail, biases, and substantial revisions. Just as recent technological progress has reduced the measurement cost for some of the data and generated entirely new types of data previously unavailable (Kolanovic and Krishnamachari, 2017; Jain, 2019), one hopes that alternative data can contribute to reducing the impact of poor data.

In the following section, we will discuss how the first two of the above problems are addressed in the macro nowcasting context. Typically, high dimensionality is

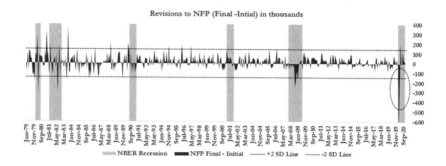

Figure 26.3
NFP revisions = Final number (released up to three months later) – the Initial number,
seem to cluster and appear at least "locally biased" in the fashion that Croushore (2011a)
postulates: with better data, expansions get revised upward and downturns downward.
Also, we can see from the figure that the biggest revisions tend to happen when the
business cycle shifts which is exactly when accurate information is the most valuable to
market participants (Jain, 2019). Data Source: FRED and Haver Analytics.

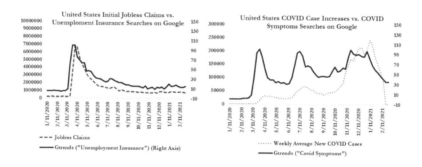

Figure 26.4 Search data seem faster than the government data. Data source: Google and
FRED.

tackled by dimension reduction and the "jagged edge" problem by ingenious data
release calendar construction and Kalman filtering.

26.2 High-dimensionality problem

A high-dimensional ($p \gg T$) problem is where the number, p, of potential
explanatory variables far exceeds the number, T, of total data points available.
This problem has been exacerbated in the last two decades as technology has
provided the means to collect an almost unlimited number of features, which
increases p. However, the number of observations, T, is limited by cost concerns
or by the inherent nature of the problem – for example, one cannot typically

generate more economic recessions. Even for the simplest linear (OLS) case[5] when p is less than T (or N) but increasing, the expected prediction error or EPE, which scales linearly with the ratio p/T, can become a serious issue[6]. The two major concerns in the high-dimensional set-up are overfitting and high variance of the resulting model.

To reduce variance, typically "simple, highly regularized approaches often become the method of choice" (Hastie et al., 2009, Chapter 18). We hope that constraining the estimated coefficients reduces the variance substantially while only minimally increasing the bias. There are many methods for doing this, and they may be categorized as:

- Parameter reduction. Unimportant dependencies are ignored in order to reduce the number of parameters to be estimated. For example, diagonalizing the within-class covariance matrix.
- Regularization. Penalty terms are added to the error minimization objective function formulation. This penalty can either eliminate features, reduce their coefficient estimates or provide a combination of elimination and reduction. For example, with either a L_1 lasso regressions penalty "selects" variables by making many coefficients *exactly* 0, and an L_2 ridge regression penalty tends to smoothly "shrink" the coefficients of the correlated variables towards each other and 0. Elastic net models are a weighted average of lasso and ridge.
- Dimensional reduction. Variables are projected on a lower-dimensional subspace. For example, principal components regression reduces the dimensionality of features via PCA and then regresses the principal components on the Y variable, Partial least squares reduces the dimension of X_p along the direction with a high correlation to Y, and supervised principal components performs dimension reduction only of variables highly correlated with Y or the outcome of interest, and then regresses.

While variance reduction methods, carefully and methodically applied, can mitigate the increased variance from the $p \gg T$ problem, they cannot *solve* it. As the number of features increases exponentially the performance of almost all variance reduction techniques declines. To illustrate the point, James et al. (2013, Chapter 6) perform a lasso on a data set with number of observations, $n = 100$, and choose the number of features, p to be 20, 50, and 1000. They find that while high regularization can help in the $p = 50$ case, when $p = 1000$ the results are poor across *all* values of the regularization penalty, λ. Even in PCA (which is conceptually similar to a ridge regression in that it shrinks all standardized coefficients towards 0), the presence of many idiosyncratic components can produce poor results. Additionally, the assessment of the significance of features or the "multiple-testing problem" becomes even more challenging in

---

[5] Requiring a simple OLS model to be unbiased is a strong condition. For instance, a regime-switching model such as one implemented by Diebold (2000) can rather better model business cycle dynamics than can OLS.

[6] Hastie et al. (2009, Section 2.5), has a good overview of high dimensionality, including the "Curse of Dimensionality" and the bigger VCV matrix estimation.

the $p \gg T$ context[7,8]. In the case of macroeconomic research, the small (about 195) cross-section of countries, as well as the limited time series of accurately recorded business cycles or other economic phenomena of interest, can be serious limitations.

Complementing machine learning methods with domain expertise and frameworks can produce better results than pure machine learning. Hastie et al. (2009) suggest the "use of scientific contextual knowledge" to figure out the most appropriate forms of regularization. The macro literature, with a long tradition of analyzing economic fluctuations[9], has provided us with a rich set of methods that combine domain and statistical expertise.

26.3 Nowcasting the big and jagged data

Nowcasting or "forecasting the present" is a term borrowed from meteorology relating to forecasting the current and near-future weather (Bok et al., 2017). Let us first define it precisely in the macro context and then elaborate and expand upon the definition to achieve a deeper understanding:

An *economic nowcast* is a current, model-based and replicable, continuously updated, time-granular, information-rich summary that approximates a key concept such as growth.

It is: *current* because its main purpose is to incorporate that latest information; *model-based and replicable* since no human judgement is used when the model is functioning; *continuously updated* for it is typically updated at daily and weekly frequencies; *time-granular* since changes in GDP are measured every day or week rather than every quarter; *information rich* as we can see the specific marginal impacts of various data releases; *summary* since there is *one* number representing the many economic data, typically via statistical techniques traditionally belonging to the family of *dynamic factor models* (DFMs); and finally it is an *approximation* since even the best measurement of the complex and dynamic economic machine is partial. As Table 26.2 shows, DFMs are indeed at the heart of many of the traditional nowcasting applications. Stock and Watson (2016) outline five main reasons why DFMs have been so popular: the observation of a few latent factors summarizing many data seems to hold empirically; this factor structure is theoretically appealing; it can accommodate lots of data; it enables real-time monitoring; and DFMs can span a wide variety of macro shocks. Moreover, the dynamic factor model structure can also handle the "jagged edge" as well

---

[7] Hastie et al. (2009, Chapter 18), discuss the family-wise error rate computation – which includes the well-known Bonferroni correction (that divides α by M, the number of multiple trials). For $p \gg T$ cases they recommend computing the false discovery rate instead of the Bonferroni correction which tends to be "too conservative" or finds too few variables significant.

[8] The time series context, where the flow of time is meaningful and one cannot arbitrarily mix past and future values, can increase the complexity.

[9] Burns and Mitchell (1946) modeled co-movements in economic data and aggregates in the 1930s and 1940s, identifying recessions and expansions; Sargent and Sims (1977) modeled common shocks via dynamic factor structure some 50 years ago; and Stock and Watson (1989) used hundreds of variables. Bok et al. (2017) and Diebold (2000) are good references for an overview.

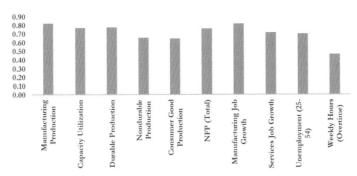

Figure 26.5 Correlation of some important economic series with the latent factor. Similar to Stock and Watson (2016) methodology, the specific economic series being correlated has not been used in making the latent factor to avoid spurious correlation. Data source FRED and calculations of the author.

as dimension reduction in a coherent state space framework to which a Kalman filter can be applied. More recently, nonlinear machine learning techniques – ranging from random forest to deep learning – have also been used effectively for nowcasting.

In order to move from many noisy variables to a few signals, one combines appropriate variable selection via domain expertise and the dimension reduction methods discussed previously. The NY Fed selects 37 variables (Table 26.2) which are "widely followed by market participants" and the Chicago Fed selected 85 indicators based on Stock and Watson (1989). The Chicago Fed's *original* monthly CFNAI Index (Chicago Fed, 2020b) listed in Table 26.2 is also a good example of a transformation summarized using principal components analysis (PCA). The basic decomposition is expressed simply as:

$$X = \Gamma F + \epsilon; \tag{26.1}$$

here[10] $X = N \times T$ stacked matrix of $N \times 1$ standardized (stationary, mean 0, and variance 1) data vectors x_t, and Γ is an $N \times R$ time-invariant matrix loading on a "few" ($R \ll N$) factors F ($R \times T$) and ϵ is typically modeled as homoskedastic mean-zero random variables with variance covariance matrix $\sigma^2 I$. In the CFNAI application, $N = 85$ and $R = 1$ and the estimation method follows an eigenvector decomposition of the sample variance covariance matrix $\frac{X^T X}{N}$ and using only the *first* principal component (the one associated with the largest eigenvalue, indicating that it explains the most variance) to generate the weights w that can subsequently be multiplied with the economic data generated at time t, x_t to provide a "less noisy summary" or economic activity index $T_t = w x_t$.

This index "closely tracks periods of economic expansion and contraction, as well as periods of increasing and decreasing inflationary pressure", thus revealing regime-switching dynamics via the common factor patterns in the data that might be hidden if we examined only the individual observations since each data series

---

[10] We are switching notation a bit from p features to N to be consistent with the literature.

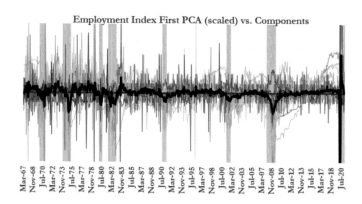

Figure 26.6 Latent factor for US employment as the first principal component (scaled) of more than 35 economic series. The less noisy principal component appears better at signaling economic cycle shifts. Data from Haver analytics.

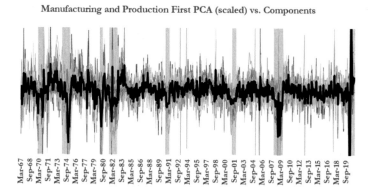

Figure 26.7 Latent factor for US manufacturing and production as the first principal component (scaled) of 25 economic series. Data from Haver analytics.

has its own idiosyncratic noise. Though the CFNAI was originally designed to track inflation (Stock and Watson, 1999), it has had more success in aligning with US expansions and recessions as Brave et al. (2019) note.

The Kalman filter that recursively generates new predictions by updating the factor estimate using the projections of the current data innovations can systematically accommodate *all* the variables into one coherent state-space framework. As an example, consider the static factor F that was invariant in (26.1) has been made dynamic into f_t and given its own transition dynamics represented by the matrix A. Once again X is an $N \times T$ stacked matrix of *measured* $N \times 1$ data release vectors x_t. This is a typical "state space" representation with f_t as the latent or unobserved factor. The matrix A can be as simple as an AR(1) process, but three lags are more typical. A benefit of this dynamic structure is that it utilizes the time-series variation in addition to the cross-sectional variation handled by the

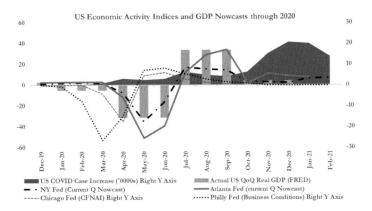

Figure 26.8 Comparing the US economic activity indices (typically mean 0) and GDP nowcasts through the recent past. Data sources in the final row of Table 26.2.

principal components.

$$x_t = \Gamma f_t + \epsilon_t \qquad (26.2)$$
$$f_t = A f_{t-1} + v_t \qquad (26.3)$$

The literature (for example Bok et al., 2017) also generally assumes that $\mathrm{Var}(v_t) = 1$ and $\mathrm{Var}(\epsilon_t) = H$, an $N \times N$ diagonal matrix, or that each ϵ_t can be decomposed into an AR(1) process with an idiosyncratic component and uncorrelated white noise. The parameters can be estimated using a Kalman smoother and EM algorithm. In the first step, the algorithm is initialized using sample PCA to estimate the factor f_t and then OLS to derive the initial parameters. The second step consists of employing a Kalman smoother that uses the entire prior sample to generate an updated estimate of the common parameter. MLE estimators that can accommodate the uncertainty of estimated factors and arbitrary missing data can also be handled.

Using the real GDP to reduce noise.

To achieve the goal of noise reduction, there are opportunities for the creative uses of many statistical learning techniques such as partial least squares, supervised principal components (dimensionality reduction), elastic net (variable selection and shrinkage) as well as unsupervised learning. A recent example is Brave et al. (2019) who utilize collapsed dynamic factor analysis, pioneered by Bräuning and Koopman (2014), that *simultaneously* models the target variable (real GDP in this case), the principal components, and the unobserved dynamic factors. The idea behind the method is that only the variables *relevant* to predicting or nowcasting the GDP should be selected instead of estimating the dynamics of the large number of other macro variables x_t. In the first step, we carry out a principal component analysis of x_t to lower the number of parameters that maximum likelihood has to estimate. In the second step, we reduce the noise related to the principal components being estimated by jointly modeling the principal components with the target variable y_t in the smaller parameter space[11]. For this example, we chose y_t to be the real GDP. However, we can also use other series related to growth or inflation. Another use for this methodology could be to improve the poorly measured growth estimate of country A if we have access to another country B's better-measured growth estimate that happens to be highly correlated to A's.

26.3.1 Nonlinearity and ML for nowcasting

An interesting strand of macroeconomics forecasting literature that has been successfully using newer machine learning techniques has emerged recently. The Y variables, as it were, remain standard – unemployment, inflation, industrial production and such,; but more and more nonlinear and computationally intensive modeling techniques are being used on X variables of both types: the traditional macro data and newer alternative data. Researchers are also successfully dealing

---

[11] Brave et al. (2019) use this technique to estimate the common components as well as to generate a monthly nowcast of the real GDP. They write y_t as the hidden or latent real GDP growth that, along with x_t, depends on two common factors μ_t and f_t as well as a long-term trend, α, and noise, η:

$$y_t = \begin{bmatrix} 1 & 1 \end{bmatrix} \begin{bmatrix} \mu_t \\ f_t \end{bmatrix} + \alpha_t + \eta_t, \tag{26.4}$$

$$x_t = \begin{bmatrix} \mathbf{1} & 1 \end{bmatrix} \begin{bmatrix} \mu_t \\ f_t \end{bmatrix} + \epsilon_t. \tag{26.5}$$

Following Bräuning and Koopman (2014), Brave et al. (2019) collapse the panel of 500 monthly time series by pre-multiplying by the factors obtained from PCA of x_t:

$$\begin{bmatrix} l' \\ \hat{f}' \end{bmatrix} x_t = \begin{bmatrix} l' & \hat{f}' \end{bmatrix} \begin{bmatrix} \mathbf{1} & \Gamma \end{bmatrix} \begin{bmatrix} \mu_t \\ f_t \end{bmatrix} + \begin{bmatrix} l' \\ \hat{f}' \end{bmatrix} \epsilon_t. \tag{26.6}$$

They set $l'\mathbf{1} = 1$, $\hat{f}'\mathbf{1} = 0$, $\hat{f}'\Gamma = 1$, and let $l'\Gamma = \gamma$, a scalar, and obtain the collapsed measurement equation describing the dynamics of $\bar{x}_t$ and $\hat{f}_t$. After this they add the real GDP as an instrument that will correct for the bias in the *estimation* of the principal components that can arise from weak factors. Equation 26.7 emphasizes treating the principal components *explicitly* as an errors-in-variables problem: note that $\hat{f}_t$ on the left indicates *estimated* f_t. Thus the overall measurement equation can be written as:

$$\begin{bmatrix} y_t \\ \bar{x}_t \\ \hat{f}_t \end{bmatrix} = \begin{bmatrix} 1 & 1 & 1 & 1 \\ 1 & \gamma & 0 & 0 \\ 0 & 1 & 0 & 0 \end{bmatrix} \begin{bmatrix} \mu_t \\ f_t \\ \alpha_t \\ \eta_t \end{bmatrix} + \begin{bmatrix} 0 \\ \bar{\epsilon}_t \\ \hat{\epsilon}_t, \end{bmatrix} \tag{26.7}$$

and the factor dynamics can be represented as:

$$\begin{bmatrix} \mu_t \\ f_t \\ \alpha_t \\ \eta_t \end{bmatrix} = \begin{bmatrix} \rho & 0 & 0 & 0 \\ 0 & \phi & 0 & 0 \\ 0 & 0 & 1 & 0 \\ 0 & 0 & 0 & 0 \end{bmatrix} \begin{bmatrix} \mu_{t-1} \\ f_{t-1} \\ \alpha_{t-1} \\ \eta_{t-1} \end{bmatrix} + \begin{bmatrix} \nu_t \\ \xi_t \\ \nu_t \\ \eta_t \end{bmatrix}. \tag{26.8}$$

Other standard assumptions, such as diagonal covariance matrices and that common factors have stationary dynamics, are also made. Finally, they perform "triangle averaging" of the various monthly GDP estimates to match the quarterly real GDP. Brave et al. (2019) report that, for understanding the business cycle, the two components μ_t and f_t, interpreted as the leading and coincident or slightly lagging indicators respectively, are quite useful. The coincident factor f_t is the most predictive, but the trend (α_t) and the irregular component (η_t) are not relevant.

Table 26.2 Some Traditional Nowcasts.

Field	NY Fed: Suspended since Sep 2021 because of pandemic induced volatility	Atlanta Fed	Chicago Fed	Philadelphia Fed
"Y" or Nowcast Variable	Current Q and Previous Q Real GDP growth rate.	Current Q and Previous Q Real GDP growth rate.	Overall Economic Activity and Inflationary Pressure with a 0 value indicating trend growth. Also estimates monthly GDP growth.	Real Business Conditions with a 0 value indicating average business conditions.
Update frequency	**Weekly**. Every Friday at 11:15 a.m.	**6-7 times a month** corresponding to major economic releases.	**Monthly** with one month lag.	**Daily** in real time.
"" or Data Used	**37 monthly and quarterly time series** "that move the market and make front page news" grouped into the following categories: Housing and construction, Manufacturing, Labor, Income, Retail and consumption, Income, Surveys, International Trade, and Other (parameter and data revisions. Local blocks are used to control for idiosyncrasies in subgroups of these variables.	**124 monthly time series from 26 data releases** are used to generate a single common latent factor that summarizes economic activity. Some series like consumer confidence are used only to generate the common factor but not to predict the GDP. GDP is further subdivided into 13 real quantity components like the BEA and the various monthly series corresponding to the 13 components are forecast and then combined using "bridge equations."	**500 monthly time series of U.S. real economic activity.** They start with the 85 time series traditionally used by CFNAI that focus on 4 major categories – production and income; employment, unemployment and hours; personal consumption and housing; and sales, orders, and inventories, and add additional series from the Conference Board, FRED MD, Atlanta Fed and NY Fed Nowcast models.	**6 series of varying frequency**: **weekly** initial jobless claims; **monthly** payroll employment, monthly industrial production, monthly real personal income less transfer payments, monthly real manufacturing and trade sales; and **quarterly** real GDP.
Methodology Summary	**Dynamic factor model** estimated by Kalman filtering and likelihood based methods that directly predict GDP. Accommodate the "jagged edge" or staggered nature of data releases. First all the macro series are summarized using PCA, and the factor loadings, as well as dynamics are initialized by OLS parameters. Then, using Kalman smoother, an updated estimate of common factors is obtained.	**Bridge equation methods** that combine **dynamic factor models'** "top down" approach (direct prediction of GDP with no subcomponents predicted like the NY Fed) with a "bottom-up" approach of predicting the 13 GDP components using bridge equations that regress the growth rate of a component on one or more related monthly series as well as their own lags.	**Collapsed dynamic factor model** This is an interesting application of jointly modeling the target variable Y_t – real GDP in this case – in a small-scale dynamic factor space. Thus, real GDP acts as an instrument or supervising variable Y_t to help correct for potential biases from many weak factors. The details are in the footnote in the previous section.	**Dynamic factor model** where business conditions are treated as the latent variable. They explicitly include different and high frequency daily data in their estimation procedure.
References	Nowcasting Report (New York Fed, 2021).	GDPNow (Atlanta Fed, 2021).	Brave–Butters–Kelly Index with 500 variables (newer), (Chicago Fed, 2021). CFNAI Index with 85 variables (older) (Chicago Fed, 2020b).	Aruoba–Diebold–Scott Business Conditions Index (Philadelphia Fed, 2021b).

with the challenge of applying these methods to time series analysis where the *order* of data is crucial. The ML techniques being used range from tree-based methods such as *random forest* (Medeiros et al., 2021; Chen et al., 2019); *gradient boosted trees* (Döpke et al., 2017) – especially XGBoost; to *elastic net* techniques (Hall, 2018) such as lasso and ridge regressions with various regularization schemes as well as time variation (Yousuf and Ng, 2021). Other successful techniques include *support vector machines* (Sermpinis et al., 2014), which resemble logistic regression, as well as others up to and including *back-propagation* (BPN)[12] and *encoder–decoder* models (Cook and Hall, 2017).

A striking conclusion of this strand is that the newer machine learning techniques seem not only to pick up on business cycle turning points with more success

---

[12] Moshiri and Cameron (2000)) is an example of the early work in the field.

than the more traditional linear techniques but also outperform the standard linear techniques at both short and long[13] forecasting horizons.

A natural question to ask is what *aspect* of machine learning lies behind this outperformance? Coulombe et al. (2022), who perform a "meta-analysis" of various machine learning techniques for macroeconomic forecasting, are investigating this question. They have designed experiments to identify which features of ML models – nonlinearity, regularization, cross-validation, or alternative loss function – are impactful for macro prediction in both data-rich and data-poor environments. Coulombe et al. (2022) report that nonlinear techniques (especially random forest and kernel ridge regression) have the most positive impact, followed by cross-validation, for picking appropriate hyper-parameters, but regularization and alternative loss functions do not seem to improve much upon existing linear methods. The effects are even more pronounced when using alternative data, predicting economic cycle turning points, and for longer-horizon forecasting.

Specifically for nowcasting, a study by the Reserve Bank of New Zealand and Bank of England researchers (Richardson et al., 2021) finds that machine learning models perform better than either a simple AR benchmark or a dynamic factor model, and also add value when ensembled with official forecasts. The study notes – *"The top-performing models – boosted trees, support vector machine regression and neural networks – can reduce average nowcast errors by approximately 20–23 percent relative to the AR benchmark. The majority of the ML algorithms also outperform the dynamic factor model."*

Meteorology, from which we have borrowed the very term 'nowcasting' is far ahead of finance in terms of both model sophistication and performance. Meteorologists use deep learning models such as convolutional LSTM (Xingjian et al., 2015) with millions of variables and achieve more than 95% predictive accuracy for three-day weather forecasts (Bauer et al., 2015). Additionally, many other fields, such as public health and epidemiology, also combine and utilize alternative data effectively for nowcasting[14].

As we learn from these fields, we should remember that the underlying phenomenon being modeled can be quite different; weather systems, though complex, do not exhibit reflexive behavior whereas humans might change their behavior based on the knowledge of how others behave as noted by Keynes (1936, Chapter 12) in his famous beauty context example, or on other financial market phenomena (Soros, 2009). Popper (1957) notes not only the philosophical differences between physical and social sciences but the negative social impact of narrow-minded and idealistic pursuits based upon these ideas. More recently, Lo and Mueller (2010) caution us against the enthusiasm for physics-like "immutable laws" in explaining economic and social phenomenon which can have unintended consequences. O'Neil (2016) reports examples of algorithms exhibiting unintended biases and causing economic and social harm in multiple fields.

---

[13] For example, Datar et al. (2020), who utilize accounting data to forecast US GDP, find that random forest algorithms are effective at forecasting 3–4 quarters ahead.

[14] Venkatramanan et al. (2021) use mobility data and machine learning to forecast influenza activity.

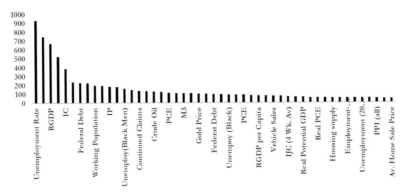

Figure 26.9 The 50 most popular FRED macro tickers sorted by number page views on a Relative Basis (with 1000 being an imaginary maximum). This seems in line with the power law in empirical data (Clauset et al., 2009). Metadata and usage statistics retrieved from FRED Data Desk team, Federal Reserve Bank of St. Louis, `https://fred.stlouisfed.org`, February 9, 2021.

26.4 Dimensions of alternative data quality

26.4.1 A crowd-sourced experiment

In practice, domain expert input is important to improving data selection, cleaning, and hypothesis generation. As a proxy for expert input we carried out an experiment with the cooperation of the St. Louis Federal Reserve Bank. We obtained the 1,000 most popular macroeconomic data releases as ranked by FRED shown in Figure 26.9. This is similar in spirit to the NY Fed's variable selection criterion of the ones that "move markets and make front page news" (Bok et al., 2017). Immediately, we can see that all macro data are not created equal: the amount of attention is distributed in a "power law" sense with the top two – unemployment rate in the first place and inflation (CPI) in the second – getting 930 and 748 views respectively on a 1–1000 imaginary relative scale. The 15th most popular release – continued unemployment insurance claims – has only 137 views and the 1000th most popular series (which happens to be the labor force participation rate specific to white males aged over 20), merely three views. We found that the top 50–100 variables obtained by this method corresponded to the variables generally used by the various nowcasts – for example, the variables obtained with this method for the employment sector were the unemployment rate, NFP, initial claims, and labor force participation.

26.4.2 The need for a hypothesis

Millions of features becoming available through new technologies (Figure 26.10) is very exciting but it brings its own set of challenges. The St. Louis Fed maintains a huge collection of macroeconomic data FRED with more than 788,000 data series *already* available. We saw in Table 26.2, that the number of variables utilized for nowcasts range from 6 to 500. A simple calculation – ignoring the

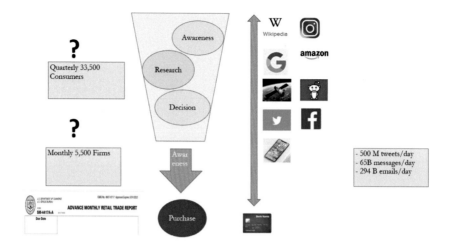

Figure 26.10 Traditional data vs. big or alternative data in the marketing consumer funnel. The consumer becomes conscious of the need in the "awareness" phase, finds information in the "research" phase, defines the purchase set in the "decision" phase, and finally procures product in the "purchase" phase. Traditional data typically have little information on the awareness and decision phases, as indicated by question marks. Alternative data typically occupy more than one spot in the funnel, as indicated by the arrow on the right stretching across all phases, support the conclusions of Jansen and Schuster (2011) that the process is more complex than the simple linear representation. The infographic of data explosion in our world is from World Economic Forum (2019).

fact that some of these series are sub-series of another – shows that even for a 500-variable nowcast, the *ex ante* probability of *any one data series* being used is 0.06%, which is orders of magnitude smaller than the probability, 99.94%, of not being used. When the statistical learning expert is making the model and considers new features, she faces the ex ante trade-off between a certain increase in variance and a certain reduction in bias. Even if a feature appears individually quite related to the 'y' variable in question, it is difficult to guess its noisy marginal predictive benefit after accounting for predictive power of other variables.

Since not all data are equally important, merely adding more data series to the large quantity already available is not an optimal way of increasing information about the state of economy. In fact, when the $p \gg T$ problem is severe, with the ratio of relevant features to the total possible being quite low (say 0.06%), the problem *worsens* as we add more features. Additionally, pure statistical techniques such as regularization cannot solve this problem for any value of the hyper-parameter.

Faced with these obstacles, we have arrived at the need for a prior hypothesis. In other words, a structure or set of constraints that can guide variable selection. This variable selection methodology should address our *ex ante* need for economic intuition or even provide a theoretically motivated assurance that this new data series *by construction* can substantially improve a *particular dimension of data*

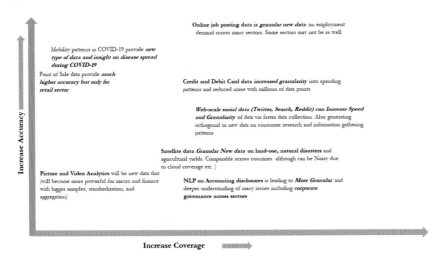

Online job posting data is *granular new data* on employment demand across many sectors. Some sectors may not be as well

Mobility patterns in COVID-19 provide *new type of data and insight on disease spread during COVID-19*

Point of Sale data provide *much higher accuracy but only for retail sector*

Credit and Debit Card data *increased granularity* into spending patterns and reduced noise with millions of data points

Web-scale social data (Twitter, Search, Reddit) can Increase Speed and Granularity of data via faster data collection. Also generating orthogonal or new data on consumer research and information gathering patterns

Satellite data *Granular New data* on land-use, natural disasters and agricultural yields. Comparable across countries although can be Noisy due to cloud coverage etc.)

Picture and Video Analytics will be new data that (will become more powerful for macro and finance with bigger samples, standardization, and aggregation)

NLP on Accounting disclosures is leading to *More Granular* and deeper understanding of many issues including corporate governance across sectors

Increase Accuracy

Increase Coverage

Figure 26.11 Sample data quality hypotheses for a few commonly used alternative datasets.

quality when compared with traditional data. For instance, we may begin with an initial hypothesis about the data set in question being beneficial for increasing timeliness if it is collected in real-time or with lowering bias if it consists of the entire population's response rather than with that of just a sample of the population, and the subsequent analysis can act as a "posterior updating" of the interaction of that data source and the economic phenomenon in question. Asset management researchers such as Lopez de Prado and Lipton (2020) recommend that we "develop theory, not trading rules", similar to the demand for structural models in macroeconomics.

Thinking through a common alternative data source – web search – to nowcast government data may illustrate the idea. Many people use web search to seek information before they act, so, by construction, search data might have the potential to be more timely than the delayed government statistics; also it is well known that the sample size of the search population is much higher than that of standard government surveys so search data are likely to be more precise. However, *ex ante* we may also be concerned about the urban and tech-centric bias as well as inherent noise in web search data, since heavy tech usage tends to be urban and trending searches can become more popular without any change in the economic phenomenon. While this is not a fully fledged theory about the economic phenomenon in question, these initial hypotheses act as informed priors that can afford a more rigorous evaluation beyond the merely statistical model fit. If subsequent analysis reports that the nowcast based on search data excels in older rural populations but not in urban tech-centers, one would be skeptical regardless of the statistical fit. Table 26.3 lists common characteristics of data that a researcher might wish to focus on and compares some traditional and alternative data sources.

26.5 Non-farm payrolls and web search case study

We identified three significant problems with macro data: that of big ($p \gg T$) data; of staggered releases; and of poor data quality. We discussed above how nowcasting addresses the first two and in this section we will present a nowcasting case study about improving the timeliness and accuracy of government data. We pick a popular alternative data set – web-search – and the government release we will focus on is non-farm payrolls (NFP) – one of the major data releases in the United States (Baumol, 2013). The question we wish to answer is:

Do search data reveal economically important, unique, and more accurate information faster than the official government employment data?

In accordance with the previous section, first we examine which data dimensions or attributes we might reasonably expect search data to have, compared with government data; second, we carefully clean the data to maximize that attribute; and finally we perform some simple analysis to test our hypothesis. In terms of data characteristics, we hypothesize, based on Figure 26.11, that search is likely to be more timely than government data, more widely applicable across multiple sectors, and more accurate – since the sample size of search data is much bigger than the government NFP surveys. A concern is the potential urban and technologically capable user-bias of search. Careful search-data cleaning will take the form of generating and matching detailed lists of job titles to the BLS categories and the final analysis will examine the statistical and economic significance of search data in nowcasting NFP revisions.

26.5.1 Background and related work

Over the last decade researchers working in this field have looked to web search activity as an attractive information option owing to its high volume (nearly 80% of people use web search to look for a new job: see Smith, 2015), continuous data generation, and real time availability. Since web-search data are naturalistically generated as a result of people addressing their true interests and concerns (Dumais et al., 2014; Goel et al., 2010), this also makes them potentially different sources of information compared with the official survey-based government data.

There are many notable papers in the field (Choi, 2009; Choi and Varian, 2012; Goel et al., 2010; Einav and Levin, 2013; Suhoy, 2009; Artola et al, 2015; Hellerstein and Middlethorp, 2012) that use web search to predict a variety of employment-related statistics across the world; moreover, there are other official macro statistics, such as consumer confidence, tourist inflows, mortgage re-financing, auto sales, and home sales. There are two main stylized facts in this literature about search data. First, there are only small improvements by using

Table 26.3 Comparing alternative and traditional data quality using some relevant categorizations.

Data Quality Dimension	Traditional Data	Alternative Data	Remarks
Timeliness	Typical lag of 1–3 months even for the US. Can be worse of developing countries. For example, for India, the current lag is more than a year (though attempts are being made to reduce it to 3 months) with the latest *official* labor market data in March 2021 is for June 2018–2019 annual survey (MOSPI, 2019).	Social data such as Twitter, Web search, Reddit can be practically instantaneously collected *and* be leading indicators. These indicators provide information on intention or the awareness, research, decision phases in Figure 26.10 that was typically unavailable previously and are likely to precede action such as purchase, which was measured by traditional indicators like retail sales.	High frequency (daily, weekly) availability can also mitigate the "jagged edge" problem with traditional data. Some alternative data sources like mobility may be delayed by up to a week and credit card data may be delayed by one to three days until the transaction settles but can still be leading indicators.
Wide coverage (high recall)	Carefully designed and covering many traditional sectors in a balanced demographic way.	Typically alternative data sources are stronger in certain sectors such as consumer behavior and spending, labor market behavior in technology-driven and gig-economy sectors, sensor data such as satellite and mobile phones to better understand mobility patterns during COVID-19 etc. Alternative sources tend to have less visibility into the government and B2B sectors.	Traditional data sources may find it difficult to measure gig-economy activity (Kässi and Lehdonvirta, 2018). One can try to combine many alternative data sources to measure government or B2B activity but the costs can be high and the sampling biases can be a big issue.
Accuracy (high precision)	The accuracy of traditional data is low with up to 50% revisions (Jain, 2019) of the most important data releases even for the United States. This issue partly springs from the high cost of information gathering in the past where it was necessary to make a trade-off between speed and accuracy via frequent small samples (50,000 people once a month) and infrequent full census (5–10 years).	Alternative data can measure cheaply and frequently at scale, but bring their own noise from automation, usage pattern changes, as well as sample biases, data collection methodology changes etc. (Jain, 2019). Careful curation and cleaning can help lower that noise in certain cases for example in inflation measurement for Argentina (Cavallo and Rigobon, 2016).	Survey responses declined during COVID-19 times (Rothbaum and Bee, 2020) making it challenging even for well designed traditional data sources. The other challenge is measuring and weighting new sectors appropriately in case of fast technological change.
Long time series	Traditional data tend to have much longer time series, typically 30 years or more, and cover multiple economic cycles.	A key issue with alternative data is the shorter time series – anywhere from 3 to 15 years which may not cover many economic cycles.	Researchers such as Adler et al. (2019) are showing creative approaches in combining official (and potentially underreported) census data to the newer and granular search data which does not have as lengthy a time series.
High granularity	Traditional data tend to offer limited granularity since they are sampling based and utilize the same set of questions. For example, the household survey for NFP polls 50,000–60,000 respondants per month.	Naturalistic alternative data can offer immense granularity on a minute-by-minute basis. For example, there are 500 million self-generated tweets per day of which even if 5,000,000 are relevant to a particular economic topic is still substantial.	While minute-by minute-granularity is interesting and extremely useful, it does not increase predictive power of economic phenomena linearly (Jain, 2019).
Low bias	Traditional data sources tend to be less biased since they tend to be carefully designed with sophisticated statistical techniques for the explicit purpose of gathering that particular data. However, these data collection are not immune to agency incompetence, government intervention and politics. Worryingly, Aruoba (2008) documents a relatively large bias – a non-zero mean deviation that may be predictable – in revisions to major economic statistics such as GDP for United States and concludes that "these are not rational forecasts."	Alternative data sources tend to be more biased since they are typically not designed for analysis purposes and may appeal to a very specific population using them – whether they are large corporations, small businesses, or individuals. Most social data, such as Web-search, Reddit, Twitter, tend to have an urban and tech-centric bias.	For traditional data bias issues, read Coyle (2014) for an account of Nigeria, Ghana, ... GDP increasing dramatically; or Cavallo and Rigobon (2016) and Cavallo (2013) about using a big data online inflation index to find misreporting on the part of the Argentinean government. Brave et al. (2020) contains good examples of potential biases in alternative data, for instance they cite potential issues with Homebase whose customers were service-oriented small businesses and hence showed a larger decline in hours worked during COVID-19 compared with the balanced BLS panel.
Low Cost	Typically free in dollar terms for users, since government agencies like the BLS for the US make the data available.	Typically more expensive with datasets costing anywhere from $30,000 to more than $100,000 and with the annual spend by asset managers continually growing (Kolanovic and Krishnamachari, 2017).	The privacy cost and risk is perceived to be low for traditional data since the user typically volunteers information and it is just a sample, but the privacy cost and risks for alternative data can be high (Jain and Seeman, 2019).

search data[15] Second, good existing data sources can overcome benefits from search data[16].

[15] Choi and Varian (2012) is a seminal paper in nowcasting or predicting near-term economic indicators such as automobile sales, unemployment claims, consumer confidence and others using web search data from Google. The paper follows several others that predict unemployment in various countries

There are three major areas of concern with the prior literature.

First, most of the predicted macro-economic variables, such as unemployment rate or unemployment insurance claims, have a high ($> 90\%$) trend component (Choi, 2009; Choi and Varian, 2012) which makes it difficult to judge the importance of the new web search data and tend not to impact the financial markets much.

Second, most studies do not explicitly control for existing experts' forecasts so it is not clear if the information is truly different and hence incremental to existing knowledge.

Third, as both government data and search data are noisy proxies for true and unknowable economic growth[17], it is unclear how to ascertain the veracity of the information content.

Using a novel data set of more than 830 million employment-related searches from a popular search engine (Bing), and leveraging the methodological innovation of nowcasting *revisions to the same official NFP number*, we address these three concerns.

We focus on NFP rather than the unemployment rate or IJC used in other studies such as Choi (2009) or Choi and Varian (2012). NFP is economically much more relevant to market participants than other statistics such as unemployment rate or

such as US, Germany and Israel (Askitas et al., 2009; D'Amuri and Marcucci, 2010). They tend to agree with Goel et al. (2010) that search data do not provide dramatic improvements in forecast performance but advance the case that small improvements are economically significant. Specifically, in the field of unemployment, they predict initial claims for jobless benefits (IJC) that are very close to a random walk. In fact, they find that the prediction error in their holdout sample increases from 3.68% to 5.95% when search variables are included, but they suggest that the search model fits better in recessions or turning points, where the mean average error decreases from 3.98% to 3.44%. While interesting, Choi and Varian's claim is weakened by the fact that their sample only has one recession and hence a limited number of turning points. Also, the difference in mean average error of 3.98% and 3.44% could just be down to sampling noise. Other work such as Clement and Stephanie (2015) also focuses on forecasting household consumption in France, which have high trending and autoregressive components, and similarly find that search data do not add much predictive power over the autoregressive or trend components.

[16] Goel et al. (2010) present empirical results in several domains and pose an important question regarding the incremental information content of search data in the presence of other data sources. They find that "the utility of search data relative to a simple autoregressive model is modest." Additionally, they find that search data add value "in absence of other data sources, or where small improvements in predictive performance are material." In our case predictions of NFP statistics by Wall Street analysts, in advance of the release of the data itself, form a very good existing data source. Most of the literature, including Choi (2009) and Choi and Varian (2012), does not explicitly control for the baseline knowledge other than controlling for previous trends. While some of the predictions of Wall Street analysts may be based on trend, it seems naive to assume that all the predictions of highly motivated professionals who are incentivized to outperform will be simply captured by a linear trend model. D'Amuri and Marcucci (2010) come closest to our idea by controlling for the survey of professional forecasters, as do Antenucci et al. (2014), but they predict the highly trending unemployment rate and jobless claims with low financial market impact. If a data release does not impact financial markets then it is quite likely that the information was already known to the participants and hence reflected in the prices as set by a standard efficient markets model (Fama, 1970).

[17] Croushore (2011b) has an example of the US real output growth for 1977 Q1 still changing as late as 2010.

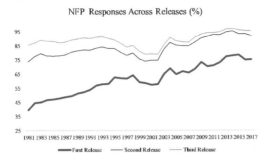

Figure 26.12 Response rates to the NFP Employment Survey for the first release vs. the second and third releases. Data Source: BLS
https://www.bls.gov/osmr/response-rates/.

IJC. Indeed, using *changes* to NFP as our variable of concern lowers the trend component from 90% to 60%.

To search the data, we use a live and realistic antagonist in the form of consensus of highly trained, professional Wall Street analysts who get ranked on how well they forecast. This is in contrast to imposing a simple auto-regressive structure or arbitrarily selecting one or two variables relevant to the labor market forecast. It seems reasonable to assume that such analysts would use all the relevant information from other government data and private vendors to make sophisticated statistical models, as well as leverage their considerable experience to forecast such an important number.

We use the novel methodology of nowcasting the *revisions* in the official NFP release itself, which typically happen 1–3 months *after* the initial NFP data are released. As more employers finally respond to the voluntary employment surveys upon which the official NFP statistics are based, the revised NFP numbers, by construction, contain more information than the initial ones. And if search data can predict the revisions in NFP 1–3 months ahead, the case for search having superior information becomes much stronger. We also verify the existence of extra information in revisions compared with the initial releases via news and noise regressions and formal Granger causality tests.

When using the Google search data, researchers have been cautious about the the lack of transparency and heavy aggregation in normalization of searches, as well as search-term relevance declining over time, as Suhoy (2009) and Chancellor and Counts (2018) report. By using the entirety of our proprietary 830 million employment search-related dataset we eliminate the normalization issue, and by cleaning and curating the employment web-search dataset with the help of experts, we better focus the dataset on the main macroeconomic problem. Encouragingly, we find that the overall performance with web search is better than what the literature has reported thus far, and it improves further when we focus on sampling areas in which information from government surveys is poor.

Figure 26.12 shows that for the establishment survey, typically only 60–70%

of the responses make it back in time for the first release. Phipps and Toth (2012) investigate which types of firms have low response rates to BLS surveys and find that four main factors are important: First, the firm's size, measured in number of employees. Second, whether a firm is a "white-collar service firm", as indicated by inclusion in three super-sectors: (1) Information, (2) Finance, and (3) Professional and Business Services. Third, whether it has multiple offices. Fourth, the size of the metropolitan service area (MSA) in terms of the population from which the official collects the employment data. The authors find that "White-collar service establishments with a larger number of multi-units have the lowest response rates." Additionally, small- and medium-sized businesses that are more sensitive to cyclical changes of the economy are also slow to respond since they might lack a specific department to respond to such surveys. MSAs with a large population are associated with lower BLS responses. Plausibly, the firms with lower initial response rates might be responsible for the most changes from revisions. White-collar service firms in MSAs are the ones quite likely to use internet search and other web-scale data for content production. Mislove et al. (2011) find that Twitter users are "more likely to live within populous counties than would be expected from census data". The potential *underrepresentation* of urban-area, service-firm bias in BLS data may thus be naturally addressed by web-scale data which tend to be biased in the direction of overrepresentation of such entities.

As has been widely documented in the literature, for instance in Goel et al. (2010) and Chancellor and Counts (2018) more recently, web searches reflect the users' *demand for information*. This demand-driven data generation is in contrast to government statistics that rely on users' voluntary *supply of information*. In troubled economic times, the former could increase while the supply of information might decline. This differential between demand and supply of information can be utilized. For instance, for NFP, we believe that this differential is potentially highest among white-collar service firms located in large MSAs in sectors such as architecture, business, finance.

26.5.2 *Government non-farm payrolls data overview*

For this case study we focus on the NFP data release, which estimates the number of jobs added to the US economy each month and is released by the Bureau of Labor Statistics (BLS). This release consists of two surveys, the first being that of around 60,000 households, and the second, known as the establishment payroll survey, is one of more than 600,000 corporate and government sites. Here we rely on the establishment survey, which further categorizes employment numbers into about 15 sectors and sub-sectors such as "Information", "Retail trade", and "Professional and business service." Changes to NFP can help signal the beginning or end of a recession if the economy starts losing or gaining more jobs when compared with the past.

Typical Wall Street consensus for NFP data release has around 80 estimates from analysts from various investment banks and other institutions, and is usually available about a week before the data release. Bloomberg tracks the previous

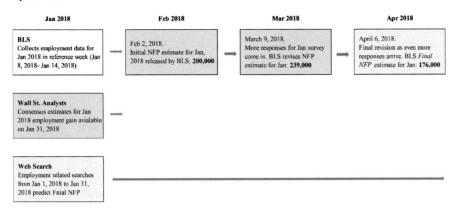

Figure 26.13 An indicative timeline of the initial NFP and final NFP data releases along with the prediction from Wall Street analysts' consensus and search-based variables.

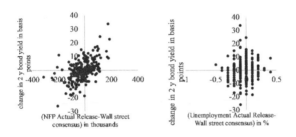

Figure 26.14 Movement in the 2-year bond yield for. NFP surprises (Actual release – Wall Street consensus) on the left and Unemployment surprises on the right. The Newey–West-corrected t-statistic for NFP is 9.4, compared with the t-statistic for unemployment at 0.32 in a horse race. Data Source: Bloomberg.

predictions of these analysts and ranks them based on their performance. We simply use the average of all the analysts' estimates.

There are two main uses of economic information. The first is to predict the short-run impact, and the second is to understand the long-run trend in order to make policy decisions (Kliesen, 2014). The short-run impact is typically a few hours or days and can be analyzed using "event studies", which involve the analysis of price impact of the incremental information on liquid financial instruments such as government bonds. The logic behind this is based on the efficient markets hypothesis of Fama (1970) according to which prices adjust as new information is revealed. Thus, the more informationally important data releases will have higher price impact whereas the less informationally important data releases will have a lower one. Taking the particular example of the 2-year bond, calculations show that the typical NFP release explains around 28% of the variance of the US 2-year bond for the day, and a one-unit change in surprise in the NFP release results in a 0.53-unit move in the 2-year bond yield, which is significant at the 0.001 level. In contrast, the surprise in the unemployment rate

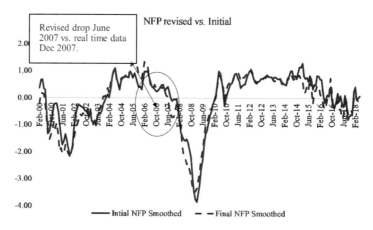

Figure 26.15 How estimating economic cycle turning points using final NFP vs. initial NFP estimates can yield different results.

accounts for *almost none* (0.8%) of the variation in the 2-year bond yield and a one-unit surprise in unemployment rate only accounts for 0.06 units of movement in the US 2-year bond yield[18].

The longer-term economic purpose of the NFP report is to help estimate the trend growth of the economy rather than just predict the one-day moves in the asset markets. As noted before, this trend growth in the US labor markets is of vital importance to the Federal Reserve for conducting monetary policy, as well as to corporations, investors, and citizens who wish to understand where the economic cycle is for making longer-term decisions (Baumol, 2013; Kliesen, 2014; Lin et al., 2019; Jain, 2019). Substantial revisions to NFP and other macroeconomic data make it difficult to truly gauge the current state of the economy and hence make optimal policy decisions in real time[19].

26.5.3 Information content of NFP revisions

Mankiw et al. (1984) suggest two possible outcomes for data revisions: news or noise. In their context, news is defined as extra information uncorrelated with the initial release that was revealed in the final numbers and noise as a forecastable

---

[18] Since both data releases occur on the same day, we also run a regression with both variables (not shown here) and find that indeed the t-statistic associated with NFP surprises is 9.4 compared with that for unemployment at 0.32. The results for IJC are similar.

[19] Specifically, for NFP, the absolute revision is about 102,000 jobs when economic times are bad compared with 57,000 jobs when good. Thus, initially NFP underestimates the extent of job losses at the starting of downturns and also consistently underestimates the job gains when the recession ends. Figure 26.15 shows two lines: in blue is represented the analysis conducted using the data available in real time; the dashed red line represents the analysis conducted with the benefit of hindsight using the revised data available 1–3 months later. As we can see, during the 2007–2008 financial crises, the initial NFP data would be delayed by a period of 6 months, compared with the Final NFP data, for determining the downturn as well as the upturn.

Table 26.4 News vs. noise regressions for NFP. Standard errors are Newey–West corrected. The β for news regressions is clearly close to 1 and for noise is significantly different than 1. A double asterisk, **, denotes statistical significance at the 1% level, whereas *** denotes it at the 0.1% level.

	News vs. Noise regressions	
	Noise hypothesis (Initial ~ Final)	News hypothesis (Final ~ Initial)
Intercept	1.73	14.89**
(NW t-stat)	(0.39)	(3.17)
Slope	0.89	0.99***
(NW t-stat)	(38.6)	(38.1)
R^2	0.87	0.87
N	492	492

Table 26.5 Granger causality tests for information in initial vs. final NFP.

Granger Causality Wald Tests				
Equation	Excluded	Chi2	df	Prob. > Chi2
NFP Initial$_{t+1}$	NFP Final$_{t+3}$	46.38	3	0.00
NFP Final$_{t+3}$	NFP Initial$_{t+1}$	5.83	3	0.12

correction that is unrelated to the final number. They test these two competing hypotheses by performing correlations of the revisions with the initial number (noise hypothesis) and final number (news hypothesis) as well as regressions. We find that both the correlation[20] and regression results correspond closer to the news hypothesis for NFP[21].

We further formalize the intuitive notion illustrated by Figure 26.15 and the analysis in Table 26.4 that the final NFP releases do contain *more* information than the initial NFP releases, rather than a mere rebasing, with a Granger (1969) causality test. The tests in Table 26.5 show that the final NFP data "Granger cause" the initial NFP data but not vice versa. In other words, knowledge of the final NFP numbers is useful in predicting the initial NFP numbers but knowing the initial NFP numbers is not useful in predicting the path of future final NFP numbers. Thus, it seems reasonable to conclude that final NFP numbers contain more forward-looking and correct information about the path of the economy than do the initial NFP numbers.

[20] The correlation between NFP revisions and the initial NFP is −0.043 with (−0.16, 0.10) as the bootstrapped 95% confidence interval, whereas final NFP and NFP revisions are correlated at +0.31, (+0.19, 0.44) as the 95% confidence interval.

[21] Like Aruoba (2008), we find a significant intercept for the news hypothesis but we believe the bias to be more "local" due to a lack of speed – in the fashion Croushore (2011a) suggests – than when we have recessions, the initial numbers being slow to reflect how bad the data are, and vice versa.

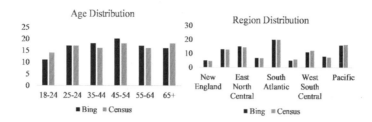

Figure 26.16 Demographic and geographic distribution of Bing searches compared with US population. Sources: MSFT and US Census.

Finance
insurance jobs in nc- sales financing jobs
Baltimore- medical billing supervisor job
description- collections jobs loisvelle
Retail
Simon outlet mall in norfoplk va job fair- small gift
shop job openings nj- Ducks sportings goods
careers- Job Description for an retail pharmacy
technician
Technology
"digital channels manager" job -Electrical
systems technology jobs alabama -altec careers
php- computer forensic jobs chandler az

Figure 26.17 Examples of hand curation of job searches into employment sectors.

26.5.4 Web search data

We collected[22] employment-related searches on MSFT's search engine Bing from January 2012 to March 2018: a total of 74 months. These searches were from both mobile and desktop devices and constrained to only be in the English language for the country region of United States. We follow the same technique Chancellor and Counts (2018) used in cleaning the data by only using four keywords: "job", "jobs", "career", and "careers" as reflecting the best trade-off for actual results and minimizing false positives, and we remove queries with co-occurring irrelevant words ("Steve Jobs"). This yields a total of around 830 million employment-related queries. Before constructing various search-related variables, we checked if search data are substantially demographically or geographically biased across states and we found that this is not so, as shown in Figure 26.16.

There are two main types of employment-related searches: broad and sector-specific. The former are typed into MSFT's Bing search engine with no specific job sector associated with them. This broad category is captured by "overall"

---

[22] Ethics Review and Data Protection: This study was undertaken in line with the Common Rule for exemption by the Microsoft Research Ethics Advisory Board under Protocol 7. Our data were gathered historically: there was no interaction with users by changing search results. No session information is used in our dataset. All data was anonymized and aggregated to county level and to national level for the final analysis. Our use and storage of this data is in agreement with Bing's End User License Agreement and Privacy Policy.

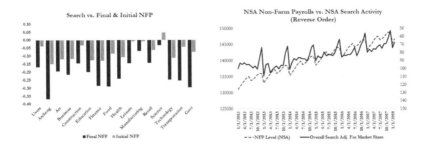

Figure 26.18 On the left, Correlation of employment-related search categories compared with initial NFP (gray) and final NFP (red) release. On the right, not-seasonally-adjusted (NSA) overall employment-related searches adjusted for Bing market share changes compared with NSA NFP levels. Data Sources: MSFT, FRED, Bloomberg.

employment-related searches. An example would be a search like "Jobs in Seattle." The second type of employment-related searches are sector-specific: their construction follows the technique used by Chancellor and Counts (2018) where job searches are carefully curated and matched to economic sectors obtained from about 15 BLS categories specified in the NFP report. Figure 26.17 provides an example of the hand-curation of searches with examples of specific searches that were grouped into various job categories corresponding to the NFP data releases. Figure 26.18 shows the plot of the "overall" employment-related searches. We note that, as the unemployment rate declined, the searches for employment also declined from January 2012 to March 2018.

To ensure our results are robust to normalization, we compute the correlations (and the regressions) *both* ways – normalized by total searches and not normalized – and find very similar results. The distribution of monthly changes in search is substantially more robust than the distribution of levels, which can change dramatically depending on the denominator. Jain (2019) discusses the normalization robustness issue in greater detail. The added intuitive benefit of relying more on the monthly changes in search is that, on this time scale, they are driven more by changes in population search behavior than by gains or losses in the search browser market share or aother normalization factors.

26.5.5 Search and NFP correlation

After seasonally adjusting both series, we performed a correlation analysis of the month-over-month *changes* in the 16 search signals; that is, the 15 curated employment sectors and the broad employment search category of "Overall" with the *changes* in initial and final NFP data releases for the period from 2012 to 2018[23]. The results are displayed in Table 26.6, which also contains the boot-

---

[23] Correlation of raw or NSA employment search first-differences with NSA NFP first-differences is 0.70, but that overestimates the economic relationship since seasonal patterns are predictable.

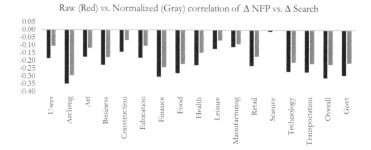

Figure 26.19 NFP (differences) compared with Search (differences) correlations, seasonally-adjusted raw differences (red) and seasonally-adjusted normalized differences (gray). These delta correlations do not change much with normalization, in contrast to levels relationships.

Table 26.6 Correlations of employment-related searches in various sectors with the initial NFP and final NFP. Median correlations are in bold and the bootstrapped 5th percentile and the 95th percentile correlations for both are displayed.

Search Category	Initial NFP 50th%ile	Initial NFP 5th%ile	Initial NFP 95th%ile	Final NFP 50th%ile	Final NFP 5th%ile	Final NFP 95th%ile
Overall	**−0.17**	−0.36	0.03	**−0.30**	−0.46	−0.12
Users	**−0.04**	−0.23	0.15	**−0.17**	−0.37	0.03
Arch.& Engg.	**−0.15**	−0.35	0.06	**−0.37**	−0.52	−0.20
Art	**−0.12**	−0.33	0.09	**−0.20**	−0.41	0.04
Business	**−0.12**	−0.32	0.09	**−0.21**	−0.39	−0.03
Construction	**−0.03**	−0.26	0.20	**−0.15**	−0.35	0.07
Education	**−0.12**	−0.33	0.09	**−0.20**	−0.41	0.04
Finance	**−0.13**	−0.31	0.07	**−0.29**	−0.44	−0.11
Food	**−0.08**	−0.25	0.09	**−0.29**	−0.47	−0.10
Healthcare	**−0.11**	−0.28	0.08	**−0.24**	−0.40	−0.07
Leisure & Hospitality	**−0.01**	−0.21	0.21	**−0.14**	−0.36	0.09
Manufacturing	**−0.01**	−0.19	0.28	**−0.07**	−0.26	0.13
Retail	**−0.06**	−0.25	0.13	**−0.14**	−0.39	0.11
Science	**0.05**	−0.19	0.28	**−0.03**	−0.31	0.23
Technology	**−0.11**	−0.30	0.10	**−0.25**	−0.42	−0.06
Transportation	**−0.04**	−0.25	0.18	**−0.25**	−0.46	−0.03
Government	**−0.07**	−0.30	0.16	**−0.30**	−0.50	−0.07

strapped 5% and 95% confidence intervals for the correlation metric. The search delta correlations follow the same pattern as the levels in the previous section and are negatively correlated with the number of jobs created. There are two potential intuitions behind the negative correlation of *net* jobs created (NFP) and the employment-related searches or demand for employment information. The first is that when economic times are good, as signalled by a low unemployment rate, it may require less effort to find jobs, thereby lowering the number of searches *per potential applicant.* The second is that in good economic times fewer people may be looking to change jobs, thus lowering the number of applicants.

We also find that the search data are more correlated across categories to the final or the revised NFP numbers that are only available 1–3 months after the initial NFP number. The stronger correlation with the final NFP numbers compared with the initial NFP numbers suggests that search data may contain forward-looking information relevant to the revised final NFP numbers. As evidence of robustness, correlations with the final NFP numbers are negative, as well as higher in magnitude. The result is consistent across the 15 hand-curated sectors in addition to the "Overall" broad category of generic job searches[24].

We now investigate the efficacy of search data in nowcasting the final NFP numbers more formally. We would like to test if search data truly contain information not completely captured by the initial NFP as well as the existing baseline estimates of the Wall Street analysts. Towards that objective we run several regressions of the form:

$$\text{Final NFP}_{t+3} = \alpha + \beta_1 \cdot \text{InitialNFP}_{t+1} + \beta_2 \cdot \text{LaggedInitialNFP}_t$$
$$+ \beta_3 \cdot \text{EmploymentSearch}_t + \beta_4 \cdot \text{SectorSpecificSearch}_t$$
$$+ \beta_5 \cdot \text{EconomistsSurvey}_t + \epsilon_t \quad (26.9)$$

26.5.6 Regression results

The results of the regressions formulated above are displayed in Table 26.7. We compute the standardized coefficients in order to understand the impact of a 1-unit move in the independent variable on the dependent variable by multiplying the actual ordinary least squares (OLS) regression coefficient by the standard deviation of the independent variable and dividing the result by the standard deviation of the dependent variable. All the t-statistics are adjusted for heteroskedasticity and autocorrelation using Newey and West (1987) standard errors. In the context of NFP, a 1-unit change is about 74,000 jobs. The first regression shows that indeed Wall Street analysts' predictions are significantly (at 0.001 level) related to the final NFP number. The second regression shows that adding "Overall" employment-related searches improves the prediction by increasing the R^2 by 4% with both Wall Street analysts and "Overall" employment-related searches being statistically significant. However, we can see that analysts have more impact (0.45

---

[24] Granger analysis of search data compared with NFP. We have only 72 monthly data points which is insufficient for serious Granger tests. We used weekly search data that increases the number of search data points. Though this is not ideal (since government data are not weekly), it did reveal some interesting patterns. Broadly speaking, search data are useful for nowcasting or predicting final NFP, initial NFP and Wall Street analysts' consensus but such data are not useful for predicting the signals based on search. We also find that Wall Street analysts' consensus is helpful for predicting the initial NFP numbers but not the final NFP number. Interestingly, it is the employment-related searches in the *second* week of the month under consideration – which typically happens to be the reference week for NFP – that has the most power. These second-week searches are helpful in predicting both the analysts' survey and the final NFP numbers, but not the other way around.

Table 26.7 Regressions to forecast final NFP numbers using various predictors. All coefficients $\beta(i)$ standardized as $\beta(i) * sd(X(i))/sd(Y)$ to enable easy comparisons across formulations. The time subscripts indicate when the data become available. Final NFP is marked as final NFP_{t+3} because the final NFP data typically become available 2–3 months after the reference month$_t$ and search data used are from the same reference month$_t$ as are the Wall Street analysts' survey$_t$. Note that previous lag of NFP is expressed at NFP_t since it becomes available earlier in the same month. Robust Newey–West t-statistics that account for serial correlation and heteroskedasticity are shown in parentheses. A single asterisk, *, denotes statistical significance at the 5% level, whereas a triple asterisk denotes it at the 0.1% level.

			Predicting final NFP_{t+3} regressions			
Initial NFP_{t+1}	Initial NFP_t (lagged)	Overall employment search$_t$	Sector-specific search$_t$	Analysts' survey$_t$	R^2	N
				0.45*** (+4.63)	0.20	73
	−0.20* (−2.15)			0.45*** (+4.63)	0.24	73
0.81*** (+9.47)				0.01 (+0.1)	0.66	73
0.79*** (+9.74)		−0.14* (−2.05)		0.02 (+0.19)	0.68	73
0.80*** (+9.44)	−0.08 (−0.97)			0.04 (0.45)	0.66	73
0.79*** (+9.69)	−0.07 (−1.11)	−0.14* (−2.38)		0.05 (0.66)	0.69	73
0.78*** (+9.55)			−0.22 (−3.89)	0.05 (0.68)	0.71	73

units or 33,000 jobs) on the final NFP prediction versus search data (−0.2 units or 14,500 jobs). The results in these and the following pairs of regressions are very similar when comparing adjusted R^2 that accounts for the increase in the number of predictors.

The third and fourth regressions show that when the noisy initial NFP number is released, the R^2 jumps from 24% to 68%, and interestingly, while the Wall Street analyst information is completely absorbed and no longer incrementally useful, the search information is still relevant with only a marginal decrease in magnitude from −0.20 to −0.14. Upon adding "Overall" employment search data, the R^2 increases by about 2%.

In the fifth and sixth regressions, we add the prior lags of NFP to account for the trending nature of the macroeconomic data. We observe a similar pattern: Wall Street analyst information is no longer useful after the initial NFP number is released but the "Overall" employment search data retains both its statistical (-2.38) and economic (-0.14 units) significance. Again, adding the "Overall" employment search data increases the R^2 by 3%, which suggests that the information in search data is not fully captured by the previous macroeconomic trends, the initial NFP numbers or Wall Street analyst consensus, and hence potentially different.

We also check the impact of focusing on the portions of employment web search, where government data are not as good, by using the architecture and engineering employment-related searches and find that those are highly significant (-0.22 units) with a t-statistic (-3.89) that is meaningful at the 1% level. Also, the R^2 increases by 5% to 71% when compared with the previous model that uses only Wall Street analysts' estimates, lagged macroeconomic trends and the noisy initial NFP release. This result suggests that web search for cyclical sectors such as architecture might contain useful information for the entire economy that is not captured by the government data.

26.5.7 Robustness

We perform an out-of-sample forecast of final NFP by keeping 30% of the sample as holdout and perform two separate next-month forecasts – the first with the initial NFP release and Wall Street analyst consensus and the second by adding "Overall" employment-related search data to the first regression to check the impact of web search data. We find that MSE decreases by 5.8% in the test sample and a similar 3.8% in the holdout set. Interestingly, results improve out-of-sample – with the MSE decreasing by 11.6% in-sample and 21.2% out-of-sample when we use architecture and engineering employment searches instead of the "Overall" employment-related searches in the model. Architecture and engineering, (as explained mentioned above), represent urban-centric, cyclical, white-collar service categories for which search data appear better suited than the BLS data for estimation purposes. While some of the 21.2% decrease in the holdout MSE could be attributed to a small sample size, the results encourage us towards doing more research on curated sectors.

In alternative regressions not included here, for reasons of space, we formulate the dependent variable directly as *Revisions* = Final NFP − Initial NFP and obtain very similar results. The variable *Revisions* can be robustly predicted by the *Overall Search* variable after controlling for its own prior lags and Wall Street analyst consensus. We also perform regressions by using the median instead of the mean and find similar and slightly better results, perhaps because medians are more stable for noisier search data.

Occasionally, using particular subsets of alternative data can potentially misrepresent how powerful the overall data are. For instance, Altenburger and Ho (2019) found that the reported results of Yelp reviews outperforming food safety inspections do not stand up to comprehensive analysis. When the entire sample of restaurants was used, rather than "Extremely Imbalanced Sampling" of restaurants with no violations or very high violations, the problem changed from one

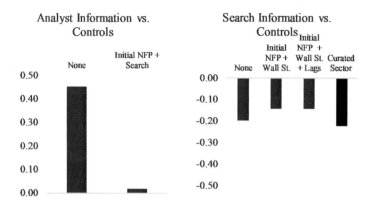

Figure 26.20 Generic employment-related web-search information (right) remains relevant after controlling for initial NFP and Wall Street analysts' information. But Wall Street analysts' information (left) is insignificant after initial NFP is released. Curated web searches (in blue) for sectors (such as architecture and engineering) with imbalance of information demand and supply are more predictive than generic employment web searches.

of classification into those extreme categories to one of score prediction for the entire sample. The results then were far less impressive. To avoid such errors, our case study uses a regression setup that analyzes *all* NFP observations, not just those at turning points. However, we obtain significant results with the "Overall" employment searches and not just with one of the sectors. While sectors such as architecture, where the information demand, via search and supply from government surveys is the highest, yield more significant results, reassuringly our results are similar across all sixteen categories[25].

26.5.8 Machine learning for NFP revisions

Thus far, the analysis has tested the *ex ante* hypotheses about *orthogonal, more accurate, and more timely* information in the search-based features. We find that NFP revisions do contain news or information that is released with a lag of up to three months, and a powerful and intuitive relationship exists between employment search and NFP that can help to meaningfully improve data timeliness and accuracy. We did not find a substantial age and geographic bias in the data, but the sector-specific results suggest search data do measure tech-centric and white-collar firms even better. How much of the difference comes from the underreporting of sectors such as architecture by the government data and how

[25] Models based on these insights were put into production and the results for NFP improved MSE by about 10% over four years which is broadly in line with the reported theoretical performance. Another conceptual and methodological verification was to use similar models to predict other volatile and crucial macroeconomic metrics, such as retail sales, for which the models outperformed Wall Street analysts by about 20% on the downturns and by about 5% on the upturns.

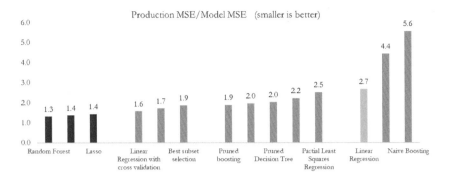

Figure 26.21 An MSE-based comparison of various simple ML techniques used to build an employment nowcast using the search NFP features in the case study. Each vertical bar in the figure is the ratio of MSE in production to the MSE of the model. The production or out-of-sample time period was about 24 months.

much from the urban and technology bias of the alternative data will need more investigation.

Having verified the hypotheses, the next stage is to use various machine learning techniques that minimize MSE to effectively incorporate these search-based features into an employment nowcast. Generally, multiple alternative data sources are also combined at this stage. As Figure 26.21 shows, for nowcasting NFP from web search, we compared various machine learning techniques to the traditional linear regression. Each vertical bar in the figure is the ratio of MSE in production to the MSE of the model and hence a test of overfitting. The actual MSE also follow the same pattern and act as a test of model fit. Random forest, ridge and lasso cluster together, with the best performance of out-of-sample MSE being 130% to 140 % of the in-sample model fit. SVM, partial least squares, decision trees, boosting and linear regression had out-of-sample MSEs that ranged from 220% to more than 400%. Adding cross-validation and stepwise variable selection improves linear regression performance but for this particular case, it still could not out-perform the best machine learning techniques. This conclusion, of higher effectiveness of random forest and elastic net compared with traditional linear regression techniques, is consistent with findings in literature cited previously in the chapter, as well as the broader machine-learning fieldwork such as Caruana and Niculescu-Mizil (2006).

Search, and most other social data like Twitter and Reddit, have "spiky" distributions – meaning they are not normal and can have a non-zero skew as well as high kurtosis. This can make asymptotic inference tricky and bootstrapping and its many variants should be used to provide a more realistic idea of the distribution. Relatedly, these data almost always have extreme observations and one must decide whether to exclude ("clean") or include them and deal with as an "influential point". This is best done in close consultation with a domain expert and can have a dramatic impact on the model. Ensembling, or combining various internally consistent models built on different philosophies that are naturally stronger in different parts of the parameter space, is a useful common practice that also increases robustness.

26.6 Conclusion and future work

This chapter presented three fundamental problems with the current macroeconomic data: high dimensionality, a staggered release schedule, and poor data quality. It provided an overview of nowcasting techniques that are used to address the problems of dimensionality and staggered release. Traditionally, nowcasting techniques were based on dynamic factor models and state-space frameworks, but more recently have started incorporating methods from machine learning. The arrival of big or alternative data with the potential for addressing the data quality problem has made macro nowcasting an exciting field. But big data is a double-edged sword – just as it promises to improve data quality, it simultaneously worsens the dimensionality problem by dramatically increasing the potential feature set. To address this, we discussed the need for an *ex ante* hypothesis to guide alternative data selection, provided some sample hypotheses for popular alternative data sources, and noted the typical data quality issues, such as improving the timeliness and increasing granularity, where alternative data are having success.

The chapter culminated in a case study that utilized a novel dataset of 830 million employment-related searches in order to nowcast NFP revisions three months ahead of government data with statistical (t-statistic of -2.4) and economic (-0.14 standardized unit impact) significance – a result that contrasts with earlier work. The case study also established that NFP revisions contain relevant information or "news" (Mankiw et al., 1984), and controlled for the information content of Wall Street analysts' expectations as well as for prior lags of NFP. After verifying the improved data quality hypotheses about the increased timeliness and accuracy, various machine-learning models were used to maximize predictive power. Random forest and elastic net models performed best, generating only half the MSE of linear regression in both the in-sample model fit and out-of-sample tests.

I suggest three related but distinct strands of development for the field. The first is an increase in the number and sophistication of statistical learning techniques being used. The second is to deal better with the more complex workflow for the insight generation process involving the data engineer, data scientist and domain expert. The third is an increased focus on privacy concerns and an increased investment in security. The first – machine learning techniques – concern specialists more, the more complex workflow concerns the entire firm, and the third – privacy and security – concerns us all.

The number of machine-learning methods is too vast to comment on specifically, but deep learning and other computationally intensive, feature-rich, and nonlinear techniques will find more success when applied towards *"feature generation"* – say via NLP of accounting statements or analysis of pictures and videos

– where the number of features (*p*) compared with the number of observations (*N* or *T*) problem is not a primary issue. That said, broader use of machine-learning methodologies has great potential, ranging from cross-validation for choosing the optimal number of parameters, to building the next generation of nonlinear regime-switching models, to generating better economic activity clusters.

Moving from the concerns of the machine-learning specialist to those of the entire firm, the more complex workflow problem that now has to account for everything from data selection and purchase frameworks to technical aspects like data pipelines and visualization is non-trivial. The different strengths and weaknesses of the players (Kolanovic and Krishnamachari, 2017; Jain, 2019) can create organizational challenges – for instance, a data scientist may find it hard to write production-level code just as data engineers may find it hard to clean the data without the scientist or the domain expert. Making fast, relevant and efficient machine-learning pipelines is a technical problem, but the tremendous activity in open-source tools has been helpful[26].

And finally, as a society – the emerging privacy and ethical concerns are a big question for us all[27]. Acquisti et al. (2016) provide a good overview of the economic value and the consequences of privacy and personal information. Kerry (2020) highlight the gap between existing laws and the data reality of today where most people simply do not and cannot read the various privacy policies. It is possible to "re-identify" individuals with incomplete datasets – for instance Rocher et al. (2019) estimate that 99.98% of the individuals can be re-identified with any dataset containing 15 demographic attributes, which is concerning and might make one argue for very strict laws and rules against data collection. However, the response to the COVID-19 pandemic demonstrated that having *some* aggregate information, for instance at the zip code level[28], could be helpful for policymakers and society. Finding the right balance, where we violate privacy the least while providing the most social benefit, is our task as citizens[29].

References

Acquisti, Alessandro, Taylor, Curtis, and Wagman, Liad. 2016. The economics of privacy. *Journal of Economic Literature*, **54**(2), 442–92.

Adler, Natalia, Cattuto, Ciro, Kalimeri, Kyriaki, Paolotti, Daniela, Tizzoni, Michele, Verhulst, Stefaan, Yom-Tov, Elad, and Young, Andrew. 2019. How search engine data enhance the understanding of determinants of suicide in India and inform prevention: observational study. *Journal of Medical Internet Research*, **21**(1), e10179.

Altenburger, Kristen M., and Ho, Daniel. 2019 (05). Is Yelp actually cleaning up the restaurant

---

[26] Kolanovic and Krishnamachari (2017) provides an overview of the industry-standard open-source tools useful for asset management.

[27] See Nature (2019, Editorial), or Lane et al. (2014).

[28] Chetty et al. (2020) show, among other things, the differential impact of COVID-19 in various populations using anonymized data in a live updateable tool that was widely reported by newspapers and utilized by academics.

[29] Jain and Seeman (2019) take the illustrative example of a fictional asset manager who invests money to obtain "extra features" that do not just violate privacy but provide no marginal predictive benefits. They suggest including potential privacy costs in assessing a dataset for purchase.

industry? A re-analysis on the relative usefulness of consumer reviews. Pages 2543–2550 of: *The Web Conference (WWW)*, https://doi.org/10.1145/3308558.3313683.

Antenucci, Dolan, Cafarella, Michael, Levenstein, Margaret C., Re, Christopher, and Shapiro, Matthew D. 2014. Using social media to measure labor market flows. Working paper, National Bureau of Economic Research, https://www.nber.org/papers/w20010.

Artola, Concha, Pinto, Fernando, and de Pedraza Garcia, Pablo. 2015. Can internet searches forecast tourism inflows? *International Journal of Manpower*, **36**(04), 103–116.

Aruoba, S. Borağan. 2008. Data revisions are not well behaved. *Journal of Money, Credit and Banking*, **40**(2–3), 319–340.

Askitas, Nikos, Zimmermann, Klaus, and Askitas, Nikolaos. 2009. Google econometrics and unemployment forecasting. *Applied Economics Quarterly*, **55**(05), 107–120.

Atlanta Federal Reserve Bank. 2021. https://www.frbatlanta.org/research/publications/wp/2014/07.aspx.

Bauer, Peter, Thorpe, Alan, and Brunet, Gilbert. 2015. The quiet revolution of numerical weather prediction. *Nature*, **525**(7567), 47–55.

Baumol, Bernard. 2013. *The Secrets of Economic Indicators: Hidden Clues to Future Economic Trends and Investment Opportunities*. Pearson Education, Inc.

Bloomberg. 2020. https://www.bloombergquint.com/global-economics/next-u-s-jobs-report-set-to-be-even-more-confusing-than-may-s.

Bok, Brandyn, Caratelli, Daniele, Giannone, Domenico, Sbordone, Argia M., and Tambalotti, Andrea. 2017. Macroeconomic nowcasting and forecasting with big data (2017–11–01). FRB of NY Staff Report No. 830.

Bräuning, F., and Koopman, S.J. 2014. Forecasting macroeconomic variables using collapsed dynamic factor analysis. *International Journal of Forecasting*, **30**(3), 572–584.

Brave, Scott, Cole, Ross, and Kelley, David. 2019. A "big data" view of the US economy: Introducing the Brave–Butters–Kelley indexes. *Chicago Fed Letter*, 422. https://www.chicagofed.org/publications/chicago-fed-letter/2019/422.

Brave, Scott A., Butters, R. Andrew, and Fogarty, Michael. 2020. The perils of working with Big Data and a SMALL framework you can use to avoid them. FRB of Chicago Working Paper No. 2020–35. Available at SSRN 3753859.

Brownstein, John S., Chu, Shuyu, Marathe, Achla, Marathe, Madhav V., Nguyen, Andre T., Paolotti, Daniela, Perra, Nicola, Perrotta, Daniela, Santillana, Mauricio, Swarup, Samarth, Tizzoni, Michele, Vespignani, Alessandro, Vullikanti, Anil Kumar S., Wilson, Mandy L., and Zhang, Qian. 2017. Combining participatory influenza surveillance with modeling and forecasting: three alternative approaches. *JMIR Public Health Surveill.*, **3**(4), e83.

Burns, Arthur F., and Mitchell, Wesley C. 1946. *Measuring Business Cycles*. National Bureau of Economic Research.

Caruana, Rich, and Niculescu-Mizil, Alexandru. 2006. An empirical comparison of supervised learning algorithms. Pages 161–168 of: *Proc. 23rd International Conference on Machine Learning*.

Cavallo, Alberto. 2013. Online and official price indexes: Measuring Argentina's inflation. *Journal of Monetary Economics*, **60**(2), 152–165.

Cavallo, Alberto, and Rigobon, Roberto. 2016. The billion prices project: using online prices for measurement and research. *Journal of Economic Perspectives*, **30**(05), 151–178.

Chancellor, Stevie, and Counts, Scott. 2018. Measuring employment demand using Internet search data. Pages 1–14 of: *Proc. 2018 Conference on Human Factors in Computing Systems*.

Chen, Jeffrey C., Dunn, Abe, Hood, Kyle, Driessen, Alexander, and Batch, Andrea. 2019. Off to the races: A comparison of machine learning and alternative data for predicting economic indicators. In: *Big Data for 21st Century Economic Statistics*. University of Chicago Press.

Chetty, Raj, Friedman, John N., Hendren, Nathaniel, Stepner, Michael et al. 2020. The economic impacts of COVID-19: Evidence from a new public database built using private sector data. Working paper, National Bureau of Economic Research. `https://www.nber.org/papers/w27431`.

Chicago Federal Reserve Bank. 2020a. `https://www.chicagofed.org/research/dual-mandate/dual-mandate`.

Chicago Federal Reserve Bank. 2020b. Background on the Chicago Fed National Activity Index. `https://www.chicagofed.org/publications/cfnai/index`.

Chicago Federal Reserve Bank. 2021. `https://www.chicagofed.org/publications/bbki/index`.

Choi, Hyunyoung. 2009. Predicting initial claims for unemployment benefits. Available at SSRN 1659307.

Choi, Hyunyoung, and Varian, Hal. 2012. Predicting the present with Google trends. *Economic Record*, **88**(s1), 2–9.

Clauset, Aaron, Shalizi, Cosma Rohilla, and Newman, Mark E.J. 2009. Power-law distributions in empirical data. *SIAM Review*, **51**(4), 661–703.

Clement, Bortoli, and Stephanie, Combes. 2015 (04). Contribution from Google Trends for forecasting the short-term economic outlook in France: limited avenues. In *A Drop of Fuel to Drive Consumption and Margins*, Institute Nationale de la Statistique et des Études Économiques. `https://www.insee.fr/en/statistiques/1408911`.

Cook, Thomas R., amd Hall, Aaron Smalter. 2017. Macroeconomic indicator forecasting with deep neural networks. Federal Reserve Bank of Kansas City Working Paper. `https://dx.doi.org/10.18651/RWP2017-11`.

Coulombe, Philippe, Leroux, Maxime, Stevanovic, Dalibor, and Surprenant, Stéphane. 2022. How is machine learning useful for macroeconomic forecasting? *Journal of Applied Economics*, **35**(5), 920–964.

Coyle, Diane. 2014. *GDP: A Brief but Affectionate History*. Princeton University Press.

Croushore, Dean D.. 2011a. Frontiers of real-time data analysis. *Journal of Economic Literature*, **49**(1), 72–100.

Croushore, Dean D. 2011b. Real-time forecasting. Pages 7–24 in: *Advances in Economic Forecasting*, Matthew L. Higgins (ed). W.E. Upjohn Institute for Employment Research.

D'Amuri, Francesco, and Marcucci, Juri. 2010. 'Google It!' Forecasting the US unemployment rate with a Google job search index. `https://mpra.ub.uni-muenchen.de/18248/`.

Datar, Srikant, Jain, Apurv, Wang, Charles C.Y., and Zhang, Siyu. 2020. Is accounting useful for forecasting GDP growth? A machine learning perspective. Available at SSRN 3827510.

Diebold, Francis. 2000. Big data dynamic factor models for macroeconomic measurement and forecasting. Pages 115–122 in: *Advances in Economics and Econometrics: Theory and Applications, Eighth World Congress (Vol. 3)*, M. Dewatripont, L.P. Hansen, and S.J. Turnovsky (eds). Cambridge University Press.

Döpke, Jörg, Fritsche, Ulrich, and Pierdzioch, Christian. 2017. Predicting recessions with boosted regression trees. *International Journal of Forecasting*, **33**(4), 745–759.

Dumais, Susan, Jeffries, Robin, Russell, Daniel, Tang, Diane, and Teevan, Jaime. 2014. Understanding user behavior through log data and analysis. Pages 349–372 in: *Ways of Knowing in HCI*, J. Olson and W. Kellogg (eds). Springer.

Einav, Liran, and Levin, Jonathan. 2013. The data revolution and economic analysis. *Innovation Policy and the Economy*, **14**(05), 1–24.

Fama, Eugene F. 1970. Efficient capital markets: A review Of theory and empirical work. *Journal of Finance*, **25**(2), 383–417.

FRED 2021. `https://fred.stlouisfed.org`.

Goel, Sharad, Hofman, Jake M., Lahaie, Sebastien, Pennock, David M., and Watts, Duncan. 2010. Predicting consumer behavior with web search. *Proc. NAS*, **107**(09), 17486–90.

Granger, C. W. J. 1969. Investigating causal relations by econometric models and cross-spectral methods. *Econometrica*, **37**(3), 424–438.

Hall, Aaron Smalter. 2018. Machine learning approaches to macroeconomic forecasting. pages 63–81 in *Federal Reserve Bank of Kansas City Economic Review*, November. DOI: 10.18651/ER/4q18SmalterHall.

Hastie, Trevor, Tibshirani, Robert, and Friedman, Jerome. 2009. *The Elements of Statistical Learning*, Second edition. Springer.

Hellerstein, Rebecca, and Middlethorp, Menno. 2012. Forecasting with internet search data. Liberty Street Economics. https://libertystreeteconomics.newyorkfed.org/2012/01/forecasting-with-internet-search-data/.

Jain, Apurv. 2019. Macro forecasting using alternative data. Pages 273–327 in: *Handbook of US Consumer Economics*, Andrew Haughwout and Benjamin Mandel (eds). Elsevier.

Jain, Apurv, and Seeman, Neil. 2019. Privacy vs. alpha: A conversation. Available at SSRN 3474983.

James, Gareth, Witten, Daniela, Hastie, Trevor, and Tibshirani, Robert. 2013. *An Introduction to Statistical Learning*. Springer.

Jansen, Bernard J., and Schuster, Simone. 2011. Bidding on the buying funnel for sponsored search and keyword advertising. *Journal of Electronic Commerce Research*, **12**(1), 1.

Kässi, Otto, and Lehdonvirta, Vili. 2018. Online labour index: Measuring the online gig economy for policy and research. *Technological Forecasting and Social Change*, **137**, 241–248.

Kerry, Cameron F., Morris, John B., Chin, Caitlin, and Lee, Nicol Turner. 2020. Bridging the gaps: A path forward to federal privacy legislation. Brookings Institute Report. https://www.brookings.edu/research/bridging-the-gaps-a-path-forward-to-federal-privacy-legislation/.

Keynes, J.M. 1936. *The General Theory of Employment, Interest and Money*. Palgrave Macmillan.

Kliesen, Kevin L. 2014. A Guide to Tracking the US Economy. *Federal Reserve Bank of St. Louis Review*. https://files.stlouisfed.org/files/htdocs/publications/review/2014/q1/kliesen.pdf.

Kolanovic, M., and R. Krishnamachari. 2017. Big data and AI strategies: machine learning and alternative data approach to investing. J.P. Morgan Quantitative and Derivative Strategy Report.

Lane, Julia, Stodden, Victoria, Bender, Stefan, and Nissenbaum, Helen (eds). 2014. *Privacy, Big Data, and the Public Good: Frameworks for Engagement*. Cambridge University Press.

Lin, Allen Yilun, Cranshaw, Justin, and Counts, Scott. 2019. Forecasting US Domestic Migration Using Internet Search Queries. Pages 1061–1072 in: *The World Wide Web Conference*. https://doi.org/10.1145/3308558.3313667.

Lo, Andrew W., and Mueller, Mark T. 2010. Warning: physics envy may be hazardous to your wealth! ArXiv:1003.2688.

Lopez de Prado, Marcos, and Lipton, Alex. 2020. Three quant lessons from COVID-19. Available at SSRN 3562025.

Mankiw, N. Gregory, Runkle, David E., and Shapiro, Matthew D. 1984. Are preliminary announcements of the money stock rational forecasts? *Journal of Monetary Economics*, **14**(1), 15–27.

Medeiros, Marcelo C., Vasconcelos, Gabriel F.R., Veiga, Álvaro, and Zilberman, Eduardo. 2021. Forecasting inflation in a data-rich environment: the benefits of machine learning methods. *Journal of Business & Economic Statistics*, **39**(1), 98–119.

Mislove, Alan, Lehmann, Sune, yeol Ahn, Yong, Pekka Onnela, Jukka, and Rosenquist, J. Niels. 2011. Understanding the demographics of twitter users. In: *Proc. 5th International AAAI Conference on Weblogs and Social Media (ICWSM)*.

Moshiri, Saeed, and Cameron, Norman. 2000. Neural network versus econometric models in forecasting inflation. *Journal of Forecasting*, **19**(04), 201–217.

MOSPI (Ministry of Statistics and Programme Implementation). 2019 http://www.mospi.nic.in/sites/default/files/press_release/Press\%20Note.pdf.

Nature. 2019. . Time to discuss consent in digital-data studies. `https://www.nature.com/articles/d41586-019-02322-z`.

Newey, Whitney, and West, Kenneth. 1987. A simple, positive semi-definite, heteroskedasticity and autocorrelation consistent covariance matrix. *Econometrica*, **55**(3), 703–08.

New York Federal Reserve Bank. 2021. `https://www.newyorkfed.org/research/policy/nowcast`.

O'Neil, Cathy. 2016. *Weapons of Math Destruction: How Big Data Increases Inequality and Threatens Democracy*. Crown.

Orphanides, Athanasios. 2001. Monetary policy rules based on real-time data. *American Economic Review*, **91**(4), 964–985.

Orphanides, Athanasios, and Williams, John C. 2007. Robust monetary policy with imperfect knowledge. *Journal of Monetary Economics*, **54**(5), 1406–1435.

Philadelphia Federal Reserve Bank. 2021a. `https://www.philadelphiafed.org/surveys-and-data/real-time-data-research`.

Philadelphia Federal Reserve Bank. 2021b. `https://www.philadelphiafed.org/surveys-and-data/real-time-data-research/ads`.

Phipps, Polly, and Toth, Daniell. 2012. Analyzing establishment nonresponse using an interpretable regression tree model with linked administrative data. *Annals of Applied Statistics*, **6**(06), 772–794.

Popper, Karl R. 1957. *The Poverty of Historicism*, Routledge.

Richardson, Adam, van Florenstein Mulder, Thomas, and Vehbi, Tuğrul. 2021. Nowcasting GDP using machine-learning algorithms: A real-time assessment. *International Journal of Forecasting*, **37**(2), 941–948.

Rocher, Luc, Hendrickx, Julien M., and De Montjoye, Yves-Alexandre. 2019. Estimating the success of re-identifications in incomplete datasets using generative models. *Nature Communications*, **10**(1), 1–9.

Rothbaum, Jonathan, and Bee, Adam. 2020. Coronavirus infects surveys, too: Survey bias and the coronavirus pandemic. US Census Bureau Working Paper, SEHSD, WP2020–10.

Sargent, Thomas, and Sims, Christopher. 1977. *Business cycle modeling without pretending to have too much a priori economic theory*. Working Paper 55. Federal Reserve Bank of Minneapolis.

Sermpinis, Georgios, Stasinakis, Charalampos, Theofilatos, Konstantinos, and Karathanasopoulos, Andreas. 2014. Inflation and unemployment forecasting with genetic support vector regression. *Journal of Forecasting*, **33**(6), 471–487.

Smith, Aaron. 2015. *Searching for Work in the Digital Era*. Report, Pew Research Center. `https://www.pewresearch.org/internet/2015/11/19/searching-for-work-in-the-digital-era/`.

Soros, George. 2009. *The Crash of 2008 and What it Means: The New Paradigm for Financial Markets*. PublicAffairs.

Stock, James H., and Watson, Mark W. 1989. New indexes of coincident and leading economic indicators. *NBER Macroeconomics Annual*, **4**, 351–394.

Stock, James, and Watson, Mark. 1999. Forecasting in action. *Journal of Monetary Economics*, **44**(01), 293–335.

Stock, J.H., and Watson, M.W. 2016. Dynamic factor models, factor-augmented vector autoregressions, and structural vector autoregressions in macroeconomics. Pages 415–525 in: *Handbook of Macroeconomics*, vol. 2. Elsevier.

Suhoy, Tanya. 2009. Query indices and a 2008 downturn: Israeli data. Research Department, Bank of Israel. Discussion Paper No. 2009.06, 01. `http://www.boi.org.il`.

Venkatramanan, Srinivasan, Sadilek, Adam, Fadikar, Arindam, Barrett, Christopher L., Biggerstaff, Matthew, Chen, Jiangzhuo, Dotiwalla, Xerxes, Eastham, Paul, Gipson, Bryant, Higdon, Dave, et al. 2021. Forecasting influenza activity using machine-learned mobility map. *Nature Communications*, **12**(1), 1–12.

World Economic Forum. 2019. `https://www.weforum.org/agenda/2019/04/how-much-data-is-generated-each-day-cf4bddf29f/`.

Xingjian, Shi, Chen, Zhourong, Wang, Hao, Yeung, Dit-Yan, Wong, Wai-Kin, and Woo, Wang-chun. 2015. Convolutional LSTM network: A machine learning approach for precipitation nowcasting. Pages 802–810 of: *Advances in Neural Information Processing Systems*.

Yahoo Finance. 2020. `https://finance.yahoo.com/news/may-jobs-payroll-surprise-masked-weaknesses-could-make-rebound-a-head-fake-183332405.html`.

Yousuf, Kashif, and Ng, Serena. 2021. Boosting high-dimensional predictive regressions with time-varying parameters. *Journal of Econometrics*, **224**(1), 60–87.

27

Nowcasting Corporate Financials and Consumer Baskets with Alternative Data

Michael Fleder[a] and Devavrat Shah[b]

Abstract

Consider getting a 360-degree view into any company's financials, as if you were the CEO – that is the promise of *alternative data*, side-channel 'alternatives' to direct corporate disclosures. Although public companies disclose financials periodically (e.g., through 10-Qs), such disclosures are infrequent and limited. This has led to an explosion in demand for alternative data sets. Despite the increasing availability of alternative data, quantitative methods for utilizing such noisy proxy signals are lacking.

This chapter provides a tutorial on quantitative methods for utilizing alternative data. Starting with datasets of anonymized consumer transactions, we focus on two problems: (i) nowcasting corporate financials and (ii) estimating the prices consumers pay for individual goods, and in what quantity. That is, first we nowcast aggregate company financials (e.g., quarterly sales) before zooming in to study customers' spending details.

First, we develop the theory for a nowcasting framework, which we then utilize to estimate quarterly sales of 34 public companies. Using consumer transactions as the primary signal, our method outperforms the Wall Street consensus (analyst) estimates, even though our method uses only transactions data as input, while the Wall St. consensus is based on various data sources including experts' input.

Next, we zoom in to study consumer spending details. We perform seemingly counterintuitive inference: given an anonymous consumer's bill total (a single number), we estimate the number and prices of products purchased. Under mild assumptions, we recover each consumer's basket. We apply our algorithm to anonymized transactions associated with consumer spending at Apple, Chipotle, Netflix, and Spotify. Using transactions data only, our algorithm (i) recovers the number of products for sale and the corresponding prices with high accuracy (ii) decomposes transaction totals into product purchases with low error (iii) identifies product launches and (iv) discovers a rumored 'secret' product of Netflix in limited release.

[a] Massachusetts Institute of Technology and Covariance Labs, New York
[b] Massachusetts Institute of Technology
Published in *Machine Learning And Data Sciences For Financial Markets*, Agostino Capponi and Charles-Albert Lehalle© 2023 Cambridge University Press.

27.1 Quant for alt data

What were my competitors' sales yesterday? Which products did their customers buy? In this chapter we show how to answer both questions with the same dataset.

The alternative data industry was born out of demand for economic information that could not be satisfied by traditional data sources. Public-company financials are lagged, infrequent, and aggregated. In contrast, alternative data are real-time, frequent, and granular. However, alternative data are not necessarily accurate. Alt data are proxies for the truth, as they contain myriad statistical biases: non-representative panels, sensor bias, etc. We show how machine learning techniques can be used to correct these issues, make alternative data more powerful, and foster broader adoption. Specifically we address the problem of nowcasting: using widely-available proxy signals to infer ground truth. As we will see, the problem of nowcasting has broad applicability to investors pricing securities, corporations studying their competition, governments looking for real-time economic measures, and intelligence analysts.

This chapter provides a tutorial focusing on (i) nowcasting corporate financials in advance of earnings and (ii) inferring consumer baskets. We tackle both problems with the same data: consumer transactions. We provide techniques for (i) precisely aggregating transactions into sales nowcasts and (ii) disaggregating transactions: inferring individual product purchases.

First, to nowcast corporate financials, we utilize a classical linear systems model to capture both the evolution of the hidden or latent state (e.g., daily sales), as well as the proxy signal (e.g., credit cards transactions). We analytically solve the problem of learning the model parameters ('system identification'), required for utilizing a linear dynamical system (LDS). This is a surprising theoretical result, because in the classical setup, system identification for LDS is irresolvable (Table 27.2). We then apply our framework to estimate quarterly sales of 34 public companies using a dataset of credit card transactions. Our method outperforms the Wall Street consensus (analyst) estimates, even though our method uses only credit card data as input, while the Wall St. consensus is based on various data sources including experts' input.

Next, we zoom in to study consumer spending details. We perform seemingly counterintuitive inference: given an anonymous consumer's bill total (a single number), we estimate the number and prices of products purchased. Formally, this corresponds to solving a system of noisy linear equations $y = Ap + \eta$, where $y \in \mathbb{R}_{>0}^{M}$ represents M transaction totals, $p \in \mathbb{R}_{>0}^{N}$ represents N product prices, $A \in \mathbb{Z}_{\geq 0}^{M \times N}$ denotes the product(s) purchased in each transaction, and η the noise (e.g., taxes). We observe only y and nothing more, with even the dimension N unknown. Effectively, we need to solve an underdetermined system of equations – a problem setup seemingly more complex than in compressed setting (Table 27.1). Nevertheless, we are able to provide an iterative algorithm that, under mild conditions, provably recovers N, p, and A. Next, we apply our algorithm to anonymized transactions associated with consumer spending at Apple, Chipotle, Netflix, and Spotify. Using transactions data only, our algorithm:

Table 27.1 Comparison of ours with relevant prior works.

Result	A	p	N	Noise	Requirements
Ordinary Least Sq.	Known. Full Rank Lai et al. (1982)	Unknown	Known	Yes	$M \gg N$
Compressed Sensing	Known (RIP Candes, 2008)	Unknown	Known	Yes	$M \gg \|p\|_0$
This Work	*Unknown (Signature)*	*Unknown*	*Unknown*	*Yes*	$M \gg \|p\|_0$

Table 27.2 System identification and tracking comparison with relevant prior works.

Result	Noisy Process ($w \neq 0$)	Latent State Inference	Finite Sample Analysis
EM or Subspace method	Yes	Yes	No
Gradient Descent (Hardt et al., 2016)	No	Yes	Yes
Regret Bound (Hazan et al., 2018)	Yes	No	No
This Work	*Yes*	*Yes*	*Yes*

(i) recovers the number of products for sale and the corresponding prices with high accuracy; (ii) decomposes transaction totals into product purchases with low error; (iii) identifies product launches; and (iv) discovers a rumored 'secret' product of Netflix in limited release.

27.2 Nowcasting company financials

27.2.1 Problem statement and model

Problem Statement.

Let $x_t \in \mathbb{R}^n$ denote the hidden state vector of actual company financials (e.g., actual daily sales) on day $t \geq 0$. Let $s_t \in \mathbb{R}^n$ denote the running summation of the hidden state from day 0 to day $t \in \mathbb{N}$, as would be included in a quarterly report (quarterly sales). Let $y_t \in \mathbb{R}^m$ denote the alt data source (e.g., a sample of daily

Table 27.3 Linear dynamical system (LDS) versus Wall Street consensus benchmark. Results aggregated across 34 public companies and 306 quarters. By 'win' we mean a test quarter for which LDS outperforms the benchmark.

Metric	LDS	Benchmark
RMSE	2.7	3.2
Median Abs Error	1.2	1.3
Wins (Total Quarters)	175	131
Win Percent	57.2%	42.8%

credit card transactions). That is x_t is never observed; s_t is infrequently observed; and y_t is always observed.

Target:

Using observations of y_t and (intermittently available) s_t, track the latent x_t and forecast future s_t.

Model.

We model the alt data y_t as a linear, noisy observation of x_t on day t. We assume x_t follows an autoregressive process and that x_t and y_t are related via the following linear dynamical system equations: for $A \in \mathbb{R}^{n \times n}$, $C \in \mathbb{R}^{m \times n}$, $\gamma_t \in \{0,1\}$, and $t \in \mathbb{N}$,

$$x_{t+1} = Ax_t + w_t \qquad y_t = Cx_t + v_t \qquad s_t = \sum_{i=0}^{t} x_i \qquad (27.1)$$

$$\gamma_t \in \{0,1\} \qquad y_t^s = \left\{ \begin{array}{ll} s_t, & \gamma_t = 1 \\ \text{missing}, & \gamma_t = 0 \end{array} \right\}$$

$$\mathbb{E}[w_t w_t'] = Q = \text{diag}(Q_{1,1}, \ldots, Q_{n,n}) \quad \mathbb{E}[w_t] = 0 \qquad A \neq 0$$
$$\mathbb{E}[v_t v_t'] = R = \text{diag}(R_{1,1}, \ldots, R_{m,m}) \quad \mathbb{E}[v_t] = 0 \qquad C \neq 0$$

where $w_t \in \mathbb{R}^n$, $v_t \in \mathbb{R}^m$ are i.i.d. random vectors with zero mean, and positive-definite covariance matrices Q, R respectively. For simplicity of exposition, we shall assume that Q, R are diagonal matrices, and w_t, v_t have Gaussian distributions. In the above, $\{\gamma_t\}$ is a binary sequence indicating the observation times of s_t (denoted y_t^s) – which could be periodic (e.g., every 90 days) or stochastic. Figure 27.2 shows a graphical model for the problem.

Model Justification:

Many alternative datasets are limited in history: perhaps just a few fiscal quarters of data at weekly frequency. That translates into just dozens of observations of alternative data, and a handful of corresponding quarterly reports. Thus, a linear model is a natural starting point for such a dynamical system. The number of key parameters (A, C, Q, R) is minimal, so we might hope to infer both the parameters and latent variables from limited data.

Alternatively, our model could have incorporated: nonlinear state transitions, nonlinear observations, time-trends, and/or explicit periodic components. Nevertheless, the immediate utility of even this simple model is validated experimentally. And deriving analytic results, even with such a simple model, is non-trivial. Thus we see our application of linear dynamical systems here as a first step; and we hope that our work provides a framework within which more complex models can be developed.

Method.

To utilize the model in Eq. (27.1), we need to: (i) solve system identification: estimate A, C, Q, R; and then (ii) use the estimated parameters to nowcast the

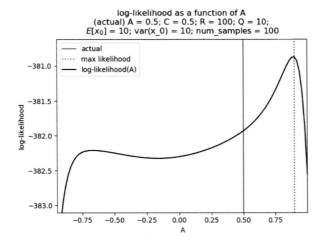

Figure 27.1 Non-convex likelihood shown here as a function of *A*. The figure also indicates that the maximum likelihood solution may be biased.

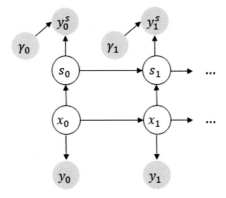

Figure 27.2 Graphical model for tracking company financials. The x_i are latent daily (say) company financials and y_i represent alt data. With frequency λ we intermittently receive a noiseless observation y_t^s of $s_t = s_{t-1} + x_t$. We indicate by γ_t whether s_t is observed ($\gamma_t = 1$) or not ($\gamma_t = 0$).

x_t and s_t. The details of our system identification and inference methods are provided in §27.2.3.

27.2.2 Contributions

In nowcasting aggregate company financials, we make three contributions. First, we provide a systematic approach for tracking company financials at high-frequency, where we combine low-frequency ground-truth aggregates with high-frequency, noisy proxy data. We utilize a variation of the classical linear dynamical systems (LDS) model with hidden state and noisy observations with Gaussian noise. Compared to the classical setting, for example that considered in Bertsekas

(1995), we are also given infrequent observations of aggregate hidden state. Our goal is the same: to develop an estimation algorithm for the hidden state. We show that our model is effective with sparse observations – for example receiving alternative data weekly but aggregate observations every three months.

Our second contribution is solving the two problems required for utilizing this LDS model. First, we solve the problem of learning the model parameters, referred to as 'system identification' in the classical literature. We show that the inclusion of infrequent aggregate state observations allows us to devise consistent estimators for all parameters, and provide a finite sample analysis of the resulting error. And for some of the parameters, the estimation is unbiased. This is surprising, because in the classical setup, system identification for LDS suffers from non-uniqueness: multiple sets of parameters give rise to identical likelihood values (Hamilton, 1994, page 387); and furthermore, the likelihood is non-convex, making optimization challenging (see Figure 27.1 and Table 27.2). In contrast, our algorithms are consistent and computationally efficient.

The next part of this contribution, of utilizing a LDS in a sparse data setting, is providing and analyzing an optimal inference algorithm. In the classical setting, Kalman filtering provides the optimal estimation procedure in terms of minimizing mean squared error. In our modified setup, this no longer holds. We develop such a method and provide an optimal estimation algorithm. We show that if ground-truth aggregate information is available with any non-zero expected frequency, then tracking estimation error remains bounded – even if the dynamics are unstable, uncontrollable, or undetectable. Furthermore, we show that the tracking estimation error decreases monotonically as ground-truth information becomes increasingly available – which is intuitively pleasing. This directly contradicts the claim General Motors Co. (GM) made in April 2018, when GM switched from monthly to quarterly reporting for its US vehicle sales; GM stated that monthly sales are not useful for investors (White, 2018). Per our analysis, having monthly versus quarterly earnings reports improves estimation of company financials; thus GM's claim is arguably incorrect.

Our third contribution is putting our model into practice. We apply our end-to-end identification and tracking algorithms to an alternative data set of real credit card transactions obtained from a hedge fund. The data set consists of typically weekly or biweekly summaries of unknown fractions of consumer spending at 34 public companies. The estimation task is to nowcast weekly sales (and hence quarterly sales) at each of the companies using both the credit card data along with historical, public, quarterly disclosures of sales. Our method outperforms a standard benchmark of Wall Street consensus estimates, beating the consensus on 57.2% of quarterly predictions as well as outperforming with respect to the mean-squared error. The model performance is significant, because we do not make use of any additional information or expert input that may have been available as input for other financial analysts' estimates.

Table 27.4 Summary of estimator properties

Estimator	Unbiased	Consistent	Error Bound	Arguments
$\hat{C}$	Y	Y	Y	$\left\{\left(s_t, \Sigma_0^t\, y_i\right)\middle|\gamma_t = 1\right\}$
$\hat{R}$	Y	Y	N	$\left\{\left(s_t, \Sigma_0^t\, y_i\right)\middle|\gamma_t = 1\right\}, \hat{C}$
$\hat{A}$	N	Y	N	$\left\{\left(s_t, y_t, y_{t+1}\right)\middle|\gamma_t = 1\right\}, y_0, \hat{C}$
$\hat{Q}$	N	Y	N	$\left\{\left(s_t, y_t, y_{t+1}\right)\middle|\gamma_t = 1\right\}, y_0, \hat{C}, \hat{R}, \hat{A}$

27.2.3 Theoretical results

System identification

The model parameters to estimate are A, C, Q and R. Table 27.4 summarizes the theoretical properties and required inputs for our estimators $\hat{A}$, $\hat{C}$, $\hat{Q}$, $\hat{R}$. The details of the system identification procedure are provided in Fleder and Shah (2020).

In summary, the goal is to estimate the parameters A, C, Q, R from a single trajectory. The intuition behind our approach is to show how aggregations of observations and (quarterly) ground truth can be related through the same parameters A, C, Q, R. Our method starts by transforming the system equations, (27.1), into a set of aggregated (quarterly) linear equations, which takes the form of multiple, multivariate regression with time-varying features (Fleder and Shah, 2020). We show that the parameters A, C, Q, R are the exact solutions of these transformed, regression problems. This enables us to apply standard results on regression (Rigollet and Hutter, 2017) with time-varying features (Hamilton, 1994).

The resulting parameter estimates are then provided as input to our inference procedure, discussed next.

Inference

We now assume the system has been identified: A, C, Q, R determined by system identification. The goal of inference is to utilize the model, (27.1), along with the identified parameters, to infer (nowcast) and forecast the latent company financials. Our inference approach is similar to Kalman filtering (Belief Propagation), but requires careful treatment. The details of the procedure and proofs of correctness are provided in Fleder and Shah (2020).

The inference procedure outputs estimates and forecasts of latent company financials. Under the model assumptions, (27.1), the nowcasts: (i) have bounded error covariance; and (ii) are optimal in terms of mean-squared error. Furthermore, the error covariance decreases monotonically as the number of available quarters increases. Next, we discuss the results of running system identification followed by inference on transactions data.

Table 27.5 Sample results for one company over nine quarters. In this instance, the LDS model outperforms the benchmark in 5/9 quarters. We highlight in bold the quarters for which the LDS model outperforms the benchmark. Overall, the mean-absolute percent error for LDS here is 9.1% versus 9.9% for the benchmark.

Quarter	LDS Abs Error(%)	Benchmark Abs Error(%)
1	22.4%	16.2%
2	9.1%	1.5%
3	**4.4%**	5.8%
4	**5.6%**	6.1%
5	18.4%	11.8%
6	**3.0%**	18.2%
7	4.6%	2.2%
8	**8.3%**	16.7%
9	**5.9%**	10.2%

27.2.4 Experiments

Here we apply our methods to real, commercially-available transactions data.

Data

We use three types of data. First, we use alternative data (the $\{y_t\}$) provided by a hedge fund. The alt data consists of aggregates of consumer credit card transactions for 34 publicly-traded retailers[1]. For each company, we are provided with credit card aggregates (in dollars) at typically weekly or biweekly frequency, but no less than monthly frequency. The credit card data is a noisy estimate of company sales, as it captures only an unknown fraction of credit card spending, and it captures no cash transactions. The 34 companies are consumer-facing and have minimal business-to-business sales. For each company, we have 7–10 quarters of data in the period from fourth-quarter 2015 to first-quarter 2019. Across all companies, we have an aggregate of 306 quarters of data – with an average and median of 9 quarters of data per company.

Second, we use the actual, quarterly year-over-year (YoY) percent change in sales – publicly available in 10-Qs and 10-Ks. These are the target values we wish to predict.

Finally, we make use of a third dataset: a benchmark for comparison. We use financial analyst consensus numbers, obtained from a Bloomberg Terminal for each company and quarter. The financial consensus numbers are an industry-standard benchmark for estimating comparable sales. Analysts may, of course, utilize machine learning models in their forecasts along with expert input.

Task

The goal is to accurately forecast YoY quarterly sales using credit card transactions as input. The point-in-time of prediction is after the fiscal quarter ends, but

---

[1] We are unable to disclose the names of the companies due to the data-use agreement.

before the quarterly number is publicly announced. For public companies, there is typically a multi-week gap between the end of a quarter, and the public release of that quarter's aggregate information. Thus, to forecast the quarterly report, we just need to run our tracking algorithm from the start to the end of the quarter and produce a nowcast. The point-in-time of the nowcast ensures that we never have access to ground truth. We then compare our estimate to the consensus number.

For each company, we use leave-one-out cross-validation on all available company data. Thus, for each of the 34 companies (with 7–10 quarters of data each), we generate 7-10 train/test splits of sizes 6–9 training quarters and 1 test quarter. The training set is used to compute per-company LDS parameters. Then, using the learned parameters, we run inference on the test sets and nowcast the quarterly sales numbers. The modeling implication is that each quarter is a separate realization of a per-company LDS with the same underlying system parameters, as opposed to a single, longer time series.

Findings

Table 27.3 tabulates the results for the 306 quarters across 34 companies. In a head-to-head comparison, ours beats the benchmark in 57.2% of quarters. To determine if this is statistically significant, we consider the following hypothesis test. Let ψ be the probability that our method outperforms the benchmark on a given quarter (assuming independence across quarters and companies). The null hypothesis is $\psi \leq 0.5$. The alternative hypothesis is $\psi > 0.5$. We will reject the null hypothesis if we observe a p-value of less than $5 \cdot 10^{-2}$. For the given sample of 306 quarters, with 175 LDS wins, we obtain a p-value of

$$\sum_{i=175}^{306} \binom{306}{i} (\frac{1}{2})^{306} = 6.9 \cdot 10^{-3}.$$

Thus we reject the null hypothesis with overwhelming statistical confidence.

Table 27.5 shows a representative sample of nowcasting error for a particular company over nine quarters. For this subset, none of the actual YoY values are zero, so we are able to compute the mean absolute percent error: 9.1% for LDS and 9.9% for the benchmark.

Significance

Perhaps the most significant statistic is the win-rate of 57.2%. Converting our nowcasts into a trading strategy is beyond the scope of this work, but we note that any statistically-significant nowcasting edge is a key ingredient for asset pricing and decision making.

Discussion

In a sense, it is quite surprising that our estimator works this well: each model is given at most nine ground-truth data points to learn the system parameters, and it uses only one alternative data source of aggregated noisy credit card transactions.

On the other hand, the Wall Street consensus is built on much longer history, and utilizes significant human analyst input as well as various other data sources.

In summary, our framework shows that one can extract value from high-frequency, noisy alternative data through a principled, systematic framework for tracking low frequency ground truth like quarterly sales.

27.3 Inferring products in anonymized transactions

27.3.1 Problem statement and model

Problem Statement.

For a given company, we assume the product catalogue contains a finite number of N products with associated prices $p \in \mathbb{R}^N_{>0}$. We treat both N and p as unknown: that is, an unknown number of products with unknown prices. Let $y \in \mathbb{R}^M_{>0}$ denote M separate, but not necessarily distinct, observed transaction totals associated with the company. Let the unknown matrix $A \in \mathbb{Z}^{M \times N}_{\geq 0}$ represent the product decomposition for each transaction in y. That is, A indicates which product(s) were purchased in each observed transaction.

Target:

Using only the observed transaction totals y as input, determine the number of products N, their associated prices p, and the decomposition A of transactions by product.

Model.

We model the observed transactions y as

$$
\begin{aligned}
y = Ap + \eta \qquad &\text{where} \qquad y \in \mathbb{R}^M_{>0} \\
p \in \mathbb{R}^N_{>0}, \qquad A \in \mathbb{Z}^{M \times N}_{\geq 0,} \qquad &M \geq N \geq 1
\end{aligned}
\tag{27.2}
$$

Each transaction corresponds to the summation of prices for the products purchased. We include a noise term η (unobserved) to account for price variations due to promotions, tax variations, etc.

Method.

To perform inference, we impose two mild conditions that enable recovery of p, A and N. First, we assume every product is purchased alone at least once – formalized as the 'Signature Condition'. Of course, we do not know which observations in y correspond to these individual purchases. Second, we place a simple constraint on price distinctiveness – formalized in the Menu Inference Theorem. These conditions allow us to perform inference using an iterative 'peeling' algorithm.

27.3.2 Contributions

As our main contribution here, we provide a simple, iterative algorithm that provably recovers N, A precisely, and p approximately, with approximation error

Table 27.6 Dataset summary of anonymized transactions. In addition to recovering product prices accurately, our method reconstructs bill totals with minimal error.

Company	Number of Transactions	Distinct Trans. Totals	Date Range	Our Findings
Netflix	2.6M	3,094	01/16-02/19	*Secret Product Ultra HD ($17.08)*
Spotify	387K	527	09/17-09/18	*Product Launch Detected ($12.99)*
Apple	197K	7,685	08/18-10/18	*iPhone XS Sales Volume*
Chipotle	133K	2.8K	09/18-02/19	*Decomposition Err MAPE < 2%*

dependent on the noise η. Our algorithm succeeds if A satisfies the Signature Condition, which requires every product to be purchased individually at least once. This requires that $M \gg \|p\|_0 = N$. It also guarantees that A has full column rank as required in traditional linear regression (or ordinary least squares). However, the Signature Condition does not necessarily require restricted iso-perimetry-like (RIP) conditions which are common in the literature on compressed sending. The algorithm recovers prices that are distinct and not multiples of each other. Indeed, no algorithm can distinguish if one or more products are sold at the same price (or as different integral multiples of the same value) without additional side information. Therefore, we require in our Theorem 27.2 for such a condition to hold. The algorithm runs in polynomial time as detailed in Fleder and Shah (2021). In summary, this simple, iterative algorithm enables solving the system of linear equations *without* knowledge of A, p or N – unlike any prior works.

We apply the algorithm to anonymized consumer transactions data[2] based on credit and debit card purchases at Netflix, Spotify, Apple and Chipotle. The details of the data in terms of number of transactions, time range and unique totals are shown in Table 27.6. We measure performance of the algorithm in terms of recovering (a) the number of key products N, (b) their corresponding prices p and (c) the decomposition (A) of transaction total into products purchased.

For (a) and (b), it is easy to verify full recovery (or not) of all products for Netflix, since Netflix has few product offerings. As shown in Table 27.7, we recover all the published offerings by Netflix. In addition, we identify two additional 'hidden' offerings which seem to agree with limited release products (Spangler, 2018). As can be seen from Table 27.7, the median error in the recovered prices is less than 0.2% (or, less than 4 cents). Accurate recovery of the number of products and their prices implies that that each transaction total is accurately modeled in terms of identifying the product purchased, i.e. performance in terms of (c). For Chipotle, which offers a larger and more complex

---

[2] We utilized anonymized debit and credit card transactions data provided by Second Measure (Second Measure, 2019) for the purpose of conducting this work.

Table 27.7 Products and their prices found by our algorithm from transaction totals for Netflix. We hand match inferred prices to Netflix's published catalog/prices from Netflix (2019); DVD.COM (2019); Spangler (2018); Rodriguez (2019). The key products are streaming basic (SB), standard (SS), and premium (SP); DS and DP refer to DVD Standard (DS) and Premier (DP) subscriptions; RS and RP refer to Netflix's HD Blu-Ray Standard (RS) and Premier (RP) subscriptions. In addition, we find two unmatched products: one at $17.08 which might be the rumored Ultra HD (Spangler, 2018), another at $18.45 is likely to be another such unknown product.

Actual Price	Inferred Price	Abs. Error	Product(s)
$7.99	$7.98	0.13%	SB; DS
$8.99	$8.98	0.11%	SB
$9.99	$9.97	0.20%	SS; RS
$10.99	$10.93	0.55%	SS
$11.99	$11.97	0.17%	SP; DP
$12.99	$12.95	0.31%	SS
$13.99	$14.00	0.07%	SP
$14.99	$14.96	0.20%	RP
$15.99	$15.99	0%	SP
$16.99	$17.08	0.53%	Rumored new product
–	$18.45	–	–

set of offerings, we reconstruct transaction totals within MAPE of 1.2% using 12 key product prices only.

Our findings enable a plethora of insights into time-varying company product catalogues. For example, for Spotify and Apple, we utilize our methods to automatically detect product launches. Specifically, we automatically detect a new Spotify product offering at a new price – starting in the month of April 2018, which matches the reported fact (Perez, 2018). Similarly, for Apple our method detects the launch of the iPhone XS Max in September 2018 from anonymized transaction totals only. Additional experiments with Spotify and Apple experiments are provided in Fleder and Shah (2021).

Collectively, these experiments verify that our method is able to recover product prices as well as decompose transaction totals accurately across different types of business: from Netflix and Spotify with few offerings, to Chipotle and Apple with extremely complex product offerings. Indeed, our method is likely to have more impactful consequences such as that represented by Table 27.8 which is the estimated sales volume by price range at Chipotle.

27.3.3 Algorithm

Here we provide a narrative description of the algorithm. A more detailed, pseudocode version of the algorithm is provided in Fleder and Shah (2021). The algorithm outputs estimators $\hat{p}$ for p, $\hat{N} = \|\hat{p}\|_0$ for N, and $\hat{A}$ for A. The estimates are such that $y \approx \hat{A}\hat{p}$. In addition to y, the algorithm takes as input parameter

$\delta > 0$ as proxy for $\|\eta\|_\infty$, and $k \in \mathbb{Z}_{>0}$ that constrains repeats of a product within a single transaction.

o Input: transaction totals $y \in \mathbb{R}^M_{>0}$; error tolerance $\delta > 0$; and $k \in \mathbb{Z}_{>0}$ constraining repeats of a product in a single transaction.

o Reduce y to distinct values, sort from smallest to largest:

$$b = (b_1, \ldots, b_M) \leftarrow \text{sorted}(\{y_1, \ldots, y_M\}).$$

o Let $\hat{p}$ denote the inferred, ordered set of product prices. Initialize $\hat{p} \leftarrow \{b_1\}$; set $\hat{A}_{1,1} = 1$ and $\hat{A}$ to be a 1×1 matrix. Going forward, we will increase the dimension of $\hat{A}$, but all the entries in the first row other than $\hat{A}_{1,1}$ will be set to 0 in that case.

o Repeat the following for each $b_i \in b$ for $i \geq 2$. Let $\hat{p} = (\hat{p}_\ell)$ be the current estimate of p and $\hat{A} \in \mathbb{Z}^{(i-1) \times |\hat{p}|}_{\geq 0}$ be current estimate of A. Find $J(i) = (a_1, \ldots, a_{|\hat{p}|}) \in [k]^{|\hat{p}|}$ as a solution to

$$\text{minimize} \quad \frac{1}{(1 + \sum_\ell a_\ell)} \left| \left(\sum_{\ell \leq |\hat{p}|} a_\ell \hat{p}_\ell \right) - b_i \right| \tag{27.3}$$

$$\text{over} \quad a_\ell \in \{0, \ldots, k\}, \, \ell \leq |\hat{p}|$$

We consider two cases:

· $|b_i - J(i)^T \hat{p}| \leq \delta \cdot (|J(i)|_1 + 1)$: add ith row, $\hat{A}_i$, to $\hat{A}$ of length $|\hat{p}|$ with $\hat{A}_i$ equal to $J(i)$ in the $|\hat{p}|$ positions. That is, the ith total can be decomposed amongst the product prices in $\hat{p}$ recovered thus far; and we record the decomposition in $\hat{A}$.

· Otherwise, b_i is not well approximated by any combination of product prices found thus far. Hence it must be a product price not yet encountered, and needs to be added to $\hat{p}$: augment $\hat{p} \leftarrow \hat{p} \cup \{b_i\}$; increase the number of columns in $\hat{A}$ by adding an all zeros column to the existing $\hat{A}$ of length $i - 1$; and then add as the ith row $\hat{A}_i$, a vector with all 0s but a single 1 in the $|\hat{p}|$ position.

o Output: after iteration through all entries in b, output $\hat{p}$, $\hat{A}$ and $\hat{N} = \|\hat{p}\|_0$.

27.3.4 Main results

Guarantee.

Here we state the result regarding correctness and robustness of the algorithm. To that end, we assume that the underlying data satisfies the constraint that every product is purchased alone at least once. This is formalized as the 'Signature Condition'.

Condition 27.1 (Signature Condition)   Given matrix $A \in \mathbb{R}^{M \times N}$, A satisfies the Signature Condition if for each $i \in [N]$, there exists $j(i) \in [M]$ such that $A_{j(i)i} = 1$ and $A_{j(i)i'} = 0$ for all $i' \neq i$.

Now we state the main result. We shall assume that $k = 1$ in the algorithm, i.e. each product is repeated at most once. The proof (see Fleder and Shah, 2021) naturally extends for $k \geq 1$.

Theorem 27.2 (Menu Inference)   *Let $A \in \{0, 1\}^{M \times N}$ satisfy the Signature Condition. Let $p \in \mathbb{R}_{>0}^N$, where for any $S_1, S_2 \subset [N]$, with $S_1 \neq S_2$*

$$\left| \sum_{i \in S_1} p_i - \sum_{i' \in S_2} p_{i'} \right| > 2\delta N \qquad (27.4)$$

for some given $\delta > 0$. Let $y = Ap + \eta$ with $\|\eta\|_\infty < \delta$. Then, with input y, δ and $k = 1$, algorithm recovers $\hat{p}$ and $\hat{A}$ so that $\|\hat{p} - p\|_\infty \leq \delta$ and $\hat{A} = A$.

Need for Conditions.

The Signature condition effectively states that each product is purchased alone, at least once. This is natural, and guarantees the full column rank of A that is needed to solve linear equation.

To understand condition (27.4), suppose $\delta = 0$, i.e. there is no noise in the data. In that case, (27.4) would indicate a difference in the summation of any two distinct subsets of product prices – indeed, if that were not the case, *no algorithm could recover A* uniquely. In that sense, our algorithm recovers A and p in a robust manner under the 'quantitative' version of the recovery condition as described in (27.4).

27.3.5 Experiments

Dataset.

We utilize a dataset of consumer transactions provided by alternative data vendor Second Measure (Second Measure, 2019). The data consists of roughly 13.4 million anonymized consumer debit and credit card transactions in the period from January 2016 through February 2019 (see Table 27.6) for four companies: Netflix, Spotify, Chipotle, and Apple. Each data point contains only: (i) the transaction total; (ii) the company name; (iii) daily-resolution timestamp; (iv) city in which the purchase was made. We discuss the results on a per-company basis, with Spotify and Apple experiments left to Fleder and Shah (2021).

Pre-processing.

We apply a simple pre-processing step to remove anomalies from the transaction data. Specifically, after sorting all transactions for a given company, we remove a small percentile of the top and bottom transactions as a robustness step. For example, with Netflix we see transaction totals ranging from \$.01 to \$182.5; however, more than 94% of transactions lie within an \$11 range. And we retain this 'majority' range.

Netflix: Inferring products and prices

Data:

The transactions data for Netflix spans the 38-month period from January 2016 through February 2019 for Boston, Chicago, Los Angeles, New York, Philadelphia, Phoenix, San Francisco, and Washington DC. The data consists of 2.6 million separate transactions, translating to $29.5 million in observed sales. Of these transactions, 3094 bill totals are distinct, ranging from single-penny transactions to $182.5. After the pre-processing described above, we are left with 1046 unique totals.

Task:

We wish to recover products prices, the number of products, and the transaction decompositions. We utilize the algorithm with $k = 1$ and δ as 1% of the transaction total.

Findings:

Table 27.7 shows the inferred product prices. Specifically, we find 11 candidate products and corresponding prices. Next, by hand, we match the first 9 of 11 inferred prices to actual Netflix product-prices. The results are shown in Table 27.7. To our knowledge, this is complete coverage of Netflix's advertised product-prices in that period. As can be seen the median error in inferred price is less than 0.2% (or less than 4 cents). Since product prices have changed over time, several of the products appear at multiple price points.

For almost all inferred products we can associate clear, low-error matches in Netflix's product catalogue. However, we also find two additional products (and associated prices) that do not correspond to publicly disclosed Netflix products (at least not in the time period of the data). The inferred product with price $17.08 matches the rumored Ultra HD product of Netflix available at $16.99 (Spangler, 2018). It seems that there might be another such product being offered at a little higher price around $18.45.

In summary, we recover the entire Netflix product price catalog accurately – from just transaction totals. In addition, we recover hidden or unadvertised products – one rumored and another completely unknown. This establishes efficacy of our algorithm.

Chipotle: Revenue attribution and reconstruction error

Data:

The transactions data for Chipotle in New York, NY covers the six-month period from September 2018 through February 2019. The data consists of 133,000 transactions, with 2800 unique transaction totals and a corresponding $1.7M in revenue. The transaction totals range from $.03 to $552.54. After pre-processing, we retain transactions in the range of $1.5–$20 and 1230 unique transaction totals.

Task:

Chipotle's menu is too complex to analyze fully: even for bill totals under $20, there are a large number of possible order combinations; and in addition, the menu contains multiple items at the same price point. Nevertheless, we hope to gain insight into Chipotle sales with two goals. First, we examine how well we are able to model the thousands of unique bill totals assuming just a small number of product price-points. Second, we would like to determine if a relatively small price range for products accounts for a large percentage of Chipotle revenue. We use $k = 2$ and δ of 0.1% of each transaction total.

Findings:

The inference algorithm identifies 12 price points within the Chipotle menu, from $1.5 to $18.65. Using these 12 products and associated prices, we examine how well each of the 1230 distinct bill totals can be decomposed into combinations of these prices. We find that with mean-absolute-percent error (MAPE) of 1.2%, all the transaction totals can be decomposed using just these 12 products/prices.

Next, we decompose each transaction total into its inferred, constituent products. Given the inferred decompositions, we examine the concentration of sales (in USD) by product price. Table 27.8 shows that a large majority of estimated sales concentrate in a narrow $2 price range – suggesting that a large fraction of Chipotle's customers order items like the 'chicken burrito' at $9.74 (Chipotle, 2019).

To benchmark the performance of our algorithm, we compare it with a simple, randomized algorithm. We consider the approach of randomly selecting X bill totals as individual item prices; and then using these selected prices, we attempt to decompose all bill totals into sums of these prices (with the previously mentioned constraints). We repeat this random selection-then-decomposition process 30 times and average the results. The resulting MAPE using the randomized algorithm is 7.5%; and if we weight the results by frequency, we see a MAPE of 22.7%. That is, this randomized benchmark has a MAPE of 5–6 times worse than our approach. To understand if this is statistically significant, we construct a hypothesis test: after each randomized trial, we compare the per-trial MAPE to our method. Let ψ be the probability that our method outperforms the randomized benchmark on a given trial. The null hypothesis is $\psi \leq 0.5$. The alternative hypothesis is $\psi > 0.5$. We will reject the null hypothesis if we observe a p-value of less than 10^{-2}. On 30/30 trials (with both unweighted and weighted errors) our method outperforms the randomized approach. We obtain a p-value of $(\frac{1}{2})^{30} < 10^{-9}$. Thus we reject the null-hypothesis with overwhelming statistical confidence. Similarly, the hypothesis test results are identical if we use a train-test split of 4 months/2 months: (i) first infer a price menu from the train set (ii) approximate totals in the test set and then (iii) run the randomized comparison on the test set – which, in fact, provides advantage to the randomized method.

Table 27.8 Inferred sales at Chipotle by product–price range using transaction data. See §27.3.5.

Price Range	Estimated Sales	Example Products
$1.5–$4	15.6%	chips, drinks, guacamole
$9–$11	79.1%	$9.74 chicken burrito, steak bowl
$11–$20	5%	double steak bowl

27.4 Conclusion

In this work, we developed quantitative methods for utilizing alternative data. We extended the theory of system identification, Kalman filtering (or Belief Propagation), and compressed sensing. Our methods come with provable performance guarantees under the modeling assumptions. In applying our algorithms, we (i) outperform an industry-standard benchmark and (ii) infer product prices, detect product launches, and take steps towards attributing sales by product – all from aggregate, anonymized, transactions data.

27.5 Relevant literature

This chapter is a summary of the doctoral thesis of Michael Fleder (2019). It has appeared with additional details in proceedings of ACM Sigmetrics conference (Fleder and Shah, 2021, 2020). This work has also featured in the MIT News (Strampel, 2021; Matheson, 2019), and is now forming the basis for a commercial startup (Covariance.AI, 2021) that is successfully applying the methodology and extensions to numerous alternative-data sources and KPIs.

References

Bertsekas, Dimitri P. 1995. *Dynamic Programming and Optimal Control*. Vol. 1. Athena Scientific.

Candes, Emmanuel J. 2008. The restricted isometry property and its implications for compressed sensing. *Comptes Rendus Mathématique*, **346**(9–10), 589–592.

Chipotle. 2019. Chipotle Online Ordering.
https://order.chipotle.com/Meal/Index/1597?showloc=1.
Accessed: 2019-05-01.

Covariance.AI. 2021. Covariance: Machine Learning with Alternative Data.
http://covariance.ai. Accessed: 2021–06–01.

DVD.COM. 2019. DVD.COM Choose a plan.
https://dvd.netflix.com/Plans?dsrc=DVDWEB_NMHOME_NMHEADER_PLANS.
Accessed: 2019–05–27.

Fleder, Michael. 2019. *Forecasting Financials and Discovering Menu Prices with Alternative Data*. PhD thesis, Massachusetts Institute of Technology.

Fleder, Michael, and Shah, Devavrat. 2020. Forecasting with alternative data. In: *Proc. ACM SIGMETRICS International Conference on Measurement and Modeling of Computer Science*.

Fleder, Michael, and Shah, Devavrat. 2021. I know what you bought at Chipotle for $9.81 by solving a linear inverse problem. In: *Proc. ACM SIGMETRICS International Conference on Measurement and Modeling of Computer Science*.

Hamilton, James Douglas. 1994. *Time Series Analysis*. Princeton University Press.

Hardt, Moritz, Ma, Tengyu, and Recht, Benjamin. 2016. Gradient sescent learns linear synamical systems. *Journal of Machine Learning Research*, **19**(1), 1025–1068.

Hazan, Elad, Lee, Holden, Singh, Karan, Zhang, Cyril, and Zhang, Yi. 2018. Spectral filtering for general linear dynamical systems. Pages 4634–4643 of *Advances in Neural Information Processing Systems*.

Lai, Tze Leung, Wei, Ching Zong, et al. 1982. Least squares estimates in stochastic regression models with applications to identification and control of dynamic systems. *Annals of Statistics*, **10**(1), 154–166.

Matheson, Robert. 2019. Model beats Wall Street analysts in forecasting business financials. `news.mit.edu/2019/model-beats-wall-street-forecasts-business-sales-1219`. Accessed: 2020–06–01.

Netflix. 2019. Netflix pick your price. `https://www.netflix.com`. Accessed: 2019-05-27.

Perez, Sarah. 2018 (April). Spotify and Hulu launch a discounted entertainment bundle for $12.99. `https://techcrunch.com/2018/04/11/spotify-and-hulu-launch-a-discounted-entertainment-bundle-for-12-99-per-month`. Accessed: 2019-06-11.

Rigollet, Philippe, and Hutter, Jan-Christian. 2021. Lecture notes in high-dimensional statistics. `http://www-math.mit.edu/~rigollet/PDFs/RigNotes17.pdf`. Accessed: 2018-05-07.

Rodriguez, Ashley. 2019. A history of Netflix US price hikes, charted. `qz.com/1524449/netflix-just-raised-prices-in-the-us-a-history-of-hikes-charted`. Accessed: 2019-05-27.

Second Measure. 2019. Data Points. `https://secondmeasure.com/datapoints`. Accessed: 2019-05-19.

Spangler, Todd. 2018 (July). Netflix testing out pricier new 'Ultra' plan at $16.99 per month. `variety.com/2018/digital/news/netflix-ultra-plan-hdr-ultrahd-test-1202865305`. Accessed: 2019-05-27.

Strampel, Kim. 2021. I know what you bought at Chipotle. `news.mit.edu/2021/i-know-what-you-bought-at-chipotle-consumer-algorithm-0202`. Accessed: 2021–03–30.

White, Joseph. 2018 (April). GM to drop monthly US vehicle sale reports. `https://www.reuters.com/article/us-usa-autos-gm/gm-to-drop-monthly-u-s-vehicle-sale-reports-idUSKCN1HA0C9`. Accessed: 2018-05-07.

28

NLP in Finance

Prabhanjan Kambadur[a], Gideon Mann[a] and Amanda Stent[b]

Abstract

The finance industry has developed advanced computational models for decision making (e.g., trading, market research) using structured data such as market fundamental data, credit card receipts, geolocation data and satellite data as input. Traditionally, the rich information present in unstructured data – such as press releases, company filings, breaking news, CEO newsletters, patent claims, government contracts, chats and voice calls – has been integrated into decision making primarily through qualitative, non-computational means. However, recent advances in natural language processing (NLP) have allowed us to create structured data – such as a time-series of topics, entities, and targeted sentiment – from unstructured data. In turn, this has led to the incorporation of information from unstructured data directly into models for decision making.

These NLP advances have had a significant impact for discretionary investors who need to scan ongoing events via content clustering and summarization. These technologies are also important to systematic, event-driven traders who derive sentiment signals on top of media streams and integrate these signals directly into their proprietary trading algorithms. Just as important, although less visibly, NLP advances support automation in the mid- and back-office through advanced information extraction over structured documents.

NLP technologies are especially useful in finance because financial applications are backed by rich structured knowledge about securities, entities and their relationships, all of which can be now be automatically extracted from unstructured data with high precision. This means that NLP for finance has clearer 'ground truth' than NLP for many other applications (such as advertising, education, or the law).

In this chapter, we showcase some novel recent finance applications of NLP for clustering and summarizing document collections, extracting data from documents, and using NLP-derived signals to predict market movements. We also

[a] Bloomberg
[b] Colby College
Published in *Machine Learning And Data Sciences For Financial Markets*, Agostino Capponi and Charles-Albert Lehalle © 2023 Cambridge University Press.

discuss opportunities for expanding the scope of current work in NLP to novel finance-related tasks such as understanding structured documents, charts, tables, and document collections.

Through the discussion, we focus on the key computational methods needed for these applications. Most of these NLP techniques rely on human supervision, either using large collections of data annotated by humans to train machine learning classifiers, or using so-called 'weak supervision' provided by subject matter experts. However, in the past five years, there have been transformative breakthroughs in NLP growing out of discoveries in deep learning. We discuss methods for text representation from bags of features to contextual word embeddings. We also cover a wide variety of machine learning models such as logistic regression, conditional random fields, convolutional neural networks, recurrent neural networks and transformer networks.

28.1 Core NLP techniques

Natural language processing (also known as human language technology) has been a core subfield of AI since its start in the 1940s and 50s. From those early days, two significant concepts stand out: the key work (Shannon, 1951) on computing the entropy (predictability) of the English language, and the Turing test (Turing, 1950) as a method for assessing intelligence. Shannon's paper set forth a lasting problem in computational approaches to language – mapping language into a discrete space, where probabilities can be assigned to sequences of characters. The Turing test opened up another avenue of research, posing the joint problems of language understanding and language generation. One of the aspects that distinguishes both of these problem formulations as machine learning (and really statistical estimation problems) is that both have an extremely high-dimensional input and output spaces. This high dimensionality means that sparsity is a core problem in natural language processing[1].

There are many techniques for addressing this problem of sparsity, but the main line of work involves annotating (or labeling) language with various levels of structure. Models constrained to a particular linguistic phenomenon can yield robust estimates of syntactic (structural), semantic (meaning-containing) and pragmatic (in-use) components of language. In turn, these building blocks can be composed to address higher-level linguistic objectives. Recently, breakthrough techniques have been proposed that leverage vast textual resources to build flexible representations of language that can be used across many different contexts and problems.

The remainder of this section will begin by covering the common building blocks used in working with natural language. Following this, we will lay out various higher-level operations that combine these building blocks into valuable outputs.

---

[1] Though this is not unique, as Fred Jelinek once stated "the problem for all science is sparsity"

28.1.1 Basic language analytics

One of the earliest projects in the study of language was on syntax: the structure of language, notably grammatical analysis. Chomsky (1956) defined this work as laying out a series of rules that could decide whether a sentence was permissible or not for a given language (grammaticality judgements). Early work in the area contained significant overlap with theory of computation, defining rule systems that required varying levels of complexity to decide. Two central examples were finite-state machines and context-free grammars (Chomsky, 1957). With an applied lens, these projects were useful in inducing a parse over linguistic structure. During the 1990s the introduction of large repositories of sentences with their associated parse trees ('tree banks') (Marcus et al., 1993), enabled a transformation of the problem into a probabilistic framework, where the goal was to find the 'most likely parse tree' for a given sentence (Charniak, 1997). Early grammars sought not only to derive the structure of sentences, but also to label sub-constituents with a phrase type (e.g., 'a noun phrase'). These days, dependency parsers are also common; with these parsers, instead of modeling phrase structure, each word is aligned to its parent – what that word 'depends' on – (De Marneffe and Nivre, 2019). Another level of synactic analysis is on part-of-speech tags – where each word is assigned to a synactic class, e.g., 'noun', 'verb', (DeRose, 1988). For languages with complex morphology, these tags may also encode lexical properties that can be resolved on the surface, e.g., gender or number, (Habash and Rambow, 2005). These syntactic representations are typically used as features for decoding semantics.

With modeling of semantics, the goal is to recover not the function a word plays within a body of text, but the meaning it carries, either in isolation or in context. Some applications of semantic modeling are particularly relevant to finance; information extraction is one, since information extraction methods can convert unstructured data (text) into structured data (facts) for consumption by trading systems or humans. Two key tasks in information extraction are named entity recognition (and linking), and relation extraction.

Some words or phrases, called 'named entities', serve to uniquely identify entities in the world (Grishman and Sundheim, 1996). Example types of named entity include people ('Alan Greenspan'), organizations ('The Federal Reserve'), companies ('GM Corporation'), and locations ('Springfield'). Automatic named entity recognition (detection) and linking (to a named entity's real-world *referent*) improves discoverability and trackability of content, and is critical to the handling of social media and news for finance. There are several complexities to named entity recognition and linking (Yosef et al., 2011). One is that the problem is open class – there are new named entities coming into being all of the time. Another complexity is ambiguity – a single name can refer to multiple entities (e.g., 'Bloomberg' may refer to Bloomberg the company, or to Michael Bloomberg); and in some cases, it is difficult to tell whether a word is a named entity or not (is 'Apple' a fruit or a company?). These complexities mean that even modern transformer-based techniques cannot reliably resolve named entities in the absence of a source of real-world knowledge such as a knowledge graph Manotumruksa et al. (2020) and up-to-date data Rijhwani and Preotiuc-Pietro (2020).

In relation/event extraction (Grishman, 2019; Gaizauskas and Wilks, 1998), the goal is to extract from text input one or more relationships holding for or between entities. As an example, given a quarterly earnings report, the system would determine the reported EPS that might occur in a sentence like: "Apple reported earnings of \$XXX/share." Relation extraction is complicated because of the many ways in which relations or events can be communicated in language; because language is often ambiguous; and because it is often useful to also extract non-entity attributes of the relationship (such as duration, or start date).

Alongside this process of increasing annotation levels for language, there has been another line of research that works in a unstructured way to make sense of individual tokens. Recall that a first step in working with words is to map linguistics sequences to numbers so that they can be manipulated in a computational fashion. As an example, traditional 'trigram' language models would take a training corpus and map each unique word in the training corpus to a unique integer. The conditioning context is then $P(w_i \mid w_{i-1}, w_{i-2})$, where each w is a unique conditioning event (i.e. a distinct integer). It is of course a relevant question as to how to treat words which aren't seen in the training data set – one common treatment is simply to map them to a distinct 'OOV' (out-of-vocabulary) word.

Over the past two decades there has been a significant amount of work to shift from representing words as a single integer to representing them as a vector of real numbers. Of course, one obvious way to do this would be to take annotations and apply them on top of the words, so that attributes like part-of-speech and number are added as linguistic features on top of the word. Alternatively, instead of an arbitrary mapping or a mapping based on layers of induced annotations, you could imagine having a set of features . This mapping of a word to a point in a multi-dimensional space is called an 'embedding' and also known as manifold learning. This shift in representation is precluded on the idea that words that are 'close in meaning' are similarly close in the embedding space. This methodology has the property that it can represent many levels of meaning on a per-word basis.

28.1.2 Higher-level linguistic analysis

Another form of natural language task is the general problem of 'text categorization', where a single label is assigned to a significant chunk of text (Sebastiani, 2002). As an example, detecting whether a particular document is on a particular topic (e.g., 'OIL'), would be an example of a text categorization task. In order to have effective text categorization, often the layers of annotations as described above are first completed, extracting for example, the relevant entities in the document or alternatively leveraging representations induced from a much larger document collection.

So far, all of the natural language tasks discussed have had as an output a fixed (small) set of labels. However, there are significant numbers of natural language tasks where the output is natural language itself. Apart from a language model prediction of next most likely word or character, summarization is another common natural language task, where the input is a large document collection and the output is a short natural language summary of those documents (Nenkova and McKeown, 2012; Lin and Ng, 2019; Sahrawat et al., 2020; Laban et al., 2020). Another example tasks is question answering, where the input is two-fold: a natural language question and a document collection from which the answer has to be extracted. The desired system response is a synthesized answer drawn from that collection (Chen and Yih, 2020; Zhu et al., 2021). Language translation, where the input is in one language and the output in another, is a further example (Sennrich and Haddow, 2017). OpenNMT (Klein et al., 2017) is a popular open source machine translation system.

Apart from the tasks of adding an annotation layer to an existing document is the problem of clustering, where the goal is to take a set of documents and naturally organize them, for example to put them into separate piles where each pile contains similar documents (Hearst and Pedersen, 1996). To some degree the problem is ill-defined, not mathematically but from an application perspective, because there may exist many divisions that are meaningful for a reader. Nonetheless, this problem is of considerable interest to readers who desire to make sense of a large document collection. There is one significant differentiation in solutions here: either a flat clustering, where each document is placed into one of a finite number of collections. Alternatively, the documents can be placed into a tree, where each document is a leaf and the documents are assorted hierarchically.

For many of the standard natural language tasks the current solutions involve supervised machine learning. As an example, for parsing, the breakthroughs in the 1990s around statistical parsing came out of the construction of a large set of parse trees that had been hand-annotated by trained linguists (Nivre et al., 2016). In order to enable this to work, there is a significant amount of standardization, training and education needed for the annotators. The 'Mechanical Turk' service deployed by Amazon was an early system that enabled end users to reach a crowd to do annotation work. There is considerable effort in generating effective ways to get accurate and complete annotations, such as improved methods for annotator selection and for annotation combination (Jamison and Gurevych, 2015; Bartolo et al., 2020; Paun et al., 2018). Because of the high cost of annotation, there has also been significant interest in weakly-supervised learning, where a machine learning model can be trained using fewer training examples (Ratner et al., 2017; Zhou, 2018).

Crucial to the application of these methods is the representation of human-readable text into a machine-readable format. In many cases, this is straightforward as the text originates in a rigidly structured electronic form (e.g., emails), and the structure that contains the text can be easily unwound. In other cases, the text is in a less easily recoverable form; for example hand-written notes, web pages, or PDFs. Even computer-generated text images, PDFs for example, pose problems in recovering structure, not necessarily at the character level but at the text paragraph level. An example of such a problem might be "is the text underneath a figure a caption or the beginning of the mainline text". Modern machine-learning techniques are proving very useful for the extraction of text and document structure from heterogeneous documents (Wang and Liu, 2020; Burdick et al., 2020).

Supporting all of this research has been a set of academic communities, associated with the larger machine-learning community, but distinct in and of itself. In particular, the Association for Computational Linguistics is the foremost community, with major conferences as the ACL (the meeting of the Association for Computational Linguistics), NAACL (North American ACL), and EMNLP (Empirical Methods for Natural Language Processing). Other conferences like ICML, and NIPS often have work that is highly relevant to natural language processing. ICDAR is a popular conference for document processing problems such as hand-writing recognition.

Alongside natural language processing are fields that draw not only from the methods but also from models trained from language collections and leverage other resources as well. For example, speech recognition heavily leverages multiple natural language processing techniques and combines with acoustic processing to recover the text from a spoken dialogue stream. wav2vec (Schneider et al., 2019) is an example open source framework for speech recognition. Another example is the emerging field of multi-modal applications, where language and another modality (typically vision) are combined to solve novel problems, e.g., visual question-answering (Antol et al., 2015) or multi-modal information extraction (Dong et al., 2020).

28.2 Mathematics for NLP

Modern NLP is largely machine learning-based, and therefore the core mathematics required for machine learning is also important for NLP. That is: calculus, linear algebra, introductory probability and statistics, and numerical optimization (Deisenroth et al., 2020).

There are, however, some unique aspects to NLP. Language is *ordered*; so-called 'bag of words' representations, that treat a document as just a collection of words, will only take you so far. Not only is language ordered, but it is also *structured*: relationships hold not only between adjacent words but between words and phrases at varying distances from each other. And finally, language has *meaning*: words and phrases have semantic relationships. These facts have implications for the machine-learning models and approaches widely used in NLP.

In this section we give a brief overview of machine-learning techniques widely used in NLP[2].

---

[2] The material in the first part of this section is taken from
`https://bloomberg.github.io/foml/\#lecture-1-black-box-machine-learning`.

28.2.1 *Introduction to supervised learning*

Although there are unsupervised and semi-supervised approaches to machine learning, the majority of applied NLP involves *supervised* approaches, in which a model is trained over annotated data (data labeled with 'ground truth').

In supervised machine learning, the computer invokes a supervised learning algorithm to obtain an approximate solution to a prediction function $f(x) = y$, where x is (features over) the input data and y is (a representation of) the label. Common supervised ML problem types include binary classification, multiclass classification, regression, and sequence tagging (binary or multiclass classification for each element in a sequence). Whether the input is characters, words, or documents, for NLP the input is typically represented as vectors of numbers (see §28.2.2).

A supervised learning algorithm 'learns' a prediction function by repeatedly providing an input x, obtaining the output y' of the current prediction function, and modifying the parameters of the prediction function based on the *loss* – a measure of how far off the predicted output is from the actual output, or $g(|y' - y|)$. The shape of the prediction function often serves as the name of the machine learning algorithm:

- linear regression: $y = b + w^T x$, where b and w are parameters to be learned
- logistic regression: $y = 1/(1 + exp(-w^T x + b))$.

Loss functions g may vary; for example, for classification tasks where no label is more important than another, the loss function may be '0/1 loss': 0 if the prediction function is correct, otherwise 1. For regression tasks, mean squared error is commonly used as a loss function. Supervised learning algorithms modify prediction functions (based on the loss) via (any variant of) gradient descent.

A prediction function (also known as a 'model') is evaluated on held-out test data; that is, data that is separate from the (x, y) pairs the prediction function is trained on. For data that is not time varying, it is possible to do a random split 80% training, 10% development (for experimenting with variations on the machine-learning algorithm, features, input representations, and so on), 10% test (for final testing). For data that is time varying (including text data, in many finance applications), it is important that the training, development and test splits be temporally separate from each other. For data that is unbalanced, it may be necessary to perform stratified sampling, i.e., sampling from each class separately. Of course, any estimate of the accuracy of a prediction function based on one sample of data is just a single point of information; to accurately assess how well a machine-learning method is performing on a data set, the data set must be large enough, and multiple rounds of training and evaluation must be performed Dietterich (1998).

The performance of a prediction function may be assessed using any number of *evaluation metrics*, including:

- accuracy: the ratio of correct predictions over total number of predictions;
- precision and recall, per class: precision is the number of true positives over

the number of true positives and false positives; recall is the number of true positives over the number of true positives and false negatives

However, it is often useful to dig deeper into the performance of a prediction function than a single number; for example, to see whether two prediction functions are performing well on different subsets of the data distribution. For example, one can look at a confusion matrix, which has all the classes (all the output labels) as row and column labels, with counts of different types of error in the cells. If one sees that different models – different prediction functions, perhaps trained using different machine learning algorithms – are performing well on different subsets of the data, one can try *ensemble* models. An ensemble is simply a collection of models that are used together. There are different ways in which models can be ensembled; some common ways include:

- bagging: train simple prediction functions on bootstrap samples of the training data, then take the average (regression) or majority vote (classification) of the simple predictors;
- boosting: train a series of individual models; in the first round, weight each training data point equally, then in subsequent rounds, base the weight of each training data point on whether the previous model(s) got the right or wrong prediction for that training data point
- stacking: train a prediction function on the outputs of all the base prediction functions.

Ensemble methods are widely used in NLP research, but ensembles of neural models are less widely used in production where explainability, transparency and speed may be more important.

The choices of data sampling, data representation (or features), machine-learning algorithm, type of prediction function, loss function, ensembling method etc. can all be made either empirically or by consideration of the nature of the phenomenon under investigation and of the data. However, it is possible to make simple mistakes that invalidate the whole experiment:

1. choosing a prediction function that is more complex than the input data warrants, leading to *overfitting* – when a prediction function has so many variables that it is fit too closely to the training data, it may not generalize to the test data;
2. leting information about labels leak into the input features (*leakage*): for example, collecting tweets that contain price move emoticons to build a market sentiment model, and including the entire text of the tweets as part of the input representation;
3. use the prediction function at 'inference time' on data the distribution of which is different from that used at 'training time' (*sample bias*); for example, using a named entity recognition model trained on broadcast news data from 1993 to tag named entities in financial text news from 2020;
4. failing to account for changes in the world (*covariate shift* and *concept drift*)

– it is rare that the same prediction function may be used forever, especially in finance, as market conditions change (Rijhwani and Preotiuc-Pietro, 2020).

28.2.2 Machine learning methods for NLP

There are three basic approaches to machine learning in NLP: text classification (label whole documents, paragraphs or sentences – for example for relation extraction); sequence tagging (label sequences of words or other units – for example for named entity recognition); and sequence generation (produce output sequences from input representations – for example for summarization). To the extent that NLP tasks can be modeled using one of these three approaches, the NLP method can operate on the text directly, without requiring intermediate representations. This is such a significant advantage that even NLP tasks that involve the prediction of linguistic representations (such as parse trees) are today typically modeled using one of these three approaches.

Features

The input to a machine-learning algorithm must be shaped as a vector of numbers; but the elements of the vectors may capture all kinds of information, including morphosyntactic, semantic and pragmatic information as outlined in the previous section. Before the arrival of deep learning, a large amount of effort would be devoted to 'feature engineering', to writing code that extracted features and converted them to numbers (Collobert et al., 2011). For example, for named entity recognition in English it might be useful to know if each word starts with a capital letter, is a number, is hyphenated, ends with a full stop, is a single letter, or starts a sentence (Chieu and Ng, 2003).

Of course, feature engineering is time consuming, may introduce bugs into an end-to-end workflow, and may slow down end-to-end workflows through repeated walks over the input data. So it is also quite common for these vectors to just capture information about frequency, or frequency in context, calculated over large text collections, as universal representations of text. To explain the various ways of constructing such vectors, we will focus on words. Here are some possible ways to construct vector representations of words useful in machine learning:

- One-hot encodings: construct an n-dimensional vector for each word; for the nth word in alphabetical order, place a unit in the ith position and zeros everywhere else. These vectors will be sparse and large.
- Raw or relative frequencies: use an n-dimensional vector as before, but the ith position contains $f_{t_i,d}/\sum_j f_{t_j,d}$, where $f_{t_i,d}$ is the frequency of occurrence of word t_i in document d.
- Term frequency inverse document frequency (tfidf): as above, but $f_{t_i,d} *$ $|D|/|d \in D : t_i \in d|$, where $f_{t_i,d}$ is the frequency of occurrence of word t_i in document d and D is the set of documents in the text collection Jones (1972).

- Fixed learned embeddings: represent each word using an n-dimensional vector, where n is small (for example, 50 to 300). Use a simple neural network to learn the numbers that should comprise the vector for each word, by having the system walk over a large collection of text. Word2vec (Mikolov et al., 2013), Glove (Pennington et al., 2014) and Fasttext (Bojanowski et al., 2016) embeddings are all examples of fixed learned embeddings. These vectors will be dense and comparatively small, but there is one vector per word, regardless of the various contexts in which the word appears.

- Contextual learned embeddings: represent each word using an n-dimensional vector, the values for which are learned using a neural network architecture such as a bidirectional long-short term memory network (Peters et al., 2018) or a transformer network (Devlin et al., 2018) (see §28.2.2). These vectors will be dense and comparatively small, and the vector for a word will vary depending on the context in which it appears.

Tfidf, fixed and contextual learned embeddings can also be calculated or learned over characters, sentences, paragraphs and documents.

Let's take a closer look at one way that fixed embeddings are learned, namely *skipgram with negative sampling* (Mikolov et al., 2013). First, walk over a large text collection word by word and store the word frequencies. Then, construct the model's vocabulary by taking the top v most frequent words. Third, construct the training data for the model: walk over the text collection word by word. For each target word that is in-vocabulary, look at the n (e.g., 2) context words before and after that word. For each pair (target word, context word), add to the training data a positive example (this step comprises 'skipgram'). Then, sample m (e.g., 4) other words at random from the vocabulary and add to the training data a negative example for each (target word, other word) pair: this step comprises 'negative sampling'. If the target word is not in the vocabulary, its data points can be replaced with a special *OOV* token. Once all the training data is constructed, create two matrices, each of width k (e.g., 50 or 300) and of length v, the size of the vocabulary; these can be initialized with random values. The first matrix is the embedding matrix and the second is the context matrix. Finally, to train, repeatedly take one positive example and its corresponding negative examples; for each example, look up the vector for the target word in the embedding matrix and that for the context word in the context matrix; and calculate the dot product of the two vectors. This, when passed through a sigmoid function, gives something like probabilities. The loss is the difference between the pseudo-probabilities and the labels (1 for each positive example, 0 for each negative example).

Now let's take a closer look at one way in which contextual embeddings are learned, the method used in *transformer* models. In these models, instead of having a *fixed* representation of a word regardless of its current context, a model predicts a vector representation of a word given its current context. This is done by training an encoder-decoder model: the encoder captures a vector representation of a word in context, and the decoder can predict a word given a context. Both are trained as so-called masked language models: mask (hide) a fraction of the words in the input, and then reward the model for predicting the correct word. These model architectures differ from those used to train fixed word embeddings in that they can product a different representation for a word depending on the words around it; they differ from model architectures like long-term sequence models in that they can learn which parts of the context to pay attention to, instead of being told by the model architecture.

Note that no matter which of these techniques is used, there is still a fixed vocabulary – for example, the out-of-the-box Glove embeddings have a vocabulary of 400,000 words. If a word occurs that is not in the vocabulary, it may be dropped, or mapped to an OOV token with a default or learned embedding (Pinter et al., 2017). For this reason, contextual embeddings (e.g., from BERT: Devlin et al., 2018) are often trained on subword units – character sequences within words. Even then, it is very likely that after the embeddings are trained one may come across character sequences never seen before (Pinter et al., 2021).

Neural networks for NLP

Deep learning (neural network learning) represents the state of the art for ML approaches to NLP. A neural network is a structured collection of artificial 'neurons', so we will start by describing an individual neuron. An individual neuron computes a logistic regression (a weighted sum) over its inputs, and then passes the output through a nonlinear *activation function* (for example, a sigmoid or a rectified linear unit). The inputs to a neuron may come directly from inputs to the network, or from the outputs of other neurons; the arrangement of connections between neurons defines the neural network architecture. Neural networks are trained using gradient descent but because some of the neurons in a network may be 'hidden', the method used is backpropagation. Common neural network architectures include the convolutional neural network, the bidirectional long-short term memory, and the encoder/decoder or transformer. We will briefly describe each.

A convolutional neural network (CNN) for NLP passes a sliding window of width *l* over the input. At each step, a *convolution filter* is applied: the vectors representing words in the window are concatenated, and a dot product is calculated between the resulting vector and a weight vector, and then the result is passed through a non-linear activation function; when the convolution has passed over the whole input, there is a final pooling layer that produces a single number. One can apply any number of convolutions over the input, and then make a final decision by passing the outputs of the pooling layers through a final softmax (Kim, 2014). In addition to being used for document classification, CNNs are often attached to other network architectures and used to learn character embeddings for the character sequences (e.g., words or subword units) input to the other network architecture.

A recurrent neural network (RNN) is constructed of one input layer, one output layer, and one or more 'hidden' inner layers. In a bidirectional recurrent neural network, there are pairs of hidden layers. The *i*th neuron in the first 'hidden' layer in each pair is connected to the neuron to the left of it as well as to the *i*th element of the input. The ith neuron in the second 'hidden' layer in each pair is connected to the neuron to the right of it as well as to the ith element of the input (Schuster and Paliwal, 1997). When training a bidirectional RNN, the gradient of the loss function decays exponentially over time, and this causes difficulties when there are long-term dependencies (as there often are with natural language). A bidirectional long-short term memory network (LSTM) is a variant on a recurrent neural network in which the neurons have special 'memory cells' that can store information, and gates that control the memory cells (Huang et al., 2015). RNNs and biLSTMs are generally applied to sequence tagging tasks.

Encoder/decoder networks are comprised of two sub-networks, one for encoding and one for decoding. The encoder constructs a compressed vector representation of the entire input (typically, the output of its last hidden state) that it passes to the decoder. The decoder processes this representation and generates the entire output. The encoder and decoder both may be RNNs. The lossiness of the data transfer between encoder and decoder is a bottleneck, so newer encoder/decoder architectures have two types of 'attention': first, the neurons in each layer in both the encoder and decoder networks can differentially pay attention to each other (instead of only listening to their neighbor as in a RNN), and second, the encoder can pass along much more context (e.g., the outputs of all its hidden layers) to the decoder, and the decoder can pay attention differentially to the various pieces of context it is given from the encoder. Encoder/decoder networks are generally applied to sequence generation tasks (Cho et al., 2014).

Finally, transformer networks are simply encoder *or* decoder networks, used for pretraining models that can produce contextual embeddings that can, in turn, be fed into downstream networks for document classification, sequence tagging and/or sequence generation tasks (Vaswani et al., 2017).

Document classification

Document classification tasks may be binary or multiclass, and single or multilabel. A binary classification task is one where each input text either is, or is not, 'in-class'. A multiclass task is one where each input text may belong to one of two, three, a hundred, or five thousand classes; obviously, the fewer the classes, in general, the more accurate the model. A multilabel task is one where each text may be labeled with one *or more* of the classes.

For document classification, NLP relied for 20 years on so-called bag of words methods. Input documents (which could be sentences, paragraphs, pages, or more) were represented typically using tfidf representations. Common techniques for document classification include naive Bayes, logistic regression and gradient boosted decision trees. Regardless of choice of machine learning algorithm, a significant amount of effort typically had to be put into defining the vocabulary and preprocessing the text. Most natural languages have large word vocabularies (even excluding digit sequences!); practitioners have used normalization (e.g., lowercasing, punctuation removal, replacement of numbers with a special token, removal of determiners, conjunctions and other function words, replacement of words with their root forms) to reduce vocabulary size.

More recently, document classification methods have relied on fixed or contextual learned embeddings and techniques like convolutional neural networks (Adhikari et al., 2019).

Sequence tagging

For sequence tagging, we need ways to model words in *context*; that is, the influence on the function or meaning of one word from the words that surround it. Consider these three examples: *she banks at Barclays; she swam between the banks of the river; she banks on getting her degree.* All three contain the word 'banks', but the function and meaning of the word varies based on its context. Sequence tagging tasks include part of speech tagging and named entity recognition. For a long time, sequence tagging was done using conditional random fields (Wallach, 2004), but today is typically done using long-short term memory (LSTM) neural architectures (Arora et al., 2021).

Sequence generation

For sequence generation ('seq2seq'), we need ways to model words in context *and* to model a very large label (model output) space. Sequence generation tasks include machine translation and summarization. Today sequence generation tasks are generally modeled using encoder/decoder architectures (Bambrick et al., 2020).

28.3 Applications

Unstructured data that exists in textual documents, such as filings, earnings calls and other transcripts, websites, emails, and instant messages, contains rich, financially relevant, information. For example, a sample of the unstructured data available to Bloomberg clients includes earnings releases, press briefings, social media, research and investigative reports, and breaking news. To assist in generating relevant and timely financial information, Bloomberg has an Editorial and Research department consisting of more than 2,700 journalists and analysts across more than 120 countries. They produce more than 5,000 stories each day, covering 75,000 securities and 10,000+ topics per year. In addition, Bloomberg is also a news aggregator, consuming global regulatory filings, financially relevant social media, Business Wire, PR Newswire, GlobeNewswire, Marketwire, The New York Times, Dow Jones, Wall Street Journal, and 100,000+ relevant websites, using predictive analytics to track the release of new information (Huang, 2018). In order to assist in the generation and consumption of this content, we have harnessed the full power of natural language processing (NLP). While we have been able to leverage academic and industrial research in NLP to help build some of our data generation and enrichment functionality, innovation has been necessary due to the real-time data distribution and unique constraints of latency, precision, and temporal-awareness required by our applications.

In this section we highlight three applications of NLP at Bloomberg and their relevance to finance.

28.3.1 Information extraction

To the average investor, market moving events are delivered during press briefings or through websites and social media. In order to be able to respond quickly to these events, it is necessary to quickly extract structured data from the unstructured content. Some examples of market moving events include: key revenue numbers from earnings releases, announcements of management changes, mergers and acquisitions, changes in credit ratings, or government policy changes. Once structured and linked data is extracted, it can be consumed better by both humans and trading applications, thereby enabling fast response times in trading strategies. In this section, we briefly discuss strategies for setting up NLP pipelines that can be used to extract structured information about mergers and acquisitions (M&A) from textual content.

Problem statement

The task described in this subsection can be expressed in plain English as "extract structured information pertaining to Mergers and Acquisitions (M&A) from textual content." First, we want to further break down this problem statement to be more precise. Note that there are different types of M&A events, such as 'acquisition', 'divestiture', 'merger', 'strategic review', 'approved', and 'termination'. Each type of M&A activity can itself consist of different sub-types. For example, an 'acquisition' event could be one of: (a) company buying another company, (b) company buying a unit of another company, (c) company buying assets of another company, (d) company buying stake in another company, and so on. Notice that there are multiple stages to each M&A event. For example, an 'acquisition' progresses through 'proposal', 'to buy', 'letter of intent', 'approval', and other such stages. Also, different M&A deals are structured differently. For example, some M&A deals are cash payments.

While all M&A event types, sub-types, and stages might be financially relevant, attempting to tackle such a large class of events at once via information extraction might be hard. Therefore, we recommend selecting specific events that are of high value to extract. For example, let us use NLP to extract the structured information pertaining to a company buying another company, and in particular, the very first stage of the acquisition – the announcement of their intent to buy the other company. First we lay out the relevant pieces of information that need to be successfully extracted to have a minimum viable product. In our case, we need to know when the M&A is first announced and the following information: (a) the acquiring company (b) the company being acquired (c) payment type and (d) deal value. Depending on the use case, knowing (a) and (b) might be sufficient, but partial or incorrect results typically yield no gains – or worse, negative gains. Second we decide on the metrics by which we measure success. In this case, suppose the information extracted is for both human and machine consumption with the intent of facilitating quick reactions to the market; that is, we need high precision and low latency. Of course, recall is also important, since low recall means that our extractions cannot be trusted to always deliver the required information. As an example, you may consider setting the precision of the model at 99%, latency at 1 second, and recall at 50%. Instead of using both precision and recall, it is possible to use different F measures, which report performance as a single number. In this case, since precision is more important, using $F_{\frac{1}{2}}$ is prudent (F_1 is the harmonic mean of precision and recall). Depending on the use case, it is possible to stipulate that confidence intervals be part of the output, but we will omit that discussion here for the sake of brevity. Similarly, if the intent is to process a large number of documents – for example, for backfilling historical data – throughput requirements can also be stipulated, as they do affect the choice of models. To summarize, our problem statement is: "given an English document, extract these data points for all mentioned M&A acquisitions events that are in the to_buy stage: the acquiring company, the company being acquired, and the value of the deal. Furthermore, ensure that the models operate under 1 second end-to-end with a precision $\geq 99\%$ and recall $\geq 50\%$".

Data

For an M&A extraction pipeline to work, we need several types of data. First, we need historical data so we can train and test our models. Second, we need real-time data in which our models can look for and extract M&A information, which can come from any one of the millions of websites and billions of social media posts around the globe. All this information has to be carefully prepared and cleaned before it is fed into the NLP models. Examples of data preparation include parsing out HTML content, OCR scans of PDF documents, and images with textual data. The process of setting up a real-time feed of potentially relevant content and then preparing it for model inference is a large-scale and complex operation, often involving ML, NLP, and speech recognition tasks to establish which websites to scrape, when to scrape these websites, how to extract text using OCR, how to recognize text in images, or transcribing audio content.

Once we have access to historical and real-time data, the next step is to establish a test and validation set to help us assess if our models meet the stipulated requirements. This step requires careful preparation of annotation tasks and for the sample data to be meticulously annotated. If all we had was a historical archive of all websites and social media, the signal-to-noise ratio would be extremely low (we would have tens of positive examples for each class of event, to thousands of negative examples). Therefore, unbiased sampling of data from this archive for annotation is a critical step. Sampling biases could easily thwart any attempt to build sophisticated models. For example, consider using a regular expression match for 'acquire' or 'buy' to identify potential content and using the returned set of documents for the annotation task. Any test set built from such a process would severely overestimate the precision of our NLP models. If we chose to deploy the same regular expression as a pre-filter in real-time, our model's recall would be significantly lower than the recall recorded on the artificial test set. Sampling correctly and correcting for sampling bias are beyond the scope of this chapter, and we refer the readers to Cortes et al. (2008) for a detailed exposition.

Assuming sampling is done correctly, the positively labeled test and valida-

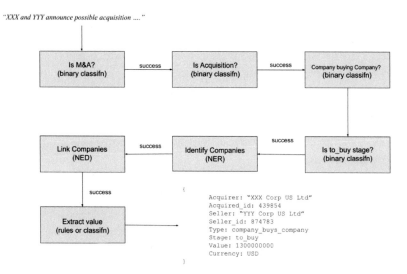

Figure 28.1 NLP pipeline to extract M&A information from documents

tion data would contain at least the following pieces of information: (a) input texts, and (b) 'ground truth' extracted M&A events with field values for the company to be acquired, acquiring company and value of the deal. For example, an annotation might look like this: {Text: "XXX and YYY announce possible acquisition. The deal is expected to be all-cash valued at 1.3B", Acquirer: "XXX Corp US Ltd", Acquirer_id: 439854, Seller: "YYY Corp US Ltd", Seller_id: 874783, Type: "company_buys_company", Stage: "to_buy", Value: 1300000000, Currency: USD}. We omit the mention of training data, as it depends on the modeling approach chosen. However, the minimal training data that would be needed is similar to the test and validation with each document and the expected structured content from the extraction clearly listed. As most documents do not contain M&A information, negative examples can be generated simply through random sampling from the historical data archive.

Modeling

There are several approaches to modelling the M&A information extraction task, and the best approach depends on factors such as the amount of training data, the quality/type of the training data, the latency needed, and the hardware available to train and run inference. For example, if low latency inference is needed, then using large transformer networks, such as BERT, might prove challenging. Similarly, if we wish to expand from the specific M&A task of a company acquiring another company to other types or subtypes of M&A activities, a modular design where a pipeline of horizontally-scalable models work together might be prudent.

A sample NLP pipeline for extracting the M&A event of interest is depicted below in Figure 28.1. The first four tasks – all accomplished using binary classification – funnel increasingly more probable M&A related documents to the final three stages where the final extraction takes place. While we have depicted these to be four separate stages, it is possible to combine all these into a single large classifier although we lose modularity and therefore extensibility to future M&A events that way. For example, to extract information on when a letter of intent has been signed, only the fourth classifier in Figure 28.1 needs to be replaced. There are several ways of gathering annotation data for this task. To get high quality training and test data, annotation tasks should be simple and so, we recommend running a separate annotation task for each stage. We also recommend that at least three annotators annotate each document. We also recommend using metrics such as Krippendorff's Alpha to measure the quality of the annotations (Artstein, 2017; Tseng et al., 2020). Krippendorff's Alpha greater than 0.6 indicates good quality annotations, while a score of 0.7 or better indicates high quality. Modular designs for NLP tasks also allow tuning of each individual task separately either through use of additional data specific to the task or better models for that task.

For each task, any number of NLP methods for document classification ranging from traditional methods such as logistic regression and SVMs (Hastie et al., 2001) to modern methods based on neural networks such as LSTMs (Hochreiter and Schmidhuber, 1997) or Transformers (Vaswani et al., 2017) can provide desirable results. To get good results, we recommend separating the test, validation and training datasets by time as the generalization performance of your classifiers might be misleading if you train on data from the future and test on the past. For our case, since we are interested in high precision classifiers, studying the precision-recall curve (CITE) allows us to pick a suitable threshold for classification. This is fundamentally a product- and application-specific decision: for example, for a fully automated low latency solution in a high impact domain like finance, precision of > 0.9 may be a business requirement even if recall is < 0.7; where a human-in-the-loop solution can be deployed (at some cost to latency), precision of 0.8 may be satisfactory.

The steps involved in extraction – named entity recognition (NER), named entity linking (NED), and slot filling (value extraction) – can be implemented in many different ways as well. NER and NED are well-studied problems in NLP. There are many off-the-shelf NER models available for a variety of languages and domains that utilize deep neural network architectures such as transformers to generate state-of-the-art results (Akbik et al., 2018, 2019; Chiu and Nichols, 2016; Lample et al., 2016). Selecting a model that is trained on the same domain as the documents from we wish to extract M&A information is important.

Building an NED model is a difficult task, which requires continuous curation of a Knowledge Graph (KG) (Reinanda et al., 2020) containing all the well-known entities, the relationships between these entities, and context about these entities. The KG serves as the database to link entities and provides rich training data for NED. For example, a Wikipedia page or a recent news clip about a company or a person is useful training data for a NED model to accurately link a mention to a well-known entity. For a full exposition of building state-of-the-art NED models, we refer the reader to Tsai et al. (2021). While public datasets such as DBPedia

(Auer et al., 2008) or YAGO2 (Hoffart et al., 2011) can be used to bootstrap a financial KG, for accurate results, a financially relevant KG is key.

Temporal accuracy of the NER and NED models is an important feature of NLP solutions for finance. The nature of language and the universe of entities both change over time. The change in language over time – introduction of new words or change in meaning of existing words – affects the accuracy of NER (Rijhwani and Preotiuc-Pietro, 2020). Similarly, in order for the NED models to link entities accurately in a document, the entities need to be current with respect to the time of the document's publication. For this, continuous curation of the KG is critical as new entities (companies and people), and new relationships (business and employer relationships) are added and deleted every day. While continuous annotation, training, and deployment of NER and NED can ensure that our M&A extraction pipeline is always up-to-date, a more subtle design is needed if backtesting for long periods of history is needed. To wit, using a NED model trained on a KG from 2021 to link entities in a document from 2018 can lead to errors as some of the entities, relationships, and context in the KG are from the future. Curating a temporally accurate KG, and training a temporally accurate NED is a subject of active research.

Finally, we have to extract the deal amount and the currency. While ML-based models can be built to extract such information from a document, rule-based approaches can also be effectively used depending on the consistency with which these values are expressed. For example, the currency can be deduced from the exchange/country in which the acquiring entity is listed. Similarly, the deal amount is a number and given a list of all the numbers in a document, we can either write rules or train a classifier to extract amounts from text and normalize them.

If an organization has a historical database of M&A transactions that have been manually curated and stored, the annotation tasks of identifying M&A documents of the right variety can be sped up using weak supervision (Zhou, 2018). For example, consider that we know company X bought company Y in March 2018. Suppose we have a list of all the documents around March 2018 that have been tagged to contain mentions of both X and Y in January, February, and March 2018. Any of these documents could contain the information of the M&A and therefore, can serve as weak supervision for some of the tasks in our pipeline and distant supervision for other tasks.

28.3.2 NSTM: Identifying key themes in news

More than 1.5 million financially relevant news articles are available to Bloomberg clients each day. Many of these stories are available for searching within 215 milliseconds, which reflects the upper bound on the time to enrich News content with topic classification, NER, NED, and other analyses, such as market sentiment and news importance. One of the main ways this volume of news is consumed is through the Bloomberg Terminal's search function. Given a search query, traditional search systems return a set of articles sorted by relevance or time.

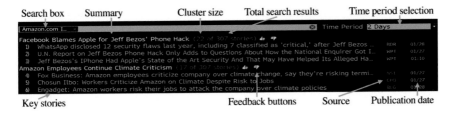

Figure 28.2 A query-based UI for NSTM showing two themes.

However, merely returning a list of relevant articles is not especially helpful to clients with short time horizons, as we are bound to return hundreds, if not thousands, of results. Many of these results are duplicates or overlap with one another, therefore adding to the complexity of getting the required answers. While some queries can be answered using text-based or table-based question answering, a majority of the queries do not have straightforward answers that can be read off of sentences and tables. In this section, we discuss a system called 'NSTM: Key News Themes' (Bambrick et al., 2020) that Bloomberg built to provide succinct overviews to user queries.

Consider how the query 'Amazon.com' over a time horizon of 2 days is answered in real-time with sub-second latency using Key News Themes in Figure 28.2. The results are neatly organized into different theme buckets, where each theme has an automatically generated summary along with exemplar news stories from each theme. Finally, both the themes and exemplar stories within each theme are ranked by their relevance to the user query to further facilitate quick consumption. The key NLP and ML techniques underlying this functionality are: (a) semantically rich document embeddings, (b) fast clustering that allows us to group documents meaningfully into themes, (c) ranking, which allows us to rank news articles in each cluster and to rank themes by relevance, and (d) summarization, allowing us to automatically decipher the meaning of each theme.

The main technical challenges in building NSTM were the lack of public datasets corresponding to this unique information composition problem, in addition to the fact that generating summaries that are accurate, informational, fluent, and concise required careful design of both the algorithmic and annotation tasks. For brevity, we omit the machine learning and software engineering challenges we faced in designing fast clustering methods to ensure our system scaled to hundreds of thousands of users in real-time. In this section, we focus on the main NLP aspects: generating document embeddings and the accurate summarization of document clusters.

Representing documents

NSTM uses clustering to meaningfully group results into themes and at the heart of any good document clustering system is the ability to represent documents in a semantically rich vector space such that similarity between documents can be accurately and efficiently computed. News articles can be modeled as a sequence of words with structure, such as headlines and sentence position, being exploited by some approaches and ignored by others. There are several unsupervised, semi-supervised, and supervised methods of learning representations of documents in vector space (Le and Mikolov, 2014; Wu et al., 2021). In NSTM, after preliminary experiments with several modeling choices, we decided to use an unsupervised Neural Variational Document Model (NVDM) (Miao et al., 2015) to compute document embeddings. NVDM is a variational inference framework for generative and conditional models of text. Our choice to use NVDM was motivated by the following two reasons. First, NVDM, like Latent Dirichlet Allocation (LDA) (Blei et al., 2003), is a generative bag-of-words (BoW) model, and such models have proven effective at capturing the rich semantic information present in documents. In NVDM, the use of cosine similarities is naturally motivated as the generative model is directly defined by the dot-product between the document embedding (z) and a shared vocabulary embedding (W). As our main operation on these document embeddings is similarity computation, we found this fact particularly useful. Second, NVDM's Variational Auto-encoder (VAE) (Kingma and Welling, 2019) framework makes inference – the task of computing a document's embedding – simpler than LDA. We trained the NVDM model on an internal corpus of 1.85M news articles, using a vocabulary of size about 200,000, and a latent dimension of 128. That is, the generated document embeddings were real vectors of size 128.

Summarization

One of the main tasks of NSTM is to generate accurate, fluent, informational, and concise summaries from clusters of news stories. Our approach to generating such summaries is bottom-up: we first generate candidate summaries from each news story, then select the best summary for each news story, and finally, we select the best story to represent a cluster and use the summary of that story. When we were building this product, there were no internal or external training datasets available for us to use. This lack of training data and the need to be able to control the generated summaries made training of sequence-to-sequence abstractive summarization tasks hard. Therefore, we settled on using a combination of two methods – OpenIE-based Tuple Extraction (Etzioni et al., 2008) and BERT-based extractive summarization – for which we collected a moderate sized dataset. In this section, we will use the sentence "Automaker ST is investing $2B in electric vehicles, atoning for the 2018 scandal" as a running example.

Open Domain Information Extraction (OpenIE) is an unsupervised approach to extracting summary candidates from sentences. First, we construct a dependency parse tree using a deep neural network, and then we extract predicate-argument n-tuples with an adapted reimplementation of Pred–Patt (White et al., 2016). Pred–Patt applies rules over Universal Dependencies dependency parses, making it language agnostic, an added benefit. (This is a use of syntax modeling, discussed in the introduction to this chapter.) The left panel of Figure 28.3 shows how OpenIE extracts candidate summaries from our running example in four stages. First, the sentence is parsed using a dependency parser. Second, predicate-argument

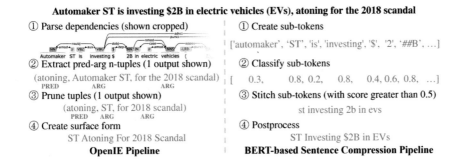

Figure 28.3 Illustrations of the symbolic OpenIE (left) and neural sentence compression (right) candidate extraction pipelines used in NSTM.

n-tuples are extracted using our modified version of Pred–Patt. For the sake of brevity, only one such *n*-tuple is shown whereas in reality, several *n*-tuples can be extracted from a single sentence. Next, the tuples are pruned. Finally, a title-cased surface form is created. This surface form is used as one of the many candidates to generate a summary from a given cluster. OpenIE offers great control over the generated summaries, is interpretable, and is fast. To complement OpenIE, and to take advantage of the advances in NLP brought about by BERT, in addition to news-specific grammatical styles, we use a BERT-based (Devlin et al., 2018) model to compress input sentences to generate summary candidates. BERT itself is pre-trained unsupervised using a masked language model on in-house news articles to generate a custom BERT pre-trained model. Next, to fine tune BERT for sentence compression, we created an in-house dataset that maps each sentence to its compressed equivalents, where the compression is achieved by deleting words in the original sentence. Our internal dataset had 10,000 examples, which were split between training (80%) and test (20%) sets. The modeling task was one of sequence labeling where, for an input sequence of words, each token (word) in the sequence is labeled with 'keep' or 'delete' symbols. The summary of a sentence is generated by combining the words labeled 'keep'. To enhance the quality of the summaries generated, we post-processed to title case the summaries and used entity recognition to ensure that an entity was either fully selected or omitted. The right panel of Figure 28.3 displays the extractive summarization of our running example in four stages. Note that, unlike in OpenIE, where each sentence can yield several *n*-tuples, each sentence yields a single summary in our BERT-based sentence compression model.

Clusters of news stories generated by NSTM might have tens, if not hundreds, of stories. Given that each news story can generate tens of summary candidates, we select one summary from potentially thousands of summaries for each cluster (theme) using a supervised learning to rank approach. For these, we collected a dataset of 33,000 summary candidates, each of which was rated GREAT, TERRIBLE, or ACCEPTABLE by our internal annotators based on readability and informativeness. Concretely, given a candidate summary *c* and the article *a*, we ask annotators to say how well *c* summarizes the content in *a*. As we use NVDM to embed both *a* and *c*, we are more likely to select summaries that reflect the whole article rather than just the headlines. From this dataset, we gather 48,000 pairwise examples ($c1 > c2 \mid a$) for training using the max margin loss in the spirit of learning to rank models. After trying several models, we realized that merely using the dot-product – that is, $a.c1 > a.c2$ implies $c1$ is a better summary of *a* than *c2* – was sufficient and had the added advantage of being faster to compute. At the end of this stage, we have one summary per new story. The summary displayed for a cluster is the summary of the news story that has the best average similarity with all the other stories in the cluster. Similarity between news stories is computed using a smoothed version of cosine similarity between the stories.

28.3.3 *Market sentiment analysis*

High quality market sentiment signals extracted from News and Social Media have been shown to produce strong signals for constructing trading strategies (Dimov, 2018; Verma, 2019). Market sentiment is different from the classical sentiment analysis in NLP (Pang and Lee, 2009) and therefore, off-the-shelf models for sentiment are insufficient to extract such a signal. Market sentiment of an article towards a company is deemed positive if the news contained in the article will likely increase the stock price of the company. Careful modeling is needed as an article can contain mentions of more than one company and the sentiments expressed in that article are different towards each of the companies. In this section, we outline a NLP pipeline that can be used to extract such a market sentiment for companies from news articles.

Problem statement

The task is to build a market sentiment analysis pipeline which, given a news article, generates a market sentiment score for all the companies mentioned in that article. That is, for each article, the sentiment analysis pipeline first extracts all the mentioned companies and links them to well-known identifiers, then predicts one of POSITIVE, NEGATIVE, or NEUTRAL labels along with a confidence score between 0 to 100 for each company salient to the article. The label is POSITIVE/NEGATIVE for a news story for a company if the news contained in the story would likely result in gains/losses for a long position holder in the company; the prediction is NEUTRAL in all other cases. The confidence score is indicative of our faith in the model's output and can serve as weighting factors when aggregating sentiment scores for a company over different intervals of time. In order to ensure that these confidence scores are comparable over time, we apply model calibration techniques (Bella et al., 2010). As precision and recall equally important and therefore, we select F_1 score as the metric for validation and testing. It is also important to identify beforehand if backtesting is a requirement as it affects both the data and modeling choices; in our case, we assume that the pipeline will be used for backtesting.

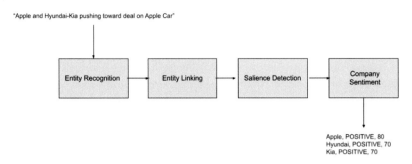

"Apple and Hyundai-Kia pushing toward deal on Apple Car"

Apple, POSITIVE, 80
Hyundai, POSITIVE, 70
Kia, POSITIVE, 70

Figure 28.4 A sample NLP pipeline for computing Market Sentiment

Data

As in the case of the M&A pipeline, several types of data are needed to build the sentiment analysis pipeline. As correctly identifying and linking companies mentioned in a news article are key to our pipeline, a financial KG, data for both NER and NED as domain-specific as possible are key (see §28.3.1). In addition to the data needed to train and test for NER and NED, we need data for two additional tasks. First, we need data to build and test a salience classifier; salience of an entity to an article tells us how much of the story is about the entity and serves as a good filter to eliminate noise from our market sentiment estimate. For example, consider the headline "X reports that Y is planning to acquire Z"; here, the story is about entities 'Y' and 'Z' but not 'X'. Salience data can be collected either as a binary label (salient and not) or as an ordinal label (high, medium, low, and not salient). While setting up an annotation task for salience is one option, weak supervision data can also be generated for this task. For example, the relative number of mentions of an entity in the news article could be one proxy measure of salience. Second, we need data to build and test a market sentiment classifier. If the annotators are finance professionals, this task can be moderately straightforward. However, if the annotators are not finance professionals, the annotation task description has to be carefully written and the annotators have to be trained as well. Along with the task description, an elaborate manual explaining the different aspects of an article that could result in a label of POSITIVE or NEGATIVE or NEUTRAL have to be outlined. In addition to providing annotated examples, training annotators to understand financial concepts that occur in the articles frequently such as 'long position investor', 'bullish', 'bearish', and 'short sale interest' could help. In short, the annotation guide for tasks such as market sentiment are as important an artifact as the models themselves.

Similar to the arguments presented in §28.3.1, sampling bias needs to be carefully addressed. For example, sampling articles for annotation non-uniformly from particular months or particular quarters, or particular years, or particular countries, or particular sectors could all induce bias due to varying language and concepts used in these articles. As even experts disagree on tasks market sentiment analysis, we recommend paying close attention to Krippendorf's alpha or other such measures of inter-annotator agreement to ensure high quality data is produced. Due to the difficult nature of assessing both salince and market moving sentiment, we recommend breaking having separate annotation tasks for salience and sentiment per company per article. When the outputs of the salience and the market sentiment annotation tasks are merged, each annotation will contain the following pieces of information: (a) the entire body of text, (b) a company that occurs in the text, (c) salience of the company, and (d) market sentiment label if that company is salient to the text. For example, an annotation might look like this: {Text: "Apple and Hyundai–Kia pushing toward deal on Apple Car", Company: "Apple Incorporated", Company_ID: "123456", Salient: "TRUE", Sentiment: "POSITIVE"}. A similar annotation would be generated for 'Hyundai' and 'Kia' companies for the same article eventually. We chose the binary classification of salience in this annotation example. Note that we have omitted to capture the 'confidence' of the annotator in this example. While confidence can be explicitly annotated and modeled, we recommend deriving the confidence measure directly from the model as that captures the confidence of the model more accurately.

Modeling

The NLP pipeline for producing a company market sentiment score is depicted in Figure 28.4. The first two steps, which are a common step in extracting value from textual content, is to identify the entities in the story using NER and then concretely linking each entity to a well-known identifier in a KG or in a financial database using NED. The NER and NED steps were discussed in detail in §28.3.1 and we omit further discussion here. The next step is salience detection, which can either be a binary classification task or an ordinal classification (regression) task depending on the annotations collected. In either case, any number of modeling approaches ranging from simple methods such as logistic regression and SVMs to more complex approaches such as random forests, gradient boosted decision trees, and deep neural networks can be used to model salience. Similarly, the market sentiment analysis task, which is a multi-class classification problem, can also be modeled using any of the above mentioned techniques. The final step is to estimate confidence of the market sentiment label, which is done through a careful calibration of probabilities emitted by the chosen model. Histogram binning (Zadrozny and Elkan, 2001), isotonic regression (Zadrozny and Elkan, 2002), Bayesian binning into quantiles (Naeini et al., 2015), and Platt scaling (Platt, 2000) are all good ways to calibrate the probabilities emitted by the market sentiment classifier.

28.4 Conclusion

Finance makes increasing use of NLP, both to extract structured information from unstructured content (via named entity recognition and linking, information extraction and document classification, among other techniques), and to synthesize natural language (for example, to translate or summarize content). In this chapter we survey the basic tasks that comprise most modern applied NLP. We provide the mathematical and ML underpinnings of the three techniques that work for all these tasks: text classification, sequence tagging and sequence generation. We then describe in detail three finance-relevant applications that use the common NLP tasks and techniques: information extraction of market moving events; summarization of key themes in news; and classification of market sentiment from social media. These applications are not an exhaustive list of the ways in which NLP is used in finance, but most other applications of NLP in finance resemble one of these. For example, prediction of market risk (Kogan et al., 2009; Wang and Hua, 2014; Rekabsaz et al., 2017) or classification of market moving news (Xie et al., 2013) is, like sentiment analysis, text classification; information extraction can be used for many types of finance data mining (Goel et al., 2020); timeline construction (Pham et al., 2009) is, like summarization, digesting of content.

Throughout this discussion we have highlighted ways in which NLP for finance is different from NLP applied to other domains. For example, it is critical in financial applications to use temporal splits of data to avoid letting the future 'leak' into the past. It is also important that NLP techniques used in finance applications be high precision and low latency: other domains, such as education, may have very different requirements.

Although in this discussion we have focused on entirely automatic methods and applications of NLP, there are many more applications of NLP in finance that are human-in-the-loop – for example, computer-assisted interactive trading (İrsoy et al., 2019), or computer-assisted research report authoring, or computer-assisted compliance-related activities (Chen et al., 2020a). Just as human experts working with AI can achieve results not possible solely through human effort or automation, so can finance experts working with NLP experts achieve results not possible through isolated expertise in each area. In addition to NLP being increasingly used by finance practitioners and researchers, NLP researchers are increasingly contributing to finance modeling (e.g., Chen et al., 2020b, 2019; Hahn et al., 2019, 2018). Continued discussion across disciplinary boundaries is necessary to make sustained, useful, forward looking progress in the application of NLP to finance.

References

Adhikari, Ashutosh, Ram, Achyudh, Tang, Raphael, and Lin, Jimmy. 2019. Rethinking complex neural network architectures for document classification. Pages 4046–4051 of: *Proc. Conference of the North American Chapter of the Association for Computational Linguistics: Human Language Technologies*.

Akbik, Alan, Blythe, Duncan, and Vollgraf, Roland. 2018. Contextual string embeddings for sequence labeling. Pages 1638–1649 of: *Proc. International Conference on Computational Linguistics*.

Akbik, Alan, Bergmann, Tanja, and Vollgraf, Roland. 2019. Pooled contextualized embeddings for named entity recognition. Page 724–728 of: *Proc. Conference of the North American Chapter of the Association for Computational Linguistics*.

Antol, Stanislaw, Agrawal, Aishwarya, Lu, Jiasen, Mitchell, Margaret, Batra, Dhruv, Zitnick, C. Lawrence, and Parikh, Devi. 2015. VQA: Visual question answering. Pages 2425–2433 of: *Proc. IEEE International Conference on Computer Vision.*

Arora, Ravneet, Tsai, Chen-Tse, and Preoţiuc-Pietro, Daniel. 2021. Identifying named entities as they are typed. Pages 976–988 of: *Proc, Conference of the European Chapter of the Association for Computational Linguistics.*

Artstein, Ron. 2017. Inter-annotator agreement. Pages 297–313 of: *Handbook of Linguistic Annotation.* Springer.

Auer, S., Bizer, C., Kobilarov, G., Lehmann, J., Cyganiak, R., and Ives, Z. 2008. DBpedia: A nucleus for a web of open data. Pages 722–735 of: *Proc. International Semantic Web Conference and the Asian Semantic Web Conference.*

Bambrick, Joshua, Xu, Minjie, Almonte, Andy, Malioutov, Igor, Perarnau, Guim, Selo, Vittorio, and Chan, Iat Chong. 2020. NSTM: Real-time query-driven news overview composition at Bloomberg. Pages 350–361 of: *Proc. Annual Meeting of the Association for Computational Linguistics.*

Bartolo, Max, Roberts, Alastair, Welbl, Johannes, Riedel, Sebastian, and Stenetorp, Pontus. 2020. Beat the AI: Investigating adversarial human annotation for reading comprehension. *Transactions of the Association for Computational Linguistics*, **8**, 662–678.

Bella, Antonio, Ferri, Cèsar, Hernández-Orallo, José, and Ramírez-Quintana, María José. 2010. Calibration of machine learning models. Pages 128–146 of: *Handbook of Research on Machine Learning Applications and Trends: Algorithms, Methods, and Techniques.* IGI Global.

Blei, David M., Ng, Andrew Y., and Jordan, Michael I. 2003. Latent Dirichlet allocation. *Journal of Machine Learning Research*, **3**, 993–1022.

Bojanowski, Piotr, Grave, Edouard, Joulin, Armand, and Mikolov, Tomas. 2016. Enriching word vectors with subword information. ArXiv:1607.04606.

Burdick, Douglas, Danilevsky, Marina, Evfimievski, Alexandre V, Katsis, Yannis, and Wang, Nancy. 2020. Table extraction and understanding for scientific and enterprise applications. *Proc. VLDB Endowment*, **13**(12), 3433–3436.

Charniak, Eugene. 1997. Statistical parsing with a context-free grammar and word statistics. Pages 598–603 of: *Proc. National Conference on Artificial Intelligence and the Conference on Innovative Applications of Artificial Intelligence.*

Chen, Chung-Chi, Huang, Hen-Hsen, Takamura, Hiroya, and Chen, Hsin-Hsi (eds). 2019. In: *Proc. First Workshop on Financial Technology and Natural Language Processing.*

Chen, Chung-Chi, Huang, Hen-Hsen, and Chen, Hsin-Hsi. 2020a. NLP in FinTech applications: past, present and future. ArXiv:2005.01320.

Chen, Chung-Chi, Huang, Hen-Hsen, Takamura, Hiroya, and Chen, Hsin-Hsi (eds). 2020b. In: *Proc. Second Workshop on Financial Technology and Natural Language Processing.*

Chen, Danqi, and Yih, Wen-tau. 2020. Open-domain question answering. Pages 34–37 of: *Proc. Annual Meeting of the Association for Computational Linguistics.*

Chieu, Hai Leong, and Ng, Hwee Tou. 2003. Named entity recognition with a maximum entropy approach. Pages 160–163 of: *Proc. Conference on Natural Language Learning.*

Chiu, Jason P.C., and Nichols, Eric. 2016. Named entity recognition with bidirectional LSTM-CNNs. *Transactions of the Association for Computational Linguistics*, **4**, 357–370.

Cho, Kyunghyun, van Merrienboer, Bart, Gülçehre, Çaglar, Bahdanau, Dzmitry, Bougares, Fethi, Schwenk, Holger, and Bengio, Yoshua. 2014. Learning phrase representations using RNN encoder–decoder for statistical machine translation. In: *Proc. Conference on Empirical Methods in NLP.*

Chomsky, N. 1956. Three models for the description of language. *IRE Transactions on Information Theory*, **2**(3), 113–124.

Chomsky, Noam. 1957. *Syntactic Structures.* de Gruyter.

Collobert, Ronan, Weston, Jason, Bottou, Léon, Karlen, Michael, Kavukcuoglu, Koray, and Kuksa, Pavel. 2011. Natural language processing (almost) from scratch. *Journal of Machine Learning Research*, **12**, 2493–2537.

Cortes, Corinna, Mohri, Mehryar, Riley, Michael, and Rostamizadeh, Afshin. 2008. Sample selection bias correction theory. *CoRR*, abs/0805.2775.

De Marneffe, Marie-Catherine, and Nivre, Joakim. 2019. Dependency grammar. *Annual Review of Linguistics*, **5**, 197–218.

Deisenroth, Marc Peter, Faisal, A. Aldo, and Ong, Cheng Soon. 2020. *Mathematics for Machine Learning*. Cambridge University Press.

DeRose, Steven J. 1988. Grammatical category disambiguation by statistical optimization. *Computational Linguistics*, **14**(1).

Devlin, Jacob, Chang, Ming-Wei, Lee, Kenton, and Toutanova, Kristina. 2018. BERT: Pre-training of deep bidirectional transformers for language understanding. ArXiv:1810.04805.

Dietterich, Thomas G. 1998. Approximate statistical tests for comparing supervised classification learning algorithms. *Neural Computation*, **10**(7), 1895–1923.

Dimov, Ivailo. 2018. *Topic tags as a key took for optimizing sentiment analysis.*

Dong, Xin Luna, Hajishirzi, Hannaneh, Lockard, Colin, and Shiralkar, Prashant. 2020. Multi-modal information extraction from text, semi-structured, and tabular data on the Web. Pages 23–26 of: *Proc. Annual Meeting of the Association for Computational Linguistics*.

Etzioni, Oren, Banko, Michele, Soderland, Stephen, and Weld, Daniel S. 2008. Open information extraction from the Web. *Communications of the ACM*, **51**(12), 68–74.

Gaizauskas, Robert, and Wilks, Yorick. 1998. Information extraction: Beyond document retrieval. *Journal of Documentation* **54**(1), 70–105.

Goel, Tushar, Jain, Palak, Verma, Ishan, Dey, Lipika, and Paliwal, Shubham. 2020. Mining company sustainability reports to aid financial decision-making. In: *Proc. AAAI Workshop on Knowledge Discovery from Unstructured Data in Financial Services*.

Grishman, Ralph. 2019. Twenty-five years of information extraction. *Natural Language Engineering*, **25**(6), 677–692.

Grishman, Ralph, and Sundheim, Beth M. 1996. Message understanding conference-6: A brief history. In: *Proc. International Conference on Computational Linguistics*.

Habash, Nizar, and Rambow, Owen. 2005. Arabic tokenization, part-of-speech tagging and morphological disambiguation in one fell swoop. Pages 573–580 of: *Proc. Annual Meeting of the Association for Computational Linguistics*.

Hahn, Udo, Hoste, Véronique, and Tsai, Ming-Feng (eds). 2018. In: *Proc. First Workshop on Economics and Natural Language Processing*.

Hahn, Udo, Hoste, Véronique, and Zhang, Zhu (eds). 2019. In: *Proc. Second Workshop on Economics and Natural Language Processing*.

Hastie, Trevor, Tibshirani, Robert, and Friedman, Jerome. 2001. *The Elements of Statistical Learning*. Springer.

Hearst, Marti A., and Pedersen, Jan O. 1996. Reexamining the cluster hypothesis: scatter/gather on retrieval results. Pages 76–84 of: *Proc> International ACM SIGIR Conference on Research and Development in Information Retrieval*.

Hochreiter, Sepp, and Schmidhuber, Jürgen. 1997. Long short-term memory. *Neural Computation*, **9**(8), 1735–1780.

Hoffart, Johannes, Suchanek, Fabian M., Berberich, Klaus, Lewis-Kelham, Edwin, de Melo, Gerard, and Weikum, Gerhard. 2011. YAGO2: Exploring and querying world knowledge in time, space, context, and many languages. Pages 229–232 of: *Proc. International Conference Companion on World Wide Web*.

Huang, Lei. 2018. Bloomberg-curated Twitter feed. `https://data.bloomberglp.com/promo/sites/12/221539_Twitter_WP.pdf`.

Huang, Zhiheng, Xu, Wei, and Yu, Kai. 2015. Bidirectional LSTM-CRF models for sequence tagging. ArXiv:1508.01991.

İrsoy, Ozan, Gosangi, Rakesh, Zhang, Haimin, Wei, Mu-Hsin, Lund, Peter, Pappadopulo, Duccio, Fahy, Brendan, Nephytou, Neophytos, and Ortiz, Camilo. 2019. Dialogue act classification in group chats with DAG-LSTMs. ArXiv:1908.01821.

Jamison, Emily, and Gurevych, Iryna. 2015. Noise or additional information? Leveraging crowdsource annotation item agreement for natural language tasks. Pages 291–297 of: *Proc. Conference on Empirical Methods in Natural Language Processing*.

Jones, Karen Sparck. 1972. A statistical interpretation of term specificity and its application in retrieval. *Journal of Documentation*, **28**(1), 11–21.

Kim, Yoon. 2014. Convolutional neural networks for sentence classification. Pages 1746–1751 of: *Proc. Conference on Empirical Methods in Natural Language Processing*.

Kingma, Diederik P., and Welling, Max. 2019. An introduction to variational autoencoders. *CoRR*, abs/1906.02691. ArXiv:1906.02691

Klein, Guillaume, Kim, Yoon, Deng, Yuntian, Senellart, Jean, and Rush, Alexander. 2017. OpenNMT: Open-source toolkit for neural machine translation. Pages 67–72 of: *Proc. Annual Meeting of the Association for Computational Linguistics*.

Kogan, Shimon, Levin, Dimitry, Routledge, Bryan R., Sagi, Jacob S., and Smith, Noah A. 2009. Predicting risk from financial reports with regression. Pages 272–280 of: *Proc. Human Language Technologies: The Conference of the North American Chapter of the Association for Computational Linguistics*.

Laban, Philippe, Hsi, Andrew, Canny, John, and Hearst, Marti A. 2020. The Summary Loop: Learning to Write Abstractive Summaries Without Examples. Pages 5135–5150 of: *Proc. Annual Meeting of the Association for Computational Linguistics*.

Lample, Guillaume, Ballesteros, Miguel, Subramanian, Sandeep, Kawakami, Kazuya, and Dyer, Chris. 2016. Neural architectures for named entity recognition. *CoRR*, **abs/1603.01360**.

Le, Quoc, and Mikolov, Tomas. 2014. Distributed representations of sentences and documents. Pages 1188–1196 of: *Proc. International Conference on Machine Learning*.

Lin, Hui, and Ng, Vincent. 2019. Abstractive summarization: A survey of the state of the art. Pages 9815–9822 of: *Proc. AAAI Conference on Artificial Intelligence*.

Manotumruksa, Jarana, Dalton, Jeff, Meij, Edgar, and Yilmaz, Emine. 2020. CrossBERT: A triplet neural architecture for ranking entity properties. Pages 2049–2052 of: *Proc. International Conference on Research and Development in Information Retrieval*.

Marcus, Mitchell P., Santorini, Beatrice, and Marcinkiewicz, Mary Ann. 1993. Building a large annotated corpus of English: The Penn treebank. *Computational Linguistics*, **19**(2), 313–330.

Miao, Yishu, Yu, Lei, and Blunsom, Phil. 2015. Neural variational inference for text processing. *CoRR*, abs/1511.06038.

Mikolov, Tomas, Sutskever, Ilya, Chen, Kai, Corrado, Greg, and Dean, Jeffrey. 2013. Distributed representations of words and phrases and their compositionality. Pages 3111–3119 of: *Proc. International Conference on Neural Information Processing Systems*.

Naeini, Mahdi P, Cooper, Gregory F. and Hauskrecht, Milos. 2015. Obtaining well calibrated probabilities using Bayesian binning. In: *Proc. Conference of Association for the Advancement of Artificial Intelligence*.

Nenkova, Ani, and McKeown, Kathleen. 2012. A survey of text summarization techniques. Pages 43–76 of: *Mining Text Data*. Springer.

Nivre, Joakim, De Marneffe, Marie-Catherine, Ginter, Filip, Goldberg, Yoav, Hajic, Jan, Manning, Christopher D., McDonald, Ryan, Petrov, Slav, Pyysalo, Sampo, Silveira, Natalia, et al. 2016. Universal dependencies v1: A multilingual treebank collection. Pages 1659–1666 of: *Proc. International Conference on Language Resources and Evaluation*.

Pang, Bo, and Lee, Lillian. 2009. Opinion mining and sentiment analysis. *Computational Linguistics*, **35**(2), 311–312.

Paun, Silviu, Carpenter, Bob, Chamberlain, Jon, Hovy, Dirk, Kruschwitz, Udo, and Poesio, Massimo. 2018. Comparing Bayesian models of annotation. *Transactions of the Association for Computational Linguistics*, **6**, 571–585.

Pennington, Jeffrey, Socher, Richard, and Manning, Christopher D. 2014. Glove: Global vectors for word representation. Pages 1532–1543 of: *Proc. Conference on Empirical Methods in Natural Language Processing*.

Peters, Matthew, Neumann, Mark, Iyyer, Mohit, Gardner, Matt, Clark, Christopher, Lee, Kenton, and Zettlemoyer, Luke. 2018. Deep Contextualized word representations. Pages 2227–2237 of: *Proc. Conference of the North American Chapter of the Association for Computational Linguistics: Human Language Technologies*.

Pham, Quang-Khai, Raschia, Guillaume, Mouaddib, Noureddine, Saint-Paul, Regis, and Benatallah, Boualem. 2009. Time sequence summarization to scale up chronology-dependent applications. Pages 1137–1146 of: *Proc. ACM Conference on Information and Knowledge Management*.

Pinter, Yuval, Guthrie, Robert, and Eisenstein, Jacob. 2017. Mimicking word embeddings using subword RNNs. Pages 102–112 of: *Proc. Conference on Empirical Methods in Natural Language Processing*.

Pinter, Yuval, Stent, Amanda, Dredze, Mark, and Eisenstein, Jacob. 2021. Learning to look inside: Augmenting token-based encoders with character-level information. ArXiv:2108.00391.

Platt, John. 2000. Probabilistic outputs for support vector machines and comparisons to regularized likelihood methods. *Advances in Large Margin Classification*, **10**(06).

Ratner, Alexander, Bach, Stephen H., Ehrenberg, Henry, Fries, Jason, Wu, Sen, and Ré, Christopher. 2017. Snorkel: Rapid training data creation with weak supervision. Page 269 of: *Proc. VLDB Endowment*.

Reinanda, R., Meij, E., and de Rijke, M. 2020. Knowledge graphs: An information retrieval perspective. *Foundations and Trends in Information Retrieval*, **14**(4), 289–444.

Rekabsaz, Navid, Lupu, Mihai, Baklanov, Artem, Dür, Alexander, Andersson, Linda, and Hanbury, Allan. 2017. Volatility prediction using financial disclosures sentiments with word embedding-based IR models. Pages 1712–1721 of: *Proc. Annual Meeting of the Association for Computational Linguistics*.

Rijhwani, Shruti, and Preotiuc-Pietro, Daniel. 2020. Temporally-informed analysis of named entity recognition. Pages 7605–7617 of: *Proc. Annual Meeting of the Association for Computational Linguistics*.

Sahrawat, Dhruva, Mahata, Debanjan, Kulkarni, Mayank, Zhang, Haimin, Gosangi, Rakesh, Stent, Amanda, Sharma, Agniv, Kumar, Yaman, Shah, Rajiv Ratn, and Zimmermann, Roger. 2020. Keyphrase extraction from scholarly articles as sequence labeling using contextualized embeddings. In: *Proc. European Conference on Information Retrieval*.

Schneider, Steffen, Baevski, Alexei, Collobert, Ronan, and Auli, Michael. 2019. wav2vec: Unsupervised pre-training for speech recognition. In: *Proc. INTERSPEECH*.

Schuster, Mike, and Paliwal, Kuldip K. 1997. Bidirectional recurrent neural networks. *IEEE Transactions on Signal Processing*, **45**(11), 2673–2681.

Sebastiani, Fabrizio. 2002. Machine learning in automated text categorization. *ACM Computing Surveys*, **34**(1), 1–47.

Sennrich, Rico, and Haddow, Barry. 2017. Practical neural machine translation. In: *Proc. Conference of the European Chapter of the Association for Computational Linguistics*.

Shannon, Claude E. 1951. Prediction and entropy of printed English. *Bell System Technical Journal*, **30**(Jan.), 50–64.

Tsai, Chen-Tse, Upadhyay, Shyam, and Roth, Dan. 2021. *Entity discovery and linking: from monolingual to crosslingual*. in submission.

Tseng, Tina, Stent, Amanda, and Maida, Domenic. 2020. Best practices for managing data annotation projects. ArXiv:2009.11654.

Turing, A. M. 1950. Computing machinery and intelligence. *Mind*, **59**(236), 433–460.

Vaswani, Ashish, Shazeer, Noam, Parmar, Niki, Uszkoreit, Jakob, Jones, Llion, Gomez, Aidan N, Kaiser, Lukasz, and Polosukhin, Illia. 2017. Attention is all you need. In: *Proc. International Conference on Neural Information Processing Systems.*

Verma, Arun. 2019. *Extracting value from social and news data.* https://www.bloomberg.com/professional/blog/extracting-value-social-news-data/.

Wallach, Hanna M. 2004. Conditional random fields: An introduction. *Technical Reports (CIS),* https://repository.upenn.edu/cis_reports/22/.

Wang, William Yang, and Hua, Zhenhao. 2014. A semiparametric Gaussian copula regression model for predicting financial risks from earnings calls. Pages 1155–1165 of: *Proc. Annual Meeting of the Association for Computational Linguistics.*

Wang, Zelun, and Liu, Jyh-Charn. 2020. PDF2LaTeX: A deep learning system to convert mathematical documents from PDF to LaTeX. Pages 1–10 of: *Proc. ACM Symposium on Document Engineering.*

White, Aaron Steven, Reisinger, Drew, Sakaguchi, Keisuke, Vieira, Tim, Zhang, Sheng, Rudinger, Rachel, Rawlins, Kyle, and Van Durme, Benjamin. 2016. Universal decompositional semantics on universal dependencies. Pages 1713–1723 of: *Proc. Conference on Empirical Methods in Natural Language Processing.*

Wu, Chuhan, Wu, Fangzhao, Qi, Tao, and Huang, Yongfeng. 2021. Hi-Transformer: hierarchical interactive transformer for efficient and effective long document modeling. ArXiv:2106.01040.

Xie, Boyi, Passonneau, Rebecca J., Wu, Leon, and Creamer, Germán G. 2013 (Aug.). Semantic frames to predict stock price movement. Pages 873–883 of: *Proc. Annual Meeting of the Association for Computational Linguistics.*

Yosef, Mohamed Amir, Hoffart, Johannes, Bordino, Ilaria, Spaniol, Marc, and Weikum, Gerhard. 2011. Aida: An online tool for accurate disambiguation of named entities in text and tables. Pages 1450–1453 of: *Proc. VLDB Endowment.*

Zadrozny, Bianca, and Elkan, Charles. 2001. Obtaining calibrated probability estimates from decision trees and naive Bayesian classifiers. Pages 609–616 of: *Proc. International Conference on Machine Learning.*

Zadrozny, Bianca, and Elkan, Charles. 2002. Transforming classifier scores into accurate multiclass probability estimates. Pages 694–699 of: *Proc. KDD 2002.*

Zhou, Zhi-Hua. 2018. A brief introduction to weakly supervised learning. *National Science Review,* **5**(1), 44–53.

Zhu, Fengbin, Lei, Wenqiang, Huang, Youcheng, Wang, Chao, Zhang, Shuo, Lv, Jiancheng, Feng, Fuli, and Chua, Tat-Seng. 2021. TAT-QA: A question answering benchmark on a hybrid of tabular and textual content in finance. In: *Proc. Annual Meeting of the Association for Computational Linguistics and the International Joint Conference on Natural Language Processing.*

The Exploitation of Recurrent Satellite Imaging for the Fine-Scale Observation of Human Activity

Carlo de Franchis[a], Sébastien Drouyer[b], Gabriele Facciolo[b],
Rafael Grompone von Gioi[b], Charles Hessel[a] & Jean-Michel Morel[b]

Abstract

Satellite image data have a financial relevance, as they allow one to access periodically and more and more frequently any point of the globe, to detect and classify human activity, or to make measurements. In this chapter, we review first the earth observation satellites, then address the problem of direct 3D measurements of the ground from space through stereovision and by synthetic aperture radar imaging, with application to automatic stockpile volumes monitoring and to the daily evaluation of crude oil stocks. We then describe the technique of earth surveillance in its main steps: creating equalized and accurately registered time series and detecting clouds. How such time series can be exploited for fine economic measurements is illustrated in a final section where we explain how to detect and count cars in parking lots.

29.1 Introduction

Satellite imaging is shifting from the photo-interpreter era to automatic monitoring and big data processing. The main focus of satellite imaging is actually akin to the organization of a steady surveillance of the earth and of human activity on earth. This is rendered possible by the launch of several civil constellations of satellites that enable a recurrent observation of any point on the globe. Hence, satellite imaging data and its applications have exploded in the past ten years with many private and public entities publishing or selling images of the Earth from space, at all resolutions from 30 cm to 10 km, and in many different modalities: optical, infrared, hyperspectral, radar imaging. All of these data have a financial relevance, as they allow one to access periodically and more and more frequently any point of the globe, to detect and classify human activity, or to make measurements. This applies to recurrent measurements of traffic, of oil and gas storage,

[a] ENS Paris-Saclay, and Kayrros, Paris
[b] ENS Paris-Saclay
Published in *Machine Learning And Data Sciences For Financial Markets*, Agostino Capponi and Charles-Albert Lehalle© 2023 Cambridge University Press.

of the volume and shape of heaps of any material from coal to garbage. This also applies to the detection of any significant change of the aspect of the ground caused by humans, like construction sites, cultivation, forest clearing, refugee camps, or by incidents or disasters such as floods, fires, earthquakes, methane leaks, etc. The periodic and frequent observation of all points of Earth from the same angle, at the same time of the day, and at many scales is essential to build an automatic perception method that delivers unsupervised measurements and alerts. A typical observation chain therefore entails a series of necessary mathematical, geometrical and image preprocessing to align the data from the different satellites and correct for many sorts of geometrical, optical and atmospheric perturbations. Alignment is supposed to deliver a stabilized "video" of the ground across dates and satellites. Once this alignment is realized, measurements can be made by the Thales triangulation method (namely stereo vision), or by other 3D reconstruction methods such as shape from shading or interferometry. But all this would be nothing but a data deluge, without a strict control of false alarms and a strategy to define the events to be detected and categorized. Here come ultimately and decisively statistical detection theory to filter alarms and machine learning methods to categorize them. We shall describe and illustrate these steps and some of their current challenges.

In this chapter, we review first the Earth observation satellites (§29.2), then address the problem of direct 3D measurements of the ground from space through stereovision (§29.3) with an application to automatic stockpile activity monitoring, and by synthetic radar imaging (§29.4) with an application to daily oil storage measurements. We then describe the technique of earth surveillance via its main steps: creating equalized and accurately registered time series (§29.5) and detecting clouds (§29.6). To what degree time series can be exploited for fine economic measurements is illustrated in a final section where we explain how to detect and count cars in parking lots (§29.7). Needless to say, this overview is still far from complete. For example, we do not address the obviously fundamental economic applications provided by meteorological satellites, or the applications of satellite imaging for the forecast of crop growth and harvests.

29.2 What is recurrent satellite imaging?

Starting during the 1950s, the development of the first Earth observation satellites was mostly driven by military needs related to the Cold War, but civilian applications followed very quickly in fields such as meteorology (Gordon, 1962), cartography (Michel et al., 2013), geology (Rosu et al., 2014), hydrology (Ruelland et al., 2008), forestry (Gumbricht, 2012) and glaciology (Marti et al., 2014). Nowadays, images from Earth observation satellites are used to solve problems ranging from geographic mapping to meteorology predictions (Joseph, 2015), including applications as diverse as the measurement of elevation changes for glaciers (Berthier et al., 2014) or rescue assistance for natural disasters (Yésou et al., 2015).

Modern Earth observation (EO) satellites are starting to provide daily high

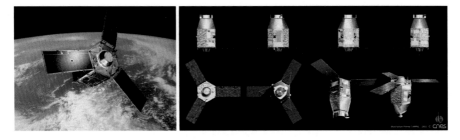

Figure 29.1 The earth observation satellite Pleiades-1A. Copyright ©Airbus DS/Master Image Films, 2007

resolution imagery of any site. This is possible thanks to large constellations and agile satellites that acquire images with different viewing angles. This trend is accelerating with the emergence of new providers, as only in the past two years, more than 150 new optical EO satellites (out of 350+ active today) have been launched. Other acquisition modalities have also become increasingly available: periodic acquisition, video modes, hyper-spectral imaging, as well as Synthetic Aperture Radar (SAR). Cloud cover is the major limitation of optical earth observation. The need for cloud-free scenes at specific dates is crucial in a number of operational monitoring applications. SAR sensors, being unaffected by clouds, provide an invaluable complement to optical satellites.

This big data explosion is pushing humans out of the analysis loop and requires automatic algorithms for analyzing this data. This is driving an AI revolution in remote sensing. An aspect of this big data revolution is the fact that data is periodically re-acquired over time and archived. Most current techniques do not explicitly model temporal evolution, which is a rich source of information. This calls for processing/analysis techniques that exploit the temporal dimension.

The ground resolution obtained by satellite imaging will always remain inferior to aerial imagery. But satellite imaging is precious in many other ways:

- It is far more cost-effective for large scale monitoring and long time series analyses;
- There are no geo-political or geographic access limitations; observing any region from space is unrestricted. In the case of SAR this continuous access is not even interrupted by meteorological events;
- The tasking time to take a new snapshot can go down to as little as 90 minutes with revisit times of a few hours depending on the constellation. With a clear trend of some constellations to deliver multiple revisits on the same day.

29.2.1 A landscape of satellites images

There are hundreds of operational EO satellites. Here we will mainly focus on the programs and providers (besides meteorological satellites) that are aimed at or can achieve a recurrent coverage of an area. We show in Table 29.1 the most common sources used for such monitoring applications.

We must first distinguish between recurrent and tasking satellites. Recurrent satellites periodically revisit all the Earth landmass; they usually have a low resolution and large swathe which allows them to cover large portions of the ground at once. In this class we find the satellites of the Landsat program, which was the first Earth observation space program for civilian purposes. Similarly, the satellites of the Copernicus program (named Sentinels) also provide periodic revisits of the whole Earth with multiple modalities. Since 2019 Planet (a private operator) has been acquiring, with the PlanetScope constellation, daily optical images over all the Earth landmass at a higher resolution than its public counterparts. This is made possible thanks to a constellation of more than 150 micro-satellites. Tasking satellites, on the other hand, acquire images on-demand over a specific area. Usually they have a much higher resolution but are more costly.

Table 29.1 List of satellites commonly used for object detection

Name	m/px	Opt. / SAR	Temporal resolution	Historical data?	Cost
Pleiades Neo	0.30	Optical	Tasking - 1 day	Catalog	+++
WorldView-3	0.31	Optical	Tasking - 1 day	Catalog	+++
WorldView-2	0.45	Optical	Tasking - 1 day	Catalog	+++
Pleiades	0.50	Optical	Tasking - 1 day	Catalog	++
Skysat	0.65	Optical	Tasking - 1 day	Catalog	++
PlanetScope	3–5	Optical	Recurrent - 1 day	Since 2016-06	+
Sentinel-2	10	Optical	Recurrent - 3 to 12 days	Since 2015-06	Free
Sentinel-1	10	SAR	Recurrent - 3 to 12 days	Since 2014-04	Free
Landsat 7, 8, 9	15	Optical	Recurrent - 8 days	Since 1999*	Free

One of the main parameters at stake is the resolution of the satellite images. Nowadays, Pleiades Neo provides the highest resolution publicly available optical satellite images at 0.30 m per pixel. At this scale, many features are visible and relatively small objects – with a width of between 1 to 2 meters - can be recognized. Other satellites – WorldView-2, Pleiades, Skysat – provide sub-meter resolutions. In general, high resolution imagery is more expensive. At the other end of the spectrum, we have Sentinel and Landsat which freely provide images at 10 meters and 15 meters per pixels respectively. Sample images are shown in Figure 29.2.

Satellite sensors can also be divided into two categories: optical and synthetic-aperture radar (SAR). Optical satellite images can be thought of as regular pictures taken from space, especially if they contain bands from the visible spectrum. These sensors are called passive as they simply measure the reflection of the Sun's emissions from Earth. They can detect multiple bands (some satellites,

(a) WorldView-3 crop of a Las Vegas street.   (b) Sentinel-2 crop of Las Vegas.

Figure 29.2 WorldView-3 and Sentinel-2 images of the same area

such as Sentinel-5P, contain thousands), each measuring a specific range in the light spectrum.

SAR satellites instead use active sensors: they send signals to the area of interest and measure what comes back. We illustrate the difference between passive and active sensors in Figure 29.3.

(a) Passive sensors (optical satellites)   (b) Active sensors (SAR satellites)

Figure 29.3 Difference between active and passive sensors in satellite images.

Due to their different mode of acquisition, each type has advantages and drawbacks. With sufficient resolution, everything that can be recognized by the human eye can be recognized in optical satellite images. As some satellites can detect bands outside the visible spectrum, some invisible gases such as methane or carbon monoxide can also be detected (using Sentinel 5-P for instance), or vegetation can be classified and monitored thanks to different ratios such as the Normalized Difference Vegetation Index (NDVI). Moreover, the different bands are acquired with a slight time lapse, so in satellite images coming from SkySat,

PlanetScope (SuperDove) or Sentinel-2 for example, this time lapse causes a parallax that can be used to detect moving objects, as shown in Figure 29.4, or clouds.

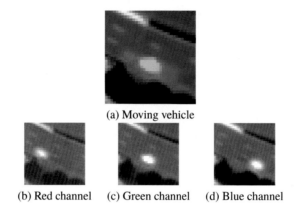

(a) Moving vehicle

(b) Red channel   (c) Green channel   (d) Blue channel

Figure 29.4 Image of a moving vehicle acquired using SkySat.

An important drawback of optical satellites is that the visible spectrum is affected by clouds and meteorological conditions, so if an area is often cloudy, a majority of images can't be processed and therefore the temporal resolution is reduced. This also adds the challenge to correctly detect clouds and shadows as explained in §29.6. On the other hand, SAR satellites tend to detect different features, some linked to topography as explained in §29.4, other linked to the nature of the material reflecting the emitted pulses. For object detection, an interesting property is that signals bounce back strongly from metallic objects, whereas water or flat areas tend to reflect a low amount of signals. As a consequence, a common use of SAR images is to detect ships near ports for instance. A great feature of SAR satellites is that they are not affected by clouds and meteorological conditions.

The most distinctive feature in satellite imaging is the difference between recurrent systematic acquisition versus the acquisition performed by tasking (done only if some customer pays for the acquisition). It is much more convenient to use providers that offer archived recurrent data (such as Sentinel, Landsat or PlanetScope), as one can get not only current images of the area of interest, but also past images (back to 2014 with Sentinel-1 for instance). Algorithms can therefore be back tested and compared to past confirmed economic and financial data.

The revisit rate is also an important factor as it will define the temporal resolution of the estimations. Most providers can approach one acquisition per day, but readers should keep in mind that optical images are often affected by clouds, so the final temporal resolution will depend on meteorological conditions. The revisit rate can also depend on the geographical position of the area to monitor. For instance, Sentinel satellites are optimized to cover the European continent;

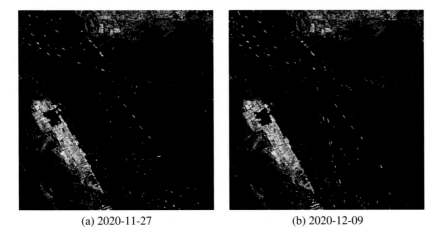

(a) 2020-11-27   (b) 2020-12-09

Figure 29.5 Sentinel-1 images of the port of Shenzhen, China

the revisit rate can be one or two days for Sentinel-1 images on the European continent, but up to twelve days in some parts of America and Asia.

As a conclusion, the optimal source will depend on different factors:

- How large the objects to be detected are;
- How precise the estimation must be;
- How often an area must be monitored;
- Whether we need to have historical data and from when;
- Whether we want to detect moving objects;
- Whether we want to detect metallic objects.

As mentioned we shall not cover all uses of EC satellites. Yet, everything ends up in two sorts of data with high economic values: measurements of volumes, or numbers of items on the one side, or simply detection and alerts. In the next section we review the problem of monitoring the 3D aspect of the ground at fine scale. In §29.4 we will address a still more particular aspect: delivering a fine-grained estimate of the volume of stored oil everywhere on Earth. Finally, in §29.7, we address another measurement of fine-grained information, namely parking occupancy.

29.3 3D monitoring from space

The crucial role of digital elevation maps

Digital elevation models (DEM) are essential to precisely evaluate changes on the ground. Indeed, high resolution imagery gives instantaneous visual knowledge. Yet, drastic visual changes make this information unstable and therefore render difficult the assessment of real changes. The irrelevant visual changes caused by the atmosphere, rain, wind, dust, changing shadows and vegetation can occur even within a short period of time. In contrast, 3D information does not depend on the appearance of the objects and is therefore stable, most objects on the ground being solid. Thus, reliably monitoring the Earth implies reconstructing it in 3D. Even when comparing optical images of the ground, accurate DEMs are needed in order to compare images obtained from two different viewpoints, which can only be compared after ortho-rectification.

In most Earth observation use cases, both civilian and military, the decisive information is not contained in each image in particular, but in the changes observed over time. For instance, geologists and glaciologists are interested in ground deformation rather than in the current configuration. In all the applications, as the displacements take place in a three-dimensional world, 3D measurements are needed to describe them.

Not only is this monitoring important for present and future observations; given that satellite imagery comprises a 40-years archive, one should also consider exploiting this archive to get a history of any region of interest. Access to such an organized archive will have obvious historical, archaeological, ecological, economic and urban applications.

Modeling the 3D geometry of real scenes is a challenging problem. The approaches that have been developed to solve it are usually grouped in two categories: active methods (Jarvis, 1983; Levoy et al., 2000) used in devices such as Lidar, and passive image-based methods (Scharstein and Szeliski, 2002; Seitz et al., 2006; Strecha et al., 2008). Passive image-based methods are the cheapest. They use multiple photographs of the scene and compute the locations of 3D points that are visible in several images. This process is commonly known as *structure from motion*, or *multi-view stereo*, or simply *stereo matching* when used with only two views (Szeliski, 2011; Furukawa and Hernández, 2015). The earliest stereo algorithms were developed in the field of *photogrammetry* to automatically construct topographic elevation maps from overlapping aerial images. In the field of *computer vision*, the topic of stereo matching has been one of the most studied problems (Marr and Poggio, 1976; Barnard and Fischler, 1982; Dhond and Aggarwal, 1989; Scharstein and Szeliski, 2002; Brown et al., 2003; Seitz et al., 2006), and continues to be one of the most active research areas.

In 1986 the French space agency, *Centre National d'Etudes Spatiales* (CNES), launched SPOT, which was the first civilian satellite to provide off-nadir viewing capabilities, enabling the acquisition of stereo images. Taking images with a resolution of 10 m per pixel, SPOT was soon followed by other satellites with stereo capabilities and increasing spatial resolutions, such as Ikonos, WorldView, QuickBird and Pleiades to name just a few. The most recent satellites of the WorldView series have a resolution of 30 cm per pixel. It is important to point out that 3D activity monitoring is of interest mainly for sub-meter imagery. Indeed, at lower resolutions the smaller geometry changes become impossible to measure.

3D remote sensing

The 3D reconstruction from satellite images is no different from the classical computer vision problem; however there are some peculiarities that make it more laborious to tackle. Besides the practical complexities related to the size of the images, it requires handling geographic coordinate systems, and special camera

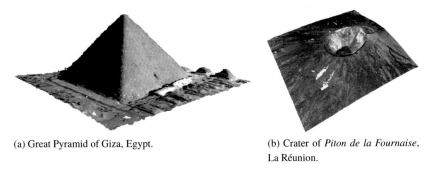

(a) Great Pyramid of Giza, Egypt.

(b) Crater of *Piton de la Fournaise*, La Réunion.

Figure 29.6 Examples of DEMs computed by S2P without human supervision.

modeling (de Franchis et al., 2015). Most of the issues come from the imperfect knowledge of the camera external parameters (position and orientation) known as pointing error (Grodecki and Dial, 2003; Fraser and Hanley, 2005) and from the scale of the structures being studied, which are usually just a few pixels across.

There are fortunately a few public modular and fully automatic 3D reconstruction pipelines (Shean et al., 2016) for satellite images. We shall illustrate the 3D reconstruction process by describing one of them, the S2P pipeline (de Franchis et al., 2014), which has open source code. The program works by cutting the input images into small tiles, performing an image-based compensation of the pointing error (de Franchis et al., 2014) and feeding these corrected pairs to standard stereo correlation algorithms. The final result is obtained by triangulation of all the disparity maps into a single 3D point cloud. This setting lets us test several state-of the art correlators and choose the best one for each situation; for urban regions, one of the best choices turns out to be the "more global matching" algorithm (Facciolo et al., 2015) (a variant of semi-global matching Hirschmüller, 2005). Figure 29.6 shows 3D models constructed without any human supervision by S2P from Pleiades images of the *Piton de la Fournaise* and of the Great Pyramid of Giza. The same procedure can also be extended to non-simultaneous sets of images (Facciolo et al., 2017; Marí et al., 2019).

It is important to note that the compensation of the pointing error does not correct the absolute position of the triangulated points. For example for Pleiades or Worldview images remnant calibration errors can lead to a shift error of the order of several meters. This of course hinders a direct comparison or fusion of DEMs obtained at different dates. Indeed, the pointing correction only aims at preserving the consistency between the images, but it cannot guarantee that the absolute position of the triangulated points is correct.

Multi-temporal fusion of DEMs and change detection

The importance of a completely automatic 3D reconstruction pipeline from satellite images is that it enables to perform diachronic analysis of Earth's relief on demand and within a reasonable time. This is particularly interesting for military and civil uses requiring the detection of significant relief changes. It is also a useful damage assessment tool after disasters (landslides, floods, earthquakes), at all stages of military operations, in the analysis of the evolution of sensitive sites, and finally for all economic and organizational applications where the monitoring of the evolution of the construction of large areas is required. Figure 29.7 shows an application of such methodology on pairs of images acquired after the 2010 earthquake of Haïti. In Figure 29.8 we can observe that once the 3D point clouds are correctly aligned it is possible to pinpoint the changes in the DSM at two different dates (two years apart). In this case, this leads to the detection and localization of several demolished buildings and of a few new built ones.

As mentioned above, the imperfect absolute geolocation of the 3D models hinders the alignment of these products and indirectly change detection. This issue can be addressed by incorporating ground control points (GCP) in the correction, which are easily identifiable points in the images with known 3D coordinates. In absence of GCPs we cannot guarantee that the absolute positions of the triangulated points are correct. Thus, if two models are generated from images at two different dates, the reconstructions might not coincide.

One alternative to using GCPs is to align the DSMs generated independently for each date before fusing (Facciolo et al., 2017) or comparing them. Another possibility consists in performing a bundle adjustment (Triggs et al., 1999) on the whole set of images. Bundle adjustment is a procedure that corrects the camera models in such a way that the image correspondences and the triangulated points are coherent.

We have observed that these two strategies, although fundamentally different, perform similarly on large images (from large-swathe commercial satellites as Pleiades or Worldview), for which a reconstruction area is usually contained in a single image (Marí et al., 2019). However, for the case of small-swathe satellites (as is the case of Planet's Skysat) dozens of smaller images are needed to cover a single area of interest. In this case a bundle adjustment of the set of images (Triggs et al., 1999) is needed to guarantee the internal consistency of the reconstruction of each date (Marí et al., 2021a,b).

Application of 3D monitoring to automatic stockpile activity monitoring

Time series of aligned DSMs can be effectively employed to measure industrial activity. As a concrete example we show here how to track stockpile volumes over an area of interest (Marí et al., 2021a). We consider a time series of SkySat L1B acquisitions covering the Richards Bay Coal Terminal (RBCT) in South Africa, which has an open-air storage area of ~1.6 km^2. The RBCT is one of the world's leading coal export terminals. Very large coal stockpiles are moved 24 hours a day to be shipped overseas. The time series comprises 43 acquisition dates, distributed non-uniformly between January and July 2020. Distance between consecutive dates oscillates between 1 and 20 days, falling below one week in most cases. For each date there are 6 to 10 SkySat L1B images, captured by the same sensor, with a difference of a few seconds, the images have a nadir Ground Sampling Distance (GSD) of ~0.72 m.

The procedure consists in first applying bundle adjustment to all the images of a single date. After that, for each date the S2P pipeline is used to reconstruct

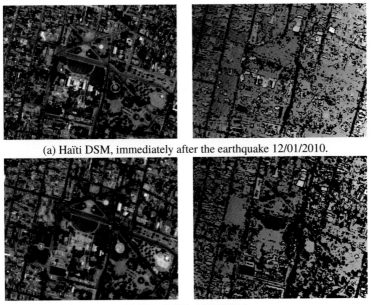

(a) Haïti DSM, immediately after the earthquake 12/01/2010.

(b) Haïti DSM, 19/07/2012.

Figure 29.7 Haïti at two distinct dates after the 12/01/2010 earthquake.

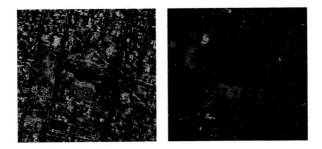

Figure 29.8 Difference between DEMs of Haïti obtained at the two dates shown in Figure 29.7. Left: difference without previous alignment of the DEMs. Right: difference after aligning the DEMs (red denotes negative difference (demolished); cyan denotes positive difference (reconstructed); black is unknown). All colors are saturated for differences larger than 5 m.

the DSM of the area of interest. Reconstructions from different pairs are natively registered in object space as a result of bundle adjustment. This allows to easily merge them into a denser and highly accurate model of the entire area. Similarly to Facciolo et al. (2017), the DSMs of different dates are then aligned by a 3D translation that maximizes the Normalized Cross Correlation between their geometry. Even if the geometry may change between different dates, this alignment serves to minimize the standard deviation of DSM heights over time. For each DSM, a Digital Terrain Model (DTM) is subtracted to consider only the heights above ground. Since our study area lies on a flat terrain, we model the DTM as a plane with height equal to the 25th percentile of the DSM heights. Without loss of generality, cloth simulation methods can be used to model non-flat DTMs (Zhang et al., 2016). We then determine a site-specific mask $\mathcal{M}_{dynamic}$ delimiting the dynamic parts of the area (Figure 29.9b). The labeling of $\mathcal{M}_{dynamic}$ is based on the point-wise standard deviation of height values over time, across the different DSMs of the area (Figure 29.9a). The normalized DSM (nDSM) containing the heights above ground in areas where changes are expected can be expressed as

$$nDSM(t) = \mathcal{M}_{dynamic}(DSM(t) - DTM(t)) \qquad (29.1)$$

where t represents the acquisition date of the time series (Figure 29.9c). Finally, the volume of the stockpiles left in nDSM(t) is computed in cubic meters as the addition of all individual cell volumes.

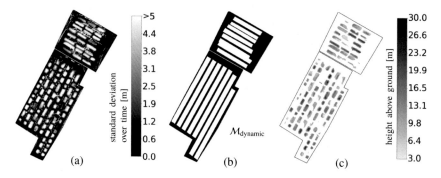

Figure 29.9 (a) Point-wise standard deviation of height values along the DSM time series. (b) Mask of dynamic parts. (c) Example of normalized digital surface model of the stockpiles (nDSM(t)). The volume of the stockpiles left in nDSM(t) is computed in cubic meters as the addition of all individual cell volumes. Experiment courtesy: Kayrros SAS.

29.4 How SAR imaging has revolutionized the world observation of oil storage

Most of crude oil and petroleum products are stored in external tanks between production and consumption. The global and local levels of storage are key indicators of the petroleum markets as they strongly reflect the demand/supply balance. However, most regions in the world do not publish official storage figures or do so infrequently and with some lag. There is therefore a great value in having near real-time independent measurements of storage tanks all over the world.

Most of these oil storage tanks are external floating roof tanks. Such tanks are made of a cylindrical wall and a floating roof. This floating roof floats freely on top of the liquid, and therefore rises and falls as the tank is filled and emptied. An example of an external floating roof tank is shown in Figure 29.10.

In this section we shall describe an automatic method for remotely measuring the volume of liquid stored in external floating roof tanks using Synthetic Aperture Radar (SAR) images.

Figure 29.10 External floating roof tank. Such a tank comprises a cylindrical wall and a floating roof. This floating roof floats freely on top of the liquid, and therefore rises and falls as the tank is filled and emptied. Illustration ©Cherezoff, Dreamstime.com, reproduced with permission.

29.4.1 Overview

The typical Areas of Interest (AOI) to be monitored are situated in a specified locus (port terminal, storage facility, refinery, etc.). A time series of images covering the AOI is used as reference against which future images will be registered, and the tanks positions are projected on the time series of radar images, thus storing the coordinates and tank dimensions that will be used to locate the tanks in future images. The routine oil storage evaluation at each site implies recurrently downloading and pre-processing each new radar acquisition over the AOI, registering it on top of the reference series, using the stored tank coordinates and dimensions to locate the tank in the new image, detecting the position of its roof, and converting it to a liquid volume. The output of the routine process is a collection of storage volume data and corresponding fill rates provided for individual tanks that can be aggregated over given geographical areas or depending on the company of owners.

29.4.2 Detailed description

Oil tanks in the world are generally first detected and mapped through analysis of optical images involving a machine learning pattern recognition algorithm. Here, classic classification deep learning is involved to recognize the patterns of tanks and groups of tanks. The longitude and latitude of the centers of the tanks of interest, as well as their dimensions, are computed from the optical images.

Once an AOI containing oil tanks is identified, all existing SAR images covering the AOI are gathered to form an image time series.

Remote sensing SAR images usually cover wide areas that are much larger than the AOI. It is thus necessary to find the part of the image that contains the AOI. The position of the AOI in the SAR image is computed thanks to the knowledge of the radar positions over time listed in the image metadata. Knowing the radar positions during the image acquisition leads to converting automatically the geographic coordinates (longitude, latitude, altitude) of a 3D point into the pixel coordinates (x, y) of its position in the SAR image (Curlander, 1982). This conversion processing is usually known as projection, localization or geocoding.

The altitude of the tank, required for projecting geographic coordinates into the SAR image, is retrieved from a Digital Surface Model (DSM) covering the AOI. Once the position of the AOI in the SAR image is computed, a rectangular crop containing the AOI is extracted. Typical dimensions of the crop range from $1\,000 \times 1\,000$ to $10\,000 \times 10\,000$ pixels, depending on the AOI dimensions and the radar resolutions.

Remote-sensing SAR instruments can operate in various modes, such as Spotlight, Stripmap, ScanSAR (Moreira et al., 2013) and TOPSAR (De Zan and Guarnieri, 2006). If the acquisition mode is ScanSAR or TOPSAR, then the SAR image subset containing the AOI may be made of two or more consecutive image strips that need to be stitched (i.e. merged) before cropping. These image strips are called bursts and the process of stitching the bursts is called de-bursting (De Zan and Guarnieri, 2006).

Radar works by emitting an electromagnetic wave and measuring its reflected echoes. The amplitude and phase of the received echoes are stored in each pixel as a complex number made of a real and an imaginary part. Some image providers distribute only amplitude images. In these images, each pixel contains only the amplitude of the signal. Other image providers distribute the full complex images. In these images, each pixel contains a real part x and an imaginary part y. The amplitude can be obtained by computing the modulus $\sqrt{x^2 + y^2}$ of the signal, pixel per pixel. The procedure of computing the amplitudes (i.e. the moduli) of the pixels of a complex SAR image is called detection.

The images of the time series must be accurately registered so that a tank has the same image coordinates in all the images of the time series. Image registration can be performed using any standard image registration algorithm such as minimizing the sum of squared differences, maximizing the cross-correlation, or maximizing the phase correlation. This step acts on an entire time series of images. All the images are processed together. The geographic coordinates and dimensions of the tanks must generally be refined by using the specific signature of the tanks in the SAR images. To this end, for example, one can use a description in which each tank is described by the position of its base center, plus its diameter and height. The tank position is a triplet of longitude λ, latitude θ and altitude z.

Altitude

If not available from external databases, the altitudes are obtained from a Digital Surface Model (DSM) of the AOI. The DSM might be outdated, especially in places where tanks were built after the DSM computation date, as tank construction often involves terrain flattening. Hence the real altitude of the ground at a tank position may differ from the DSM measured altitude by several meters. This can be a cause of mismatch when projecting geographic coordinates into radar images.

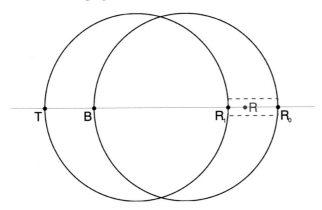

Figure 29.11 How an external floating roof tank is seen in a SAR image. Points T and B are the image peaks caused by the double bounce reflections of the tank top (T) and base (B). Point R corresponds to the floating roof. It can range from R_0 (tank empty) to R_1 (tank full).

Radius and height

If not available from external databases, the radius of each tank can be estimated by fitting a circle on its top rim in a top view optical image. The height of a tank is the vertical distance between the top and the bottom of the tank. It should not be confused with the altitude of the ground at the tank position. Typical tank heights range between 10 and 25 meters. If not available from external databases, the height can be measured directly from optical slant view images, or derived from the sunlight shadow in optical images, or measured in the SAR images as explained below.

Floating roof tank signature in SAR images

Floating roof tanks are made of metal thus they strongly reflect radar waves and are easily visible in SAR images. Moreover, a well-known double bounce accumulation occurs at the cylinder base and top points facing the radar. All the radar echoes reflected on the tank wall facing the radar bounce down to the ground and bounce again back to the radar. They all travel the same total distance, hence the radar sees them as coming from the same point which appears very bright in the SAR image (see Figure 29.11). This point is named B, for "Base", in Figure 29.11. The same phenomenon occurs for the top rim, as there is usually a metallic walkway around the top that creates an orthogonal intersection with the tank wall. The radar echoes that reflect on the upper part of the tank wall, above the walkway, bounce down to the walkway then bounce again back to the radar. The resulting bright point is named T, for "Top". The same phenomenon happens again on the opposite side of the tank, at the orthogonal intersection formed by the floating roof and the internal face of the tank wall. The resulting bright point is named R, for "Roof". The position of R varies between R_0 (tank empty) and R_1 (tank full) according to the tank filling rate. Figure 29.12 illustrates the three

608 C. de Franchis et al.

608   C. de Franchis et al.

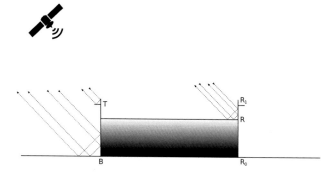

Figure 29.12 All radar echoes reflected on the tank wall facing the radar bounce down to the ground and bounce again back to the radar. They all travel the same total distance, hence the radar sees them as coming from the same point which appears very bright in the image.

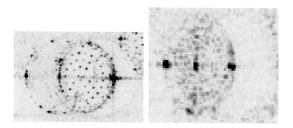

Figure 29.13 Floating roof tanks in SAR images from COSMO-SkyMed. Left: Spotlight image, resolution is 1 m per pixel. Right: Stripmap image resolution is 3 m per pixel. High amplitude values are represented by black pixels, low amplitude values are represented by white pixels. The three black dots visible in each one of these images correspond, from left to right, to the strong reflections of the tank top rim (point T), the tank base (point B) and the tank floating roof (point R).

occurrences of this double bounce phenomenon and shows the three points B, T and R.

The tank base and top are seen as bright circles when the horizontal and vertical resolutions are equal. If the horizontal and vertical resolution differ, then the tank base and top rims are seen as ellipses (as for instance in Figure 29.13). The position of R varies horizontally in the dashed window between R_0 (tank empty) and R_1 (tank full) according to the tank filling rate.

Projection function

Given a radar image u, the precise radar positions over time listed in the image metadata allow to define a projection function P_u. This function converts the 3D coordinates λ, θ, z of a world point into the 2D coordinates of its pixel position in image u (Curlander, 1982). For most SAR sensors, such as TerraSAR-X/TanDEM-X, COSMO-SkyMed, or Sentinel-1, this projection function is accurate up to a tenth fraction of a pixel. It is thus safe to assume that the mismatches we may observe between the projected coordinates of a tank and its actual SAR image signature coordinates come from inaccuracies of the tank geographic coordinates.

Problem formulation.

For a given tank, the problem is to refine the longitude λ, latitude θ and altitude z of the tank base as well as the tank height h and radius r, using all the available SAR images of a registered time series. For that one can maximize the following energy:

$$
E(\lambda, \theta, z, r, h) =
$$
$$
\sum_{\alpha \in [0, 2\pi]} u(P(\lambda + r \cos \alpha, \theta + r \sin \alpha, z) + u(P(\lambda + r \cos \alpha, \theta + r \sin \alpha, z + h),
$$

$$(29.2)$$

where α spans over a regular sampling of $[0, 2\pi]$. The longitude λ and latitude θ in the formula above are expressed in meters using for instance a Universal Transverse Mercator (UTM) coordinates system.

The average image u is obtained by averaging the images of the time series pixel per pixel. The average image is less noisy than a single SAR image. The average projection function is P. To evaluate P on the point λ, θ, z one needs to evaluate all the projection functions of the time series on that point and then average the resulting series of pixel coordinates. The energy defined above in (29.2) is the sum of the average SAR amplitudes at the projections of points sampled on the tank base and the tank top rim. It should be maximal when the values of λ, θ, z, r, and h are such that the projections of the tank base and top rim pass through the bright spots B and T (see Figure 29.11).

For each tank of the AOI, its refined coordinates λ, θ, z and dimensions r, h and the average projection function P of the time series are used to compute the image coordinates of points R_0 and R_1 in the registered time series (see Figure 29.11), where R_0 is the position of the floating roof when the tank is empty and R_1 is the position of the floating roof when the tank is full. As the images are registered, the pixel coordinates of these points are the same in all the images of the time series.

The actual position of the roof image bright spot R between R_0 and R_1 is directly linked to its current height h_R in the tank:

$$
h_R = \frac{|R - R_0|}{|R_1 - R_0|} h, \tag{29.3}
$$

where h denotes the total tank height computed in A15 when optimizing the energy E. By detecting for each tank the bright spot R between R_0 and R_1, one can compute its current roof height. Liquid volumes in each tank are then computed using the current height h_R of their floating roof and their radius r:

$$
V = \pi r^2 h_R. \tag{29.4}
$$

The tanks volumes can then be aggregated per tanks product types, or per tank terminal, per city, country, region, or aggregated over the whole world.

29.5 Creating a movie of the earth: the challenges of registration and equalization

The recent proliferation of earth observation satellites has opened the way to the analysis of long image time series with denser temporal repetition. Given the numerous satellites observing the earth, any point on the ground is present in multiple and long image time series. It becomes indispensable to build more complete time series obtained by fusing the output of different imaging devices (with roughly comparable resolution). For example, the Landsat-8 and Sentinel-2 were built with very similar specifications, thus they are very good candidates for fusion, but this can be carried with many more satellites. Furthermore, given this wealth of images, it is crucial to design automatic tools to process them.

Coppin and Bauer noted in 1996 that there were two outstanding requirements for satellite time series analysis: multitemporal image registration and radiometric calibration (Coppin and Bauer, 1996). This is still true today. Although more that two decades of research in these fields has provided us with accurate tools, most of them are tailored to the processing of only a few images, if not only a pair of images. The community's current problem is then to scale those tools for the automatic processing of increasingly longer time series.

Besides facilitating human visual scene interpretation, registration and relative radiometric normalization in time series are preparation steps needed by many remote sensing applications. Registration is a fundamental requirement of most time series analysis tools. Indeed, only features precisely located in space permit an accurate analysis of their temporal evolution. Radiometric normalization is also used in many applications (Canty and Nielsen, 2008): to prepare data for change detection (Paolini et al., 2006; Jianya et al., 2008); tracking vegetation indices over time (Du et al., 2002); for supervised and unsupervised land cover classification; for multi-temporal satellite image mosaicing (Rahman et al., 2015).

29.5.1 Overview of the existing methods

Registration and radiometric normalization of multi-temporal satellite image time series are particularly difficult, because the image's content may vary greatly from one acquisition to another. Moreover, when fusing images from different satellites, the task is even more intricate than expected due to the differences between sensors and acquisition modes. In particular, resolutions are different, dynamic ranges are different, color bands are different. Still more challenging, the pixels where the ground is visible and stable can vary in each image due to solar time changes, modifications of the satellite viewpoint, atmospheric perturbations, clouds, and human actions.

According to Du et al. (2002), there are four main reasons for the variations of sensor measurements of the same scene at different dates:

(1) changes in satellite sensor calibration over time;
(2) differences in illumination and observation angles;
(3) variation in atmospheric effects; and

(4) changes in target reflectance.

In the case of multi-satellite fusion, a fifth reason is the difference of sensors.

Image registration methods are generally divided into two categories: feature-based methods and area-based methods (Zitová and Flusser, 2003). The latter includes correlation methods, among which the normalized cross correlation and other Fourier-based correlation methods, dating back to the 1970s (Anuta, 1970). The particularly successful phase correlation method (Kuglin and Hines, 1975) normalizes the correlation image in the Fourier domain by its modulus, thus keeping only the phase information. This makes it invariant to global linear variations in contrast and brightness (X. *et al.*, 2019). The sub-pixel displacement can be estimated from the cross correlation image (Stone et al., 2001; Foroosh et al., 2002; Ye et al., 2020) or by estimating the slope of a plane in the phase of its Fourier transform (Leprince et al., 2007; Dong et al., 2018). Accuracy and robustness of correlation algorithms can be further improved by frequency weighting (Leprince et al., 2007).

These methods only register two images. Moreover, in spite of all these improvements and advances, image registration often fails due to occlusions or extreme contrast changes, which in turn result in severe misalignment. A way around this is to leverage the availability of all images of the time series. The problem of registering several images at once is called multi-image alignment. It consists in bringing a group of images into a common reference. In addition to being a critical first step in many remote sensing applications, it is also the cornerstone of many image fusion techniques such as high dynamic range imaging, super-resolution, burst deblurring and burst denoising (Aguerrebere et al., 2018). There is much to gain with the registration on several reference images (Aguerrebere et al., 2016; Rais et al., 2014). Methods in Govindu (2004) and Farsiu et al. (2005) use this over-determined system of equations by taking advantage of the fact that the shift between two distant images is the sum of the shifts of all adjacent images between them (Aguerrebere et al., 2018).

Radiometric correction methods are also classified in two categories: the first encompasses techniques that aim at mapping the sensors' values to physical measurements like ground radiance or reflectance (Yuan and Elvidge, 1996). These methods require external measurements. The second category includes methods whose objective is simply to normalize the sensors' values, so that the relative changes remaining are only due to changes in the target reflectance. These methods use only information from the time series itself.

In Yuan and Elvidge (1996), a review of seven relative radiometric normalization techniques, the authors noticed that all used affine models whose parameters were found by regression. The assumption is that among the different sources of change between the same channel of two images at two different dates, linear effects dominate nonlinear effects (Du et al., 2002). Hall et al. (1991) noted that the affine nature of the transformation linking two images can be destroyed by the heterogeneity in the atmospheric properties across the scene and the non-linearity in calibration differences between sensors. From our own experiments, the former is a far more serious issue. The most important and difficult parts of relative radiometric correction are the selection of pseudo-invariant features (PIFs) (Du et al., 2002; Canty et al., 2004), and the regression method used to find the affine model parameters (Syariz *et al.*, 2019).

29.5.2 Algorithms

Phase correlation

The phase correlation method is used to estimate translations in pairs of images.

Let us denote $U = \mathcal{F}\{u\}$ the discrete Fourier transform of image u. We call c the phase-correlation image between u and a second image v. Its discrete Fourier transform is given by

$$C = \frac{U \cdot V^*}{|U \cdot V^*|},\qquad(29.5)$$

where * denotes the complex conjugate and $|\cdot|$ the norm. The shift is given by the position of the maximum in $c = \mathcal{F}^{-1}\{C\}$. Some weights can be applied to C to reduce the effect of the borders and of the noise (Leprince et al., 2007). Since this estimation only gives integer translations, a second refinement step is needed to obtain a shift at sub-pixel precision. Assuming the maximum position at position $(x^{\max}, y^{\max})$, we fit a one-dimensional sinc function for the x-axis using the two neighboring points $x^{\max} - 1$ and $x^{\max} + 1$:

$$\arg\min_{a_x,b_x,c_x} \sum_{x=x^{\max}-1}^{x^{\max}+1} \|c(x, y^{\max}) - c_x \cdot \text{sinc}(b_x \cdot (x - a_x))\|^2,\qquad(29.6)$$

and similarly for the y-axis. Finally, the position of the maximum is $\hat{\tau} = (a_x, a_y)$.

Relative radiometric normalization

As an example we now consider the fusion of Landsat-8 and Sentinel-2 images.

Sentinel-2 visible bands have a resolution of 10 m, while Landsat-8 panchromatic band is at 15 m and its visible blue, green and red bands are at 30 m resolution. To create a time series with images from both satellites, the pixels' size must be made uniform. We thus use the panchromatic band to pansharpen Landsat-8 visible bands to 15 m/pixel and then upsample them to 10 m/pixel by interpolation. The time series is then registered with sub-pixel accuracy using the phase correlation method (Foroosh et al., 2002). We sample the image after translation by spline interpolation (Briand and Davy, 2019). The overall image series processing chain then works as follows:

(1) Apply the ground visibility detector (Grompone von Gioi et al., 2020) described in §29.6, remove exceedingly cloudy images from the time series;
(2) Apply quality metrics and select reference images as local maxima of the global quality score;
(3) Find stable pixels by comparing the gradient angles in the pairs of source and reference images;

(4) Correct the spectral values based on relevant pixels and the two closest reference images;

(5) Apply a final tone-mapping step.

We assume that affine effects dominate between the measured radiometric values of two images from different dates and possibly from different sensors. This can be modeled as:

$$v_c = a_c \cdot u_c + b_c, \tag{29.7}$$

where c is the channel, u and v are images at two different dates, and (a_c, b_c) are the parameters of our affine model for this specific channel. This model is of course valid only on persistent features between the two images. We thus need to find first the parts of the ground that are visible in both images. To this end we use the angle of the images' gradients, as described in §29.6. Persistent features are found as regions where the gradients' angles are similar in the two images. This criterion has the advantage of being insensitive to affine changes of the pixels' intensities. Denoting by u_D the input image restricted to its persistent pixels, we thus want

$$\arg\min_{a_c, b_c} \|v_{c,D} - a_c \cdot u_{c,D} - b_c\|, \tag{29.8}$$

with $\|\cdot\|$ a certain norm. An affine model is computed for each channel independently, and with the two closest reference images in the time series. Final correction parameters are obtained by linear interpolation.

Figure 29.14 shows the aspect of a time series without radiometric normalization. The image is recomposed by concatenating horizontal or vertical bands from the inputs. After the relative radiometric normalization, seams between the bands are much less visible, even though bands come from images taken at different dates and with different sensors. The first and last images were captured on Feb. 18, 2018 and Sep. 21, 2019, respectively, over Versailles (France). Their size is 2×5 km.

29.6 Ground Visibility and cloud detection

Assessing ground visibility is an important step in optical satellite images analysis. Indeed, the presence of clouds and haze concealing the surface of the Earth often causes detection errors in automatic image analysis. This task is usually addressed as a cloud detection problem, where the image pixels are classified into classes such as ground, clouds, cirrus, snow, haze, cloud shadows, etc. (Chandran and Jojy, 2015; Mahajan and Fataniya, 2019). Satellite cloud detection often exploits spectral bands specifically designed for cloud detection (Irish, 2000; Zhang et al., 2001; Irish et al., 2006; Scaramuzza et al., 2012; Chandran and Jojy, 2015; Taravat et al., 2015; Hollstein et al., 2016). The inter-band delay in push-broom satellites also allows cloud detection by parallax analysis of the color bands (Shin and Pollard, 1996; Panem et al., 2005; Manizade et al., 2006; Wu et al., 2016; Sinclair et al., 2017; Frantz et al., 2018).

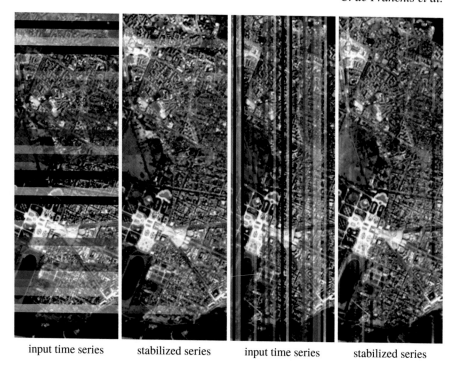

input time series   stabilized series   input time series   stabilized series

Figure 29.14 Composite image, formed by concatenating horizontal (left side) or vertical (right side) bands from each image of the time series. The time series contains a total of 39 images, captured by Sentinel-2 (correction levels 1C and 2A) and Landsat-8. This algorithm automatically normalizes the spectral values in the series using a few well chosen references. Experiment courtesy: Kayrros SAS.

Instead of detecting clouds, an alternative approach is to detect ground visibility, that is, the parts of each image where the ground is visible (Dagobert et al., 2019, 2020; Grompone von Gioi et al., 2020). This can be done by comparing the corresponding parts of a time series and selecting matching regions. The rationale is that the ground has persistent patterns that are observed repetitively in the time series, while clouds are changing phenomena appearing differently each time. This approach is based on the following three hypotheses:

1. the images in the time series are well registered (as described in §29.5);
2. the region of interest changes slowly in the observed time series;
3. each part of the region of interest is visible at least twice in the time series.

Under these hypotheses, any pattern observed more than once at the same position must correspond to a visible region of the ground.

The method works as follows. Given a set of N registered images, the ground visibility masks for the N images are all initialized as *not visible*; then the $\frac{N(N-1)}{2}$ pairs of images are compared and when a local match is found, the corresponding parts are marked as *visible* in both ground visibility masks.

Given two images u and v and a set of pixels R, we would like to know whether both images are similar in the region R. To this aim, the following distance is used:

$$s_{u,v}(R) = \sum_{\omega \in R} \frac{|\text{Angle}(\nabla u(\omega), \nabla v(\omega))|}{\pi}, \qquad (29.9)$$

namely the sum of all normalized gradient angle errors in R (Patraucean et al., 2013; Grompone von Gioi and Patraucean, 2015). Using only the image gradient *orientation* renders this distance invariant to illumination changes. This distance can take values between zero and $|R|$ (the size of the region). Zero corresponds to a perfect match, while $|R|$ corresponds to the case where the gradient orientation in the two images are opposite at every pixel of R. For a given region R, we need to decide whether the distance $s_{u,v}(R)$ is small enough, indicating whether the corresponding parts of the images are similar or not. This decision is handled with an *a contrario* formulation.

The *a contrario* theory (Desolneux et al., 2000, 2008) is a statistical framework used to set detection thresholds automatically in order to control the number of false detections. It is based on the non-accidentalness principle (Witkin and Tenenbaum, 1983; Lowe, 1985) which informally states that an observed structure is meaningful only when the relation between its parts is too regular to be the result of an accidental arrangements of independent parts. A stochastic background model $\mathcal{H}_0$ needs to be defined, where the structure of interest is not present and can only arise as an accidental arrangement. We also need to define a family of events of interest T. Then, given a candidate, one measures how well the image corresponds to the candidate event, and evaluates the probability of observing such a good agreement by chance in the *a contrario* model. When this probability is small enough, there exists evidence to reject the null hypothesis and declare the event meaningful. The number of tests N_T is included as a correction term, as it is done in the statistical multiple hypothesis testing framework (Gordon et al., 2007), to obtain an estimate of the number of false detections.

For comparing images by gradient, a natural background model $\mathcal{H}_0$ is that the gradient orientations at each pixel are independent random variables, uniformly distributed in $[-\pi, \pi)$. Following the *a contrario* framework, the NFA associated to a candidate region match is defined as

$$\text{NFA}(u, v, R) = N_T \cdot \mathbb{P}\Big[S_{\mathcal{H}_0}(R) \leq s_{u,v}(R)\Big], \qquad (29.10)$$

where $S_{\mathcal{H}_0}(R)$ is a random variable corresponding to the distance $s_{U,V}(R)$ for random images U and V whose gradient orientation follow $\mathcal{H}_0$. It can be shown (Desolneux et al., 2000, 2008) that the NFA gives an upper bound to the mean number of false detections under $\mathcal{H}_0$. Thus, the smaller the NFA, the more unlikely the match is to be observed by chance in the background model $\mathcal{H}_0$, and the more meaningful. The *a contrario* approach prescribes accepting as valid detections the candidates with NFA $< \varepsilon$ for a predefined value ε. In many practical applications, including the present one, the value $\varepsilon = 1$ is adopted. Indeed, it allows for less than one false detection on an image or on a set of images, which is usually quite tolerable.

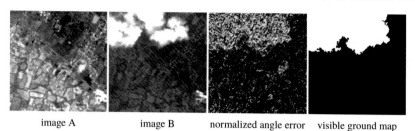

image A   image B   normalized angle error   visible ground map

Figure 29.15 Comparison of two images to evaluate ground visibility. In the normalized gradient orientation error map between the two images, black corresponds to zero error (same gradient orientation in both images) while white corresponds to the worst error (the gradient orientations have opposite direction). In the two ground visibility maps, black pixels correspond to *visible* ground while white corresponds to *not visible* ground. In this simple example with only two images, a common mask can be produced, and the zones that do not match are declared as not visible. However, when more images are included in the series, zones that are not at first detected as visible in the first image might match with other images in the series and therefore be declared as visible. This shows the relevance of the third hypothesis of the method: each zone should be visible at least twice in the time series.

The *a contrario* model described is used to decide whether a couple of images are similar or not in a given region. But trying every possible connected region in the image domain is obviously impossible as it would take too much time. Instead, a greedy algorithm is used for performing the comparison of all pairs of images in a reduced time, see Figure 29.15.

Figure 29.16 shows the result of the ground visibility detection on a time series of ten Sentinel-2 images. The first and third rows show the ten input images and the second and fourth rows show the resulting ground visibility masks; white corresponds to *not visible* and black to *visible*. At first glance one can see the correct result for images 2, 3, 4, 7 and 8; the images are fully covered by clouds and so indicate the complete white masks. The other images need more careful observation. In most of the cases these masks are mainly black, correctly indicating the general ground visibility of those images. The white spots correspond mostly to the presence of clouds or to zones were the topography does not satisfy the slow change hypothesis, essentially zones covered by water. But there are also *not visible* parts which are due to the lack of repetitive structure; this is the case in some small fields where no relevant texture can be matched from one image to another.

In general, the method produces good results when its functioning hypotheses are satisfied. Among them, a demanding one is that the target zone changes slowly relative to the sampling frequency, which is not valid, for example, over the sea.

29.7 Detecting and monitoring cars from space

Detecting and counting objects in satellite images bring valuable insights for estimating economic activity. For instance, counting the number of ships or trucks transiting between two countries can give an indication on the magnitude of the economic exchange between those two countries (Vachon et al., 2012; Drouyer and de Franchis, 2019). Monitoring the number of oil tanks, wind turbines or

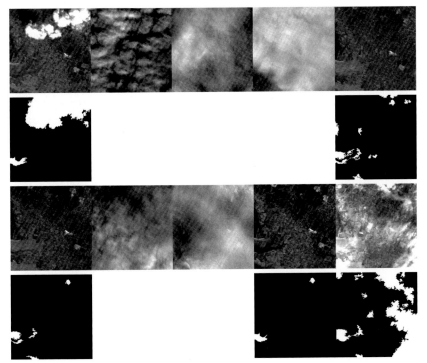

Figure 29.16 Ground visibility computed in a time series of ten images from the Sentinel-2 constellation. First and third rows: input images. Second and fourth rows: resulting ground visibility maps, where black pixels correspond to *visible* while white corresponds to *not visible*. Experiment courtesy: Kayrros SAS.

solar panels can help predict energy output (Tadros et al., 2020; Mandroux et al., 2021). Monitoring parking occupancy in front of stores can constitute an alternative consumption measure on a national, regional or store level (Drouyer and de Franchis, 2020; Drouyer, 2020a).

Data obtained from satellite images – also known as alternative data – provide a complementary edge to core financial data. In most cases, they give an early indication on some aspects of these core financial data: economic activity, sales estimations, import and export. But they can also provide additional or more detailed insights, especially on areas and countries where official figures are not considered reliable.

We saw in §29.2.1 that there are many satellite image sources with different resolutions, different sensors, different revisit rates and different pricing. Acquiring these images is the easiest step as they must be processed in order to extract the target data. Again, many different techniques can be applied, from simple thresholds to advanced deep learning models. The combination of satellite image source and computer vision techniques used depends on the objective that the alternative data must achieve, notably the target accuracy, the temporal resolution and the cost.

In this section, we will explore several strategies through a case study on vehicle detection. Vehicle counting and detection can be a great proxy for estimating economic activity. First, monitoring storage parking lots near car factories is a way to estimate car production. But, above all, it is an indirect estimation of the number of people in an area. For instance, monitoring parking lots in front of stores allows to estimate how popular they are, and monitoring parking lots in front of offices can be an indication of their activity.

This study will be done using two different approaches as they apply on different resolutions with distinct precision/cost trade-offs; high resolution satellites images (<1 m/px) and medium resolution images (3 m/px).

29.7.1 Vehicle detection on high resolution satellite images

Vehicles in high resolution satellite images are easily discernable by a human, as shown in Figure 29.17. As a consequence, if we manually annotate an image, we can get a precise estimate of the number of vehicles. The challenge lies in developing algorithms that achieve a similar performance.

On these high resolution images, there is an important variability in how vehicles can appear. Vehicles have different shapes: cars, pickups and trucks appear differently, but even two cars of the same size can have different features. Vehicles are also painted in different colors; black, white, grey, blue, red,. . . And, even between slightly different resolutions, similar cars can appear differently.

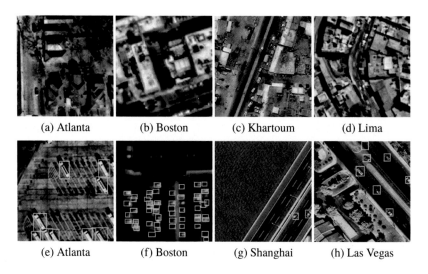

| (a) Atlanta | (b) Boston | (c) Khartoum | (d) Lima |
| (e) Atlanta | (f) Boston | (g) Shanghai | (h) Las Vegas |

Figure 29.17 Examples of crops from the VehSat dataset (Drouyer, 2020b). The second line also contains examples of detection using the faster R-CNN algorithm (Ren et al., 2015).

The variability in the aspects of a single object hints that using a machine learning method is better suited than a hand crafted algorithm. We shall use deep learning as it achieves state of the art results for detecting and localizing objects (Jiao et al., 2019). However, several difficulties must be addressed for achieving a good performance.

A large number of images must be gathered to constitute a dataset. Although

Method	WV3	WV2	Pleiades	SkySat
FRCNN (Ren et al., 2015)	75.8%	70.2%	51.9%	48.3%
FRCNN corr	0.97	0.96	0.76	0.91
FRCNN-Spe	66.6%	66.8%	46.2%	43.0%
CS	70.3%	52.8%	31.2%	23.7%
YOLOv2 (Redmon and Farhadi, 2017)	39.6%	47.1%	13.4%	9.1%

Table 29.2 Comparison of the Average Precision per source satellite for each method. FRCNN: Faster R-CNN algorithm. FRCNN corr: in this line, instead of displaying the Average Precision, we display the correlation between the number of vehicles detected in the patches and the number of vehicles in the ground truth. FRCNN-Spe: Faster R-CNN algorithm specialized for each satellite. CS: Crowdsourcing.

several public datasets already exist, images for a specific detection problem might not be publicly available, making this process expensive. In our case, we were able to obtain images through publicly available datasets and we generated 4544 crops of 70×70 meters areas.

The dataset must be correctly annotated. Using a multi-pass annotation system is ideal for improving the quality of the annotation, especially if using crowdsourcing services. In our case, annotators draw a line from the back to the front of each vehicle, allowing us to obtain a bounding box. A second annotator corrects the annotations in a second pass. The performance in the first pass is far from perfect: the F1 score is 0.79 for the highest resolution WorldView-3 images, whereas, for instance, it is 0.53 for Pleiades images. In total, 36851 vehicles were annotated.

The dataset must be diverse: it must contain images from different geographic areas, under different atmospheric conditions, ideally using different satellite instances if the source is a fleet of satellites. As can be seen in Figure 29.17, the aspect of different cities is very variable, so vehicles appear in different environments and different lighting conditions. In our case, our dataset is comprised of images from eight different places and acquired using four different satellites.

As with any machine learning method, the dataset must be divided into three training, validation, testing non-overlapping subsets. The training subset represents 60% of our dataset, the validation and testing subsets represent each 20% of our dataset. This repartition is quite common, but proportions might differ especially if the dataset is small.

A deep learning method must be chosen, ideally through a benchmark. We chose to compare two state-of-the-art methods at the time of the experiment: Faster R-CNN (Ren et al., 2015) and YOLOv2 (Redmon and Farhadi, 2017). It appears that, as shown in Table 29.2, Faster R-CNN achieves the best performance on our dataset.

We show in Figure 29.17 several examples of vehicle detections. Although the dataset is relatively small compared to deep learning standards, the algorithm's

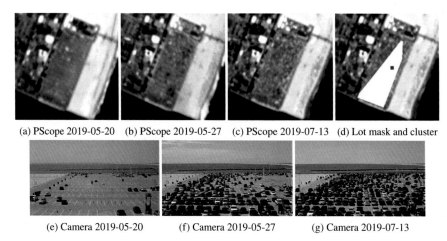

(a) PScope 2019-05-20   (b) PScope 2019-05-27   (c) PScope 2019-07-13   (d) Lot mask and cluster

(e) Camera 2019-05-20   (f) Camera 2019-05-27   (g) Camera 2019-07-13

Figure 29.18 Extract of the image series in the validation site, from low occupancy to near full occupancy. In (d), the parking lot mask is depicted in white and only covers a part of the lot in accordance to the camera field of view, the cluster of pixels used for estimating histograms in Figure 29.19(a) is depicted in red. PScope = Planetscope.

performance is adequate for counting vehicles. However, the performance decreases as the resolution decreases.

We described in this section a general pipeline for object detection on high resolution satellite images using deep learning. However, although allowing a great precision, this evaluation has several limitations. First, high resolution satellite images are very expensive, especially if this detection must occur regularly. Second, obtaining a large dataset might also be expensive and might not be feasible if we want to detect rare objects/instances. Finally, annotating the data might be difficult because of the size of the target objects and the way they appear.

Using simpler methods and lower resolution images can therefore often provide more satisfying results. Next, we study vehicle detection on PlanetScope satellite images.

29.7.2 *Parking occupancy estimation on PlanetScope satellite images*

We described in §29.7.1 a general pipeline for training object detectors on high resolution satellite images that we applied for vehicle detection. This approach worked because of two conditions. First, vehicles were easily discernible. Second, we were able to manually annotate a ground truth.

We will now study a parking occupancy estimator on PlanetScope satellite images. Some images of a validation parking lot are shown in Figure 29.18. In contrast with high resolution images, the resolution of PlanetScope images is 3 m/px, so we are unable to visually recognize or discern vehicles. Instead of detecting each object individually, another way to look at the problem is to estimate a density map where we evaluate the number of objects for each pixel in an image.

In our case, we can classify each pixel in our parking lot as occupied or not, as the area of cars ($2{\times}5$ m/px) is similar to the area of one pixel ($3{\times}3$ px). Such

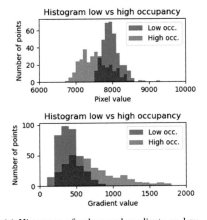

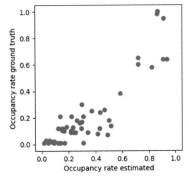

(a) Histogram of values and gradients on low and high occupancy areas.

(b) Relationship between the estimated occupancy rate (using brightness + gradient features) and the ground truth occupancy rate.

Figure 29.19 Histograms of pixel values depending on two parking lot conditions: low and high occupancy. The value attributed to the pixel is the average of its three RGB channels. Low/high occ. = low/high occupancy.

a classification leads to what is known as semantic segmentation where each pixel in an image is assigned one or multiple classes. Deep learning might again be a solution there, if we had access to a reliable ground truth on hundreds of thousands of parking lots. However, obtaining such a ground truth directly from the images is challenging as we are unable to recognize cars. The ideal is to estimate our semantic segmentation without the need of annotated data.

We were able to get satellite images from a validation parking lot in Ocean City, MD, USA. We were able to gather the camera records of this lot, with an image taken every two minutes, as shown in Figure 29.18. This allowed us to establish an approximate parking occupancy ground truth for this particular lot over several months.

Analyzing these images also led us to the following observation. Once PlanetScope images have been registered and equalized, as described in §29.5, we observe that some features in occupied areas tend to have a greater variance than in unoccupied areas. For instance, empty areas tend to appear as homogeneous and grey, whereas occupied areas will tend to appear darker but also with white spots (depending on the color of the cars).

When measuring a specific area in our parking lot where the occupancy is known – see Figure 29.18(d) – we can confirm that the distribution of grey levels in our parking lot fall into two different Gaussian distributions depending on whether or not the pixel is occupied, as shown in Figure 29.19(a). One can therefore use Gaussian mixtures models to find those two distributions. Each pixel can then be classified as occupied if it belongs to the Gaussian distribution with the highest variance and unoccupied otherwise.

We can do this with different features or a combination of them. We explore this subject more deeply in Drouyer (2020a) with additional features. We show in Figure 29.19(b) the relationship between the estimated occupancy rate and the ground truth occupancy rate. We achieve a correlation of 0.83, far from perfect but can be used in large scale analysis to estimate global trends. We show some images and results in Figure 29.20.

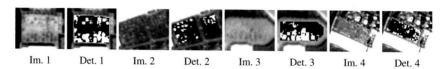

Im. 1   Det. 1   Im. 2   Det. 2   Im. 3   Det. 3   Im. 4   Det. 4

Figure 29.20 Examples of images and associated detections for several parking lots (using brightness + gradient features). Im. = Image. Det. = Detection. Experiment courtesy: Kayrros, Inc.

References

Aguerrebere, C., Delbracio, M., Bartesaghi, A., and Sapiro, G. 2016. Fundamental limits in multi-image alignment. *IEEE Transactions on Signal Processing*, **64**(21), 5707–5722.

Aguerrebere, C., Delbracio, M., Bartesaghi, A., and Sapiro, G. 2018. A practical guide to multi-image alignment. Pages 1927–1931 of: *Proc. ICASSP*.

Anuta, P.E. 1970. Spatial registration of multispectral and multitemporal digital imagery using Fast Fourier Transform techniques. *IEEE Transactions on Geoscience Electronics*, **8**(4), 353–368.

Barnard, Stephen T., and Fischler, Martin A. 1982. Computational stereo. *ACM Computing Surveys*, **14**(4), 553–572.

Berthier, Etienne, Vincent, C, Magnússon, E., Gunnlaugsson, A. P., Pitte, P., Le Meur, E., Masiokas, M., Ruiz, L., Palsson, F., Belart, J.M.C., and Wagnon, Patrick. 2014. Glacier topography and elevation changes derived from Pléiades sub-meter stereo images. *The Cryosphere*, **8**, 2275–2291.

Briand, Thibaud, and Davy, Axel. 2019. Optimization of image B-spline interpolation for GPU architectures. *Image Processing On Line*, **9**, 183–204.

Brown, M.Z., Burschka, D., and Hager, G.D. 2003. Advances in computational stereo. *IEEE Transactions on Pattern Analysis and Machine Intelligence*, **25**(8), 993–1008.

Canty, Morton J, and Nielsen, Allan A. 2008. Automatic radiometric normalization of multitemporal satellite imagery with the iteratively re-weighted MAD transformation. *Remote Sensing of Environment*, **112**(3), 1025–1036.

Canty, Morton J., Nielsen, Allan A., and Schmidt, Michael. 2004. Automatic radiometric normalization of multitemporal satellite imagery. *Remote Sensing of Environment*, **91**(3-4), 441–451.

Chandran, Geethu, and Jojy, Christy. 2015. A survey of cloud detection techniques for satellite images. *International Research Journal of Engineering and Technology*, **2**(9),2485–2490.

Coppin, Pol R., and Bauer, Marvin E. 1996. Digital change detection in forest ecosystems with remote sensing imagery. *Remote Sensing Reviews*, **13**(3-4), 207–234.

Curlander, John C. 1982. Location of spaceborne SAR imagery. *IEEE Transactions on Geoscience and Remote Sensing*, **20**(3), 359–364.

Dagobert, T., Morel, J. M., de Franchis, C., and Grompone von Gioi, R. 2019. Visibility detection in time series of Planetscope images. Pages 1673–1676 of: *Proc. IEEE International Geoscience and Remote Sensing Symposium*.

Dagobert, Tristan, Grompone von Gioi, Rafael, Morel, Jean-Michel, and de Franchis, Carlo. 2020. Temporal repetition detector for time series of spectrally limited satellite imagers. *Image Processing On Line*, **10**, 62–77.

de Franchis, Carlo, Meinhardt-Llopis, Enric, Michel, Julien, Morel, Jean-Michel, and Facciolo, Gabriele. 2014. An automatic and modular stereo pipeline for pushbroom images. *ISPRS Annals of the Photogrammetry, Remote Sensing and Spatial Information Sciences*, **2**(3), 49.

de Franchis, Carlo, Meinhardt-Llopis, Enric, Michel, Julien, Morel, Jean-Michel, and Facciolo, Gabriele. 2014. Automatic sensor orientation refinement of Pléiades stereo images. Pages 1639–1642 of: *Proc. Geoscience and Remote Sensing Symposium (IGARSS)*.

de Franchis, Carlo, Meinhardt-Llopis, Enric, Greslou, Daniel, and Facciolo, Gabriele. 2015. Attitude refinement for orbiting pushbroom cameras: a simple polynomial fitting method. *Image Processing On Line*, **5**, 328–361.

De Zan, Francesco, and Guarnieri, Andrea Monti. 2006. TOPSAR: Terrain observation by progressive scans. *IEEE Transactions on Geoscience and Remote Sensing*, **44**(9), 2352–2360.

Desolneux, A., Moisan, L., and Morel, J.-M. 2000. Meaningful alignments. *International Journal of Computer Vision*, **40**(1), 7–23.

Desolneux, Agnes, Moisan, Lionel, and Morel, Jean-Michel. 2008. *From Gestalt Theory to Image Analysis: a Probabilistic Approach*. Springer.

Dhond, U.R., and Aggarwal, J.K. 1989. Structure from stereo – a review. *IEEE Transactions on Systems, Man, and Cybernetics*, **19**(6), 1489–1510.

Dong, Y., Long, T., Jiao, W., He, G., and Zhang, Z. 2018. A novel image registration method based on phase correlation using low-rank matrix factorization with mixture of Gaussian. *IEEE Transactions on Geoscience and Remote Sensing*, **56**(1), 446–460.

Drouyer, S. 2020a. Parking occupancy estimation on Planetscope satellite images. In: *Proc. IEEE International Geoscience and Remote Sensing Symposium (IGARSS)*.

Drouyer, S. 2020b. VehSat: A large-scale dataset for vehicle detection in satellite images. In: *Proc. International Geoscience and Remote Sensing Symposium (IGARSS)*.

Drouyer, S., and de Franchis, C. 2019. Highway traffic monitoring on medium resolution satellite images. Pages 1228–1231 of: *Proc. International Geoscience and Remote Sensing Symposium (IGARSS)*.

Drouyer, Sébastien, and de Franchis, Carlo. 2020. Parking occupancy estimation on Sentinel-1 images. *ISPRS Annals of the Photogrammetry, Remote Sensing and Spatial Information Sciences*, **2**, 821–828.

Du, Yong, Teillet, Philippe M, and Cihlar, Josef. 2002. Radiometric normalization of multitemporal high-resolution satellite images with quality control for land cover change detection. *Remote Sens. Environ.*, **82**(1), 123–134.

Facciolo, G., de Franchis, C., and Meinhardt-Llopis, E. 2017. Automatic 3D reconstruction from multi-date satellite images. In: *Proc. CVPR Workshops*.

Facciolo, Gabriele, de Franchis, Carlo, and Meinhardt, Enric. 2015. MGM: A significantly more global matching for stereovision. In: *Proc. BMVC*.

Farsiu, S., Elad, M., and Milanfar, P. 2005. Constrained, globally optimal, multi-frame motion estimation. Pages 1396–1401 of: *Proc. IEEE Workshop on Statistical Signal Processing Proceedings*.

Foroosh, Hassan, Zerubia, Josiane B, and Berthod, Marc. 2002. Extension of phase correlation to subpixel registration. *IEEE Trans. Image Process.*, **11**(3), 188–200.

Frantz, David, Haß, Erik, Uhl, Andreas, Stoffels, Johannes, and Hill, Joachim. 2018. Improvement of the Fmask algorithm for Sentinel-2 images: separating clouds from bright surfaces based on parallax effects. *Remote Sens. Environ.*, **215**, 471 – 481.

Fraser, Clive S, and Hanley, Harry B. 2005. Bias-compensated RPCs for sensor orientation of high-resolution satellite imagery. *Photogrammetric Engineering & Remote Sensing*, **71**(8), 909–915.

Furukawa, Yasutaka, and Hernández, Carlos. 2015. Multi-view stereo: A tutorial. *Foundations and Trends in Computer Graphics and Vision*, **9**(1-2), 1–148.

Gordon, A., Glazko, G., Qiu, X., and Yakovlev, A. 2007. Control of the mean number of false discoveries, Bonferroni and stability of multiple testing. *Annals of Applied Statistics*, **1**(1), 179–190.

Gordon, A. H. 1962. Satellite meteorology. *Nature*, **195**(4847), 1161–1162.

Govindu, V.M. 2004. Lie-algebraic averaging for globally consistent motion estimation. Pages 684–691 of: *Proc. IEEE Computer Society Conference on Computer Vision and Pattern Recognition*, vol. 1.

Grodecki, Jacek, and Dial, Gene. 2003. Block adjustment of high-resolution satellite images described by rational polynomials. *Photogrammetric Engineering & Remote Sensing*, **69**(1), 59–68.

Grompone von Gioi, R., and Patraucean, V. 2015. *A contrario* patch matching, with an application to keypoint matches validation. Pages 946–950 of: *Proc. ICIP*.

Grompone von Gioi, R., Hessel, C., Dagobert, T., Morel, J. M., and de Franchis, C. 2020. Temporal repetition detection for ground visibility assessment. *ISPRS Annals of the Photogrammetry, Remote Sensing and Spatial Information Sciences*, **V-2-2020**,829–835 .

Gumbricht, T. 2012. Mapping global tropical wetlands from earth observing satellite imagery. *Center for International Forestry Research.* https://doi.org/10.17528/cifor/004014.

Hall, Forrest G, Strebel, Donald E, Nickeson, Jamie E., and Goetz, Scott J. 1991. Radiometric rectification: toward a common radiometric response among multidate, multisensor images. *Remote Sens. Environ*, **35**(1), 11–27.

Hirschmüller, Heiko. 2005. Accurate and efficient stereo processing by semi-global matching and mutual information. Pages 807–814 of: *Proc. Computer Vision and Pattern Recognition*.

Hollstein, André, Segl, Karl, Guanter, Luis, Brell, Maximilian, and Enesco, Marta. 2016. Ready-to-use methods for the detection of clouds, cirrus, snow, shadow, water and clear sky pixels in Sentinel-2 MSI images. *Remote Sensing*, **8**(8).

Irish, Richard R. 2000. *Landsat 7 automatic cloud cover assessment*. In *Proc. SPIE 4049, Algorithms for Multispectral, Hyperspectral, and Ultraspectral Imagery VI*. https://doi.org/10.1117/12.410358.

Irish, Richard R., Barker, John L., Goward, Samuel N., and Arvidson, Terry. 2006. Characterization of the Landsat-7 ETM+ automated cloud-cover assessment (ACCA) algorithm. *Photogrammetric Engineering & Remote Sensing*, **72**(10), 1179–1188.

Jarvis, R. A. 1983. A perspective on range finding techniques for computer vision. *IEEE Transactions on Pattern Analysis and Machine Intelligence*, **5**(2), 122–139.

Jianya, Gong, Haigang, Sui, Guorui, Ma, and Qiming, Zhou. 2008. A review of multi-temporal remote sensing data change detection algorithms. *Int. Archives of the Photogrammetry, Remote Sensing and Spatial Information Sciences*, **37**(B7), 757–762.

Jiao, Licheng, Zhang, Fan, Liu, Fang, Yang, Shuyuan, Li, Lingling, Feng, Zhixi, and Qu, Rong. 2019. A survey of deep learning-based object detection. *IEEE Access*, **7**, 128837–128868.

Joseph, George. 2015. *Building Earth Observation Cameras*. CRC Press.

Kuglin, C. D., and Hines, D. C. 1975. The phase correlation image alignment method. Pages 163–165 of: *Proc. IEEE International Conference on Cybernetics and Society*.

Leprince, S., Barbot, S., Ayoub, F., and Avouac, J.P. 2007. Automatic and precise orthorectification, coregistration, and subpixel correlation of satellite images, application to ground deformation measurements. *IEEE Transactions on Geoscience and Remote Sensing*, **45**(6), 1529–1558.

Levoy, Marc, Ginsberg, Jeremy, Shade, Jonathan, Fulk, Duane, Pulli, Kari, Curless, Brian, Rusinkiewicz, Szymon, Koller, David, Pereira, Lucas, Ginzton, Matt, Anderson, Sean, and Davis, James. 2000. The digital Michelangelo project. Pages 131–144 of: *Proc. SIGGRAPH '00*.

Lowe, D. 1985. *Perceptual Organization and Visual Recognition*. Kluwer Academic Publishers.

Mahajan, Seema, and Fataniya, Bhavin. 2019. Cloud detection methodologies: variants and development – a review. *Complex & Intelligent Systems*, **6**, 251–261.

Mandroux, N., Drouyer, S., and Grompone von Gioi, R. 2021. Wind turbine detection on Sentinel-2 images. In: *Proc. IEEE International Geoscience and Remote Sensing Symposium (IGARSS)*.

Manizade, K. F., Spinhirne, J. D., and Lancaster, R. S. 2006. Stereo cloud heights from multi-spectral IR imagery via region-of-interest segmentation. *IEEE Transactions on Geoscience and Remote Sensing*, **44**(9), 2481–2491.

Marí, R., de Franchis, C., Meinhardt-Llopis, E., and Facciolo, G. 2019. To bundle adjust or not: A comparison of relative geolocation correction strategies for satellite multi-view stereo. In: *Proc. ICCV Workshops*.

Marí, R., de Franchis, C., Meinhardt-Llopis, E., and Facciolo, G. 2021a. Automatic stockpile volume monitoring using multi-view stereo from SkySat imagery. Submitted to *IEEE International Geoscience and Remote Sensing Symposium (IGARSS)*.

Marí, R., de Franchis, C., Meinhardt-Llopis, E., and Facciolo, G. 2021b. A generic bundle adjustment methodology for indirect RPC model refinement of satellite imagery. Image Processing On Line, **11**, 344-373.

Marr, D, and Poggio, T. 1976. Cooperative computation of stereo disparity. *Science*, **194**(4262), 283–287.

Marti, Renaud, Gascoin, Simon, Houet, Thomas, Laffly, Dominique, and René, Pierre. 2014. Evaluation du modèle numérique d'élévation d'un petit glacier de montagne généré à partir d'images stéréoscopiques Pléiades: cas du glacier d'Ossoue, Pyrénées françaises. *Revue Française de Photogrammétrie et de Télédétection*, **208**, 57–62.

Michel, Pausader, Jean-Philippe, Cantou, Claire, Tinel, and Delphine, Fontannaz. 2013. Potential of Pleiades VHR data for mapping applications. Pages 4313–4316 of: *Proc. IEEE International Geoscience and Remote Sensing Symposium (IGARSS)*.

Moreira, Alberto, Prats-iraola, Pau, Younis, Marwan, Krieger, Gerhard, Hajnsek, Irena, and Papathanassiou, Konstantinos P. 2013. A tutorial on synthetic aperture radar. *IEEE Geoscience and Remote Sensing Magazine*, **1**(1), 6–43.

Panem, C., Baillarin, S., Latry, C., Vadon, H., and Dejean, P. 2005 (July). Automatic cloud detection on high resolution images. 4 pp. in: *Proc. IEEE International Geoscience and Remote Sensing Symposium (IGARSS)*.

Paolini, Leonardo, Grings, Francisco, Sobrino, José A, Jiménez Muñoz, Juan C., and Karszenbaum, Haydee. 2006. Radiometric correction effects in Landsat multi-date/multi-sensor change detection studies. *International Journal of Remote Sensing*, **27**(4), 685–704.

Patraucean, V., Grompone von Gioi, R., and Ovsjanikov, M. 2013. Detection of mirror-symmetric image patches. Pages 211–216 of: *CVPRW*.

Rahman, M. M., Hay, G. J, Couloigner, I., Hemachandran, B., and Bailin, J. 2015. A comparison of four relative radiometric normalization (RRN) techniques for mosaicing H-res multi-temporal thermal infrared (TIR) flight-lines of a complex urban scene. *ISPRS J. Photogramm.*, **106**, 82–94.

Rais, M., Thiebaut, C., Delvit, J.M., and Morel, J.M. 2014. A tight multiframe registration problem with application to Earth observation satellite design. Pages 6–10 of: *Proc. IEEE IST*.

Redmon, Joseph, and Farhadi, Ali. 2017. YOLO9000: better, faster, stronger. Pages 7263–7271 of: *Proc. CPVR*.

Ren, Shaoqing, He, Kaiming, Girshick, Ross B., and Sun, Jian. 2015. Faster R-CNN: towards real-time object detection with region proposal networks. In: *Proc. CoRR*.

Rosu, Ana-Maria, Pierrot Deseilligny, Marc, Delorme, Arthur, Binet, Renaud, and Klinger, Yann. 2014. Measurement of ground displacement from optical satellite image correlation using the free open-source software MicMac. *ISPRS Journal of Photogrammetry and Remote Sensing* **100**, 48–59.

Ruelland, D., Dezetter, A., Puech, C., and Ardoin-Bardin, S. 2008. Long-term monitoring of land cover changes based on Landsat imagery to improve hydrological modelling in West Africa. *International Journal of Remote Sensing*, **29**(12), 3533–3551.

Scaramuzza, P. L., Bouchard, M. A., and Dwyer, J. L. 2012. Development of the Landsat data continuity mission cloud-cover assessment algorithms. *IEEE Transactions on Geoscience and Remote Sensing*, **50**(4), 1140–1154.

Scharstein, Daniel, and Szeliski, Richard. 2002. A taxonomy and evaluation of dense two-frame stereo correspondence algorithms. *International Journal of Computer Vision*, **47**(1-3), 7–42.

Seitz, Steve M., Curless, B., Diebel, J., Scharstein, Daniel, and Szeliski, Richard. 2006. A comparison and evaluation of multi-view stereo reconstruction algorithms. Pages 519–528 of: *Proc CVPR*, vol. 1.

Shean, David E., Alexandrov, Oleg, Moratto, Zachary M., Smith, Benjamin E., Joughin, Ian R., Porter, Claire, and Morin, Paul. 2016. An automated, open-source pipeline for mass production of digital elevation models (DEMs) from very-high-resolution commercial stereo satellite imagery. *ISPRS Journal of Photogrammetry and Remote Sensing*, **116**, 101–117.

Shin, D., and Pollard, J. K. 1996. Cloud height determination from satellite stereo images. Pages 4/1–4/7 of: *Proc. IEE Colloquium on Image Processing for Remote Sensing*.

Sinclair, Kenneth, van Diedenhoven, Bastiaan, Cairns, Brian, Yorks, J., Wasilewski, Andrzej, and McGill, Matthew. 2017. Remote sensing of multiple cloud layer heights using multi-angular measurements. *Atmospheric Measurement Techniques Discussions*, **2**, 1–23.

Stone et al., H.S. 2001. A fast direct Fourier-based algorithm for subpixel registration of images. *IEEE Transactions on Geoscience and Remote Sensing*, **39**(10), 2235–2243.

Strecha, Christoph, Von Hansen, W., Van Gool, Luc, Fua, Pascal, and Thoennessen, U. 2008. On benchmarking camera calibration and multi-view stereo for high resolution imagery. Pages 1–8 of: *CPVR*.

Syariz, Muhammad Aldila, et al. 2019. Spectral-consistent relative radiometric normalization for multitemporal Landsat 8 imagery. *ISPRS J. Photogramm.*, **147**, 56–64.

Szeliski, Richard. 2011. *Computer Vision*. Springer.

Tadros, A., Drouyer, S., Grompone, von Gioi, R., Facciolo, G., and Carvalho, L. 2020. Circular-shaped object detection in low resolution satellite images. *ISPRS Annals of the Photogrammetry, Remote Sensing and Spatial Information Sciences*, **2**, 901–908.

Taravat, Alireza, Proud, Simon, Peronaci, Simone, Del Frate, Fabio, and Oppelt, Natascha. 2015. Multilayer perceptron neural networks model for meteosat second generation SEVIRI daytime cloud masking. *Remote Sensing*, **7**(2), 1529–1539.

Triggs, B., McLauchlan, P.F., Hartley, R.I., and Fitzgibbon, A.W. 1999. Bundle adjustment – a modern synthesis. In: *Proc. International Workshop on Vision Algorithms*.

Vachon, Paris W, Wolfe, John, and Greidanus, Harm. 2012. Analysis of Sentinel-1 marine applications potential. Pages 1734–1737 of: *Proc. IEEE International Geoscience and Remote Sensing Symposium (IGARSS)*.

Witkin, A. P., and Tenenbaum, J. M. 1983. On the role of structure in vision. Pages 481–543 of: *Human and Machine Vision*, Beck, J., Hope, B., and Rosenfeld, A. (eds). Academic Press.

Wu, Teng, Hu, Xiangyun, Zhang, Yong, Zhang, Lulin, Tao, Pengjie, and Lu, Luping. 2016. Automatic cloud detection for high resolution satellite stereo images and its application in terrain extraction. *ISPRS Journal of Photogrammetry and Remote Sensing*, **121**, 143–156.

X., Tong, et al. 2019. Image registration with Fourier-based image correlation: a comprehensive review of developments and applications. *IEEE JSTAR*, **12**(10), 4062–4081.

Ye et al., Z. 2020. Robust fine registration of multisensor remote sensing images based on enhanced subpixel phase correlation. *Sensors*, **20**(15), 4338.

Yésou, Hervé, Escudier, Aurélie, Battiston, Stéphanie, Dardillac, Jean-Yves, Clandillon, Stephen, Uribe, Carlos, Caspard, Mathilde, Giraud, Henri, Maxant, Jérôme, Durand, Arnaud, Fellah, Kader, Studer, Mathias, Huber, Claire, Philippoteaux, Laurent, de Fraipont,

Paul, and Fontannaz, Delphine. 2015. Exploitation de l'imagerie Pléiades en cartographie réactive suite à des catastrophes naturelles ayant affecté le territoire français en 2013. *Revue Française de Photogrammétrie et de Télédétection*, **209**, 39–45.

Yuan, Ding, and Elvidge, Christopher D. 1996. Comparison of relative radiometric normalization techniques. *ISPRS Journal of Photogrammetry and Remote Sensing*, **51**(3), 117–126.

Zhang, W., Qi, J., Wan, P., Wang, H., Xie, D., Wang, X., and Yan, G. 2016. An easy-to-use airborne lidar data filtering method based on cloth simulation. *Remote Sensing*, **8**(6), 501. `https://doi.org/10.3390/rs8060501`.

Zhang, W. D., He, M. X., and Mak, M. W. 2001. Cloud detection using probabilistic neural networks. Pages 2373–2375: *Proc. IEEE International Geoscience and Remote Sensing Symposium (IGARSS)*.

Zitová, B., and Flusser, J. 2003. Image registration methods: a survey. *Image and Vision Computing*, **21**(11), 977–1000.

Part VII

Biases and Model Risks of Data-Driven Learning

30

Introduction to Part VII
Towards the Ideal Mix between Data and Models

Mathieu Rosenbaum[a]

Abstract

With the rise of machine learning methods, a recent opposition has emerged in the financial community and beyond between data-driven and model-driven approaches. It seems that all classical stochastic approaches are now outdated and only data-based ones, whatever that means, are acceptable.

Of course the reality is more subtle since data can only describe what already happened, whereas "good models", relying on stationary assumptions, can go beyond the anecdote of past data.

30.1 What are we exactly talking about?

The two extreme views are on the one hand model free approaches, where all the knowledge about the dynamics is learnt from data in an agnostic way, and on the other hand more stochastic models where *a priori* random dynamics are postulated with some degrees of freedom through parameters to be estimated/calibrated (possibly infinite-dimensional). The choice of these parameters is then based on the adequacy of the model to data given the chosen values. So we see that actually the distinction between data-driven and model-driven modelling may not be really relevant as actually all financial models use data. The discussion is on how much of the model should be dictated by data and how much should be fixed a priori by the financial engineer. In some sense, this is the 2021 version of the debate between parametric versus semi-parametric versus non-parametric approaches in statistics and finance or even in some sense between stochastic control and reinforcement learning in optimization.

30.2 What is a good model?

Before going deeper into this debate we should probably ask ourselves what a good model is. Traditionally, in finance, a model is said to be good if:

[a] CMAP, Ecole Polytechnique
    Published in *Machine Learning And Data Sciences For Financial Markets*, Agostino Capponi and Charles-Albert Lehalle© 2023 Cambridge University Press.

- it fits the data reasonably well: the main "stylized facts" have to be reproduced;
- it is parsimonious: one should not need too many parameters;
- it is tractable: one should be able to do computations in the model, at least numerical ones, and to simulate it;
- one can understand the model and give an interpretation for it: if a model works very well there must be a simple underlying explanation why it is able to capture the phenomena with great accuracy;
- it is useful for solving a financial problem: risk management, algorithmic trading, portfolio optimization, derivatives hedging. . .

These elements remain essentially valid nowadays except that one is more willing to accept models that are less parsimonious living in high-dimensional parameter spaces. Furthermore, having a nice interpretation is not considered so important in some situations. As was actually already the case before the machine learning era, in some cases, having a "black box" type approach may be reasonable while in others it can be very dangerous. A good illustration of this is provided for example by Chapter 33 "Black-box model risk in finance". Note also that using overparametrized models is common practice in some areas of finance for a long time already. For example, some methods to approximate the implied volatility surface with a very large number of parameters proved to be quite efficient in specific contexts.

30.3 Simulating what has never been observed

One of the most important aspects and use of random modelling is scenari generation. It is obviously very different to try to somehow reproduce the past and to anticipate things that have never been seen before. To generate never observed yet plausible trajectories, there is a need to understand the mechanisms at stake in our system of interest so that these mechanisms can be encoded into a reasonable model, and lead to meaningful new and original dynamics. In particular, one wants to comprehend what stress tests correspond to in terms of economic meaning. As a matter of fact, there will be another financial crisis but it will be different from the previous ones. Therefore, if one hopes to be able to be prepared for it as well as possible, it is necessary to go beyond what is encoded in past data. In addition, the time series aspect of financial data is a big challenge for purely data based approaches as they often rely on importing methods coming from static worlds (imaging, textual data etc.. . .). A large part of financial dynamics are fondamentally non stationary, therefore it is sometimes hard to learn anything.

30.4 Being pragmatic

There is in fact no such thing as an opposition between data driven and model driven approaches. They are complementary and the freedom we let to data to dictate the dynamics should just depend on the application in mind.

For example, in asset price modelling for risk management of derivatives, financial theory tells us that prices must be driven by Brownian motion and that volatility is rough. These ingredients should be put *a priori* in the modelling so that the set of admissible models becomes smaller and thus the task of the financial engineer easier. Similarly, there are excellent order book models that are intuitive, robust and parsimonious. Consequently, trying to tackle the very high-dimensional problem of order book modelling with a purely data based approach is not only very complex but probably a waste of time. Obviously, there is always an intricate trade-off between adding numerical/mathematical/technical complexity and increasing significantly the performance as illustrated in Chapter 31, "Pricing Model Complexity: The Case for Volatility-Managed Portfolios".

In contrast, other problems such as anonymizing data, mimicking sample paths, some types of backtesting or the highlight of subtle financial factors as in the Chapter 32 "Bayesian Deep Fundamental Factor Models" are clearly very well suited in principle for data-driven approaches as one does not wish to put *a priori* ideas in such contexts.

So, as we know very well, all models are wrong but some are useful. One should just use an approach based on the applicability and relevance of the model to address the considered problem. It can be very much data-driven, or almost completely set in stone *a priori* depending on the context. Any good financial engineer should actually be able to master a wide range of types of approaches, from the most "mathematical" to the most "statistical" ones. Reducing finance to machine learning on past data would be a big mistake, and ignoring the revolutionary tools of artificial intelligence and what they can bring to the financial world would be just as much of one.

Generative Pricing Model Complexity: The Case for Volatility-Managed Portfolios

Brian Clark[a], Akhtar Siddique[b] and Majeed Simaan[c]

Abstract

AI/ML models are used for many financial applications ranging from portfolio selection to efficient credit allocation. However, the drawback to applying these models in practice is that performance (i.e., predictive power) is generally inversely related to model complexity. In this chapter, we formalize the problem that a risk manager faces when trading off the benefits of model performance with the drawbacks of complexity. Because it is difficult to define a single metric that quantifies complexity across all settings, we use several metrics to capture model complexity. We use the case of a volatility-managed portfolio to show the impact of model complexity on performance by constructing a mean-variance-complexity efficient frontier. We find that for any level of risk, there is a positive relation between model complexity and portfolio return. Moreover, we show that model complexity affects the classic mean-variance efficient frontier. For low levels of complexity, the risk–return tradeoff is an inverted u-shape, while at high levels of complexity the risk-return tradeoff is positive.

31.1 Introduction

Machine learning (ML) models have been shown to be of value for many financial applications ranging from efficient asset allocation to efficient credit allocation (see, e.g., Khandani et al., 2010, Butaru et al., 2016, Sirignano et al., 2016, and Gu et al., 2020). However, one criticism of the models is that performance (i.e., predictive power) tends to be inversely related to model complexity (see, e.g., Mullainathan and Spiess, 2017, Athey, 2018, Hind, 2019)[1]. In some applications, nonetheless, this is not an issue. For example, if a trading strategy requires an accurate forecast of volatility but the risk manager is not concerned with

[a] Rennsselaer Polytechnic Institute
[b] Office of the Comptroller of the Currency
[c] Stevens Institute of Technology
   Published in *Machine Learning And Data Sciences For Financial Markets*, Agostino Capponi and Charles-Albert Lehalle © 2023 Cambridge University Press.

[1] A similar argument follows for portfolio optimization. Due to estimation risk, decision-maker could end up maximizing error than risk-reward tradeoffs (Michaud, 1989). On the other hand, a naive approach that corresponds to minimal complexity could yield better out-of-sample performance (DeMiguel et al., 2009).

understanding the factors that drive volatility, there are several ML models that are suitable for the job (see, e.g., Liu, 2019; Sun and Yu, 2020), and generally, more complex models will be preferable. Even deep learning models, which are notoriously difficult to interpret, are suitable if the portfolio manager is only interested in minimizing the forecast error – and is not concerned with understanding the factors that are driving volatility.

There are other applications, however, where a risk manager or other stakeholders such as external regulators would benefit from the increased predictability of ML models *if and only if* they could interpret or explain the results. For example, consider a bank manager seeking an asset allocation strategy to maximize performance (e.g., risk-adjusted return). She is faced with two choices: (1) implement a simple linear model the results of which her stakeholders (e.g., shareholders, executives, etc.) can easily interpret; or (2) a complex ML model which will make asset allocation more efficient, but is practically a black box.[2] Which should she choose?

The benefit of the first choice is that traditional linear econometric models are built with the primary intention of being able interpret the factors that drive the results (Mullainathan and Spiess, 2017). As such, managers can develop a risk management strategy to exploit the firm's core competencies while hedging the risk factors that drive potentially harmful outcomes. A linear model makes the risk mapping process straightforward. However, it may also be limited in terms of identifying the true risk factors – especially if the true risk drivers exhibit complex, non-linear relations. Clearly, this comes at the cost of a loss in the predictive power of the model. For instance, a bank may grant loans to undeserving customers or fail to finance well-qualified potential customers. Ultimately, the manager must balance this tradeoff when selecting a suitable model for their task.

In this chapter, we examine the problem a risk manager faces when trading off the benefits of model accuracy versus model complexity. Conceptually, we view our framework as a generic ML problem in which model complexity is a risk factor. The enhancement of our framework compared to the usual model selection problem is that our framework compares model performance based on two dimensions: (1) predictive or forecasting power; and (2) the usability or complexity of the model. The former is straightforward. For example, Gu et al. (2020) compare the predictive power of several ML models to predict asset returns. The best model is defined as the one with the best forecasting power (e.g., highest out of sample R^2 or lowest MSE). Other papers such as Khandani et al. (2010) and Butaru et al. (2016) use profitability metrics based on trading

---

[2] Regulatory requirements have also been a significant driver for explanability. The "Right to explain" incorporated in the European Union General Data Protection Regulation (GPDR) has been viewed as one such requirement. The drafters of the GPDR inserted nonbinding recitals in the text to explain the regulation. Recital 71 states that *automated processing* "should be subject to suitable safeguards, which should include information specific to the data subject and the right to obtain human intervention, to express his or her point of view, to obtain an explanation of the decision reached after such assessment and to challenge the decision."

off Type I and Type II errors which are also based on commonly used statistics derived from the confusion matrix (for the case of binary classification problems).

We should also note that the topic of complexity/interpretability is related to the problem of estimation risk that has been studied in the literature on optimal portfolio choice. Brown (1979), Chen and Brown (1983), etc. investigated the impact of estimation risk on optimal portfolios. Intuitively, with estimation risk the portfolio is riskier than if the true parameters were known with certainty. Hence, the optimal portfolio with estimation risk tends to be closer to the Global Minimum Variance Portfolio. However, to our knowledge, no one has studied how different choices in algorithms may induce different estimation risk. In application of machine learning to portfolio optimization, that becomes a natural problem to investigate.

Quantifying model complexity presents a more substantial challenge. Model complexity is very much related to the concept of interpretability or explainability because more complex models tend to be more difficult to explain. Interpretability in machine learning is examined in several ways and typically falls into two categories. The first evaluates interpretability in the context of an application: if the system is useful in either a practical application or a simplified version of it, then it must be somehow interpretable (see, e.g., Ribeiro et al., 2016, Lei et al., 2016, Doshi-Velez and Kim, 2017). The second evaluates interpretability via a quantifiable proxy: a researcher might first claim that some model class – e.g. sparse linear models, rule lists, gradient boosted trees – are interpretable, and then present algorithms to optimize within that class (see, e.g., Bussmann et al., 2020, Sachan et al., 2020, and Bücker et al., 2020 for explainability tools for credit risk applications; Hind, 2019, Arrieta et al., 2020, and Li et al., 2020 for general surveys).

In our framework, we seek to classify models based on how difficult they are to interpret from the perspective of model stakeholders (e.g., risk managers, external regulators, model validation staff, etc.). That is, we consider a model to be a nexus of functions that could potentially include *any* ML alogorithm or econometric model. In other words, the risk manager is not simply tuning a set of hyper-parameters to optimize the bias–variance tradeoff inherent to a single algorithm such as an elastic net regression or a deep learning neural network, but rather is choosing the "best" algorithm among a suite of alternatives. This difference creates several practical challenges that mimic the complex reality of a model selection framework in a regulated financial institution. Most notable is that we need model-agnostic measures of both complexity and model accuracy. The latter is straightforward. For example, confusion-matrix-based metrics such as κ, precision, or recall are acceptable for classification problems or metrics such as R^2 or MSE^{-1} are acceptable for regression settings and are regularly used by modelers at banks to compare performance of different algorithms and are also common in the literature (see, e.g., Khandani et al., 2010, Butaru et al., 2016, and Gu et al., 2020).

The real challenge is to quantify the complexity of a model using a single metric that allows for a fair comparison across different classes of models (e.g., regression, neural networks, random forests, boosted regression, etc.). Unfortunately, there no widely accepted view of what complexity means in the financial industry, let alone across different fields. As such, we start by defining what is meant by model interpretability our context. For tractability, we restrict ourselves to a setting that meets the following criteria:

1. The model user benefits from increased model accuracy.
2. The model user benefits from being able to interpret the model.
3. Models that perform better are more difficult (or costly) to interpret because they are inherently more complex.

The first condition is obvious. The second condition requires some elaboration. Start with the obvious question of, "why would a model user benefit from being able to interpret the model results?" While stewards of traditional econometric modeling may find this question obvious (and a necessary condition for building a robust model), it is not always the case that one cares about interpretation. Note that the ML algorithms that we test in this chapter have been developed without the goal of interpretation in mind. For example, while deep learning models can read X-rays, drive cars, and perform complex facial recognition tasks, we do not necessarily care *why* they make correct decisions; rather that they *do* make correct decisions. Having said that, there are many counter-examples where being able to interpret a model is imperative to the user. For example, banks that deny customers credit must legally provide them with a rationale explaining their decision. This requires that a bank be able to identify and convey specific risk factors that drive their decisions – i.e., they must be able to interpret their models to reduce their legal risk exposure. However, it is also easy to imagine a scenario when a bank is using a similar set of credit data to identify customers to whom they want to market additional products and services.[3] In this case, they may want to have some level of interpretation of the model results but their legal exposure is less so they would likely be willing to accept a more complex model that performs better in terms of accuracy. Finally, the third condition connects the first two. Since models that perform better are generally more difficult to interpret, one can see the natural tradeoff that a risk manager faces when selecting a model. Practically speaking, this is the quintessential ML tradeoff – *how does one balance model performance and the need to understand the black box*? This is the issue we attempt to quantify in this chapter. In this case, they may want to have some level of interpretation of the model results but their legal exposure is less so they would likely be willing to accept a more complex model that performs better in terms of accuracy. Finally, the third condition connects the first two. Since models that perform better are generally more difficult to interpret, one can see the natural tradeoff that a risk manager faces when selecting a model. Practically speaking, this is the quintessential ML tradeoff - *how does one balance model performance and the need to understand the black box*? This is the issue we attempt to quantify in this chapter.

---

[3] This is commonly known a cross-selling.

Assuming that we can use well-established statistics to compare performance across models, the next step to quantify model complexity. A robust model complexity metric needs to satisfy several conditions. For example, the metric must be quantifiable, model-agnostic, and ideally measured on a continuous scale. However, it is difficult to image a single metric that quantifies model complexity across all settings. Therefore, we instead focus on the quality of observable risk management decisions a model user makes based on the interpretation of the model. We build on the functional decomposition proposed by Molnar et al. (2019). In particular, we implement three complexity metrics, which we ultimately combine them into a single metric in our empirical framework:

1. Interaction Strength (IAS)
2. Main Effect Compexity (MEC)
3. Number of Features (NF)

The intuition behind the complexity measure is the following. An ML predictive function can be approximated using first-order with respect to the Accumulated Local Effects (ALE) (Apley and Zhu, 2020). Given this approximation, we are able to compute residual which corresponds to the IAS. For a linear ML model, such residual is zero and, hence, the IAS. Whereas for a support vector regression (SVR) with radial kernel the IAS is significant. Recall that an SVR model with radial kernel emulates a plain vanilla neural net with one hidden layer. Clearly, the IAS for such models is not zero. On the other hand, the MEC denotes the complexity associated with the ALE functions to describe such first-order approximation. For a linear model such measure is one, by definition. On the other hand, for a non-linear model the MEC describes the average number of segments from a stepwise regression needed to describe the complexity associated with the ALE function. Finally, the NF corresponds to the number of features used in the model. That is, the higher the NF the more complex the interpretation behind the model is.

The financial risk management application we focus on is related tactical asset allocation that rotates between equity and Treasury bond exchange-traded funds (ETFs). We focus on a simple paradigm to link between ML complexity and financial performance. Specifically, we motivate our asset allocation problem according to the volatility managed portfolios idea from Moreira and Muir (2017). Given this financial problem, we examine how model complexity is associated with out-of-sample performance. The main premise behind our investigation is whether model complexity is associated with lower or higher estimation risk in asset allocation. It is well known that estimation risk hinders portfolio performance *ex post* (Best and Grauer, 1991). At the same time, higher complexity denotes a higher model risk in which the decision-making process could overfit on historical data.

Overall, our chapter makes the following observations:

1. As expected, we find that higher complexity which is associated with higher in-sample overfitting corresponding to higher in-sample statistical performance.

We measure statistical performance by the ability of each model to predict next-month realized volatility. At the same time, we observe that complexity is also associated with higher out-of-sample R^2. In particular, a one standard deviation increase in the complexity measure corresponds to a 2% increase in out-of-sample. However, this magnitude is much smaller than in-sample, which is 8%.

2. Among the seven ML models explored in this chapter, random forest is deemed to be the only algorithm that yields consistent results in terms of positive economic performance–complexity tradeoff. Altogether, we find that higher complexity is associated with higher portfolio turnover, out-of-sample volatility, and expected shortfall. While this increased complexity is associated with an out-of-sample portfolio mean return, its economic magnitude is relatively small. Hence, this observation undermines whether complexity mitigates estimation risk and, thus, leads to better out-of-sample performance. It is worthwhile mentioning, nevertheless, that this inverse relationship between complexity and out-of-sample performance decays as complexity decreases beyond certain level. This result could be attributed to the persistent performance of the random forest algorithm.

3. Taking into consideration risk-reward-complexity tradeoff, we propose an efficient set that associates risk-adjusted performance with complexity. In particular, we propose a mean-variance-complexity (MVC) efficient frontier. we find that model complexity affects the classic risk–return efficient frontier. For low levels of complexity, the risk–return tradeoff is an inverted ∪-shape; at high levels of complexity the risk–return tradeoff is positive.

4. The corresponding MVC efficient frontier supports the following recommendations:

   1. The use of mean absolute error (MAE) rather than mean-squared error (MSE) for tuning. In our experiment, squared deviations are larger than absolute deviations. Given non-stationarity and potential outliers inherent in financial data, we expect the MSE to amplify such errors due to transient events that eventually lead to an increase in estimation risk.

   2. In terms of algorithm choice, our final results advocate for either a simple model such as Lasso or elastic net, or a highly complex one such as random forest. However, a choice in between corresponds to an inverse relationship between risk-adjusted performance and complexity. In other words, our findings suggest that when it comes to model complexity, risk managers could be off with more extreme levels of complexity (high or low) rather than intermediate levels.

The remainder of this chapter proceeds as follows. In §31.2, we define the model complexity measures and the idea behind them according to Molnar et al. (2019). This is supported by a number of numerical illustrations. In §31.3 we discuss the financial problem of the chapter that focuses on the idea of volatility-managed portfolios and their relation to model complexity. Finally, §31.4 covers the main empirical investigation and results of the chapter, and §31.5 concludes.

31.2 Quantifying model complexity

In this section we discuss the methodology we use to quantify the complexity of different machine learning algorithms. The analysis below is built on the functional decomposition proposed by Molnar et al. (2019). The idea behind this relies on Accumulated Local Effects (ALE) (Apley and Zhu, 2020). In line with Molnar et al. (2019), we discuss the main idea behind the functional decomposition to compute ML complexity. The discussion below is presented at a high level with numerical illustration to provide better understanding of the main idea behind the complexity measure.

31.2.1 The main idea

Let $f : \mathbb{R}^{N \times d} \to \mathbb{R}$ denote a prediction function that maps the data matrix $\mathbf{x} \in \mathbb{R}^{d \times N}$ into a column vector $\hat{\mathbf{y}} \in \mathbb{R}^{N \times 1}$. Here d and N denote, respectively, the number of features and the number of observations. According to Equation (3) of Molnar et al. (2019), the fitted value is approximated as

$$f(\mathbf{x}) = f_0 + \sum_{j=1}^{d} f_{j,\text{ALE}}(\mathbf{x}_j) + \text{IAS}(\mathbf{x}). \qquad (31.1)$$

A couple of comments are in order. First, the intercept f_0 denotes the average of the fitted values, i.e.

$$f_0 = \frac{1}{N}\hat{\mathbf{y}}^{\top}\mathbf{1}, \qquad (31.2)$$

where $\mathbf{1}$ is a column vector of 1s. Next, the second component corresponds to the first-order sensitivity of the fitted function to each feature in the data. Theoretically, for a given j, the function $f_{j,\text{ALE}}(\mathbf{x}_j)$ is computed as the conditional expectation of the $f(\mathbf{x})$ for given value of $\mathbf{x}_{j,n}$. Here $\mathbf{x}_{j,n}$ denotes the nth observation of feature $\mathbf{x}_j$ for $n = 1, \ldots, N$. Given this definition, $f_{j,\text{ALE}}(\mathbf{x}_j)$ computes the sensitivity of conditional expectation over a range of values $\mathbf{x}_{j,n}$. In the following, we will discuss the further the implementation and approximation of ALEs. Third, $\text{IAS}(\mathbf{x})$ captures the interaction strength. Since the function is approximated using an intercept and first-order approximation, the residual of $f(\mathbf{x})$ is attributed to interactions among the features. For example, for a linear regression model, we expect IAS to be zero. On the other hand, for non-linear models such as SVM or random forest, we expect such term to be non-zero. Below we provide further details regarding each component.

31.2.2 ALE functions

According to Molnar (2019), ALE functions "describe how features influence the prediction of a machine learning model on average. ALE plots are a faster and unbiased alternative to partial dependence plots (PDPs)." Therefore, to understand ALE, it is worth reviewing PDPs first before we move to the formal definition of ALE. By design, PDPs denote the conditional expectation of the prediction function for a given feature

$$\text{PDP}\left(\mathbf{x}_j\right) = \int f(\mathbf{x})D\left(\mathbf{x}_{-j}\right)d\mathbf{x}_{-j}, \tag{31.3}$$

where $f(\mathbf{x})$ is the prediction function and $D\left(\mathbf{x}_{-j}\right)$ is the joint density function of the the data with the exclusion of feature j. Specifically, Equation (31.3) denotes the conditional expectation of the fitted function for different values of feature j.

In practice, we do not know the joint density function and, hence, the conditional expectation is unknown. Instead, PDP rely on numerical methods to become operational. The implementation is similar to a bootstrap approach with N_b observations. In particular, for feature j and observation $n = 1$, the methodology copies the training data and replaces the original values of feature j with this single value. Given the "new" data, denoted by $\tilde{\mathbf{x}}$, it applies the mapping function to derive a series of N values. By taking the average over these N fitted values, we find a single number which we denote by $\text{PDP}_n(\mathbf{x}_j)$. Repeating the previous steps N_b times for $n = 1,\ldots,N_b$, the procedure results in the PDP values.

By design, PDP is intuitive as it assesses the sensitivity of the prediction function to a given feature. Its implementation is straightforward and relies on basic bootstrapping. However, its interpretation could imply both advantages and disadvantages. On the one hand, the implementation of PDP presumes that the features are independent, making it simple to understand how it works. On the other hand, the independence assumption can cause misleading interpretations, resulting in a bias. This is bias stems from the fact that if features are correlated, there are limits on the potential values that feature j can take. For instance, consider a simple case in which the features correspond to the height and weight of an individual. Clearly, there is a lower bound on how much a person can weigh for a given height. In terms of financial application, a similar intuition follows suit. For a well-diversified portfolio, one cannot consider different portfolio returns without considering market movement.

Given the above disadvantage of PDP, this is where ALE comes into the picture. In particular, ALE tells us how the model prediction changes within a given window of feature j. In this regard, ALE has two advantages over PDP. First, it computes conditional expectation by taking into account correlated variables. Second, the expectation is applied on value changes rather than actual levels. By design, this mitigates concerns regarding the omitted variable bias and results in a more reliable sensitivity interpretation.

Put formally, ALE of feature j is given by

$$f_{j,\text{ALE}}(\mathbf{x}_j) = \int_{x_{\min}}^{x_{j,n}} \int f'\left(\mathbf{z}_j\right) D\left(\mathbf{x}_{-j} \mid \mathbf{x}_j\right) d\mathbf{x}_{-j}d\mathbf{z}_j - \text{constant}. \tag{31.4}$$

A few comments follow Equation (31.4). First, note that $f'\left(\mathbf{z}_j\right)$ corresponds to the sensitivity of the prediction function to small changes in feature j. Second, the ALE function considers the conditional expectation and, hence, the co-dependence structure among the features. This is reflected by the conditional density, $D\left(\mathbf{x}_{-j} \mid \mathbf{x}_j\right)$. Third, for a given window of feature j, the ALE computes the conditional expectation within that interval. Finally, note that the ALE considers the marginal effect by subtracting a constant. In this regard, the main decomposition of the prediction function from Equation (31.1) is approximated by the total prediction average plus the marginal contribution of each feature. Note that this decomposition holds when IAS = 0. We devote the next subsection to this component.

31.2.3 Interaction strength (IAS)

Given the intuition behind the ALE approximation from (31.1), we discuss the IAS metric proposed by Molnar et al. (2019). Specifically, Molnar et al. (2019) define IAS as an approximation error measured for a given loss function L, where

$$\text{IAS} = \frac{\mathbb{E}\left(L\left(f, f_{\text{ALE}}\right)\right)}{\mathbb{E}\left(L\left(f, f_0\right)\right)} \geq 0. \tag{31.5}$$

Note that f_0 denotes the mean prediction (see Equation (31.1)) and f_{ALE} is the approximation of the prediction function based on the ALE components. Additionally, the loss function L denotes the distance between the prediction values and the one approximated using the ALE values only. If the prediction model is linear, all prediction values are captured using the ALE components only. In fact, if we measure this distance/loss using the second norm, then IAS equals one minus the R-squared measure, such that

$$\text{IAS} = \frac{\sum_{i=1}^{N}\left(\hat{\mathbf{y}}_i - f_{\text{ALE}}(i)\right)^2}{\sum_{i=1}^{N}\left(\hat{\mathbf{y}}_i - f_0\right)^2} = 1 - R^2, \tag{31.6}$$

where $\hat{\mathbf{y}}_i$ is the fitted/predicted value for observation $i = 1, \ldots, N$, while $f_{\text{ALE}}(i)$ denotes the ALE approximation of the same observation. By construction, if the predicted value $\hat{\mathbf{y}}_i$ is a linear map of the underlying features, then its variability is attributed solely to the ALE first-order approximation. This denotes an R-squared of 1 and, hence, IAS = 0. Otherwise, there is a positive loss such that IAS > 0.[4]

To demonstrate the principle behind the IAS measure, we introduce a simple example using the Bike Rentals data from Fanaee-T and Gama (2014), which is also implemented in the empirical analysis of Molnar et al. (2019). For a given model, the response variable is defined as the number of rentals and the feature space consists of three numerical variables (temperature, humidity, and wind speed). To demonstrate the ALE approximation, we consider the first attribution of one variable on the response one. In this case, we look into the relationship between wind speed, $\mathbf{x}_j$ and the number of rentals, $\hat{\mathbf{y}}$. It is important to indicate that this rather the relationship between variable j and the corresponding first-order approximation, $f_{j,\text{ALE}}(\mathbf{x}_j)$. In Figure 31.1, we illustrate the ALE approximation for four different ML algorithms – both linear and non-linear. The panels correspond

[4] According to Molnar et al. (2019), feature effects such as PDP or ALE are more reliable when the interaction strength is lower. In particular, the authors note that the "effects can vary greatly for individual instances and even have opposite directions when the model includes feature interactions."

to the following algorithms, respectively, Lasso, decision tree, support vector regression with radial kernel (ksvm), and random forest (rf). In all cases, the hyperparameters are tuned using a 10-fold cross validation with three repetitions. The ML implementation is carried using Kuhn *et al.* (2008), whereas the ALE approximation is conducted using Apley (2018).

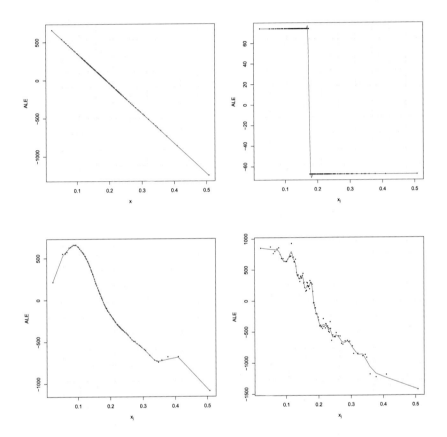

Figure 31.1 *Accumulated Local Effects.* The figure provides an empirical illustration of the Accumulated Local Effects by Apley and Zhu (2020). In all cases, a given model is trained using the same data and features. The data corresponds to the Bike Rental from Fanaee-T and Gama (2014). The response variable denotes the number of rentals, whereas the explanatory variable is the wind speed. The panels below the ALE approximation for different ML algorithms. Panel (a) is Lasso, where the first norm is penalized. Panel (b) is a decision tree. Panel (c) is a support vector regression with radial kernel. Finally, Panel (d) corresponds to random forest. In all cases, the hyperparameters are tuned using 10-fold cross validation with three repetitions. The ML implementation is carried using Kuhn *et al.* (2008), whereas the ALE approximation is conducted using Apley (2018).

Given the implementation of Apley (2018), we are able to visualize the relationship between $\mathbf{x}_j$ and $\hat{\mathbf{y}}$. The range of $\mathbf{x}_j$ is the same of the one given in the data. In all cases, the dots denote the values computed using ALE, whereas the smoothed line is derived using a local regression with span of 10%. The latter line is derived for a reason that will be discussed shortly. For Lasso, we observe that $\mathbf{x}_j$ feeds into the model and is not eliminated as part of the norm-regularization. Since the model is linear, we observe a downward sloping linear line. For decision tree, we observe that the values are truncated into parts: low and high wind speed. For ksvm and rf, we observe a more non-linear relationship between the two.

The ALE approximation does not return N values for given $\mathbf{x}_j$. In order to measure the IAS, we use the fitted local regression to map the $\mathbf{x}_j$ values into $f_{j,\mathrm{ALE}}(\mathbf{x}_j)$ for $j = 1,\ldots,d$. As a result, this allows us to explain the fitted values using the intercept and the first-order approximation $f_{j,\mathrm{ALE}}(\mathbf{x}_j)$. According to (31.1), the equation can be seen as a regression in which we try to explain the fitted values using the first-order approximation. If there is a perfect match between the two, then it indicates the ML algorithm does not rely on any interaction between the variables and, hence, the first-order approximation enables us fully to understand the relationship between regressions and the response variables. Clearly, under such a scenario the regression R^2 is 1, indicating that IAS = 0.

For a non-linear model – such as ksvm – we do not expect the first-order approximation to fully explain the fitted values. The regression results in an $R^2 < 1$, leading to a non-zero component corresponding to interactions. For this reason, Molnar et al. (2019) define the IAS as $1 - R^2$. Consistent with the panels from Figure 31.1, the values of IAS are, respectively, 0.00, 0.11, 0.04, and 0.22. This indicates that, in terms of interaction, the rf algorithm corresponds to the highest strength, followed by decision tree and ksvm. Obviously, the IAS for the Lasso algorithm is zero given the fact that the fitted value is a linear map of the feature space.

31.2.4 Main effect complexity (MEC)

Given the above discussion, we proceed to discussing the idea behind the Main Effect Complexity (MEC) proposed in Algorithm 2 of Molnar et al. (2019). In principle, the proposed metric measures the shape complexity of the $f_{j,\mathrm{ALE}}(\mathbf{x}_j)$ for $j = 1,\ldots,d$. This metric is denoted by MEC_j. To gain a single number for each model, the authors propose the following weighted measure:

$$\mathrm{MEC} = \sum_{j=1}^{d} w_j \times \mathrm{MEC}_j, \tag{31.7}$$

with

$$w_j = \frac{V_j}{\sum_{j=1}^{d} V_j}, \tag{31.8}$$

and with w_j corresponding to the weight attributed to the complexity of the $f_{j,\mathrm{ALE}}(\mathbf{x}_j)$. In particular, V_j measures the variability of the ALE approximation for feature j. For instance, from Figure 31.1, we note that such variability for the rf algorithm is larger than that for the lasso model. Nonetheless, such weighting is conducted within the model for different features. On the other hand, MEC_j is defined as the number of parameters needed to approximate the curve with piece-wise linear models.

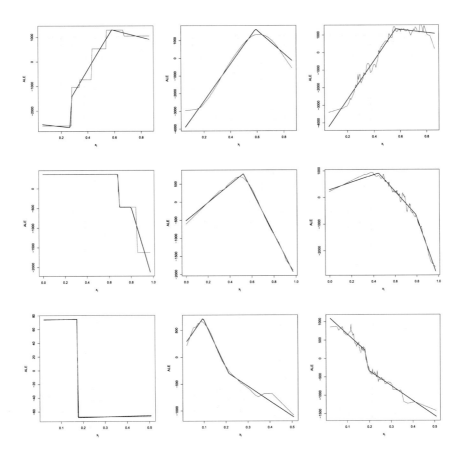

Figure 31.2 *Main Effect Complexity.* The figure provides an empirical illustration of the Accumulated Local Effects (ALE) by Apley and Zhu (2020). The red line denotes the first-order ALE approximation. The black line represents the curve fitted using a piece-wise linear regression. The more complex the model is the more parameters needed to fit the piece-wise regression in order to achieve a minimum level of 95% R-squared. In all cases, a given model is trained using the same data and features. The data corresponds to the Bike Rental from Fanaee-T and Gama (2014). The response variable denotes the number of rentals, whereas the explanatory variable is either temperature ($j = 1$), wind speed ($j = 2$), or humidity ($j = 3$). The panels below illustrate the ALE approximation for different ML algorithms. Panel (a) is Decision Tree. Panel (b) is a support vector regression with radial kernel, whereas Panel (c) corresponds to random forest. The panels after follow the same order in terms of algorithms, however, for different features. For each algorithm, the hyperparameters are tuned using 10-fold cross validation with three repetitions. The ML implementation is carried using Kuhn *et al.* (2008), whereas the ALE approximation is conducted using Apley (2018).

To demonstrate the idea behind MEC, we consider similar ALE functions from

Figure 31.1. Figure 31.2 demonstrates the fitted piece-wise linear regression needed to achieve an $R^2 = 95\%$.[5]. Starting with the wind speed variable. We make the following observations. First, the decision tree in Panel (a) indicates that three lines are needed in order to represent the corresponding ALE function. In this case, we observe that the piece-wise regression attains a perfect match of the ALE function. For the ksvm and rf, i.e. Panels (b) and (c), we also need three lines to represent the ALE function. Hence, in all cases the corresponding MEC_j value is 3. On the other hand, for the humidity measure, we observe more heterogeneity. In this case, the decision tree needs four segments to fit the ALE function, while the rf needs three lines. On the other hand, we note that the ksvm needs only two.

Table 31.1 *Measuring Complexity.* This table reports the complexity results for seven different machine learning algorithms. NF denotes the number of features used in the model. IAS denotes the interaction strength computed with respect to the first-order ALE approximation from 31.1. MEC denotes the main effect complexity according to Equation (31.7).

	Model	MEC	IAS	NF
1	lasso	1.00	0.00	3
2	glmnet	1.00	0.00	3
3	svmLinear	1.00	0.00	3
4	svmRadial	2.08	0.04	3
5	rf	2.30	0.22	3
6	rpart	4.30	0.11	3
7	rpart2	4.13	0.11	3

In order to compute the final MEC for each algorithm, we weight each MEC_j for $j = 1, 2, 3$ with the corresponding variance of the ALE function. In Table 31.1, we report the MEC measure for each seven different algorithms investigated by Molnar et al. (2019). We start our discussion based on the MEC results. We note that the MEC measure for the three linear models (Lasso, glmnet, and svmLinear) is 1. This is expected given the fact that one segment is needed from the piece-wise regression to represent a linear function. On the other hand, we note that for the other non-linear models the values range between 2.08 and 4.13. The reason we get non-integers as a final measure is due to the weighting scheme. As Figure 31.2 illustrates, a different number of segments is needed for the same algorithm depending on the feature of interest. At the same time, the approximated ALE function for each feature exhibits different variability. For instance, the variability for decision tree algorithm is largest for the first feature which is about 84% – see Panel (a) from Figure 31.2. In this case, $MEC_1 = 4$. At

[5] We follow the same suggestion as Molnar et al. (2019) for the minimum level needed to achieve a good fit of the ALE approximation.

the same time, $\text{MEC}_2 = 5$ while $\text{MEC}_3 = 3$. Since $w_2 = 15\% > w_3 = 1\%$, the final MEC measure is closer to 5 than 3.

For the other metrics, we report the interaction intensity (IAS) in Table 31.1. For all linear models, IAS is zero. On the other hand, the ksvm has an IAS of 4%, whereas the tree-based models exhibit higher IAS. For instance, the random forest (rf) has the highest IAS among the seven algorithms. This is expected given the fact that the rf bootstraps multiple trees in finding the optimal model. In addition, we refer to Algorithm 1 from Molnar et al. (2019) to report the number of features (NF) used in the model. In all cases, the seven algorithms use the full feature space in the final model. In their empirical example, Molnar et al. (2019) use nine features including categorical variables. The NF result depends on the model as Table 1 from their paper illustrates.

31.3 Portfolio problem

We consider a simple portfolio problem for two assets, high and low risk. This is similar to the problem studied by Boudt et al. (2020) and Clark et al. (2020) that corresponds to tactical asset allocation. In their analysis, the authors focus on two exchange-traded funds (henceforth ETFs): equity and Treasury bonds, corresponding to high and low risk assets, respectively.

31.3.1 Volatility managed portfolio

Due to estimation risk concerns associated with mean returns (Merton, 1980), the portfolio literature focuses mostly on the global minimum variance portfolio (GMVP) that depends solely on the estimation of the covariance matrix (see, e.g., Ledoit and Wolf, 2003). Let $\hat{\Sigma}_{t+1|t}$ denote the forecast of the covariance matrix at month t over the subsequent month $t + 1$. The GMVP, hence, at time t, is given by

$$\mathbf{w}_t = \frac{\hat{\Sigma}_t^{-1}\mathbf{1}}{\mathbf{1}^\top\hat{\Sigma}_t^{-1}\mathbf{1}}, \tag{31.9}$$

with $\mathbf{1}$ is a column vector of 1s. In the case of two assets, Equation (31.9) takes the following form:

$$\mathbf{w}_t = \begin{bmatrix} \dfrac{\hat{\sigma}_{2,t+1|t}^2 - \hat{\rho}_{t+1|t}\hat{\sigma}_{1,t+1|t}\hat{\sigma}_{2,t+1|t}}{\hat{\sigma}_{1,t+1|t}^2 + \hat{\sigma}_{2,t+1|t}^2 - 2\hat{\rho}_{t+1|t}\hat{\sigma}_{1,t+1|t}\hat{\sigma}_{2,t+1|t}} \\ \dfrac{\hat{\sigma}_{1,t+1|t}^2 - \hat{\rho}_{t+1|t}\hat{\sigma}_{1,t+1|t}\hat{\sigma}_{2,t+1|t}}{\hat{\sigma}_{1,t+1|t}^2 + \hat{\sigma}_{2,t+1|t}^2 - 2\hat{\rho}_{t+1|t}\hat{\sigma}_{1,t+1|t}\hat{\sigma}_{2,t+1|t}} \end{bmatrix}, \tag{31.10}$$

where $\hat{\sigma}_{i,t+1|t}^2$ denotes the volatility forecast for asset $i = 1, 2$ at month t for the subsequent month. At the same time, $\hat{\rho}_{t+1|t}$ denotes the forecast of the correlation between the two at time t. One can think about the volatility forecast in line with GARCH models, whereas the correlation forecast in a similar manner as dynamic conditional correlations (DCC).

If we assume that the correlation between the two assets is zero, then the portfolio problem simplifies to

$$\mathbf{w}_t = \begin{bmatrix} \dfrac{1/\hat{\sigma}^2_{1,t+1|t})}{(1/\hat{\sigma}^2_{1,t+1|t})+(1/\hat{\sigma}^2_{2,t+1|t})} \\[2ex] \dfrac{1/\hat{\sigma}^2_{2,t+1|t})}{(1\hat{\sigma}^2_{1,t+1|t})+(1\hat{\sigma}^2_{2,t+1|t})} \end{bmatrix} \tag{31.11}$$

The portfolio weights from (31.11) provide a simple intuition. If the forecast of Asset 1 is relatively higher (lower) compared to Asset 2, then the portfolio tilts less (more) weights to Asset 1 relative to Asset 2. Note that in either portfolio definition, the portfolio weights sum to 1. In fact, the result from (31.11) is similar to the definition of the volatility-managed portfolio by Moreira and Muir (2017). In their formation, the weight allocated to asset i is

$$\mathbf{w}_{i,t} = \frac{c}{\hat{\sigma}^2_{i,t+1|t}} \tag{31.12}$$

for some constant c. Clearly, if we put

$$c = \frac{1}{(1/\hat{\sigma}^2_{1,t+1|t}) + (1\hat{\sigma}^2_{2,t+1|t})} \tag{31.13}$$

then we get the result from Equation (31.11).

In this chapter, we focus on a formulation similar to that in Moreira and Muir (2017) for an arbitrary volatility forecasting problem. This provides a parsimonious framework for estimating the underlying parameters and, hence, constructing our portfolio. Specifically, our analysis focuses on forecasting the volatility of the high risk asset. Given the definition of (31.12) and an arbitrary value of c, the portfolio construction is straightforward. For instance, if one chooses historical volatility as the estimate for the SPY volatility, which is approximately 20% according to our sample, then the weight allocated to the SPY is roughly 25% for $c = 0.01$. Since the weight allocated to the other asset is the complement, then the portfolio formation suffices to derive forecasts for the high risk asset. In this research, we set $c = 0.01$ for all ML algorithms. In order to derive the optimal forecast, we refer to the appeal of ML in the discussion below.

31.3.2 The appeal of ML in portfolio

In typical GARCH formulation, the volatility forecast is written as an autoregressive function of the previous volatility as well as the return residuals. Motivated by this, we reformulate a similar principle from the ML point of view. Let $\mathbf{x}_t$ denote the features available at time t, while y_{t+1} denotes the response variable. In GARCH analysis, the feature space is given by the realized variance, σ_t^2, and the return residual, ϵ_t. On the other hand, the response variable is σ_{t+1}^2. Using maximum likelihood estimation, one yields the optimal forecast for $\hat{\sigma}_{t+1|t}$. In principle, the GARCH model finds the optimal mapping function $f : \mathbf{x}_t \to \hat{\sigma}_{t+1|t}$. In similar fashion, for a time series of realized volatilities, the ML provides greater flexibility in mapping the feature space into forecast.

While volatility is autoregressive, other important features for the same asset or other assets may provide first-order importance in determining future volatility. In a similar way, risk factors that can be mapped through the process could provide greater perspective on future volatility, e.g. VIX. We discuss these features more in §31.4. For now, we stress the appeal of ML into the portfolio formation problem. For a given ML algorithm a, there is an optimal predictive function that maps the current feature space $\mathbf{x}_t$ into $\hat{\sigma}_{t+1|t}(a)$. We denote the forecast as a function of the ML algorithm index a to study their complexity and portfolio performance. Clearly, for a given algorithm a, the portfolio weights at time t are given by

$$\mathbf{w}_t(a) = \begin{bmatrix} c \left[\hat{\sigma}^2_{t+1|t}(a) \right]^{-1} \\ 1 - c \left[\hat{\sigma}^2_{t+1|t}(a) \right]^{-1} \end{bmatrix}. \tag{31.14}$$

Hence, for a fixed c and ML algorithm a, the portfolio weights are given by $\mathbf{w}_t(a)$. For brevity, we suppress the subscript from $\hat{\sigma}^2_{t+1|t}(a)$ to denote the volatility forecast of the high risk asset, SPY. Note that the first and second elements of the vector $\mathbf{w}_t(a)$ correspond to the weight allocated to the high and low risk assets, respectively. Additionally, it is evident that the portfolio weights in this formulation requires only one volatility forecast for the high risk asset volatility.

The main idea behind this illustration is that each ML algorithm yields a sequence of volatility forecast. Given which, we can evaluate both the statistical and economic performance of each ML algorithm. For the former, we compare the forecast volatility with the realized one and report conventional measures such as R-squared or MSE. For the latter, the portfolio formation from (31.14) allows us to evaluate it out-of-sample (OOS) performance, e.g. portfolio volatility and value-at-risk. At the same time, given the complexity associated with each algorithm, we can "price" the complexity in terms of OOS portfolio performance. We devote the next section to discussion.

31.4 Empirical investigation

In this section, we discuss the data collection, forecasting, portfolio formation and performance evaluation. Relying upon the performance evaluations, we derive similar complexity measures to those considered in §31.2. Eventually, we examine the portfolio performance results in relationship with model complexity.

31.4.1 Data

Similar to Boudt et al. (2020) and Clark et al. (2020), our empirical investigation relies on several indicators for forecasting the next month's market volatility. In particular, we focus on the future volatility of S&P 500 ETF (henceforth SPY), which corresponds to the main response variable, i.e. $\hat{\sigma}^2_{t+1|t}(a)$. Hence SPY ETF is the focus of our analysis. Additionally, we refer to five different securities to construct the main feature space:

1. IEF: iShares 7–10 Year Treasury Bond ETF
2. XLF: Financial Select Sector SPDR Fund
3. GLD: Gold Shares ETF by State Street
4. VIX: CBOE Volatility Index

For each security, we compute the following: monthly price change (return): five features; realized volatility: five features; change in monthly volume – except for VIX: four features; difference between return and its 12-month moving average (MA): five features; ratio between volatility and its 12-month moving average (MA): five features; and difference between volume change and its 12-month moving average (MA): four features. The final feature space, hence, consist of 28 variables, i.e. $\mathbf{x}_t \in \mathbb{R}^{28}$.

In Figure 31.3, we present the correlation coefficients between the feature space and the response variable, i.e. the next month realized volatility. We group the correlation matrices with respect to each of the six categories mentioned above.

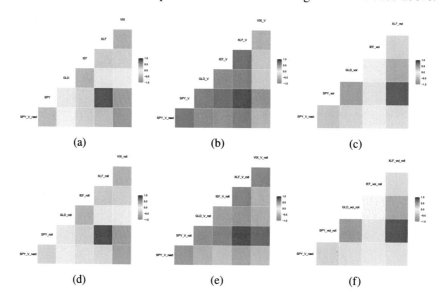

Figure 31.3 *Correlation between Features and Future Realized Volatility.* The parts report the correlation between the future realized volatility and the followings features, respectively: (a) lagged returns; (b) lagged volatility; (c) lagged volume; (d) lagged MA returns; (e) lagged MA volatility; and (f) lagged MA volume.

31.4.2 Training and testing

The training period is between Jan 2006 and Dec 2015, and the testing period is between Jan 2016 and Dec 2020, included. Given the training period, we find the optimal mapping function $f_a : \mathbf{x}_t \rightarrow \hat{\sigma}^2_{t+1|t}(a)$ for all months in the training period. Given this mapping function, we form forecast for the following month volatility using current previous data. For instance, our first forecast is constructed for Jan 2016. This is based on recent $\mathbf{x}_t$ from Dec 2015. In the subsequent month, we repeat the process until the end of the sample. We note that in portfolio selection literature it is common to perform rolling window estimation. Under a rolling a window, there would be a time series of complexity measures. To keep the results comprehensive and the computations not too demanding, we do not pursue a rolling window in this chapter.

For tuning, we consider a sliding window of size T, whereas the holdout window is H. Given the training the sample of 120 months, for instance, we consider the first T months for in-sample and the next H months for validation. The window is rolled over the next the month the procedure is repeated. The end of this procedure results in a series of validation observations. Based on which we find the optimal hyperparameters that minimize/maximize a given performance metric, returning the optimal mapping function. Our analysis considers different specifications. For each model, we tune the hyper-parameters based on three different criteria: RMSE, R-squared, or MAE. Additionally, we consider different time slices for tuning. For the initial sample window, T, we consider either 24, 36, 48, or 60 months. For the horizon, H, which corresponds for the holdout window, we consider either 3, 6, 9, or 12 months. In total this results in $3 \times 4 \times 4 = 48$. Hence, for 7 algorithms this experiment corresponds to 336 final performance and complexity metrics.

31.4.3 Performance

We report the complexity measures discussed in §31.2. In order to compute an algorithm-specific complexity measure, we aggregate the three measures in the following way. First, we standardize the values of each measure with respect to our sample, such that each measure has mean zero and standard deviation of one. Second, we consider the sum of the three values to come up a single measure for each algorithm-specification. In this case, for specific algorithm and experiment specification, we define complexity as

$$COM = Z_{\mathrm{MEC}} + Z_{\mathrm{IAS}} + Z_{\mathrm{NF}}, \tag{31.15}$$

with Z_v denoting the standardized score of measure v.

Our results consider both statistical and economic performance. For the statistical performance, we compute both the in-sample and out-of-sample coefficient of determination denoted by R_{in}^2 and R_{out}^2, respectively. In terms of economic performance, we evaluate algorithms using the portfolio results. For each algorithm, we form a portfolio based on the corresponding forecast $\hat{\sigma}_{t+1|t}^2(a)$ and evaluate its performance out-of-sample. We use common metrics used in the literature. Those are the annual mean return and volatility. In terms of risk-adjusted performance, we compute the annualized Sharpe ratio. For downside risk, we consider both value-at-risk (VaR) and expected shortfall (ES). The former is computed as the mean return minus the 5% percentile, whereas the latter is computed as the portfolio mean return conditioned on that the return drops below the empirical 5% percentile of the testing period. In addition, we compute the portfolio turnover

(TO) as the average portfolio monthly turnover. To take into account transaction cost, we penalize portfolio returns for turnover. In particular, for a k change in position, we subtract $k \times 20$ basis points from the portfolio return at each period when k is non-zero.[6]

31.4.4 Results and discussion

Table 31.2 shows the complexity and portfolio performance results for each of the seven different algorithms. Panels (a) and (b) shows the mean and standard deviation complexity and performance metric by algorithm. The algorithms we test are as follows:

1. *lasso*: Lasso regression.
2. *glmnet*: Elastic net regression.
3. *svmLinear*: Support vector machine with a linear kernel.
4. *svmRadial*: Support vector machine with a radial kernel.
5. *rf*: Random forest.
6. *rpart*: CART decision tree with complexity pruning.
7. *rpart2*: CART decision tree with maximum tree depth.

Overall, the linear models tend to be less complex as the LASSO and elastic net regressions are the least complex, followed by the linear support vector machines (SVM). The most complex models are the random forest and radial SVM models. Indeed, these results are consistent with our priors.

In terms of performance, we plot the relations between complexity and performance in Figure 31.4. We measure performance using the six common metrics discussed above. The linear regression models tend to perform better as they become more complex as shown by the positively sloped $R^2 in$ and R^2_{out} metrics in panels (a) and (b) of Figure 31.4, where each color represents an algorithm and the points represent each of the 48 different cases we tested. There is a substantial amount of variability both within and across algorithms. To test the statistical importance of these relations, we run several regressions and report the results in Tables 31.3 through 31.5.

Tables 31.3 and 31.4 show the results of linear regressions of performance on complexity. Table 31.3 includes model fixed effects and Table 31.4 does not. Overall, the coefficients on complexity are positive and significant in both tables which suggests that measures of risk and return are generally increasing with model complexity. Inspecting the plots in Figure 31.4, it appears that there is a nonlinear relation between performance and complexity. To test for this possibility, we include a quadratic term for complexity and show the results in Table 31.5. There is a convex relation between each of the performance metrics and complexity; complexity increases model performance (R_2) and risk up to a point until it actually starts to decline. However, it is difficult to draw any strong conclusions from the above results because we are only comparing one

[6] By definition of portfolio turnover, k is non-negative.

Table 31.2 *Complexity and Portfolio Performance.* This table reports different statistics and measures. The first four measures correspond to the complexity approximation discussed in §31.2. The measures MEC, IAS, and NF are computed with respect to Molnar et al. (2019) and they correspond, respectively, to the main effect complexity, the interaction strength, and the number of features used in the model. COM is a single metric that into account all three measures to provide a single complexity assessment of each algorithm – see Equation (31.15). The statistics R^2_{in} and R^2_{out} denote the statistical performance of each model. The former is in-sample, where the fitted values are regressed against the true in-sample values. On the other hand, the latter is out-of-sample, where the true values from the testing sample are regressed against the forecasts. The testing period dates between Jan 2016 and Dec 2020, included, whereas the training period dates between Jan 2006 and Dec 2015. The fianl six columns correspond to the portfolio performance based on the testing period forecasts. For each algorithm, we form a portfolio based on the corresponding forecasts and evaluate its performance out-of-sample. The Mean and Std denote the annual mean return and volatility; SR stands for the annualized Sharpe ratio; ES is the expected shortfall computed as mean portfolio returns conditioned on that the return drops below the empirical 5% percentile of the testing period. Finally, TO stands for the average portfolio monthly turnover. In all cases, portfolio returns are penalized for turnover by taking into considerations transaction cost of 20 basis points.

	Model	Complexity Measures						Portfolio Performance					
		MEC	IAS	NF	COM	R^2_{in}	R^2_{out}	Mean	Std	SR	VaR	ES	TO
					Panel (a) Average by Algorithm								
1	lasso	1.00	0.00	15.27	−1.33	0.71	0.21	0.10	0.08	1.33	0.47	0.70	0.37
2	glmnet	1.00	0.00	5.17	−2.17	0.65	0.12	0.09	0.06	1.45	0.41	0.46	0.22
3	svmLinear	1.00	0.00	28.00	−0.28	0.75	0.24	0.12	0.12	0.99	0.51	1.24	0.50
4	svmRadial	2.12	0.15	28.00	1.91	0.68	0.19	0.09	0.07	1.27	0.43	0.66	0.48
5	rf	3.88	0.14	28.00	3.21	0.92	0.26	0.09	0.07	1.32	0.41	0.58	0.39
6	rpart	0.87	0.12	0.56	0.33	0.13	0.03	0.09	0.06	1.48	0.34	0.37	0.08
7	rpart2	3.15	0.15	2.12	0.53	0.42	0.12	0.09	0.07	1.23	0.42	0.52	0.33
					Panel (b) Standard Deviation by Algorithm								
1	lasso	0.00	0.00	5.22	0.43	0.02	0.08	0.01	0.01	0.05	0.01	0.09	0.05
2	glmnet	0.00	0.00	3.27	0.27	0.04	0.04	0.00	0.01	0.06	0.04	0.09	0.09
3	svmLinear	0.00	0.00	0.00	0.00	0.00	0.00	0.00	0.00	0.00	0.00	0.00	0.00
4	svmRadial	0.20	0.03	0.00	0.33	0.12	0.02	0.00	0.01	0.08	0.02	0.07	0.07
5	rf	0.69	0.01	0.00	0.51	0.01	0.06	0.00	0.00	0.12	0.01	0.05	0.04
6	rpart	1.46	0.17	1.09	1.73	0.23	0.05	0.00	0.01	0.17	0.05	0.10	0.15
7	rpart2	0.28	0.20	1.59	2.05	0.15	0.02	0.00	0.00	0.06	0.02	0.03	0.13

dimension at a time to complexity. In other words, the relation between risk, return, and model complexity is difficult to fully capture without examining all three dimensions simultaneously. For this reason, we next explore model complexity and performance from the perspective of a risk manager who is choosing amongst the set of portfolios estimated in this section in an attempt to capture the relation between the classic risk-return tradeoff and model complexity.

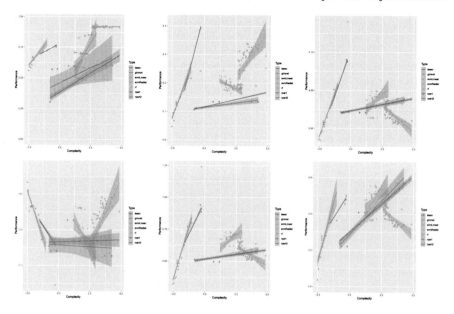

Figure 31.4 *Performance vs. Complexity.*

Table 31.3 *Panel Regression.* This table reports the results of panel regression in which the dependent is a performance measure. The main regressor is the combined complexity measure. In all cases, the regression control for model fixed effects.

			Dependent variable:			
	R^2_{in}	R^2_{out}	Mean	Std	ES	TO
	(1)	(2)	(3)	(4)	(5)	(6)
COM	0.081***	0.023***	0.001***	0.002***	0.033***	0.068***
	(0.004)	(0.002)	(0.0002)	(0.0002)	(0.003)	(0.003)
Constant	0.789***	0.235***	0.078***	1.332***	0.727***	0.454***
	(0.010)	(0.006)	(0.001)	(0.012)	(0.010)	(0.008)
Observations	336	336	336	336	336	336
Adjusted R^2	0.925	0.819	0.881	0.939	0.949	0.902

Notes: *p<0.1; **p<0.05; ***p<0.01

31.4.5 Additional results

The previous analysis is built on 336 different specifications. For each, there is time series of out-of-sample portfolio returns. One can view these results as a set of different portfolios/assets.[7] Given which, we address the following the issues.

---

[7] We note that the median correlation among these portfolios is 89% , ranging between 58% and 100%. While all portfolios share the same assets, then differ in the weight allocated to the risky asset. If the

Table 31.4 *Panel Regression without Algorithm Fixed Effects.* This table reports the results of panel regression in which the dependent is a performance measure. The main regressor is the combined complexity measure.

	R^2_{in} (1)	R^2_{out} (2)	Mean (3)	Std (4)	ES (5)	TO (6)
			Dependent variable:			
COM	0.073*** (0.005)	0.025*** (0.002)	0.0001 (0.0002)	0.001*** (0.0004)	0.023*** (0.007)	0.050*** (0.003)
Constant	0.605*** (0.011)	0.169*** (0.004)	0.095*** (0.001)	0.075*** (0.001)	0.646*** (0.015)	0.338*** (0.006)
Observations	336	336	336	336	336	336
R^2	0.384	0.403	0.0004	0.025	0.037	0.485
Adjusted R^2	0.382	0.402	−0.003	0.022	0.034	0.483

Notes: *p<0.1; **p<0.05; ***p<0.01

Table 31.5 *Panel Regression without Algorithm Fixed Effects with Second-Order Complexity.* This table reports the results of panel regression in which the dependent is a performance measure. The regressors are the combined complexity measure and the complexity measure squared.

	R^2_{in} (1)	R^2_{out} (2)	Mean (3)	Std (4)	ES (5)	TO (6)
			Dependent variable:			
COM	0.088*** (0.005)	0.031*** (0.002)	0.001*** (0.0002)	0.003*** (0.0004)	0.051*** (0.006)	0.063*** (0.002)
COM^2	−0.018*** (0.002)	−0.007*** (0.001)	−0.001*** (0.0001)	−0.002*** (0.0002)	−0.032*** (0.002)	−0.016*** (0.001)
Constant	0.696*** (0.015)	0.203*** (0.005)	0.099*** (0.001)	0.085*** (0.001)	0.807*** (0.017)	0.417*** (0.007)
Observations	336	336	336	336	336	336
R^2	0.497	0.552	0.182	0.331	0.371	0.717
Adjusted R^2	0.494	0.549	0.177	0.327	0.367	0.715

Notes: *p<0.1; **p<0.05; ***p<0.01

Consider a risk averse decision-maker who is aware of complexity. In order to

ML algorithm is unaffected by different specification, then we have the same portfolio. However, in other cases as Figure 31.4 illustrates, there is a large heterogeneity among portfolios depending on the algorithm.

find the optimal allocation among these assets, she chooses the portfolio that yields the highest risk-reward-complexity trade-off.

To formalize the above problem, consider the following optimization problem:

$$\min_{\xi} \quad \xi^\top \Omega \xi$$

$$\text{s.t.} \quad \xi^\top \mathbf{1} = 1$$

$$\xi^\top \eta \geq l \qquad\qquad (31.16)$$

$$\xi^\top \gamma \leq u$$

with $\Omega \in \mathbb{R}^{336 \times 336}$ denoting the covariance matrix of the portfolio returns, $\eta \in \mathbb{R}^{336}$ the vector of mean returns of the portfolio returns, and $\gamma \in \mathbb{R}^{336}$ is the ML complexity associated with each portfolio. The objective is to minimize the portfolio variance subject to three constraints. The first is that that portfolio weights sum to 1. The second corresponds to the mean target denoted by l. The third is the maximum complexity level (denoted by u) that the agent is willing to bear.

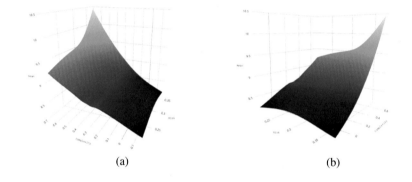

(a)   (b)

Figure 31.5 *Mean-Variance-Complexity Efficient Frontier.* This figure illustrates the mean-variance-complexity efficient frontier correspond to the optimization problem from (31.16). The surface is computed using a range of values for mean and complexity targets. The z-axis denotes the mean return, whereas the x- and y-axes denote risk and complexity. In Panel (a) (respectively (b)), complexity is on the bottom left-hand (right-hand) side axis.

The optimization problem is (31.16) is a typical mean-variance optimization with an additional constraint. For instance, for $u \to \infty$, then the optimal portfolio corresponds to the conventional mean-variance portfolio (Markowitz, 1952). We solve the optimization numerically using gradient descent methods with linear constraints. We consider a range of 100 values for each of l and u, resulting in 10,000 combinations. For each single combination, we compute solve for efficient portfolio denoted by ξ^*. In all cases, we set the initial guess as the equally portfolio. This results in a series of optimal portfolios, each of has a mean, variance, and complexity. Based on this, we interpolate the mean returns on the other two. This allows to derive a mean-variance-complexity surface. We summarize this surface in Figure 31.5.

A couple of comments follow from Figure 31.5. First, either panel indicates that there is a positive-trade-off between complexity and mean return. This trade-off is stronger when the agent is willing to take more risk. Second, it is only when complexity is high enough we observe a positive risk-reward trade-off. For instance, either perspective illustrates an inverse U-shaped function between mean and risk when complexity is low. However, when complexity is high, we observe the conventional positive risk-reward trade-off.

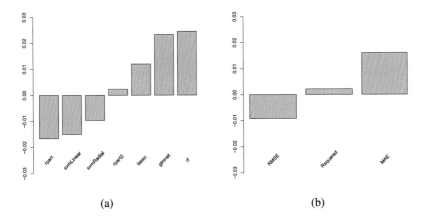

(a)   (b)

Figure 31.6 *Mean-Variance-Complexity Efficient Portfolio Weights.* This figure illustrates the average weight allocated to each of the 336 portfolios. The summary below provides an average per algorithm in Panel (a) and per metric in Panel (b).

In Figure 31.6, we demonstrate the average portfolio weight from (31.16) for each algorithm. By design the weights are attained to maximize the mean-variance-complexity trade-off. Hence, to better understand the contribution of each algorithm to this trade-off, we illustrate the average weight with respect to the algorithm type. In addition, we also report the average by metric. Overall, Figure 31.6 provides two important takeaways. First, consistent with Figure 31.4, we note that the weight allocated to rf, LASSO, and glmnet are positive; whereas, the weights attributed to support vectors is negative. Second, in terms of tuning metric, we observe that the mean-absolute error (MAE) yields higher trade-off. One potential explanation is fact that MAE takes the absolute deviation rather than the squared one. In our analysis, squared deviation are larger than absolute deviation since we consider the natural log of realized volatility. Hence, we expect the MSE to amplify such errors due to outliers.

31.5 Concluding remarks

We explore the relation between model performance and model complexity in the the context of a volatility managed portfolio. In constructing a mean-variance-complexity efficient frontier, we show how model complexity affects the traditional mean-variance efficient frontier. We find that for any level of risk, there is a positive relation between model complexity and portfolio return. Moreover, we show that model complexity affects the classic mean-variance efficient frontier. For low levels of complexity, the risk-return tradeoff is an inverted u-shape, while at high levels of complexity the risk-return tradeoff is positive.

Among the seven ML models explored in this chapter, random forests is deemed as the only algorithm that yields consistent results in terms of positive performance-complexity tradeoff. Future research should take into considerations a number of aspects. First, the robustness of our findings over a longer testing period and larger number of ML algorithms. Second, estimation risk could be potentially mitigated by tuning ML algorithms on a more regular basis. Due to computational constraints, the investigation in this chapter was restricted to a single period tuning. Hence, whether higher tuning frequency decreases estimation risk and, hence, out-of-sample performance is of an empirical investigation. We leave both for future research.

Implementation

The implementation of this chapter was conducted using R (R Core Team, 2020) available at `https://github.com/simaan84/pricing_model_complex`.

Acknowledgements

We are grateful for useful comments from the editors Charles-Albert Lehalle and Agostino Capponi. We also thank the R and open-source community. To mention a few, the production of our results greatly benefited from the following packages: Kuhn *et al.* (2008); Apley (2018); Muggeo *et al.* (2008); Ryan and Ulrich (2019); Sievert (2018); Wickham (2016). The opinions expressed in this chapter do not represent those of the Office of the Controller of the Currency and do not, explicitly or implicitly, establish any expectations or procedures.

References

Apley, Dan. 2018. *ALEPlot: Accumulated Local Effects (ALE) plots and Partial Dependence (PD) plots*. R package version 1.1.

Apley, Daniel W., and Zhu, Jingyu. 2020. Visualizing the effects of predictor variables in black box supervised learning models. *J. Royal Statistical Society: Series B (Statistical Methodology)*, **82**(4), 1059–1086.

Arrieta, Alejandro Barredo, Díaz-Rodríguez, Natalia, Del Ser, Javier, Bennetot, Adrien, Tabik, Siham, Barbado, Alberto, García, Salvador, Gil-López, Sergio, Molina, Daniel, Benjamins, Richard, et al. 2020. Explainable artificial intelligence (XAI): Concepts, taxonomies, opportunities and challenges toward responsible AI. *Information Fusion*, **58**, 82–115.

Athey, Susan. 2018. The impact of machine learning on economics. Pages 507–547 of: *The Economics of Artificial Intelligence: An Agenda*. University of Chicago Press.

Best, Michael J., and Grauer, Robert R. 1991. On the sensitivity of mean-variance-efficient portfolios to changes in asset means: some analytical and computational results. *Review of Financial Studies*, **4**(2), 315–342.

Boudt, Kris, Cela, Muzafer, and Simaan, Majeed. 2020. In search of return predictability: Application of machine learning algorithms in tactical allocation. Pages 35–73 of: *Machine Learning for Asset Management: New Developments and Financial Applications*, Emmanuel Jurczenko (ed). Wiley.

Brown, S. J. 1979. The effect of estimation risk on capital market equilibrium. *Journal of Financial and Quantitative Analysis*, **14**, 215–220.

Bücker, Michael, Szepannek, Gero, Gosiewska, Alicja, and Biecek, Przemyslaw. 2020. Transparency, auditability and eXplainability of machine learning models in credit scoring. ArXiv:2009.13384.

Bussmann, Niklas, Giudici, Paolo, Marinelli, Dimitri, and Papenbrock, Jochen. 2020. Explainable machine learning in credit risk management. *Computational Economics*, **57**, 203–216.

Butaru, Florentin, Chen, Qingqing, Clark, Brian, Das, Sanmay, Lo, Andrew W., and Siddique, Akhtar. 2016. Risk and risk management in the credit card industry. *Journal of Banking & Finance*, **72**, 218–239.

Chen, S., and Brown, S. J. 1983. Estimation risk and simple rules for optimal portfolio selection. *Journal of Finance*, **38**, 1087–1093.

Clark, Brian, Feinstein, Zachary, and Simaan, Majeed. 2020. A machine learning efficient frontier. *Operations Research Letters*, **48**(5), 630–634.

DeMiguel, Victor, Garlappi, Lorenzo, and Uppal, Raman. 2009. Optimal versus naive diversification: How inefficient is the 1/N portfolio strategy? *Review of Financial Studies*, **22**(5), 1915–1953.

Doshi-Velez, Finale, and Kim, Been. 2017. *Towards a rigorous science of interpretable machine learning*. ArXiv:1702.08608.

Fanaee-T., Hadi, and Gama, Joao. 2014. Event labeling combining ensemble detectors and background knowledge. *Progress in Artificial Intelligence*, **2**(2–3), 113–127.

Gu, Shihao, Kelly, Bryan, and Xiu, Dacheng. 2020. Empirical asset pricing via machine learning. *Review of Financial Studies*, **33**(5), 2223–2273.

Hind, Michael. 2019. Explaining eXplainable AI. *XRDS: Crossroads, The ACM Magazine for Students*, **25**(3), 16–19.

Khandani, Amir E., Kim, Adlar J., and Lo, Andrew W. 2010. Consumer credit-risk models via machine-learning algorithms. *Journal of Banking & Finance*, **34**(11), 2767–2787.

Kuhn, Max, et al. 2008. Building predictive models in R using the caret package. *J. Stat Softw*, **28**(5), 1–26.

Ledoit, Olivier, and Wolf, Michael. 2003. Improved estimation of the covariance matrix of stock returns with an application to portfolio selection. *Journal of Empirical Finance*, **10**(5), 603–621.

Lei, Tao, Barzilay, Regina, and Jaakkola, Tommi. 2016. Rationalizing neural predictions. ArXiv:1606.04155.

Li, Xiao-Hui, Cao, Caleb Chen, Shi, Yuhan, Bai, Wei, Gao, Han, Qiu, Luyu, Wang, Cong, Gao, Yuanyuan, Zhang, Shenjia, Xue, Xun, et al. 2020. A survey of data-driven and knowledge-aware eXplainable AI. *IEEE Transactions on Knowledge and Data Engineering*, **34**(1), 29–49.

Liu, Yang. 2019. Novel volatility forecasting using deep learning–long short term memory recurrent neural networks. *Expert Systems with Applications*, **132**, 99–109.

Markowitz, Harry. 1952. Portfolio selection. *Journal of Finance*, **7**(1), 77–91.

Merton, Robert C. 1980. On estimating the expected return on the market: An exploratory investigation. *Journal of Financial Economics*, **8**(4), 323–361.

Michaud, Richard O. 1989. The Markowitz optimization enigma: is 'optimized' optimal? *Financial Analysts Journal*, **45**(1), 31–42.

Molnar, Christoph. 2019. *Interpretable Machine Learning*. Lulu.com.

Molnar, Christoph, Casalicchio, Giuseppe, and Bischl, Bernd. 2019. Quantifying model complexity via functional decomposition for better post-hoc interpretability. Pages 193–204 of: *Proc. Joint European Conference on Machine Learning and Knowledge Discovery in Databases*.

Moreira, Alan, and Muir, Tyler. 2017. Volatility-managed portfolios. *Journal of Finance*, **72**(4), 1611–1644.

Muggeo, Vito M., et al. 2008. Segmented: an R package to fit regression models with broken-line relationships. *R News*, **8**(1), 20–25.

Mullainathan, Sendhil, and Spiess, Jann. 2017. Machine learning: an applied econometric approach. *Journal of Economic Perspectives*, **31**(2), 87–106.

R Core Team. 2020. *R: A Language and Environment for Statistical Computing*. R Foundation for Statistical Computing, Vienna, Austria.

Ribeiro, Marco Tulio, Singh, Sameer, and Guestrin, Carlos. 2016. Model-agnostic interpretability of machine learning. ArXiv:1606.05386.

Ryan, Jeffrey A. and Ulrich, Joshua M. 2019. *quantmod: Quantitative Financial Modelling Framework*. R package version 0.4-15.

Sachan, Swati, Yang, Jian-Bo, Xu, Dong-Ling, Benavides, David Eraso, and Li, Yang. 2020. An explainable AI decision-support-system to automate loan underwriting. *Expert Systems with Applications*, **144**, 113100.

Sievert, Carson. 2018. Interactive Web-based Data Visualization with R, plotly, and shiny. `https://plotly-r.com/`.

Sirignano, Justin, Sadhwani, Apaar, and Giesecke, Kay. 2016. Deep learning for mortgage risk. ArXiv:1607.02470.

Sun, Hao, and Yu, Bo. 2020. Forecasting financial returns volatility: a GARCH-SVR model. *Computational Economics*, **55**(2), 451–471.

Wickham, Hadley. 2016. *ggplot2: Elegant Graphics for Data Analysis*. Springer.

Bayesian Deep Fundamental Factor Models

Matthew F. Dixon[a] and Nicholas G. Polson[b]

Abstract

Bayesian deep fundamental factor models are developed to automatically capture non-linearity and interaction effects in factor modeling. Uncertainty quantification provides interpretability with interval estimation, ranking of factor importances and estimation of interaction effects. With no hidden layers we recover a linear factor model and for one or more hidden layers, uncertainty bands for the sensitivity to each input naturally arise from the network weights. As evidence of superior ability to capture outliers, we use fundamental factor data representing 3290 assets in the Russell 1000 index, over monthly periods spanning December 1989 to December 2017. Our Bayesian deep fundamental factor model with 49 factors generates L_∞ test errors which are 30% lower than quadratic Lasso regression and 40% lower than OLS regression. We also measure the performance of a portfolio constructed from the top predicted asset returns over 2017 and generate information ratios which are approximately 1.5 times greater than quadratic Lasso regression and 5 times greater than OLS regression. Furthermore, we compare sector and factor tilts of the portfolios and demonstrate how factor sensitivity with confidence bands provides interpretability in our model.

32.1 Introduction

In this chapter, we present a Bayesian framework for deep fundamental factor (DFF) models. The key aspect is a methodology for interpretability in a Bayesian deep learner under mild restrictions on the network architecture which builds on the point-wise estimation framework of Dixon and Polson (2020). Our method explicitly identifies interaction effects and ranks the importance of the factors, both with uncertainty intervals. In the case when the network contains no hidden layers, we recover Bayesian linear regression and the only source of uncertainty in the factor sensitivities is from the parameters. For one or more hidden layers, we demonstrate how uncertainty intervals arise from the network weights *and* the distribution of the factor loadings. By conditioning on a single asset (or

[a] Illinois Institute of Technology
[b] ChicagoBooth, University of Chicago
    Published in *Machine Learning And Data Sciences For Financial Markets*, Agostino Capponi and Charles-Albert Lehalle © 2023 Cambridge University Press.

averaging over a portfolio of assets) at any period in time, we demonstrate how the uncertainty intervals of the factor sensitivities can be attributed solely to the parameter uncertainty. This then renders the model useful for assessing the model risk in a practical setting where portfolios are constructed and hedged by their factor exposures.

Deep learning

Deep learning applies hierarchical layers of hidden variables to construct non-linear predictors which scale to high dimensional input space. The deep learning paradigm for data analysis is algorithmic rather than probabilistic. Deep learning has been shown to 'compress' the input space by projecting the input variables into a lower dimensional space using auto-encoders, as in deep portfolio theory (Heaton et al., 2017). A related point-wise estimation approach introduced by Fan et al. (2017), referred to as *sufficient forecasting*, provides a set of sufficient predictive indices which are inferred from high-dimensional predictors. Their approach uses projected principal component analysis under a semi-parametric factor model and has a direct correspondence with deep learning.

Non-linearity

While high-dimensional data representation is one distinguishing aspect of machine learning over linear regression, it is not alone. Bayesian deep learning resolves predictor non-linearities and interaction effects without over-fitting through an implicit bias-variance tradeoff. As such, it provides a highly expressive regression model for complex data which relies on compositions of simple non-linear functions rather than additive functions.

32.1.1 Why Bayesian deep learning?

Artificial neural networks have a long history in financial modeling. Most recently, the literature has been extended to include deep neural networks (see, for example, Feng et al., 2018; Heaton et al., 2017; Chen et al., 2019). It is well-known that shallow neural networks are furnished with the universal representation theorem, which states that any shallow[1] feedforward neural network can represent all continuous functions (Hornik et al., 1989), *provided there are enough hidden units*. It has recently been shown that deep networks can achieve superior performance versus linear additive models, such as linear regression, while avoiding the curse of dimensionality (Poggio, 2016).

Feng et al. (2018) show that pointwise deep neural networks provide powerful expressability[2] when combined with regularization, however, their use in factor modeling presents some fundamental obstacles, one of which we shall address in this chapter, namely interpretability. Neural networks have been presented

---

[1] While there is no consensus on the definition of a shallow network, we shall refer to a network with one hidden layer as shallow.

[2] Expressability is a measure of the generality of the class of functions represented by the network. In the context of classification, expressibility is measured by VC dimension.

to the investment management industry as 'black-boxes'. As such they are not viewed as interpretable and their internal behavior can't be reasoned on statistical grounds. We add to the literature by introducing a method of interpretability for Bayesian deep factor models. In particular, the method characterizing the effect of parameter uncertainty on the factor sensitivities. One important caveat is that we do not attempt to solve the causation problem in economic modeling.

Bayesian deep learning has been shown to be a powerful tool for probabilistic inference (Graves, 2011; Kingma and Welling, 2013; Blundell et al., 2015; Gal, 2016; Polson and Sokolov, 2017). It is well known that pointwise deep networks are prone to overfitting – when fitted by pointwise estimation deep networks are also often incapable of correctly assessing the uncertainty in the training data and so make overly confident decisions about the correct prediction. Using cross validation to tune penalty terms on the weights may remedy this but at the expense of more computation. In the Bayesian framework, all weights in a deep network are represented by probability distributions over possible values, rather than learning a fixed value for each weight. In doing so, regularisation is made implicit via the prior on the weights and a compression cost in the loss function (Polson and Sokolov, 2017). For example, if the weights have a Gaussian prior, this results in L_2 regularisation of the weights. If, on the other hand, the weights have a Laplace prior, then L_1 regularisation is applied.

A second benefit is the rich representation of the forward map simply by model averaging over a ensemble of networks, where each network has its weights drawn from a shared, learned probability distribution (Blundell et al., 2015). For this reason, relatively simple architectures can result in highly accurate predictions in contrast to pointwise estimation which may require more complex architectures.

32.1.2 Connection with fundamental factor models

Barra factor models (see Rosenberg and Marathe, 1976; Carvalho et al., 2012) are appealing because of their simplicity and their economic interpretability, generating tradable portfolios. Under the assumption of homoscedasticity, factor realizations can be estimated in the Barra model by ordinary least squares regression[3]. OLS linear regression exhibits poor expressability and relies on the Gaussian errors being orthogonal to the regressors. Generalizing to non-linearities and incorporating interaction effects is a harder task.

Asset managers seek novel predictive firm characteristics to explain anomalies which are not captured by classical capital asset pricing and factor models. Recently a number of independent empirical studies, have shown the importance of using a higher number of economically interpretable pointwise predictors related to firm characteristics and other common factors (Moritz and Zimmermann, 2016; Harvey et al., 2015; Gu et al., 2018; Feng et al., 2018). Gu et al. (2018) analyze a dataset of more than 30,000 individual stocks over a 60-year period from 1957 to

---

[3] The Barra factor model is often presented in the more general form with heteroscedastic error but is not considered here.

2016, and determine over 900 baseline signals. Moritz and Zimmermann (2016); Chen et al. (2019); Gu et al. (2018) highlight the inadequacies of OLS regression in variable selection over high-dimensional datasets – in particular the inability to capture outliers. In contrast, deep neural network can explain more structure in stock returns because of their ability to fit flexible functional forms with many covariates. In particular, Feng et al. (2018) demonstrate the ability of a three-layer deep neural network to effectively predict asset returns from fundamental factors. Our work contributes to the growing body of research on the importance of deep learning for factor modeling by not only providing interpretability, as in Dixon and Polson (2020), but by providing uncertainty quantification. In contrast to Chen et al. (2019), we do not attempt to enforce the no-arbitrage constraint but share similiar goals in attempting to rank the importance of the factors, albeit with the additional benefit of having uncertainty intervals.

32.1.3 Overview

The benefits of our Bayesian fundamental factor framework can be summmarized as follows:

- **Automatically capture non-linearity** in asset returns without the need for combinatorially intractable augmentation of linear models with pairwise interaction terms, or feature engineering;
- **Identify** and **assess** the material impact of what is *not* being captured by linear and quadratic returns models, e.g. higher order effects, outliers etc;
- **Rank factor sensitivities** for economic interpretation of the model; and
- **Uncertainty quantification** of asset returns and factor sensitivities to characterize the model risk.

In this chapter we shall demonstrate using Russell 1000 index stocks that Bayesian deep fundamental factor models result in material differences over linear models such as OLS and Lasso. While the latter can be extended to capture high-order effects, the modification becomes combinatorially intractable with larger datasets – there are $\binom{n}{2}$ pairwise interaction terms for n-dimensional data. Additionally, these terms are inadequate for capturing higher-order effects which may be needed to capture outliers. Moreover such additive models are prone to the multi-collinearity problem. We shall further identify the most important factors in predicted portfolio returns and characterize the uncertainty due to model parameters at any point in time.

The rest of the chapter is outlined as follows. Section 32.2 introduces the linear Barra fundamental factor model and discusses the experimental design and various use cases in asset management. Section 32.3 introduces Bayesian deep learning and variational algorithms for training the network. Section 32.4 provides the connection with Bayesian deep learning and fundamental factor models. Section 32.5 introduces a sensitivity based interpretability approach which characterizes the uncertainty in factor sensitivities in our Bayesian model.

Section 32.6 demonstrates the application of our framework to Bayesian NN-based factor models. Finally, Section 32.7 concludes with a brief discussion and directions for future research.

32.2 Barra fundamental factor models

Rosenberg and Marathe (1976) introduced a cross-sectional fundamental factor model to capture the effects of macroeconomic events on individual securities. The choice of factors are microeconomic characteristics – essentially common factors, such as industry membership, financial structure, or growth orientation (Nielsen and Bender, 2010). We present the original model in slightly non-standard notation so as to later describe our deep fundamental factor model in Section 32.4.

The model setup is as follows: given observations of N asset prices, $\{\mathbf{p}_t\}_{t=0}^T$ and K fundamental factor loadings at times $\{B_t\}_{t=1}^T$, a common version of the Barra fundamental factor model expresses the linear relationship between the loadings and the N asset returns:

$$\mathbf{r_t} = F_{\theta_t}(B_t) + \epsilon_t = B_t\theta_t + \epsilon_t, \quad t = 1,\ldots,T, \tag{32.1}$$

where $B_t = [\mathbf{1} \mid \boldsymbol{\beta}_1(t) \mid \cdots \mid \boldsymbol{\beta}_K(t)]$ is the $N \times K + 1$ matrix of known factor loadings (betas): $\beta_{i,k}(t) := (\boldsymbol{\beta}_k)_i(t)$ is the exposure of asset i to factor k at time t. The factors are asset specific attributes such as market capitalization, industry classification, style classification. The term $\theta_t = [\alpha_t, f_{1,t}, \ldots, f_{K,t}]$ is the $K + 1$ vector of unobserved factor realizations at time t, including the intercept α_t. $\mathbf{r_t}$ is the N-vector of asset returns at time t, i.e. $\mathbf{r_t} = \mathbf{p}_t/\mathbf{p}_{t-1} - \mathbf{1}$. The errors are assumed independent of the factor realizations $\rho(f_{i,t},\epsilon_{j,t}) = 0$, for all i, j, t with homoschedastic Gaussian error, $\mathbb{E}[\epsilon_{j,t}^2] = \sigma^2$. This form of the Barra model is conventionally used for risk modeling, however, we shall adopt a slight variant of this model for a stock selection strategy based on predicted returns, as presented in Dixon and Polson (2020). The other application where the Barra model is used is for attribution and shall not be considered here. For completeness, we summarize the three main applications of this model:

1. **Risk**: The covariance of the asset returns is

$$\mathbb{E}[(\mathbf{r}_t(B_t) - \bar{\mathbf{r}}_t)(\mathbf{r}_t(B_t) - \bar{\mathbf{r}}_t)^\top]$$
$$= \mathbb{E}[(F_\theta(B_t) - \mathbb{E}[F_\theta(B_t)])(F_\theta(B_t) - \mathbb{E}[F_\theta(B_t)])^\top] + \sigma^2 I_N,$$

   for $t = 1,\ldots,T$ and $\bar{\mathbf{r}}_t = \mathbb{E}[\mathbf{r}_t] = \mathbb{E}[F_\theta(B_t)]$ is the *unconditional* expectation of the asset returns[4].

2. **Prediction**: The next period returns are projected on to factor loadings by the *conditional* expectation operator:

$$\hat{\mathbf{r}}_{t+1} = \mathbb{E}[\mathbf{r}_{t+1} \mid B_t] = \mathbb{E}[F_\theta(B_t) \mid B_t], \quad t = 1,\ldots,T - 1, \tag{32.2}$$

---

[4] A common requirement is to then apply Euler's method to decompose the asset return covariance from factor covariances. See, for example, Zivot and Wang (2003) for further details.

to predict the next period asset returns from the factor loadings.

3. **Attribution**: The current period returns are projected on to factor loadings by the *conditional* expectation operator:

$$\hat{\mathbf{r}}_t = \mathbb{E}[\mathbf{r}_t \mid B_t] = \mathbb{E}[F_\theta(B_t) \mid B_t], \quad t = 1,\ldots,T, \qquad (32.3)$$

to identify the relative importance of the factors in explaining realized asset returns.

Attribution and prediction are practically identical in terms of modeling methodology and hence can be treated as one. Under the empirical distribution of the factor loadings, risk modeling requires estimating the sample covariance of the asset returns from history which only simplifies to estimating sample covariances of factor loadings when the model is linear. The extension to sample covariance matrices of a non-linear function of the factor loadings requires estimating $N \times N$ covariance matrices rather than $K \times K$ factor covariance matrices. Furthermore, how best to subsequently perform an Euler decomposition, given the non-linearity in the factors, shares the same difficulties as delta-gamma models used for market risk models (Britten-Jones and Schaefer, 1999) and is an open question left to future work.

32.2.1 Prediction with the Barra model

Prediction involves applying the Barra model, trained in the previous period, to a new observation B_t. The fitted map $F_{\hat{\theta}_{t-1}}(B_t) = B_t \hat{\theta}_{t-1}$ gives the next period returns

$$\hat{\mathbf{r}}_{t+1} = F_{\hat{\theta}_{t-1}}(B_t), \quad t = 2,\ldots,T-1. \qquad (32.4)$$

Figure 32.1 illustrates the experimental design for the predictive factor model: each factor model is fitted over period t and then tested over period $t + 1$. Each supervised training set is a factor loading matrix B_t and asset return N-vector $\mathbf{r}_t$. Similarly, each supervised test set is the pair $(B_{t+1}, \mathbf{r}_t)$. Note, for simplicity, that the estimation universe of N assets is assumed to be fixed from $t = 1,\ldots,T$.

The form of the statistical model (32.4) uses point-wise estimation of each asset return from its K factor loadings. Our goal is to cast the Barra model into a Bayesian deep learning framework. This framework has the benefit of providing uncertainty quantification in addition to representing the relationship between asset returns and factors in a more general way amenable to, for example, capturing outliers. In the next two sections, we shall present Bayesian deep learning and the training algorithm, before describing how we implement these techniques to train and predict asset returns using a Bayesian deep fundamental factor model in Section 32.4.

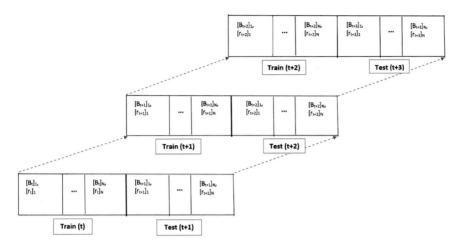

Figure 32.1 The experiment is designed so that the factor model is fitted over period t and then tested over period $t + 1$. Each training set is a factor loading matrix B_t and asset return N-vector $\mathbf{r}_t$.

32.3 Bayesian deep learning

Given i.i.d. data $\mathcal{D} := (X, Y) = \{\mathbf{x}_i, \mathbf{y}_i\}_{i=1}^n$, deep feedforward neural networks learn a parameterized forward map

$$\hat{Y} = F_\theta(X), \tag{32.5}$$

where F_θ is a superposition of L univariate *semi-affine* functions, $\sigma_{\theta^{(\ell)}}^{(\ell)}, \ell \in \{1, \ldots, L\}$:

$$F_\theta(X) = (\sigma_{\theta^{(L)}}^{(L)} \circ \cdots \circ \sigma_{\theta^{(1)}}^{(1)})(X), \tag{32.6}$$

and the unknown parameters $\theta = (W, b)$ are a set of weight matrices $W = (W^{(1)}, \ldots, W^{(L)})$ and a set of bias vectors $b = (b^{(1)}, \ldots, b^{(L)})$. Any weight matrix $W^{(\ell)} \in \mathbf{R}^{m \times n}$, can be expressed as n column m-vectors $W^{(\ell)} = [\mathbf{w}_{,1}^{(\ell)}, \ldots, \mathbf{w}_{,n}^{(\ell)}]$. We denote each weight as $w_{ij}^{(\ell)} := \left[W^{(\ell)} \right]_{ij}$.

The ℓth semi-affine function is itself defined as the composition of the activation function, $\sigma^{(\ell)}(\cdot)$, and an affine map:

$$\sigma_{\theta^{(\ell)}}^{(\ell)}(Z^{(\ell-1)}) := \sigma^{(\ell)} \left(W^{(\ell)} Z^{(\ell-1)} + b^{(\ell)} \right), \tag{32.7}$$

where $Z^{(\ell-1)}$ is the output from the previous layer, $\ell - 1$. The activation functions, $\sigma^{(\ell)}(\cdot)$, e.g. $\sigma^{(\ell)}(\cdot) = \max(\cdot, 0)$, are critical to non-linear behavior of the model. Without them, F_θ would be a linear map and, as such, would be incapable of capturing interaction effects between the inputs. This is true even if the network has many layers.

32.3.1 Deep probabilistic models

Polson and Sokolov (2017) show that the response Y can be viewed as a random variable being generated by a probability model $p(Y \mid \hat{Y}_\theta)$ for a deep learner $\hat{Y}_\theta(X) \equiv F_\theta(X)$. Given the network weights and biases θ, which in the context of a fundamental factor model no longer carry direct interpretability[5], the negative log-likelihood defines $\mathcal{L}$ as

$$\mathcal{L}(Y, \hat{Y}_\theta(X)) = -\log p(Y \mid \hat{Y}_\theta(X)).$$

which is a Gauss least squares error term, $\mathcal{L}(\mathbf{y}_i, \hat{\mathbf{y}}_\theta(\mathbf{x}_i)) = \|\mathbf{y}_i - \hat{\mathbf{y}}_\theta(\mathbf{x}_i)\|_2^2$. The procedure to obtain estimates $\hat{\theta}$ of the deep learning model parameters from $\mathcal{D}$ is described shortly. To control the predictive bias-variance trade-off we add a regularization term and optimize

$$\mathcal{L}_\lambda(Y, \hat{Y}_\theta(X)) = -\log p(Y \mid \hat{Y}_\theta(X)) - \log p(\theta \mid \lambda).$$

Probabilistically this is a negative log-prior distribution over parameters, namely

$$-\log p(\theta \mid \lambda) =\propto \lambda\phi(\theta),$$
$$p(\theta \mid \lambda) \propto \exp(-\lambda\phi(\theta)).$$

If this prior is a zero mean multivariate Gaussian with covariance $\frac{1}{2\lambda}$ then $\phi(\cdot)$ is the L_2 norm. Deep predictors are regularized maximum a posteriori (MAP) estimators, where

$$p(\theta \mid \mathcal{D}, \lambda) \propto p(Y \mid \hat{Y}_\theta(X))p(\theta \mid \lambda)$$
$$\propto \exp\left(-\log p(Y \mid \hat{Y}_\theta(X)) - \log p(\theta \mid \lambda)\right).$$

Training requires finding the optimal forward map as a solution of a highly nonlinear optimization

$$Y = F_{\hat{\theta}}(X) + \epsilon, \quad \text{where } \hat{\theta} := \arg\max_\theta \log p(\theta \mid \mathcal{D}),$$

and the log-posterior is optimised given the training data, $\mathcal{D}$ with

$$-\log p(\theta \mid \mathcal{D}, \lambda) = \sum_{i=1}^n \mathcal{L}(\mathbf{y}_i, \hat{\mathbf{y}}_\theta(\mathbf{x}_i)) + \lambda\phi(\theta).$$

Deep learning has the key property that $\nabla_\theta \log p(Y \mid \hat{Y}_\theta(X))$ is computationally inexpensive to evaluate using tensor methods for very complicated architectures and fast implementation on large datasets. `TensorFlow` and `TPUs` provide a state-of-the-art framework for a plethora of architectures. From a statistical perspective, one caveat is that the posterior of θ is highly multi-modal and providing good hyper-parameter tuning can be expensive. This is clearly a fruitful area of research for state-of-the-art stochastic Bayesian MCMC algorithms to provide more efficient algorithms (Polson and Sokolov, 2017). Note for ease of notation, throughout the remainder of this chapter, we shall drop the dependency on λ and refer to $\hat{Y} := \hat{Y}_\theta(X)$.

---

[5] A method for interpreting the deep factor model is presented in Section 32.5.1.

32.3.2 Variational approximation

Bayesian inference for neural networks estimates the posterior distribution of the fitted weights and biases given the training data, $p(\hat{\theta} \mid \mathcal{D})$. This distribution answers predictive queries about unseen data by taking expectations: the predictive distribution of an unknown response $\mathbf{y}'$ given a test feature vector $\mathbf{x}'$, is of the form $p(\mathbf{y}' \mid \mathbf{x}') = \mathbb{E}_{p(\theta\mid\mathcal{D})}[p(\mathbf{y}' \mid \mathbf{x}', \theta)]$. Each possible configuration of the weights, weighted according to the posterior distribution, makes a prediction about the unknown response given the test feature vector $\mathbf{x}'$. Thus taking an expectation under the posterior distribution on parameters is equivalent to using an ensemble of an uncountably infinite number of neural networks. Unfortunately, this is intractable for neural networks of any practical size.

The variational inference relies on approximating the posterior $p(\theta \mid \mathcal{D})$ with a more computationally tractable variation distribution $q(\theta \mid \mathcal{D}, \omega)$ parameterized by ω. Then q is found by minimizing the Kullback–Leibler divergence between the approximate distribution and the posterior:

$$\mathrm{KL}(q \parallel p) = \int q(\theta \mid \mathcal{D}, \omega) \log \frac{q(\theta \mid \mathcal{D}, \omega)}{p(\theta \mid \mathcal{D})} d\theta.$$

Since $p(\theta \mid \mathcal{D})$ is not necessarily tractable, we replace minimization of $\mathrm{KL}(q \parallel p)$ with maximization of the evidence lower bound (ELBO):

$$\mathrm{ELBO}(\omega) = \int q(\theta \mid \mathcal{D}, \omega) \log \frac{p(Y \mid X, \theta)p(\theta)}{q(\theta \mid \mathcal{D}, \omega)} d\theta.$$

Note that ELBO is also referred to as the 'variational free energy'. The evidence is then

$$\log p(\mathcal{D}) = \mathrm{ELBO}(\omega) + \mathrm{KL}(q \parallel p).$$

Note that the evidence does not depend on ω, thus minimizing $\mathrm{KL}(q \parallel p)$ is the same as maximizing $\mathrm{ELBO}(\omega)$. Also, since $\mathrm{KL}(q \parallel p) \geq 0$, which follows from Jensen's inequality, we have $\log p(\mathcal{D}) \geq \mathrm{ELBO}(\omega)$. Hence the reasoning behind the evidence lower bound nomenclature. The resulting maximization problem $\mathrm{ELBO}(\omega) \to \max_{\omega}$ is solved using stochastic gradient descent.

To calculate the gradient, it's convenient to write the ELBO as

$$\mathrm{ELBO}(\omega) = \int q(\theta \mid \mathcal{D}, \omega) \log p(Y \mid X, \theta) d\theta - \int q(\theta \mid \mathcal{D}, \omega) \log \frac{q(\theta \mid \mathcal{D}, \omega)}{p(\theta)} d\theta.$$

In this form it's easy to see that it's composed of a log likelihood term, and a prior-dependent term which penalizes complexity. ELBO thus enforces the bias-variance trade-off between satisfying the complexity of the data $\mathcal{D}$ and satisfying the simplicity prior $p(\theta)$. Note also that ELBO is given an information theoretic interpretation as a minimum description length cost (Hinton and Van Camp, 1993; Graves, 2011).

The gradient of the first term

$$\nabla_{\omega} \int q(\theta \mid \mathcal{D}, \omega) \log p(Y \mid X, \theta) d\theta = \nabla_{\omega} \mathbb{E}_q \log p(Y \mid X, \theta)$$

is not an expectation and thus cannot be estimated using Monte Carlo methods. A 'reparameterization trick' is used to represent the gradient $\nabla_\omega \mathbb{E}_q \log p(Y \mid X, \theta)$ as an expectation of some random variable, so that Monte Carlo techniques can be used to estimate it. This involves using the identity $\nabla_x f(x) = f(x)\nabla_x \log f(x)$ to obtain $\nabla_\omega \mathbb{E}_q \log p(Y \mid X, \theta)$.

Thus, if we select $q(\theta \mid \omega)$ so that it is easy to compute its derivative with respect to ω and generate samples from it, the gradient can be efficiently estimated using Monte Carlo simulation. More precisely, to evaluate the expectations, a backpropagation-like algorithm is obtained for variational Bayesian inference in neural networks – Bayes by Backprop – which uses unbiased estimates of gradients of ELBO to learn a distribution over the weights of a neural network.

32.3.3 Bayes by backprop

We follow Blundell et al. (2015) and present Bayes by Backprop in more detail. Suppose, for simplicity, that the variational posterior is a diagonal Gaussian distribution, then a sample of the weights and biases θ can be obtained by sampling a unit Gaussian, shifting it by a mean μ and scaling by a standard deviation σ. We parameterise the standard deviation pointwise as $\sigma = \log(1 + \exp(\rho))$ and so σ is always non-negative. The variational posterior parameters are $\omega = (\mu, \rho)$. Thus the transform from a Gaussian noise vector ϵ and the variational posterior parameters that yields a posterior of the weights and biases $q(\theta \mid \mathcal{D}, \phi)$ is $\theta = \mu + \log(1 + \exp(\rho)) \circ \epsilon$, where $\circ$ is pointwise multiplication. Each step of the optimisation proceeds as follows:

1. Sample $\epsilon \sim \mathcal{N}(0, I)$.
2. Let $\theta = \mu + \log(1 + \exp(\rho)) \circ \epsilon$.
3. Let $\omega = (\mu, \rho)$.
4. Let $f(\theta, \omega) = \log q(\theta \mid \mathcal{D}, \omega) - \log p(\theta)p(\mathcal{D} \mid \theta)$.
5. Calculate the gradient with respect to the mean

$$\Delta_\mu = \frac{\partial f(\theta, \omega)}{\partial \theta} + \frac{\partial f(\theta, \omega)}{\partial \mu}. \tag{32.8}$$

6. Calculate the gradient with respect to the standard deviation parameter ρ

$$\Delta_\rho = \frac{\partial f(\theta, \omega)}{\partial \theta} \frac{\epsilon}{1 + \exp(-\rho)} + \frac{\partial f(\theta, \omega)}{\partial \rho}. \tag{32.9}$$

7. Update the variational parameters:

$$\mu \leftarrow \mu - \alpha \Delta_\mu \tag{32.10}$$

$$\rho \leftarrow \rho - \alpha \Delta_\rho. \tag{32.11}$$

Note that the $\frac{\partial f(\theta,\omega)}{\partial \theta}$ term of the gradients for the mean and standard deviation are shared and are exactly the gradients found by the usual backpropagation algorithm on a neural network. Thus, remarkably, to learn both the mean and the standard deviation we must simply calculate the usual gradients found by backpropagation, and then scale and shift them as above.

Scale mixture prior

There are many types of scale mixture priors which are used in Bayesian modeling to generalize the class of priors to more flexible models. Following Blundell et al. (2015), we implement a slightly more general case of the above algorithm, using a scale mixture of two Gaussian densities as the prior, each density is zero mean, but has different variances:

$$p(\boldsymbol{\theta}) = \prod_j \pi \mathcal{N}(\boldsymbol{\theta}_j | 0, \sigma_1^2) + (1 - \pi)\mathcal{N}(\boldsymbol{\theta}_j | 0, \sigma_2^2), \tag{32.12}$$

where $\boldsymbol{\theta}_j$ is the jth parameter of the network, $\mathcal{N}(x | \mu, \sigma^2)$ is the Gaussian density evaluated at x with mean μ and variance σ^2 and σ_1^2 and σ_2^2 are the variances of the mixture components. The first mixture component of the prior is given a larger variance than the second, $\sigma_1 > \sigma_2$, providing a heavier tail in the prior density than a plain Gaussian prior. The second mixture component has a small variance $\sigma_2 \ll 1$ causing many of the weights to *a priori* tightly concentrate around zero. The prior resembles a spike-and-slab prior (Chipman, 1996), where instead all the prior parameters are shared among all the weights. This renders the prior more practical for optimisation by stochastic gradient descent as it avoids prior parameter optimisation based on training data.

32.4 Bayesian deep fundamental factor models

For any realization of $\boldsymbol{\theta} = (W_0, b_0) = \boldsymbol{\theta}_0$, our deep network provides a pointwise non-linear cross-sectional fundamental factor model of the predictive form

$$\mathbf{r}_{t+1} = F_{\boldsymbol{\theta}_0}(B_t) + \boldsymbol{\epsilon}_t, \tag{32.13}$$

which is a non-linear generalization of the Barra linear model. In fact, without activation, this deep network recovers the linear model. Hence, we seek to predict returns $\mathbf{r}_{t+1}$ at time $t + 1$ using the beta covariates B_t, which are by construction measurable at time t. This model can hence be mapped on to a growing body of research on returns prediction using neural networks. One appealing aspect of this approach is that stationarity of the factor realizations is not required since we only predict one period ahead and do not rely on unconditional expectations of variances which must be estimated from historical data.

32.4.1 *Probabilistic prediction*

Given a new observation of the factor loadings, B_t, we estimate the predictive distribution $p(\hat{\mathbf{r}}_{t+1} \mid B_t, \mathcal{D}_{t-1})$ of the asset returns given by the deep probabilistic predictor, $\hat{\mathbf{r}}_{t+1} := \hat{\mathbf{r}}_{t+1}(B_t) = F_{\boldsymbol{\theta}_{t-1} \in \hat{\Theta}_{t-1}}(B_t)$. This distribution is given by averaging the conditional probability $p(\hat{\mathbf{r}}_{t+1} \mid B_t, \mathcal{D}_{t-1}, \boldsymbol{\theta}_{t-1})$ over the fitted posterior distribution of $\boldsymbol{\theta}_{t-1} \in \hat{\Theta}_{t-1}$. More precisely the predictive distribution of the asset returns is given by:

$$p(\hat{\mathbf{r}}_{t+1} \mid B_t, \mathcal{D}_{t-1}) = \mathbb{E}_{\boldsymbol{\theta}_{t-1} \mid \mathcal{D}_{t-1}}[p(\hat{\mathbf{r}}_{t+1} \mid B_t, \mathcal{D}_{t-1}, \boldsymbol{\theta}_{t-1})]$$

$$= \int_{\boldsymbol{\theta} \in \hat{\Theta}_{t-1}} p(\hat{\mathbf{r}}_{t+1} \mid B_t, \mathcal{D}_{t-1}, \boldsymbol{\theta}) p(\boldsymbol{\theta} \mid \mathcal{D}_{t-1}) d\boldsymbol{\theta}$$

$$\approx \int_{\boldsymbol{\theta} \in \hat{\Theta}_{t-1}} p(\hat{\mathbf{r}}_{t+1} \mid B_t, \mathcal{D}_{t-1}, \boldsymbol{\theta}) q(\boldsymbol{\theta} \mid \mathcal{D}_{t-1}, \hat{\omega}_{t-1}) d\boldsymbol{\theta}, \quad (32.14)$$

which is found by subsampling $\boldsymbol{\theta}_{t-1}$ from the cheap variational distribution, $q(\boldsymbol{\theta}_{t-1} \mid \mathcal{D}_{t-1}, \hat{\omega}_{t-1})$ using the fitted variational parameters $\hat{\omega}_{t-1}$ found by the Bayes by Backprop algorithm, so that any predictive asset returns sample $\hat{\mathbf{r}}_{t+1}^{(j)}$, drawn from the distribution of predicted returns with density $p(\hat{\mathbf{r}}_{t+1} \mid B_t, \mathcal{D}_{t-1})$, is

$$\hat{\mathbf{r}}_{t+1}^{(j)} = \frac{1}{|I_j|} \sum_{i \in I_j} F_{\boldsymbol{\theta}_{t-1}^{(i)}}^{(j)}(B_t), \quad j = 1, \ldots, M, \quad (32.15)$$

where I_j indexes a set of random subsamples $\{\boldsymbol{\theta}_{t-1}^{(i)}\}_{i \in I_j}$.

32.5 Deep network interpretability

Once the neural network has been trained, a number of important issues surface around how to interpret the model parameters. This aspect is by far the most prominent issue in deciding whether to use neural networks in favor of other machine learning and statistical methods for estimating factor realizations, sometimes even if the latter's predictive accuracy is inferior. In fact, as mentioned earlier, it's a widely held misnomer in the finance industry that neural networks are 'black-boxes'. In this section, we shall revisit Dixon and Polson (2020) and present the method for interpreting multi-layer perceptrons which imposes minimal restrictions on the neural network design.

Dixon and Polson (2020) use a gradient-based technique for determining the importance of the input variables. The method is directly consistent with how coefficients are interpreted in linear regression, i.e. as *model* sensitivities and is thus appealing to asset managers. In this approach, the model sensitivities are the partial derivatives of the fitted model output with respect to each input. This method is in the same vein as Horel and Giesecke (2019) who develop statistical tests for the significance of the factors in a deep network, also using their partial derivatives. Their test statistic is based on a weighted distribution of the square of the neural network partial derivative, with respect to each input. Without a bound on the gradient, everywhere, the integral for their test statistic is unbounded and hence the test statistic could be unbounded.

The scope of their study is limited to a single-hidden layer network and treats the asymptotic distribution of the network as a Gaussian process. Dixon and Polson (2020) describe an approach for an arbitrary number of layers and do not rely on asymptotic approximations, which may be limited when the network has a small number of neurons. They derive closed form expressions for the Jacobian and

Hessian – the off-diagonals provide sensitivities to interaction terms. In contrast to Horel and Giesecke (2019), we characterize the dispersion of the empirical sensitivity distribution – confidence intervals of the distributions can be found by Bootstrap sampling (see, for example, Wood, 2005). Such an approach is only useful in practice if the variance of these distributions are bounded, which was shown in Dixon and Polson (2020).

To evaluate fitted model sensitivities analytically, we require that the function $\hat{Y} = F_\theta(X)$ is continuous and differentiable everywhere. Furthermore, for stability of the interpretation, we shall require that $F_\theta(x)$ is a Lipschitz continuous[6]. That is, there is a positive real constant K such that for all $x_1, x_2 \in \mathbb{R}^p$, $|F_\theta(x_1) - F_\theta(x_2)| \leq K|x_1 - x_2|$. Such a constraint is necessary for the first derivative to be bounded and hence amenable to the derivatives, with respect to the inputs, providing interpretability.

Provided that the values of the weights and biases are finite, each semi-affine function is Lipschitz continuous everywhere. For example, the function $\tanh(x)$ is continuously differentiable with derivative, $1 - \tanh^2(x)$, is globally bounded. With finite weights, the composition of $\tanh(x)$ with an affine function is also Lipschitz. Clearly $\text{ReLU}(x) := \max(\cdot, 0)$ is not continuously differentiable and one can't use the approach described here.

32.5.1 Factor sensitivities

To see the main idea, let us briefly revisit interpretability in the Barra factor model

$$\hat{r}_{t+1} = F_{\theta_t}(B_t) := \theta_0(t) + \theta_1(t)\beta_1(t) + \cdots + \theta_K(t)\beta_K(t), \qquad (32.16)$$

where, for a fixed time period t, the factor coefficients model are the factor 'sensitivities', $\theta_j(t) = \partial_{\beta_j(t)}\hat{r}_{t+1}$. Thus, for scaled factor loadings, we can interpret θ_t as the relative importance of the factor. In a feedforward neural network, the interpretation of θ_t is no longer meaningful, since it represents edge weights and biases in the neural network. However, the sensitivities of the neural network, as a network sub-graph, now provide the interpretability. As the network is non-linear, these sensitivities are no longer constant, but depend on the input. We can use the chain rule to obtain the model sensitivities:

$$\partial_{\beta_j(t)}\hat{r}_{t+1} = \partial_{\beta_j(t)}F_{\theta_t}(B_t) = \partial_{\beta_j(t)}\sigma^{(L)}_{\theta_t^{(L)}} \circ \cdots \circ \sigma^{(1)}_{\theta_t^{(1)}}(B_t). \qquad (32.17)$$

Alternatively, we can write the Jacobian in the matrix form. Denoting $\theta_t = (W_t, b_t)$, where $W_t = (W^{(1)}, \ldots, W^{(L)})$ and $b_t = (b^{(1)}, \ldots, b^{(L)})$:

$$\partial_{B_t}\hat{r}_{t+1} = W^{(L)}J(I^{(L-1)}) = W^{(L)}D(I^{(L-1)})W^{(L-1)} \cdots D(I^{(1)})W^{(1)}, \qquad (32.18)$$

where $D_{ii}(I) := \sigma'(I_i), D_{ij} = 0, i \neq j$ is a diagonal matrix and $I := I(X) = WX + b$ is an affine transformation. For example, with one hidden layer, $\sigma(x) := \tanh(x)$

---

[6] If Lipschitz continuity is not imposed, then a small change in one of the input values could result in an undesirable large variation in the derivative

and $\sigma^{(1)}_{W^{(1)},b^{(1)}}(X) := \sigma(I^{(1)}) := \sigma(W^{(1)}X + b^{(1)})$:

$$\partial_{\beta_j(t)}\hat{\mathbf{r}}_{t+1} = \sum_i \mathbf{w}^{(2)}_{,i}(1 - \sigma^2(I^{(1)}_i))w^{(1)}_{ij} \quad \text{where } \partial_\beta\sigma(\beta) = (1 - \sigma^2(\beta)). \quad (32.19)$$

In matrix form, with general σ, the Jacobian[7] of σ with respect to B_t is $J = D(I^{(1)})W^{(1)}$ of σ,

$$\partial_{B_t}\hat{\mathbf{r}}_{t+1} = W^{(2)}J(I^{(1)}) = W^{(2)}D(J^{(1)})W^{(1)}. \quad (32.20)$$

Bounds on the factor sensitivities are given by the product of the weight matrices

$$\min(W^{(2)}W^{(1)},0) \leq \partial_{B_t}\hat{\mathbf{r}}_{t+1} \leq \max(W^{(2)}W^{(1)},0). \quad (32.21)$$

For convenience, worked expressions for the Jacobian of a two hidden layer network are provided in §32.A.

Interaction terms

The pairwise interaction effects are readily available by evaluating the elements of the Hessian matrix. For an L-layer network, we define the (i,j)th element of the Hessian as

$$\partial^2_{\beta_i(t)\beta_j(t)}\hat{\mathbf{r}}_{t+1} = \sum_{\ell=1}^{L-1} H_{i,j,\ell}, \quad H_{i,j,\ell} := W^{(L)}D^{(L-1)}W^{(L-1)}\cdots\partial_{\beta_j(t)}D^{(\ell)}W^{(\ell)}\cdots w^{(1)}_i. \quad (32.22)$$

where it is assumed that the activation function is at least twice differentiable everywhere, e.g. $tanh(x)$, softplus etc. For convenience, worked expressions for the Hessian of a two hidden layer network are provided in §32.A.

Probabilistic factor sensitivities

In our Bayesian deep fundamental factor framework, the predictive distribution of the factor sensitivities (and interaction terms) is found by simply replacing the predictor with the closed form derivatives in the Bayesian model averaging formula (32.14). The distribution of the sensitivities of the predictive model are:

$$p(\partial_{B_t}\hat{\mathbf{r}}_{t+1} \mid B_t, \mathcal{D}_{t-1}) = \mathbb{E}_{\theta_{t-1}|\mathcal{D}_{t-1}}[p(\partial_{B_t}\hat{\mathbf{r}}_{t+1} \mid B_t, \mathcal{D}_{t-1}, \hat{\theta}_{t-1})]$$

$$\approx \int_{\theta\in\hat{\Theta}_{t-1}} p(\partial_{B_t}\hat{\mathbf{r}}_{t+1} \mid B_t, \mathcal{D}_{t-1}, \theta)q(\theta \mid \mathcal{D}_{t-1}, \hat{\omega}_{t-1})d\theta,$$

and so any predictive asset return sensitivity sample $\partial_{B_t}\hat{\mathbf{r}}^{(j)}_{t+1}$ is

$$\partial_{B_t}\hat{\mathbf{r}}^{(j)}_{t+1} = \frac{1}{|I_j|}\sum_{i\in I_j}\partial_{B_t}F^{(j)}_{\theta^{(i)}_{t-1}}(B_t), \quad j = 1,\dots,M,$$

where I_j indexes a set of random subsamples $\{\theta^{(i)}_{t-1}\}_{i\in I_j}$.

Note, for pointwise consistency, that the identical samples $\{\theta^{(i)}_{t-1}\}_{i\in I_j}$ are used to both estimate the distribution of the factor sensitivities as the predictor in (32.15). Therefore, for any sample, the factor sensitivities are the derivatives of the predictors under the identical realizations of θ.

---

[7] When σ is an identity function, the Jacobian $J(I^{(1)}) = W^{(1)}$.

32.6 Applications: Russell 1000-Factor modeling

This section presents the application of our Bayesian framework to the fitting of a fundamental factor model. The Barra factor model often includes many more explanatory variables than used in our experiments below, but the purpose, here, is to illustrate the application of our framework to a larger dataset.

For completeness, we provide evidence that our Bayesian deep neural network factor model generates positive and higher information ratios than linear and quadratic regressions when used to sort portfolios from a larger universe, using 49 factors from Bloomberg – 19 of which are fundamental factors and the remainder are GICS sector dummy variables (see repository documentation for a description of the factors), over an approximately 30-year period of monthly updates between January 1989 to December 2017, and with a coverage universe of 3290 stocks from the Russell 1000 index. All stocks with missing factor exposures are removed the estimation universe for the date of the missing factors only. To avoid excessive turn-over in the estimation universe over each consecutive period, we include all dropped symbols in the index over the 12 consecutive monthly periods.

Training and testing is alternated each period. For example, in the first historical month of the data, the model is fitted to the factor exposures and monthly excess monthly returns over the next period. Cross-validation is performed using this cross-sectional training data, with approximately 1000 symbols. Once the model is fitted and tuned, we then apply the model to the factor exposures in the next period, $t + 1$, to predict the excess monthly returns over $[t + 1, t + 2]$. The process is repeated over all periods. Note, for computational reasons, we can avoid cross-validation over every period and instead stride the cross-validation, every other say 10 periods, relying on the optimal hyper-parameters from the last cross-validation periods for all subsequent intermediate periods. In practice we find that performing cross-validation in the first period only is adequate.

We use the `Blitz` module with `torch` to implement the Bayesian Deep Network. All predictions are performed by averaging over an ensemble size of 100. The Bayes' by Backprop algorithm uses an Adam optimization with a learning rate of 0.01 and training is performed with a batch size of 16. The BNN is trained for up to 300 epochs and early stopping is used with a "patience" of 10 and minimum absolute difference of 10^{-2}. Patience defines the number of iterations that must consecutively satisfy the threshold constraint on the absolute difference between current and previous losses. So, in our case, since the patience is set to 10 and the minimum loss difference is 10^{-2}, then fifty consecutive loss evaluations on mini-batch updates must each be within 10^{-2} of the previous update before the training terminates. In each iteration, the ELBO cost function is sampled ten times.

We choose the standard deviations of the zero mean Gaussian priors to be $\sigma = (0.4, 0.1)$ and the mixture weight $\pi = 0$. The variational parameters are initialized as $\omega = (\mu, \rho) = (0, -7.0)$. Three-fold cross-validation over $\{50, 100, 200\}$ hidden units per layer, $\{1, 2, 3\}$ hidden layers, prior mixture $\{0.0, 0.1, 0.2, 0.3, 0.4, 0.5, 0.6, 0.7, 0.8, 0.9, 1.0\}$ and tanh or smooth ReLu (i.e. softplus) is performed. We find that the optimal architecture has two hidden layers, 100 units per layer, $\pi = 0.7$ and smooth ReLU activation.

The OLS model is implemented using Python `StatsModels` module. The linear regression is an OLS model with an intercept. The Lasso model includes quadratic powers in the factors and pairwise interaction terms. The Lasso model uses cross-validation to optimize the L_1 regularization parameter and iterative fitting, with 50 alphas, along the regularization path.

To evaluate the relative advantage of the BNN in comparison to linear and quadratic regressions, we construct an equally weighted, long-only, portfolio of n stocks with the highest predicted monthly returns in each period. The portfolio is reconstructed each period, always remaining equal but varying in composition. This is repeated over the most recent 10-year period in the data. We then estimate the information ratios from the mean and volatility of the excess monthly portfolio returns, using the Russell 1000 index as the benchmark.

Figure 32.2 compares the out-of-sample performance of Bayesian neural networks, Lasso, and OLS regression using the L_∞ norm of stock return errors over the coverage universe. We observe the ability of the neural network to capture outliers, with the L_∞ norm of the error in the BNN being an order of magnitude smaller than in the OLS model at two dates, 2000–01–01 and 2015–10–01. The Lasso model exhibits several mid-size outliers which aren't present in the BNN model and is outperformed by BNN. The average L_∞ norms over all periods is shown in parenthesis and is 40% smaller for BNNs than for OLS and 30% smaller than for Lasso. The out-of-sample MSEs for BNNs, Lasso, and OLS are 0.028, 0.029, and 0.254 – the latter decreases to 0.028 with these dates removed.

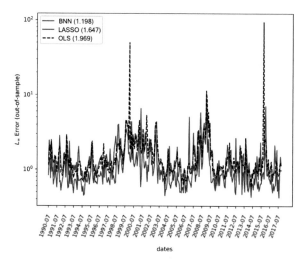

Figure 32.2 The out-of-sample stock returns error, under the L_∞ norm, is compared between OLS, Lasso, and a two-hidden layer Bayesian deep network (with ReLU activation) applied to a coverage universe of 3290 stocks from the Russell 1000 index over the period from January 1990 to December 2017. The average L_∞ error is shown in parenthesis.

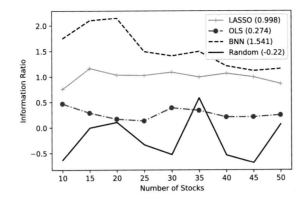

Figure 32.3 The information ratios of an equally weighted portfolio selection strategy which selects the *n* stocks from the universe with the highest predicted monthly returns. The information ratios are evaluated for various portfolios whose number of stocks are shown by the *x*-axis.

Figure 32.3 compares the information ratios using the BNN, Lasso, and OLS models to identify the stocks with the highest predicted returns over the period from Feb 17 to Dec 17. Note that Jan 17 is used for training and we choose not to use returns from the previous year to test Jan 17. The information ratios are evaluated for equally weighted portfolios with varying numbers of stocks. Also shown, for control, are randomly selected portfolios, without the use of a predictive signal. The mean information ratio for each model, across all portfolios, is shown in parentheses. We observe that the information ratio of the portfolio returns, using the BNN (with ReLU), is approximately 1.5 times greater than Lasso regression and approximately 5 times greater than OLS regression. We also observe that the information ratio of the baseline random portfolio is on average negative, thus establishing a negative information ratio baseline for this period.

Figure 32.4 shows the sector tilts of equally weighted portfolios constructed from the predicted top performing 50 stocks, in each monthly period tested in 2017. The sectors are ranked by their time averaged ratios, but their tilts vary each month as the portfolios turn-over. Note that the factor symbols are defined in §32.B.

Financials is the most dominant sector, with almost 20% time averaged representation. This is followed by Consumer Discretionary. Note that three of the least representative sectors are excluded: Energy, Communication Services, and Real Estate. The sector tilts across the BNN and the Lasso model are found to be broadly comparable on average.

Figure 32.5 shows the corresponding (scaled) factors, averaged over the portfolio, for a subset of the factors with non-trivial differences in tilts between Lasso and BNNs. In comparison with Lasso, we observe that BNNs favor assets with higher Book to Price, Earning to Price, and Cash Flow Volatility to Total Assets.

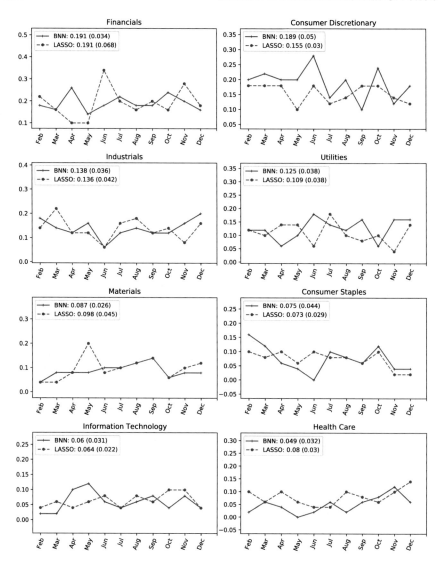

Figure 32.4 The sector tilts are shown over time for each sector, in descending order. The mean and std. devs. of the sector ratios, from Feb to Dec 2017, are shown in parentheses.

Lasso favors stocks with higher Total Asset Growth, Sales Volatility to Total Assets, and Rolling CAPM Beta.

Figure 32.6 shows the MAP of the BNN predicted portfolio monthly returns against the realized portfolio returns, for the same portfolio, from Feb to Dec 2017, and are found to closely track each other. The uncertainty bands represent an envelope of one standard deviation about the MAP and are shown to vary widely. For example, the uncertainty is smaller in July and August than April or December. Figure 32.7 shows the MAP of the top BNN predicted factor sensitivities evaluated for the same portfolios as in Figure 32.6 over 2017. Each

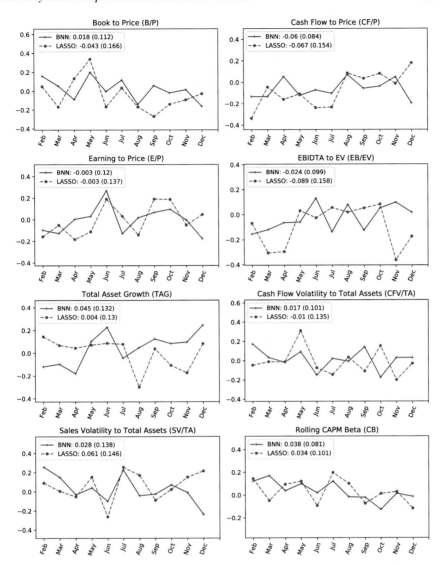

Figure 32.5 The (scaled) factors, averaged over the portfolio, are shown from Feb. to Dec. 2017, for a subset of the factors.

graph shows the MAP for a separate factor together with the uncertainty bands which represent an envelope of one standard deviation about the MAP. We note that the MAP of some factor sensitivities exhibit more variability over time, e.g. Cash Flow to Price (CF/P), whereas others transition more gradually, e.g. Log Sales (S).

To illustrate the utility of factor sensitivities with uncertainty bands, Figure 32.8 compares the distribution of the top ten most positive and negative predicted factor sensitivities from February to April using the interpretable BNN. In each

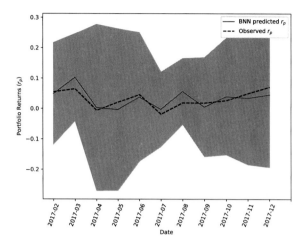

Figure 32.6 The MAP of the BNN predicted portfolio monthly returns (red solid line) is compared with the realized portfolio returns (black dashed line) over 2017. The uncertainty bands represent an envelope of one standard deviation about the MAP.

plot, the sensitivities are sorted in descending order from left to right by their median values.

We note that the most prominent predicted factors vary over each test period. For example, Cash Flow/Price (CF/P) followed by sub-industry SI5 and SI6 are the three most important positive factors and Rolling CAPM Beta (CB), followed by the Communications (SE9) and Financial (SE7) sectors are the most important negative factors in February 2017. Log Sales (S), sub-industry SI9, and Earnings growth (EG) feature most prominently as positive predictive factors where as Industry I3, Earnings volatility in total assets (EaV/TA), and sub-industry SI2 are the most prominent negative predictive factors in March, 2017. Then sub-industry SI9, Market Cap (MC), and Financials (I4) are the most prominent positive predictive factors and Health-Care (SE6), Industry I3, and sub-industry SI3 are the most negative in April, 2017. We also observe substantial variability in the shape of the sensitivity distributions across factors. We observe substantial differences in uncertainty across these prominent predictive factor sensitivities. For example, the interquartile range for sub-industry SI6 in February is much lower than the interquartile range of Health-Care (SE6) in April.

32.7 Discussion

In this chapter, we introduce a Bayesian deep learning framework for fundamental factor modeling which generalizes the linear fundamental factor models. Our framework provides interpretability by ranking the factor importances or interaction effects and providing uncertainty quantification. In the case when the network contains no hidden layers, our approach recovers a Bayesian linear fundamental factor model and the framework therefore allows the impact of non-linearity in

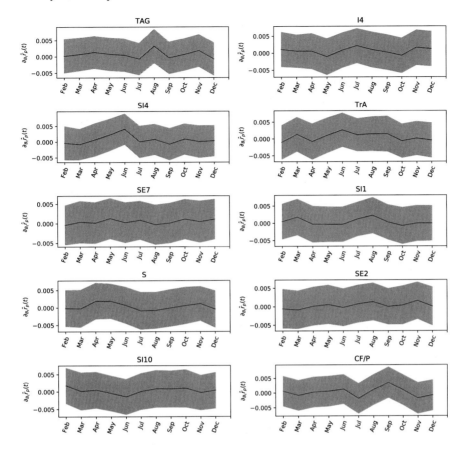

Figure 32.7 The MAP of the top BNN predicted factor sensitivities (red solid line) evaluated for the same portfolio as in Figure 32.6 over 2017. Each figure shows the MAP for a separate factor together with the uncertainty bands which represent an envelope of standard deviations about the MAP.

factors to be assessed. The Bayesian NN is observed to better capture outliers, with a 30% reduction in L_∞ normed error compared to quadratic Lasso regression. We further show that this translates into superior portfolio performance, the Bayesian NN is observed to generate information ratios which are a factor of 1.5 times higher than quadratic Lasso regression. Factor sensitivity plots are provided to demonstrate how the relative importance of factors evolve over time together with the degree of uncertainty in their estimation due to parameter uncertainty in the network. Such information can be used to inform hedging strategies and other portfolio risk management techniques, by for example, limiting exposure of the portfolio to certain negative factors or even those with a high degree of uncertainty in their factor sensitivities. With a preliminary Bayesian deep learning methodology for fundamental factor modeling etched out, we turn to various important considerations which take us a step further towards adoption in practice.

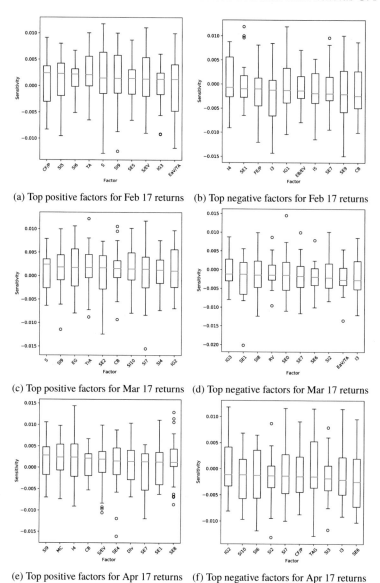

(a) Top positive factors for Feb 17 returns   (b) Top negative factors for Feb 17 returns

(c) Top positive factors for Mar 17 returns   (d) Top negative factors for Mar 17 returns

(e) Top positive factors for Apr 17 returns   (f) Top negative factors for Apr 17 returns

Figure 32.8 The distribution of factor model sensitivities for the predicted portfolio monthly returns over three separate periods: February, March, and April, 2017.

Dimensionality reduction:

A common practice in portfolio risk management is to identify the subset of factors which explain the majority of asset return variability (see for example Chapter 3 of McNeil et al. (2005)). More precisely, how can the covariance of the portfolio's asset returns be substantially reconstructed using a subset of factors or, alternatively, how can the portfolio variance be approximated with a lower dimensional factor vector? Such a question, of course, is well suited to principle component analysis, in which such factors are not fundamental but principal components and there is typically a linear relationship between these and the asset returns. Singular value decomposition of the covariance matrix provides a guaranteed ordering of principal components through monotonically decaying eigenvalues.

However, there is no such guarantee provided by fundamental factors because they are not orthogonal, although this does not completely prohibit factor reduction nonetheless. The sensitivity of the portfolio returns to each factor is given by a weighted combination of the deep learner's factor sensitivities: $\partial_{B_t}\hat{P}_t = \mathbf{w}^\top \partial_{B_t}\hat{\mathbf{r}}_t$, where $\mathbf{w}$ denotes the portfolio weights[8]. By ranking $|\partial_{B_t}\hat{P}_t|$, we can potentially eliminate the lowest ranking factors and estimate the portfolio variance σ_P^2 with and without the full set of factors to determine the proportion of portfolio variance attributed to the most important factors. The caveat is that there is no guarantee that a subset of factors will substantially explain the portfolio variance. Such low ranking factors may be highly correlated, and hence in aggregate, still non-trivially contribute to the portfolio variance.

Portfolio construction:

The strategy developed in the previous section is not intended as an exhaustive representation of the strategies that are suitable for exploiting the predictive signal from the deep learner. In our experimental setup, we arbitrarily choose the top 50 performing stocks in each monthly period leading to substantial portfolio turnover. Over a 30-month period, the average monthly turn over is 81% and our portfolio performance does not account for transaction costs.

As factor sensitivities change in each period, we broadly envisage more elaborate strategies, which attempt to reduce the turn-over, by swapping or downsizing long portfolio positions with the least exposure to the portfolio's overall most important factor sensitivities. Conversely, we may look to the lowest rank sensitivities and eliminate or downsize those positions with the largest exposure to them. The details of how such an approach would perform in real market conditions remains an open question. However, because confidence intervals in the portfolio returns' factor sensitivities are a linear combination of the confidence intervals of the asset's factor sensitivities, a portfolio manager could select those factors which not only contribute the most to the portfolio returns and variance but also those with the narrowest confidence intervals. Such an approach becomes especially attractive in market regimes which are dominated by at most a few fundamental factors, e.g. a 'momentum regime', and therefore becomes more tractable. Portfolios which track a smaller investment universe are likely more amenable to this approach as its more conceivable that a few factors may dominant the portfolio returns and variance.

Future work shall address how to develop the framework for portfolio risk modeling and the best approach for estimating the covariance matrix of asset returns from the network is an open question as the transformation to a lower-

---

[8] In our numerical experiments we choose, for simplicity, equal weights across a portfolio of 50 assets, however the weights could be adjusted based on, say, the prediction and its confidence.

dimensional factor covariance matrix is no longer feasible. Further work is also required to build on linear risk decomposition methods (e.g. Euler's method) by incorporating the non-linear factor sensitivities.

32.A Appendix: Gradients of two-layer feedforward networks

For convenience, we provide the first- and second-order derivatives of feedforward network with two hidden layers. For simplicity we can choose tanh as the activation function. The Jacobian of the model output is:

$$\frac{\partial \hat{Y}}{\partial X} = W^{(3)} D^{(2)} W^{(2)} D^{(1)} W^{(1)}.$$

Since $\sigma(x) = \tanh(x) \Rightarrow \sigma'(x) = 1 - \tanh^2(x)$, we can express the Jacobian as

$$\frac{\partial \hat{Y}}{\partial X} = W^{(3)} D^{(2)} W^{(2)} D^{(1)} W^{(1)},$$

where

$$D^{(2)}_{i,i} := D^{(2)}_{i,i}(I^{(2)}_i) = 1 - \tanh^2(I^{(2)}_i), \qquad D^{(2)}_{i,j \neq i} = 0,$$
$$D^{(1)}_{i,i} := D^{(1)}_{i,i}(I^{(1)}_i) = 1 - \tanh^2(I^{(1)}_i), \qquad D^{(1)}_{i,j \neq i} = 0.$$

The Hessian takes the form:

$$\frac{\partial \hat{Y}}{\partial X_i X_j} = W^{(3)} W^{(2)} \left(\frac{\partial D^{(2)}}{\partial X_j} D^{(1)} + \frac{\partial D^{(1)}}{\partial X_j} D^{(2)} \right) w^{(1)}_i,$$

where applying the chain rule we have

$$\frac{\partial D^{(2)}}{\partial X_j} = D'(I^{(2)}) W^{(2)} D^{(1)} \operatorname{diag}(w^{(1)}_j)$$

$$\frac{\partial D^{(1)}}{\partial X_j} = D'(I^{(1)}) \operatorname{diag}(w^{(1)}_j).$$

For $\sigma(x) := \tanh(x)$, we note that

$$\frac{\partial \sigma}{\partial x} = 1 - \tanh^2(x)$$

and

$$\frac{\partial^2 \sigma}{\partial x^2} = -2 \tanh(x)(1 - \tanh^2(x)).$$

Substituting in these derivatives gives terms for D' and D:

$$D'_{i,i}(I^{(2)}_i) = -2 \tanh(I^{(2)}_i)(1 - \tanh^2(I^{(2)}_i)), \qquad D'_{i,j \neq i}(I^{(2)}_i) = 0$$

and

$$D'_{i,i}(I^{(1)}_i) = -2 \tanh(I^{(1)}_i)(1 - \tanh^2(I^{(1)}_i)), \qquad D'_{i,j \neq i}(I^{(1)}_i) = 0.$$

32.B Description of Russell 1000-Factor model

Table 32.B.1 A short description of the factors used in the Russell 1000 fundamental factor model.

ID	Symbol	**Value Factors**
1	B/P	Book to Price
2	CF/P	Cash Flow to Price
3	E/P	Earning to Price
4	S/EV	Sales to Enterprise Value (EV). EV is given by $EV = \text{Market Cap} + \text{LT Debt} + \max(\text{ST Debt} - \text{Cash}, 0)$, where LT (ST) stands for long (short) term
5	EB/EV	EBIDTA to EV
6	FE/P	Forecasted E/P. Forecast Earnings are calculated from Bloomberg earnings consensus estimates data. For coverage reasons, Bloomberg uses the 1-year and 2-year forward earnings.
7	DIV	Dividend yield. The exposure to this factor is just the most recently announced annual net dividends divided by the market price. Stocks with high dividend yields have high exposures to this factor.
		Size Factors
8	MC	Log (Market Capitalization)
9	S	Log (Sales)
10	TA	Log (Total Assets)
		Trading Activity Factors
11	TrA	Trading Activity is a turnover based measure. Bloomberg focuses on turnover which is trading volume normalized by shares outstanding. This indirectly controls for the Size effect. The exponential weighted average (EWMA) of the ratio of shares traded to shares outstanding: In addition, to mitigate the impacts of those sharp shortlived spikes in trading volume, Bloomberg winsorizes the data: first daily trading volume data is compared to the long-term EWMA volume(180-day half-life),then the data is capped at 3 standard deviations away from the EWMA average.
		Earnings Variability Factors
12	EaV/TA	Earnings Volatility to Total Assets. Earnings Volatility is measured over the last 5 years/Median Total Assets over the last 5 years
13	CFV/TA	Cash Flow Volatility to Total Assets. Cash Flow Volatility is measured over the last 5 years/Median Total Assets over the last 5 years
14	SV/TA	Sales Volatility to Total Assets. Sales Volatility over the last 5 years/Median Total Assets over the last 5 years
		Volatility Factors
15	RV	Rolling Volatility which is the return volatility over the latest 252 trading days
16	CB	Rolling CAPM Beta which is the regression coefficient from the rolling window regression of stock returns on local index returns
		Growth Factors
17	TAG	Total Asset Growth is the 5-year average growth in Total Assets divided by the Average Total Assets over the last 5 years
18	EG	Earnings Growth is the 5-year average growth in Earnings divided by the Average Total Assets over the last 5 years
		GSIC sectorial codes
19–24	(I)ndustry	{10, 20, 30, 40, 50, 60, 70}
25–35	(S)ub-(I)ndustry	{10, 15, 20, 25, 30, 35, 40, 45, 50, 60, 70, 80}
36–45	(SE)ctor	{10, 15, 20, 25, 30, 35, 40, 45, 50, 55, 60}
46–49	(I)ndustry (G)roup	{10, 20, 30, 40, 50}

References

Blundell, Charles, Cornebise, Julien, Kavukcuoglu, Koray, and Wierstra, Daan. 2015. Weight uncertainty in neural networks. ArXiv:1505.05424.

Britten-Jones, Mark, and Schaefer, Stephen M. 1999. Pages 115–143 in: *Nonlinear Value-At-Risk*. Springer.

Carvalho, Carlos M., Lopes, Hedibert, Aguilar, Omar, and Mendoza, M. 2012. Dynamic stock selection strategies: a structured factor model framework. Pages 69–90 in: *Bayesian Statistics 9*, José M. Bernardo et al. (eds). Oxford University Press.

Chen, Luyang, Pelger, Markus, and Zhuz, Jason. 2019. Deep learning in asset pricing. Tech. Rept. Stanford University.

Chipman, Hugh. 1996. Bayesian variable selection with related predictors. *Canad. J. Statist.*, **24**(1), 17–36.

Dixon, Matthew, and Polson, Nick. 2020. Deep fundamental factor models. *SIAM Journal on Financial Mathematics*, **11**(3), SC–26–SC–37.

Fan, Jianqing, Xue, Lingzhou, and Yao, Jiawei. 2017. Sufficient forecasting using factor models. *Journal of Econometrics*, **201**(2), 292–306.

Feng, Guanhao, He, Jingyu, and Polson, Nicholas G. 2018. Deep learning for predicting asset returns. ArXiv:1804.09314.

Gal, Yarin. 2016. *Uncertainty in Deep Learning*. PhD thesis, University of Cambridge.

Graves, Alex. 2011. Practical variational inference for neural networks. Pages 2348–2356 of: *Advances in Neural Information Processing Systems*.

Gu, Shihao, Kelly, Bryan T., and Xiu, Dacheng. 2018. Empirical asset pricing via machine learning. Chicago Booth Research Paper 18-04.

Harvey, Campbell R., Liu, Yan and Zhu, Heqing. 2015. . . . and the cross-section of expected returns. *Review of Financial Studies*, **29**(1), 5–68.

Heaton, J. B., Polson, Nicholas G., and Witte, Jan H. 2017. Deep learning for finance: deep portfolios. *Applied Stochastic Models in Business and Industry*, **33**(1), 3–12.

Hinton, Geoffrey E., and Van Camp, Drew. 1993. Keeping the neural networks simple by minimizing the description length of the weights. Pages 5–13 of: *Proceedings of the 6th Annual Conference on Computational Learning Theory*.

Horel, E., and Giesecke, K. 2019. Significance tests for neural networks. ArXiv:1902.06021.

Hornik, K., Stinchcombe, M., and White, H. 1989. Multilayer feedforward networks are universal approximators. *Neural Netw.*, **2**(5), 359–366.

Kingma, D.P., and Welling, M. 2013. Auto-encoding variational Bayes. ArXiv:1312.6114.

McNeil, Alexander J., Frey, Rüdiger, and Embrechts, Paul. 2005. *Quantitative Risk Management: Concepts, Techniques and Tools*. Princeton University Press.

Moritz, Benjamin, and Zimmermann, Tom. 2016. Tree-based conditional portfolio sorts: the relation between past and future stock returns. Available from SSRN 2740751.

Nielsen, Frank, and Bender, Jennifer. 2010. The fundamentals of fundamental factor models. Tech. Rept. 24. MSCI Barra Research Paper. Available from SSRN 1707661.

Poggio, T. 2016. Deep learning: mathematics and neuroscience. Pages 9–12 in: *Brain-Inspired Intelligent Robotics: The Intersection of Robotics and Neuroscience*. A sponsored supplement to *Science*.

Polson, Nicholas G., and Sokolov, Vadim. 2017. Deep learning: a Bayesian perspective. *Bayesian Anal.*, **12**(4), 1275–1304.

Rosenberg, Barr, and Marathe, Vinay. 1976. Common factors in security returns: microeconomic determinants and macroeconomic correlates. Research Program in Finance Working Papers 44. University of California at Berkeley.

Wood, Michael. 2005. Bootstrapped confidence intervals as an approach to statistical inference. *Organizational Research Methods*, **8**(4), 454–470.

Zivot, Eric, and Wang, Jiahui. 2003. Factor models for asset returns. Chapter 15 of *Modeling Financial Time Series with S-PLUS* Springer.

33

Black-Box Model Risk in Finance

Samuel N. Cohen[a], Derek Snow[b] and Lukasz Szpruch[c]

Abstract

Machine learning models are increasingly used in a wide variety of financial settings. The difficulty of understanding the inner workings of these systems, combined with their wide applicability, has the potential to lead to significant new risks for users; these risks need to be understood and quantified. In this chapter, we will focus on a well studied application of machine learning techniques, to pricing and hedging of financial options. Our aim will be to highlight the various sources of risk that the introduction of machine learning emphasises or de-emphasises, and the possible risk mitigation and management strategies that are available.

Acknowledgements.

We are grateful to Katia Babbar for helpful comments and suggestions on a draft of this chapter. We acknowledge the support of the Alan Turing Institute under EPSRC grant no. EP/N510129/1

33.1 Introduction

Traditionally, the tractability of pricing and hedging methods was arguably more critical than their accuracy, and the limits of computation determined what methods were useful. The Black–Scholes formula is concise, simple to understand, can be implemented on a handheld calculator (Lo, 2019); these features were critical to its wide adoption. Similarly, the Heston model benefits from convenient (fast) Fourier transformation methods (see, for example, Gatheral, 2006) and the SABR model from a convenient approximation (see Hagan et al., 2002; Obój, 2008), which have formed a key part of their attractiveness. While many more sophisticated and accurate models have been developed, computational bottlenecks have impeded their wide adoption.

In recent years, machine learning models in finance have become streamlined; in just a few lines of packaged code, modellers can develop state-of-the-art models with online computing power and open-source software (Snow, 2020; Dixon et al.,

---

[a] University of Oxford and The Alan Turing Institute. samuel.cohen@maths.ox.ac.uk.
[b] The Alan Turing Institute. dsnow@turing.ac.uk.
[c] University of Edinburgh and The The Alan Turing Institute. l.szpruch@ed.ac.uk.
Published in *Machine Learning And Data Sciences For Financial Markets*, Agostino Capponi and Charles-Albert Lehalle © 2023 Cambridge University Press.

2020). However, the risks of blindly using machine learning solutions, without understanding their inner workings and inherent drawbacks, are significant.

In this chapter, we seek to give an overview of the key issues which arise when using machine learning in finance, and some remedies which have been suggested. Rather than concentrating on developing a particular algorithm, we take a higher-level view of the risks and challenges which arise in these contexts. We wish to highlight that machine learning is not a panacea for financial markets, instead it provides tools which allow practitioners to shift between different sources of risk, some of which have not been a primary concern in the past.

We will focus on those risks which are a core part of machine learning – the risks inherent in data and in the modelling algorithms used. We will not discuss what The World Economic Forum calls the erosion of "human financial talent" where humans lose the skill to challenge and disagree with machine learning systems (McWaters et al., 2019), although this is potentially a significant concern in many financial applications.

There are two broad uses of machine learning in finance. The first is to remove computational barriers and enable use of advanced models in day to day business operations. When used in this manner, 'machine learning' is providing the next generation of computational tools, which are used to speed up and improve traditional modelling. For example, when calibrating an option pricing model one often needs to price many derivatives many times, using a variety of potential parameter values – this is a task that can be improved by using a machine-learnt approximation for the pricing operator. Hutchinson et al. (1994) trained a neural network on simulated data to learn the Black–Scholes option pricing formula. A number of efficient algorithms have recently been developed to approximate parametric pricing operators with flexible modelling assumptions (for example, see Horvath et al., 2020; Jacquier and Oumgari, 2019; Ferguson and Green, 2018; McGhee, 2018; Sabate-Vidales et al., 2018, 2020). This in turn can eliminate the calibration bottlenecks commonly found in using realistic pricing models.

The second application involves a more fundamental change in the approach to modelling and working with data, where traditional, low-dimensional, hand-crafted models are replaced with abstract over-parameterized models. These models may be used to represent the statistical features of underlying assets, to determine prices, hedges and risk properties of portfolios in terms of market observables, as well as a combination of these tasks. This application of machine learning depends in a far more significant way on the historical data available, leading to various challenges: for example, it becomes hard to understand what is driving the price of a derivative, and the data modelling and preprocessing steps might introduce an additional set of risks, which can be a cause of unease for regulators and risk managers.

In this chapter, for the sake of concreteness, we focus our attention on the challenge of pricing and hedging derivatives, which we outline in §33.2, and principally on the use of one machine learning method (deep neural networks) in this challenge. In §33.3 we discuss issues connected with the sources of data that are used as inputs into machine learning algorithms, while in §33.4 we are concerned with the risk associated with the way probabilistic modelling is incorporated within machine learning.

33.2 A practical application of machine learning

To get acquainted with a neural network solution, we will take a closer look at the problem of pricing and hedging an exotic derivative, by trading in a financial market. Here we give a non-technical description; for a technical primer on neural networks see, for example, the work of Bengio et al. (2017). The inputs to our problem will be a combination of historical market data and commonly accepted handcrafted models (depending on the precise approach), the latter may be used to generate additional simulated data for training purposes. The key outputs are prices, hedging strategies and risk assessments for exotic options and portfolios of exotic options and other assets.

33.2.1 How to use neural networks for derivative modelling

The precise role of machine learning in options pricing and the data used to support it can vary significantly. If we consider one particular class of machine learning methods – neural networks – in Figure 33.1 we present one way of classifying some applications of this method, looking at whether they principally are concerned with the processing and generation of data, or with building models for financial markets, and how these contribute to different outputs.

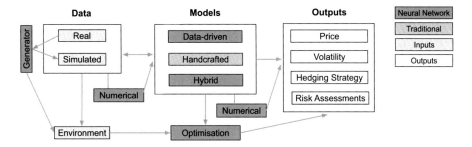

Figure 33.1 *Neural Network Pricing and Hedging.* This figure illustrates five broad roles for neural networks as part of developing pricing, volatility, and hedging models. Neural networks can learn directly from real data (Data-driven), can be used as a numerical or computational tool (Numerical), can enhance handcrafted models (Hybrid), can generate novel simulated data (Generator), and can be used in reinforcement learning models to develop strategies in a dynamic way (Optimisation).

Many of the use-cases for neural networks boil down to their ability to learn complex, high-dimensional, non-linear relationships; for example, to solve partial differential equations in high dimension, to develop data-driven models with large feature sets, and to find optimal policies in large state-spaces via reinforcement learning. The effectiveness of neural nets in high-dimensional settings suggests they have ability to overcome the computational curse of dimensionality[1]. We divide the neural network use-cases into five broad modes of application: Data-driven models, Hybrid models, Numerical approximations, Online Optimisation methods, and Generator models.

(a) **Numerical approximations** are based on traditional parametric models, and exploit neural networks to, for example, approximate pricing or hedging functionals in the form of solutions to parametric families of PDEs. This approximation can significantly speed up calibration problems. In these applications, the neural network is not trained against real-world historical data, but is used purely to approximate complex functions, in an efficient way. The key difficulty in this area is the calibration of hyperparameters and network architectures, and the implementation to industrial standards.

- Some applications depend on solving a high-dimensional and/or non-linear PDE, even if underlying model is simple. For example to price and hedge path-dependent options or to compute XVAs[2]. Neural networks can be used as a function approximation tool which works well in high dimension, and are particularly efficient at solving PDEs when blended with Monte Carlo simulation (Barucci et al., 1997; Beck et al., 2021; Sabate-Vidales et al., 2020; Sirignano and Spiliopoulos, 2018; Han et al., 2018; Gnoatto et al., 2020).

- Some problems, in particular in calibration of handcrafted models (for example, Heston or SABR models), require repeated calculation of various option prices prices under a variety of parameters. By providing an efficient means of approximating this calculation for a range of parameter choices, neural networks speed up the process of calibration, allowing a more efficient use of data (Andreou et al., 2010; Bayer et al., 2019; Sabate-Vidales et al., 2018). These methods often depend on simulating option values from the handcrafted model, under a range of parameter values.

(b) **Data-driven models** rely on real market data to approximate pricing and hedging functions. These models disregard handcrafted models in their entirety and simply use historical, synthetic or simulated data of any type to learn new relationships and features (Ghaziri et al., 2000; Montesdeoca and Niranjan, 2016).

(c) **Hybrid models** rely on historical, simulated, or synthetic data to approximate pricing and hedging functions and also constrain or impose knowledge onto the architecture of an otherwise unconstrained neural network.

- Some models first leverage a handcrafted model to estimate prices and then build a data-driven model to learn the difference or residuals between

---

[1] It is widely conjectured that neural nets are effective in situations where the problem at hand admits an accurate low dimensional representation, however, this representation is not known a priori (Fefferman et al., 2016).

[2] This is similar to the now classic use of regression in the Longstaff–Schwarz approach to pricing American options (Longstaff and Schwartz, 2001), where neural networks can give a more flexible approximating class of functions.

the observed price and the handcrafted model estimate (Lajbcygier and Connor, 1997).

- Other models constrain a universal neural network by adding domain knowledge to the architecture to learn more realistic relationships that increases the interpretability or efficiency of the model; for example, forcing monotonous relationships towards one direction by adding penalties to the loss function (Garcia and Gençay, 2000; Dugas et al., 2009; Gierjatowicz et al., 2020).

(d) **Online Optimization methods**: A number of option types, for example American options, benefit from learning optimal stopping rules using neural networks in a reinforcement learning framework; others may benefit from learning a value function or a hedging strategy that benefits from temporal optimal control, e.g. a model that takes evolving market frictions into account in an environment or control system (Buehler et al., 2019; Kolm and Ritter, 2019).

(e) **Generator models** can take any data as input and generate new data that has the same statistical properties. Data can be generated by applying a calibrated 'handcrafted' model (or from a range of handcrafted models) or from a machine learning generative model. Alternatively, they may be learn from data observed in one situation to generate representative data in a related setting. The first of these uses (where the generated data matches statistical properties of historical observations) is called 'synthetic data', and is a subset of 'simulated data', which includes scenarios that were not present in the historical data. The generated data's purpose is principally to aid the performance of machine learning pipelines, for example to provide an environment to train further models with reinforcement learning. It's worth noting that the generator, and hence the simulated data, should be seen as a statistical model[3] for our observations. This approach can be viewed as model-based data boosting (Buehler et al., 2020; Ni et al., 2020; Mariani et al., 2019).

Using a combination of these approaches, we can now build an abstract pipeline for learning to price and hedge options.

(1) Using historical data and current market data, build up a collection of training trajectories of the assets under consideration, as well as a representation of the state of the market. As we typically only have one trajectory of past data, one often needs to augment historical observations with models or simulated data. We have discussed two main approaches to this:

    (i) Design and train *generative* models to provide additional realistic data, or provide a rich parametric class with which to work.

    (ii) Train *handcrafted* models (possibly using neural networks as a *numerical* tool to speed up calibration) from which to simulate.

(2) Using this data, either

---

[3] Here the network architecture, loss function and training method of the generator are all modeling choices.

(i) Use reinforcement learning to *optimize* hedging strategies (and thus determine initial prices), using the simulator as a training environment.

(ii) Learn *data-driven* pricing relationships and hedging strategies by observing prices in historical data.

(iii) Learn a *hybrid* model that first trains on simulated data, and then transfers this learning over to real data for efficient training.

(iV) Use a further *numerical* approach, to solve the PDEs arising from the (possibly high dimensional) models. Equivalently, one can ensure the calibrated model generates trajectories with probabilities from a risk-neutral measure, and use Monte Carlo simulation to estimate prices.

33.2.2 Black-box trade-offs

The pipeline we have outlined above gives a very flexible approach to modelling and pricing of financial derivatives, however this is not a 'free lunch'. Neural networks are notoriously opaque as a modelling tool, and are often implemented simply as a 'black-box' approach to function approximation (the hybrid models discussed above being a partial exception to this). An important practical question is whether the potential disadvantages of black-box hedging can be justified by increased performance, and whether the risks associated with this approach can be distinguished and quantified.

An industry standard for assessing the quality of a new model is to compare it with simpler benchmarks such as Black–Scholes delta-hedging with the presence of transaction costs (Davis et al., 1993; Whalley and Wilmott, 1999). There is preliminary evidence that suggests, at least in simple, constant volatility settings, these benchmark models have performance close to that of reinforcement learning agents (Mikkilä, 2020). If that is true, these benchmarks should be preferred because they have easy to explain analytical solutions.

Ruf and Wang (2021) have shown, on an out-of-sample test set, that a simple fixed hedging strategy that hedges calls by $0.9 \times \delta_{BS}$ and puts by $1.1 \times \delta_{BS}$, where δ_{BS} is the delta under the classic Black–Scholes model, outperformed 14 out of 16 models, including all the supervised neural network models with 1-day rebalancing, and outperformed all models with 2-day rebalancing. It should be noted, that these tests were performed in a simple one period setting, with no transaction costs. These results do not directly extend to a large basket of derivatives; as a result, more tests are needed. However, these results suggest that more complex black-box models may fail to outperform simpler ones.

It is also possible that improved feature selection and simple models might be a better solution than a direct application of neural networks. Ruf and Wang (2021) compared a neural network model with a linear regression model to estimate the hedge ratio, on simulated and real data. Predictors in the linear model included standard model sensitivities under the Black–Scholes model: moneyness, Delta, Vega, implied volatility, time to maturity and Vanna. They conclude that the classical option sensitivities already contain the non-linearities necessary to build an effective hedging strategy for common options, in a financially significant and efficient way. They further showed that this linear regression model outperformed their neural network model.

Their approach diverges from Buehler et al. (2019), who do not use option sensitivities as variables, but instead rely on the belief that the Greeks indirectly present themselves as non-linear functions that the agent has access to via the market state, in the form of hedging instrument prices. Future experiments should test hedging performance on a basket of derivatives, in a multi-period setting, with dynamic volatility, transaction costs, and environmental feedback. Such a real-life setup could benefit the deep hedging approach (Buehler et al., 2019), as this has the capacity for direct feedback from the environment and online training.

33.3 The role of data

33.3.1 Data risks

We now focus our attention on how the use of machine learning methods highlights risks associated with the underlying financial data. One can identify three primary sources of risk: biases in the training data, erroneous data or erroneous preprocessing, and legal and regulatory data risks. We will not focus on legal and regulatory risks. As we are moving the dial from handcrafted towards data-driven models, the data risk increases significantly. On the one hand, handcrafted models are more robust to biases and errors in the data, and the risk of using inadequate models is easier to detect. On the other other hand, for data-driven models the training data becomes an integral part of the model, making them more sensitive to data risk.

Biased data

A key issue in financial data is that the majority of data is backward-looking, and there is no guarantee that future behaviour of financial markets will be represented by historical observations. We typically only have one trajectory of historical data – we cannot see what *might* have happened in different scenarios – which makes it difficult to build a clear view of the range of likely outcomes in the future. Any recent changes in the true underlying state of a system that are not incorporated in a model's training dataset will also lead to biased predictions.

These risks are not particular to machine learning – they are well known issues in financial markets; however, the use of machine learning models, which often depend on observed data in a more significant way than traditional handcrafted models, means understanding and managing these risks is critical for the success of these approaches. Here we summarize some key forms of bias that historical data could exhibit:

(a) **Backward looking**: most data reflect prices and signals obtained in the past. This means that data could reflect a state of affairs that no longer applies, e.g. an options model might only have access to data from a low volatility period, or from an old regulatory regime. Financial markets are reactive and don't follow universal laws – for example, the increase of high-frequency trading has changed the nature of many financial markets (see, for example, the discussion in MacKenzie (2018)). Restricting to only the very recent past, and projecting a model's predictions only into the near future, can mitigate this concern somewhat, but often results in significantly less data being available for training.

(b) **Spurious correlations**: for some financial domains it is prudent to record and collect attributes that have some theoretical basis. For example, it is questionable whether an option pricing model should contain sentiment features. It is often difficult to identify intuitively unreasonable relationships within a black-box model, and the increase in dimensionality of the models being used results in a vast increase in the range of potential relationships that could be inferred (Fan and Zhou, 2016). Spurious correlations are well known in finance, but the increased use of machine learning techniques can exacerbate this problem.

(c) **Sample disparity**: biases in the sampling procedure might lead to data that doesn't fairly represent the state of the market. For example, a firm may wish to use the same algorithm for trading over multiple exchanges and geographic locations, each of which has subtly different conventions and data. This introduces biases within the data which can be magnified through the use of machine learning models, particularly when a model is trained in one setting then deployed for use elsewhere.

(d) **Imbalanced inputs**: some evidence shows that even when your sample accurately reflects the true state of the market, it remains imbalanced – rare events may be significant, but are only infrequently represented in historical samples. Many data-driven models are known to favour the performance of the majority outcome, to limit overall model errors (Provost, 2000). Within a financial setting, this might correspond to a model which performs well in low-volatility regimes, even when high-volatility periods are observable (infrequently) within the training data. A related idea is the use of stressed periods for calculating risks within the Basel accords – these infrequent periods are significant to overall performance, and need to be explicitly taken into account.

(e) **Insufficient data**: one often has insufficient data to use machine-learning models well. The calibration of neural networks requires significant quantities of data, which are often not available for training in a financial context (Gu et al., 2018). The richer the context in which an algorithm is to be run, and the more finely tuned its behaviour needs to be, the larger the quantity of data needed. It is worth emphasising that this is not to say that the data in finance is 'small', but that often it is not 'large' in the directions needed – we may have enormous datasets due to high-frequency observations of a large number of asset's order books, but these will be of little use in determining good models over long time periods.

Data errors and preprocessing

A further concern in many applications is that data may display subtle errors, which need to be addressed before it can be effectively used. This is a common concern in many applications of machine learning, and data-cleaning methodologies form a key part of the implementation of these methods.

(a) Observed financial data can fail to satisfy fundamental economic constraints, which can be subtle. For example, as discussed in Cohen et al. (2020), historic options price data, for both listed and OTC contracts, may be inconsistent with no-arbitrage constraints, particularly in emerging markets. If such data is naïvely used when training a trading system, it is plausible that the system would learn to exploit this apparent arbitrage opportunity. Given these errors could arise due to multiple sources (for example, stale quotes being listed as live in historical data), this can lead to significant error in the resulting learnt behaviour.

(b) When working with time-series data, it is critical to respect information flow when e.g splitting data into training, validation, and test sets; engineering features; or normalising data. Errors in this process can 'leak' information from the future, leading to unrealistic performance.

(c) Financial data often has a particular concern around precise timekeeping, which may not be reflected in the accuracy of the data given. Particularly when working with very high-frequency data, failing to take into account latency and other implementation issues can have a significant effect, which may not be well reflected or available in historic data (see, for example, the effect of latency in Cartea and Sánchez-Betancourt (2021)). This is particularly the case with the increased attention being given to non-market data sources (for example signals from online news sources), where historic time-stamping may be of low quality.

(d) Financial data is often heavy tailed and not stationary, making it difficult to detect and exclude erroneous data. Typical methods (such as Winsorizing) have the potential to introduce significant bias, particularly when considering extreme events.

33.3.2 Data solutions

There are many process improvements that can be implemented to decrease data risks, for example, performing data quality monitoring, documenting and reviewing the manipulation of input data, and educating and training individuals involved in data manipulation tasks. Another key approach, to fix biased and limited data, is to generate synthetic or simulated data which is free from (or even corrects for) these issues. We will outline two key approaches – the top-down approach of synthetic data generation, and the bottom-up approach of agent based modelling.

Synthetic data generation

Synthetic data generation (SDG) is a top-down data generation solution. It can help to address some of the data biases and errors listed above. It does so by augmenting the quantity and quality of historical data, but it does not attempt to provide a simulator which can model feedback effects for an agent's interventions in a market.

At a high level, a synthetic data generator attempts to build a probabilistic model which would generate observations similar to historical data. Generative models such as generative adversarial networks (GANs) and variational autoencoders (VAEs) have demonstrated great success in seemingly high dimensional setups (Wan et al., 2017; Lin et al., 2020). If used correctly, SDGs could allow for a more comprehensive approach to future-proofing and validating machine learning pipelines; ameliorating some structural deficiencies in data and amending distributional biases (Louizos et al., 2016).

In a financial context, Takahashi et al. (2019) have shown that (GANs) can be used to generate synthetic data that matches most known stylized features of returns; Ni et al. (2020) have shown how mathematically principled feature extraction methods such as signature models can be used to efficiently implement conditional GANs for generic time series data. Related ideas, but combined with VAEs, are presented in Buehler et al. (2020). Henry-Labordere (2019) developed efficient algorithms building upon optimal transport theory and highlighted an interesting application of data generators for detecting anomalies. Algorithms based on restricted Boltzmann machines have been developed in Kondratyev and Schwarz (2019), who coined the term 'market generators'. An alternative approach is to learn the underlying dynamics of the system, allowing a path to evolve through time – this is the approach taken by neural-SDE models (Gierjatowicz et al., 2020). Fu et al. (2019) have shown how conditional GANs can be used to produce synthetic data for different market scenarios. Koshiyama et al. (2020) have shown how these methods can be used to validate trading strategies.

SDGs still pose a form of modelling risk, since a generator is only as good as the data from which it constructs its generating function; building an SDG involves choosing a metric, a loss function, and a training algorithm for parameter selection. As such, SDGs introduce model risks within the data used to train downstream models, and these risks may be difficult to identify depending on the use-case.

At the present time, research in this area lacks standardised benchmarks and theoretical guarantees. Most off-the-shelf methods are not built with financial applications in mind, and are therefore likely to generate simulated data which exhibits arbitrage or other economically unrealistic phenomena. Moreover, many of these models remain black-box and are not easily interpretable.

The key benefit, however, is that these methods are expressive and work in high dimensions. This is the main difference when comparing with traditional methods using handcrafted features. Synthetic data can also be used to generate data according to expert opinions and known facts, e.g., can be conditioned to form the observed volatility smiles. And SDGs generally offer a more accurate and robust oversampling method than traditional methods like SMOTE (Synthetic Minority Oversampling Technique) that simply repeat existing records (Chawla et al., 2002). They also provide a convenient solution for missing data imputation and outlier treatment (Xu and Veeramachaneni, 2018).

Synthetic data generation tools can be used as part of a larger solution to address some of the most common upstream data errors. They can be used side-by-side with federated learning techniques to improve the quality of single standing resources, by pooling data across, departments, subsidiaries, companies, or data-providers (Goetz and Tewari, 2020).

Deep generative models for synthetic data generation remains a new field, and although they have potential to alleviate some of the known issues of neural network models, it is clear that they have the potential to introduce further risks. Overall, as with other methods, they can be seen as shifting risks away from the quantity and quality of data, by including probabilistic modelling (with its associated risks) at a very early stage in the analysis pipeline.

Market simulator engine

Agent-based model (ABM) simulators, unlike SDGs, are a bottom-up data solution and date back to the 1990s. Notable early models include those by Levy et al. (1994) and the Santa Fe Artificial Stock Market (Palmer et al., 1994; Arthur et al., 1996). ABMs model markets as evolving systems of competing, autonomous interacting agents (LeBaron, 2000).

The development of ABMs has seen multiple waves of interest. The first wave of market simulators in the 1990s was a deliberate move away from classical economic theories to advance financial market knowledge, the second wave was a reaction to the failure of economic models in foreseeing the financial crises of 2008, the third wave was a call to understand high frequency trading and the flash-crashes in 2010 and 2013, and the fourth and current wave combines the concerns with the past, but emphasises the use of simulators to train machine learning agents.

A key advantage of ABMs is that, as bottom-up models, they attempt to learn the feedback effects of agents acting within the market. This has the advantage that these effects can be modelled, but makes training much more difficult – usually involving explicit modelling decisions, and requiring more data to train. Again we see that the issues of historical data not containing counterfactual histories, or being too limited for our purposes, are being addressed, but doing so introduces increases our reliance on statistical models, rather than on observed data.

Modeling feedback is important for training environments to be realistic. For example when hedging or trading strategies are trained and tested on historical data, the success of the model still cannot be reliably demonstrated, even when using holdout sets for validation. Training environments with appropriately modelled feedback can, at least partially, mitigate this issue. Such training environments are also critical for deploying on-line reinforcement learning solutions as they allow pre-training of these systems before implementation in the real market. This is critical in applications where the costs and risks of exploration are significant.

Agent-based modelling has, in recent years, allowed for the design of high-fidelity simulated markets (Belcak et al., 2020; Byrd et al., 2020). These artificial markets can run millions of in-silico trials to test counterfactual theories, research emergent phenomena, and train and test algorithms.

A current trend is that quantitative funds are looking to establish risk management systems that develop scenarios with no historical precedent[4]. With a simulator, one can perform training and backtesting for trading, execution, and placement algorithms under various conditions. Causal assessments can be performed for market impact and market slippage. Lastly, simulators can also be used as a means of generating synthetic data, given that financial data of sufficient granularity is often highly proprietary and/or expensive to access.

Standardized cleaning and preprocessing methodologies

Issues surrounding data quality often are specific to the particular use-case. The increasing use of varied data sources, often with little standardization, will inevitably result in the preprocessing of financial data becoming more important.

Some approaches, for example the no-arbitrage constraints for option books in Cohen et al. (2020), rely on preprocessing data to conform with prescribed characteristics. In this case, given the no-arbitrage constraints restrict the range of possible option prices significantly, imposing these requirements has the potential to address errors coming from a variety of sources. These methods can also be run on data coming from a SDG or ABM, in order to ensure that the simulated data is economically reasonable.

An approach which can serve to highlight potential concerns, is to look at the sampling frequency and periods of data. By comparing the results of using different but comparable datasets, it is possible to gauge the stability of calibration and models, and hence to identify causes for concern.

More generally, learning from other areas of machine learning, the development of common examples, codebases and resources, in an open-source manner, has the potential to improve the identification and processing of data errors. A significant risk is that inappropriate methods for dealing with data errors will be separately developed, implemented and used, without sufficient oversight or criticism. The use of well-developed, understood, and standardized tools is a key part of modern machine learning, and the development of preprocessing tools appropriate to finance should be seen in this light.

As part of this, the development of publicly discussed use-cases, with realistic data, would allow for new methods to be evaluated in a consistent manner, and for best-practice to be developed. While this is the case in other areas of machine learning, there is still much scope for improvement when it comes to financial data and problems.

---

[4] In 2017, Jane Street published a technical presentation of their own exchange, motivated to train and test new algorithms and models. It has been reported to handle messages in the rate of 500k/second with latencies in the single digit microseconds (Nigito, 2021).

33.4 The role of models

33.4.1 Model risks

As we have discussed, the use of machine learning changes, but does not eliminate, the use of classical mathematical modelling. Classical models may appear explicitly in machine learning methods (for example, in a hybrid pricing model), or may be subtly incorporated in the simulations used to support more explicitly data-driven approaches. Typically, however, machine learning methods aim to construct models from flexible ('non-parametric') families, combining the classical tasks of model selection and calibration into a single step. In this section, we will discuss the risks which arise from these modelling decisions, in a machine learning context.

It is worth noting that the importance of model risk depends strongly on how machine learning methods are used. Using machine learning tools for numerical procedures typically introduces little additional model risk, as one can often verify the solutions using other techniques. For example, when using a neural network to estimate option prices, for the sake of quickly calibrating the parameters of a classical model, it is straightforward to verify (using traditional PDE or Monte Carlo methods) that the calibrated model gives the correct prices of those options – the neural network is only serving as a numerical tool.

Conversely, end-to-end deep reinforcement learning, for example of a hedging strategy, exposes users to risks in multiple forms: models with too many parameters risk overfitting to available data, leading to both poor performance and a misunderstanding of a model's accuracy; the common use of synthetic and simulated data hides an additional layer of model risk in the training environment; complex models are more exposed to reward hacking, poisoning attacks, and other adversarial concerns and are typically less interpretable than simpler models.

The problem of calculating a price for a financial derivative which is consistent with the market can be seen as equivalent to finding a map that takes market data (e.g. prices of underlying assets, interest rates, prices of liquid options) and returns the no-arbitrage price of the derivative. One way to do this is to select a martingale model (to prevent arbitrage) that can be calibrated to market data, by which we mean that the model matches the observed prices of liquid assets.

While this is a dominating approach in the industry, the introduction of a model necessarily introduces model risk, and there are infinitely many models that can fit market data. In the robust finance paradigm, see Hobson (1998); Cox and Obłój (2011), one takes a conservative approach and, instead of computing a single price, one constructs pricing intervals that are consistent with market data. Without imposing further constraints, the class of all calibrated models might be too large, and consequently, the corresponding pricing intervals too wide to be of practical use (Eckstein et al., 2021). It is therefore natural to consider a smaller search space of models (e.g. SDEs with continuous coefficients) and use data and machine learning to select an appropriate model (i.e. the coefficients of the SDE).

This approach has been recently applied in Gierjatowicz et al. (2020). The key idea is to use SDEs to describe the model dynamics but, instead of fixing its coefficients, to allow the drift and diffusion to be given by an overparametrized neural network. These 'neural-SDE' models not only provide a systematic framework for model selection, but can also produce robust estimates on the derivative prices.

A concern for model risk is not new in finance, but the use of machine learning methods can be seen as typically emphasising some risks over others. In Table 33.1, we present an overview of the typical distinctions between handcrafted and machine learning perspectives on model risk.

Table 33.1 Comparing Typical Risks Between Handcrafted and Machine Learning Methods

Risk	Handcrafted	Machine Learning
Structural Risk	Lower-dimensional models which are easy to calibrate, but fail to capture all aspects of the market's behaviour. Generally a higher bias than variance and more prone to underfitting.	High-dimensional models which require large amounts of data to calibrate, but can capture fine detail when fitted well. Can often incorporate new sources of information in a convenient manner. Generally a higher variance than bias and more prone to overfitting.
Model Sensitivity	Few parameters and model inputs. Model outputs vary smoothly with calibration and input. Well understood sensitivities to erroneous inputs.	High-dimensional parameters and data inputs. Model outputs can vary sharply with inputs. Sensitivities to erroneous inputs can vary significantly.
Adversarial Attacks	Reasonably robust calibration and not susceptible to data poisoning attacks. Calibration can be easily monitored by users. Adversarial defences not a key part of most models.	Susceptible to attacks, require robust training and adversarial defences, but these can be incorporated as a key part of the model. Not easily monitored by users.
Model Drift	Models naturally incorporate economic intuition and underpinnings. Few parameters to update online, but do not often incorporate updating as a core part of the model.	Model based on data patterns which may change over time. Many parameters need to be updated dynamically, which can lead to unstable behaviour. Model updating can be included as a core part of the approach.

Structural risk

Within a machine-learning paradigm, one usually combines the stages of model selection and calibration. Given data on a supposed relationship or phenomenon, one aims to directly fit a model to this data with which to predict, simulate and build understanding.

For our example of pricing and hedging of options, we can focus on the task of pricing an option given historical market data. Our data consist of historical observations of market data, and we aim to build a function which can take new observations and provide us with prices in the future. To do this, some basic modelling assumptions are unavoidable:

- Does the price of an option depend only on the current market state, on the recent past, or on a long history of market observations. Equivalently, what are the inputs to the pricing function that I wish to find?
- Do I wish to make conditional predictions (say of an option price given a stock price) or do I wish to give simulations of both simultaneously?
- Does the relationship between market observations and prices remain stable through time? If not, how do I choose training periods which are representative of the situations where I will apply my function in the future?
- If the observed prices are not perfectly predicted by market data, so I have noisy observations, are the noises independent, or are they correlated between times and assets?

In each case, the answer given to these questions will be incorporated in our machine learning model, and introduces model risk at a structural level.

These general concerns are common to both classical and machine learning methods, however the increased flexibility of machine learning methods may suggest that (as one can include more observations in a model), they would be less present in a machine learning approach.

Even after these general concerns are addressed, machine learning methods introduce risks similar to the 'model risk' of classical mathematical finance. Within the paradigm of machine learning, models are not chosen explicitly but implicitly, through the choice of training data, training algorithm and the often *ad hoc* choice of a large parametric model (e.g. a neural network and its architecture). Unlike handcrafted methods that are explicitly specified, or hybrid approaches relying on feature engineering, neural networks construct an internal representation of features to capture and approximate functions.

With neural networks, model specification is not in the direct control of the modeller. Due to this flexibility in feature specification, a larger space of plausible models are explored than in traditional or many other machine learning approaches. The cost of this flexibility is that the model selected may not be the 'best' available. Since the fitting of traditional models typically involve solving some convex optimisation problem, a best model can be identified due to the existence of a unique minimum. Neural networks fitting techniques are typically non-convex and many good solutions can be found.

Adding fuel to the fire, neural networks are known to be sensitive to initialisation conditions (McMormack and Doherty, 1993). Moreover, many sources of randomness are often injected into the training phase of neural networks, this includes the use of dropout (where some neurons are randomly set to zero for network regularisation), early stopping (where the process of gradient descent stops when the performance on a validation set stops improving), and stochastic gradient descent (where random selections of observations are used to fit the network). These additional factors introduce uncertainty in the output of neural network models. The injection of noise during training is critical to the performance of these methods, and it leads to, so called, implicit regularisation (Neyshabur, 2017). That means that stochastic gradient descent methods select regularised solutions, even though regularisation is not explicitly incorporated at the training stage (Heiss et al., 2019). In this sense, the model selection step of classical approaches is replaced by the choice of training algorithm, which has a less easily understood connection with model performance.

Drawing from interpretability research by Lipton (2018), any model's transparency can be broken down in *simulatibility*, *decomposability*, and *algorithmic transparency*. With simulatibility, a human should be able to step through each of the operations in a reasonable time; with decomposability, each part of the model has an intuitive explanation that is understood in isolation; with algorithmic transparency, there are theoretical guarantees about the behaviour of the algorithm, for example certainty of convergence. Going down this checklist it is clear that neural networks lack simulatibility and decomposability because the parameters in the hidden layers do not have an intuitive explanation. Moreover, for non-convex problems stochastic gradient descent is not guaranteed to converge. Instead, one can show that the weights of neural networks are represented by Monte Carlo samples from optimal distribution over the parameter space. This perspective allows one to establish convergence guarantees, but does not help with the issue of interpretability (Hu et al., 2019; Jabir et al., 2019).

Model sensitivity

A key selling point of neural networks is their ability to work with high dimensional inputs. However, this comes with a well documented issue of sensitivity, where the learnt relationships vary wildly with small perturbations to the underlying inputs.

Models are known to be fragile when using high dimensional inputs. The reasons are numerous: given the randomness involved in training neural networks, some inputs may spuriously be considered important. This is a particular issue when only limited data is available, or simulated data (from a low dimensional model) is used as training data – simulated data will typically not explore a full range of market conditions (as it is constrained by the model from which it's generated), and so the neural net will not learn to provide good answers when novel conditions are encountered. Secondly, when many inputs are used within a model, there is an increased probability that some variables might not be available when a model is put into practice.

Since the model specification of neural networks is implicit, the modeller and end-user of these methods will often no longer understand how the model has been fit, significantly increasing model specification risk. Consequently, it is not clear how we can quantify sensitivity of the model. The field is therefore largely left with developing more interpretable model alternatives (Nakagawa et al., 2019) or using post-hoc explanations to assess and visualise what models have learned (Li et al., 2020). This however also comes with risk as many post-hoc explanation are not robust and may lead to false sense of security (Anders et al., 2020).

Robustness and adversarial attacks

The competitive nature of financial markets often leads to particular concerns for machine learning models. As models are used in increasingly automated ways, they need to be able to respond to the pressures placed on them by competitive forces, who have strong incentives to identify and exploit potential weaknesses of a model.

For example, we could consider our challenge of managing an options portfolio, but in a context where market price impact reduces the efficiency of trading. A classic model for order execution with market impact, Almgren and Chriss (2001), yields deterministic policies for executing a large buy or sell order, which may have the undesirable effect of 'information leakage' (revealing your strategy to other market participants) when used in an illiquid market. In the more complex situation of managing a portfolio, one could consider building a neural network model to perform this task (for optimal execution, a model of this type is given in work by Ning et al. (2018)). The additional randomness of the neural network model would arguably assist in preventing information leakage, when compared to the traditional model. Nevertheless, it is *a priori* unclear whether this additional randomisation would be sufficient, or whether further precautions against information leakage would be needed.

Adversarial attacks can be grouped into many categories, for example, attacks can either be intentional or unintentional. Behzadan and Munir (2018) splits them into attacks on model confidentiality, integrity (does the model behave as intended?), and availability (can the model be disabled by an external actor?). Attacks could also be split into the components that are susceptible to the attack, for reinforcement learning this includes the environment, the observation channel, the reward channel, the decision making system, and the online training system.

We can consider various way in which a financial reinforcement learning agent can be attacked, with a simple description and illustrative example. These classifications have been adapted from the adversarial threat a matrix developed by MITRE in collaboration with Microsoft, IBM, NVIDIA, and Bosch (Kumar et al., 2020).

In Table 33.2, we present examples of adversarial attacks against a trading system. We first list those which are internal to the company, many of which can arise inadvertently in building and implementing machine learning methods and then follow with examples of attacks that an adversary can exploit without having direct access to a trader's codebase. The examples are our own, and are purely illustrative.

These intentional and unintentional attack examples are hypothetical, and relate to problems seen in other machine learning domains. Nonetheless, these examples have significant implications for financial model risk management. A substantial level of compounded risks could exist where multiple of these susceptibilities overlap.

Although there is a need to test and benchmark the robustness and resilience of trading agents with private systems and historical data, these agents ultimately

Table 33.2 Examples of Adversarial Attacks in Finance

Failures	Description	Example
Reward Hacking	When training, the stated reward differs from a true reward.	A learning agent was trained to create a perfect hedge, however transaction costs were poorly modelled, leading to poor performance.
Side Effects	A reinforcement learning system disrupts the environment by advancing its goal.	A model has learned an order execution strategy for an illiquid asset, but by executing this strategy, changes the dynamics of the order book significantly, leading to increased risk.
Distributional Shifts	The system is trained on one environment, but unable to adapt to changes.	A pricing model was trained on data during normal times, and is unable to react to the higher correlations between assets during crises.
Natural Adversarial Examples	Even without being attacked, the system fails from natural errors.	A pricing model was trained individually for each strike and maturity, resulting in arbitrageable prices being offered in the market.
Common Corruptions	The system is not able to deal with common corruptions.	A pricing model failed due to a halt on trading being placed on a closely related underlying instrument.
Incomplete Testing	The system is not tested on the right environment nor over multiple periods.	A pricing model is tested only on one exchange, but is deployed in multiple locations with differing market behaviours.
Poisoning attack	Contaminate training phase.	Contaminated data is introduced into a pricing model, for example when using sentiment analysis based on social media.
Model stealing	Recover the entire model.	A proprietary model is trained and can be queried online by counterparties. By repeated queries it is possible that the inputs can be matched with the outputs, to reverse engineer the original model.
Model inversion	Recover hidden features.	A pricing model is trained using proprietary trading data on market impact. The fitted model is then made public, without the underlying data. By repeated queries, it may be possible to extract the training data used (Fredrikson et al., 2015).
Reprogramming system	Repurpose system for other use.	An online pricing model is used to identify expected future market volatility.
Adversarial example in physical domain	Fool a system by changing some interface component.	An adversary determines that a pricing model has sensitivity to the volumes deep in the order book – by posting to this part of the book, they influence the model's behaviour.
Exploit software dependencies	The use of traditional software exploits.	The model relies on code dependencies; these dependencies are exploited by modifying the code to introduce nonsensical values, leading to a trading halt. (The 2016 NPM/left-pad debacle illustrates this external dependency risk, where a disgruntled developer deleted a tiny piece of code that 'broke' the internet (Collins, 2016).)

have to move to the real world, where a slight distributional shift could impair performance. In other areas of machine learning, in addition to internal testing, models can be subjected to public audits. However, in finance the competitive risks from revealing private models are significant, leading to a far lower level of transparency.

Model drift

A good model not only fits historical data well, but also captures changes in the environments in which it is deployed. The challenge of updating models exists in both handcrafted and machine learning models, and reflects the basic challenge that finance does not operate according to stable physical laws, but arises from the interactions of many agents.

The challenge of changing market behaviour can be significant: the overwhelming belief is that the value of a derivative and its underlying are kept in line due to no-arbitrage. However, during the 2007-08 financial crisis, these relationships were observed to break down, as arbitrage calculations did not account for counterparty creditworthiness. As a result, a theoretical arbitrage opportunity was observable in the market, but was not available in practice (Baba and Packer, 2009).

Handcrafted models, typically, require updates of few parameters to capture the shift of the data distribution. For overparameterised models, this may not be the case, and a small change in the data may require a significant change in the model. For example: fraud detection models lose their discriminatory power against maliciously evolving strategies, hedging strategies have to evolve as market conditions are changing.

Off-line machine learning suffers from a lack of robustness to distribution shifts, and hence a lack of on-line monitoring can significantly impair its performance (Sugiyama and Kawanabe, 2012). This has become particularly clear in recent years in other applications of machine learning. For example, in the airline industry it was quickly realised that the standard machine learning pricing models that study flight patterns, fuel costs, and user behaviour became useless during the covid-19 pandemic, with data scientists choosing to fall back on traditional macroeconomic modelling (McCartney, 2020).

On the other hand, online learning approaches have the promise of being able to dynamically and naturally adapt to new situations (Zeng and Klabjan, 2019; Soleymani and Paquet, 2020). This comes with significant issues, however, as these methods require training at a meta-level: the rate at which they adapt to new information needs to be tuned and adjusted, with rapid adjustment speeds typically associated with increased volatility in performance.

33.4.2 Model solutions

Any given model provides only a crude approximation to reality; the risk of using an inadequate model is often hard to detect and quantify. While modern data science techniques are opening the door to more data-driven model selection mechanisms, this comes with new risks, as described previously. In this section, we argue that by combining old and new approaches, it is possible to regain control over newly emerging risks (e.g. lack of interpretability) while improving over classical models currently favoured by industry. We base our presentation on a few hybrid modelling approaches which have recently emerged in the research literature.

A natural idea is to incorporate prior knowledge/modelling into deep learning. This can be achieved through incorporating modelling constraints during the training. However, as the number of constraints increases, and hence the search space of possible network parameters decreases, stochastic gradient descent algorithms struggle to find good solutions, so bespoke machine learning methods need to be developed.

Machine learning as a numerical tool

As mentioned above, using machine learning as a numerical tool introduces only modest model risks, while potentially providing significant speed and accuracy benefits. In Sabate-Vidales et al. (2018, 2020), the authors developed deep learning algorithms for solving parametric families of (path-dependent) partial differential equations (P)PDEs that arise in pricing and hedging. The key idea in these works is to use a probabilistic representation of the (P)PDE, and learn both the solution and its gradient simultaneously. An advantage of this approach is that the gradient of the solution to the (P)PDE provides access to the hedging strategies. While this method is of interest in its own right, it can also be used as a control variate for unbiased Monte Carlo pricing. In other words, by combining deep learning with standard Monte Carlo pricing, one can remove the bias due to approximation with neural nets and easily compute confidence intervals (which are, in general, hard to obtain for large networks). This approach has been tested on several models and (path dependent) payoffs. We stress that while the literature on deep learning for PDEs is growing rapidly, for finance applications it is critical to approximate parametric *families* of PDEs, where parameters correspond to the possible values of calibrated coefficients of the model. A similar observation has been made in Horvath et al. (2020).

Another interesting approach, one that combines ideas emerging from ML and classical modelling has been put forward in Lyons et al. (2019). The key idea here is to lift both modelling and pricing into the signature space. Intuitively, signatures provide efficient basis functions for representing functionals defined on the path space (e.g exotic derivatives or non-Markovian models) and play a similar role to polynomials on Euclidean space. In particular, the signature expansion of a path represents the values of integrals against that path, and so can capture the effect of dynamic trading and hedging. The classical idea of replicating an option via trading in the market then reduces to regressing the option payoff on the signature of the underlying and other vanilla securities.

It has been shown that one can effectively represent many exotic derivatives using this signature expansion, and consequently obtain the prices of derivatives in terms of the expectation (under the pricing measure) of the signature expansion terms. Consequently, one only needs to calibrate expected signatures to market data, which in some settings can be done efficiently. The advantage of using signatures when compared with recursive neural networks is that the computational cost does not increase with the number of time points in a time-series.

The idea of model selection using signatures has been proposed in Arribas et al. (2020). Here, one still works with the familiar SDEs type model but aims to learn (possibly non-Markovian) coefficients from data.

Expert knowledge

A viable approach to controlling the risk of non-transparent model specifications is to develop algorithms and training methods that embed expert knowledge into the architecture or training stage of machine learning. A handful of papers have attempted to embed financial domain knowledge into their models. These methods can offer regularisation, efficiency, consistency, and stability benefits.

Drawing from the review by Ruf and Wang (2020), methods that adjust the *architectural* design of neural networks include models that incorporate a homogeneity hint by training a neural network in two parts, the first part controls for moneyness, and the other for time-to-maturity (Garcia and Gençay, 1998). Other methods restrict the shape of outputs (Dugas et al., 2001) or enforce no-arbitrage conditions such as the convexity of a neural network pricing function and monotonicity (Zheng et al., 2019).

Approaches that impart expert knowledge at the *training* stage include data augmentation, which involves the generation of synthetic data to help with neural network training (Yang et al., 2017), adjustments in the penalty terms of the loss function to promote no-arbitrage (Itkin, 2019; Ackerer et al., 2019), as well as the development of bespoke training algorithms for neural networks for options hedging, including the use of the extended Kalman filter, sequential Monte Carlo, and evolutionary algorithms (Niranjan, 1996; de Freitas et al., 2000; Palmer, 2019).

Benchmarks

A safe and efficient transition toward using machine learning in finance is only possible when models and methods are well understood and tested on reliable data sets. In other areas of machine learning, standard benchmarks and data sets are a common way to proof-test new methodologies. For example, recent advances in computer vision or reinforcement learning were significantly accelerated due to the emergence of challenging benchmarks, such as ImageNET (Deng et al., 2009) or ALE (Bellemare et al., 2013). These benchmarks have enabled open, systematic cross-validation of various AI solutions.

In machine learning, the term 'benchmarking' has been used to refer to the evaluation and comparison of machine learning models, particularly regarding their ability to learn patterns from benchmark datasets (Olson et al., 2017). This process can be thought of as a check to validate the improvement of a new method, but also more broadly to identify the respective advantages and disadvantages of each method. Comparisons can be made across a wide range of metrics, for example accuracy in detecting signals, interpretability, and computational complexity.

Currently, in finance, various algorithms and machine learning methods are tested on disparate data sets, which are often only accessible to a small community or at high cost. A consequence of this is that very little comparison of methods is done, and we have little understanding of the appropriateness or optimality of these methods. In addition, evaluating new AI techniques on real-world applications often requires expert domain knowledge and consideration of scalability and the cost of development.

A key difficulty, in financial applications, is that a more open approach to benchmarking will often involve revealing details of each participant's methodologies. While this is reasonable within the academic community, within industry it is clear that confidentiality is needed, both regarding algorithms and, in some cases, their performance. For this reason, it is important to build our understanding of which problems can be discussed and benchmarked in a public way, and which related data science problems provide insight for those cases where confidentiality is needed.

The typical datasets which the benchmarking literature has well studied come from real-world data and simulated data with known underlying patterns. As alluded to before, in finance there are relatively few datasets that have been made publicly available, and often these contain only a small sample of the data that would be needed in practice. There is therefore a clear opportunity for Finance to benefit from synthetic data generators. Synthetic data has been used in other fields[5] but has not yet flourished in the financial literature.

Benchmarking has its own problems, many of which are not new to machine learning. There has been an increasing concern that published research findings are misleading due to the number of studies addressing the same question and datasets (Ioannidis, 2005). Benchmarking has a similar problem, in that a lot of models are prodding the same unchanging datasets leading to a lack of generalisation. Studies reveal that the accuracy of state-of-the-art deep learning models can drop from 4%-10% when moving to a new test-set, highlighting the risk of overfitting (Recht et al., 2018). For this reason, the regular evaluation and updating of benchmarks remains important for future development.

Adversarial defenses

In order to be reliably implemented, algorithms must be robust with respect to a variety of objectives (e.g. safety, accuracy). Summarizing the range of adversarial challenges outlined above, we see that machine learning pipelines should come with robustness guarantees against: (i) shifts in data distribution (distributional robustness), (ii) intentional input manipulations (adversarial robustness) and (iii) intentional feature manipulation to 'game' the system (strategic robustness).

Recent work (Huang et al., 2017) has begun to address these issues for neural

---

[5] For example, the Open Graph Benchmark, released in 2020, has become a popular repository of challenging and realistic benchmark datasets to help facilitate scalable, robust, and reproducible graph machine learning research (Hu et al., 2020).

network based models. Drawing on adversarial machine learning and distributionally robust optimisation (Rahimian and Mehrotra, 2019; Cohen et al., 2019a; Wicker et al., 2020), it is possible to certifiably train models to provably ensure robustness, by providing guaranteed bounds on the probability of the model output (decision) satisfying a combination of objectives.

Data-driven models cannot automatically guarantee model robustness (Kwiatkowska, 2019). An adversarial defence is anything that decreases the efficacy of adversarial attacks. There are a range of techniques that can be used to provide adversarial defences; they can generally be classified into adversarial training methods, randomisation-based schemes, denoising methods, and provable defences.

- Adversarial learning techniques simply train a neural network using adversarial samples. It is one of the most effective defences against attacks as revealed in benchmark studies (Madry et al., 2018). These can be thought of as a preprocessing technique.
- Randomisation schemes can also protect against perturbation in inputs. These generally involve some transformation, such as random resizing, or can also be achieved by adding a noise layer to the neural network (Liu et al., 2018).
- Denoising inputs in the prediction phase can help to rectify or remove adversarial perturbations. This denoising can be done with generative adversarial networks or autoencoders and can be thought of as a postprocessing technique (Xie et al., 2019).
- Provable defences are unlike the above approaches in that they are theoretically proven, rather than purely being experimentally validated. These methods can certify a level of robustness before the prediction stage (Balunovic and Vechev, 2019).

The defences listed here can only verify and protect a system against a limited number of attacks. Security vulnerability attacks will have to be dealt with using domain expertise, rather than relying on generalist defence mechanisms. Adversarial defences will not protect against a badly developed model, and appropriate fail-safe mechanisms and human oversight remain a critical part of implementation.

Explainability

Explainability allows for human oversight of machine learning to be carried out effectively, ensuring that model risk is understood and controlled. Understanding the causes behind performance is a common part of risk management – for example, the 'Profit and Loss attribution test', which forms part of the Fundamental Review of the Trading Book (BIS, 2019), requires a bank's hypothetical profits using front-office pricing models to be explained against their back-office risk models and factors, as part of the validation of those risk models.

The understanding of models and their risks is a significant challenge in finance. The 2007–2008 financial crises demonstrated that copulas, especially those proposed by Li (2000), were underpowered for modelling the risks of CDOs, but yet still were too large and complex to be understood and critiqued by users. In contrast, machine learning models are overpowered, have shiny user-interfaces, but are even more obscure. Machine learning has been promoted in much-cited papers as a method for systemic risk analysis, with only limited discussion of the risks of using machine learning and its lack of interpretability (Kou et al., 2019; Aziz and Dowling, 2019).

Neural networks are not inherently explainable, as input features become entangled and compressed into a single value via repeated non-linear transformations of a weighted sums (Ras et al., 2022). Explainability can be improved by *prima facie* selecting a more interpretable 'white box' model: that is, adopting models which intrinsically are easier to query and understand. Neural network models can be designed to be more interpretable through joint training (Hendricks et al., 2016; Iyer et al., 2018) or including attention mechanisms (Bahdanau et al., 2016; Devlin et al., 2019; Anderson et al., 2018).

Although these solutions apply for neural networks in general, they do not necessarily apply in a reinforcement learning framework. In this setting, rule-based (Verma et al., 2018; Hein et al., 2017), or hierarchical (Shu et al., 2017) methods are available. The purpose of rule-based methods is to present the policies in high level human-readable language, e.g. IF-THEN sequences. Hierarchical methods divide policies into simpler sub-tasks, each of which are separately more interpretable than a flat policy, and are therefore useful to explain individual decisions, i.e. they provide 'local' interpretability.

The above interpretable models generally forgo some performance for enhanced comprehensibility. As a result, as performance is often the primary concern, explainability techniques which can be applied to a black-box model need to be identified. These techniques can be grouped under the name 'post-hoc' explainability.

The types of post-hoc explanation methods are broad and include perturbation analysis, gradient analysis, example based explanations, and surrogate-modelling for local and global explanations (Adadi and Berrada, 2018). Different applications and tasks require a different balance between explainability and performance.

'Deep' reinforcement learning is based on neural network models, and adds an additional layer of incomprehensibility to the modelling process (Mnih et al., 2013). Reinforcement learning models are complex, but it is often possible to use interpretable surrogate models as a means of simplifying and representing their actions; this is often easier than developing inherently interpretable models (Puiutta and Veith, 2020). A range of surrogates are available for this purpose, and include genetic programming techniques (Hein et al., 2018), causal DAGs (Madumal et al., 2020) and the use of tree-based models to approximate predictions (Coppens et al., 2019). However, when using surrogate models for explainability, it is wise to keep the underlying model as simple as possible, in order to make it easier for a surrogate model to reproduce its outputs.

Monitoring and control

Models not only have to be validated on historical data, i.e. benchmarked, they also have to be monitored and controlled when running 'live'. In machine learning, this is related to 'concept drift', which refers to data distributions changing over time, leading to faulty predictions (Žliobaitė et al., 2016). The hope is that, with online learning incorporated in the approach, models can self-diagnose and self-correct when this occurs, but this is not always the case. Continuous recalibration may not be possible in all settings, due to regulatory requirements and the cost of recalibration (Cohen et al., 2019b). A good survey of concept and data drift and how to deal with it can be found in Gama et al. (2014). The importance of monitoring, recalibrating and updating systems, and ensuring sufficient human control, is a key part of the implementation of most automated systems in practice.

In a financial setting, we might also want to base the criteria for drift on the execution of other methods (for example, handcrafted strategies) that are run in parallel as 'controls' for performance. This allows one to study those occasions in which the performance of controls differed significantly from the model, highlighting points of concern.

Machine learning models need more extensive monitoring procedures than handcrafted approaches due to the various risks they come with. Nevertheless, the promise of improved performance, the flexibility of modelling, and the speed advantages associated with embracing these new technologies means that there is no doubt about their broad incorporation into many parts of the finance industry.

References

Ackerer, Damien, Tagasovska, Natasa, and Vatter, Thibault. 2019. Deep smoothing of the implied volatility surface. Available at SSRN 3402942.

Adadi, Amina, and Berrada, Mohammed. 2018. Peeking inside the black-box: a survey on explainable artificial intelligence (XAI). *IEEE Access*, **6**, 52138–52160.

Almgren, Robert, and Chriss, Neil. 2001. Optimal execution of portfolio transactions. *Journal of Risk*, **3**, 5–40.

Anders, Christopher, Pasliev, Plamen, Dombrowski, Ann-Kathrin, Müller, Klaus-Robert, and Kessel, Pan. 2020. Fairwashing explanations with off-manifold detergent. Pages 314–323 of: *International Conference on Machine Learning*.

Anderson, Peter, He, Xiaodong, Buehler, Chris, Teney, Damien, Johnson, Mark, Gould, Stephen, and Zhang, Lei. 2018. Bottom-up and top-down attention for image captioning and visual question answering. Pages 6077–6086 of: *Proc. IEEE Conference on Computer Vision and Pattern Recognition*.

Andreou, Panayiotis C., Charalambous, Chris, and Martzoukos, Spiros H. 2010. Generalized parameter functions for option pricing. *Journal of Banking & Finance*, **34**(3), 633–646.

Arribas, Imanol Perez, Salvi, Cristopher, and Szpruch, Lukasz. 2020. Sig-SDEs model for quantitative finance. In: *Proc. First ACM International Conference on AI in Finance*. Article 7, 1–8. DOI: https://doi.org/10.1145/3383455.3422553.

Arthur, W, Brian, Holland, John H., LeBaron, Blake, Palmer, Richard, and Tayler, Paul. 1996. Asset pricing under endogenous expectations in an artificial stock market. Pages 15-44 in: *The Economy as an Evolving Complex System II*, W. B. Arthur, S. Durlauf, and D. Lane (eds). Addison-Wesley. Available at SSRN 2252.

Aziz, Saqib, and Dowling, Michael. 2019. Machine learning and AI for risk management. Pages 33–50 of: *Disrupting Finance*. Palgrave Pivot.

Baba, Naohiko, and Packer, Frank. 2009. Interpreting deviations from covered interest parity during the financial market turmoil of 2007–08. *Journal of Banking & Finance*, **33**(11), 1953–1962.

Bahdanau, Dzmitry, Cho, Kyunghyun, and Bengio, Yoshua. 2016. Neural machine translation by jointly learning to align and translate. ArXiv:1409.0473v7.

Balunovic, Mislav, and Vechev, Martin. 2019. Adversarial training and provable defenses: Bridging the gap. In: *International Conference on Learning Representations*.

Barucci, Emilio, Cherubini, Umberto, and Landi, Leonardo. 1997. Neural networks for contingent claim pricing via the Galerkin method. Pages 127–141 of: *Computational Approaches to Economic Problems*. Springer.

Bayer, Christian, Horvath, Blanka, Muguruza, Aitor, Stemper, Benjamin, and Tomas, Mehdi. 2019. On deep calibration of (rough) stochastic volatility models. ArXiv:1908.08806v1.

Beck, Christian, Becker, Sebastian, Cheridito, Patrick, Jentzen, Arnulf, and Neufeld, Ariel. 2021. Deep splitting method for parabolic PDEs. *SIAM Journal on Scientific Computing*, **43**(5), A3135–A3154.

Behzadan, Vahid, and Munir, Arslan. 2018. The faults in our pi stars: Security issues and open challenges in deep reinforcement learning. ArXiv:1810.10369v1.

Belcak, Peter, Calliess, Jan-Peter, and Zohren, Stefan. 2020. Fast agent-based simulation framework of limit order books with applications to pro-rata markets and the study of latency effects. ArXiv:2008.07871v2.

Bellemare, Marc G., Naddaf, Yavar, Veness, Joel, and Bowling, Michael. 2013. The arcade learning environment: An evaluation platform for general agents. *Journal of Artificial Intelligence Research*, **47**, 253–279.

Bengio, Yoshua, Goodfellow, Ian, and Courville, Aaron. 2017. *Deep Learning*. MIT Press.

BIS. 2019 (Dec). MAR32 – Internal models approach: backtesting and P&L attribution test requirements. `https://www.bis.org/basel_framework/chapter/MAR/32.htm?inforce=20220101`, [Accessed Feb. 1, 2021].

Buehler, Hans, Gonon, Lukas, Teichmann, Josef, Wood, Ben, Mohan, Baranidharan, and Kochems, Jonathan. 2019. Deep hedging: hedging derivatives under generic market frictions using reinforcement learning. *Swiss Finance Institute Research Paper*.

Buehler, Hans, Horvath, Blanka, Lyons, Terry, Perez Arribas, Imanol, and Wood, Ben. 2020. A data-driven market simulator for small data environments. Available at SSRN 3632431.

Byrd, David, Hybinette, Maria, and Balch, Tucker Hybinette. 2020. Abides: towards high-fidelity market simulation for AI research. Pages 11–22 of: *Proc. ACM SIGSIM Conference on Principles of Advanced Discrete Simulation*. DOI: https://doi.org/10.1145/3384441.3395986.

Cartea, Álvaro, and Sánchez-Betancourt, Leandro. 2021. The shadow price of latency: Improving intraday fill ratios in foreign exchange markets. *SIAM Journal on Financial Mathematics*, **12**(1), 254–294.

Chawla, Nitesh V., Bowyer, Kevin W., Hall, Lawrence O., and Kegelmeyer, W. Philip. 2002. SMOTE: synthetic minority over-sampling technique. *Journal of Artificial Intelligence Research*, **16**, 321–357.

Cohen, Jeremy M., Rosenfeld, Elan, and Kolter, J. Zico. 2019a. Certified adversarial robustness via randomized smoothing. Pages 1310–1320 of: *Proc. ICML*. `http://proceedings.mlr.press/v97/cohen19c.html`.

Cohen, Samuel N., Henckel, Timo, Menzies, Gordon D., Muhle-Karbe, Johannes, and Zizzo, Daniel J. 2019b. Switching cost models as hypothesis tests. *Economics Letters*, **175**, 32–35.

Cohen, Samuel N, Reisinger, Christoph, and Wang, Sheng. 2020. Detecting and repairing arbitrage in traded option prices. *Applied Mathematical Finance*, **27**(5), 345–373.

Collins, Keith. 2016. How one programmer broke the Internet by deleting a tiny piece of code. *Quartz Magazine*, `https://qz.com/646467`, [Accessed Feb. 1, 2021].

Coppens, Youri, Efthymiadis, Kyriakos, Lenaerts, Tom, Nowé, Ann, Miller, Tim, Weber, Rosina, and Magazzeni, Daniele. 2019. Distilling deep reinforcement learning policies in soft decision trees. Pages 1–6 of: *Proc. IJCAI 2019 Workshop on Explainable Artificial Intelligence.*

Cox, Alexander M.G., and Obłój, Jan. 2011. Robust pricing and hedging of double no-touch options. *Finance and Stochastics*, **15**(3), 573–605.

Davis, Mark H.A., Panas, Vassilios G., and Zariphopoulou, Thaleia. 1993. European option pricing with transaction costs. *SIAM Journal on Control and Optimization*, **31**(2), 470–493.

de Freitas, João F.G., Niranjan, Mahesan, and Gee, Andrew H. 2000. Hierarchical Bayesian models for regularization in sequential learning. *Neural Computation*, **12**(4), 933–953.

Deng, Jia, Dong, Wei, Socher, Richard, Li, Li-Jia, Li, Kai, and Fei-Fei, Li. 2009. Imagenet: A large-scale hierarchical image database. Pages 248–255 of: *2009 IEEE Conference on Computer Vision and Pattern Recognition.*

Devlin, Jacob, Chang, Ming-Wei, Lee, Kenton, and Toutanova, Kristina. 2019. BERT: Pre-training of deep bidirectional transformers for language understanding. Pages 4171–4186 of: *Proc. Conference of the North American Chapter of the Association for Computational Linguistics: Human Language Technologies, Volume 1.* https://aclanthology.org/N19-1423.

Dixon, Matthew F., Halperin, Igor, and Bilokon, Paul. 2020. *Machine Learning in Finance.* Springer.

Dugas, Charles, Bengio, Yoshua, Bélisle, François, Nadeau, Claude, and Garcia, René. 2001. Incorporating second-order functional knowledge for better option pricing. Pages 472–478 in: *Advances in Neural Information Processing Systems.*

Dugas, Charles, Bengio, Yoshua, Bélisle, François, Nadeau, Claude, and Garcia, René. 2009. Incorporating functional knowledge in neural networks. *Journal of Machine Learning Research*, **10**(6), 1239–1262.

Eckstein, Stephan, Guo, Gaoyue, Lim, Tongseok, and Obloj, Jan. 2021. Robust pricing and hedging of options on multiple assets and its numerics. *SIAM Journal on Financial Mathematics*, **12**(1), 158–188.

Fan, Jianqing, and Zhou, Wen-Xin. 2016. Guarding against spurious discoveries in high dimensions. *Journal of Machine Learning Research*, **17**(1), 7123–7156.

Fefferman, Charles, Mitter, Sanjoy, and Narayanan, Hariharan. 2016. Testing the manifold hypothesis. *Journal of the American Mathematical Society*, **29**(4), 983–1049.

Ferguson, Ryan, and Green, Andrew. 2018. Deeply learning derivatives. ArXiv:1809.02233.

Fredrikson, Matt, Jha, Somesh, and Ristenpart, Thomas. 2015. Model inversion attacks that exploit confidence information and basic countermeasures. Pages 1322–1333 of: *Proc. 22nd ACM SIGSAC Conference on Computer and Communications Security.*

Fu, Rao, Chen, Jie, Zeng, Shutian, Zhuang, Yiping, and Sudjianto, Agus. 2019. Time series simulation by conditional generative adversarial net. ArXiv:1904.11419.

Gama, João, Žliobaitė, Indrė, Bifet, Albert, Pechenizkiy, Mykola, and Bouchachia, Abdelhamid. 2014. A survey on concept drift adaptation. *ACM Computing Surveys*, **46**(4), 1–37.

Garcia, René, and Gençay, Ramazan. 1998. Option pricing with neural networks and a homogeneity hint. Pages 195–205 of: *Decision Technologies for Computational Finance.* Springer.

Garcia, René, and Gençay, Ramazan. 2000. Pricing and hedging derivative securities with neural networks and a homogeneity hint. *Journal of Econometrics*, **94**(1-2), 93–115.

Gatheral, Jim. 2006. *The Volatility Surface: A Practitioner's Guide.* Wiley.

Ghaziri, H., Elfakhani, S., and Assi, J. 2000. Neural networks approach to pricing, options. *Neural Network World*, **1**(2/00), 271–277.

Gierjatowicz, Patryk, Sabate-Vidales, Marc, Siska, David, Szpruch, Lukasz, and Zuric, Zan. 2020. Robust pricing and hedging via neural SDEs. Available at SSRN 3646241.

Gnoatto, Alessandro, Reisinger, Christoph, and Picarelli, Athena. 2020. Deep xVA solver – a neural network based counterparty credit risk management framework. Available at SSRN 3594076.

Goetz, Jack, and Tewari, Ambuj. 2020.   Federated learning via synthetic data. ArXiv:2008.04489.

Gu, Shihao, Kelly, Bryan, and Xiu, Dacheng. 2018. Empirical asset pricing via machine learning. Tech. Rept. National Bureau of Economic Research.

Hagan, Patrick S., Kumar, Deep, Lesniewski, Andrew S., and Woodward, Diana E. 2002. Managing smile risk. *The Best of Wilmott*, **1**, 249–296.

Han, Jiequn, Jentzen, Arnulf, and E, Weinan. 2018. Solving high-dimensional partial differential equations using deep learning. *Proc. National Academy of Sciences*, **115**(34), 8505–8510.

Hein, Daniel, Hentschel, Alexander, Runkler, Thomas, and Udluft, Steffen. 2017. Particle swarm optimization for generating interpretable fuzzy reinforcement learning policies. *Engineering Applications of Artificial Intelligence*, **65**, 87–98.

Hein, Daniel, Udluft, Steffen, and Runkler, Thomas A. 2018. Interpretable policies for reinforcement learning by genetic programming. *Engineering Applications of Artificial Intelligence*, **76**, 158–169.

Heiss, Jakob, Teichmann, Josef, and Wutte, Hanna. 2019. How implicit regularization of neural networks affects the learned function – Part I. ArXiv:1911.02903.

Hendricks, Lisa Anne, Akata, Zeynep, Rohrbach, Marcus, Donahue, Jeff, Schiele, Bernt, and Darrell, Trevor. 2016. Generating visual explanations. Pages 3–19 of: *European Conference on Computer Vision*. Springer.

Henry-Labordere, Pierre. 2019. Generative models for financial data. Available at SSRN 3408007.

Hobson, David G. 1998. Robust hedging of the lookback option. *Finance and Stochastics*, **2**(4), 329–347.

Horvath, Blanka, Muguruza, Aitor, and Tomas, Mehdi. 2020. Deep learning volatility: a deep neural network perspective on pricing and calibration in (rough) volatility models. *Quantitative Finance*, **21**(1) 11–27.

Hu, Kaitong, Ren, Zhenjie, Siska, David, and Szpruch, Lukasz. 2019. Mean-field Langevin dynamics and energy landscape of neural networks. *Ann. Inst. H. Poincaré Probab. Statist.*, **57**(4), 2043–2065.

Hu, Weihua, Fey, Matthias, Zitnik, Marinka, Dong, Yuxiao, Ren, Hongyu, Liu, Bowen, Catasta, Michele, and Leskovec, Jure. 2020. Open graph benchmark: Datasets for machine learning on graphs. ArXiv:2005.00687.

Huang, Xiaowei, Kwiatkowska, Marta, Wang, Sen, and Wu, Min. 2017. Safety verification of deep neural networks. Pages 3–29 of: *International Conference on Computer-Aided Verification*. Springer.

Hutchinson, James M., Lo, Andrew W., and Poggio, Tomaso. 1994. A nonparametric approach to pricing and hedging derivative securities via learning networks. *Journal of Finance*, **49**(3), 851–889.

Ioannidis, John P.A. 2005. Why most published research findings are false. *PLoS Medicine*, **2**(8), e124.

Itkin, Andrey. 2019. Deep learning calibration of option pricing models: some pitfalls and solutions. ArXiv:1906.03507.

Iyer, Rahul, Li, Yuezhang, Li, Huao, Lewis, Michael, Sundar, Ramitha, and Sycara, Katia. 2018. Transparency and explanation in deep reinforcement learning neural networks. Pages 144–150 of: *Proc. AAAI/ACM Conference on AI, Ethics, and Society*.

Jabir, Jean-François, Šiška, David, and Szpruch, Łukasz. 2019. Mean-field neural odes via relaxed optimal control. ArXiv:1912.05475.

Jacquier, Antoine Jack, and Oumgari, Mugad. 2019. Deep PPDEs for rough local stochastic volatility. Available at SSRN 3400035.

Kolm, Petter N, and Ritter, Gordon. 2019. Dynamic replication and hedging: A reinforcement learning approach. *Journal of Financial Data Science*, **1**(1), 159–171.

Kondratyev, Alexei, and Schwarz, Christian. 2019. The market generator. Available at SSRN 3384948.

Koshiyama, Adriano, Firoozye, Nick, and Treleaven, Philip. 2020. Generative adversarial networks for financial trading strategies fine-tuning and combination. *Quantitative Finance*, **21**(5) 797–813.

Kou, Gang, Chao, Xiangrui, Peng, Yi, Alsaadi, Fawaz E., and Herrera-Viedma, Enrique. 2019. Machine learning methods for systemic risk analysis in financial sectors. *Technological and Economic Development of Economy*, **25**(5), 716–742.

Kumar, Ram Shankar Siva, Nyström, Magnus, Lambert, John, Marshall, Andrew, Goertzel, Mario, Comissoneru, Andi, Swann, Matt, and Xia, Sharon. 2020. Adversarial machine learning-industry perspectives. Pages 69–75 of: *2020 IEEE Security and Privacy Workshops (SPW)*.

Kwiatkowska, Marta Z. 2019. Safety verification for deep neural networks with provable guarantees. In: *Leibniz International Proceedings in Informatics, LIPIcs*.

Lajbcygier, Paul R., and Connor, Jerome T. 1997. Improved option pricing using artificial neural networks and bootstrap methods. *International Journal of Neural Systems*, **8**(04), 457–471.

LeBaron, Blake. 2000. Agent-based computational finance: Suggested readings and early research. *Journal of Economic Dynamics and Control*, **24**(5-7), 679–702.

Levy, Moshe, Levy, Haim, and Solomon, Sorin. 1994. A microscopic model of the stock market: cycles, booms, and crashes. *Economics Letters*, **45**(1), 103–111.

Li, David X. 2000. On default correlation: A copula function approach. *Journal of Fixed Income*, **9**(4), 43–54.

Li, Yimou, Turkington, David, and Yazdani, Alireza. 2020. Beyond the black box: an intuitive approach to investment prediction with machine learning. *Journal of Financial Data Science*, **2**(1), 61–75.

Lin, Zinan, Jain, Alankar, Wang, Chen, Fanti, Giulia, and Sekar, Vyas. 2020. Using GANs for Sharing Networked Time Series Data: Challenges, Initial Promise, and Open Questions. Pages 464–483 of: *Proc. ACM Internet Measurement Conference*.

Lipton, Zachary C. 2018. The mythos of model interpretability: in machine learning, the concept of interpretability is both important and slippery. *Queue*, **16**(3), 31–57.

Liu, Xuanqing, Cheng, Minhao, Zhang, Huan, and Hsieh, Cho-Jui. 2018. Towards robust neural networks via random self-ensemble. Pages 369–385 of: *Proc. European Conference on Computer Vision*.

Lo, Andrew W. 2019. *Adaptive Markets: Financial Evolution at the Speed of Thought*. Princeton University Press.

Longstaff, Francis A., and Schwartz, Eduardo S. 2001. Valuing American Options by simulation: a simple least-squares approach. *Review of Financial Studies*, **14**, 113–147.

Louizos, Christos, Swersky, Kevin, Li, Yujia, Welling, Max, and Zemel, Richard. 2016. The variational fair autoencoder. In: *Proc. 4th International Conference on Learning Representations*.

Lyons, Terry, Nejad, Sina, and Arribas, Imanol Perez. 2019. Nonparametric pricing and hedging of exotic derivatives. *Applied Mathematical Finance*, **27**(6), 457–494.

MacKenzie, Donald. 2018. Material signals: A historical sociology of high-frequency trading. *American Journal of Sociology*, **123**(6), 1635–1683.

Madry, Aleksander, Makelov, Aleksandar, Schmidt, Ludwig, Tsipras, Dimitris, and Vladu, Adrian. 2018. Towards deep learning models resistant to adversarial attacks. *Proc. 4th International Conference on Learning Representations*.

Madumal, Prashan, Miller, Tim, Sonenberg, Liz, and Vetere, Frank. 2020. Explainable reinforcement learning through a causal lens. pages 2493–2500 of: *Proc. AAAI Conference on Artificial Intelligence*.

Mariani, Giovanni, Zhu, Yada, Li, Jianbo, Scheidegger, Florian, Istrate, Roxana, Bekas, Costas, and Malossi, A. Cristiano I. 2019. PAGAN: Portfolio Analysis with Generative Adversarial Networks. Arxiv:1909.10578.

McCartney, Scott. 2020. Coronavirus has upended everything airlines know about pricing. *Wall Street Journal*, Aug 5.

McGhee, William A. 2018. An artificial neural network representation of the SABR stochastic volatility model. *Journal of Computational Finance*, **25**(7), 1–27.

McMormack, C., and Doherty, James. 1993. Neural network super architectures. Pages 301–304 of: *Proc. International Conference on Neural Networks*, vol. 1.

McWaters, R.J., Blake, M., and Galaski, R. 2019. *Navigating uncharted waters: a roadmap to responsible innovation with AI in financial services* Part of the *Future of Financial Services Series*. World Economic Forum.

Mikkilä, Oskari. 2020. *Optimal Hedging with Continuous Action Reinforcement Learning.* Master's Thesis, Tampere University

Mnih, Volodymyr, Kavukcuoglu, Koray, Silver, David, Graves, Alex, Antonoglou, Ioannis, Wierstra, Daan, and Riedmiller, Martin. 2013. Playing Atari with deep reinforcement learning. ArXiv:1312.5602.

Montesdeoca, Luis, and Niranjan, Mahesan. 2016. Extending the feature set of a data-driven artificial neural network model of pricing financial options. Pages 1–6 of: *2016 IEEE Symposium Series on Computational Intelligence (SSCI).*

Nakagawa, Kei, Ito, Tomoki, Abe, Masaya, and Izumi, Kiyoshi. 2019. Deep recurrent factor model: interpretable non-linear and time-varying multi-factor Model. ArXiv:1901.11493.

Neyshabur, Behnam. 2017. Implicit regularization in deep learning. ArXiv:1709.01953.

Ni, Hao, Szpruch, Lukasz, Wiese, Magnus, Liao, Shujian, and Xiao, Baoren. 2020. Conditional Sig-Wasserstein GANs for Time Series Generation. ArXiv:2006.05421.

Nigito, Brian. 2021. How to Build an Exchange: Jane Street. `https://www.janestreet.com/tech-talks/building-an-exchange`, [Accessed Jan. 25, 2021].

Ning, Brian, Lin, Franco Ho Ting, and Jaimungal, Sebastian. 2018. Double deep q-learning for optimal execution. ArXiv:1812.06600.

Niranjan, Mahesan. 1996. Sequential tracking in pricing financial options using model based and neural network approaches. *Advances in Neural Information Processing Systems*, **9**, 960–966.

Oblój, Jan. 2008. Fine-tune your smile: Correction to Hagan et al. *Wilmott Magazine*, **35**, 102–104.

Olson, Randal S., La Cava, William, Orzechowski, Patryk, Urbanowicz, Ryan J., and Moore, Jason H. 2017. PMLB: a large benchmark suite for machine learning evaluation and comparison. *BioData Mining*, **10**(1), 1–13.

Palmer, Richard G., Arthur, W. Brian, Holland, John H., LeBaron, Blake, and Tayler, Paul. 1994. Artificial economic life: a simple model of a stockmarket. *Physica D: Nonlinear Phenomena*, **75**(1–3), 264–274.

Palmer, Samuel. 2019. *Evolutionary Algorithms and Computational Methods for Derivatives Pricing.* PhD thesis, University College London.

Provost, Foster. 2000. Machine learning from imbalanced data sets 101. Pages 1–3 of: *Proc. AAAI Workshop on Imbalanced Data Sets*, vol. 68. AAAI Press.

Puiutta, Erika, and Veith, Eric. 2020. Explainable reinforcement learning: a survey. In *Machine Learning and Knowledge Extraction*, Holzinger, A., Kieseberg, P., Tjoa, A., Weippl, E. (eds). Lecture Notes in Computer Science, vol. 12279.

Rahimian, Hamed, and Mehrotra, Sanjay. 2019. Distributionally robust optimization: A review. ArXiv:1908.05659.

Ras, Gabrielle, Xie, Ning, van Gerven, Marcel, and Doran, Derek. 2022. Explainable deep learning: A field guide for the uninitiated. *Journal of Artificial Intelligence Research*, **73**, 329–396.

Recht, Benjamin, Roelofs, Rebecca, Schmidt, Ludwig, and Shankar, Vaishaal. 2018. Do CIFAR-10 classifiers generalize to CIFAR-10? ArXiv:1806.00451.

Ruf, Johannes, and Wang, Weiguan. 2020. Neural networks for option pricing and hedging: a literature review. *Journal of Computational Finance*, **24**(1), 1–46.

Ruf, Johannes, and Wang, Weiguan. 2021. Hedging with neural networks. *Journal of Business and Economic Statistics*, DOI: 10.1080/07350015.2021.1931241.

Sabate-Vidales, Marc, Siska, David, and Szpruch, Lukasz. 2018. Unbiased deep solvers for parametric PDEs. ArXiv:1810.05094.

Sabate-Vidales, Marc, Šiška, David, and Szpruch, Lukasz. 2020. Solving path dependent PDEs with LSTM networks and path signatures. ArXiv:2011.10630.

Shu, Tianmin, Xiong, Caiming, and Socher, Richard. 2017. Hierarchical and interpretable skill acquisition in multi-task reinforcement learning. ArXiv:1712.07294.

Sirignano, Justin, and Spiliopoulos, Konstantinos. 2018. DGM: A deep learning algorithm for solving partial differential equations. *Journal of Computational Physics*, **375**, 1339-1364.

Snow, Derek. 2020. Machine learning in asset management – Part 2: Portfolio construction – weight optimization. *Journal of Financial Data Science*, **2**(2), 17–24.

Soleymani, Farzan, and Paquet, Eric. 2020. Financial portfolio optimization with online deep reinforcement learning and restricted stacked autoencoder – DeepBreath. *Expert Systems with Applications*, **156**, 113456.

Sugiyama, Masashi, and Kawanabe, Motoaki. 2012. *Machine Learning in Non-Stationary Environments: Introduction to Covariate Shift Adaptation*. MIT Press.

Takahashi, Shuntaro, Chen, Yu, and Tanaka-Ishii, Kumiko. 2019. Modeling financial time-series with generative adversarial networks. *Physica A*, **527**, 121261.

Verma, Abhinav, Murali, Vijayaraghavan, Singh, Rishabh, Kohli, Pushmeet, and Chaudhuri, Swarat. 2018. Programmatically interpretable reinforcement learning. Pages 5045–5054 of: Proc. ICML. `https://proceedings.mlr.press/v80/verma18a.html`.

Wan, Zhiqiang, Zhang, Yazhou, and He, Haibo. 2017. Variational autoencoder based synthetic data generation for imbalanced learning. Pages 1–7 of: *2017 IEEE Symposium Series on Computational Intelligence*.

Whalley, A.E., and Wilmott, Paul. 1999. Optimal hedging of options with small but arbitrary transaction cost structure. *European Journal of Applied Mathematics*, **10**(2), 117–139.

Wicker, Matthew, Laurenti, Luca, Patane, Andrea, and Kwiatkowska, Marta. 2020. Probabilistic safety for Bayesian neural networks. Pages 1198–1207 of: *Conference on Uncertainty in Artificial Intelligence*.

Xie, Cihang, Wu, Yuxin, Maaten, Laurens van der, Yuille, Alan L., and He, Kaiming. 2019. Feature denoising for improving adversarial robustness. Pages 501–509 of: *Proc. IEEE/CVF Conference on Computer Vision and Pattern Recognition*.

Xu, Lei, and Veeramachaneni, Kalyan. 2018. Synthesizing tabular data using generative adversarial networks. ArXiv:1811.11264.

Yang, Yongxin, Zheng, Yu, and Hospedales, Timothy. 2017. Gated neural networks for option pricing: Rationality by design. Pages 52-58 of: *Proc. 31st AAAI Conference on Artificial Intelligence*.

Zeng, Yaxiong, and Klabjan, Diego. 2019. Online adaptive machine learning based algorithm for implied volatility surface modeling. *Knowledge-Based Systems*, **163**, 376–391.

Zheng, Yu, Yang, Yongxin, and Chen, Bowei. 2019. Gated deep neural networks for implied volatility surfaces. ArXiv:1904.12834.

Žliobaitė, Indrė, Pechenizkiy, Mykola, and Gama, Joao. 2016. An overview of concept drift applications. Pages 91–114 in: *Big Data Analysis: New Algorithms for a New Society*, Nathalie Japkowicz, Jerzy Stefanowski (eds). Springer.

Index

Printed in the United States
by Baker & Taylor Publisher Services